608705

MARINE CONSERVATION

The land-sea coastal realm from the tropics to polar regions, where the majority of marine conservation issues lie. See Chapters 2 and 4 for physical and biological/ecological characterization. Illustration © Robert L. Smith, Jr.

Cover image: created in Photoshop by Robert L. Smith, Jr., from separate pictures of the walrus and the ice floe, Bering Sea. Photographs © Ray & McCormick-Ray.

MARINE CONSERVATION

SCIENCE, POLICY, AND MANAGEMENT

G. Carleton Ray and Jerry McCormick-Ray
Department of Environmental Sciences
University of Virginia
Charlottesville, Virginia, USA

Illustrations by Robert L. Smith, Jr.

WILEY Blackwell

Registered office: John Wiley & Sons, Ltd, The Atrium, Southern Gate, Chichester, West Sussex, PO19 8SQ, UK

Editorial offices: 9600 Garsington Road, Oxford, OX4 2DQ, UK
The Atrium, Southern Gate, Chichester, West Sussex, PO19 8SQ, UK
111 River Street, Hoboken, NJ 07030-5774, USA

For details of our global editorial offices, for customer services and for information about how to apply for permission to reuse the copyright material in this book please see our website at www.wiley.com/wiley-blackwell.

Library of Congress Cataloging-in-Publication Data

Ray, G. Carleton.
Marine conservation : science, policy, and management / G. Carleton Ray and Jerry McCormick-Ray.
pages cm
Includes bibliographical references and index.
ISBN 978-1-118-71444-7 (cloth) – ISBN 978-1-4051-9347-4 (pbk.) 1. Marine resources conservation–Textbooks.
I. McCormick-Ray, Jerry. II. Title.
GC1018.R39 2014
333.91'6416–dc23
2013014734

A catalogue record for this book is available from the British Library.

Wiley also publishes its books in a variety of electronic formats. Some content that appears in print may not be available in electronic books.

Set in 9/11 pt Photina by Toppan Best-set Premedia Limited
Printed and bound in Malaysia by Vivar Printing Sdn Bhd

1 2014

To Sally Lyons Brown for her vision and support, and to Raymond F. Dasmann and F. Herbert Bormann who continue to inspire us.

CONTENTS

CONTRIBUTORS

Mark A. Albins
Marine Fish Laboratory
Auburn University
Fairhope, Alabama, USA

Alan B. Bolten
Department of Biology
Archie Carr Center for Sea Turtle Research
University of Florida
Gainesville, Florida, USA

Karen A. Bjorndal
Department of Biology
Archie Carr Center for Sea Turtle Research
University of Florida
Gainesville, Florida, USA

Robert L. Brownell, Jr.
NOAA Southwest Fisheries Center
Pacific Grove, California, USA

Claudio Campagna
Wildlife Conservation Society
Buenos Aires, Argentina

Randolph M. Chambers
Department of Environmental Science and Policy
College of William and Mary
Williamsburg, Virginia, USA

Mark R. Christie
Department of Zoology
Oregon State University
Corvallis, Oregon, USA

Philip J. Clapham
National Marine Mammal Laboratory
National Oceanic and Atmospheric Administration
Seattle, Washington, USA

Barry Clark
Zoology Department
University of Cape Town
South Africa

Paul K. Dayton
Scripps Institution of Oceanography
University of California, San Diego
San Diego, California, USA

Hon. Earl D. Deveaux
[Retired, Minister of the Environment, The Bahamas]
Nassau, Bahamas

Robert J. Diaz
Department of Biological Sciences
Virginia Institute of Marine Sciences
College of William and Mary
Gloucester Point, Virginia, USA

David B. Eggleston
Department of Marine, Earth and Atmospheric Sciences
North Carolina State University
Raleigh, North Carolina, USA

R. Michael Erwin
[Retired, Department of Environmental Sciences, University of Virginia]
Weaverville, North Carolina, USA

James A. Estes
Center for Ocean Health
University of California, Santa Cruz
Santa Cruz, California, USA

Valeria Falabella
Wildlife Conservation Society
Buenos Aires, Argentina

Michael Garstang
[Retired, Department of Environmental Sciences, University of Virginia]
Charlottesville, Virginia, USA

J. Frederick Grassle
Institute of Marine and Coastal Sciences
Rutgers University
New Brunswick, New Jersey, USA

Allan Heydorn
[Retired CEO, World Wildlife Fund, South Africa]
Stellenbosch, South Africa

Mark A. Hixon
University of Hawai'i at Manoa
Honolulu, Hawai'i, USA

Edward D. Houde
Center for Environmental Science
University of Maryland
Solomons, Maryland, USA

Gary L. Hufford
[Retired National Weather Service]
National Oceanic and Atmospheric Administration
Eagle River, Alaska, USA

Brian J. Huntley
[Retired Director, Kirstenbosch National Botanical Garden]
Capetown, South Africa

Yulia V. Ivashchenko
National Marine Mammal Laboratory
National Oceanic and Atmospheric Administration
Seattle, Washington, USA

Steven Kohl
Department of the Interior
Fish and Wildlife Service
Washington, D.C., USA

Igor Krupnik
Arctic Studies Center
National Museum of Natural History
Smithsonian Institution
Washington, D.C., USA

Craig A. Layman
Department of Applied Ecology
School of Agriculture
North Carolina State University
Raleigh, North Carolina, USA

Romuald N. Lipcius
Department of Fisheries Science
Virginia Institute of Marine Science
College of William and Mary
Gloucester Point, Virginia, USA

Thomas R. Loughlin
[Retired, NOAA National Marine Mammal Laboratory]
Redmond, Washington, USA

John C. Ogden
Florida Institute of Oceanography
St. Petersburg, Florida, USA

James E. Perry
Department of Coastal and Ocean Policy
Virginia Institute of Marine Science
College of William and Mary
Gloucester Point, Virginia, USA

Eleanor Phillips
The Nature Conservancy
Northern Caribbean Office
Nassau, The Bahamas

James H. Pipkin
[Retired, U.S. Department of State Special Negotiator for Pacific Salmon (1994 to 2001); U.S. federal Commissioner on the bilateral Pacific Salmon Commission (1999 to 2002); Counselor to the U.S. Secretary of the Interior (1993 to 1998); and Director of the Interior Department's Office of Policy Analysis (1998 to 2001)]
Bethesda, Maryland, USA

Robert Prescott-Allen
[Retired, International Union for Conservation of Nature and Natural Resources, author of The Wellbeing of Nations, co-author of Blueprint for Survival, World Conservation Strategy, and Caring for the Earth: a Strategy for Sustainable Living]
Victoria, British Columbia, Canada

Brandon J. Puckett
Department of Marine, Earth and Atmospheric Sciences
North Carolina State University
Raleigh, North Carolina, USA

Sam Ridgway
[Retired, U.S. Navy Marine Mammal Program, and Professor of Comparative Pathology, Veterinary Medical Center, University of California, San Diego]
San Diego, California, USA

Rutger Rosenberg
Kristineberg Marine Research Station
University of Gothenburg
Sweden

Brian R. Silliman
Division of Marine Science and Conservation
Nicholas School of the Environment
Duke University
Durham, North Carolina, USA

N. A. Sloan
Gwaii Haanas National Park Reserve, National Marine Conservation Area Reserve, and Haida Heritage Site
Haida Gwaii, British Columbia, Canada

Paul Snelgrove
Ocean Sciences Centre and Biology Department
Memorial University of Newfoundland
St. John's, Newfoundland, Canada

Amber J. Soja
NASA National Institute of Aerospace
Hampton, Virginia, USA

William T. Stockhausen
Alaska Fisheries Science Center
National Marine Fisheries Service, NOAA
Seattle, Washington, USA

Lori A. Sutter
Virginia Institute of Marine Science
College of William and Mary
Gloucester Point, Virginia, USA

Richard M. Warwick
Plymouth Marine Laboratory
Plymouth, Devon, UK

Kirk O. Winemiller
Texas A&M University
Department of Wildlife and Fisheries Sciences
College Station, Texas, USA

Victoria Zavattieri
Wildlife Conservation Society
Buenos Aires, Argentina

PREFACE

We are at a time in history when science allows us better to understand our global environment, and when human societies are beginning to recognize the urgency of marine conservation and the need for sustainable use of marine resources. As John A. Moore (1993) has put it: "We have reached a point in history when biological knowledge is the *sine qua non* for a viable human future . . . A critical subset of society will have to understand the nature of life, the interaction of living creatures with their environment, and the strengths and limitations of the data and procedures of science itself. The acquisition of biological knowledge, so long a luxury except for those concerned with agriculture and the health sciences, has now become a necessity for all."

During the past century, humans have acquired the ability to intrude, exploit, and better understand the last, previously unexplored portion of Earth—the contiguous global oceans. The rates and magnitude of change brought on by the Marine Revolution (Ray, 1970) followed 5–10,000 years of the Agricultural Revolution and two centuries of the Industrial Revolution, with dangers of repeating errors of the past. Observation of the quickening pace of change and the way that humans behave and manage themselves, and increasing knowledge of the way marine ecosystems function have made apparent major ecosystem instabilities and management incongruences. Approaches deemed feasible when marine conservation was emerging only a half-century ago no longer fulfill needs of the 21st century. That the world has become "hot, flat, and crowded" (Friedman, 2008) makes clear the need for new marine conservation approaches.

Our previous book, *Coastal-Marine Conservation: Science and Policy* (Blackwell Science, 2004) called attention to the fundamental role natural history and ecosystem-based science play in conservation policy and management planning. That is, conservation must be informed by the natural histories of organisms *together with* the hierarchy of scale-related linkages and ecosystem processes. This book continues that focus on a whole-systems approach to marine conservation, taking account of major advances in marine ecosystem understanding to guide marine conservation practice. Our objective is to expose students and other readers to the broad range of overlapping issues (Chapter 2) in the context of present conservation mechanisms that have been devised to achieve marine conservation goals (Chapter 3). Achieving these goals depends on understanding basic marine ecosystem science (Chapter 4) and the natural histories of marine organisms (Chapter 5), that is, how organisms make a living in dynamic and often stressful environments. In that process, we call attention to emergent and unexpected properties that are changing coastal and marine systems—climate change, ocean acidification, dead zones, and loss of biodiversity—that challenge the resilience of coast-ocean systems, hence also governance and human well-being. We present seven "real-world" case studies that exemplify coastal and marine conservation in action, each presenting a central issue or issues in the context of its biogeographic and social setting. Each combines theoretical ("pure") and applied science, and each concludes with challenges to governance that are not yet fully resolved.

A final synthesis chapter looks to the future, to transition coastal and marine conservation from the *being* of traditional, fragmented, protection, and management to the *becoming* of ecosystem-based approaches, intertwined in a social-ecological system, that propel marine biodiversity and society into the future. Overall, this book is an attempt to provide a framework for thoughtful, critical thinking in order to incite innovation in the new Anthropocene Era of the 21st century.

References, scientific terms, Latin names, and units. This book provides readers with a window into a massive literature on conservation science, policy, and management as a context for understanding the present state of knowledge of marine ecosystems, their life, and their current conservation and management. The language of science is enormous and similar terms often have different, even contradictory, meanings among disciplines. We have attempted to explain these terms by defining some of them in the text. We do not include a glossary, as definitions can be accessed in science dictionaries or through search engines on the Internet. We use the International System of Units (SI units) and metric measurements (e.g., m = meters, mt = tonnes, km = kilometers, nmi = nautical miles, etc.) throughout the text.

Species are referred to by their vernacular ("common") names (blue crab, herring, porpoise, etc.) with Latin names for proper identification. Care must be taken with vernacular names because for the great majority of species these names are not standardized (mammals, birds, and some fishes are notable exceptions). For example, "cod" is a common name for a valuable Atlantic fish of the cod family (Gadidae), but "cod" in Australia refers to groupers of the sea bass family (Serranidae), and for some species of the Southern Ocean "cod" refers to ice fishes of the family Nototheniidae; similarly, "rockfish" may refer to a number of fishes from a half dozen families of fishes; and, the "Dover sole" of the north eastern Pacific is not the highly valued Dover sole of the eastern Atlantic. Therefore, scientific names are essential for identification, and are given with the vernacular the first time the species is mentioned in each chapter, or if far separated.

Acknowledgements and permissions. We first wish to thank the Curtis and Edith Munson Foundation and The Henry Foundation of Washington, D.C., for generous support for the writing and publication of this book in full color, as well as other donors to the University of Virginia Global Biodiversity Fund. We remain deeply grateful to the late Sally Lyons Brown and the W. L. Lyons Brown Foundation for encouragement and support of the 2004 edition, without which the writing of the present book would have been far more arduous.

We especially wish to thank all co-authors and contributors of boxes for volunteering their time and expertise and for their patience during the four-year preparation of this book; these persons are identified accompanying their contributions, and their affiliations are given under Collaborating Authors following this Preface. Acknowledgements to persons and publications who provided photographs and figures are given as those materials appear. Chapters that list no authors, and photographs acknowledged as "the authors" are ours.

We are especially indebted to persons who have influenced our thinking during past decades: Frederick B. Bang, Raymond F. Dasmann, Francis H. Fay, J. Frederick Grassle, Starker Leopold, John A. Moore, Kenneth S. Norris, John C. Ogden, Fairfield Osborn, C. Richard Robins, William E. Schevill, William A. Watkins, and Sir Peter Scott. Others have been especially helpful by providing information specific to individual chapters: David Argument, Pat Bartier, Peter Berg, Charles Birkeland, Peter Boveng, Michael Braynen, Michael Cameron, Eric Carey, Mark H. Carr, Roger Covey, Jon Day, Terri Dionne, Catlyn Epners, D. Fedje, Lynn Gape, Robert Ginsburg, E. Gladstone, Samuel H. Gruber, Bruce P. Hayden, William J. Hargis, Nick Irving, Paula Jasinski, Victor S. Kennedy, Bjorn Kjerfve, Casuarina Lambert McKinney, Jeffrey Lape, Ian G. Macintyre, Dennis Madsen, Karen McGlathery, Pericles Maillis, John D. Milliman, Mary Morris, Roger Newell, Rick Parish, David Pollard, Cliff Robinson, Yvonne Sadovy, Rodney V. Salm, Neil E. Sealey, Kenneth Sherman, Herman H. Shugart, Richard Starke, Ann Swanson, Brent and Robin Symonette, Hillary Thorpe, Brian Walker, Douglas Wartzok, Maire Warwick, and Pat Wiberg. Mark Hixon, Mark Albins, and Mark Christie acknowledge support from the U. S. National Science Foundation for their research on reef-fish larval connectivity, as reviewed in Chapter 5 (grants OCE-00-93976 and OCE-05-50709), and the lionfish invasion in Chapter 6 (grants OCE-08-51162, OCE-12-33027, and a Graduate Research Fellowship). Norm Sloan (Ch. 10) also expresses thanks to those who contributed to the 2006 technical overview of Gwaii Haanas Marine: Pat Bartier, John Broadhead, Lyle Dick, Catlyn Epners, Daryl Fedje, Debby Gardiner, John Harper, Anna-Maria Husband, Greg Martin, Mary Morris, Trevor Orchard, Marlow Pellatt, Cliff Robinson, Ian Sumpter, and Ian Walker. Finally, we thank students who have taken our marine conservation courses at the University of Virginia for their often-insightful comments.

We also wish to thank the authors, illustrators, journals, and publishers who have given permission to use their work, as noted in captions to figures and tables.

We are grateful for the encouragement of funding organizations and agencies for which we have been granted support or have served an advisory or consultancy role, most particularly: ICSU (International Council for Science), IUCN (International Union for the Conservation of Nature and National Resources), NASA (National Aeronautics and Space Administration), National Geographic Society, National Oceanic and Atmospheric Administration, National Science Foundation, New York Aquarium (Wildlife Conservation Society), Office of Naval Research, UNESCO, U.S. Man and the Biosphere Program, Bahamas National Trust, U.S. Marine Mammal Commission, and U.S. National Park Service.

We are indebted to Ward Cooper, Izzy Canning, Kelvin Matthews, Delia Sandford, Carys Williams, Kenneth Chow and Audrie Tan of Wiley Blackwell for their efficiency and positive encouragement, as well as Ian Sherman and his co-workers who worked with us on the 2004 edition, without whom this new edition would not have been possible. We finally wish to thank Ruth Swan, project manager for Toppan Best-set, and Mark Ackerley, freelance copyeditor, for their diligent and thorough proofing, and freelancer Elizabeth Paul for her arduous permission seeking.

G. Carleton Ray and Jerry McCormick-Ray

REFERENCES

Friedman TL (2008) *Hot, Flat, and Crowded: Why We Need a Green Revolution—and How it Can Renew America*. Farrar, Straus and Giroux, New York.

Moore JA (1993) *Science as a Way of Knowing: The Foundation of Modern Biology*. Harvard University Press, Cambridge, Massachusetts.

Ray C (1970) Ecology, law, and the "marine revolution." *Biological Conservation* **3**, 7–17.

ABOUT THE COMPANION WEBSITE

This book is accompanied by a companion website:

www.wiley.com/go/ray/marineconservation

The website includes:

- Powerpoints of all figures from the book for downloading
- PDFs of tables from the book

CHAPTER 1

IN PURSUIT OF MARINE CONSERVATION

There is a tide in the affairs of men
Which, when taken at the flood, leads on to fortune . . .
On such a full sea are we now afloat;
And we must take the current when it serves
Or lose our ventures.

William Shakespeare *Julius Caesar*

Open-ocean systems may seem not to be so disturbed at their surface, but signs of ecological disruption are apparent. The lone walrus on our cover is a metaphor for Planet Earth's fragmented habitats, disrupted ecosystems, and diminished biodiversity. As oceans change, tropical reefs die, polar regions lose sea ice, and marine life that we hardly know is increasingly becoming vulnerable to extinction. Nowhere is this change more apparent than in the land-sea coastal realm (Frontispiece), where the majority of humanity lives, ecosystems are most productive, and biodiversity is greatest.

During the rise of human civilizations, societies have inherited the economics of resource exploitation from an ocean perceived as "limitless." Fisheries, shipping, and coastal settlement as old as civilization, have increasingly expanded to force conservation into defense of species and spaces. And as the ecosystems upon which species depend have changed, scientists have become increasingly involved. Modern science, which had moved from studies in natural history to environmental modeling and statistics to better understand marine systems, is returning to natural history, recognizing that it forms the basis for environmental and evolutionary science itself (Box 1.1). The advancing state of knowledge and the increasing need for sustainable ecosystems are forcing marine conservation science to become more proactive and to expand its scope to encompass whole regional seas. Recognition of depleted fisheries, coastal catastrophes, and consequences of natural events tied to human activities have led to new ways of thinking about how marine conservation may modify society's relentless pursuit of ocean wealth.

The past decades' tendency to compartmentalize marine conservation issues has changed. Marine conservation is now forced to embrace the totality of issues together, because the oceans are interconnected, dynamic, and complex. Knowing how marine life makes a living is fundamental in the vast, bioenergetic marine environment undergoing continual change. And the dynamic features of the global ocean and of the coastal realm make the pursuit of marine conservation different from that for the land.

1.1 THE EMERGENCE OF MODERN MARINE CONSERVATION

Modern marine conservation arose after World War II when the oceans took on greater political, economic, and social importance. The oceans became viewed as a "supplier" to meet expanding human wants for food, resources, and wealth. Humans rapidly began to acquire the ability to explore and exploit this last, previously unavailable portion of Earth—the oceans—to fish and seek petroleum and minerals facilitated by new technology that allowed humans to invade, and also better to understand the oceans to their utmost depths. We call this era of emerging ocean importance the "Marine Revolution" (Ray, 1970). It followed the Industrial Revolution of about two centuries before, which had expanded the human footprint with the invention of the steam engine, electric power, industrialization, and urbanization. And the Industrial Revolution followed the Agricultural Revolution, circa 10,000 to 5000 BP, that transformed landscapes into patches of farmland on such massive scales as to alter Earth processes, including climate (Ruddiman, 2005). Each successive revolution promoted human well-being and population growth as it also depleted natural resources, and as land resources became depleted and consumption grew, societies looked to the oceans for food, energy, and economic benefits. Today, human activities are globally pervasive, marked by resource shortages and the need to conserve what remains in the new age of the Anthropocene (Crutzen and Stoermer, 2000; Steffen *et al.*, 2007).

The economic value that humans place on coastal and marine systems and their workings no doubt arose during the earliest of human cultures. The need for conservation that scientists and writers called attention to focused on overexploited commercial species as early as the 18th and 19th centuries with the squandering of Steller sea cows, fur seals, and others. George Perkins Marsh's *Man and Nature* (1864) was first to link culture with nature, science with society, and landscape with history, and spearheaded nature conservation by leading to forest conservation and establishment of the first

Marine Conservation: Science, Policy, and Management, First Edition. G. Carleton Ray and Jerry McCormick-Ray.

Box 1.1 The importance of studying nature outdoors

Paul K. Dayton
Scripps Institution of Oceanography, University of California, San Diego, USA

> The most basic rules of the world—the ones we all live by—are ecological rules. You can't study them or even perceive them very well in a classroom or laboratory. It is imperative to go out on the mountainside, watch the rain fall over a valley, dig into the earth beneath a fallen tree, or wade a creek for cobbles with sources upstream. The best work in the natural disciplines all starts with observations in nature.
>
> Kenneth S. Norris, in Dayton (2008)

Ken Norris wrote this, in late 1960, making a pitch to the University of California Regents to create a natural reserve system. He was successful and the UC Natural Reserve System has grown into the best such system in the world. But to what avail are patches of nature if people do not immerse themselves in those natural systems?

In the past few decades the powerful tools of molecular biology and capacity of modern computers have joined with technical advances that allow us to monitor and analyze the world around us with unprecedented precision. These new and powerful tools have seduced would-be ecologists into the comfortable idea that they can do good ecology in the laboratory or at a computer terminal without bothering to actually study nature. Indeed, the tools are so complicated that there has been strong selection for ecologists to become increasingly specialized with a laser-like focus. We have thus deprived ourselves of a sense of place of nature that comes from personal experiences, smelling, feeling, and seeing important if episodic relationships. Many ecologists and especially universities have lost respect for the broad view of nature, the understanding of the components and processes of the whole natural world or "natural history" of the systems we study. These specialists fail to perceive the critical relationships and ecosystem workings that their powerful machines were not designed to study. Deprived of personal experience in nature, many forget natural history and accept habitats and systems that are a pale shadow of their former selves and substitute simplistic models for understanding of nature.

Here we are concerned with the conservation of these habitats. We understand that we are reducing populations and losing species, and we are disrupting the important relationships that define our ecosystems. As populations decline, the relationships that define the ecosystems are lost long before the species go extinct, and it is precisely these relationships that we most need to protect. The damage to these relationships and ecosystems is often so persuasive that it may be impossible to understand what has been lost because generations of biologists have reduced expectations of what is natural. This sliding baseline of reality is exacerbated by the lack of personal experience in nature. Without a deep understanding of the history of their systems, ecologists can be beguiled by short-term events or introduced, inappropriate imposters that replace and mask the traces of the natural systems we hope to study and protect. The natural relationships simply disappear, leaving no conspicuous evidence of what has been lost. This loss is paralleled by the loss of human cultures and languages with the passing of elders; we, too, have lost the ecological cultural wisdom of the ages as well as the evolutionary wisdom found in intact ecosystems.

Conservation biologists face extremely difficult problems much more complex than most realize. For example, we need to understand ecosystem stability, recoverability, and resilience. How do we define stability, and what processes maintain it? What spatial and temporal scales are optimal for the analyses of trends? How do we define ecosystem stress? How can we understand when "natural" disturbances ratchet into new "stable states" that resist recovery? What relationships are most critical, what processes define strong and weak interactions, and how do we evaluate the most critical interactions? How do we define multispecies relationships important to ecosystem resilience? Can we predict thresholds in these relationships?

Sustainable ecosystem-based management is an ecological mantra, but how does "single-species management" morph into ecosystem-based management? What do we need to protect and how can we prioritize the relationships? People perturb all ecosystems, but how do we evaluate cumulative effects and understand how much is too much? That is, all ecological relationships have thresholds defined in the context of ongoing natural interactions, but which thresholds are most critical and how do we measure them?

The above questions focus on difficult science that cannot be done without a very deep sense of place that only comes from intimate familiarity with the natural world. But consider also the great importance of social values in addition to the natural sciences. The scientific focus is on important relationships critical for management, but how do we evaluate the value of species? Do we also need to protect weak interactions? Ecologists lose credibility when they claim that every species and interaction is critical to the ecosystem, because this assertion simply is not true. Most systems are comprised of many populations that can be altered without much ecosystem effect. There are many rare and very obscure species with no discernible interactions, and there are charismatic species such as pandas or leatherback sea turtles with roles that are hard to evaluate. Thus, we are asked whether some species are expendable, and we must learn to shift seamlessly from our scientific value systems to cultural value systems that define human values. It is very hard to argue for aesthetic or cultural values for nature without having an intimate understanding of the natural world. If you have not experienced first hand the awe and wonder of nature, it is very hard to communicate it!

Finally, you went into biology because you love nature, and this involves regular contact with nature. The intuitive sense of place so very important to ecological understanding must come from personal experience—smelling, feeling, and seeing the important lessons nature offers an open and prepared mind. It is easy to be seduced by the demands of everyday life and to forget to visit nature and fuel your passion and sense of self as well as a sense of place necessary for your science.

U.S. Commissioner of Fish and Fisheries. But only since the 1940s did conservation become an ethic among the wider public. Aldo Leopold's *Sand County Almanac* (1960), Fairfield Osborn's *Our Plundered Planet* (1947) and *Limits of the Earth* (1953), Raymond Dasmann's *A Different Kind of Country* (1968) and *No Further Retreat* (1971), and others inspired a conservation movement that saw the founding of governmental agencies and non-governmental organizations dedicated to wildlife management and environmental protection. Rachel Carson's *Silent Spring* (1962)—on the *New York Times'* bestseller list for 31 weeks—served as an indictment of the pesticide industry and helped to catalyze ecological awareness and action. However, opposition to ocean abuse—a major feature of the Marine Revolution—has been relatively new.

Little had been said for the marine world until Rachel Carson's *The Sea Around Us* (1951) and, especially, Jacques-Yves Cousteau and Frédéric Dumas' The *Silent World* (1953) made the oceans and their life familiar to the public. Cousteau and Dumas' invention of the "Aqualung" (self-contained underwater breathing apparatus or scuba) allowed anyone in reasonably good health to explore and find value in the sea and marine life "up-close and personal." This self-conscious awareness of the sea's value, beyond only "resources," had immense, global impact. Under a new sense of urgency, Marine Protected Areas began to be established and charismatic species to be protected. Whales, sea turtles, and others that had suffered from over-exploitation, and dolphins and killer whales that were displayed in oceanaria became icons of the ocean's value.

The immediate responses for ocean protection were based on practices that had long proved appropriate for terrestrial environments, namely protection of *species*—overwhelmingly charismatic ones deemed threatened or endangered—and protection of *spaces* that served as habitats for unique, endemic, or threatened plants and animals, or as scenic inspirations. Marine conservation had finally joined an era of environmental concern that reached a climax, fervently expressed on Earth Day, 1970, that aroused the necessary social and political will to make transformational change (Graham, 1999): "In 1965 the environment was not a leading issue. Five years later it was the national problem Americans said they worried about most, second only to crime. Earth Day 1970, celebrated just as that crescendo in public concern was reaching its peak, became the lasting symbol of past frustrations and future hopes." Increased awareness of coastal impacts and recognition of failures to conserve marine resources brought on a quickening pace of change. The public opposed the ruthless slaughter of marine mammals, impacts of polluted water, and shores tarnished by oil spills. The result was a suite of environmental legislation, particularly in the U.S., that set standards that became adopted globally. U.S. legislation centered on species protection, coastal zone management, fisheries management, curbing ocean dumping, and establishment of marine sanctuaries. Marine Protected Areas became institutionalized, albeit operationally stalled by difficulties of designating environmentally or legally defensible boundaries, sizes, and locations, compounded by jurisdictional conflicts, established national priorities, and deficiencies of international ocean law. Internationally, the first effort (mid-1970s) specifically directed towards marine conservation became the *Marine Programme* of the *International Union for the Conservation of Nature and Natural Resources* (IUCN), which persists to this day. This program helped direct efforts towards regional-seas agreements organized and promoted by the United Nations Environmental Program (UNEP). Conservation focus remained on charismatic marine species—whales, seals, walruses, albatrosses, sea turtles, etc.—and natural areas of high biodiversity (coral reefs) and/or scenic beauty, which served to promote marine conservation to the vast majority of humankind that had little direct experience *in* the sea.

However, these programs lacked appropriate mechanisms for addressing new and emergent issues, which made obvious the enormity of the task confronting marine conservation. A cadre of non-governmental organizations (NGOs) began to expand, each with its own interests and goals. At this same time, marine ecology was advancing, generated by new technologies for undersea exploration; satellites allowed "world views" of the coasts and oceans, computers analyzed large data sets, and models revealed insights into system-level phenomena. A principal finding was that change is a fundamental property of ecosystems, at all scales from local to global, and that such change responds to ecological and social domains beyond protected-area boundaries. That is, "protection" of valued or threatened species and spaces—presumably isolated from harm—would not suffice. Marine boundaries are continuously on the move.

From about 1980 to the turn of the 21st century, human-caused ocean change deepened, grew wider, and became more complex, along with the public recognition that "biodiversity" was seriously under threat (Wilson and Peter, 1988). Conservation gradually began to take on a new role—that of protecting biodiversity "hot spots" and restoring diminished natural systems in a shrinking world dominated by human needs.

Additionally, a host of independent initiatives arose, but too many individually directed and often-conflicting laws, regulations, agreements, and treaties added up to challenge conservation—a "tyranny of small decisions" (Odum, 1982). By protecting one part of a whole system, another part unexpectedly reacts, often resulting in consequences that no one wanted or intended, including species depletion and ecological degradation.

We are now at a time in history when science allows better opportunities to understand our global environment and to more clearly recognize the limits of the oceans and the urgency of marine conservation. The need for a comprehensive "systems" approach to protect species and spaces has become increasingly apparent. Coherent national ocean policies are being called for, and international policies are being formulated, but the challenge of implementing comprehensive conservation policy remains. But as Graham (1999) warned: "A generation later, the political and economic ground has shifted . . . The public's sense of crisis has been replaced with enduring support for improving pollution control and conservation, but also with a frequent reluctance to pay the public costs of increased protection or to change everyday habits."

1.2 DEFINING "MARINE CONSERVATION"

Marine conservation is an elusive concept to grasp. What exactly is it? "Conservation," as defined in *Webster's Third New International Dictionary*, is "deliberate, planned, or thoughtful preserving, guarding, or protecting . . . planned management of a natural resource to prevent exploitation, destruction, or neglect . . . wise utilization of a natural product . . . a field of knowledge concerned with coordination and plans for the practical application of data from ecology, limnology, pedology, or other sciences that are significant to preservation of natural resources." These definitions presume a basic understanding of natural-resource science and illustrate that conservation is an issue-directed activity towards which science can provide a guide to inform decision-makers at all levels. However, solutions to sector-based conservation problems have proved elusive for reasons that are not always straightforward, not for want of a plethora of laws, regulations, agreements, organizations, and procedures that have been adopted, but for their applications in a society divided by priorities. Many difficulties also relate to recognizing the differences between land and sea and their respective conservation needs.

The oceans are not like the land. Physically, the three-dimensional ocean is driven by interactions of fluid dynamics, light, nutrients, and temperature. Biologically, ocean volume exceeds the land by almost two orders of magnitude, being dominated by small, non-charismatic microbes and plankton that support larger invertebrates and fishes and a few highly developed, charismatic air-breathing reptiles, birds, and mammals. Phyletic diversity and total biomass in the sea far exceeds that of the land, although large plants are few and restricted to shallow, nearshore waters. Functionally, marine ecosystems are continuous and connected across huge spatial extents, as exhibited by planktonic larvae, billfishes, sharks, sea turtles, and whales. Yet the ocean has boundaries to which species respond. Many widely distributed species exhibit taxonomic and genetic differences in biogeographic patterns and in metapopulations (Ch. 5). Ocean boundaries can move, and can change unexpectedly and unpredictably over decadal time scales or less, and at spatial scales rarely known for terrestrial environments. Such boundary changes are difficult to know, often being observed through natural history and genetic studies of species. Furthermore, the distributions and behaviors of species depend not only on the physical environment, but also on species that can affect and change environments. Species–environment feedbacks modify ecosystems and create conditions that support many other species. Many marine species are opportunistic, depending on chance or changes in response to highly dynamic marine systems. Furthermore, species and environments are interdependent and may coevolve. Such relationships are particularly difficult to observe in the moving fluid of the marine environment. Thus, defining species–environmental interdependencies under conditions of continual change and lack of natural-history knowledge for most of them remains a critical conservation arena. As Levin (2011) put it: "Sustainable management requires that we relate the macroscopic features of communities and ecosystems to the microscopic details of individuals and populations." But how?

1.3 MARINE CONSERVATION'S SCOPE

The rise of ecology, globalization, and the ubiquity of human activities makes obvious the fact that by the later 20th century humans had so altered global ecosystems that the rapidly decreasing number of natural spaces on Earth left to defend may soon be few. This raises the ambiguous issue of "scope." Does scope simply mean size, as established through spatially designated protected or managed areas, i.e., that the larger the boundaries or percentage of protected areas designated means that *more* is protected? Conversely, should preference be given to those species that we believe to be "charismatic"? Or does scope imply a greater suite of procedures, regulatory or otherwise, which translates to *how* conservation is conducted? Answers are not as simple as they may seem.

Currently, marine conservation draws public support and legislative action more from emotional and personal preferences and less from scientifically based information on marine system processes. Hardly anyone would not wish to save a whale, but what about its food supply of very small copepods and krill? And how do ocean processes operating over huge scales support those foods? Clearly, marine conservation is drawn into a large spatial context, as well as being subject to socio-economic conflicts. If marine conservation is to be about biodiversity maintenance, resource sustainability, and human well-being—and all at once—it should become fundamentally hierarchical, from protecting the rarest and most valued (in human and ecological terms) species and spaces, to sustainable use, and to enable the resilience of ecosystems; that is, conservation needs to become "systemic" in its approaches. The crisis is this: as the increasing human population demands ever more marine natural resources, the environmental deficit also grows (Ch. 13; Bormann, 1990). The objective of marine conservation, then, is to slow and eventually stop the ecological cascades resulting from social/ecological imbalances by protecting, restoring, and sustainably using resilient ocean systems and their living components as Earth's last frontier. This objective requires a better understanding of the living and physical components that marine conservation aims to address holistically. As Franklin (1993) has said in another context: "We must see the larger task—stewardship of all the species on all of the landscape with every activity we undertake as human beings—a task without temporal and spatial boundaries."

1.4 ADAPTING MARINE CONSERVATION TO THE 21ST CENTURY

The 21st century is much different from preceding centuries. The Earth is now "hot, flat, and crowded" (Friedman, 2008) and marine issues are converging, thus requiring new approaches. As this century advances, a systems approach is needed for improving society's ability to take effective action through improved understanding of the physical and biological worlds under an accelerating pace of environmental change (Forrester, 1991). Such an approach requires identifying and understanding the components in the system, and how system behavior arises from their interactions over time (Sweeney and Sterman, 2000).

Marine management institutions that arose in the 20th century are today challenged by the interactions among resources, the environment, critical habitats, and conflicts among institutions that undermine their mandated goals. The organisms that institutions aim to protect inhabit a dynamic world in which feedbacks and complex interdependencies sustain them. While the history of ecology is firmly grounded in natural history, understanding ecological patterns and being able to conserve resources requires understanding dynamics (Levin, 2011). This understanding requires a process that starts with a problem to be solved, and advances with better knowledge about the situation and the wealth of information available (Forrester, 1991). For conservation to advance, this wealth of information needs to: (i) place conservation issues in the context of environmental-social systems; (ii) connect species natural history to interconnected natural and human systems; and (iii) place ecosystem resilience in the forefront of conservation action (Walker and Salt, 2012). These goals relate to the art of systems thinking, which involves the ability to represent and assess dynamic complexity. Implicit in thinking about systems is the ability to have good science and quantitative data in order to see relationships between the issue to be addressed and the conservation tools to address it.

Marine conservation is confronted by an overwhelming array of complex issues and an astonishing amount of information. Categories of issues confronting marine conservation are introduced in Chapter 2 to help sort out this complexity. While solutions to many issues are being sought (Ch. 3), most of them have been addressed singly, as if in isolation. Yet some issues are emergent, have arisen suddenly and unexpectedly to catch both science and society unprepared, notably climate change, ocean acidification, and anoxia. Such issues relate to the nature and properties of the ocean's ecological systems, the natural histories of marine species, and their interactions, which requires relating dynamics and linkages of organisms to each other and to their environment. A conceptual level of ecosystem understanding helps make these connections real (Chs. 4, 5).

The book introduces seven case studies that exemplify pursuit of marine conservation. They illustrate an array of attempts to address specific conservation issues in geo-social-ecological contexts. Implicit in each case study is the relationship of social and ecologic systems to each other and to the task of conservation.

Some questions to consider along the way:

- How can marine conservation be framed to protect, restore, and accommodate both a dynamic marine environment and expanding human needs?
- How does systems thinking relate the environmental debt to social well-being and economics?
- How can a focus on "charismatic" iconic species be expanded to encompass biodiversity protection?
- How big, how many, and where should Marine Protected Areas be placed to maximize benefits for marine conservation?
- What lessons can be learned from real-world cases that can be extrapolated to other situations?
- How do 21st century needs fit within 20th century mandates?

Ecosystem approaches to marine conservation focus on issues holistically, rather than repeating fragmented approaches that fail to account for unexpected changes that arise from complex system behavior. Maintaining the status quo through sector-based decisions (e.g., fishing, coastal development, water quality, and energy) needs reconsideration, which requires thinking differently about solutions in order to better fit future policies with procedures. Successful alternatives are being sought (Ch. 13) to protect and sustain biodiversity and the species that both serve society's needs and refresh human minds. As complex systems defy intuitive solutions, it is time to explore new frontiers for marine conservation practice.

Marine conservation itself is now at a crossroads, transitioning from "protection" and sector-based regulations to a wider context. That marine conservation has lagged behind its terrestrial counterpart gives it the potential to be innovative by devising a "best mix" of old ways to new ones, taking historic successes and failures into account. Aware that the oceans are no longer "out of sight, out of mind" to most people, as in the recent past, and armed with "science as a way of knowing," as John Moore put in the title of his seminal book (1993), marine conservation should be capable of avoiding future pitfalls. Humans are not to be faulted for lack of caring. Rather, future progress lies in perceiving connectedness and feedbacks to and from the environment and human societies, leading to the hopeful well-being of both.

REFERENCES

Bormann FH (1990) The global environmental deficit. *BioScience* **40**, 74.
Carson R (1951) *The Sea Around Us*. Oxford University Press, Oxford, UK.
Carson R (1962) *Silent Spring*. Houghton Mifflin Company, Boston, Massachusetts.

Cousteau J-Y, Dumas F (1953) *The Silent World: A Story of Undersea Discovery and Adventure* (ed. Dugan J). Harper Brothers (HarperCollins), New York.

Crutzen PJ, Stoermer EF (2000) The "Anthropocene". *Global Change Newsletter* **41**, 17–18.

Dasmann RF (1968) *A Different Kind of Country*. The MacMillan Company, New York.

Dasmann RF (1971) *No Further Retreat: The Fight to Save Florida*. The MacMillan Company, New York.

Dayton P (2008) Why nature at the University of California? *Transect* **26**(2), 7–14.

Forrester JW (1991) System dynamics and the lessons of 35 years. In *The Systemic Basis of Policy Making in the 1990s* (ed. De Greene KB). MIT Press, Cambridge, Massachusetts.

Franklin JF (1993) Preserving biodiversity: Species, ecosystems, or landscape. *Ecological Applications* **3**, 202–205.

Friedman TL (2008) *Hot, Flat, and Crowded*. Farrar, Straus and Giroux, New York.

Graham M (1999) *The Morning after Earth Day*. The Brookings Institute, Washington, D.C.

Leopold A (1960) *A Sand County Almanac and Sketches Here and There*. Oxford University Press, New York.

Levin SA (2011) Evolution at the ecosystem level: On the evolution of ecosystem patterns. *Contributions to Science* **7**.

Marsh GP (1864) *Man and Nature: Or, Physical Geography as Modified by Human Action*. President and Fellows of Harvard College, Cambridge, Massachusetts.

Moore JA (1993) *Science as a Way of Knowing: The Foundation of Modern Biology*. Harvard University Press, Cambridge, Massachusetts.

Odum WE (1982) Environmental degradation and the tyranny of small decisions. *BioScience* **32**, 728–729.

Osborn F (1947) *Our Plundered Planet*. Little, Brown and Company, Boston, Massachusetts.

Osborn F (1953) *Limits of the Earth*. Little, Brown and Company, Boston, Massachusetts.

Ray GC (1970) Ecology, law, and the "marine revolution." *Biological Conservation* **3**, 7–17.

Ruddiman WF (2005) *Plows, Plagues, and Petroleum: How Humans Took Control of Climate*. Princeton University Press, Princeton, New Jersey.

Steffen W, Crutzen PJ, McNeill JR (2007) The Anthropocene: are humans now overwhelming the great forces of nature. *Ambio* **36**, 614–621.

Sweeney LB, Sterman JD (2000) Bathtub dynamics: initial results of a systems thinking inventory. *System Dynamics Review* **16**(4), 249–286.

Walker B, Salt D (2012) *Resilience Practice: Building Capacity to Absorb Disturbance and Maintain Function*. Island Press, Washington, D.C.

Wilson EO, Peter FM (1988) *Biodiversity*. National Academy Press, Washington, D.C.

CHAPTER 2

MARINE CONSERVATION ISSUES

. . . man has greatly reduced the numbers of all larger marine animals, and consequently indirectly favored the multiplication of the smaller aquatic organisms which entered into their nutriment. This change in the relations of the organic and inorganic matter of the sea must have exercised an influence on the latter. What that influence has been, we cannot say, still less can we predict what it will be hereafter; but its action is not for that reason the less certain.

George Perkins Marsh (1864) *Man and Nature: Or Physical Geography as Modified by Human Action.*

2.1 IGNITING MARINE CONSERVATION CONCERN

Issues attract conservation concern for changes threatening marine biological richness and ecosystem function. Marine ecosystems sustain the largest species on Earth (blue whale), the fastest swimmers (mako shark, marlins), the most bizarre (octopus), most serene (kelp forests, coral reefs), most intriguing (dolphins, orcas, sea horses), most fearsome (great white shark), and most tasty (shellfish, salmon). Depletion of some species, overabundance of others, ill health, and degradation of habitats are *primary* issues for concern, followed by *secondary issues* that illustrate the concentration of human activities impinging on marine ecosystems. *Tertiary issues* focus on fundamental changes in marine ecosystems that are global in scope and propelling marine ecosystems toward unexpected and unintended outcomes. These issues, largely hidden beneath the undulating waves, contrast with a seemingly resilient ocean undergoing change, with major social and economic consequences.

2.2 PRIMARY ISSUES: LOSS OF MARINE BIODIVERSITY

Scientific evidence makes clear that marine ecosystems are losing some of their largest, most charismatic and most productive species. Overabundance of nuisance and toxic species, ill health and pandemics, abnormal behaviors, and deteriorating critical habitats highlight biological changes in the marine environment. This set of issues focuses conservation concern on the ethical and ecological loss of species and marine biological diversity, moving marine environments increasingly toward biological homogenization with consequences for ecosystem integrity and function.

2.2.1 Species extinctions and depletions

Many of the largest and most charismatic marine species, the icons of the oceans, are being depleted worldwide and/or risk extinction. The IUCN 2008 *Red List of Threatened Species* documents about 1500 marine species (Polidoro *et al.*, 2008; Fig. 2.1). Documented extinctions of less obvious species are few (e.g., sediment fauna; Snelgrove *et al.*, 1997), but ramifications could be significant (Emmerson *et al.*, 2001). The ability of scientists to anticipate extinction is elusive, and understanding the causes is a central problem in biology (Ludwig, 1999).

Of the more than 120 species of marine mammals, at least a quarter is presently depleted (Polidoro *et al.*, 2008), and a few have gone extinct. The Steller sea cow (*Hydrodamalis gigas*) was wastefully hunted to extinction 27 years after its discovery in 1741 (Stejneger, 1887; Fig. 2.2); its living Sirenian relatives, the dugongs (*Dugong dugon*) and manatees (*Trichechus* spp.), face potential extinction. Whaling drastically reduced the great whales and recovery of some is slow. The North Atlantic gray whale (*Eschrichtius robustus*) population went extinct in the 18th century, but the relatively rare, iconic blue whale (*Balaenoptera musculus*) appears to be recovering. Right whales (*Eubalaena glacialis*) remain at risk in the North Atlantic and North Pacific (the latter was victim of illegal whaling, Box 3.1), but the Southern Hemisphere population is rapidly recovering (FAO, 2011). The sperm whale (*Physeter macrocephalus*) of *Moby Dick* fame has recovered to 32% of pre-whaling levels (Whitehead, 2002). A declining population of the iconic orca or "killer" whale (*Orcinus orca*) in Washington State is in danger of extinction due to reduced prey and toxic pollution (Wiles, 2004). The Gulf of California porpoise (*Phocoena sinus*) and all river dolphins (family Platanistidae) are greatly depleted and near extinction; the Chinese Yangtze River dolphin (*Lipotes vexillifer*) is considered extinct (Turvey *et al.*, 2007). The seriously depleted Mediterranean (*Monachus monachus*)

Marine Conservation: Science, Policy, and Management, First Edition. G. Carleton Ray and Jerry McCormick-Ray.

and Hawaiian monk (*M. schauinslandi*) seals may be following the now extinct Caribbean monk seal (*M. tropicalis*) that was last reliably sighted in the 1950s near Jamaica. International protection of fur seals (*Callorhinus* and *Arctocephalus* spp.) and sea otters (*Enhydra lutris*) prompted their recovery from near-extinction during the 19th century's fur and oil exploitation, although some are currently declining for unknown reasons (Ch. 7). Atlantic walruses (*Odobenus rosmarus rosmarus*) remain depleted to this day, following a centuries-long period of exploitation; the Pacific subspecies (*O.r. divergens*) recovered following the collapse of Bering Sea whaling, but appears now to be declining (Ch. 7).

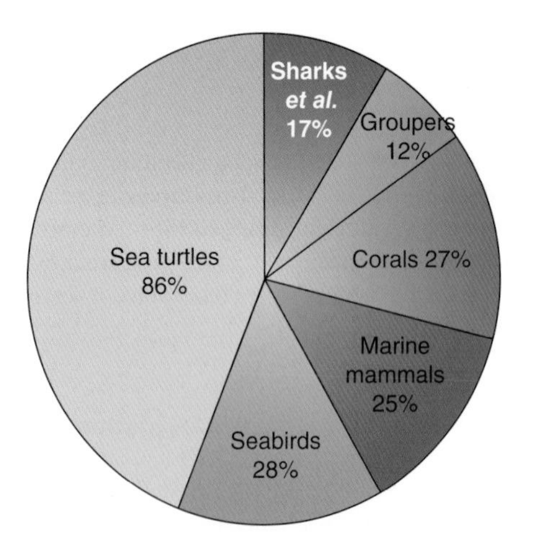

Fig. 2.1 Percent marine species in taxonomic groups are listed in the *Red List of Threatened Species* as Critically Endangered, Endangered, and Vulnerable to extinction (IUCN, 2012). The number of marine species assessed for extinction lags far behind those on land. Percents of Red-Listed species of sharks and rays, groupers, reef-building corals, seabirds, marine mammals, and sea turtles have been calculated from data in Polidoro *et al.* (2008).

Many seabirds are in serious decline. Some 312 species (albatrosses, penguins, puffins, auks, etc.) in 17 families are vulnerable to extinction due to their dual dependence on land and sea, which subjects them to both terrestrial development and marine fishing activities (Ballance, 2007). Of particular concern are petrels and albatrosses that migrate over great ocean distances to feed and return to land to breed. Coastal pollution and climate change increase the threat.

Sea turtles are also threatened with extinction due to dual dependence to breed on sandy beaches and long-life ocean feeding (NRC, 2010a). Their sea migrations cover whole ocean basins (Ch. 8) where fisheries bycatch is an especially serious form of mortality. All seven species of these air-breathing reptiles face direct and indirect human impacts: loggerhead (*Caretta caretta*); green (*Chelonia mydas*); hawksbill (*Eretmochelys imbricata*); Kemp's ridley (*Lepidochelys kempii*); olive ridley (*Lepidochelys olivacea*); leatherback (*Dermochelys coriacea*); and flatback (*Natator depressus*).

Fishes are by far the most diverse and numerous of vertebrates, and the list of threatened and depleted species is long and growing. Many of the largest are targeted by commercial and sports fisheries, and examples are many. The largest and fastest tuna and billfish are depleted as a result of high market

Fig. 2.2 Extinct Steller sea cow (*Hydrodamalis gigas*) as conceived from existing sources. This herbivorous marine mammal, exploited to extinction, was the largest member of the order Sirenia, a group that includes dugongs (*Dugon dugon*) and manatees (*Trichechus* spp.). All four extant species of this group are listed by IUCN as Vulnerable to extinction. Illustration © R. L. Smith, Jr.

Fig. 2.3 Atlantic bluefin tuna (*Thunnus thynnus*), is the largest of tuna (4 m long and weighing up to nearly 900 kg), a prime target for game and longline fishing, and a favorite for sushi. This highly migratory, top predator has declined more than 80% since the 1970s and is listed by IUCN as "Endangered" (IUCN, 2012). Illustration © R. L. Smith, Jr.

value that encourages overfishing. The largest of them, the Atlantic bluefin tuna (*Thunnus thynnus*; Fig. 2.3), is subject to intense fishing pressure and may be on the path to extinction (IUCN, 2012). The depleted great white shark (*Carcharodon carcharias*) has a low reproductive potential (Smith *et al.*, 1998). Other sharks (e.g., scalloped hammerhead (*Sphyrna lewini*), thresher shark (*Alopias vulpinus*), etc.) have declined more than 75% just in the last 15 years (Baum *et al.*, 2003); coastal sand tiger sharks (*Carcharias taurus*) of the Atlantic, Caribbean, and Gulf of Mexico are threatened by poor water quality, fishing, and fisheries bycatch (Meadows, 2009). Sawfishes (*Pristis* spp.) and some species of skates and rays (order Rajiformes) are threatened worldwide due to fisheries bycatch and gill-net fishing. The 5.5 m shallow-water smalltooth sawfish (*P. pectinata*) is in a critical state. Estuarine fishes that travel between land and sea to breed and feed (salmons, sturgeons, anguillid eels) are particularly vulnerable; natural populations of Atlantic salmon (*Salmo salar*) are seriously depleted, as are southerly northwest Pacific populations of five species of salmon (*Oncorhynchus* spp.). Groupers as a whole, especially the tropical West Atlantic Nassau grouper (*Epinephelus striatus*), are much depleted (Ch. 8). Deep-living ocean fish are also especially vulnerable; the slow-growing, late-to-mature orange roughy (*Hoplostethus atlanticus*), which lives below 200 m in the deep sea, is especially vulnerable to fishing, due to a low reproductive rate, and is greatly depleted.

Invertebrates are particularly difficult to assess due to their overwhelming numbers, variety, and lack of high conservation priority. Iconic corals and some shellfish are approaching extinction from a variety of causes. Tropical corals, especially the historically abundant Caribbean reef-building elkhorn (*Acropora palmata*) and staghorn (*A. cervicornis*) corals are now much reduced (Ch. 8). Two rare endemic coral species of the Galápagos Archipelago (*Tubastraea floreana* and *Rhizopsammia wellingtoni*) are declining, presumably due to climate change. Abalone, in particular white (*Haliotis sorenseni*) and black (*Haliotis cracherodii*) abalones of the Northwest Pacific, as well as the perlemoen (*Haliotis midae*) of South Africa (Ch. 11), are prized food items and key members of coastal ecosystems, and face high risk of extinction.

2.2.2 Overabundant species

Conversely to depletion, some species are flourishing beyond expected levels. Overabundance reflects a species' ability to dominate a natural community and become a nuisance or harmful. This situation is often the result of an unnatural (deliberate or accidental) transfer of a species (termed alien, exotic, invasive) into a new location, where it can thrive with few natural controls, and outcompete native species. Even relatively uncommon species in their native environments can prove successful in changed environments or when their predators are absent, reproducing in such massive numbers that they can deplete their own food resources (e.g., sea urchin "barrens"; VanBlaricom and Estes, 1988). And some native species may thrive, for example, the common reed (*Phragmites*) in North American wetlands (Box 2.1). Others may transform ecosystems into monocultures, later to crash and leave barren seascapes.

Increasingly, exotic species introduced by human activities into new locations are transforming environments; coastal waters appear to be particularly vulnerable (Preisler *et al.*, 2009). About 329 marine invasive species are documented for 84% of the world's 232 marine ecoregions (Molnar *et al.*, 2008). Most are benign, but some can transform marine habitats, displace native species, alter community and ecosystem structure through nutrient cycling and sedimentation patterns, damage fisheries, and clog ship hulls and power plants. A particularly severe invasion is that of the lionfish (*Pterois volitans*) in The Bahamas (Ch. 8). The social and economic consequences are major national and global concerns (Vitousek *et al.*, 1996).

Sea plants globally have invaded new environments in unprecedented numbers. A fast-growing exotic alga (*Caulerpa* sp.) is transforming parts of the Mediterranean Sea's benthos into dense, single-species cover; *Caulerpa* has also invaded southern California and Australia. Sea lettuce (*Ulva prolifera*) formed a massive green tide on the popular tourist beaches of Brittany, France (June 2008), that killed dogs, a horse, and a clean-up worker. This alga reappeared in 2011 to rot en masse, releasing massive amounts of hydrogen sulfide gas (H_2S) that killed 36 wild boars (Hu *et al.*, 2010). Sea lettuce also blooms massively in the East China and Yellow seas; another green alga (*Enteromorpha prolifera*) covered 13,000–30,000 km^2 in the Yellow Sea (Sun *et al.*, 2008).

Exotic species can disrupt flows of energy and materials and biogeochemical pathways important to nutrient recycling, thus altering whole ecosystems. Such species may also alter evolutionary routes important to biodiversity, habitat stability, and ecological biomass (Crooks, 2009). For example, the European intertidal common periwinkle (*Littorina littorea*) that invaded New England shores changed mud flats and salt marshes into rocky shores by grazing on stabilizing algae and marsh grass (Williamson, 1996).

Natural phytoplankton blooms described as "red tides" (dinoflagellates; Fig. 2.4), "green films" (cyanobacteria), "brown tides" (chrysophytes), and micro-planktonic algae (dinoflagellates, blue-green algae, diatoms) are increasingly discoloring coastal waters and some are toxic, e.g., harmful algal blooms (HABs; Anderson, 2004). Blooms may remain localized or

Box 2.1 Invasion of common reed (*Phragmites*) in North American wetlands

Randolph M. Chambers
College of William and Mary, Williamsburg, Virginia, USA

Wetlands are often sites of invasion by non-native species of plants (Zedler and Kercher, 2004). In the U.S., one of the most abundant, conspicuous, and notorious invaders is common reed, *Phragmites australis*, a grass that grows in dense stands up to 3–4 m tall, effectively blocking the growth of other potential plant competitors (Meyerson *et al*., 2009). *Phragmites* is considered a "cryptic invader" (Saltonstall, 2002) because an invasive genotype, introduced from Europe around the advent of the Industrial Revolution (Saltonstall *et al*., 2004), can displace the native subspecies. The native typically is a minor component of wetland communities, so the expansive growth of exotic *Phragmites* monocultures eliminates many other species as well.

The negative consequences of invasion and subsequent expansion of *Phragmites* into both tidal and non-tidal wetland environments include loss of wetland biodiversity and shifts in ecosystem structure and function (Chambers *et al*., 1999). With significant assimilation and storage of water and nutrients (Mozdzer and Zieman, 2010), a *Phragmites*-dominated wetland exhibits patterns of energy flow through food webs and nutrient cycling different from that of the native plant community. *Phragmites*-dominated wetlands tend to be drier than those they displace, with consequences for fish use of these habitats (Osgood *et al*., 2006). Further, bird use of *Phragmites* wetlands tends to include more generalist species than wetland specialists.

Exotic *Phragmites* invades open space in wetlands via both seed and rhizome dispersal (McCormick *et al*., 2010). During the 20th century, U.S. *Phragmites* invasion was tied to human activities in wetlands (Bart *et al*., 2006). Shoreline development that extended from uplands to wetland borders created nitrogen-enriched habitat into which *Phragmites* could establish (Silliman and Bertness, 2004). Some researchers suspect that eutrophication of waterways in North America creates conditions that encourage the introduction and spread of *Phragmites*. Interestingly, however, those same nutrient-rich conditions have been cited as a possible cause of *Phragmites* die-back in some parts of Europe.

Owing in part to the "no net loss" policy of wetland mitigation in the U.S., created wetlands provide additional open space for invasion and expansion of *Phragmites*, as *Phragmites* is one of the first species to arrive and thrive in these sites (Havens *et al*., 2003). At present, even undisturbed, pristine wetlands are susceptible to invasion, perhaps due to nitrogen enrichment via atmospheric deposition. *Phragmites* now occurs in wetlands from all 48 of the conterminous United States. Along the middle Atlantic seaboard, a broad invasion "front" appears to be working south through Virginia and the Carolinas. Some wetlands are taken over by *Phragmites* quickly, whereas others seem more resistant to invasion. *Phragmites* has also spread northward into eastern Canada, where yet another invasive species (purple loosestrife, *Lythrum salicaria*) is considered a bigger threat to native biodiversity.

Efforts to stop *Phragmites* expansion using controlled burning, chemical spraying, and physical removal have been largely unsuccessful. Without chronic application of these methods every growing season, *Phragmites* stands tend to recover more quickly than other species. Bio-control methods are under development but run the risk of non-specific actions by the control agents; a number of rhizome-boring insects have been introduced accidentally from Europe, but their North American impacts on *Phragmites* and potentially on other species have not been assessed (Tewksbury *et al*., 2002). From a management perspective, most invaded wetlands cannot be restored to a pre-*Phragmites* condition. Many coastal wetlands that once were restricted to tidal flows have been re-opened, allowing extended flooding by anoxic saltwater sufficient to kill *Phragmites* and encourage re-establishment of natives. However, managers often cannot exercise this option and must accept ecological changes brought on by a new wetland dominant.

Managing non-native *Phragmites* invasion is also complicated by the presence of the native, less aggressive genotype of *Phragmites* that is losing ground. How can *Phragmites* be managed to maintain the native and kill the exotic? Recent research has demonstrated that hybridization between native and non-native genotypes is possible (Meyerson *et al*., 2009), further limiting the available options for control of the abundant invader. Despite the negative impacts of having such an aggressive species in wetlands of North America, non-native stands are significant sinks for nutrients and may be important in mitigation of polluted, non-point source runoff to waterways. Because of rapid, extensive root and rhizome growth, *Phragmites* may also serve to stabilize shorelines in the face of coastal erosion and rising sea level. Additionally, in European and Asian wetlands where it has grown for centuries, *Phragmites* is used for thatching roofs and for paper production; this practice also occurs in portions of the U.S. by immigrants. Because of these positive qualities valued by humans, the new invasion and overabundance of exotic *Phragmites* cannot be easily categorized as either a bane or a blessing. Appropriate policy and management decisions regarding the invasion and spread of *Phragmites* must be considered within site-specific social and ecological contexts.

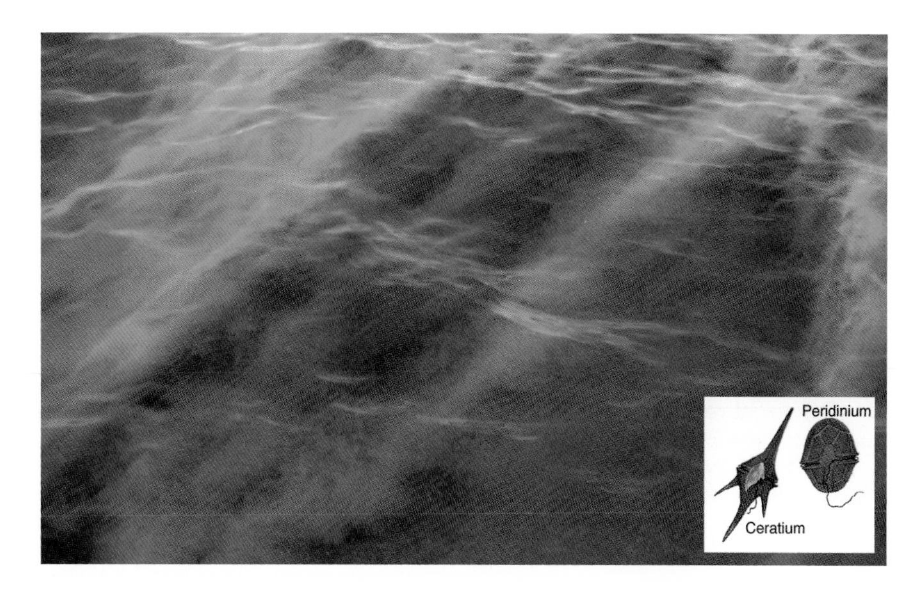

Fig. 2.4 Algae bloom, popularly known as "red tide." Small inserted picture illustrates two microscopic toxic dinoflagellates that cause red tide blooms. Photograph © Ray & McCormick-Ray. Dinoflagellates from U.S. Public Health Service online.

cover thousands of square kilometers for weeks; some occur at the same time and place each year and others are unpredictable, as for example: *Alexandrium fundyense* in the Gulf of Maine (Anderson *et al.*, 2005); *Karenia brevis* in the Gulf of Mexico (Steidinger *et al.*, 1998; Vargo, 2009); cyanobacteria in the Baltic Sea (Kononen, 1992; Bianchi *et al.*, 2000); and others (Pitcher and Pillar, 2010). HABs produce toxins, noxious gases, or anoxic water that kill marine life, and are becoming more frequent. Some produce a neurological biotoxin (domoic acid) that causes amnesic shellfish poisoning that affects people and a variety of sea life from fish to blue whales (Grant *et al.*, 2010). Some dinoflagellate HABs (e.g., *Alexandrium* sp., *Gymnodinium* sp., *Pyrodinium* spp., etc.) produce saxitoxin, also a neurotoxin that caused massive humpback whale mortality in 1987 (Geraci *et al.*, 1989). A toxic dinoflagellate (*Noctiluca scintillans*) bloom stretched more than 20 miles along the California coast in 1995 (Anderson, 2004); another killed more than 1600 New Zealand sea lion pups (*Phocarctos hookeri*) at Auckland Island in 1998. And the first known toxic dinoflagellate bloom (*Gymnodinium* sp.) in the Arabian Sea in 1999 killed fish, closed aquaculture facilities, and caused significant economic impact (Heil *et al.*, 2001). Toxic algae not only affect sea life, but also alter marine food-chain structure and habitats, and are linked to public health, seafood safety, and aquaculture, causing human deaths and illnesses and threatening coastal areas (Stommel and Watters, 2004).

Dense aggregations of jellies ("jellyfish") are increasing in severity and frequency worldwide (Parsons and Lalli, 2002; Graham and Bayha, 2007; Richardson *et al.*, 2009). Overabundant jelly animals (pelagic cnidarians, ctenophores) may cause severe threats to ecosystem function on massive scales (Graham *et al.*, 2003), with most notable blooms occurring in the Far East and East Asian marginal seas (Uye, 2008; Dong *et al.*, 2010). Jellies are a natural feature of healthy pelagic ecosystems; in the Far East three species (*Aurelia aurita*, *Cyanea nozakii*, *Nemopilema nomurai*) naturally form large blooms. However, the population of the giant jellyfish *N. nomurai* (2 m maximum bell diameter, 200 kg wet weight) in Southeast Asia increased 250% between 2000 and 2003 with 300–500 million medusae being observed in 2005 (Uye, 2008). In Japan, moon jellyfish (*Aurelia* sp.) clog power plant intake lines (Purcell, 2005). The American comb jelly (*Mnemiopsis leidyi*) that invaded the Black Sea bloomed in the late 1980s to reach concentrations of 300–500 animals per m^3, with a biomass in some regions of over a billion tons (Mills, 2001); it also spread into other European seas, including the central Baltic, causing concern for fisheries. Overfishing, eutrophication, climate change, translocation, and habitat modification may contribute to blooms of jellies. Such abundance reduces food for fishes, alters food webs, and collapses fisheries to impact fishermen and national economies.

2.2.3 Ill health

Diseases of sea life, expressed as lesions, deformities, and infections, are collectively referred to as "ill health." Region-wide epidemic diseases of a wide variety of taxa have caused massive die-offs. Such phenomena appear to be increasingly frequent globally (Harvell *et al.*, 1999). Examples are numerous. Corals worldwide exhibit "bleaching" due to loss of zooxanthellae (Fig. 2.5a). Caribbean corals exhibit microbial infections in epidemic proportions described as "white pox," "black line," and fungal diseases (Goreau *et al.*, 1998; Fig. 2.5b). Reef-building Caribbean corals are also infected by the bacterium *Vibrio* sp. (Cervino *et al.*, 2004). High mortalities of Caribbean sea fans (*Gorgonia ventalina*) caused by a worldwide terrestrial fungus (*Aspergillus sydowii*) carried on airborne dust from Africa (Weir-Brush *et al.*, 2004) were related to ocean warming and nutrient enrichment (Ellner *et al.*, 2007). Sponges worldwide are exhibiting significantly more diseases, with decimated populations throughout the Mediterranean and Caribbean seas (Webster, 2007). A "wasting disease" caused by a slime mold (*Labyrinthula macrocystis*) extirpated North Atlantic eelgrass (*Zostera marina*) in 1931–2, and 10 other species are

Fig. 2.5 Examples of diseased marine species. (a) Bleached fire coral (*Millepora* sp.). Photograph © Ray & McCormick-Ray. (b) Blackline coral disease (*Montastrea* sp.). Photograph © Ray & McCormick-Ray. (c) Green sea turtle with viral tumors, fibropapillomatosis, Andros Island, Bahamas. Photograph © Karen Bjorndal. (d) California sea lion (*Zalophus californianus*) with poxvirus (parapox). Reproduced with permission of The Marine Mammal Center, Sausalito, California. Disease patterns in the ocean are diverse, making it difficult to discern a clear increasing trend (Lafferty *et al.*, 2004).

at elevated risk of extinction with three more qualifying as endangered (Short *et al.*, 2011). Only recently have some species shown signs of slow recovery (Godet *et al.*, 2008).

Vertebrates are also affected. Fishes exhibit a wide variety of well-studied diseases, some caused by humans (Noga, 2000). Sea turtles are infected by a herpes virus that causes multiple cutaneous masses called fibropapillomatosis, associated with heavily polluted coastal areas, areas of high human density, or where agricultural runoff and/or biotoxin-producing algae occur (Fig. 2.5c; Aguirre and Lutz, 2004). Marine mammals, e.g., seals and polar bears, also exhibit epidemic diseases, including a highly contagious, incurable, and often deadly disease called canine distemper virus (CDV) caused by a morbillivirus (de Swart *et al.*, 1995), which is a leading cause of death in unvaccinated dogs. In 1987, many freshwater Baikal seals (*Phoca sibirica*) died from CDV. Other significant morbillivirus species include dolphin morbillivirus (DMV), porpoise morbillivirus (PMV; Saliki *et al.*, 2002), and in pinnipeds, phocine distemper virus (PDV). PDV killed more than 23,000 harbor seals (*Phoca vitulina*) in Europe in 1988 and 30,000 in 2002 (Härkönen *et al.*, 2006) and has been reported for sea otters (*Enhydra lutris*) in the North Pacific Ocean (Goldstein *et al.*, 2009). DMV and PMV are now considered the same species, renamed cetacean morbillivirus (CMV). Viruses have also caused mortalities among striped dolphins (*Stenella coeruleoalba*), endangered Mediterranean monk seals (*Monachus monachus*), and fin whales (*Balaenoptera physalus*). Viral infections and pollutants were implicated in the deaths off U.S. mid-Atlantic shores of more than 700 bottlenose dolphins (*Tursiops truncatus*) in 1987–8, and in excess of 500 harbor seals in New England waters in 1979–80. This massive mortality caused by an influenza virus carried by birds killed 3 to 5% of the 10,000 to 14,000 seals along the New England coast (Geraci *et al.*, 1982). Viruses also infect California sea lions (Fig. 2.5d).

Ill health brings into question: what is normalcy? Are diseases in the ocean increasing (Lafferty *et al.*, 2004), are they new, or are they re-emergent (Harvell *et al.*, 1999)? Much remains to be known about the "normal state" of health for most marine species. Nevertheless, the magnitude of such phenomena and extent are difficult to ignore.

2.2.4 Abnormal behaviors

Although normal behaviors of most marine species are poorly known, changes in species distributions and behavior such as altered breeding times and places are being increasingly reported. For example, some migratory waterfowl that normally feed on shallow-water vegetation consume farm crop residues and no longer migrate. Expanding numbers of gulls opportunistically feed in garbage dumps and around fishing boats. California sea lions are choosing docks and piers rather than natural shores to rest, and Florida manatees seek the warm-water effluents of power plants during cold winters. Some cetaceans are hybridizing with other species (Zornetzer and Duffield, 2003), a phenomenon apparently unique among mammals (Willis *et al.*, 2004), but that may be normal for Cetacea.

Increasing interactions with humans are proving to be aggressive, mutualistic, positive, or learned. Shark attacks on humans are not common, but raise much public concern and speculation. Sharks' decreasing numbers do not translate into reduced attacks on humans, possibly because of increased numbers of swimmers and divers in nearshore waters (West, 2011). Sharks are not alone; dolphin (*Tursiops* sp.) interactions with humans in Monkey Mia in western Australia have turned aggressive (Orams *et al.*, 1996; Orams, 1997), betraying the illusion of their friendly behavior toward humans.

2.2.5 Critical habitat degradation

Marine life depends on habitats, which are increasingly being modified, fragmented, and lost. Such changes worldwide are seriously threatening many species (Sih *et al.*, 2000). At the interface of land and sea, coastal habitats include salt marshes (Box 2.2), estuaries (Ch. 6), mangroves, reefs, and seagrasses (Ch. 7) that are particularly under severe threat worldwide, being increasingly exposed to poor water quality and erosion. Islands and sandy beaches are disappearing, exacerbated by interactions between human activities, tsunamis, hurricanes, and global warming. Most notably in the Indian Ocean, the 1200 islands and atolls composing the island nation of the Maldives are threatened by inundation due to sea-level rise. Loss of coastal habitats and islands is reducing critical ecosystem services that provide social benefits (Barbier *et al.*, 2011).

Estuaries are among the most productive of all ecosystems and vital to fisheries yet face worldwide decline (Lotze *et al.*, 2006). Deteriorating estuarine health is commonly due to poor water quality, depletion of native species (e.g., shellfish, estuarine fishes), and monocultures of invasive species. In the U.S., estuaries are typically over-enriched with nutrients (Bricker *et al.*, 2008). Once diverse and productive, estuaries and coastal seas have lost more than 90% of their formerly important species' populations and more than 65% of their associated seagrass and wetland habitats.

Seagrasses that provide key ecological services are in a global crisis (Orth *et al.*, 2006; Fourqurean *et al.*, 2012). An estimated 29% of their known global areal extent has disappeared since being first recorded in 1879, and losses have accelerated worldwide since 1980 at an annual rate of 110 km^2 (Waycott *et al.*, 2009). Fourteen percent of all seagrass species are at risk of extinction, with nearly one-quarter (15 species) in serious trouble (Short and Wyllie-Echeverria, 1996; Short *et al.*, 2011). Loss of seagrass habitat is attributable to a broad spectrum of anthropogenic and natural interactions—disease, destructive fishing practices such as dredging, nutrient pollution, natural dieback, etc.—affecting dependent fishes, invertebrates, waterfowl, dugongs, manatees, green turtles, and others.

Hard-bottom reefs (oyster, coral) are globally threatened. Temperate oyster reefs have been intensively depleted over a long period, those remaining being only vestiges of their former extents (Ch. 6; Beck *et al.*, 2009). Tropical coral reefs are threatened worldwide (Ch. 8; Box 2.3).

2.3 SECONDARY ISSUES: HUMAN ACTIVITIES

Secondary issues focus on human activities as agents of coastal change. Thirty-eight percent of the world's 6.5 billion people occupy only 7.6% of Earth's total land area—the narrow coastal fringe (UNEP/GPA, 2006). Fishing is the major agent of change, followed by chemical pollution, eutrophication, and invasive species (NRC, 1995). Such resource extraction, additions of novel substances, and physical alterations have historical roots imbedded in the social fabric of the global society. Expanding this level of coastal impact is the physical alteration of watersheds, new dam construction, wetland filling and/or drainage, and coastal armoring. These human activities act cumulatively over time to physically and functionally alter the coastal system on which so many species and a large portion of the global economy depend.

2.3.1 Extractions: over-harvesting natural coastal resources

Human civilizations extract many benefits from the oceans. With increasing technological advances driven by expanding human needs with increasing intensity, activities and impacts are moving ever deeper into the unknown realm of deep-ocean basins.

2.3.1.1 Overfishing

The limits of ocean bounty have been reached, and in some cases exceeded. Whaling drastically reduced the great whales when the International Whaling Commission stopped it in 1982 (Fig. 2.6; Ch. 3). But the seas continued to bring hope of meeting global food shortages (Idyll, 1978), and global fisheries in the 1950s extracted <20 million metric tons (mt) annually. By the late 1980s, expanding fisheries reached maximum global capacity (Pauly, 2008) and have since been declining (Fig. 2.7a); by 2004, 366 fisheries had collapsed, nearly one of four (Mullon

Box 2.2 Salt marshes under global siege

Brian R. Silliman
Division of Marine Science and Conservation, Nicholas School of the Environment, Duke University, North Carolina, USA

Salt marshes are hugely productive intertidal grasslands that form in low-energy, wave-protected shorelines along continental margins. For over 8000 years, humans have benefited greatly from salt marshes and relied on them for direct provisioning of materials (Davy *et al.*, 2009). For example, starting roughly 2000 years ago and to this day, marsh grasses are still purposely planted and protected by the Dutch so as to act as buffers against storm surges and as natural-engineering tools to reclaim shallow seas and build up sea barriers to facilitate greater human reclamation and development (Davy *et al.*, 2009). Indeed, over 40% of the land in present-day Netherlands was once estuarine intertidal mud habitat and was reclaimed with the help of the engineering services of salt marsh plants (Davy *et al.*, 2009). Besides this poignant service, salt marshes provide many other valuable benefits to humans, including water filtration, buffering of storm waves and surges, carbon sequestration and burial, critical habitat for both adult and juvenile fishes and birds, grasses for building houses and baskets, land for grazing ungulates and development, and for scientific and educational opportunities.

Despite this list of abundant and valuable critical services, salt marshes are under global siege from an impressive portfolio of human-generated threats (Gedan *et al.*, 2009). Salt marsh coverage, as well as the structure of these ecosystems, continues to deteriorate drastically due to human-induced changes. The critical ecosystem services these systems support are likewise endangered. No longer can marshes be viewed in scientific, conservation, social, and political circles as one of the most resilient and resistant ecological communities. And no longer can they be championed as systems that can and should be used to buffer human impacts (e.g., absorption of nutrients in wastewater and terrestrial runoff). These systems are in desperate need of protection from human influence. Most of these threats are currently underestimated or even overlooked by coastal conservation managers because marsh preservation practitioners have historically worried most about stopping reclamation efforts (Silliman *et al.*, 2009a). Current threats to salt marshes include human-precipitated species invasions, small- and large-scale eutrophication and accompanying plant species declines, runaway grazing by snails, geese, crabs, and nutria that denude vegetated marsh substrate over vast extents, climate-change induced effects including sea-level rise, increasing air and sea surface temperatures, increasing CO_2 concentrations, altered hydrologic regimes, and a wide range of pollutants, including nutrients, synthetic hormones, metals, organics, and pesticides (Silliman *et al.*, 2009a).

Already about 50% of the value of services marshes provide have been lost as salt marsh ecosystems have been degraded or lost (Gedan *et al.*, 2009). On some coasts, such as the West Coast of the U.S., this number rises above 90%, for both marsh area and their services (Bromberg and Silliman, 2009). Without proper conservation action, it is now predicted that this key coastal community will become a non-significant, ecosystem-service-generating habitat in <100 years (Silliman *et al.*, 2009a). Key to saving salt marsh ecosystems and their services is recognizing a wide variety of threats and abating them through up-to-date conservation strategies (Silliman *et al.*, 2009b) and providing justification of these conservation measures by both describing and valuing all of the critical services marshes provide.

One of the most important and effective acts that conservation practitioners can begin to do to ensure the long-term protection and persistence of salt marsh habitats is to champion the use of Marine Protected Areas in marsh management. This has been done widely for reefs, kelps, mangroves, and seagrasses, but not marshes and their surrounding waters. These protected areas must: (i) include associated marine habitats, such as seagrass beds and oyster reefs; (ii) incorporate extensive areas of undisturbed terrestrial border to buffer marshes from excessive eutrophication via runoff and allow for their landward migration as sea-level rises; (iii) account for the inclusion of positive interactions (Halpern *et al.*, 2007) at all levels of biological association (e.g., between species—trophic cascades; and across ecosystems—nursery benefits); and (iv) be large, numerous, and appropriately spaced (See Halpern *et al.*, 2007 for discussion). Around the world, coral and rocky reef conservation practitioners and scientists lead the field of marine conservation in this effort. Salt marsh conservationists and ecologists are far behind this work and, thus, should look to these fields for lessons-learned and guidance when establishing Marine Protected Areas for temperate coastal areas whose intertidal zone is dominated by salt marshes. Because of the conservation prestige associated with the designation of a site as a Marine Protected Area, using this method as a means to preserve marshes will also raise public awareness as to the critical role marshes play in the ecology and economy of local human communities.

Box 2.3 People and coral reefs

John C. Ogden
Florida Institute of Oceanography, St. Petersburg, Florida, USA

The complex and diverse assemblages of organisms that compose shallow coral reef ecosystems (<30 m deep) cover a total area of the global tropical ocean approximately the size of the state of Nevada. Coral reefs are ecologically linked to other coastal ecosystems, notably seagrasses, mangroves, and the open ocean, considerably extending their total area and influence. About 90% of shallow coral reefs occur in the Indo-Pacific and most of the remainder is in the Wider Caribbean, including Florida, Bermuda, and the Gulf of Mexico. Corals also form diverse assemblages in deep, cold waters at all latitudes, generally encrusting on the rocky surfaces of deep-sea mounts, but they do not form reefs there. Advances in fishing technology have opened seamounts to trawling and there is increasing international political action to protect deep-water corals.

As the marine ecosystem with the most species, shallow coral reefs are often compared to tropical rain forests. Reefs contain far fewer described plant and animal species than tropical forests, but support almost twice as many phyla (the basic forms of life) as all terrestrial ecosystems combined. This rich biodiversity and the brilliant display of form and color make coral reefs attractive to scientists who explore the origin, maintenance, and functioning of biological diversity. These same characteristics are also attractive to people who draw joy, fascination, and inspiration from the myriad life forms arrayed in intricate patterns in warm, transparent waters.

Coral reefs provide significant and economically important ecosystem services to human societies and are critical to future economic development of ecotourism in many developing countries. Fishing on coral reefs contributes approximately 10% of the world's estimated annual fishery production of more than 80 million metric tons and this protein is mostly locally consumed by coastal communities. As governments look for economic development opportunities, tourism is potentially more economically important than fishing in most places, but fishing will remain an important cultural activity. In an era of projected climate warming, a healthy, robust reef, with growth pacing sea-level rise, protects shorelines and productive coastal lagoons from the erosion of open-ocean waves, storms, and occasional tsunamis. Finally, a significant scientific effort is devoted to prospecting for new drugs and chemical compounds, active, for example, in cancer and AIDS therapies.

Modern coral reefs are the largest biogenic structures in nature. They originated millions of years ago, but reefs of today represent only approximately 6000–10,000 years of post-Pleistocene growth, during the most recent period of sea-level rise. Reefs are essentially cemented piles of coral rock, formed by the limestone skeletons of stony corals and other carbonate-producing organisms (notably calcareous algae), and sand. The living part of the reef is a thin veneer over the reef surface. Corals live in all seas, but the hermatypic, or reef-building, corals are restricted to shallow (<100 m) waters >20°C. Water temperatures much higher than 30°C can be lethal. Hermatypic corals are distinguished by a symbiotic relationship with single-celled, photosynthetic algae called zooxanthellae, which live within the cells of the individual polyps that make up the coral colony and give corals their muted colors of green, red, brown, and yellow. Most corals actively feed by capturing zooplankton with specialized stinging cells on their tentacles, which provides their zooxanthellae with essential nutrients. In return, the coral receives oxygen and photosynthetic products from the algal cells. Healthy, growing coral reefs require high light, wave action, low nutrient levels, and a narrow range of temperature and salinity in order to survive.

Coral reefs have evolved in association with episodic natural disturbances such as tropical storms, sedimentation from terrestrial runoff, and occasional harmful variations in seawater temperature and salinity. However, with over one-half of the world's population now living in the world's coastal zones, coral reefs are increasingly being destroyed by chronic human disturbances such as poor land-use practices, including coastal deforestation; sediment runoff; pollution; diseases; the direct and indirect effects of fishing and aquaculture; destruction of habitat by mining, dredging, and coral collecting; and global climate change.

Coral diseases have been particularly damaging. For example, the abundant and ubiquitous elkhorn coral (*Acropora palmata*), a major Caribbean reef builder, was so dramatically reduced by white band disease over the past five decades that it was listed as threatened under the U.S. *Endangered Species Act*. The impact of this and other diseases has stimulated a major research effort to understand the role of the many bacteria associated with coral mucus. Some coral diseases have been linked to human enteric bacteria from sewage, but these bacteria are ubiquitous in tropical seas, and the conditions that cause some species of bacteria to become pathogenic are not yet fully understood.

Coral bleaching is a special case of disturbance. When corals are subjected to unusually elevated or prolonged periods of seasonal high water temperatures (>30°C), often in concert with other stresses such as sedimentation, they expel their symbiotic algae and bleach, turning pale or white. Bleaching is not necessarily fatal, but can kill corals if it persists. While coral bleaching has been known for a long time, since 1990 bouts of bleaching have

(Continued)

increased in severity and global extent, notably in 1998, 2005, and 2010, reducing many formerly vibrant coral reefs to algae-covered rubble piles. These massive and globally coherent temperature events coincide with increasing scientific certainty that the atmosphere and the oceans are warming and that the cause is the well-documented increase in atmospheric carbon dioxide (CO_2), a key greenhouse gas, originating in the steadily increasing combustion of fossil fuels since the start of the mid-19th century Industrial Revolution.

Increased atmospheric CO_2 has another insidious impact on the global oceans. The rising level of CO_2 in the atmosphere has caused increased absorption of CO_2 in the oceans, where it has already reduced the pH by about 0.1 units (i.e., increased ocean acidity by about 30%). This, in turn, reduces the available level of carbonate ion required by the myriad organisms which build skeletal structures from calcium carbonate. Based on models and archaeological records, by the end of this century the amount of CO_2 in the oceans will exceed that of any time in the last 300 million years. Ocean chemistry is complex and we cannot predict at this point what will happen, but what we know so far makes changes in global energy policy ever more urgent.

Taken in sum, human disturbances have already extracted a heavy toll from global coral reefs. Recent estimates of the status of reefs suggest that by 1992, about 10% of the world's coral reefs had been severely damaged; by 2000 this increased to about 25%; and, by 2050, it has been projected that three-quarters of the world's coral reefs may be damaged or destroyed. If reefs are to be saved, an international effort must begin the difficult task of defining a global energy policy, reducing greenhouse gas emissions by phasing out fossil fuels, and implementing alternative energy sources. In addition, within the defined large marine ecosystems of the world it will be critical to implement international management and governance of coral reefs at the geographic scale of the ecosystem processes that sustain them.

Regionally and locally, disturbances to reefs may be identified, and reasonable, almost common-sense, actions for controlling such disturbances as pollution and land runoff may be implemented. Active reef restoration by transplanting corals from coral nurseries is being explored and may prove to be useful in some cases. An ecosystem-based management (EBM) approach combined with modern geographic information systems and modeling holds great promise. Finally, large geographic-scale ocean-use planning incorporating large management and conservation areas, far larger than the relatively small Marine Protected Areas already in place, are required for integrated protection and sustainable human use of these remarkable structures.

In this first century of what some are calling the Anthropocene, the human footprint has extended to almost all parts of the global ocean. Similar to the canary in the mine that warned early Welsh coal miners of the presence of poisonous gases, the sharp decline of coral reefs may be our warning that "business as usual" must, through policy and human behavior, be changed to new models that stress living sustainably with our environment, whose services sustain the biodiversity upon which our future lives will depend.

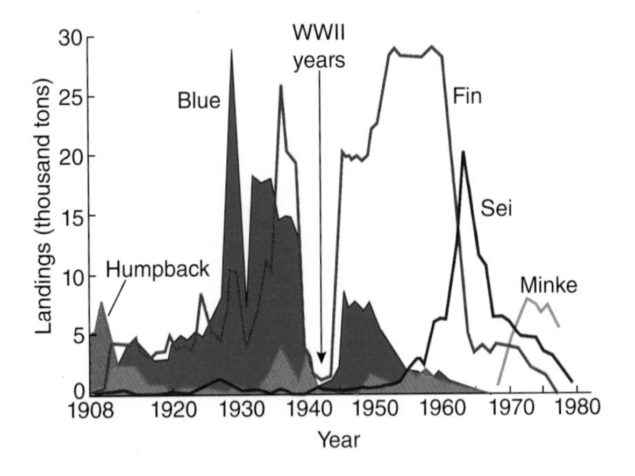

Fig. 2.6 Landings of Southern Ocean baleen whales by commercial whaling, 1908–82. The two largest whales (humpback, *Megaptera novaeangliae*, 11.5–15.0 m; blue, *Balaenoptera musculus*, 24–27 m) were depleted first, then smaller whales (fin, *B. physalus*, 18–22 m; sei, *B. borealis*, 12–16 m; minke, *B. acutorostrata*, 7–10 m). World War II interrupted whaling and some whale populations temporarily increased. The moratorium signed in 1982 by the International Whaling Commission prohibited further whaling, and some populations are rebounding. From Jennings *et al.* (2001, p. 14). Reproduced with permission of John Wiley & Sons.

et al., 2005). And due to illegal, unreported, and unregulated (IUU) fishing and related activities, often encouraged by corrupt practices, fisheries statistics underestimate total removal (FAO, 2012), threatening efforts to secure long-term sustainable fisheries (Fig. 2.8). The ever-growing demand for fish drives many fisheries to exploit natural populations beyond the capacity of fishes to replenish themselves and for ecosystems to recover from loss. Most fisheries are fully exploited (Fig. 2.9), overfishing continues, collapses are accelerating (Worm *et al.*, 2006), and major ecological disturbances are evident (Pauly and Christensen, 1995; Jackson *et al.*, 2001). Few areas of the world remain unexploited, and few are protected from fishing; recovery when and if it occurs can be slow. To offset the declining capture fisheries, increasing aquaculture production aims to fill the ever-rising need for more food (Fig. 2.10; FAO, 2010).

The drama in fisheries over-exploitation has historical roots, exacerbated by improved efficiencies. Fisheries have moved from one location to another in a *slash-and-burn* pattern of exploitation, exploiting areas where the harvestable biomass is greatest, fishes are most accessible, or both (Law, 2000). North Atlantic fish depletions began in the late 1800s with plaice (*Pleuronectes platessa*, a European flatfish) followed by Atlantic herring (*Clupea harengus*), Atlantic cod (*Gadus morhua*), and other fisheries in the 1900s (Holt, 1969). Four centuries of

Fig. 2.7 Major fisheries extractions from global oceans. (a) World fisheries landings increase since 1950. Total marine fish landings (right ordinate; data from FAO, 2012) and individual landings (left ordinate) for five high seas marine fish species that dominate global landings: Peruvian anchovy (*Engraulis ringens*); Alaskan pollock (*Theragra chalcogramma*); Atlantic herring (*Clupea harengus*); Atlantic cod (*Gadus morhua*); Japanese pilchard (*Sardinops melanostictus*). Based on data in FAO Fisheries and Aquaculture Information and Statistics Service—06/08/2012 online (2012): www.fao.org/fishery/statistics/global-production/en. (b) Tons of albacore tuna (*Thunnus alalunga*) being loaded from a factory fishing ship (arrow) into truck in Manta, Ecuador (May 2012), where the tuna industry is the largest city employer. Photograph © Charles Clarkson.

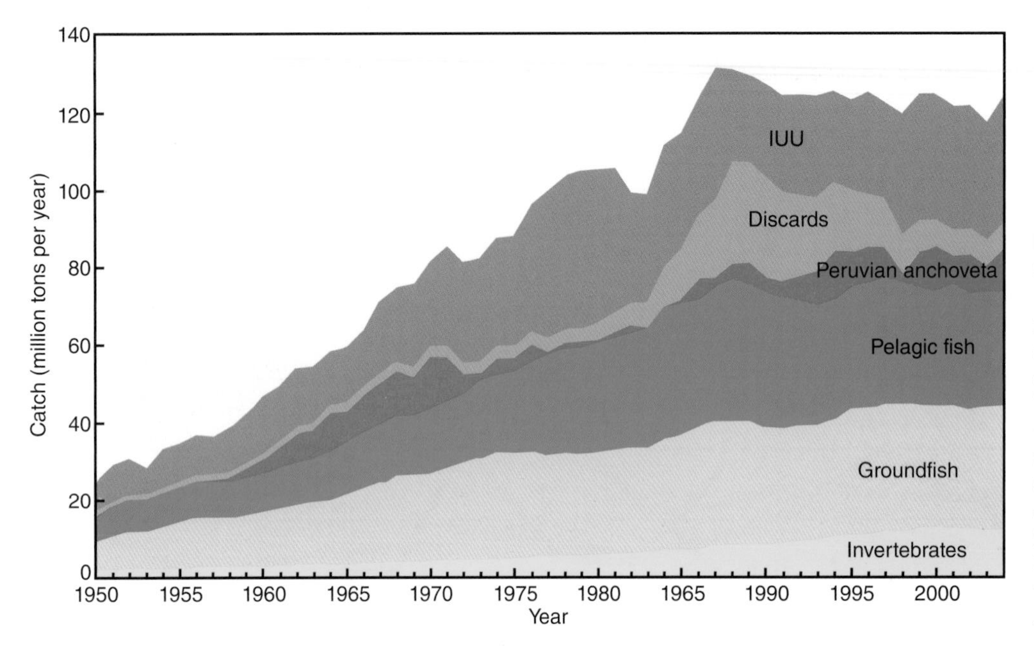

Fig. 2.8 Total extraction of marine fisheries from 1950–2004, which accounts for over-reporting by China, discards of bycatch and other discards, and Illegal Unreported or Unregulated (IUU) extractions. Such IUU fishing occurs in virtually all capture fisheries, valued at an estimated $10–23.5 billion per year (FAO, 2010). From Pauly 2008. Reproduced with permission from Sea Around Us.

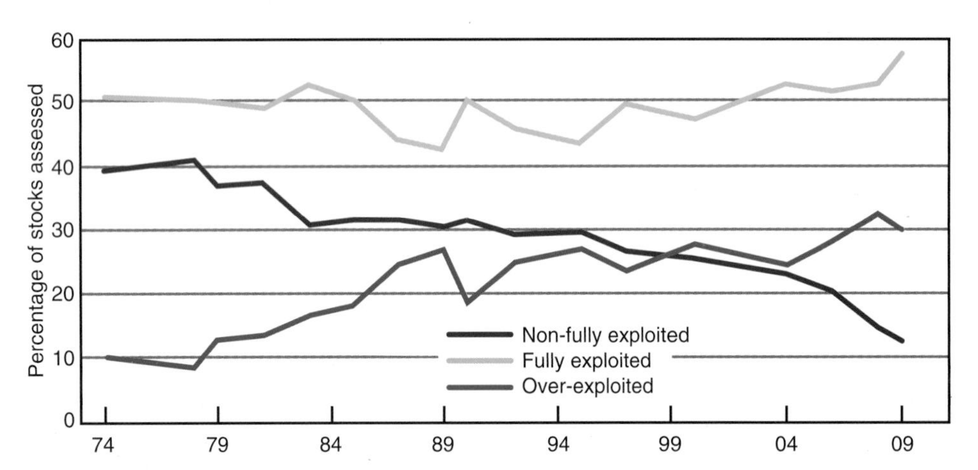

Fig. 2.9 Trends in world marine fisheries indicate that ocean fisheries production is approaching global capacity. Global assessment of fisheries stocks since 1974 indicated that the proportion of non-fully exploited stocks gradually decreased as the percentage of over-exploited stocks increased, especially late 1970s and 1980s. From Food and Agriculture Organization of the United Nations (2012). *The State of World Marine Fishery Resources*. FAO Fisheries and Aquaculture Department. Food and Agriculture Organization of the United Nations, Rome. www.fao.org/docrep/016/i2727e/i2727e00.htm

overfishing Atlantic cod that intensified after World War II forced its collapse in the 1990s and devastated coastal economies (Hutchings, 2005). With technological improvements (Thurstan *et al.*, 2010) and larger, more numerous vessels with greater efficiency after the War, fishing efforts rapidly expanded. Today, about 4.3 million large fishing vessels and increasing numbers of small, unaccounted fishing boats are actively pursuing fish, accompanied by much under- and over-reporting (FAO, 2010). These vessels are being equipped with efficient fish-finding devices and navigational aids that pinpoint fishing grounds. Lloyd's 2005 register of ships greater than 100 tons accounted for about 1200, with increasing numbers of flag-state registrations being listed as *unknown* (Gianni and Simpson, 2005), and contributing increasingly to widespread and profitable IUU fishing impacts (Flothmann *et al.*, 2010). Vessels are using 2 km-long drift nets and longlines that extend tens of kilometers with thousands of hooks that fish indiscriminately. Factory ships meet fishing vessels at sea to efficiently process loads of fish for markets (Fig. 2.7b). This massive removal of more fish is accompanied by increasing numbers of countries failing to report landings, inappropriately reporting catches, and/or engaging in illegal

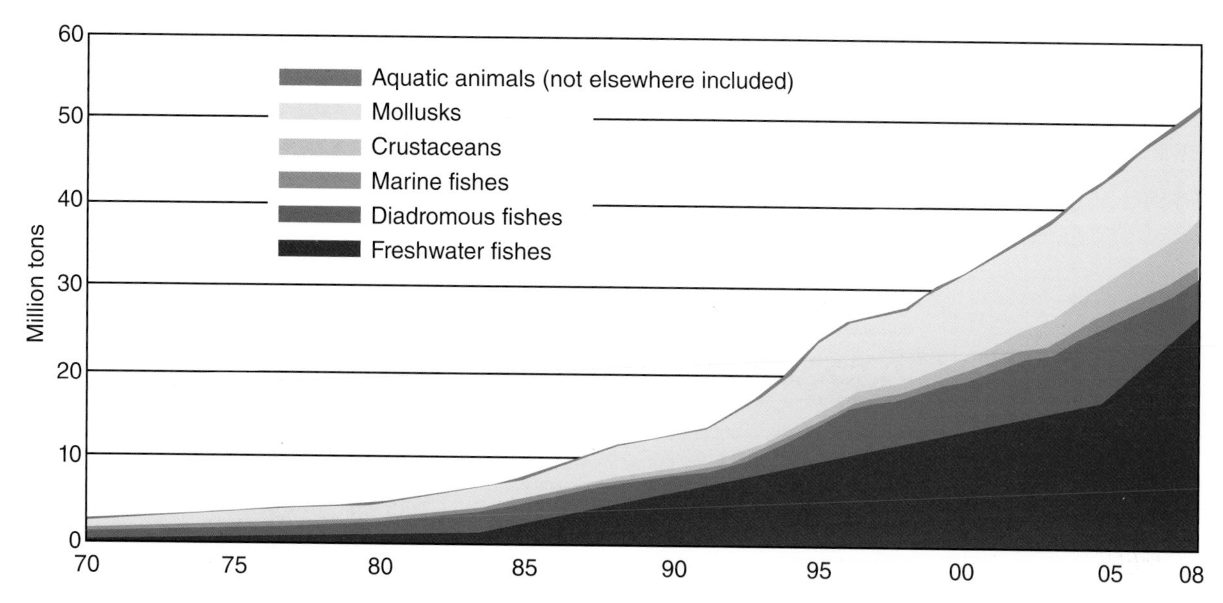

Fig. 2.10 World aquaculture production reflects increasing market demand since 1970 for major fisheries groups, especially mollusks (oysters, mussels, clams) and crustacea (shrimp, lobster, crabs), which supplements capture fisheries and accounts for almost half of total food fish. Aquaculture is the fastest-growing animal-food-producing sector and a major source of income and livelihood for hundreds of millions of people. From Food and Agriculture Organization of the United Nations (2012). *The State of World Marine Fishery Resources*. FAO Fisheries and Aquaculture Department. Food and Agriculture Organization of the United Nations, Rome. www.fao.org/docrep/016/i2727e/i2727e00.htm

trade. Fishing now extends into the deep sea where impacts are poorly known (Davies *et al.*, 2007). Ocean fisheries production is approaching global capacity (Garcia *et al.*, 2005; NRC, 2006; FAO, 2012). Furthermore, highly efficient fishing methods have greatly increased captures of non-targeted species (bycatch; discards), including cetaceans, seals, sea turtles, sharks, tuna, sea birds, and juvenile fishes. The amount of bycatch, estimated at about 20% of the total, is difficult to estimate due to lack of reporting (Fig. 2.8); discards in tropical shrimp trawl fisheries alone may be orders of magnitude greater than the retained catch (Zeller and Pauly, 2005).

Benthic trawling is an expansive and variably damaging operation that adds another dimension to fishery impacts (Gray *et al.*, 2006). Extensive trawling occurs on the northeast Atlantic shelf (Kaiser *et al.*, 1998), southern North Sea (Rijnsdorp *et al.*, 1998), and elsewhere; some areas are completely dredged three to four times per year. Trawling and dredging can reduce biodiversity, cause serious ecological impacts (Thrush and Dayton, 2002), and can restructure benthic environments (NRC, 2002a). Dredges are used in estuarine waters to harvest clams, oysters, conch, and crabs, and in offshore waters to harvest sea scallops, surf calms, and quahog clams. Trawling grounds of 24 countries encompass more than 57% of global continental-shelf area, covering 8.8 million km^2 (Burke *et al.*, 2001). The effects are increased turbidity, alteration of benthic habitats, crushing, burying, and smothering biota, and subjecting non-targeted sessile species to predation. In comparison to forest clear-cutting, modern trawling is globally 150 times more expansive (Table 2.1).

Marine ecosystems are further affected by extractions of targeted species for aquarium and ornamental species trade, a $25 billion-per-year worldwide industry that is growing at 14% per year (Padilla and Williams, 2004). This trade involves 1471 species of fish and more than 500 invertebrate species, including 140 stony corals and millions of individuals. Over-harvesting of precious corals follows a sequence similar to fisheries: i.e., exploration, discovery, exploitation, depletion (Wabnitz *et al.*, 2003). And while trade is legitimate, illegal trade is a growing ecological and management concern. Between 2003 to 2006, for example, up to 1.05 million Caribbean queen conch (*Strombus gigas*), a marine mollusk valued for its shell in the ornamental trade, were illegally harvested, with a conservative estimate of more than $2.6 million in value (Daves, 2009). This adds to the total impact of biomass removal, with consequences for biodiversity and marine ecosystem function (Donaldson *et al.*, 2010), and involves complex issues of management, ecosystem impact, and social justice (FAO, 2010).

2.3.1.2 Minerals

The marine environment is a distinct geological province (Mangone, 1991), a hotbed for mineral extraction with active sources of mineralization occurring along plate boundaries (Rona, 2008; Rona, 2003). Salt, magnesium, and bromine are recovered from seawater; rock, coral, calcareous marls, shells, sand, gravel, and lime used for coastal protection, beach replenishment, industrial construction, etc., are commonly extracted from beaches and the seabed (UN-ISA, 2004). Phosphorite is mined from salt marshes for fertilizer and offshore

Table 2.1 Comparison of impacts: forest clear-cutting and benthic trawling. Watling & Norse 1998. Reproduced with permission of John Wiley & Sons.

Impact on:	Forest clear-cutting	Bottom trawling/dredging by fishing gear
Substrate	Exposes soils to erosion; compresses soils; loss of nutrients	Overturns, moves, buries boulders/cobbles; homogenizes sediments; eliminates microtopography; leaves long-lasting grooves; alters nutrient flux
Roots and infauna	Saprotrophs (that decay roots) are stimulated then eliminated	Infauna crushed and buried; others become susceptible to scavenging
Biogenic structures	Removes above-ground logs; buries structure-forming species; simplifies habitat	Removes, damages, displaces structure-forming species/habitat, e.g., seagrass, oyster beds; simplifies benthic habitat
Cascading effects	Eliminates most late-succession species; encourages pioneer species	Eliminates most late-succession species; encourages pioneer species
Biogeochemistry	Releases large carbon pulse to atmosphere by removing and oxidizing accumulated organic material; eliminates arboreal lichens that fix nitrogen	Releases large carbon pulse to water column and atmosphere by removing and oxidizing accumulated organic material; increases oxygen demand
Recovery time to original structure	Decades to centuries	Years to centuries; variable, depends on size, duration, frequency
Typical return time	40–200 years	40 days to 10 years
Global area affected per year	~0.1 million km^2 of net forest and woodland loss	8.8 million km^2
Latitudinal range	Subpolar to tropical	Subpolar to tropical
Ownership	Private and public	Public
Scientific documentation (publications)	Many	Well documented
Public awareness	Substantial; visual impact	Very little; obscure, out-of-sight
Legal status	Modify activity to lessen impacts; prohibit or favor alternate logging methods and preservation	Fisheries management; restricted activity for few areas

coal deposits are extracted by tunneling from shore out to sea. Throughout Southeast Asia, large quantities of tin and its mineral cassiterite are dredged from shallow offshore waters. Shores also contain many high-value, low-volume minerals, e.g., platinum, gold, silver, titanium, zirconium, chromium, and rare-earth minerals; gold-bearing sands and gravels buried in fluvial channels are dredged in the shallow offshore waters of Alaska, New Zealand, and the Philippines. Diamonds are dredged from the seabed in water up to 200 m deep in Namibia and South Africa. Mining has a long history of dumping waste into nearshore areas, with very damaging effects, as for example at Placer Dome's Marcopper mine in the Philippines (Coumans, 2003).

Mining contributes significantly to the global economy, with risks and conflicts over its potentially lucrative future. As many mineral resources are located on continental margins and in the deep sea, geopolitics and uncertainties over national jurisdictions come into play (UN-ISA, 2004). Once resolved, greater certainties over ownership will open doors to deep-sea mining for manganese nodules; nearly 70% of the deep-sea floor at depths below 4000 m contains these small, golf-ball-size nodules. The deep sea also contains gas hydrates (350–5000 m depth), cobalt (1000–3000 m along undersea mountain ranges), and massive amounts of sulfides (500–4000 m near plate boundaries; Bollmann *et al.*, 2010). The extraction of almost all minerals produces considerable amounts of waste, much of which is toxic, and generates fine sediment plumes that may contain heavy metals that settle and suffocate benthic organisms at varying distances from the source.

2.3.1.3 Fossil fuel extractions: oil and gas

Oil and natural gas are key fossil fuel resources that power industrial societies. World oil consumption in 2007 reached about 3.9 billion tons a year and is expected to increase at least another 50% by 2030. As prices rise and deposits dwindle, energy demand will escalate offshore oil extractions for the greatest consumers: USA, China, and Russia (Bollmann *et al.*, 2010). The ecological consequences of accidents are acute, chronic, and highly variable (Peterson *et al.*, 2003).

Oil and gas lie under vast layers of ocean-floor sediments. Of a worldwide estimate of 157 billion tons of easily obtainable oil, 26% comes from offshore sources, an extraction that is growing more strongly than onshore (Bollmann *et al.*, 2010; Fig. 2.11). By 2012, 17 giant offshore fields are expected, each containing at least 500 million barrels of recoverable oil, with drilling operations moving into deeper water depths and into more hazardous areas (Robelius, 2007). Within five years,

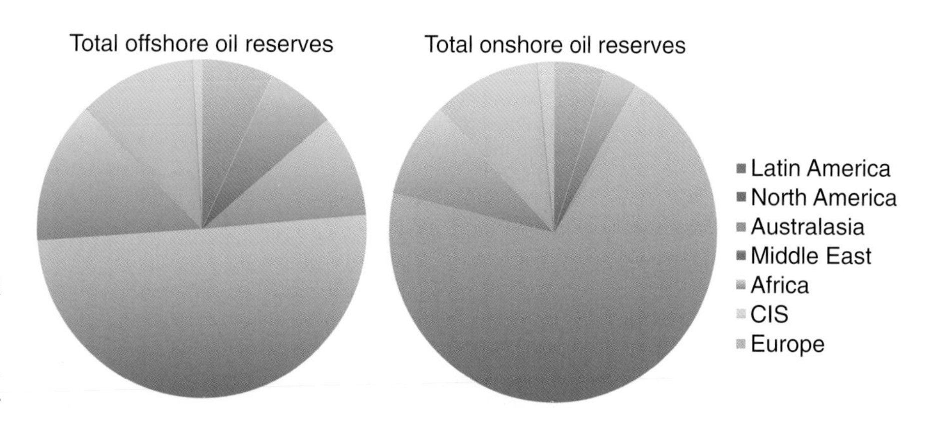

Fig. 2.11 Relative comparisons of conventional onshore and offshore oil reserves by geographic region in 2007. From onshore and offshore combined estimated total of 157 gigatons (Gt) of oil reserves, Bollmann estimates that 26% (4 Gt) or more comes from offshore and deep-water sources. An estimated 30% of undiscovered gas and 13% of undiscovered oil in the marine areas occur north of the Arctic Circle, mainly in Russian waters. CIS = Commonwealth of Independent States. Based on data from Bollmann *et al.* (2010).

Cambridge Energy Research Associates expects global deep-water production to extract the equivalent production of Saudi Arabia—an estimated 10 million barrels a day from the most productive offshore areas: North Sea, Atlantic Ocean off Brazil, West Africa, Gulf of Mexico, Arabian Gulf, and seas off Southeast Asia. The Gulf of Mexico now produces 91% of offshore oil/gas extraction, followed by the relatively shallow North Sea (average depth 40 m) whose production is declining, followed by the Arabian Gulf, and Southeast Asia seas. The offshore areas of India, South China Sea, and Caspian Sea (off Kazakhstan) are emerging as major fields. In the U.S. over the past two decades, about one-quarter of its total natural gas production came from offshore sources, yielding almost 30% of the total oil production (Office of Oil and Gas, 2005). An estimated 30% of undiscovered gas and 13% of undiscovered oil is in the Arctic Ocean, a vulnerable region. Extraction of oil and gas in high-latitude seas is subject to sea ice and harsh winter storms, where spill clean-up presents technological challenges yet to be resolved. Similarly, drilling in areas prone to earthquakes, hurricanes, and tsunamis is technologically hazardous, as in areas of the Western Pacific and elsewhere.

2.3.1.4 Extracting ocean energy

Coastal waters potentially provide plentiful and predictable renewable, clean energy (Pelc and Fujita, 2002). Wind, waves, and currents contain about 300 times more energy than humans are currently consuming, and power plants are converting ocean energy into electricity (Bollmann *et al.*, 2010). Worldwide, about 40 offshore wind energy projects have been implemented, mostly in the United Kingdom, Denmark, the Netherlands, and Sweden, with facilities becoming bigger and venturing into deeper waters. However, although renewable energy generators may be relatively environmentally clean, they have their own set of problems, for example, adding human activity, vibration, and underwater noise to the marine environment, with poorly known consequences for marine life.

2.3.2 Introductions: adding novelty to marine ecosystems

Humans continually add new products to the ocean's chemical soup. These include synthetic chemicals, toxic metals, trash, radioactive materials, pathogens, pesticides, exotic species, artificial heat, noise and light, nutrients, disease agents, and endocrine-disrupting compounds, added slowly and chronically, or suddenly and in concentrated forms, but rarely in accord with natural rhythms. Watersheds and groundwater transfer pollutants and materials to coastal waters, ships discharge litter, waste, exotic species, and chemicals, accidentally or deliberately at sea (ocean dumping, bilge cleaning, antifouling paints). And from the air, storms, winds, and rain deliver debris, chemicals, nutrients, microbes, and others.

2.3.2.1 Adding nutrients to coastal waters

High crop yields and green lawns benefit from fertilizer applications. About 143 million tons are globally applied to land each year (FAO, 2005), and its use is increasing (Bumb and Baanante, 1996), adding high concentrations of nitrogen and phosphorous (Fig. 2.12), essential chemicals for life. Increasing quantities of anthropogenic fixed nitrogen are entering rivers, groundwater, and atmosphere (Duce *et al.*, 2008), directly from wastewater treatment plants, sewage systems, and stormwater overflows, or indirectly via groundwater contamination, precipitation, and land runoff from farms, septic systems, lawns, streets, and roads.

2.3.2.2 Adding toxic petroleum and related byproducts

Roughly 1.4 billion liters of petroleum and related hydrocarbons enter the oceans annually, chronically in low doses and catastrophically in high doses (NRC, 2002c). Tanker accidents exceeding one thousand barrels account for most of the world's oil spills, but chronic exposures from daily releases (accidental, illegal) occur regularly. A city of five million people, for example, might annually release roughly the equivalent of the tanker *Exxon Valdez* oil spill, the difference being that a city's input is chronic whereas an oil spill is acute.

"Petroleum" is a broad term that describes naturally occurring and refined compounds of oil and natural gas, with toxic qualities that vary with degree of industrial refinement (NRC, 1985). Petroleum from natural (e.g., submarine seeps) and anthropogenic (tanker accidents, ship deballasting operations, etc.) sources contains varying degrees of benign and toxic properties. Unrefined crude oil is a widely varying hydrocarbon

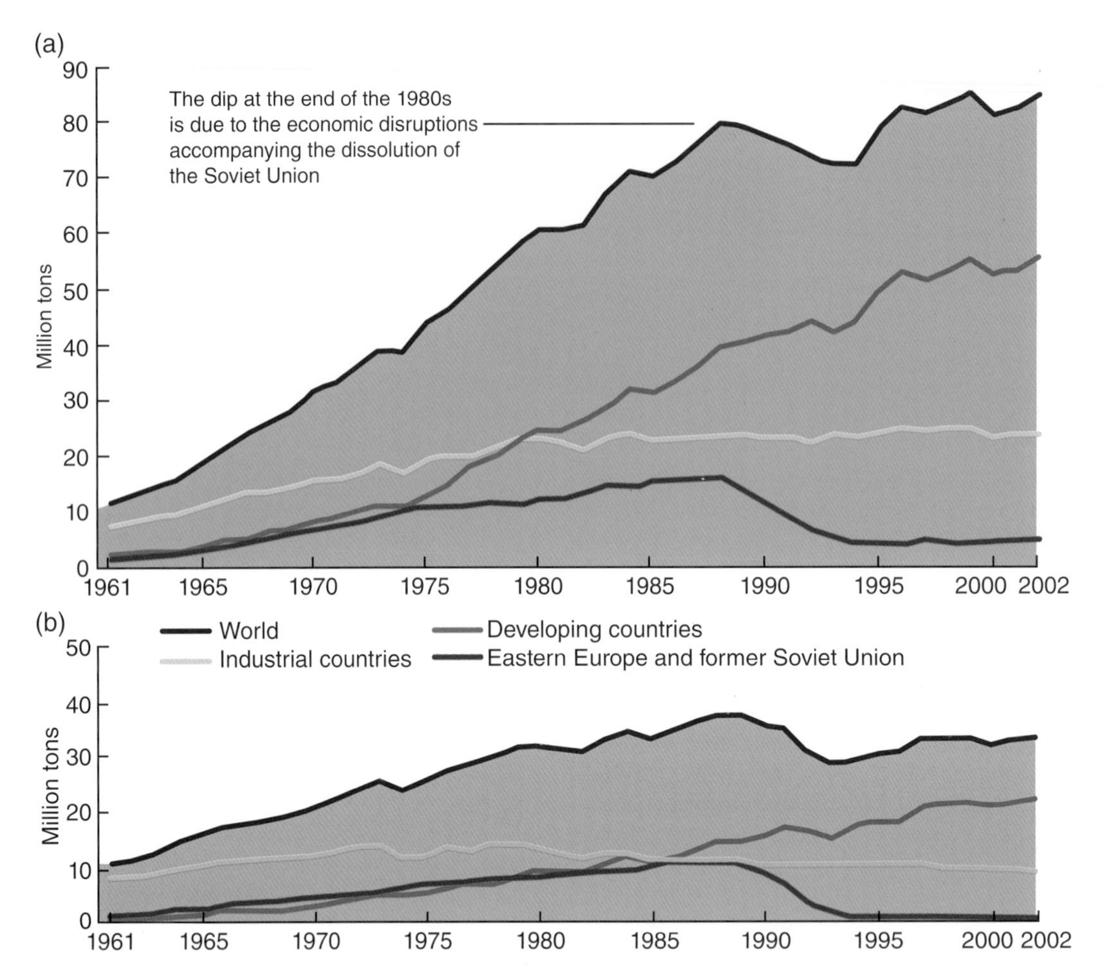

Fig. 2.12 Regional, global, and country trends in use of fertilizers, 1961–2001. Nitrogen and phosphorous are key fertilizers used globally to increase food production and are degrading water quality. (a) Increased applications of synthetic production of nitrogen fertilizer and other uses of nitrogen. (b) A steady threefold increase in use of phosphate occurred until 1990, then banning use decreased its levels to about equal to 1980 applications. Nitrogen and phosphate pollution enriches freshwater ecosystems, creates hypoxia in coastal marine ecosystems, and contributes to nitrous oxide emissions contributing to global climate change and air pollution in urban areas. From Millennium Ecosystem Assessment (2005), www.wri.org.

compound that contains a complex mixture of some toxic chemicals and heavy metals with a variety of sticky and persistent properties. Refined petroleum, in contrast, contains high concentrations of soluble, highly toxic components (e.g., benzene, toluene, xylene) in acute exposures that can quickly evaporate. Due to oil pollution, toxic arsenic concentrations are increasing in the ocean (Wainipee *et al.*, 2010).

Polynuclear aromatic hydrocarbons (PAHs), mostly from incomplete combustion of fossil fuels, pose potential hazards to humans and sea life. PAHs are ubiquitous compounds from such human activities as industrial effluents, domestic sewage, oil spills, bilge water, and creosote on dock pilings that are released into the air, transported in particulates, and precipitate to the ocean surface. The heavier, more persistent PAHs, e.g., carcinogenic benzo(a)pyrene that can concentrate in organisms, especially shellfish, have greater carcinogenic potential. Bottom-dwelling fish exposed to chronic PAHs in harbors have lesions and deformities.

The risk of oil spills is increasing worldwide. For several decades, oil companies have been venturing into deeper waters, and waters subject to harsh weather and environmental conditions. A major oil field offshore of Russia's eastern Sakhalin Island (Sea of Okhotsk) is subject to severe earthquakes, storms, and sea ice, where spills are a potential threat to the dwindling western North Pacific population of endangered gray whales (*Eschrichtius robustus*) and to indigenous people. In the Arctic today, major oil companies are seeking leases and building infrastructures to tap into rich oil deposits, venturing into hazardous conditions in areas that are also critical habitat for marine mammals.

Large-volume spills have increased since the 1950s (Table 2.2). Spills from large, out-of-control wells are infrequent, but they are massive. The earliest in the U.S. occurred off Oregon coast when the tanker *Mandoil II* released about 314,000 barrels into the Pacific Ocean. In 1979–80, the exploratory well *Ixtoc*, with a subsurface drill depth reaching 3600 m,

Table 2.2 Major ocean oil spills. Compiled from Clark (1997); Irwin *et al.* (1997); World Almanac Books (1998); NOAA (2012), online Incident News.

Date	Location	Source	Amount in barrels (x 1000)	Cause
1967	Atlantic Ocean (Scilly Isles, UK)	*Torry Canyon*	860	Grounding
1968	Pacific Ocean (Oregon)	Mandoil II	314	Tanker collision
1969	Pacific Ocean (Santa Barbara, Calif)	*Alpha Well 21 Platform*	103	Well blowout
1970	Pacific Ocean (Bermuda)	*Chrissi*	235	Tanker structure
1976	Atlantic Ocean (Nantucket, Mass.)	*Argo Merchant*	183	Grounding
1977	Pacific Ocean (Hawaii)	*Hawaiian Patriot*	95	Tanker fire
1978	Atlantic Ocean (Brittany, France)	*Amoco Cadiz*	1 600	Grounding
1979	Atlantic Ocean (West Indies)	*Atlantic Empress*	2 124	Tanker Collision
1979	Gulf of Mexico (Mexico)	IXTOC I	3 500	Wellhead blowout
1980	Niger Delta (Nigeria)	Funiwa No 5 Well	200	Well blowout
1983	Atlantic Ocean (South Africa)	*Castillio de Bellver*	1 870	Tanker fire
1988	Nova Scotia (Canada)	*Odyssey*	968	Tanker explosion
1989	Prince William Sound (Alaska)	*Exxon Valdez*	275	Grounding
1991	Arabian Gulf (Kuwait)	Iraq bombing	11 000	Gulf War
1991	Atlantic Ocean (Angola coast)	*ABT Summer*	1 920	Tanker explosion
1991	Mediterranean Sea (Italy)	*MT Haven*	1 140	Tanker fire
1993	Atlantic/North Sea (Shetland Islands)	*MV/Braer*	623	Heavy weather
1994	Arctic (Russia)	*Usinsk*	765	Pipeline rupture
2002	Atlantic O. (Galicia coast, Spain)	*The Prestige*	576	Aging tanker
2005	Gulf of Mexico (US)	Hurricane *Katrina*	190	Hurricane
2010	Gulf of Mexico (US)	*Deepwater Horizon* oil rig	4 900	Platform explosion

released about 3.3 million barrels of crude oil for 290 days into Bay of Campeche, Mexico (Jernelöv and Lindén, 1981). In 1989, the *Exxon Valdez* released about 275,000 barrels of crude oil into Prince William Sound (Alaska) that spread more than 900 km westward to contaminate shores and subtidal sediments. Many species have still not recovered, and recovery for orcas has been especially slow (Matkin *et al.*, 2008). In 2005, Hurricane Katrina spread 190,000 barrels of crude and refined oil products into the Mississippi River and Gulf of Mexico, followed in 2010 by the *Deepwater Horizon*, the world's largest oil spill outside of war. The *Deepwater Horizon* platform exploded and caught fire on April 20 2010 in the northern Gulf of Mexico. It released about 4.9 million barrels during 86 days from its ultra-deep 1500 m well (Cleveland *et al.*, 2011) that spread across about 75,000 km^2 to pollute offshore waters, lagoons, sandy beaches, and barrier islands. The oil affected a variety of sea life (some endangered, Campagna *et al.*, 2011), wetlands, shellfish reefs, and deep-benthic habitat; these impacts are still being felt, for example on marshes and other wetlands, further exacerbated by multiple human causes (Silliman *et al.*, 2012). Concentrated toxins and tons of methane gas contaminated one of the world's most productive ecosystems, probably for years to come. Scientists most often lack critical data to predict ecological consequences (Bjorndal *et al.*, 2011), but as the *Exxon Valdez* disaster proves, the persistence of toxic and sublethal exposures may affect wildlife long after the acute phase ends (Peterson *et al.*, 2003; Chen and Denison, 2011).

2.3.2.3 Adding persistent bioaccumulative toxins (PBTs)

Coastal waters receive many highly toxic substances in common use, and that are extremely persistent. Some are endocrine disrupters that can cause reproductive or nervous system dysfunction, behavioral abnormalities, and birth defects. Many are carcinogenic and more are lethal or sublethal in effect. PBTs are primarily synthetic chemicals designed and manufactured to meet a wide range of industrial, agricultural, residential, and personal needs (Table 2.3). Integral to PBTs are persistent organic pollutants (POPs), which include trace metals and organo-metal compounds. PBTs released into the environment generally resist physical, chemical, and metabolic breakdown and accumulate in sediments near industrial or urbanized areas, where organisms feed. The toxin concentrates in the fatty tissues of marine organisms; over time the body-load can reach a magnitude greater than ambient seawater, such that the organisms themselves become toxic (Fig. 2.13). Among the most contaminated are marine mammals, especially orcas (Hickie *et al.*, 2007); one dead individual washed up on a Pacific Northwest shore had a PCB body burden of such magnitude to be declared toxic waste!

PBTs are ubiquitous and persist despite commercial bans and regulated use. They are emerging in the Arctic in high enough concentrations to affect wildlife and people (Muir and de Wit, 2010). They include PAHs, e.g., halogenated aromatic hydrocarbons (e.g., organochlorines), polychlorinated biphenyls

Table 2.3 Acute and persistent anthropogenic toxins harmful to sea life.

TOXIN TYPE	SOURCES	BIOLOGICAL IMPACT
Heavy metals:	**Industrial concentrations of naturally occurring. Ubiquitous.**	**Interacts with biomolecules; impact varies with marine species and organ; liver is great accumulator.**
Cadmium	Industrial production into products; from mines, rivers, atmosphere, dredging	Collects in inshore mud flats, bacterial films, and organic matter
Copper	Electrical industry alloys; electrical wiring, algicides, acid mine drainage	Essential in biochemical processes; catalyst in hemoglobin formation; highly toxic to invertebrates
Mercury	Power plants; pulp/paper mills, fungicides, fossil fuel combustion, mercurial catalysts, weathering	Natural bacteria convert mercury to methylmercury that concentrates in fish, bioaccumulates; endocrine disruptor; toxic if eaten
Tin	Organotin production for pesticide, PVC stabilizer, biocide, etc. increased from 5000 tons in 1995 to 35,000 tons in 1985; now ~ 50,000 tons	Damages marine life worldwide. Tributyltin is extremely toxic to oysters. Sediment microorganisms convert metallic tin into methyltin
Zinc	Sewage/industrial discharge. Used in alloys, paints, cosmetics, etc. to coat steel/iron against corrosion, roofing	Catalytic activity essential in biochemical processes. Most toxic to aquatic microscopic organisms; larvae
PBT, POP: ***Persistent***	**Industrial organic synthetics**	**Many are lipophilic. Can be carcinogenic, mutagenic, endocrine disrupter**
DDT, DDE, DDD	Organo-pesticides, extensively used	Sediment; lipophilic, bioaccumulate, hormonally active
Chlordane	Ubiquitous pesticide 1948 to 1978	Bioaccumulate in sea life (marine mammals, sea turtles)
Octachlorostyrene	Waste byproduct from chlorine production prior to 1970	Suspected endocrine disruptor. Concentrates in sediments, sea life, marine mammals, dead cormorants
Mirex	Chlorinated pesticides, used extensively	Persistent. Concentrates and chronically toxic to sea life
Polyaromatic hydrocarbons (PAHs)	Crude oil, refined oil, aerosols; sewage; surface runoff. Widespread	Low water solubility. Concentrates in sediment, biomagnify: benzo(a)pyrene is a carcinogen and endocrine disrupter.
Polychlorinated biphenyls (PCBs)	Dielectric fluids in electrical equipment (transformers, hydraulic systems); sewage; widely distributed	Highly persistent; concentrates in sediments, liver, gonad, sea turtles
Dioxins (2,3,7,8-TCDD)	Unintentional byproducts of combustion and industrial chemical processes; from incineration, pulp/paper production	Among the most toxic: occur in seafood: affect food webs, carcinogenic, endocrine disrupter
Recent concerns		
Polybrominated diphenyl ethers (PBDEs)	Flame retardants used in plastics, foams, fabrics etc., important to human safety; exponential increase since 1970s	Persistent; disperse globally to deep ocean water, polar regions; in sediment, soil; concentrate/biomagnifies in sea life; fetal toxicity; teratogenicity; mass mortality; marine mammals; seabirds
Pharmaceuticals/ personal care products (PPCPs)	Prescription, over-counter therapeutic drugs, veterinary drugs, fragrances; cosmetics; in sewage, from industry	Mostly unknown. Antibiotic resistant; subtle, insidious effects; bioaccumulative potential

(PCBs) from plastics and electrical equipment, and pesticides being added at more than 5.0 billion pounds to land each year (Fishel, 2007). Pesticides can cause unexpected and unintentional harm to many invertebrate marine species that are essential to food-web transfer. The organochlorines are of most concern, e.g., DDT (dichloro diphenyl trichloroethane) and its derivatives DDE (dichloro diphenyl ethane) and DDD (dichloro diphenyl dichloroethane); these were banned or phased out from use internationally by the Stockholm Conference (Ch. 3), but persist in the environment to affect immune and nervous systems of organisms. PCBs are stored in sediments and bioaccumulate and biomagnify up food chains to cause tumors, fetal death, and birth defects in various species. Dioxins, the byproducts of chlorinated substances used to bleach paper pulp, are extremely toxic, persistent, and can cause abnormalities and tumors. Newer PBTs in current use and entering aquatic systems include brominated flame retardants and numerous pharmaceutical and personal care products (PPCPs) for which the impact is little known (SCCWRP, online). PBTs and hundreds or thousands of other toxics entering coastal/marine waters are mostly not monitored, and generate concerns for their cumulative and increasing impact on coastal ecosystems (Dachs and Méjanelle, 2010) and on people. Due to bioaccumulation, the highly toxic organic form of mercury (methyl mercury) raises serious concerns about seafood consumption safety for humans (Fitzgerald *et al.*, 2007), as first witnessed in Minimata (Japan) in 1956 that caused severe neurological damage to many people.

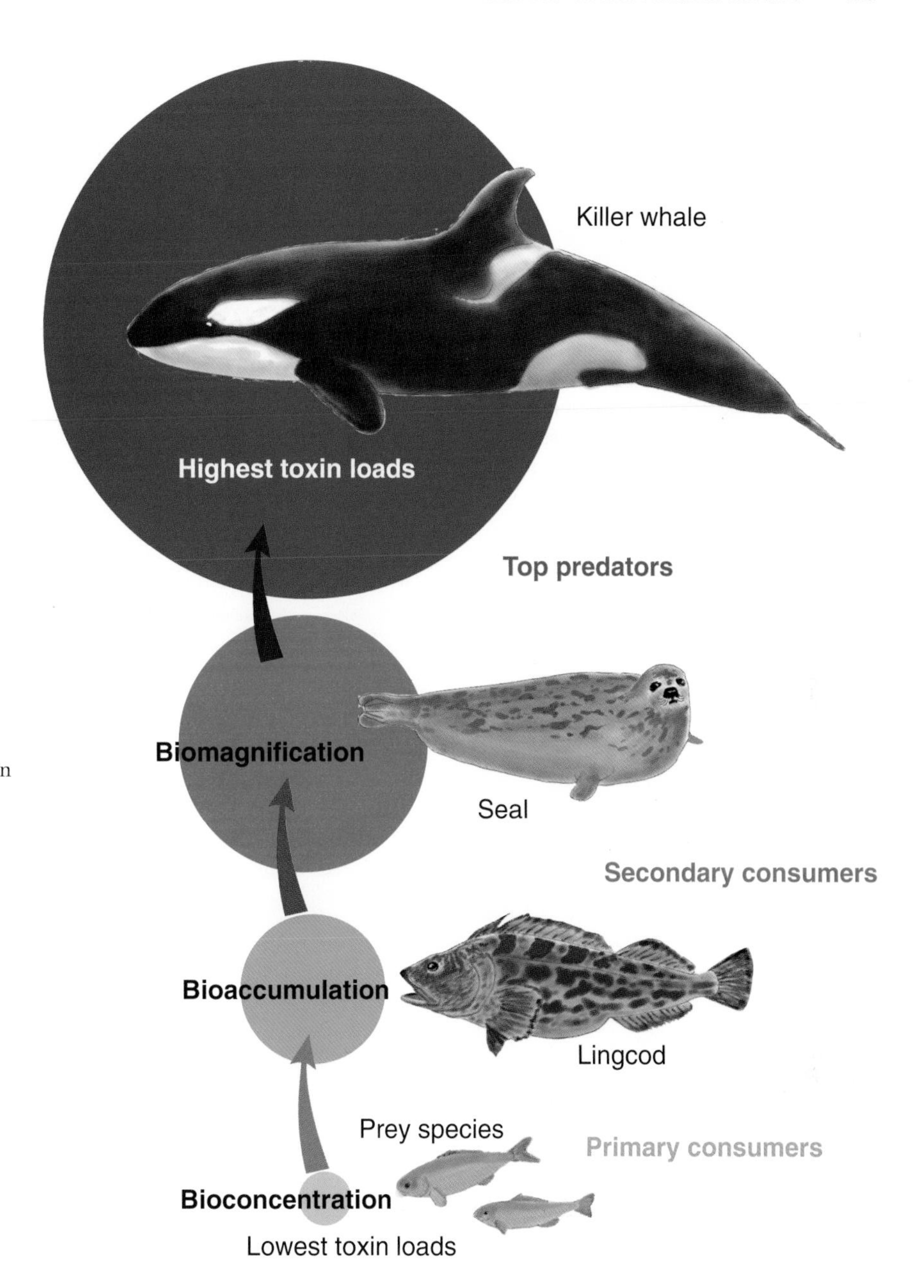

Fig. 2.13 Biotoxicity. Toxic accumulations in organisms are made worse by human activities. Biotoxicity results from feeding on contaminated food that becomes transferred and accumulated up the food chain: from bioconcentration (concentration of toxin in an organism, not through food web) in primary consumers; to bioaccumulation (biological sequestering of toxin at higher concentration than environment) in secondary consumers; to biomagnification (biological transfer of a toxin amplified up through trophic levels of food web). Organisms at high trophic levels (e.g., orcas, humans) are often exposed to high concentrations of toxic chemicals, e.g., persistent organic pollutants (POPs). USGS, online definitions. Data from Ross and Birnbaum (2003).

2.3.2.4 Adding litter and persistent plastics

Marine litter occurs in all oceans, accumulating along shores and in water, both in densely populated regions and remote places far from obvious sources and human activities (Fig. 8.28). Most litter is accidentally or carelessly released, although many commercial, fishing, metropolitan, and military disposal operations purposefully introduce tens of thousands of tons of litter annually into the seas. The North Pacific Subtropical Gyre, called the "Garbage Patch" (Kaiser, 2010), draws attention to the poorly quantified deaths of thousands, even millions, of marine mammals, seabirds, sea turtles, and fishes every year (NOAA, online). Introduced materials profoundly alter feeding behavior of animals. Sea turtles, seabirds, and marine mammals are especially likely to ingest foreign objects, and each year many thousands may be injured, receive inadequate nutrition, or die due to digestive-system blockage.

Plastics are a dominant component of ocean litter. Globally, its production increased from 1.5 million tons in 1950 to 230 million tons in 2009, resulting in durable and serious toxic and mechanical effects (Gregory, 2009; Hirai *et al.*, 2011). Plastic litter ranges in size from micro-millimeter particles (human body washes, cosmetics) to large objects, including abandoned fishing gear (e.g., buoys, fishing line, nets). The amount entering the oceans from all sources is staggering. Trawls regularly retrieve plastics and other debris, even from the deep sea. Especially serious are lost or discarded fishing gear (Fig. 2.14a). Tens of thousands of plastic fish traps lost at sea continue to "ghost fish" for long periods of time. Floating plastic sheets may cling to plants, corals, and intertidal

Fig. 2.14 Unintentional impacts by humans on coastal animals. (a) Steller sea lion (*Eumetopias jubatus*) skeleton entangled in discarded fishing gear on Amak Island, southeastern Bering Sea, 1974. The skeleton shows signs of having been consumed by a predator, probably a grizzly bear. Photograph © Ray & McCormick-Ray. (b) Porpoise (*Phocoena phocoena*) with propeller wounds found stranded in San Francisco Bay, California. Courtesy of The Marine Mammal Center, Sausalito, California.

animals. And thousands of exposed animals are trapped, smothered, strangled, entangled, killed, or deformed, while others suffer malnourishment and die slowly. Furthermore, small plastic fragments consumed by fish, larvae, and small organisms may contain toxic chemicals (PCBs, PAHs, DDTs, PBDEs, alkylphenols, bisphenol) that are endocrine disrupters (chemicals that interfere with endocrine or hormone systems of animals and people), and that can also bioaccumulate and biomagnify through food webs.

2.3.2.5 Adding harmful species

Humans regularly introduce pathogenic microbes and exotic species into new locations, where they may become extremely abundant and invasive. Some pathogenic and antibiotic-resistant microorganisms are introduced from human sewage systems, despite treatment. The common sewage bacterium *Serratia marcescens*, for example, is devastating Caribbean elkhorn corals (Sutherland *et al.*, 2011). Ships, too, have introduced pathogenic bacteria through routine discharge of ballast water, e.g., cholera (*Vibrio cholerae*) and other species that have caused severe economic impact. Ballast water discharged from ships is also responsible for introducing the Japanese toxic dinoflagellate (*Gymnodinium catenatum*) into Australian waters, zebra mussels and European shore crabs into North America, and comb jellies into the Black Sea (see Section 2.2.2). Precautionary measures of pumping bilge water into holding tanks and providing treatment at ports have been proposed, but are expensive and difficult to monitor; violations are frequent.

Entrepreneurs add cultivated, aquaculture fish in massive numbers to new locations. Worldwide demand for fish has dramatically spurred mariculture development since the early 1990s (Naylor and Burke, 2005). For some species (e.g., Atlantic salmon), global production has roughly quadrupled and now surpasses wild salmon fisheries. Successful Atlantic salmon farming encouraged aquaculturists to begin farming numerous other marine finfish, including depleted wild species, e.g., Atlantic cod (*Gadus morhua*), Atlantic halibut (*Hippoglossus hippoglossus*), Pacific threadfin (*Polydactylus sexfilis*), mutton snapper (*Lutjanus analis*), bronzini (*Morone labrax*), and bluefin tuna (*Thunnus thynnus*). Most are raised in coastal waters, but some are held in ocean net pens or cages. Serious local pollution often accompanies these facilities (Levin *et al.*, 2001; Christensen *et al.*, 2003; Krkošek *et al.*, 2006).

2.3.2.6 Adding noise, heat, light pollution, collisions

The underwater world is getting noisier, for example for marine mammals that depend on sound for an acoustical image of their world (Weilgart, 2007; Box 2.4). Supertankers, huge fish-factory ships, cruise ships, military exercises, dredging and construction, oil exploration and production, fish finders,

Box 2.4 Noise pollution: a threat to dolphins?

Sam Ridgway
University of California, San Diego, USA

The significance of human-made sound in disturbing or injuring cetaceans has been considered only recently (Popper *et al.*, 2000). Earlier studies of dolphin hearing were motivated by the discovery of the animal's sonar. Audiograms, plots of hearing threshold at different sound frequencies, have been done on several species of the cetacean superfamily Delphinoidea (narwhals, white whales, all dolphins, and the porpoises). The first detailed audiogram of the bottlenose dolphin showed especially good underwater hearing with a threshold of 42 dB re 1 μPa (10^{-14} W m^2) at 60 kHz. Studies also showed sensitivity to sound frequencies from about 60 Hz to 150 kHz, almost eight times the frequency span of human hearing (humans are slightly more sensitive to sound pressure in air, but our frequency range is limited to about 20 kHz).

Sensitive ears connected to a massive auditory central nervous system are fundamental to the dolphin's echolocation and communication. It is reasonable to ask how the animal, with such excellent hearing, avoids damaging its own ears with the loud sounds it produces during echolocation. The dolphin ear, anatomically only a few centimeters away from its sound production mechanism, processes high-frequency echolocation pulses up to 230 dB re 1 μPa in peak-to-peak amplitude. Using intense pulses and sensitive ears, dolphins can detect echoes (as quiet as a human whisper) from small objects at 100 m and more. Because the dolphin's pulses are very brief—on the order of 40 μs, and 25,000 of these would equal a second of sound, although the total energy within each pulse is minuscule. Anatomical structures, including highly reflective air sinuses that attenuate sound, probably help the animal avoid damaging its own ears.

On a comparative basis, the baleen whale auditory system does not appear as specialized as that of dolphins. The acoustic centers of the baleen whale brain are smaller than those of dolphins for whom the auditory nerve is the largest cranial nerve; the trigeminal nerve of the baleen is larger. Unlike dolphins, whose sense of hearing predominates, baleen whales appear to rely most on the sense of touch. Although we have made no audiograms, observations show that baleen whales usually produce low-frequency sounds often as low as 15 Hz.

If, as anatomical study suggests, the baleen whale ear is specialized for low frequencies, then the inference is that the animal's hearing is adapted to protection from considerable acoustic interference such as that which occurs from natural ocean background noise in the part of the acoustic spectrum below 1000 Hz. It is unlikely that baleen whales will be captured and trained for audiograms as have the delphinoids; nonetheless, physiological methods could be used to obtain audiograms on beached or entrapped whales.

The question arises: can baleen whales detect calls of other whales by means other than audition? The arrays of vibrissae about their heads suggest that baleen whales may use these adaptations to sense low-frequency vibrations, including the calls of other baleen whales. Uses of tactile detection like these may explain the large trigeminal nerve in baleen whales. Until audiograms can be measured on baleen whales, we are left to speculate about their hearing thresholds and frequency sensitivities. The absence of definitive audiograms compounds the problem of determining what levels of human-generated sound may damage baleen whale hearing.

Dolphins have evolved robust mechanisms to protect their ears and body tissues from loud natural sounds such as lightning strikes, earthquakes, pounding surf, volcanic eruptions, whale calls, and even their own echolocation pulses. Year after year, these adaptations are eroded as oceanic shipping raises the ambient background noise in the oceans. Intensified technology also introduces loud noise for purposes of improved sonar, oil exploration, and acoustic communication modems. These animals should not be continuously exposed to the equivalent of a boiler factory or even a loud discotheque. Increasing production of intrusive noise in the sea poses a serious threat to marine life. Science and technology must take action together in order to protect marine mammals such as dolphins and whales from dangerous noise.

depth sounders, speed boats, small watercraft, and submersibles that explore the deepest parts of oceans create a noisy presence. Also, the increasing presence of ships and recreational watercraft may result in collisions with sea life, inflicting serious injury or mortality. A major source of mortality for North Atlantic right whales (*Eubalaena glacialis*) has been collisions with large ships, a situation that has now been brought under partial control. But many smaller marine mammals are frequently killed or injured by collisions with recreational or small fishing boats, for example sirenians, seals, and porpoises (Fig. 2.14b).

Thermal pollution from industrial discharges changes ambient water temperatures, which can increase species' metabolism and affect ecosystems. Large amounts of heated water of more than 10°C above ambient sea temperature are released from industrial sources (e.g., power plants) into local waters. Elevated water temperatures also result from roads and parking lot drainage into urban and stormwater drains that ultimately discharge into nearby waters. Heat can differentially affect species' distributions and reproductive behavior; when levels are above a species' physiological tolerance, the impact can be lethal, most notably for tropical corals sensitive to temperatures >30°C. Elevated water temperatures also reduce oxygen saturation, exacerbating hypoxia.

The increasing extent and intensity of artificial night lighting near shores and at sea has substantial effects on wild species (Longcore and Rich, 2004). Artificial light from urban and seaside development disrupts marine bird and sea turtle navigation and orientation; sea turtle hatchlings emerging from sandy nests to scurry to the sea are disoriented and subject to predation where urban lights add a glow to dark shorelines. Lighted fishing fleets (Fig. 12.9), offshore oil platforms, and cruise ships bring artificial light to the world's oceans, potentially disrupting behavior patterns of many ocean species.

2.3.3 Physical alterations: structural changes of coastal systems

Humans constitute a massive geophysical force, globally transforming coastal land and sea for human uses, which causes major ecological change (Airoldi *et al.*, 2005; 2009). Globally, humans have altered almost a third of coastal lands within 100 km of the coast (Burke *et al.*, 2001) by physically transforming productive coastal-marine areas into urban/industrial complexes and by armoring the coasts against erosion and sea-level rise. Artificial structures now occupy more than half of the coasts of many nations; e.g., 1 km of coastline was developed every day in Europe between 1960 and 1995 (Airoldi and Beck, 2007). Along the northern Mediterranean coastline, urbanization, harbor, and port development now cover nearly 90% (e.g., French Riviera, Athens, Barcelona, Marseille, Naples, north Adriatic shorelines), with a projected increase of 10–20% expected for most shores.

2.3.3.1 Reclaimed land

Reclamation is most obvious in estuaries, lagoons, and shallow coastal waters, where millions of tons of dredged material that may contain toxic contaminants are dumped and potentially contaminate groundwater. Artificial structures often replace or are constructed adjacent to some of the richest coastal farmlands and fishery areas. Since the Agricultural Revolution 5000–10,000 BP, coastal waters have been reclaimed for coastal city-state development and trade. Extensive salt marshes of England's Wash were reclaimed as early as 900 AD; only half remain intact today. The Zuiderzee of the Netherlands was closed off from the sea for agriculture, which caused significant loss of estuary-dependent species (Hood, 2004); brackish lakes (polders) emerged behind high dikes that now protect more than half of the Dutch population from rising seas. The short supply of land along Japan's coasts is augmented by artificial islands, which have expanded human occupation, industries, and sea farming.

Physical coastal structures intensely serve the needs of gigantic petrochemical complexes, harbors, steel mills, power plants, urban centers, tourism, recreation, international trade, and shipbuilding facilities as they alter the coastal system (Fig. 2.15). Offshore oil and gas extraction from fixed platforms requires onshore facilities and a variety of shore-stabilizing structures (e.g., seawalls, dikes, etc.), which alter tidal conditions and habitats (Hood, 2004). Artificial structures may enhance dispersal of nonindigenous species by creating a system of "novel" habitats favorable for invasive species in coastal waters (Glasby and Connell, 1999).

2.3.3.2 Obstructed water flow

Man-made dams pervade present-day river systems worldwide (Freeman *et al.*, 2003), impeding water flows into coastal systems. In northern Europe, six dams obstruct the Rhine watershed on its route to the North Sea. The Nile, the world's longest river, drains more than 3 million km^2 of watershed of nine countries and once flowed directly into the southeastern Mediterranean Sea. Construction of Egypt's Aswan High Dam and seven other large dams drastically reduced the Nile's flow into the Mediterranean Sea, fundamentally affecting fisheries and coastal ecology. In South America, 29 large dams on the Paraná River affect the 2.5 million km^2 Pantanal wetland that straddles four countries; several more are planned to control floods and generate electric power. Dams built for flood control on Laos' upper Mekong River and its tributaries have destroyed major freshwater fisheries and the spawning and nursery habitats of the river's unique and endangered giant fishes, even before the flow enters the 2200 km-long lower Mekong, which empties into the South China Sea. Laos is planning to construct a dozen more hydropower projects on the lower Mekong, and Cambodia is constructing one more near its mouth (Stone, 2011). The world's largest dam is China's Three Gorges Dam on the Yangtze River, a river in which 45,600 large and small dams intercept its flow path into East China Sea, which supports a major fisheries area.

For centuries, dams have been constructed across rivers to produce energy, mitigate against destructive floods and droughts, and to provide water to farms and cities. The number of large dams increased seven-fold worldwide during the major dam-building period between 1950 and the mid-1980s. These obstructions block migratory fauna, impact fisheries, imperil

Fig. 2.15 Baltimore Harbor in Maryland has been transformed from a natural harbor at the fall line of the Patuxent River into a major port observed here in 1977, hard armored to stabilize its shore for concentrated development that the city supports today. Photograph © Ray & McCormick-Ray.

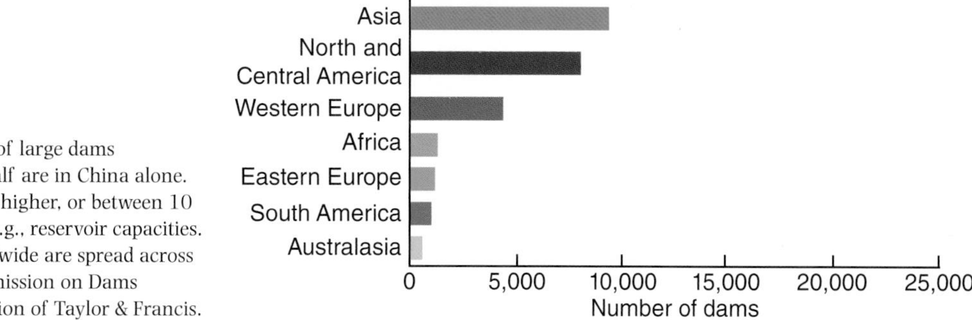

Fig. 2.16 Regional distribution of large dams worldwide in 2000; more than half are in China alone. A large dam is defined as 15 m or higher, or between 10 and 15 m depending on criteria, e.g., reservoir capacities. Today, >45,000 large dams worldwide are spread across 150 countries. From World Commission on Dams (2000). Reproduced with permission of Taylor & Francis.

obligate riverine species, and impede sediment flow important to coastal beaches, shores, and estuaries. An estimated 800,000 large dams existed in 1997, 45,000 of them higher than a five-story building, with more than half built in China (Fig. 2.16). Dams are a primary reason for declines or extirpations of diadromous fish (shad, herrings) in the eastern United States, salmons (*Oncorhynchus* spp.) of the Pacific Northwest, and scores of other fish species elsewhere (Kocovsky *et al.*, 2009). Although many stream-dependent species may persist in dam-altered basins, populations restricted to fragments of their former ranges are reduced in abundance (Freeman *et al.*, 2003). Also due to dam construction, estuaries at the mouths of three of the world's largest and most famous rivers (Nile, Mississippi, Yangtze) are starved of sediments and their deltas are rapidly disappearing and becoming less productive.

2.3.3.3 Ports and coastal mega-urban centers

Many of the world's largest cities and ports occupy coastal systems (Martínez *et al.*, 2007; Table 2.4), especially estuaries and protected regional seas. Rapid growth in international trade and increased vessel traffic has spurred development of ports important to national economies. Ports serve major cities through active loading, unloading, storage, and transport of materials in complex port operations, where fires, explosions, and toxic releases cause serious impacts and chronic degradation (Darbra and Casal, 2004; Table 2.5). U.S. ports are economic development engines that hold remarkable autonomy (Giuliano, 2007). In England's North Sea, vessels that support oil fields cause most accidents (Sii *et al.*, 2003). Port developments along the Pacific and Indian Ocean shipping routes are expanding into megacities that degrade waterways. The Pearl River Delta of southern China that includes Hong Kong is rapidly developing industrial, municipal, and agricultural activities over 8000 km^2 of delta, causing serious toxic organic pollution (Fu *et al.*, 2003). This Port's annual traffic volume of 40,000 ocean-going ships and 200,000 coastal vessels maneuvers among ferries, barges, and recreational and fishing boats and experienced 2012 accidents between 2001 and 2005 (Yip, 2008).

Ports in developed nations have long been subject to chronic toxic pollution sequestered in sediments, in which fish diseases are reported (Murchelano and Wolke, 1991); e.g., high levels of PCBs are reported in catfish, blue crabs, and speckled trout in the Houston Ship Channel, where about 40% of the U.S.

Table 2.4 World's major ports and megacities occur on major coastal waters. Concrete and steel separate land and sea but unite global trade.

World's largest cities/ ports (country)	Area (km²)	Population (million)	Traffic port container (TEUs**) (millions)	River system and coastal water	Port max. draft (m)
Singapore* (Singapore)	1200	6.4	29.9	Straits of Malacca	12.5
Shanghai* (China)	6340	24.8	28.0	Yangtze Delta, East China Sea	7.0–12.5
Hong Kong* (China)	1104	7.1	24.5	Pearl R. estuary, South China Sea	Deep
Shenzhen* (China)	72.6	9.6	21.4	Pearl R. estuary, South China Sea	3.0
Busan* (South Korea)	—	3.7	13.4	Korean Straits	16.0
Dubai Ports (U. Arab Emir)	—	1.0	11.8	Persian Gulf	7.2
Ningbo* (China)	2462	2.3	11.2	Yangtze R. Delta, East China Sea	Deep
Guangzhou* (China)	—	25.2	11.0	Pearl R. estuary, South China Sea	3.0
Rotterdam* (Netherlands)	105	2.8	10.8	Rhine R./Meuse R., North Sea	24.0
Qingdao* (China)	—	3.9	10.0	Yellow Sea	17.5
Hamburg* (Germany)	—	2.6	9.7	Elbe R., North Sea	7.0–17.0
Kaohsiung* (Taiwan)	—	2.8	9.7	Taiwan Strait	8.2
Antwerp (Belgium)	204	0.5	8.7	Scheldt Estuary, North Sea	11.5
Tianjin* (China)	107	9.7	8.5	Haihe R., Bohai Bay, Yellow Sea	18.0
Port Kelang (Malaysia)	573	0.6	8.0	Kelang R., Strait of Malacca	13.3
Los Angeles* (USA)	30	18.1	7.9	Pacific Ocean	16.2
Long Beach (USA)			6.4	Pacific Ocean	12.0
Bremen/Bremerhaven (Ger.)	400	0.1	5.5	Geest R., Weser estuary, North Sea	14.5
Tanjung Pelepas (Malaysia)	3.1 M	0.9	5.5	Pulai R. estuary, Strait of Malacca	15.0–19.0
New York/New Jersey Metropolitan area (USA)	3900	18.9	5.3	Hudson estuary, New York Bight	—
Laem Chabang (Thailand)	10	0.006	5.1	Gulf of Thailand	14.0
Xiamen* (China)		2.4	5.0	Jiulong River, Taiwan Straits	14.0
Dalian* (China, PR of)	15	3.6	4.5	Yellow Sea	23.0
Tanjung Priok (Indonesia)	6	—	4.0	Java Sea	3.0–12.0
Nhava Sheva (India)	—	0.2	4.0	Arabian Sea	12.0–13.5
Tokyo* (Japan)	—	34.3	3.7	Tokyo Bay, Pacific Ocean	10.5

*Based on population criteria of world's 26 megacities (>1 million inhabitants), the area of central city and linked neighboring communities. Source: Thomas Brinkhoff online www.citypopulation.de **TEU (or teu) "Twenty-Foot Equivalent Unit" the standard linear measurement used in quantifying container traffic flows.

chemical production and oil refineries is concentrated (Howell *et al.*, 2008). Dredging of ever-deeper channels to accommodate increasingly large cargo and container ships releases toxics to the water column, while increasing ship traffic also pollutes the air. Emerging industrial nations lack the capacity to deal with contaminated sediments.

2.4 TERTIARY ISSUES: EMERGENT AND UNINTENDED CONSEQUENCES

Humans have for centuries, if not millennia, been foremost agents of environmental change. Today, the world's coastal and ocean ecosystems are experiencing unprecedented rates of biological and ecological change from historic and systemic consequences that are most difficult to address. Observations of ocean change draw attention to regional and global issues and a mounting environmental debt from historical abuses that is threatening the natural resiliency and sustainability of many coastal and marine systems as well as local economies. Tertiary issues thus call attention to multiple sources that cumulatively are moving marine ecosystem performance toward undesirable change. We consider only a few of the most notable.

2.4.1 Degraded coastal water quality

On a global scale, nutrient loads from human activities are entering coastal systems to over-enrich (e.g., nutrient pollution) and degrade water quality (Nixon, 1995). Nutrient enrichment increases water turbidity, degrades habitat, alters

Table 2.5 Vessels and port routine operations and episodic-accidental events degrade coastal habitats and sea life through chronic exposures.

EPISODIC events	ROUTINE operations
Vessel impacts	
Collisions with sea life	Underwater noise
Vessel collisions	Air emissions
Toxic and cargo spills	Ballast water release/hull fouling of exotic species
Sewage release	Hull coating toxic release
Ocean dumping	Lights
Oily waste water	
Port impacts	
Dredging maintenance, toxin releases	Storm water runoff—sediment, toxins
Port expansion/land reclamation	Vessel wake erosion
Ship construction	Cargo-handling air emissions
Explosions/accidental spillage	Habitat alteration (estuary, sea grass)
Ground transport collisions/spillage	Shoreline stabilization/alteration
	Lights: timing; intensity
	Altered currents/hydrodynamics
	Artificial habitats encourage exotic species
	Dust generation
	Noise

food-web structure, reduces biodiversity, and initiates harmful algal blooms (Howarth, 2008). These effects can alter marine food webs critical to commercial fisheries and lead to ecological changes detrimental to estuaries, coral reefs, and other habitats important to marine biodiversity and ocean productivity.

Human activities linked to coastal degradation concentrate mostly in bay and estuarine habitats (Vitousek *et al.*, 2009), with consequences for human health (Emch *et al.*, 2008; de Magny *et al.*, 2010). Ill health affects species inhabiting overdeveloped and simplified environmental systems, where high stress conditions may compromise their immunity to microbial infection (Harvell *et al.*, 1999). High daily loads of wastewater that empty into coastal waters add bacteria, viruses, and numerous other forms of human pathogens into marine reservoirs. Evidence indicates that marine organisms host a diversity of parasites and pathogens (Stewart *et al.*, 2008), some of which are pathogenic to a variety of sea life (Munn, 2006). Viruses are abundant and play enormous roles in ocean processes (Ch. 5), and bacteria from sewage effluents are contributing to antibiotic-resistant bacterial lesions in fishes (Al-Bahry *et al.*, 2009) and to coral diseases (Sutherland *et al.*, 2011). Also, "sick seas" may be sickening marine mammals (Lowenstine, 2004), alarming the general public (Noga, 2000).

2.4.1.1 Expanding anoxic bottom waters

Over-enrichment leads to eutrophication and low oxygen concentration (hypoxia), lack of oxygen (anoxia), and to "dead zones" (Box 2.5) in which all organisms that require dissolved

Box 2.5 Hypoxia

Robert J. Díaz
College of William and Mary, Gloucester Point, Virginia, USA

Rutger Rosenberg
University of Gothenburg, Sweden

Low dissolved oxygen environments (known as hypoxic or dead zones) occur in a wide range of aquatic systems and vary in frequency, seasonality, and persistence. While there have always been naturally occurring hypoxic habitats, anthropogenic activities related primarily to organic and nutrient enrichment related to sewage/industrial discharges and runoff from agricultural lands have led to increases in hypoxia and anoxia in both freshwater and marine systems. A consequence of this over-enrichment has been a rapid rise in the areas affected by hypoxia over the last 50 years. No other environmental variable of such ecological importance to estuarine and coastal marine ecosystems as dissolved oxygen has changed so drastically, in such a short time. Currently there are over 500 hypoxic areas or dead zones around the world related to human activities (Fig. B2.5.1).

By the early 1900s dissolved oxygen (DO) was a topic of interest in research and management, and by the 1920s it was recognized that a lack of DO was a major hazard to fishes. It was not obvious, however, that DO would become critical in estuarine and shallow coastal systems until the 1970s and 1980s when large areas of low dissolved oxygen started to appear with associated mass mortalities of invertebrate and fishes. From the middle of the 20th century to today, there have been drastic changes in dissolved oxygen concentrations and dynamics in many marine coastal waters. Prime examples would be the northwest continental shelf of the Black Sea, the Baltic Sea, the Gulf of Mexico continental shelf off Louisiana and Texas, and the East China Sea.

There is a similarity of faunal response across systems to varying types of hypoxia that range from beneficial to mortality. Consequences of low DO are often sublethal and can affect growth, immune responses, and reproduction. When a system becomes hypoxic, mobile fauna have to contend with two simultaneous problems: (i) loss of habitat as they are forced to migrate into higher DO waters; and (ii) increased risk of negative species interactions and

(*Continued*)

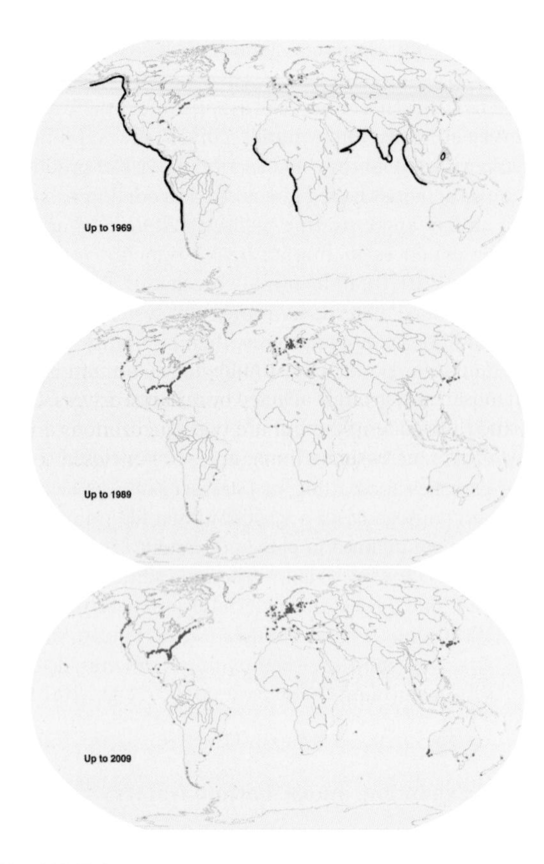

Fig. B2.5.1

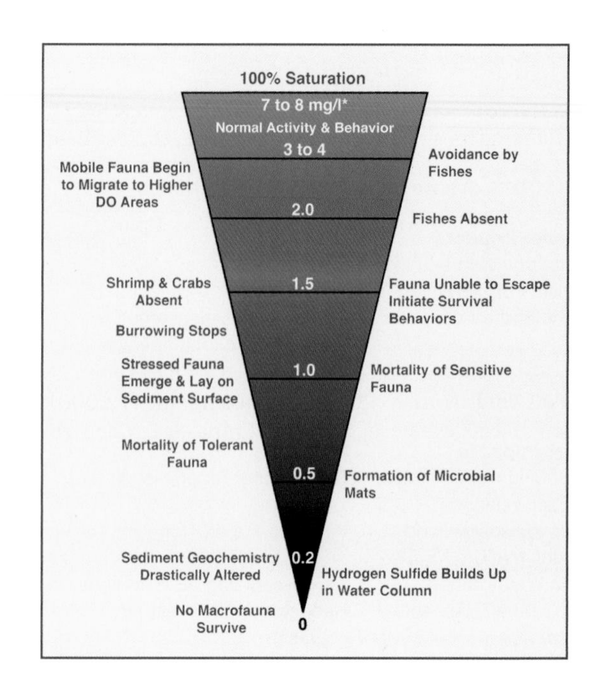

Fig. B2.5.2

predation. Sessile fauna initiate a graded series of behaviors to survive and will eventually die as DO declines or extends through time (Fig. B2.5.2).

Climate change, whether from global warming or from microclimate variation, will have consequences for eutrophication-related oxygen depletion. Climate change may make systems more susceptible to development of hypoxia through direct effects on water column stratification, solubility of oxygen, metabolism, and mineralization rates. This will likely occur primarily through warming, which will lead to increased water temperatures and a subsequent decrease in oxygen solubility, and an increase in organism metabolism and remineralization rates. Indirect effects on the quality and quantity of organic matter produced will also be important. All factors related to climate change will progressively lead to an onset of hypoxia earlier in the season and possibly extending it through time. Earlier warmer surface waters would also extend and enhance water column stratification intensity.

Much of how climate change will affect hypoxia in estuarine and coastal systems will depend on coupled land-sea interactions with climate drivers. The future pervasiveness of hypoxia will also be linked to land management practices, expansion of agriculture to feed a growing global population, and production of biofuels. Climate change will affect physical and biological processes of water column stratification, organic matter production, nutrient discharges, and rates of oxygen consumption. Land management will affect the nutrient budgets and concentrations of nutrients applied to land through agriculture. If in the next 50 years humans continue to modify and degrade coastal systems as in previous years, human population pressure will likely continue to be the main driving factor in the persistence and spreading of coastal dead zones. The expansion of agriculture for production of crops to be used for food and biofuels will result in increased nutrient loading, expand eutrophication effects, and contribute to greenhouse gases. Overall, climate drivers will tend to magnify the effects of an expanding human population.

Climate-related changes in wind patterns are of great concern for coastal systems as they control Ekman transport and upwelling/downwelling strength, which would affect stratification strength and delivery of deep-water nutrients into shallow coastal areas. Even relatively small changes in wind and current circulations could lead to large changes in the area of coastal seabed exposed to hypoxia, particularly along the Pacific coast. Changes in the pattern of upwelling in Pacific coastal waters off the Oregon and Washington coasts, due to shifts in winds that affected California current systems, appeared to be responsible for the recent development of severe hypoxia over a large area of the inner continental shelf.

oxygen are depleted or perish, thus leading to deterioration of the structure and function of the ecosystem. Marine hypoxia/anoxia is a source of greenhouse gases, methane (CH_4), and nitrous oxide (N_2O) (Naqvi *et al.*, 2010). Benthic hypoxic/anoxic areas have spread exponentially since the 1960s, with few reverses (Diaz and Rosenberg, 2008). A dead zone was first observed in the northern Gulf of Mexico in the mid-1970s (Rabalais *et al.*, 2002), which is maintained year-round by the eutrophic Mississippi and Atchafalaya rivers, pouring nutrients into the system (Turner *et al.*, 2008). This hypoxic zone reached an average size of 13,600 km^2 between 1985–2010 that increased to as much as 26,515 km^2 in 2011 (Rabalais and Turner, 2011), surpassing the 12,000 km^2 hypoxic/anoxic waters of the East China Sea (Chen *et al.*, 2007).

2.4.1.2 Regional change

Rapidly deteriorating coastal seas are losing biodiversity. The semi-enclosed Mediterranean Sea, dominated by coastal developments that house about 150 million people and 200 million tourists seasonally, exhibits poor health (El-Sayed, 2008) and loss of biodiversity (Bianchi and Morri, 2000). Overfishing has removed indigenous fishes (Tudela, 2004), and the opening of the Suez Canal in 1869 connected two biogeographical provinces (Atlanto-Mediterranean and Indo-Pacific) and introduced more than 300 Red Sea species into the Mediterranean's eastern portion (Spanier and Galil, 1991). Red Sea jellyfish (*Rhopilema nomadica*) now proliferate along the Levantine coast, profoundly affecting indigenous biota and causing significant ecologic and economic consequences (Galil, 2000). Introductions of non-native toxic fishes are now public health concerns for ciguatera and tetrodotoxin (Bentur *et al.*, 2008).

In the U.S., air, sediments, and organisms in the urban, intensely developed San Francisco Bay are polluted. High mercury concentrations are found in fish-consuming birds (Ackerman *et al.*, 2008). Concentrations of the unregulated and widely used flame retardant, polybrominated diphenyl ethers, which dramatically increased during the last decade, occur in humans; levels in harbor seals (*Phoca vitulina richardsi*) are amongst the highest known (She *et al.*, 2002). The more than 210 invasive species, including the proliferating European green crab (*Carcinus maenas*) and cordgrass (*Spartina* sp.), are reducing native populations, impacting ecological transfers, and altering ecosystem function (Grosholz and Ruiz, 2009). And wetland habitat has been destroyed by reduction of freshwater inflows from diking and filling (Nichols *et al.*, 1986). Total impacts are decreasing estuarine-dependent fish in the Bay (Feyrer *et al.*, 2007).

The northern Gulf of Mexico is another region, among many, exhibiting change. A spectacular number of jellies (*Phyllorhiza punctata*) erupted in summer 2000. The Gulf received the disastrous *Deep Horizon* Gulf oil spill in 2010, as well as increased nutrient-enriched runoff from the Mississippi River (Milly and Dunne, 2001), powerful hurricanes (e.g., Ivan, Katrina, Rita), and devastating 27 m-high waves (Stone *et al.*, 2005) that severely impacted its coasts, wetlands, and numerous oil and gas platforms (Turner, 1997). In 2002, the Gulf of Mexico and the East China Sea had the largest dead bottom areas in the world (Rabalais *et al.*, 2002).

The East China Sea, the Yellow Sea, and Bohai Sea in Southeast Asia are undergoing major modification, exhibiting eutrophy, hypoxia, and fishery collapses due to overfishing. The North Sea, Wadden Sea, and Baltic Sea are among others exhibiting major impacts. Growth in industrial fishing, human populations, industrialization, agro-industries, and water use make clear that as much as 41% of the world's coastal-marine ecosystems are "strongly affected by multiple drivers" (Halpern *et al.*, 2008).

In sum, no regional sea is unaffected by human influences, with most being strongly affected by multiple drivers. Overfishing and nutrient enrichment are two synchronous anthropogenic effects increasingly impacting semi-enclosed seas since World War II (Caddy, 1993, 2000). Within these ecosystems, synergistic effects interfere with wildlife, feeding hierarchies, setting off an initial increase in productivity of benthic/demersal and pelagic food webs due to "bottom-up" and "top-down" impacts. These impacts lead toward a progressive predominance of short-lived pelagic species (nuisance species). Biomass removal from the most productive ocean ecosystems (upwelling areas, temperate continental-shelf systems, etc.) removes primary production from that system (Pauly and Christensen, 1995), changes energy flow and species dominance, and causes profound, poorly explored impacts that are raising concern. The removal of fishes from Large Marine Ecosystems (LMEs, Ch. 4), areas that produce the major portion of global fisheries yield, affects ecosystem structure, function, and fish yields (Worm *et al.*, 2006). Removing apex predators from pelagic food webs, i.e., removing the largest and oldest individuals, causes an ecological cascade that can restructure food webs (Myers and Worm, 2003; Kitchell *et al.*, 2006). The excessive removal of cod, for example, has disrupted food-web energy flow, which caused a progressive decline of the benthic system (Frank *et al.*, 2005). Removal of large sharks has resulted in increased nuisance prey species, for example over-abundance of cownose rays that feed on scallop beds, thereby forcing closure of a century-long scallop fishery (Myers *et al.*, 2007). Evidence makes clear that historical overfishing of large marine vertebrates, i.e., whales, sea turtles, manatees, seals, fishes, and others, has affected the functional performance of ecosystems. For example, overfishing productive coastal seas that have complex food webs, high biomass, and large animals may simplify biological systems, amplifying boom-and-bust fishing cycles. As these systems become increasingly disturbed and overfished, energy that goes into fisheries production increasingly favors microorganisms and jellies (Jackson *et al.*, 2001; Mills, 2001). Some LMEs appear to have increasing abundance of jellies (Brotz *et al.*, 2012), but their global rise is difficult to substantiate (Condon *et al.*, 2012), being masked by normal fluctuations (Arai, 2001) and exacerbated by climate change (Gibbons and Richardson, 2009). Jelly increases may be due to combinations of overfishing, eutrophy, introduced species, and habitat modification that restructure pelagic ecosystems into less desirable states (Richardson *et al.*, 2009). The overall impact reduces biodiversity, species recovery potential, ecosystem stability, and water quality (Worm *et al.*, 2006).

2.4.2 Global ocean change

Changes occurring in the global ocean relate to climate, global cycles, and the stability of the Earth's cybernetic processes.

2.4.2.1 Climate and ocean warming

Warming of the climate system is unequivocal (IPCC, 2007; Hansen *et al.*, 2012), and change can be abrupt (NRC, 2002b). Global climate temperature increased considerably during the 20th century, especially since the 1970s (Jones and Moberg, 2003; Vose *et al.*, 2005; Gleason *et al.*, 2008), although the timing and severity of warming, its regional impacts, and the magnitude of feedback processes remain uncertain (Millero, 2007). However, subpolar ice and mountain glaciers are declining in mass worldwide (Dyurgerov, 2003; Oerlemans, 2005; Ch. 7). The world ocean is a significant temperature sink with a close connection to climate (Ch. 4).

The world ocean is warming, with robust certainty (Levitus *et al.*, 2005, 2009; Lyman *et al.*, 2010; Fig. 2.17a). Its upper 700 m has warmed an average of 0.18°C over the last 40 years (IPCC, 2007) and is predicted to increase within decades. There may be significant warming below 700 m depths as well (Trenberth and Fasullo, 2010). Rates of change are fastest in the Arctic Ocean and adjacent seas (Spielhagen *et al.*, 2010), threatening the survival of ice-dependent species (Ch. 7); loss of sea ice set summer records in 2007 (Perovich *et al.*, 2008), which was exceeded in 2012 (National Snow and Ice Data Center, online). Sea-ice loss is helping to open an area rich in oil and gas and cruise boat opportunities. This loss is also bringing promise of new, shorter shipping routes and initiating a race among Arctic nations to claim jurisdiction over extensive natural resources on continental shelves (Ch. 3). Polar sea-ice melt and diminishing ice caps on Greenland and Antarctica are most striking examples of global climate change.

2.4.2.2 Sea-level rise

A model analysis indicates a close link between global temperature and the rate of sea-level rise (Vermeer and Rahmstorf, 2009). Melting of Greenland and Arctic continental ice sheets and the expansion of seawater volume from ocean warming contribute to global sea-level rise. Throughout the 20th century, global sea level has risen at an accelerating rate

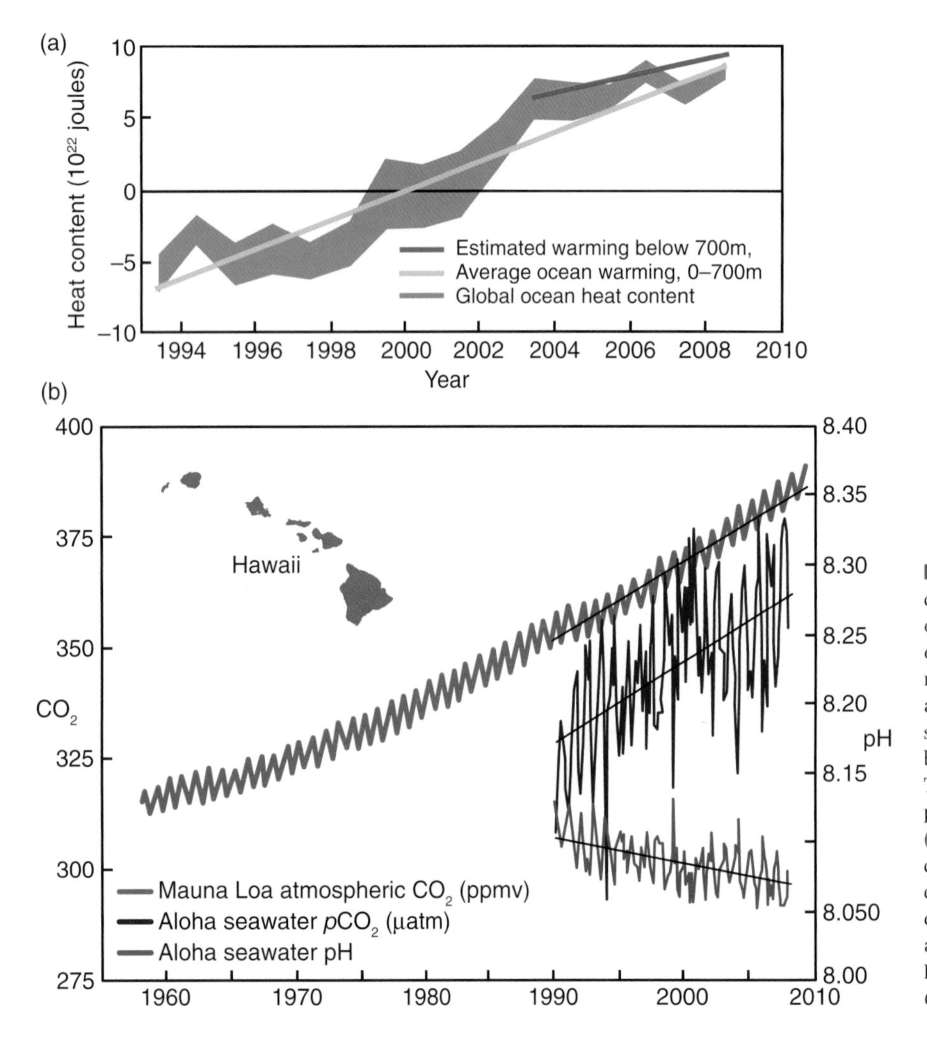

Fig. 2.17 Trends in ocean warming, carbon dioxide (CO_2) concentrations, and ocean acidification. (a) Increasing global ocean heat content encompassing error range with respect to mean (blue), showing an average (0.64 watts m^{-2}) increase from surface to 700 m depths (yellow); estimates below 700 m also show increases (red). Trenberth 2010. Reprinted with permission of MacMillan Publishers Ltd. (b) Increasing trend in atmospheric CO_2 concentrations (red), surface ocean carbon dioxide (pCO_2, purple), and decreasing ocean surface acidity (pH, green) measured at Mauna Loa, Hawaii. From Doney SC, Balch WM, Fabry VJ, Feely RA (2009) *Oceanography* **22**, 16–25, with permission.

(Church and White, 2006), and since 1950 it has risen by an average of 1.7 ± 0.3 mm $year^{-1}$ (Domingues *et al.*, 2008). Rising sea levels and the potential for stronger storms threaten the world's coastal regions, flooding many of the largest and most densely populated cities, ports, and low-lying regions and islands, e.g., Hurricane Sandy that devastated the northeastern U.S. in October 2012. Low-lying Bangladesh and island nations such as Kiribati, the Maldives, and The Bahamas are further subjected to increases in hurricanes, typhoons, and tsunamis resulting from climate change.

2.4.2.3 Altered water cycle

Growing human demands for water and groundwater sources (Konikow and Kendy, 2005) are depleting and polluting/degrading water ecosystems (Rosegrant *et al.*, 2002), worsened by climate change (Vörösmarty *et al.*, 2000), and impacting the terrestrial water cycle on a global scale (Hoff, 2002). Humans withdraw about 10% of the globally available freshwater from rivers and groundwater each year (Oki and Kanae, 2006), with 50% from groundwater for drinking (Llamas and Martínez-Santos, 2005) and increasingly for irrigation and non-consumptive uses, both of which have increased in the last 50 years. Large-scale changes in freshwater flux have potentially important implications for ocean circulation, climate (Peterson *et al.*, 2002), and the hydrologic cycle (Rahmstorf, 1995). Terrestrial water storage associated with dam building (Section 2.3.3.2; Chao *et al.*, 2008) and groundwater mining contributes to changes in global water cycling, through processes yet to be understood (Domingues *et al.*, 2008). Climate change and groundwater depletions together are globally affecting the global hydrologic cycle and decreasing water quality and yields, which damage ecosystems, cause land subsidence, and increase pumping costs. Higher temperatures will increase the snowmelt season, turn snowfall into rainfall, and substantially change the timing and volume of spring flood, altering hydrologic conditions important to coastal ecosystems, especially estuaries.

2.4.2.4 Altered biogeochemical cycling: sinks, sources, and transformation

The oceans play a critical role in global biogeochemical cycles (Schlesinger, 1997; Sarmiento and Gruber, 2006) that human activities are significantly influencing (Galloway *et al.*, 2004; Duce *et al.*, 2008; Doney, 2010). The role of oceans directs attention to key biogenic elements of carbon and nitrogen (Jickells, 1998) and to heat-trapping gases ("greenhouses gases": carbon dioxide, CO_2; methane, CH_4; nitrous oxide, N_2O; fluorinated gases, etc.). Heavily populated coastlines release reactive nitrogen, which increased nearly 80% between 1860 and 1990 (JOCI, 2008) to fertilize and degrade coastal water, China's releases being globally significant (Zhang, 2002). And marine vessels are contributing a large portion of the world's greenhouse gases; international shipping in 2007 contributed an estimated 2.7% to global CO_2 emissions (Buhaug *et al.*, 2009). These releases are expected to increase as ships become larger and more numerous.

The rise in CO_2 concentrations influences the global carbon cycle and increases ocean acidification (Feely *et al.*, 2004). Oceans have absorbed an estimated third of the CO_2 released by human activities between 1800 and 1994 (Sabine *et al.*, 2004). And along with increases in carbon dioxide emissions, the powerful and persistent nitrous oxide gas (N_2O) is on the rise, with methane gas also playing an increasingly important role (Heimann, 2010). These greenhouse gases, with water vapor and chlorofluorocarbons (CFCs), are increasing in the atmosphere and increasing global temperatures (Millero, 2007; Doney, 2010); their interplay with the ocean is an active field of research.

2.4.2.5 Ocean acidification

The input of gigatons of carbon dioxide into the oceans is acidifying ocean water, i.e., lowers its pH, where pH is a measure of acidity (Fig. 2.17b; NRC, 2010b). Ocean acidification results when carbon dioxide combines with water to instantly form bicarbonate (HCO_3^-) and hydrogen ions—the H+ that increases acidity that causes major impacts on marine organisms (Feely *et al.*, 2004). The impact of ocean acidity on sea life is varied. Evidence reveals that in sufficient concentration it can lower net calcification (Ries *et al.*, 2009). Ocean acidity impairs such animals as clams, oysters, corals, and more, interfering with their capacity to extract calcium carbonate to build their skeletons or shells. When hydrogen ion concentrations are high enough and carbonate concentrations are driven down, an organism's calcium carbonate shell begins to dissolve, which increases the energetics required by organisms to extract carbonate from surrounding water. Early signs of ocean acidification and its effects on organisms (Orr *et al.*, 2005; Yamamoto-Kawai *et al.*, 2009), are affecting calcium-carbonate shells and skeletons for a broad range of marine species, e.g., corals, microscopic protozoa, certain algae (Guinotte and Fabry, 2008). This effect is supported in a controlled study carried out at Woods Hole Oceanographic Institution (WHOI, 2010; Fig. 2.18; Ries *et al.*, 2009).

The regions with greater capacity for carbon dioxide absorption are cold, high-latitude surface waters. They are first to experience the impact, then the tropics (Feely *et al.*, 2004). This effect is, however, far from clear due to regional differences and the diversity of tolerances among species and their complex life histories. How these impacts translate into ecosystem change is unknown, but one thing is certain: acidification is massive and rapid and the more carbon dioxide emitted, the worse it is going to get—"an experiment we would not choose to do" (Kerr, 2010). Quick and aggressive emissions reductions are key to minimizing this apparently irreversible acidification process.

2.5 THE CHALLENGE FOR THE 21ST CENTURY

Primary issues awaken us to diminishing biological richness, loss of ocean habitat, proliferation of undesirable nuisance and harmful species, ill health, and effects on critical ecosystems (Millennium Ecosystem Assessment, 2005). Secondary

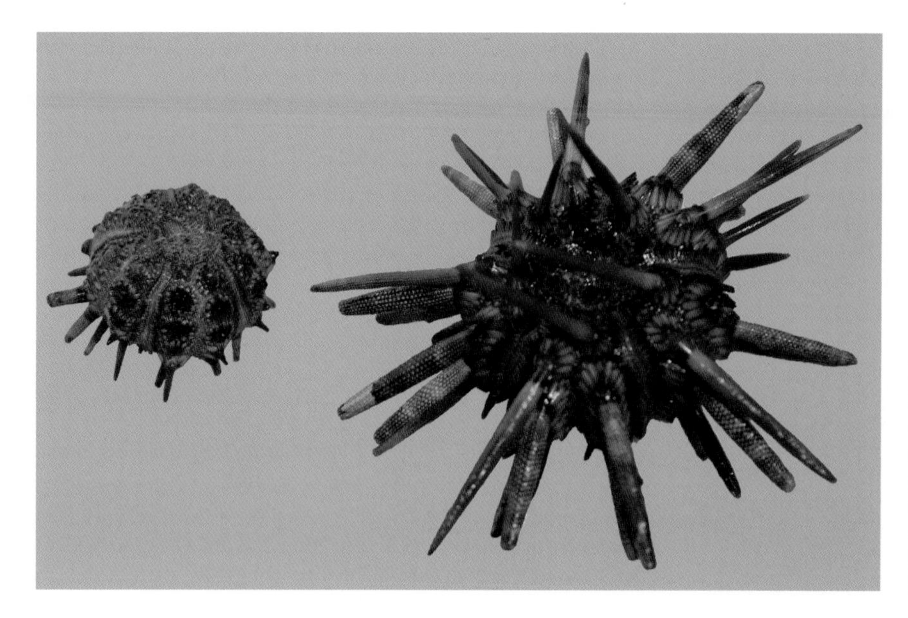

Fig. 2.18 Impact of ocean acidity on sea life, as studied by Justin Ries at Woods Hole, Massachusetts. Right: sea urchin grown under present-day ocean acidity has normal spines and appears healthy. Left: sea urchin grown under higher CO2 concentrations in more acidic seawater conditions is substantially damaged. Photograph by Tom Kleindinst © Woods Hole Oceanographic Institution.

issues highlight the concentration of human activities in the coastal fringe—removing resources, adding foreign substances, and physically altering coastal systems through human activities that are expanding in magnitude, duration, and intensity. Tertiary issues draw attention to emergent phenomena, with global losses of ecological services that are moving Planet Earth beyond a safe operating space for humanity—its planetary boundaries (Rockström *et al.*, 2009), including climate change, biodiversity loss, and human interference in the nitrogen cycle, whose boundaries have already been exceeded.

Increasing world population and resource demands challenge individuals and society to confront ecosystem limits through decision-making power to protect, restore, and conservatively use the benefits and services that coastal and marine ecosystems provide. The dynamic interactions between land, sea, air, and human influences are moving local, regional, and global ecosystems into an unpredictable future burdened with environmental debts from historical abuses. Marine conservation is equipped with an arsenal of advancing and evolving social mechanisms (Ch. 3) and increasing scientific understanding (Chs. 4, 5) needed to address the accelerating pace of this intensifying *Anthropocene* era (Steffen *et al.*, 2007; Ch. 13). The range of issues presented here are global in scope, demanding a multi-scale, global system focus.

REFERENCES

Ackerman JT, Eagles-Smith CA, Takekawa JY, Jill D, Bluso JD, Adelsbach TL (2008) Mercury concentrations in blood and feathers of prebreeding Forster's terns in relation to space use of San Francisco Bay, California, USA, habitats. *Environmental Toxicology and Chemistry* **27**, 897–908.

Aguirre AA, Lutz PL (2004) Marine turtles as sentinels of ecosystem health: is fibropapillomatosis an indicator? *EcoHealth* **1**.

Airoldi L, Abbiati M, Beck MW, Hawkins SJ, Jonsson PR, Martin D, Moschella PS, *et al.* (2005) An ecological perspective on the deployment and design of low-crested and other hard coastal defense structures. *Coastal Engineering* **52**, 1073–1087.

Airoldi L, Beck MW (2007) Loss, status and trends for coastal marine habitats of Europe. *Oceanography and Marine Biology: Annual Review* **45**, 345–405.

Airoldi L, Connell SD, Beck MW (2009) Chapter 19: The loss of natural habitats and the addition of artificial substrata. In *Marine Hard Bottom Communities* (ed. Wahl M), Ecological Studies **206**, Springer-Verlag Berlin Heidelberg, 269–280.

Al-Bahry SN, Mahmoud IY, Al-Belushi KIA, Elshafie AE, Al-Harthy A, Bakheit CK (2009) Coastal sewage discharge and its impact on fish with reference to antibiotic resistant bacteria and enteric pathogens as bio-indicators of pollution. *Chemosphere* **77**, 1534–1539.

Anderson DM (2004) The growing problem of harmful algae. *Oceanus Magazine* **43**, 1–5. www.whoi.edu/oceanus

Anderson D, McGillicuddy D, Townsend D, Turner J (2005) The ecology and oceanography of toxic *Alexandrium fundyense* blooms in the Gulf of Maine. *Deep-Sea Research* Part II **52**, 2365–2368.

Arai MN (2001) Pelagic coelenterates and eutrophication: a review. *Hydrobiologia* **451**, 69–87.

Ballance L (2007) Understanding seabirds at sea: why and how? *Marine Ornithology* **35**, 127–135.

Barbier EB, Hacker SD, Kennedy C, Koch EW, Stier AC, Silliman BR (2011) The value of estuarine and coastal ecosystem services. *Ecological Monographs* **81**, 169–193.

Bart DM, Burdick D, Chambers R, Hartman J (2006) Human facilitation of *Phragmites australis* invasions in tidal marshes: a review and synthesis. *Wetland Ecology and Management* **14**, 53–65.

Baum JR, Myers A, Kehler DG, Worm B, Harley SJ, Doherty PA (2003) Collapse and conservation of shark populations in the Northwest Atlantic. *Science* **299**, 389.

Beck MB, Brumbaugh RD, Airoldi L, Carranza A, Coen LD, Crawford C, *et al.* (2009) *Shellfish reefs at risk: a global analysis of problems and solutions*. The Nature Conservancy, Arlington VA, 1–52.

Bentur Y, Ashkar J, Lurie Y, Levy Y, Azzam ZS, Litmanovich M, Golik M, *et al.* (2008) Lessepsian migration and tetrodotoxin poisoning due to *Lagocephalus sceleratus* in the eastern Mediterranean. *Toxicon* (Oxford) **52**, 964–968.

Bianchi CN, Morri C (2000) Marine biodiversity of the Mediterranean Sea: situation, problems and prospects for future research. *Marine Pollution Bulletin* **40**, 367–376.

Bianchi, TS, Engelhaupt E, Westman P, Andren T, Rolff C, Elmgren R (2000) Cyanobacterial blooms in the Baltic Sea: natural or human induced? *Limnology and Oceanography* **45**, 716–726.

Bjorndal KA, Bowen BW, Chaloupka M, Crowder LB, Heppell SS, Jones CM, *et al.* (2011) Better science needed for restoration in the Gulf of Mexico. *Science* **331**, 537–538.

Bollmann M, Bosch T, Colijn F, Ebinghaus R, Kortzinger A, Latif M, *et al.* (2010) *World Ocean Review 2010*. Maribus gGmbH, Hamburg, Germany, 240pp.

Bricker SB, Longstaff B, Dennison W, Jones A, Boicourt K, Wicks C, Woerner J (2008) Effects of nutrient enrichment in the nation's estuaries: A decade of change. *Harmful Algae* **8**, 21–32.

Brinkhoff T (online) Principal agglomerations of the world. www.citypopulation.de

Bromberg K, Silliman BR (2009) Patterns of salt marsh loss within coastal regions of North America: pre-settlement to present. In *Human Impacts on Salt Marshes: A Global Perspective* (eds Silliman BR, Grosholz T, Bertness MD). University of California Press.

Brotz L, Cheung WWL, Kleisner K, Pakhomov E, Pauly D (2012) Increasing jellyfish populations: trends in Large Marine Ecosystems. *Hydrobiologia* **690**, 3–20.

Buhaug Ø, Corbett JJ, Endresen Ø, Eyring V, Faber J, Hanayama S, Lee DS, *et al.* (2009) *Second IMO GHG Study 2009*. International Maritime Organization (IMO), London, UK, April.

Bumb BL, Baanante CA (1996) World trends in fertilizer use and projections to 2020. International Food Policy Research Institute, 2020, Brief 38, October.

Burke LA, Kura Y, Kassem K, Revenga C, Spalding M, McAllister D (2001) *Coastal Ecosystems*. World Resources Institute, Washington, D.C., 1–93.

Caddy JF (1993) Toward a comparative evaluation of human impacts on fishery ecosystems of enclosed and semi-enclosed seas. *Reviews in Fisheries Science* **1**, 57–95.

Caddy JF (2000) Marine catchment basin effects versus impacts of fisheries on semi-enclosed seas. *ICES Journal of Marine Science* **57**, 628–640.

Campagna C, Short FT, Polidoro BA, McManus R, Collette BB, Pilcher NJ, Sadovy Y, Stuart SN, Carpenter KE (2011). Gulf of Mexico oil blowout increases risks to globally threatened species. *Bioscience* **61**, 393–397.

Cervino JM, Hayes RL, Polson SW, Polson SC, Goreau TJ, Martinez RJ, Smith GW (2004) Relationship of *Vibrio* species infection and elevated temperatures to yellow blotch/band disease in Caribbean corals. *Applied and Environmental Microbiology* **70**, 6855–6864.

Chambers RM, Meyerson LA, Saltonstall K (1999) Expansion of *Phragmites australis* into tidal wetlands of North America. *Aquatic Botany* **64**, 261–273.

Chao BF, Wu YH, Li YS (2008) Impact of artificial reservoir water impoundment on global sea level. *Science* **320**, 212–214.

Chen C-C, Gong G-C, Shiah F-K (2007) Hypoxia I: the East China Sea: one of the largest coastal low-oxygen areas in the world. *Marine Environmental Research* **64**, 399–408.

Chen J, Denison MS (2011) The deepwater horizon oil spill: environmental fate of the oil and the toxicological effects on marine organisms. *Journal of Young Investigators* **21**, 84–95.

Christensen PB, Glud RN, Dalsgaard T, Gillespie P (2003) Impacts of longline mussel farming on oxygen and nitrogen dynamics and biological communities of coastal sediments. *Aquaculture* **218**, 567–588.

Church JA, White NJ (2006) A 20th century acceleration in global sea-level rise. *Geophysical Research Letters* **33**, L01602.

Clark RB (1997) *Marine Pollution*. Fourth Edition. Clarendon Press, Oxford.

Cleveland C, Hogan CM, Saundry P (2011) Deepwater Horizon oil spill. In *Encyclopedia of Earth* (ed. Cleveland CJ). Environmental Information Coalition, National Council for Science and the Environment, Washington, D.C. www.eoearth.org/article/Deepwater_Horizon_oil_spill?topic=50364

Condon RH, Graham WM, Duarte CM, Pitt KA, Lucas CH, *et al.* (2012) Questioning the rise of gelatinous zooplankton in the world's oceans. *BioScience* **62**, 160–169.

Coumans C (2003) Will rising closure costs sink Placer Dome? Environmental Mining Council of BC, February 14, 1–4. Yates St., Victoria, BC. *Mining's Problem with Waste*. Mining Watch Canada. Ottawa, Ontario K1R 6K7, Canada, 1–4.

Crooks JA (2009) Chapter 16. The role of exotic marine ecosystem engineers. In *Biological Invasions in Marine Ecosystems* (eds Rilov G, Crooks JA). *Ecological Studies* **204**, 287–304.

Dachs J, Méjanelle L (2010) Organic pollutants in coastal waters, sediments, and biota: a relevant driver for ecosystems during the Anthropocene? *Estuaries and Coasts* **33**, 1–1.

Darbra R-M, Casal J (2004) Historical analysis of accidents in seaports. *Safety Science* **42**, 85–98.

Daves NK (2009) CITIES gives hope to the Queen conch. *Endangered Species Bulletin* **34**, 14–15.

Davies AJ, Roberts JM, Hall-Spencer J (2007) Preserving deep-sea natural heritage: emerging issues in offshore conservation and management. *Biological Conservation* **138**, 299–312.

Davy A, Figueroa E, Bakker J (2009) Human modification European salt marshes. In *Human Impacts on Salt Marshes: A Global Perspective* (eds Silliman BR, Grosholz T, Bertness MD). University of California Press, Berkeley, California.

de Magny GC, Long W, Brown CW, Hood RR, Huq A, Murtugudde R, Colwell RR (2010) Predicting the distribution of vibrio spp. in the Chesapeake Bay: a vibrio cholerae case study. *EcoHealth* online.

De Swart RL, Harder TC, Ross PD, Vos HW, Osterhaus ADME (1995) Morbilliviruses and morbillivirus diseases in marine mammals. *Infectious Agents and Disease* **4**, 125–130.

Diaz RJ, Rosenberg R (2008) Spreading dead zones and consequences for marine ecosystems. *Science* **321**, 926.

Domingues CM, Church JA, White NJ, Gleckler PJ, Wijffels SE, Barker PM, Dunn JR (2008) Improved estimates of upper-ocean warming and multi-decadal sea-level rise. *Nature* **453**, 1090–1093.

Donaldson A, Gavriel C, Harvey BJ, Carolsfeld J (2010) Impacts of Fishing Gears other than Bottom Trawls, Dredges, Gillnets, and Longlines on Aquatic Biodiversity and Vulnerable Marine Ecosystems. Fisheries and Oceans (DFO) Canadian Science Advisory Secretariat, Research Document 2010/011, Canada, vi +84pp.

Doney SC (2010) The growing human footprint on coastal and open-ocean biogeochemistry. *Science* **328**, 1512–1516.

Doney SC, Balch WM, Fabry VJ, Feely RA (2009) Ocean acidification. *Oceanography* **22**, 16–25.

Dong Z, Liu D, Keesing JK (2010) Jellyfish blooms in China: dominant species, causes and consequences. *Marine Pollution Bulletin* **60**, 954–963.

Duce RA, LaRoche J, Altieri K, Arrigo KR, Baker AR, Capone DG, Cornell S, *et al.* (2008) Impacts of atmospheric anthropogenic nitrogen on the open ocean. *Science* **320**, 893–897.

Dyurgerov M (2003) Mountain and subpolar glaciers show an increase in sensitivity to climate warming and intensification of the water cycle. *Journal of Hydrology* **282**,164–176.

Ellner SP, Jones LE, Mydlarz LD, Harvell CD (2007) Within-host disease ecology in the sea fan *Gorgonia ventalina*: modeling the spatial immunodynamics of a coral-pathogen interaction. *The American Naturalist* **170**, E143–E00.

Emch M, Feldacker C, Islam MS, Ali M (2008) Seasonality of cholera from 1974 to 2005: a review of global patterns. *International Journal of Health Geographics* **7**, 31.

El-Sayed MK (2008) Chapter 6. marine environment. In *Arab Environment: Future Challenges* (eds Tolba MK, Saab NW). Report of the Arab Forum For Environment and Development, UNEP.

Emmerson MC, Solan M, Emes C, Paterson DM, Raffaelli D (2001) Consistent patterns and the idiosyncratic effects of biodiversity in marine ecosystems. *Nature* **411**, 73–77.

FAO (2005) *Current World Fertilizer Trends and Outlook to 2009/10*. Food and Agriculture Organization of the United Nations, Rome.

FAO (2010, 2011, 2012) *The State of World Fisheries and Aquaculture*. Food and Agriculture Organization of the United Nations, Rome. Online.

Feely RA, Sabine CL, Lee K, Berelson W, Kleypas J, Fabry VJ, Millero FJ (2004) Impact of anthropogenic CO_2 on the $CaCO_3$ system in the oceans. *Science* **305**, 362–366.

Feyrer FF, Nobriga ML, Sommer TR (2007) Multidecadal trends for three declining fish species: habitat patterns and mechanisms in the San Francisco estuary, California USA. *Canadian Journal of Aquatic Science* **64**, 723–734.

Fishel FM (2007) *Pesticide Use Trends in the U.S.: Global Comparison*. University of Florida IFAS Extension, doc P-I 143, EDIS website at www.edis.ifas.ufl.edu

Fitzgerald WF, Lamborg CH, Hammerschmidt CR (2007) Marine biogeochemical cycling of mercury. *Chemical Reviews* **107**, 641–662.

Flothmann S, von Kistowski K, Dolan E, Lee E, Meere F, Album G (2010) Closing loopholes: getting illegal fishing under control. *Science* **328**, 1235–1236.

Fourqurean JW, Duarte CM, Kennedy H, Marbà N, Holmer M *et al.* (2012) Seagrass ecosystems as a globally significant carbon stock. *Nature Geoscience*, published online, 20 May.

Frank KT, Petrie B, Choi JS, Leggett WC (2005) Trophic cascades in a formerly cod-dominated ecosystem. *Science* **308**, 1621–1623.

Freeman MC, Pringle CM, Greathouse EA, Freeman BJ (2003) Ecosystem-level consequences of migratory faunal depletion caused by dams. *American Fisheries Society Symposium* **35**, 255–266.

Fu J, Mai B, Sheng G, Zhang G, Wang X, Peng P, Xiao X, *et al.* (2003) Persistent organic pollutants in environment of the Pearl River Delta, China: An Overview. *Chemosphere* **52**, 1411–1422.

Galil BS (2000) A sea under siege—alien species in the Mediterranean. *Biological Invasions* **2**, 177–186,

Galloway JN, Dentener FJ, Capone DG, Boyer EW, Howarth RW, Seitzinger SP, Asner GP, Cleveland CC, Green PA, Holland EA, Karl DM, Michaels AF, Porter JH, Townsend AR, Vörösmarty CJ (2004) Nitrogen cycles: past, present and future. *Biogeochemistry* **70**, 153.

Garcia SM, Moreno IL, Grainger R (2005) Global trends in the state of marine fisheries resources 1974–2004. In *Review of the State of World Marine Fishery Resources*. FAO Fisheries Technical Paper 457. Food and Agriculture Organization of the United Nations, Rome.

Gedan BK, Silliman BR, Bertness MD (2009) Centuries of human-driven change in salt marsh ecosystems. *Annual Review of Marine Science* **1**, 117–141.

Geraci JR, St. Aubin DJ, Barker IK, Webster RG and others (1982) Mass mortality of harbor seals: pneumonia associated with influenza A virus. *Science* **215**, 1129–1131.

Geraci JR, Anderson DM, Timperi RJ, St. Aubin DJ, Early GA, Prescott JH, Mayo CA (1989) Humpback whales (*Megaptera novaeangliae*) fatally poisoned by dinoflagellate toxin. *Canadian Journal of Fisheries and Aquatic Sciences* **46**, 1895–1898.

Gianni M, Simpson W (2005) The Changing Nature of High Seas Fishing: how Flags of Convenience Provide Cover for Illegal, Unreported and Unregulated Fishing. An independent report funded by Australian Government Department of Agriculture, Fisheries and Forestry, International Transport Workers Federation and WWF International, October, 1–83.

Gibbons KJ, Richardson AJ (2009) Patterns of jellyfish abundance in the North Atlantic. *Hydrobiologia* **616**, 51–65.

Giuliano G (2007) *Port-Related Trade: Regulation and Response in Los Angeles*. Presentation Center for Transportation Studies, November 29, University of Virginia, 2007.

Glasby TM, Connell SD (1999) Urban structures as marine habitats. *Ambio* **28**, 595–598.

Gleason KL, Lawrimore JH, Levinson DH, Karl TR, Karoly DJ (2008) A revised U.S. climate extreme index. *Journal of Climate* **21**, 2124–2137.

Goldstein T, Mazet JAK, Gill VA, Doroff AM, Burek KA, Hammond JA (2009) Phocine distemper virus in northern sea otters in the Pacific Ocean, Alaska, USA. *Emerging Infectious Diseases* **15**, 925–927.

Goreau TJ, Cervino J, Goreau M, Hayes R, Hayes M, *et al.* (1998) Rapid spread of diseases in Caribbean coral reefs. *Revista de Biologia Tropical (San Jose)* **46** (Suppl. 5), 157–171.

Godet L, Fournier J, van Katwijk MM, Olivier F, Le Mao P, Retière C (2008) Before and after wasting disease in common eelgrass *Zostera marina* along the French Atlantic coasts: a general overview and first accurate mapping. *Diseases of Aquatic Organisms* **79**, 249–255.

Graham WM, Bayha KM (2007) Biological invasions by marine jellyfish. In *Biological Invasions, Ecological Studies* **193**, Springer, 239–255.

Graham WM, Martin DL, Felder DL Asper VL, Perry HM (2003) Ecological and economic implications of a tropical jellyfish invader in the Gulf of Mexico. *Biological Invasions* **5**, 53–69.

Gray JS, Dayton P, Thrush S, Kaiser MJ (2006) On effects of trawling, benthos and sampling design. *Marine Pollution Bulletin* **52**, 840–843.

Gregory MR (2009) Review: Environmental implications of plastic debris in marine settings—entanglement, ingestion, smothering, hangers-on, hitch-hiking and alien invasions. *Philosophical Transactions of the Royal Society B* (2009) **364**, 2013–2025.

Grant KS, Burbacher TM, Faustman EM, Gratttan L (2010) Domoic acid: Neurobehavioral consequences of exposure to a prevalent marine biotoxin. *Neurotoxicology and Teratology* **32**,132–141.

Grosholz ED, Ruiz GM (2009) Chapter 17. Multitrophic effects of invasions in marine and estuarine systems. In *Biological Invasions in Marine Ecosystems* (eds Rilov G, Crooks JA). *Ecological Studies* **204**, Springer-Verlag Berlin, 305–324.

Guinotte JM, Fabry VJ (2008) Ocean acidification and its potential effects on marine ecosystems. *Annals of the New York Academy of Sciences* **1134**, 320–342.

Halpern BS, Walbridge S, Selkoe KA, Kappel CV, Micheli F, D'Agrosa C, *et al.* (2008) A global map of human impact on marine ecosystems. *Science* **319**, 948–952.

Halpern BS, Silliman BR, Olden J, Bruno J, Bertness MD (2007) Incorporating positive interactions in aquatic restoration and conservation. *Frontiers in Ecology and the Environment* **5**, 153–160.

Hansen J, Sato M, Ruedy R (2012) Perception of climate change. *Proceedings of the National Academy of Sciences USA* (PNAS) Early Edition, 1–9. www.pnas.org/cgi/doi/10.1073/pnas.1205276109

Härkönen T, Dietz R, Reijnders P, Teilmann J, Harding K, Hall A, Brasseur S, Siebert U, Goodman SJ, Jepson PD, Rasmussen TD, Thompson P (2006) A review of the 1988 and 2002 phocine distemper virus epidemics in European harbour seals. *Diseases of Aquatic Organisms* **58**, 116–130.

Harvell CD, Kim K, Burkholder JM, Colwell RR, Epstein PR, Grimes DJ, *et al.* (1999) Emerging marine diseases—climate links and anthropogenic factors. *Science* **285**, 1505–1510.

Havens KJ, Berquist H, Priest WI (2003) Common reed grass, *Phragmites australis*, expansion into constructed wetlands: Are we mortgaging our wetland future? *Estuaries* **26**, 417–422.

Heil CA, Glibert PM, Al-Sarawi MA, Faraj M, Behbehani M, Husain M (2001) First record of a fish-killing *Gymnodinium* sp. bloom in Kuwait Bay, Arabian Sea: chronology and potential causes. *Marine Ecological Progress Series* **214**, 15–23.

Heimann M (2010) How stable is the methane cycle? *Science* **327**, 1211–1212.

Hickie BE, Ross PS, MacDonald RW, Ford JB (2007) Killer whales (*Orcinus orca*) face protracted health risks associated with lifetime exposure to PCBs. *Environmental Science and Technology* **41**, 6613–6619.
Hirai H, Takada H, Ogata Y, Yamashita R, Mizukawa K, Saha M, *et al.* (2011) Organic micropollutants in marine plastics debris from the open ocean and remote and urban beaches. *Marine Pollution Bulletin* **62**, 1683–1692.
Hoff H (2002) The water challenge: Joint Water Project LOICZ. *Global Change Newsletter* **46**.
Hood WG (2004) Indirect environmental effects of dikes and estuarine channels: thinking outside of the dike for habitat restoration and monitoring. *Estuaries* **27**, 273–282.
Holt S (1969) The food resources of the ocean. *Scientific American* **221**, 2–15.
Howarth RW (2008) Coastal nitrogen pollution: a review of sources and trends globally and regionally. *Harmful Algae* **8**, 14–20.
Howell NL, Suarez MP, Rifai HS, Koenig L (2008) Concentrations of polychlorinated biphenyls (PCBs) in water, sediment, and aquatic biota in the Houston Ship Channel, Texas. *Chemosphere* **70**, 593–606.
Hu C, Li D, Chen C, Ge J, Muller-Karger FE, Liu J, Yu F, He M-X (2010) On the recurrent *Ulva prolifera* blooms in the Yellow Sea and East China Sea. *Journal of Geophysical Research*, **115**, C05017.
Hutchings JA (2005) Life history consequences of overexploitation to population recovery in Northwest Atlantic cod (*Gadus morhua*). *Canadian Journal of Fisheries and Aquatic Science* **62**, 824–832.
Idyll CP (1978) *The sea against hunger*. Thomas Y. Crowell, New York City.
IPCC (Intergovernmental Panel on Climate Change) (2007) *Climate Change 2007: Synthesis Report*. Summary for Policymakers. www.ipcc.ch/pdf/assessment-report/ar4/ar4_syr_spm.pdf
Irwin RJ, VanMouwerik M, Stevens L, Seese MD, Basham W (1997) *Environmental Contaminants Encyclopedia*. National Park Service, Water Resources Division, Fort Collins, Colorado. Distributed within the Federal Government as an Electronic Document (Projected public availability).
IUCN (2012) *IUCN Red List of Threatened Species*. www.redlist.org
Jackson JBC, Kirby MX, Berger WH, Bjorndal KA, Botsford LW, Bourque BJ, *et al.* (2001) Historical overfishing and the recent collapse of coastal ecosystems. *Science* **293**, 629–638.
Jennings S, Kaiser MJ, Reynolds JD (2001) *Marine Fisheries Ecology*. Blackwell Science, Oxford UK.
Jernelöv A, Lindén O (1981) Ixtoc I: Case study of the world's largest oil spill. *Ambio* **10**, 299–306.
Jickells TD (1998) Nutrient biogeochemistry of the coastal zone. *Science* **281**, 217–222.
Jones PD, Moberg A (2003) Hemispheric and large-scale surface air temperature variations: an extensive revision and an update to 2001. *Journal of Climate* **16**, 206–223.
JOCI (2008) Changing Oceans, Changing World. Joint Ocean Commission Initiative. September. Meridian Institute Washington, D.C. www.jointoceancommission.org
Kaiser MJ, Edwards DB, Armstrong PA, Radford K, Lough NEL, Flatt RP, Jones HD (1998) Changes in megafaunal benthic communities in different habitats after trawling disturbance. *ICES Journal of Marine Science* **55**, 353–361.
Kaiser J (2010) The dirt on ocean garbage patches. *Science* **328**, 1506.
Kerr RA (2010) Ocean acidification unprecedented, unsettling. *Science* **328**, 1500–1501.
Kitchell JF, Martell SJD, Walters CJ, Jensen OP, Kaplan IC, Watters J, *et al.* (2006) Billfishes in an ecosystem context. *Bulletin of Marine Science* **79**, 669–682.
Kononen K (1992) Dynamics of the toxic cyanobacterial blooms in the Baltic Sea. *Finnish Marine Research* **261**, 3–36.
Konikow LF, Kendy E (2005) Groundwater Depletion: A global problem. *Hydrogeology Journal* **13**, 317–320.
Kocovsky PM, Ross RM, Dropki DS (2009) Prioritizing removal of dams for passage of diadromous fishes on a major river system. *River Research and Applications* **25**, 107–117.
Krkošek M, Lewis MA, Morton A, Frazer LN, Volpe JP (2006) Epizootics of wild fish induced by farm fish. *Proceedings of the National Academy of Sciences* **103**, 15506–15510.
Law R (2000) Fishing, selection, and phenotypic evolution. *ICES Journal of Marine Science* **57**, 659–668.
Lafferty KD, Porter JW, Ford SE (2004) Are diseases increasing in the ocean? *Annual Review of Ecology, Evolution, and Systematics* **35**, 31–54.
Levin PS, Zabel RW, Williams JG (2001) The road to extinction is paved with good intentions: negative association of fish hatcheries with threatened salmon. *Proceedings of the Royal Society of London B* **268**, 1153–1158.
Levitus S, Antonov JI, Boyer TP, Locarnini RA, Garcia HE, Mishonov AV (2009) Global ocean heat content 1955–2008 in light of recently revealed instrumentation problems. *Geophysical Research Letters* **36**, L07608.
Levitus S, Antonov J, Boyer T (2005) Warming of the world ocean, 1955–2003. *Geophysical Research Letters* **32**, L02604.
Llamas MR, Martínez-Santos P (2005) Intensive groundwater use: silent revolution and potential source of social conflicts. *Journal of Water Resources Planning and Management*, September-October, 337–341.
Longcore T, Rich C (2004) Ecological Light Pollution. *Frontiers in Ecology and the Environment* **2**, 191–198.
Lotze HK, Lenihan HS, Bourque BJ, Bradbury RH, Cooke RG, Kay MC, *et al.* (2006) Depletion, degradation, and recovery potential of estuaries and coastal seas. *Science* **312**, 1806–1809.
Lowenstine LJ (2004) Sick sea mammals: a sign of sick seas? In 55th Annual Meeting of the American College of Veterinary Pathologists (ACVP) and 39th Annual Meeting of the American Society of Clinical Pathology (ASVCP) (eds ACVP, ASVCP). American College of Veterinary Pathologists and American Society for Veterinary Clinical Pathology, Middleton WI, USA.
Ludwig D (1999) Is it meaningful to estimate a probability of extinction? *Ecology* **80**, 298–310.
Lyman JM, Good SA, Gouretski VV, Ishii M, Johnson GC, Palmer MD, Smith DM, Willis JK (2010) Robust warming of the global upper ocean. *Nature* **465**, 334–337.
Mangone GJ (1991) *Concise Marine Almanac*. Second edition. Van Nostrand Reinhold Co., New York.
Marsh GP (1864) *Man and Nature: Or Physical Geography as Modified by Human Action*. Belknap, Harvard (1965).
Martínez ML, Intralawan A, Vázquez G, Pérez-Maqueo O, Sutton P, Landgrave R (2007) The coasts of our world: ecological, economic, and social importance. *Ecological Economics* **63**, 254–272.
Matkin CO, Saulitis EL, Ellis GM, Olesiuk P, Rice SD (2008) Ongoing population-level impacts on killer whales *Orinus orca* following the "Exxon Valdez" oil spill in Prince William Sound, Alaska. *Marine Ecological Progress Series* **356**, 269–281; doi: 10.3354/meps07273.
McCormick MK, Kettenring KM, Baron HM, Whigham DF (2010) Extent and mechanisms of *Phragmites australis* spread in brackish wetlands in a subestuary of the Chesapeake Bay. *Wetlands* **30**, 67–74.
Meadows D (2009) Conserving species before they need the ESA. *Endangered Species Bulletin* **34**, summer US FWS.
Meyerson LA, Viola DV, Brown RN (2009) Hybridization of invasive *Phragmites australis* with a native subspecies in North America. *Biological Invasions*. Open access paper, Springer.
Mills C (2001) Jellyfish blooms: are populations increasing globally in response to changing ocean conditions? *Hydrobiologia* **451**, 55–68.

Millennium Ecosystem Assessment (2005) *Ecosystems and Human Well-being: Biodiversity Synthesis*. World Resources Institute, Washington, D.C.

Millero F (2007) Marine inorganic carbon cycle. *Chemical Reviews (Washington, D.C.)* **107**, 308–341.

Milly PCD, Dunne KA (2001) Trends in evaporation and surface cooling in the Mississippi River basin. *Geophysical Research Letters* **28**, 1219–1222.

Mozdzer TJ, Zieman JC (2010) Ecophysiological differences between genetic lineages facilitate the invasion of non-native *Phragmites australis* in North American Atlantic coast wetlands. *Journal of Ecology* **98**, 451–458.

Molnar JL, Gamboa RL, Revenga C, Spalding MD (2008) Assessing the global threat of invasive species to marine biodiversity. *Frontiers in Ecology and the Environment* **6**, 485–492.

Muir DCG, de Wit (2010) Trends of legacy and new persistent organic pollutants in the circumpolar Arctic: overview, conclusions, and recommendations. *Science of the Total Environment* **408**, 3044–3051.

Mullon C, Fréon P, Cury P (2005) The dynamics of collapse in world Fisheries. *Fish and Fisheries* **6**, 111–120.

Munn CB (2006) Viruses as pathogens of marine organisms from bacteria to whales. *Journal Marine Biological Association UK* **86**, 453–467.

Murchelano RA, Wolke RE (1991) Neoplasms and nonneoplastic liver lesions in winter flounder, *Pseudopleuronectes americanus*, from Boston Harbor, Massachusetts. *Environmental Health Perspectives* **90**, 17–26.

Myers RA, Worm B (2003) Rapid worldwide depletion of predatory fish communities. *Nature* **423**, 280–283.

Myers R, Baum JK, Shepherd TD, Powers SP, Peterson CH (2007) Cascading effects of the loss of apex predatory sharks from a coastal ocean. *Science* **315**, 1846–1850.

Naylor R, Burke M (2005) Aquaculture and ocean resources: raising tigers of the sea. *Annual Review of Environment and Resources* **30**, 185–218.

Naqvi SWA, Bange HW, Farías L, Monteiro PMS, Scranton MI, Zhang J (2010) Marine hypoxia/anoxia as a source of CH_4 and N_2O. *Biogeosciences* **7**, 2159–2190.

Nichols FH, Cloern JE, Luoma SN, Peterson DH (1986) The modification of an estuary. *Science* **231**,567–573.

Nixon S (1995) Coastal marine eutrophication: a definition, social causes, and future concerns. *Ophelia* **41**, 199–219.

NRC (National Research Council) (1985) *Oil in the Sea. Inputs, Fates, and Effects*. National Academy Press, Washington, D.C., 1–601.

NRC (1995) *Understanding Marine Biodiversity*. Committee on Biological Diversity in Marine Systems, National Research Council. National Academy Press, Washington, D.C.

NRC (2002a) *Effects of trawling & dredging on seafloor habitat*. Committee on Ecosystem Effects of Fishing. National Academy Press, Washington, D.C., 126pp.

NRC (2002b) *Abrupt climate change: inevitable surprises*. National Research Council. National Academy Press, Washington, D.C.

NRC (2002c) *Oil in the sea III: inputs, fates, and effects*. Committee on Oil in the Sea, National Research Council. National Academy Press, Washington, D.C.

NRC (2006) *Dynamic changes in marine ecosystems*. National Academies Press, Washington, D.C.

NRC (2010a) *Assessment of sea turtle status and trends*. National Research Council. National Academy of Sciences, Washington, D.C.

NRC (2010b) *Ocean acidification: a national strategy to meet the challenges of a changing ocean*. Committee on the Development of an Integrated Science Strategy for Ocean Acidification Monitoring, Research, and Impacts Assessment; National Research Council. National Academy of Sciences, Washington, D.C. www.nap.edu/catalog/12904.html

NOAA National Ocean Service (2012) Incident News. Office of Response and Restoration. www.incidentnews.gov

Noga EJ (2000) Review article: skin ulcers in fish: pfiesteria and other etiologies. *Toxicologic Pathology* **28**, 807.

Office of Oil and Gas (2005) Overview of U.S. Legislation and Regulations Affecting Offshore Natural Gas and Oil Activity. US Energy Information Administration, Washington, D.C., online.

Oerlemans JH (2005) Extracting a climate signal from 169 glacier records. *Science* **308**, 675–677.

Oki T, Kanae S (2006) Global hydrological cycles and world water resources. *Science* **313**, 1068.

Orams MB (1997) Historical accounts of human-dolphin interaction and recent developments in wild dolphin based tourism in Australasia. *Tourism Management* **18**, 317–326.

Orams MB, Greg GJ, Baglioni Jr. AJ (1996) "Pushy" behavior in a wild dolphin feeding program at Tangalooma, Australia. *Marine Mammal Science* **12**, 107–117.

Orr JC, Fabry VJ, Aumont O, Bopp L, Doney SC, Feely RA, *et al.* (2005) Anthropogenic ocean acidification over the Twenty-First century and its impact on calcifying organisms. *Nature* **437**, 681–686.

Orth RJ, Carruthers TJB, Dennison WC, Duarte CM, Fourqurean JW, Heck Jr. KL, *et al.* (2006) A global crisis for seagrass ecosystems. *BioScience* **56**, 987–996.

Osgood DT, Yozzo DJ, Chambers RM (2006) Patterns of habitat utilization by resident nekton in *Phragmites* and *Typha* marshes on the Hudson River Estuary, New York. *American Fisheries Society Symposium* **51**, 151–173.

Padilla DK, Williams SL (2004) Beyond ballast water: aquarium and ornamental trades as sources of invasive species in aquatic ecosystems. *Frontiers in Ecology and the Environment* **2**, 131–138.

Pauly D, Christensen V (1995) Primary production required to sustain global fisheries. *Nature* **374**, 255–257.

Pauly D (2008) Global fisheries: a brief review. *Journal of Biological Research-Thessaloniki* **9**, 3–9.

Parsons TR, Lalli CM (2002) Jellyfish population explosions: revisiting a hypothesis of possible causes. *La Mer* (Paris) **40**, 111–121.

Pelc R, Fujita RM (2002) Renewable energy from the ocean. *Marine Policy* **26**, 471–479.

Perovich DK, Richter-Menge JA, Jones KF, Light B (2008) Sunlight, water, and ice: extreme Arctic sea ice melt during the summer of 2007. *Geophysical Research Letters* **35**, L1150.

Peterson BJ, Holmes RM, McClelland JW, Vörösmarty CJ, Lammers RB, Shiklomanov AI, *et al.* (2002) Increasing river discharge to the Arctic Ocean. *Science* **298**, 2171–2173.

Peterson CH, Rice SD, Short JW, Esler D, Bodkin JL, Ballachey BE, Irons DB (2003) Long-term ecosystem response to the EXXON Valdez oil spill. *Science* **302**, 2082–2086.

Pitcher G, Pillar S (2010) Harmful algal blooms in eastern boundary upwelling systems. *Progress in Oceanography* **85**, 1–4.

Polidoro BA, Livingstone SR, Carpenter KE, Hutchinson B Mast RB, Pilcher NJ, Sadovy de Mitcheson Y, Valenti SV (2008) Status of the world's marine species. In *The 2008 Review of The IUCN Red List of Threatened Species* (eds Vié J-C, Hilton-Taylor C, Stuart SN). IUCN, Gland, Switzerland, 55–65.

Popper AN, DeFerrari HA, Dolphin WF, Edds-Walton PL, Greve GM, McFadden D, *et al.* (2000) *Marine Mammals and Low-Frequency Sound: Progress Since 1994*. National Academy Press, Washington, D.C., 1–146.

Preisler RK, Wasson K, Wolff WJ, Tyrrell MC (2009) Chapter 33. Invasions of estuaries vs. the adjacent open coast: a global perspective. In *Biological Invasions in Marine Ecosystems* (eds Rilov G, Crooks JA). *Ecological Studies* **204**, 587–617.

Purcell J (2005) Climate effects on formation of jellyfish and ctenophore blooms: a review. *Journal of Marine Biological Associations of the United Kingdom* **85**, 461–476.

Rabalais NN, Turner RE, Wiseman Jr. WJ (2002) Gulf of Mexico, a.k.a. "The Dead Zone." *Annual Review of Ecology and Systematics* **33**, 235–263.

Rabalais NN, Turner RE (2011) *Forecast: Summer Hypoxic Zone Size, Northern Gulf of Mexico*. Online. LUMCON, nrabalais@lumcon.edu

Rahmstorf S (1995) Bifurcations of the Atlantic circulation in response to changes in the hydrologic Cycle. *Nature* **378**, 145–149.

Richardson AJ, Bakun A, Hays GC, Gibbons MJ (2009) The jellyfish joyride: causes, consequences and management responses to a more gelatinous future. *Trends in Ecology and Evolution* **24**, 312–322.

Ries, JB, Cohen AL, McCorkle DC (2009) Marine calcifiers exhibit mixed responses to CO_2-induced ocean acidification. *Geology* **37**, 1131–1134.

Rijnsdorp AD, Buys AM, Storbeck F, Visser EG (1998) Micro-scale distribution of beam trawl effort in the southern North Sea between 1993 and 1996 in relation to the trawling frequency of the sea bed and the impact on benthic organisms. *ICES Journal of Marine Science* **55**, 403–419.

Robelius F (2007) *Giant oil fields— the highway to oil*. Acta Universitatis Upsaliensis. Digital Comprehensive Summaries of Uppsala Dissertations from the Faculty of Science and Technology. Uppsala, 1–168.

Rockström J, Steffen W, Noone K, Persson Å, Chapin III FS, Lambin EF, *et al.* (2009) A safe operating space for humanity. *Nature* **461**, 472–475.

Rona PA (2003) Resources of the sea floor. *Science* **299**, 673–674.

Rona PA (2008) The Changing vision of marine minerals. *Ore Geology Reviews* **33**, 618–666.

Rosegrant MW, Ximing C, Cline SA (2002) *Global Water Outlook to 2025: Averting an Impending Crisis.* International Food Policy Research Institute Washington, D.C. and the International Water Management Institute Colombo, Sri Lanka, September.

Ross PS, Birnbaum LS (2003) Integrated human and ecological risk assessment: a case study of persistent organic pollutants (POPs) in humans and wildlife. *Human and Ecological Risk Assessment* **9**, 303–324.

Sabine CL, Feely RA, Gruber N, Key RM, Lee K, Bullister JL, Wanninkhof R, Wong, *et al.* (2004) The oceanic sink for anthropogenic CO_2. *Science* **305**, 367–371.

Saliki JT, Cooper EJ, Gustavson JP (2002) Emerging morbillivirus infections of marine mammals: development of two diagnostic approaches. *Annals of the New York Academy of Science* **969**, 51–59.

Saltonstall K (2002) Cryptic invasion by a non-native genotype of the common reed, *Phragmites australis*, into North America. *Proceedings of the National Academy of Sciences* (USA) **99**, 2445–2449.

Saltonstall K, Peterson PM, Soreng R (2004) Recognition of *Phragmites australis* subsp. *americanus* (Poaceae: Arundinaceae) in North America: evidence from morphological and genetic analyses. *Sida* **21**, 683–692.

Sarmiento JL, Gruber N (2006) *Ocean Biogeochemical Dynamics*. Princeton Univ. Press, Princeton, NJ.

SCCWRP (online) Project: emerging contaminant effects on coastal fish. www.sccwrp.org/ResearchAreas/Contaminants/Contaminants OfEmergingConcern/EffectsOnBiota/EmergingContaminantEffectsOn CoastalFish.aspx

Schlesinger WH (1997) *Biogeochemistry: an analysis of global change.* Academic Press, San Diego, CA.

She J, Petreas M, Winkler J, Visita P, McKinney M, Kopec D (2002) PBDEs in the San Francisco Bay Area: measurements in harbor seal blubber and human breast adipose tissue. *Chemosphere* **46**, 697–707.

Short FT, Wyllie-Echeverria S (1996) Natural and human-induced disturbance of seagrasses. *Environmental Conservation* **23**, 17–27.

Short FT, Polidoro B, Livingstone SR, Carpenter KE, Bandeira S, Bujang JS, *et al.* (2011) Extinction risk assessment of the world's seagrass species. *Biological Conservation* **144**, 1961–1971.

Sih A, Jonsson BG, Luikart G (2000) Habitat loss: ecological, evolutionary and genetic consequences. *Trends in Ecology and Evolution* **15**, 132–134.

Sii HS, Wang J, Ruxton T (2003) A statistical review of the risk associated with offshore support vessel/platform encounters in UK waters. *Journal of Risk Research* **6**, 163–177.

Silliman BR, Bertness MD (2004) Shoreline development drives invasion of *Phragmites australis* and the loss of plant diversity on New England salt marshes. *Conservation Biology* **18**, 1424–1434.

Silliman BR, Bertness MD, Thomsen MS (2009a) Top-down control and human intensification of consumer pressure in U.S. southern salt marshes. In *Human Impacts on Salt Marshes: A Global Perspective* (eds Silliman BR, Grosholz T, Bertness MD). University of California Press, Berkeley, California, 103–114.

Silliman BR, Grosholz ED, Bertness MD (2009b) *Human Impacts on Salt Marshes: A Global Perspective*. University of California Press, Berkeley, California.

Silliman BR, van de Koppel J, McCoy MW, Diller J, Kasozi GN, Earl K, Adams PN, Zimmerman AR (2012) Degradation and resilience in Louisiana salt marshes after the BP Deepwater Horizon oil spill. *Proceedings of the National Academy of Sciences* **109**, 11234–11239.

Smith SE, Au DW, Show C (1998) Intrinsic rebound potentials of 26 species of Pacific sharks. *Marine and Freshwater Research* **49**, 663–78.

Snelgrove P, Blackburn TH, Hutchings PA, Alongi DM, Grassle JF, *et al.* (1997) The importance of marine sediment biodiversity in ecosystem processes. *Ambio* **26**, 578–583.

Spanier E, Galil BS (1991) Lessepsian migration: a continuous biogeographical process. *Endeavour* **15**, 102–106.

Spielhagen RF, Werner K, Sørensen SA, Zamelczyk K, Kandiano E, Budeus G, Husum K, *et al.* (2010) Enhanced modern heat transfer to the Arctic by warm Atlantic water. *Science* **331**, 450–453.

Steffen W, Crutzen PJ, McNeill JR (2007) The Anthropocene: are humans now overwhelming the great forces of nature? *Ambio* **36**, 614–621.

Steidinger KA, Vargo GA, Tester PA, Tomas CR (1998) Bloom dynamics and physiology of *Gymnodinium breve* with emphasis on the Gulf of Mexico. In *Physiological Ecology of Harmful Algal Blooms* (eds Anderson DM, Cembella AD, Hallegraeff GM). NATO ASI Series, Springer-Verlag, Berlin, 133–153.

Stewart JR, Gast RJ, Fujioka RS, Solo-Gabriele HM, Meschke JS, Amaral-Zettler LA, *et al.* (2008) The coastal environment and human health: microbial indicators, pathogens, sentinels and reservoirs. *Environmental Health* **7**(Suppl. 2), S3.

Stejneger L (1887) How the great northern sea cow (*Rytina*) became exterminated. *American Naturalist* **21**, 1047–1054.

Stone GW, Walker ND, Hsu SA, Babin A, Liu B, *et al.* (2005) Hurricane Ivan's impact on the northern Gulf of Mexico. *EOS*, Transactions, American Geophysical Union **86**, 497–508.

Stone R (2011) Mayhem on the Mekong. *Science* **333**, 814–815.

Stommel EW, Watters MR (2004) Marine neurotoxins: ingestible toxins. *Current Treatment Options in Neurology* **6**, 105–114.

Sun S, Wang F, Li C, Qin S, Zhou M, Ding L, Pang S, *et al.* (2008) Emerging challenges: massive green algae blooms in the Yellow Sea. *Nature Precedings*.

Sutherland KP, Shaban S, Joyner JL, Porter JW, Lipp EK (2011) Human pathogen shown to cause disease in the threatened elkhorn coral *Acropora palmata*. *PLoS ONE* **6**, e23468.

Tewksbury L, Casagrande R, Blossey B, Häfliger P, Schwarzländer M (2002) Potential for biological control of *Phragmites australis* in North America. *Biological Control* **23**, 191–212.

Thurstan RH, Brockington S, Roberts CM (2010) The effects of 118 years of Industrial fishing on UK bottom trawl fisheries. *Nature Communications* **1**, 15. www.nature.com/naturecommunications

Trenberth KE (2010) The ocean is warming, isn't it? *Nature* **465**, 304.

Trenberth KE, Fasullo JT (2010) Tracking earth's energy. *Science* **328**, 316–317.

Thrush SF, Dayton PK (2002) Disturbance to marine benthic habitats by trawling and dredging: implications for marine biodiversity. *Annual Review of Ecology and Systematics* **33**, 449–473.

Tudela S (2004) Ecosystem effects of fishing in the Mediterranean: an analysis of the major threats of fishing gear and practices to biodiversity and marine habitats. *Studies and Reviews* **74**. General Fisheries Commission for the Mediterranean, Food and Agriculture Organization of the United Nations, Rome, 1–44.

Turner RE (1997) Wetland loss in the Northern Gulf of Mexico: multiple working hypotheses. *Estuaries* **20**, 1–13.

Turner RE, Rabalais NN, Justic D (2008) Gulf of Mexico hypoxia: alternate states and a legacy. *Environmental Science and Technology* **42**, 2323–2327.

Turvey ST, Pitman RL, Taylor BL, Barlow J, Akamatsu T, Barrett LA, Zhao X, Reeves RR, Stewart BS, Wang K, Wei Z, Zhang X, Pusser LT, Richlen M, Brandon JR, Wang D (2007) First human-caused extinction of a cetacean species? *Biological Letters* **3**, 537–540.

UNEP/GPA (2006) *The State of the Marine Environment: Trends and Processes*. UNEP/GPA, The Hague.

UN-ISA (International Seabed Authority) (2004) *Marine Mineral Resources. Scientific Advances and Economic Perspectives*. United Nations Division for Ocean Affairs and the Law of the Sea, Office of Legal Affairs, and the International Seabed Authority.

Uye S-I (2008) Blooms of the giant jellyfish *Nemopilema nomurai*: a threat to the fisheries sustainability of the East Asian Marginal Seas. *Plankton Benthos Research* **3** (Suppl.), 125–131.

VanBlaricom GR, Estes JA, eds (1988) *The Community Ecology of Sea Otters*. Berlin, Springer-Verlag.

Vargo GA (2009) A brief summary of the physiology and ecology of *Karenia brevis* Davis (Hansen G and Moestrup comb. nov.) red tides on the West Florida shelf and of hypotheses posed for their initiation, growth, maintenance, and termination. *Harmful Algae* **8**, 573–584.

Vermeer M, Rahmstorf S (2009) Global sea level linked to global temperature. *Proceedings of the National Academy of Sciences of the USA (Washington, D.C.) (PNAS)* **106**, 21527–21532. www.pnas.org_cgi_doi_10.1073_pnas.0907765106

Vitousek PM, D'Antonio CM, Loope LL, Westbrooks R (1996) Biological invasions as global environmental change. *American Scientist* **84**, 468–478.

Vitousek PM, Naylor R, Crews T, David MB, Drinkwater LE, Holland E, *et al.* (2009) Nutrient imbalances in agricultural development. *Science* **324**, 1519–1520.

Vörösmarty CJ, Green P, Salisbury J, Lammers RB (2000) Global water resources: vulnerability from climate change and population growth. *Science* **289**, 284–288.

Vose RS, Easterling DR, Gleason B (2005) Maximum and minimum temperature trends for the globe: an update through 2004. *Geophysical Research Letters* **32**, L23822.

Wabnitz C, Taylor M, Green E, Razak T (2003) *From Ocean to Aquarium*. UNEP-WCMC, Cambridge, UK.

Wainipee W, Weiss DJ, Sephton MA, Coles BJ, Unsworth C, Court R (2010) The effects of crude oil on arsenate adsorption on geothite. *Water Research* **44**, 5673–5683.

Watling L, Norse EA (1998) Disturbance of the sea bed by mobile fishing gear: a comparison to forest clearcutting. *Conservation Biology* **12**, 1180–1197.

Waycott M, Duarte CM, Carruthers TJB, Orth RJ, Dennison WC, Olyarnik S, *et al.* (2009) Accelerating loss of seagrasses across the globe threatens coastal ecosystems. *Proceedings of the National Academy of Science* **106**, 12377–12381.

Webster NS (2007) Sponge disease: a global threat? *Environmental Microbiology* **9**, 1363–1375.

Weilgart LS (2007) The impacts of anthropogenic ocean noise on cetaceans and implications for management. *Canadian Journal of Zoology* **85**, 1091–1116.

Weir-Brush JR, Garrison VH, Smith GW, Shinn EA (2004) The relationship between gorgonian coral (Cnidaria: Gorgonacea) diseases and African dust storms. *Aerobiologia* **20**, 119–126.

West JG (2011) Changing patterns of shark attacks in Australian Waters. *Marine and Freshwater Research* **62**, 744–754.

Whitehead H (2002) Estimates of the current global population size and historical trajectory for sperm whales. *Marine Ecological Progress Series* **242**, 295–304.

WHOI (Woods Hole Oceanographic Institution) (2010) Ocean acidification: a risky shell game. *Oceanus Magazine* **48**. www.whoi.edu/oceanus

Wiles GJ (2004) *Washington state status report for the killer whales*. Washington Department of Fish and Wildlife, Olympia, 106pp.

Williamson M (1996) *Biological invasions*. Chapman and Hall, London.

Willis PM, Crespi BJ, Dill LM, Baird RW, Hanson MB (2004) Natural hybridization between Dall's porpoises (*Phocoenoides dalli*) and harbour porpoises (*Phocoena phocoena*). *Canadian Journal of Zoology* **82**, 828–834.

World Almanac Books (1998) The World Almanac and Book of Facts. K-111 Reference Corporation, Mahwah, New Jersey.

World Commission on Dams (2000) *Dams and Development*. The Report of the World Commission on Dams. Earthscan Publications Ltd, London and Sterling, VA.

Worm B, Barbier EB, Beaumont N, Duffy JE, Folke C, Halpern BS, *et al.* (2006) Impacts of biodiversity loss on ocean ecosystem services. *Science* **314**, 787–790.

Yamamoto-Kawai Y, McLaughlin FA, Carmack EC, Nishino S, Shimada K (2009) Aragonite undersaturation in the Arctic Ocean: effects of ocean acidification and sea ice melt. *Science* **326**, 1098–1100.

Yip TL (2008) Port traffic risks—A study of accidents in Hong Kong Waters. *Transportation Research Part E* **44**, 921–931.

Zedler JB, Kercher S (2004) Causes and consequences of invasive plants in wetlands: opportunities, opportunists, and outcomes. *Critical Reviews in Plant Sciences* **23**, 431–452.

Zeller D, Pauly D (2005) Good news, bad news: global fisheries discards are declining, but so are total catches. *Fish and Fisheries* **6**, 156–159.

Zhang J (2002) Biogeochemistry of Chinese estuarine and coastal waters: nutrients, trace metals and biomarkers. *Regional Environmental Change* **3**, 65–76.

Zornetzer HR, Duffield DA (2003) Captive-born bottlenose dolphin x common dolphin (*Tursiops truncatus* × *Delphinus capensis*) intergeneric hybrids. *Canadian Journal of Zoology* **81**, 1755–1762.

CHAPTER 3

MARINE CONSERVATION MECHANISMS

If the misery of our poor be caused not by the laws of nature, but by our institutions, great is our sin.

Charles Darwin

3.1 THE TOOLKIT

Roman law declared that air, running water, the sea, and its shores were to be shared among all people. Ocean policies have evolved from this basic tenet into *freedom of the seas*, for use as a global commons. The development of nations, government, laws, and policies have increasingly claimed control of this global commons, with uses of the coastal foreshore evolving from traditional claims of property. Usage rights triggered debates of ownership claimed by the state, community, or individual as the value of marine assets intensified. The littoral zone and nearshore waters became battlegrounds that required resolution through agreements, laws, and regulations. As conflicts increased and resources diminished in the 20th century, calls for marine conservation intensified. Today, marine environmental change is forcing society to confront a paradox: protect and restore diminishing marine assets *and* preserve traditions as economic and social uses intensify. Society's response to the growing loss of marine and coastal assets is tied to culture, economics, and social benefits in an arsenal of tools—the toolkit.

The purpose of this chapter is to present an overview of the legal and social mechanisms that have achieved widespread use for conservation and management of coastal and ocean systems. As public awareness of the importance of biotic services grows, marine environmental policy is slowly moving from sector-based actions to ecosystem-based approaches. Thus, the resultant conservation toolkit of socially approved tools is evolving with conservation practice by means of science, public participation, governance, law, administrative processes, politics, and dedicated action of governments, individuals, communities, and committed organizations.

3.2 BIOLOGICAL CONSERVATION

Wildlife protection originated in Roman law. It evolved into the modern era of the early 20th century when protestors sought to protect decimated individual and groups of species from industrial slaughter, ushering in the modern era of conservation. Under Roman law, *ferae naturae* were regarded as property of no one, like the air or oceans. Yet, species could become the property of anyone who captured or killed them (Bean and Rowland, 1997). In early American law: "The wild bird in the air belongs to no one, but when the fowler brings it to the earth and takes it into his possession it is his property" (Blumm and Ritchie, 2005). "Wildlife" as a conservation objective under law was not yet conceived.

3.2.1 Species conservation

In the early 20th century, outraged protestors sought to protect species from massive industrial exploitation for fur, oil, and fashion. Henry Wood Elliot, who studied the northern fur seal (*Callorhinus ursinus*) in the Pribilof Islands, Alaska (Ch. 7), in the late 1800s, revealed unsustainable exploitation and sought protection. Government officials and commercial interests failed to act, forcing Elliott to seek an international mechanism that became the first international wildlife treaty, the *North Pacific Fur Seal Convention* (*Fur Seal Treaty*) of 1911 (NOAA, 2006). This Treaty initiated the restoration of fur seal and sea otter populations. Concurrently in 1910, the *U.S. Lacey Act* prohibited trade in wildlife, fish, and plants illegally taken in violation of state law. This Act is considered the first example of the U.S. federal government using its power to preserve species by addressing illegal trade in wildlife (Anderson, 1995).

Protection for other migratory species soon followed (Table 3.1). The international *Migratory Bird Treaty* (1918) protected shorebirds, waterfowl, and others by forbidding market hunting for the feather trade. But species and their habitat were better secured by governments, as agreed upon at the *Convention on Nature Protection and Wildlife Preservation in the Western Hemisphere* in 1940 (56 Stat. 1354, 161 U.N.T.S. 193). This Convention sought to protect not only migratory birds and natural landscapes, but also encouraged spatial-area protection through government establishment of national parks, national reserves, nature monuments, and strict wilderness reserves. The acknowledged unregulated depletion of highly migratory whales in international waters brought international pressure to the commercial whaling industry that agreed to provide for the proper conservation of whale stocks and thus make possible the orderly development of the whaling industry by

Marine Conservation: Science, Policy, and Management, First Edition. G. Carleton Ray and Jerry McCormick-Ray.

Table 3.1 Examples of marine species protection: national laws, regional councils, international agreements.

Protection	Instrument (date in force)	Intended action
Fur seals	*North Pacific Fur Seal Treaty* (1911)	First international treaty for wildlife protection. Ended fur seal harvest; protect sea otters.
Birds	*Migratory Bird Treaty* (1918)	Regulates taking, selling, transporting, and importing migratory birds.
Whales	*International Convention on the Regulation of Whaling* (1946)	Regulates harvest and industry; moratorium on commercial whaling established, 1985–1986.
Marine mammals	*U.S. Marine Mammal Protection Act* (1972)	Established concept of "optimum sustainable population" placing a moratorium on taking most marine mammals.
Antarctic seals	*Convention on Conservation of Antarctic Seals* (1972)	Adopted standards for conservation of Antarctic seals.
Polar bears	*Agreement on Conservation of Polar Bears, Oslo* (1973)	Limits hunting to sustainable levels.
Dolphins	*Agreement to Reduce Dolphin Mortality in the Eastern Tropical Pacific Tuna Fishery* (1992)	Regulates dolphin "bycatch" to the lowest possible level, with an objective of zero take.
Endangered species	*Convention on the International Trade in Endangered Species of Wild Fauna and Flora* (CITES: 1973)	Prohibits or controls all trade of listed species; however, few marine taxa in active fisheries are on the CITES list
Endangered species	*U.S. Endangered Species Act* (1973)	Forbids jeopardizing listed endangered or threatened species or adversely modifying critical habitats.
Endangered migratory species	*"Bonn" Convention on the Conservation of Migratory Species of Wild Animals* (l979)	Forbids take of listed migratory species of wild animals, including sea turtles, birds, and marine mammals.
Sea turtles	*Inter-American Convention for the Protection and Conservation of Sea Turtles* (2000)	Protects sea turtles and their nesting habitats in the Americas.
Fisheries	*Convention for the Establishment of an Inter-American Tropical Tuna Commission* (IATTC: 1949)	Regulates catch of tunas in the tropical Pacific Ocean.
Fisheries	*International Convention for the Conservation of Atlantic Tuna* (ICCAT 1966)	Regulates catch of tunas in the Atlantic Ocean.
Fisheries	*Convention for the Conservation of Salmon in the North Atlantic Ocean* (1982)	For regulation of catch of salmon and for their conservation in the Atlantic Ocean.
Fisheries	*Convention for the Prohibition of Fishing with Long Driftnets in the South Pacific* (1989)	Prohibits use of longlines in designated areas.
Fisheries	*Convention for the Conservation of Anadromous Stocks in the North Pacific Ocean* (1992)	Principally directed towards regulation and conservation of salmons.

means of the *International Agreement on the Regulation of Whaling* (1946). Conservation of marine mammals as a whole became fully established in 1972 with the U.S. *Marine Mammal Protection Act*, the first instrument to legally recognize species as "wildlife" (Ray and Potter, 2011).

Wildlife laws stir strong passions and intense controversy, not only for marine mammals but also for such commercially valued, endangered, migratory species such as Atlantic salmon, bluefin tuna, and sea turtles, today debated as to whether they are to be classified as resources or wildlife (Bean and Rowland, 1997). This resource/wildlife issue was contentious at the *Convention on International Trade in Endangered Species of Wild Fauna and Flora in the 2010* (CITES), which determined that threatened species of commercial fish are not considered wildlife but are traded commodities that require different instruments (Doukakis *et al.*, 2009).

Instruments that protect species now have wide social and ethical appeal. Species protection is established in social traditions, legal regulation, and in numerous international agreements that attempt to curb the crisis of species and biodiversity loss. The deep personal attachment towards individual species (e.g., polar bears, whales, flamingoes, butterflyfish, corals, and even oysters) is used to catalyze action around symbolic icons, which when faced with extinction heighten public attention and conservation action (McCay, 1998). Public outcries had prompted Congress to pass the *Endangered Species Act* in 1973 (ESA, Section 3.5.1.7), and more than 160 countries to agree on the *Convention of International Trade in Endangered Species of Wild Fauna and Flora* (CITES 1973), intended to prevent trade of species threatened with extinction. It brought bookkeeping of depleted species with IUCN's (2010) Red List of Threatened Species. However, such lists lack important information needed for action, especially information on the status and trends for a host of individual species critically in need of management, most importantly wide-ranging species of sharks, oceanic fishes, sea turtles, marine mammals, and others (Wallace *et al.*, 2011).

Many highly mobile, valued marine organisms venture outside of national boundaries and into international waters, causing management problems. These "transboundary" species also cross nations' Exclusive Economic Zones and are designated "straddling species" (Fig. 12.10b). To protect such species requires knowledge of their entire life-history distribution, and mandates negotiation between coastal nations, distant-water nations, and contiguous coastal nations (Caddy and Seijo, 2005). In particular, all seven species of sea turtles are listed as endangered or threatened worldwide, and each have different movement patterns (Fig. 8.16), which requires different conservation options and close working relationships that need to be addressed through diplomatic channels, capacity building, and scientific exchange. Conservation efforts have mainly focused on turtle nesting beaches, mostly in protected areas that have proved successful for some populations. However, outside of protected areas regional and international mechanisms often remain vague and ineffectual; e.g., the voluntary, non-binding United Nations Food and Agriculture Organization (FAO) *Guidelines to Reduce Sea Turtle Mortality in Fishing Operations* (2005) is based on a series of "soft-law" fishery instruments.

Protection of large, commercial, oceanic fishes depends on international cooperation. The largest and most commercially valuable of all fishes is bluefin tuna (*Thunnus thynnus*), which has declined more than 80% since 1970 as overfishing and international trade continue (Ch. 2) under soft-law instruments. The Atlantic population's life-history pattern encompasses international and coastal waters and is monitored by the International Commission for the Conservation of Atlantic Tunas (ICCAT). Their critical spawning occurs in the Gulf of Mexico. The species is managed as a bycatch issue under the U.S. National Marine Fisheries Service (NMFS), and the international *Code of Conduct for Responsible Fisheries* attempts to control bycatch. For greatly depleted sharks (Ch. 2), FAO's Committee on Fisheries adopted IPOA-SHARKS (1999), which addresses not only bycatch but also de-finning, which continues to decimate shark populations worldwide. Such international soft-law instruments fail to curb depletions.

Whales are highly migratory species with a long history of exploitation. The International Whaling Commission (IWC) since its formation in 1946 has been responsible for setting catch quotas, but whales continued to be depleted largely due to lack of information on natural history and habitat and a need for better management procedures (Schevill, 1974; Box 3.1). In 1972, the U.S. passed the *Marine Mammal Protection Act* that placed a moratorium on US commercial whaling, and the IWC followed suit in 1986. Since then, the disputed international moratorium is negotiated annually by the IWC, with species being treated unequally with different degrees of success. Under the earlier IWC moratorium, the eastern North Pacific gray whale (*Eschrichtius robustus*) had recovered fully by the end of the 20th century to become the first marine mammal to be removed from the U.S. *Endangered Species Act* list (Section 3.5.1.7), but in recent years it has declined steadily for unknown reasons (Keller and Gerber, 2004). The most

Box 3.1 The era of excess: Soviet illegal whaling and the failure of the IWC

Phillip J. Clapham
National Oceanic and Atmospheric Administration, Seattle, Washington, USA

Yulia V. Ivashchenko
National Oceanic and Atmospheric Administration, Seattle, Washington, USA

Robert L. Brownell, Jr.
NOAA Southwest Fisheries Center, Pacific Grove, California, USA

The signing, in 1946, of the *International Convention for the Regulation of Whaling* presaged what was supposed to be a new era of management in the exploitation of whales. Faced with the industry's excessive catches in previous decades—more than 150,000 blue whales were killed in the 1930s alone—the whaling nations brokered an agreement whose principal aim was to provide for the proper conservation of whale stocks and thus make possible the orderly development of the whaling industry. The Convention established the International Whaling Commission (IWC), whose Scientific Committee was to oversee research on whales and recommend quotas that, in theory, would allow sustainable whaling into the future.

What began as a good idea inevitably fell victim to the desire to maintain profits. Faced with the uncertainties about whale numbers that typify the perpetually difficult field of cetacean research, the whalers consistently gave the benefit of the doubt to their revenues rather than to their resources. Despite mounting evidence of declines in some whale populations, the whaling industry continued to take record numbers of animals; in the decade following implementation of the Convention, over a quarter million fin whales were killed in the Southern Hemisphere, together with tens of thousands of whales of other species. Regulation, such as it was, came too little and too late for many populations, and by the early 1960s it was becoming increasingly clear that the IWC was failing to fulfill the objectives of the Convention (Clapham and Baker, 2008).

(Continued)

As it turned out, the situation was far worse than anyone could have imagined at the time. In addition to the known and publically reported excesses of the industry in the years following the Second World War, one nation had been conducting a huge campaign of illegal whaling that was to devastate already over-exploited stocks of whales, and bring at least one population to the brink of extinction. The USSR, driven by an absurd system of domestic industrial planning, made more than 150,000 unreported catches in the Southern Hemisphere alone, and another 26,000 in the North Pacific (Ivashchenko *et al.*, 2011; Ivashchenko *et al.*, 2013).

In the USSR of Joseph Stalin and his successors, everything was about reaching and exceeding the production targets of the endless plans—monthly, annual, five-year, ten-year, and regional—that were created for every aspect of Soviet industry. High production numbers meant much-coveted bonuses, awards, and recognition for workers and managers alike. In the beginning, when whales were plentiful, it was easy for Soviet whalers to exceed their production targets, which were supposed to be based upon a rational, scientific assessment of the abundance of the resource. In reality, however, the following year's targets were often set at the level achieved the previous year, regardless of what industry scientists said about the ability of whale populations to withstand these ever-increasing catches (Ivashchenko *et al.*, 2011). The result was an utterly predictable collapse of populations, as Soviet whalers took more and more animals and ranged ever further in the oceans in their attempts to fulfill their quotas. Nor was the resource even utilized well: all that counted was the number of whales killed, and in their zeal to take more and more animals, carcasses were sometimes left to rot before the factory ships of the Soviet whaling fleets could process them.

The Soviets took everything, regardless of size, age, or protected status. Among their many excesses, a few examples stand out as particularly egregious. In Antarctic waters south of Australia and Oceania, Soviet factory fleets killed nearly 25,000 humpback whales in just two seasons (1959/60 and 1960/61), an unparalleled catch which caused the immediate crash of the populations concerned, and precipitated the closure of shore whaling stations in Australia and New Zealand. In the eastern North Pacific, Soviet fleets killed 529 right whales in the 1960s, almost finishing off an already small population that was struggling to come back from the excesses of 19th century whaling (Ivashchenko and Clapham, 2012); today, the right whales remaining in this region are estimated to number only about 30 animals (Wade *et al.*, 2011), and they may not survive.

Sperm whales in the North Pacific were hit particularly hard, not only by the USSR but also by Japanese land station and pelagic whaling operations. The USSR alone killed almost 160,000 sperm whales in the period 1948–79 (Ivashchenko *et al.*, 2013). The devastation wreaked on sperm whales led the Soviet scientist Alfred Berzin to note that as a result of the catches, some breeding areas for sperm whales became deserts.

The truth about Soviet whaling was finally revealed following the collapse of the Soviet Union (Yablokov, 1994; Berzin, 2008). In subsequent years, the true catch record—an essential component of assessments of the status of whale populations today—was reconstructed through release of previously secret data; in many cases, these records had been saved and hidden for years, sometimes at considerable personal risk, by biologists working on the factory fleets (Clapham and Ivashchenko, 2009).

In 1972, following years of disagreements and obstruction—not least by the USSR—the IWC finally passed an "International Observer Scheme" aimed at independent monitoring of catches through placement of foreign observers on factory ships. In a final irony, Soviet vessels were monitored by inspectors from Japan (and vice versa), a country that was buying large quantities of whale meat from the USSR. We now know that this "monitoring" was weak at best, and that Japanese inspectors were in some cases fully aware of continued illegal catches, while in others they were taken off the factory ship's flensing deck and treated to "hospitality" involving food and much vodka before illegal whales were processed.

Today, the IWC is revisiting the issue of catch monitoring as the organization discusses a possible resumption of commercial whaling. Once again the discussion is characterized by assurances from the whaling nations that the widespread violations that occurred are part of a history that has been put to bed, and that things are different today—and therefore that a truly independent system is not needed. It remains to be seen whether the IWC will agree, and in so doing ignore the grave lessons of a shameful and excessive past.

abundant of all baleen whales, the minke whale (*Balaenoptera acutorostrata*), remains hunted by Japan under the IWC clause of "scientific whaling" and by Norway as a traditional activity. The North Pacific right whale (*Eubalaena glacialis*), depleted to near-extinction by illegal commercial whaling, now numbers only a few dozen. The depleted southern right whale (*Eubalaena australis*) has shown encouraging signs of recovery in both South Africa and Argentina. A notable success story concerns the once heavily exploited Arctic bowhead whale (*Balaena mysticetus*) that today has recovered and supports subsistence harvest by native Iñupiat people. And, while Antarctic whale populations may also be recovering, some are vulnerable to expansion of a global fishery for their major food item, krill (*Euphausia superba*), currently under current weak management (FAO, 2011).

Among pinnipeds, all species of fur seals have made remarkable recoveries. Since passage of the *Fur Seal Convention*, North Pacific fur seals (*Callorhinus ursinus*) have recovered fully, although recently are showing decline (Ch. 7). And since their decimation in the 18th to early 20th centuries, Antarctic and sub-Antarctic fur seals (*Arctocephalus* spp.) and Pacific walruses (*Odobenus rosmarus divergens*, Ch. 7) have all expanded almost throughout their historic ranges. So successful has been the recovery of some *Arctocephalus* spp. that the Antarctic Treaty Consultative Meeting (ACTM) has been prompted to revise their legal protection (Jabour, 2008). However, Pacific walruses are again declining, this time due to climate change and sea-ice loss (Ch. 7).

Like marine mammals, highly migratory, transboundary, threatened and endangered species have distributions that rarely coincide with a nation's political/jurisdictional boundaries, thus requiring some form of inter-jurisdictional cooperation for protection. To meet life-history demands, the vast majority of species depend on multiple habitats, where threats at any one life-history stage may threaten their survival. For fish, numerous international agreements are intended to regulate, restrict, or prohibit certain commercial fishing methods and reduce fishing effort, but unfortunately, non-commercial species are relatively neglected while very high profits from fish exploitation prompt over-exploitation.

3.2.2 Habitat conservation

Association of species with habitat is traditional practice for subsistence users, sportsmen, and others, but is poorly documented in lists of threatened or endangered marine species. However, listing highlights the need for habitat protection, with a prescription for how that species can be restored or how the threatening activity can be changed. This approach is widely endorsed by national and international agencies and organizations, and is explicit or implicit in numerous laws and agreements (Table 3.2). Thus, listing species under the U.S. ESA (Section 3.5.1.7) legally forces its habitat to be preserved, unlike IUCN's

Table 3.2 International instruments for ocean/habitat,wildlife regional protection. Examples (Nowlan, 2001; UNEP, 2005; Prideaux, 2003).

Region or habitat type to be protected	Instrument	Purpose
Oceans: continental shelf; high seas; territorial sea, contiguous zone.	*UN Convention on Law of Sea* (UNCLOS III: 1982)	Ocean jurisdiction and conservation.
Antarctic/Southern Ocean	*Antarctic Treaty* (1959); *Convention for the Conservation of Antarctic Seals* (1972); *Protocol on Environmental Protection to the Antarctic Treaty* (1991)	Specially protected land and ice habitats areas for scientific research. Extends Antarctic Treaty to the Antarctic Convergence; adopts ecosystem management.
Arctic Ocean/Northeast Atlantic	*OSPAR Convention* (1998); *Arctic Environmental Protection Strategy* (l991); *Global Program of Action for Protection of Marine Environment from Land- based Activities* (l995)	Protect Arctic ecosystems: protect, enhance, restore natural resources; recognize traditional-cultural needs of indigenous peoples; review state of Arctic environment; address pollution.
Mediterranean	*Barcelona Convention* (1976); *Protocol Concerning Specially Protected Area and Biological Diversity in the Mediterranean* (1995)	For establishment of a network of protected areas for habitats and wildlife.
West/Central Africa	*Abidjan Convention* (1984)	Protect marine, coastal environment.
East Africa	*Nairobi Convention* (1985) *Protocol Concerning Protected Areas and Wild Fauna and Flora in the Eastern African Region* (1985)	For establishment of a network of protected areas for habitats and wildlife.
South Pacific	*Noumea Convention* (1990); *Protocol for Conservation and Management of Protected Marine and Coastal Areas of the South-East Pacific* (1989)	For establishment of a network of protected areas for habitats and wildlife.
Caribbean	*Cartagena Convention* (1983); *Protocol Concerning Specially Protected Areas and Wildlife* (1990)	For establishment of a network of protected areas for habitats and wildlife.
Wetlands	*Ramsar Convention* (1971)	Protect wetlands and waterfowl habitat.
High-seas fishery	*UN General Assembly Resolution 61/05* (2010)	Manage deep-sea fishes on high seas (seamounts, deep-sea habitat)
Regional seas for migratory species	*Convention on Conservation of Migratory species of Wild Animals* (1983)	A global environmental treaty to conserve and manage migratory species and habitats throughout range; facilitated through Multilateral Agreements.
Coastal waters	*Convention on Biological Diversity, Agenda 21*	UNEP Program to Protect Marine Environment from Land-based Activities Implement *Agenda 21*.

Red List of Threatened Species that carries no such regulatory function, leaving habitat protection to individual nations.

Habitat conservation is advocated to protect, restore, or manage natural areas that species depend upon for their survival. Species and/or higher taxa targeted for protection have become synonymous with protection of wetlands, seagrass beds, seashores, mangrove forests, coral reefs, and other habitats. Habitat protection is often extended to protect communities, i.e., multiple species that depend on these habitats at one or more stages of their life histories. The *Ramsar Convention* is the only international agreement specific to a particular habitat type (wetlands), and it came into being to protect birds that use wetlands as major habitats, rather than to protect wetlands per se.

Preserving habitat or potential habitat (a habitat lacking a target species, but capable of supporting it in future) is essential for recovery of depleted or endangered species. Designation of such areas requires that the area be of sufficient size to restore or sustain the species in the ecosystem. Habitats have also become recognized as important for conservation science as control areas for studies in natural history, community interactions, ecosystem function, responses to environmental change, and for their esthetic and recreational values. Habitat protection thus has moved species and wildlife conservation into spatial conservation, and into jurisdictional arenas granted by social tradition, constitution, and law, with established rules of conduct administered through systems of governance.

3.2.3 Biodiversity conservation

Although species and habitat protection are critical for conservation, that form of protection is insufficient. Experience with species- and habitat-specific laws, agreements, and other mechanisms supports the need to manage units larger than individual species and their habitats. The rate of species loss has forced recognition of a problem much larger than loss of single species, i.e., loss of biodiversity and genetic uniqueness important to ecological function. Species require well-functioning ecosystems in which to live and ecosystems depend on complexes of diverse species to maintain their functions. Hence *biodiversity conservation* and *ecosystem-based management* (EBM; Section 3.6.6; Ch. 13) have become new catchwords in the vocabulary of natural resources management.

Biodiversity, literally the diversity of life, includes species and their genetic variability, habitats, and ecosystems (Ch. 5). In 1992, 159 nations met in Rio de Janeiro, Brazil, at the "Rio Summit" to discuss conservation of biodiversity and natural resources. The results were the *Rio Declaration on Environment and Development* and the *United Nations Convention on Biological Diversity* (CBD). Some 190 countries consequently pledged to reduce the rate of biodiversity loss by 2010—the year designated by the UN as *International Year of Biodiversity*. That year has passed with some progress, but biodiversity continues to decline (Millennium Ecosystem Assessment, 2005).

Biodiversity conservation is a broad-spectrum approach to protection, restoration, and sustainable use for species, habitats, and ecosystems. It uses many of the same mechanisms as for species and habitat protection, but involves a broader, more holistic, ecosystem-based perspective encompassing species, habitats, ecosystems, and humans. Biodiversity research has increased dramatically in recent years, but less than 10% of that research has been devoted to marine biodiversity (Hendriks *et al.*, 2006). Terrestrial predominance creates a severe imbalance that percolates through national and international programs. Efforts are underway to promote conservation of marine biological diversity through a series of conventions and other international instruments, most notably, the UN Law of the Sea (LOS). LOS offers a critical international framework for enforcing conservation and sustainable use of marine biodiversity, and for increasing marine scientific research in international, oceanic areas. However, challenges remain, especially that no nation can effectively prevent ecological harm perpetuated by others. Construction of large-scale dams (e.g., Egypt's Assam High Dam, China's Three Gorges Dam, etc.) can affect ecological processes beyond national borders, but current mechanisms among nations cannot effectively conserve biodiversity or prevent introduction of alien (exotic) species across international waters.

Marine biodiversity conservation is in its infancy. A current deterrent to its progress is lack of information on numbers, distributions, and life histories of the vast majority of marine species; the *Census on Marine Life* attempts to confront this problem (Box 5.1). The inertia of some governments, notably the U.S., to become signatory to the CBD and LOS deprives the Convention of necessary commitment and resources. Yet the ocean provides reliable goods and services to humanity, with many potential tools for conserving marine biodiversity, e.g., through sustainable fisheries management, pollution control, maintenance of essential habitats, and creation of marine spatial reserves (Worm *et al.*, 2006). The goal to reduce the rate of species loss can be targeted by tracking the net balance of species improvement and/or depletion of species listed on IUCN's *Red List of Threatened Species* (Sachs *et al.*, 2009; Walpole *et al.*, 2009). And from this list, the loss and rate of habitat and ecosystem change can be inferred, but only after the fact.

3.3 SPATIALLY EXPLICIT CONSERVATION

Efforts to protect or restore any species and/or its habitat require spatially explicit approaches, especially through "protected area" designation. Preserving habitats, e.g., reefs, seagrass beds, estuaries, lagoons, the benthos, seashores, sea ice, open waters, seamounts, and more, as protected areas, aims to sustain valued species, recover depleted or endangered species, and protect ecosystems. Protected areas can support scientific research needed on natural history, community interactions, ecological function, and also can serve as control areas to investigate responses to environmental change. Where spatial extent is important, designated protected areas often fall within the jurisdiction of multiple authorities, some by tradition, constitution, law, or social contract, with separate established rules of conduct and governance.

During past centuries, Marine Protected Areas (MPAs) were established to perpetuate traditional uses such as hunting

Fig. 3.1 Cultural Marine Protected Area at Manono, Western Samoa, honoring Samoans who defeated and expelled the Tongans in battle 25 generations ago, or about the year 1250. As they left, the Samoan chief cried out, "*Malietoa melietau*" (*Brave heroes, well fought!*). The sign indicates the location of the battle. Photograph © Ray & McCormick-Ray.

or fishing, or for cultural reasons (Fig. 3.1). In the early to mid-20th century, governments established MPAs almost solely for species and habitat conservation or for scenic or cultural values; e.g., Glacier Bay National Monument in Alaska (1925), Fort Jefferson National Monument in Florida (1935), Green Island in Queensland, Australia (1938). The modern era of MPA establishment arose with the establishment of the Exuma Cays Land and Sea Park, The Bahamas (1959, Ch. 8). In 1962, the International Union for the Conservation of Nature and Natural Resources (IUCN) hosted the First World Conference on National Parks and Reserves. A prescient recommendation for the first time encapsulated several concepts that remain relevant today: a land-sea approach, no-take fishery reserves, the need for research, and habitat protection (Box 3.2). Shortly thereafter, the need for MPA guidelines became evident. The total known list of marine parks and reserves by the early 1970s was only 125 (Björklund, 1974), making evident that MPA establishment initially was based on expediency, opportunism, and pragmatism that lacked scientifically defensible choices. In 1968, the decade-long (1962–73) International Biological Programme (IBP) held an international conference on *Man and the Biosphere* under UNESCO (United Nations Educational, Scientific, and Cultural Organization) sponsorship, which recommended development of an international network of *biosphere reserves* (Ch. 11). A MAB task force in 1974 proposed five general selection criteria for terrestrial research and conservation: biogeographic representativeness, diversity, naturalness, uniqueness, and effectiveness. These criteria were presented to the 1975 International Conference on Marine Parks and Reserves in Tokyo (Japan), which expanded concepts for MPA selection. In 1977–8, the first international effort to promote MPAs occurred when the International World Wildlife Fund and IUCN together initiated a global program, *The Seas Must Live*, to inspire the lagging need for marine conservation. The seas still seemed "healthy" then, as only a few species (some marine mammals, sea turtles, and fewer than a dozen fish) were recognized as endangered. Following the WWF/IUCN effort, in 1981 UNEP's Mediterranean Action Plan, the first of several such regional plans,

Box 3.2 First World Conference on National Parks: Recommendation 15

The First World Conference on National Parks, held in Seattle in 1962, was a gathering of more than 60 countries and represented the first international exchange of ideas on protected areas. The following Recommendation was adopted:

> WHEREAS it is recognized that the oceans and their teeming life are subject to the same dangers of human interference and destruction as the land, that the sea and land are ecologically interdependent and indivisible, that population pressures will cause man to turn increasingly to the sea, and especially to the underwater scene for recreation and spiritual refreshment, and that the preservation of unspoiled marine habitat is urgently needed for ethical and esthetic reasons, for the preservation of rare species, for the replenishment of stocks of valuable food species, and for the provision of undisturbed areas for scientific research.
>
> THE FIRST WORLD CONFERENCE ON NATIONAL PARKS invites the Governments of all those countries having marine frontiers, and other appropriate agencies, to examine as a matter of urgency the possibility of creating marine parks or reserves to defend underwater areas of special significance from all forms of human interference, and further recommends the extension of existing national parks and equivalent reserves with shorelines, into the water to the 10 fathom depth or the territorial limit or some other appropriate off-shore boundary.

Source: Adams AB, ed (1962). First World Conference on National parks, U.S. Dept. Interior, Washington DC

finally provided comprehensive MPA criteria and guidelines that became widely adopted, modified, and are still applicable today (www.unepmap.org).

By the 1990s, marine species depletions and habitat deterioration rose to obvious crises. Bolstered by increasing information about species, natural history, and ecological structure and function, efforts to develop MPAs intensified. Scientific evidence increasingly revealed that marine communities plus environments constitute a viable unit needed for restoring depleted habitats and species' populations. Furthermore, the mobility of species and the openness of marine systems emphasized that the scope of marine conservation must be large, even regional. Scientists urged MPA establishment in the open ocean (e.g., Lubchenco *et al.*, 2003) for both biodiversity protection and no-take fishery reserves, urging adoption of very large MPAs (e.g., LORs, Ch. 12). IUCN (2005) proposed expansion of regional programs. By that time, the First Conference of the Parties of the *Convention on Biological Diversity* in 1994 listed 1306 MPAs (Kelleher *et al.*, 1995) with coral reefs attaining international status via the *International Coral Reef Initiative* to become the marine equivalent to tropical forests for biodiversity conservation. The CBD called for effectively managed networks of MPAs of ecologically representative areas by 2012, and to effectively conserve at least 10% of each of the world's marine and coastal ecological regions (Toropova *et al.*, 2010). This important goal sparked efforts towards global ocean conservation.

By the first decade of the 21st century, more than 5880 MPAs had been established worldwide, and since 2003 have increased in number by >150% (Table 3.3a,b). Although covering only 1.17% of ocean space, MPAs' total cover is >4.2 million km^2 (Toropova *et al.*, 2010; UNEP, 2009) with sizes from the smallest of 0.4 ha (Echo Bay Provincial Park, 1971, Canada) to the largest at 544,000 km^2 (Chagos Marine Reserve, 2010, Indian Ocean). The latter surpassed the 408,250 km^2 Phoenix Islands Protected Area in Kiribati. Australia's Great Barrier Reef Marine Park in 1975 that covered 340,000 km^2 was expanded to include Great Barrier Reef Coast Marine Park (GBR Coast MP, 2004) to help protect the Great Barrier Reef and environs, a designated World Heritage site. The Great Australian Bight Marine Park designated in 1998 protects 19,395 km^2 of Australia's ocean space. The 11 largest MPAs of more than 100,000 km^2 also include the U.S. Papahānaumokuākea Marine National Monument designated in 2008 that protects 360,000 km^2 of the northwest ocean of the Hawaiian Islands (also a UNESCO World Heritage Site).

3.4 GOVERNANCE: POLICY, STRATEGY, TACTICS

Coastal and marine issues are being addressed through a variety of governing mechanisms that involve established procedures, instruments of government, and the private sector. Inspiration and motivation for collective conservation action is captured in a vision statement, followed by agreed-upon goals to guide conservation action, as carried out in policy, strategy, and tactics. While policy is intended to set a course or principle of action through governing mechanisms, establishing policy is an art, with science and persuasion playing key roles in negotiating among stakeholders with divergent views. And as governance involves government, control, and authority, good governance involves law, science, economics, and the sovereign power of nations.

Optimally, policy goals are clear and agreed upon, options and criteria are defined, and information is complete (Fiorino, 1995). However, uncertainty forces environmental policy into the arena of politics, social norms, and spirit of the times, and when policies collide, as among public trust, property rights, and depletion of shared resources, the interaction between policy and politics can result in a series of compromises with unintended consequences. Hence, environmental policies may owe little to environmental paradigms and a lot to political expediency, often taking the form of piecemeal efforts that lack focus on ultimate outcomes, and tending to happen as much by default as by intended action.

Science provides rational arguments to help narrow uncertainty, but policy decisions are not based solely on technical information. As norms and politics play strongly into policy decisions, the core assumptions of science and policy are fundamentally different: science is empirical and requires expert interpretation, whereas policy establishes a standard among collective interests (stakeholders) that hold varied beliefs, values, and ideals (Wagner, 2001). A paradigm conflict that requires resolution thus exists between science-based assessment and interest-based policy-making (Cahn, 2002).

Once a desired policy outcome is established, the government or group creates a framework for priority action, a strategy that sets tactical targets (Fig. 3.2). Strategy is a plan of action, a military term defined as "the art of defeating the enemy in the most economical and expeditious manner" (Morison, 1958). Strategy provides incentives, establishes institutional capacity with clear accountability, exposes errors and inefficiencies, and identifies true costs. It involves inventory, research, and monitoring as well as cooperation (networking), programs (projects), and resources (money, facilities, etc.). A successful conservation strategy includes preventative action, precautionary approaches, public participation, research, and monitoring so as to incorporate feedbacks that improve the ability to lead and to adjust without losing sight of the goal. As conservation strategy is a plan of action to achieve an overall goal, as for biodiversity protection, sustainable use, and ecosystem health, it is least glamorous and most difficult to carry out, and too often the "missing link" in conservation programs.

The need for conservation strategy was foreseen in the 1970s with publication of the World *Conservation Strategy* (IUCN, 1980). Its goals were to maintain essential ecological processes and life-support systems, preserve genetic diversity, and ensure sustainable utilization of species and ecosystems by emphasizing "processes" and "systems." A decade later, its revision—*Caring for the Earth: A Strategy for Sustainable Living* (IUCN *et al.*, 1991)—contained nine "Principles" and included a chapter on Oceans and Coastal Areas with 12 recommended actions. Finally, the *Global Biodiversity Strategy* of 1992 (WRI *et al.*, 1992) was agreed upon at the tenth meeting of the CBD at Nagoya, Japan (2010), that established a new Strategic Plan for 2012–20 (Box 3.3). These documents identify aspirations

Table 3.3 Summary of 5878 globally established Marine Protected Areas that cover ~1% of global ocean surface. From Toropova C, *et al.* (2010) *Global Ocean Protection: Present Status and Future Possibilities*. Brest, France; Agence des aires marines protégées, Gland, Switzerland, Washington, D.C. and New York, USA; IUCN WCPA, Cambridge, UK; UNEP-WCMC, Arlington, USA; TNC, Tokyo, Japan; UNU, New York, USA; World Conservation Strategy, pp. 1–96.

(a) MPAs in marine realms/provinces with approximate MPAs area sizes and percentage covered.

Marine *realms and provinces*	**Shelf area (km²)**	**Marine area under some form of protection (km²)**	**Percentage marine area protected (within coastal belt)**
Southern Ocean	792,253	28,330	4
Tropical Eastern Pacific	254,137	27,558	11
Temperate Australasia	1,025,333	56,288	5
Temperate Northern Pacific	3,029,022	74,156	2
Temperate Southern Africa	284,261	7,225	3
Western Indo-Pacific	2,233,848	39,119	2
Temperate Northern Atlantic	4,178,449	66,113	2
Arctic	7,636,248	372,132	5
Eastern Indo-Pacific	150,287	29,448	20
Temperate South America	1,704,401	6,052	0.4
Central Indo-Pacific	5,881,372	421,679	7
Tropical Atlantic	2,162,800	138,764	6
Totals	**29,332,411**	**1,266,864**	**4**

(b) Oceanic MPAs: ocean zones, off-shelf, and regional seas. MPA coverage of the off-shelf, bathyal, and abyssal areas breaks down to 1.32% and 0.67%, respectively. The total global MPA area coverage is mostly of a relatively few very large MPAs and many very small sites.

Region	**Type**	**Area (km²)**	**MPA (km²)**
Pelagic			
Atlantic Ocean	Boundary and equatorial currents, gyres, transitional, Gulf of Mexico, Caribbean Sea, SE U.S.A shelf	57,982,554	87,253
Pacific Ocean		86,073,399	1,136,277
Indian Ocean		82,816,824	628,926
Southern Subtropical Front		21,837,584	345,893
Antarctic Ocean		33,003,858	618,246
Semi-enclosed seas			
Indonesian Through-Flow	Complex: straits and seas	3571343	42,895
Mediterranean Sea		1840,859	4,382
Red Sea		229,964	2
Sea of Japan/East Sea		740,969	2
South China Sea		1,586,354	7
Black Sea		292,027	0
Bathyal	Whole oceans, plates, ridges, regions	830,60,170	1,093,774
Abyssal	Whole regional ocean basins	23,7436,097	1,586,537

for a comprehensive approach, being largely goal-setting instruments for nations and conservation groups to help develop coastal-marine conservation strategies.

Tactics is the art of carrying out strategy through deployment and maneuvers of targeted actions. Tactics are unequivocally the most costly portion of conservation, involving real-world applications guided by legal (regulation, zoning, resource quotas, MPAs, etc.) and non-legal measures (e.g., partnerships, agreements, community action, etc.). Government agencies, national to local, have the authority to carry out the greater portion of marine conservation tactics, often with collaboration of international agencies, NGOs, conservation groups, and private citizens. Specifically, government tactics are carried out with legislative authority or decree with enforcement responsibility, public accountability, and taxpayer support for subsidies, direct funding, and public relations.

3.4.1 Law

Laws create the framework for solving environmental problems (Salzman and Thompson, 2010). Normative law is the

Fig. 3.2 Conceptual linkages in policy, strategy, and tactics, highlighting the central role of strategy for marine conservation.

oldest code of law known to civilization, with deep roots in basic truths that govern all people everywhere and that are often taken for granted; e.g., taking a person's property or stealing is wrong; lying is mostly wrong. Environmental laws relate not only to scientific uncertainty about complex issues, but also to conflicts between financial interests, misaligned natural and political boundaries, different concepts about issues, and to clashes between competing interests.

U.S. environmental law has provided some of the most important legal innovations of the modern age through its creation of national parks, environmental assessments, and public access to information, many of which are being applied worldwide. These environmental legal innovations stem from Roman law, which is acceptable to three-fourths of the civilized world because of its equity, universal adaptability, and its applications to government (Burdick, 2004). In Roman law, the Institutes of Justinian declared that air, running water, the seas, and its shores are a commons for all people to use, a declaration that became known as the *Public Trust Doctrine*, a common-law principle passed from Romans to England and to the English Empire. In Britain, the King held public-trust authority as a benefit to all English subjects, and which was passed after the U.S. Revolution to give authority to states. State courts thus became the chief enforcers of what has remained a common-law doctrine of property (Ruhl and Salzman, 2006). Continental Europe inherited civil law from the Roman Empire, which was further developed through a code of laws established by Napoleon Bonaparte. Worldwide, common law, civil law, customary law, Muslim law, and mixed law are different forms of law. As globalization is connecting societies, the surge of international agreements and regulatory regimes now being created means that environmental-policy innovations are being transported into countries with different legal and cultural traditions. The effectiveness and enforceability of such policy innovations in some regions may be more aspirational than legally obligatory (Yang and Percival, 2009). Environmental issues and conflicts thus require resolution among different forms of law.

Box 3.3 Key elements of the new *Global Biodiversity Strategic Plan* 2011–2020

The vision: "Living in Harmony with Nature" where "By 2050, biodiversity is valued, conserved, restored and wisely used, maintaining ecosystem services, sustaining a healthy plant and delivering benefits essential for all people."

The mission: to "take effective and urgent action to halt the loss of biodiversity in order to ensure that by 2020 ecosystems are resilient and continue to provide essential services, thereby securing the planet's variety of life, and contributing to human well-being, and poverty eradication. To ensure this, pressures on biodiversity are reduced, ecosystems are restored, biological resources are sustainably used and benefits arising out of utilization of genetic resources are shared in a fair and equitable manner; adequate financial resources are provided, capacities are enhanced, biodiversity issues and values mainstreamed, appropriate policies are effectively implemented, and decision-making is based on sound science and the precautionary approach."

Five strategic goals, with twenty Aichi targets: These are both aspirations for achievement at the global level, and a flexible framework for establishing national or regional targets.

- Goal A: Address the underlying causes of biodiversity loss by mainstreaming biodiversity across government and society.
- Goal B: Reduce the direct pressures on biodiversity and promote sustainable use.
- Goal C: To improve the status of biodiversity by safeguarding ecosystems, species and genetic diversity.
- Goal D: Enhance the benefits to all from biodiversity and ecosystem services.
- Goal E: Enhance implementation through participatory planning, knowledge management and capacity building.

Implementation: The *Strategic Plan* will be implemented primarily through activities at the national (i.e., *National Biodiversity Strategies* and *Action Plans*), or subnational level with supporting action at the regional and global levels.

Source: COP (2010) Tenth meeting of the Conference of the Parties to Convention on Biological Diversity in Nagoya, Aichi Prefecture, Japan

Global environmental law is emerging through the United Nations with a set of legal principles developed from regulatory systems to protect the environment and to manage natural resources. The United Nations Environment Program (UNEP) role in international governance and policy, through its Division of Environmental Law and Conventions, plays a key role in the development and facilitation of international environmental law (UNEP, online). International sources of environmental law are documented in UNEP's *Register of International Treaties and Other Agreements in the Field of the Environment* (UNEP, 2005). The International Court of Justice is the primary judicial organ of the United Nations, and is charged with resolving various disputes between nations (ICJ, online). A nation can recognize the compulsory jurisdiction of the Court, or it can choose to be exempt from compulsory jurisdiction for certain classes of cases. This partial exemption is controversial, but is upheld by the 15-member Court elected by the United Nations General Assembly and the Security Council.

3.4.2 Science

Science is a highly organized, self-correcting process that requires iterative communication among peers. As it attempts to narrow uncertainty, science provides an objective understanding of the environment, a measurement of change, and information for policy-makers, generating defensible evidence and dispelling myth through rational evaluation, e.g., that the ocean is so large that *the solution to pollution is dilution*. Lack of scientific understanding has led to fatal consequences across the socio-economic spectrum, especially for those for whom fish is a dietary mainstay (Mergler *et al.*, 2007), as in the case of Japan's Minamata methylmercury poisoning in May 1956 that killed more than 100 people (Harada, 1995). Since World War II, science has increasingly formed a basis for modern thought, judicial fairness, and democratic equity—a way of knowing about the natural world through an objective lens of rational analysis (Moore, 1993) that requires rigorous assessment and re-assessment to avoid fallacious interpretation. As such, science forms a sound basis for conservation action. But lacking complete scientific information and much complexity, marine conservation most often requires preemptive, precautionary approaches to guide policy, management, and use (Ch. 13).

3.4.3 Economics

Economics plays a central role in social, political, and environmental issues, and environmental economics plays a key role in marine environmental policy and conservation (Costanza *et al.*, 1999). Ecological economics is about the sustainability of healthy marine environments in an era of globalization that is shrinking and reshaping the increasingly interconnected and interrelated world: i.e., generating economic growth while increasing social disparity and decreasing resources. Economic disparity increases ineffective responses and leads to a cluster of risks involving national fragility, organized crime, corruption, and a growing illicit trade of goods estimated at U.S. $1.3 trillion in 2009 (WEF, 2011; Smith *et al.*, 2003). The World Economic Forum in Switzerland declared that the global financial crisis that began in 2008 reduced global economic resilience, increased geopolitical tension, and heightened social concerns (WEF, 2011). Economic concerns cause politically corrupt nations to adopt less protective measures for conservation priorities and species diversity than other nations (Smith *et al.*, 2003). Some call for new indicators of economic progress to be geared to the economy that actually exists (Daly and Cobb, 1994; Cobb *et al.*, 1995).

3.4.4 Sovereign power of nations

What makes a nation sovereign is its claim on the environment and its rights over lands, adjacent ocean space, and resources, as defined under its constitution. Nations hold legal power to determine norms of behavior and conditions of life for humans and non-humans, to conserve and manage resources and environments mostly under binding rules of law, and to enter into international agreements. International agreements are largely voluntary, lack power, and considered "soft" mechanisms, as they depend on consent or consensus among nations and parties for implementation through bottom-up procedures.

Nations, on the other hand, traditionally engage in "hard" measures, e.g., command-and-control regulatory power, to achieve conservation goals. National policies are typically administered hierarchically in a top-down, vertical flow of power, from national government to provinces to local authorities and stakeholder (e.g., native) groups. But when integration among powers is required, governments usually operate horizontally through institutions and agencies that carry out assigned duties.

Most governments struggle to balance social, economic, and political interests as they also enact protective measures to safeguard natural resources, wildlife, and the environment. When policies collide, differing values attached to resources by different subsets of society often result in contentious debate, as occurs among public trust, property rights, and depletion of shared resources. And because policies are frequently formed under conditions of uncertainty, urgency, small budgets, and other constraints, interest groups can pursue goals to fit a favorable outcome for themselves (Kamieniecki, 2006). Hence, policies for sustainability, biodiversity protection, social equity, and resource conservation must compete with policies for economic growth, pursuit of wealth, energy acquisition, and resource consumption.

Land and sea jurisdictions among almost all nations fall under distinctly different private and public domains. Use of coastal intertidal areas and shorelands immediately landward of the water's edge is usually a customary right of citizens, but varies considerably among the world's dominant legal systems. Under the English *Public Trust Doctrine*, the government holds uses of coastal waters, submerged soils, and their resources for public benefit. Because the doctrine is grounded in property ownership and "best use," it is intimately connected with the economy, family structure, and the political system. Under "nuisance law," legislative bodies can declare that certain activities constitute public nuisances. Scandinavian "rights of common access" run counter to "property rights," allowing public access to lands, beaches, and intertidal areas. Some societies do not base laws on forms of property ownership (e.g., Polynesian and Inuit cultures). Confusion and conflict can result when these traditions coexist with western law, being resolved through highly variable social and legal mechanisms and traditions that are currently evolving.

3.5 POLICY INSTRUMENTS FOR MARINE CONSERVATION

Policy instruments are about decision-making tools on such issues as pollution, species protection, climate change, fisheries, transboundary situations, coastal zone management, biodiversity, sustainability, and environmental conservation. Marine environmental governance incorporates not only many policy instruments, but also financing mechanisms, rules, procedures, and social norms that directly or indirectly affect marine conservation.

3.5.1 U.S. national environmental policy

U.S. environmental policy is ultimately about politics and government (Fiorino, 1995). Policy includes a comprehensive set of mechanisms covering all major human-caused environmental issues. Policy shifts with public attitude and private-interest persuasion, and through conflicting goals and compromises that can have unintended consequences. Policy is ultimately determined by power politics, and when power (or priority) is granted to one person, institution, or resource, another may step in to limit it, owing less to environmental paradigms and more to political expediency. Environmental policy thus is a paradox (Smith, 2004), i.e., it needs to protect the environment from users while also protecting user interests.

The right to fish and hunt came with the *Public Trust Doctrine* inherited from British customary (common) law. Natural resources are held in public trust by state authority that protects the common shared resource for the benefit of citizens to fish and hunt. Entrepreneurs challenged that basic right in court in 19th century disputes over oysters (McCay, 1998; Ch. 6). As common law lacked judicial clarity, regulatory law soon outpaced it to become the primary tool for resource and environmental protection. De Tocqueville's *Democracy in America* (1831) stated: "scarcely any political question arises in the United States which is not resolved, sooner or later, into a judicial question" (Bodansky, 1998). The fundamental function of courts has been toward democratization, but public concern and the 20th century environmental movement are moving courts in a different direction, bringing dramatic changes in law and public values (Coglianese, 2001) that reshape environmental policy. Environmental protection may be implemented through the *Public Trust Doctrine* with the need to protect ecosystem services important to the public good (Ruhl and Salzman, 2006), but is vulnerable to the will of a concerted minority that can manipulate a diffuse majority in the public-trust arena (Sax, 1970).

U.S. environmental policies thus struggle to balance social, economic, and political interests against safeguarding natural resources, wildlife, and the environment. The Legislative Branch of Congress passes environmental laws, the Executive Branch administers federal environmental regulations through a variety of federal agencies (Smith, 2004; Salzman and Thompson, 2010), and the Judiciary Branch adjudicates the allocation of resources. Federal regulation is one of the basic tools that government uses to implement public policy (Copeland, 2008) and volunteerism is not generally seen as a reliable solution (Smith, 2004). Congress has the constitutional authority to regulate commerce, become party to international treaties, and to regulate spending. Individual constitutions of states provide mechanisms to carry out environmental policies and programs critical to marine conservation. In cases of overlap or conflict, federal law takes precedence.

Some major U.S. environmental legislation illustrates legislative action for marine conservation. Almost all others are pursuant to specific issues, as for example, the *Harmful Algal Bloom and Hypoxia Research and Control Act* (1998) that established a Department of Interior Inter-Agency Task Force to assess the economic and ecological impacts of algal blooms and hypoxia. This Act was amended in 1998 and 2004 to establish a program for prevention and control of harmful algal blooms and hypoxia.

3.5.1.1 *National Environmental Policy Act* (NEPA, 42 U.S.C. 4321 et seq.)

NEPA was enacted four months prior to Earth Day in 1970, over public concern over oil spills, raw sewage, and industrial pollutants contaminating lands, water, and air (Brooks, 2009). Prior to NEPA, the first federal environmental legislation involved air pollution, the *Air Pollution Control Act* (1955) and Clean Air Act (1963), but existing laws did not provide environmental protection the public demanded. Through passage of NEPA, the U.S. established its first national policy to balance environmental concerns with social-economic requirements. While it does not preserve the environment, it recognizes a rapidly changing world with diminishing natural resources. NEPA set in place procedural requirements for all federal government agencies to enhance the general welfare needs by requiring these agencies to prepare Environmental Assessments (EAs) and Environmental Impact Statements (EISs) about environmental effects of proposed actions. These gave decision-makers and the public the opportunity—not the needed action—to consider alternatives that would minimize or avoid environmental impacts. To monitor environmental policy and strategy occurring on federal land, NEPA established the President's Council on Environmental Quality (CEQ) and created the Environmental Protection Agency (EPA) to implement strategy and tactics.

3.5.1.2 *Clean Air Act* (CAA 1970, 42 U.S.C. §7401 et seq.)

This Act is a comprehensive federal law for air emissions from both stationary and mobile sources, and is administered by the EPA. It resulted in a major shift in the federal government's role in air pollution control and addressed public health and welfare risks posed by certain widespread air pollutants. This Act has four basic programs: the National Ambient Air Quality Standards (NAAQS) to control air pollutants, e.g., ozone, nitrogen dioxide, particulate matter, lead, etc.; National Emission Standards for Hazardous Air Pollutants to control release of known toxins; New Source Performance Standards to regulate new sources of industrial pollution; and State Implementation Plans, authorizing states to implement CAA rules and regulations. EPA establishes emission standards that require the maximum degree of reduction in emissions of hazardous air pollutants.

3.5.1.3 *Clean Water Act* (CWA 1972, 33 U.S.C. §1251 et seq.)

This Act forms the cornerstone of U.S. surface-water quality protection, regulating pollutant discharges into waterways and establishing industrial water quality standards. A permitting process of the U.S. Army Corps of Engineers and EPA establishes the basic structure for regulating discharge of dredged or fill materials into waters and wetlands. CWA employs a variety of regulatory and non-regulatory tools to achieve the broader goal of restoring and maintaining the chemical, physical, and biological integrity of the nation's waters for "the protection and propagation of fish, shellfish, and wildlife and recreation in and on the water." In recent times, CWA programs have shifted from a program-by-program, source-by-source, pollutant-by-pollutant approach to more holistic watershed-based strategies, giving equal emphasis to protecting healthy waters and restoring impaired ones (Ch. 6).

3.5.1.4 *Coastal Zone Management Act* (CZMA 1972, P.L. 92–583, 86 Stat. 1280, 16 U.S.C. §1451–1464, Chapter 33)

CZMA established national policy to preserve, protect, develop, and where possible, restore or enhance the resources of the Nation's coastal zone. Administered by NOAA's Office of Ocean and Coastal Resource Management (OCRM), it is largely voluntary and depends on cooperation and incentives among federal and state levels of government to achieve goals, encouraging coastal states to develop and implement coastal zone management plans (CZMPs) through federal grants. Ambitious CZMPs attempt to balance competing land and water uses while also protecting sensitive resources, encompassing two national programs: the National Coastal Zone Management Program and the National Estuarine Research Reserve System.

3.5.1.5 *Marine Protection, Research, and Sanctuaries Act* (1972, 33 U.S.C. §1401–1445; 16 U.S.C. §1431 et seq.; also 33 U.S.C. 1271)

Titles I and II of this Act, also referred to as the *Ocean Dumping Act*, generally prohibit: (i) transportation of material from the United States for the purpose of ocean dumping; (ii) transportation of material from anywhere for the purpose of ocean dumping by U.S. agencies or U.S.-flagged vessels; and (iii) dumping of material transported from outside the United States into U.S. territorial seas. Title III created the National Marine Sanctuary Program managed under a National Marine Sanctuary System whereby the Secretary of Commerce may designate any discrete area of the marine environment of special national significance for conservation, recreation, and/or of ecological, historical, scientific, archaeological, educational, or esthetic quality as a Sanctuary. Fishing is permitted. Amendments in 1980 allow removal if the Governor of the affected state finds it unacceptable or if both houses of Congress disapprove.

Broadening the scope of Sanctuaries, Presidential Executive Order (EO) 13158 (May 2000) encouraged Marine Protected Area (MPA) designation. It defined MPA as "any area of the marine environment that has been reserved by federal, state, territorial, tribal, or local laws or regulations to provide lasting protection for part or all of the natural and cultural resources within." A national system of MPAs is developed jointly by the Departments of Commerce and Interior to strengthen management, protection, and conservation of existing protected areas, establish new ones, and reduce harm of federally approved or funded activities. This scientifically based system encompasses Department of Interiors' National Estuarine Research Reserves, national seashores, national parks, national monuments, critical habitats, national wildlife refuges, NOAA's National Marine Sanctuaries, fishery management zones where use of specific types of fishing gear is restricted, state conservation areas, state reserves, and others. As of 2012, 1700 MPAs of some form cover approximately 41% of all U.S. coastal waters, with <8% of these as "no-take" fishery areas (National Marine Protected Areas Center, online). Legally, "When a nation declares a marine protected area or an exclusive fishing zone, is it exercising rights as the proprietor of marine systems or is it exercising regulatory authority?" (Osherenko, 2006).

3.5.1.6 *The Marine Mammal Protected Act* (MMPA 1972, 16 U.S.C. §1361–1421h et seq.)

The MMPA set precedent by: establishing "optimum sustainable population" (OSP) for marine mammals as significant functioning elements of ecosystems by: requiring a science-based ecosystem approach for management; placing a moratorium on the taking and importation of marine mammals and marine-mammal products; calling for application of the "precautionary principle"; and by establishing a Marine Mammal Commission. Under the Commission, a Committee of Scientific Advisors reviews activities of, and makes recommendations to, agencies responsible for marine-mammal management that are also required to report results directly to Congress (Ray and Potter, 2011). Two agencies are responsible: the U.S. Department of Interior's Fish and Wildlife Service (FWS) for walrus, polar bear, sea otter, and West Indian manatee; Department of Commerce's NOAA National Marine Fisheries Service (NMFS) for all others, i.e., seals, whales, dolphins, and porpoises. The MMPA superseded all other Acts pertaining to marine mammals, most significantly the *Fur Seal Act* that supported the international *North Pacific Fur Seal Convention* of 1911, which prohibited taking North Pacific fur seals (*Callorhinus ursinus*) and sea otters (*Enhydra lutris*) except by Alaska natives for subsistence purposes or by others under permit from NOAA. Although these Acts were abrogated by the MMPA, Alaska subsistence hunters are still allowed to take marine mammals.

MMPA became the first legislation anywhere in the world to mandate an ecosystem approach to marine resource management. The controversial management concept of "optimum sustainable population" was intended to replace simplistic "maximum sustainable yield" approaches, as applied to fisheries and marine mammals alike. The primary objective of marine-mammal management is to maintain the health and stability of the marine ecosystem and when consistent with the primary objective, i.e., to obtain and maintain optimum sustainable populations of marine mammals. The ecosystem approach has been incorporated in other U.S. statutes such as the *Magnuson-Stevens Fishery Conservation and Management Act* (Section 3.5.1.8), in legislation in other countries, and in international agreements such as CCAMLR (Section 3.5.2.4).

3.5.1.7 *Endangered Species Act* (ESA1973, 7 U.S.C. §136; 16 U.S.C. §460 et seq.)

ESA takes a very different approach than MMPA, focusing on species already in danger of extinction, requiring habitat protection and species restoration. This law is one of the most powerful environmental laws for the preservation of endangered species ever enacted by any nation (Mueller, 1994) and many Americans approve of the need to save numerous species from extinction. Scientific questions about what constitutes a *species* and its *habitat* are critical. ESA defines a species as "any subspecies of fish or wildlife or plants, and any distinct population segment of any species of vertebrate fish or wildlife which interbreeds when mature"; endangered species are "any species which is in danger of extinction throughout all or a significant portion of its range"; threatened species are "any species which is likely to become an endangered species within the foreseeable future throughout all or a significant portion of its range." With the exception of recognized insect pests, all animals and plants are eligible, and species listed are protected without regard to commercial or sport value. A species is *fully recovered* when it no longer requires protection, and then may be delisted. Often, *fully recovered* cannot be known, due to lack of a baseline (Jackson *et al.*, 2011). For example, only 39 of approximately 1800 species protected under ESA have been removed from the list, but only 15 of those are considered to be fully recovered (Keller and Gerber, 2004); thus longer-term protection is advised. Furthermore, under the ESA, the U.S. federal government is required to designate *critical habitat* for any listed species, defined as: "specific areas within the geographical area occupied by the species at the time of listing and/or specific areas outside the geographical area occupied by the species if the agency determines that the area itself is essential for conservation." However, ESA mechanisms for habitat protection are often insufficient to prevent species depletion or to protect the habitat itself. Enforcement is subject to lawsuits regarding property rights. The U.S. Supreme Court has argued that by placing limits on federal legislative authority to protect habitat, private citizens' rights may be affected.

The joint FWS/NMFS Office of Protected Resources administers the Act. Listing requires good science, yet documentation of threatened species' natural history and habitat is too often rudimentary at best, with narrowly focused wildlife agencies frequently lacking technical skills, information, and necessary budgets (Carden, 2006). For marine systems, natural-history information is especially difficult to acquire, and opponents to environmental action can use scientific uncertainty to obfuscate evidence (Oreskes and Conway, 2010). While the public strongly supports the ESA, political debates contest the extent to which the nation's natural resources should be protected, and how best to utilize them.

3.5.1.8 *Magnuson-Stevens Fishery Conservation and Management Reauthorization Act* (2006, P.L. 109–479)

This is the primary law governing marine fisheries management in U.S. federal waters, having exclusive fishery management authority over all fish, but leaving to states the management of fisheries within their jurisdiction. The Act aims to conserve and manage fishery resource, supported by strong ecological considerations.

This Act has a long history of change. It began with the *Fishery Conservation and Management Act* (1976) that established 200 nmi fishery conservation zones and created eight regional fishery management councils to oversee management and to promote fisheries conservation. Amendments in 1996 phased out foreign fishing within the U.S. Exclusive Economic Zone (EEZ), and focused on rebuilding overfished fisheries, protecting essential fish habitat, and reducing bycatch. The Act thereby aided the domestic fishing industry, and intended to promote conservation. The U.S. claims exclusive fishery conservation and management authority over all anadromous and catadromous species throughout their migratory ranges and beyond the EEZ, except within a foreign nation's waters. To improve shark conservation, the *International Fisheries Agreement Clarification Act* (Public Law No: 111–348, 2011) amended the *High Seas Driftnet Fishing Moratorium Protection Act* and the *Magnuson-Stevens Fishery Conservation and Management Act*.

3.5.1.9 National ocean policy

The need for a comprehensive U.S. ocean policy was made clear by the need to coordinate 140 federal laws and 18 federal implementing agencies that are responsible for managing U.S. coastal-marine systems, and to address conflicts arising among them (ICOSRM, 2008). Marine species, for example, are utilized for economic yield (*Fisheries Act*), for optimum sustainable population (MMPA), and may be listed when threatened by extinction (ESA). In the case of marine mammals, conflicts can arise in carrying out these policies. Comprehensive policy is also important because ocean-dependent industries generate billions of dollars every year, contributing 2.5 times more to the U.S. economy than the agricultural industry. And deciding whether commercial priorities take precedence over environmental and species sustainability can only be resolved in the policy/political arena under a comprehensive policy.

Until the late 1940s when industries began to expand offshore, offshore oil and gas production was mostly unregulated and land ownership beneath states' navigable waters was controversial (Office of Oil and Gas, 2005). In 1953, Congress passed the *Submerged Lands Act* (SLA) that established the federal government's title to submerged lands over most of the continental margin, giving most states jurisdiction over any natural resources within 3 nmi of the coastline. The SLA led to the *Outer Continental Shelf Lands Act* (OCSLA, 1953), the cornerstone of offshore legislation that defines the outer continental shelf (OCS) as any submerged land outside state jurisdiction. However, ocean policy remained confused and uncoordinated.

"Who owns the coast?" (McCay, 2008). Customs and policies inherited from traditional practices and historical policies have divided the coastal system into separate domains (Armstrong and Ryner, 1980), with different management regimes that follow separate policies for submerged land, the water column, pollution, and resource use. Thus, the coastal land-sea zone lacks legal "distinctiveness" for management. Furthermore, marine species that utilize both land and sea fall mostly within the public domain as "common-pool" resources, where a number of people lacking incentives for conservation have access (NRC, 2002). Considering fish and other marine organisms as *wildlife* rather than *resources/commodities* is an option, protection under the ESA being another. However, increasing species listing is already over-loading administrative capacity, with the result that some potentially endangered species cannot get listed (e.g., Pacific walrus, Ch. 7; Woody, 2011). Furthermore, those advocating property rights resist protection of critical habitat, as required under ESA.

National ocean policy remains a work in progress. The *Oceans Act* of 2000 (P.L. 106–256) established the U.S. Commission on Ocean Policy to make recommendations for a coordinated and comprehensive national ocean policy. It and the Pew Oceans Commission created the bipartisan collaborative Joint Ocean Commission Initiative that called for a national ocean policy to improve federal coordination, considered climate change and acidification, protection of ocean resources, development of a unifying policy framework, and acceding to law-of-the-sea (UNCLOS) principles (JOCI, 2008). In July 2010, Presidential Executive Order EO 13547 gave ocean policy a boost by calling for stewardship of the ocean, coasts, and Great Lakes, to protect, maintain, and restore their health and biological diversity, and to increase scientific understanding of coastal and ocean ecosystems as parts of globally interconnected air, land, ice, and water systems. The National Ocean Council created by this EO recommended implementing the nation's first national ocean policy for stewardship, but remains challenged both by lack of ecosystem understanding and by the laws, authorities, and governance structures intended to manage the nation's coasts and oceans (CEQ, 2010).

3.5.2 International governance and cooperation

Policy-making at the international level is an integral part of international governance and cooperation, but among the greatest failures of international cooperation is the inability to manage ocean resources (Weaver, 2010). Analysis and assessment of data and self-interests of nations feed into foreign policy decisions, under mechanisms defined by international law (Box 3.4; Table 3.2). Primary tools include treaties, which are binding (hard law) and involve norms or principles made explicit and documented, and cooperative agreements that are non-binding (soft law). Differences between hard and soft law can be debated, but soft law is gaining influence, for example, the *Rio Declaration on Environment and Development* and agreements on the deep seabed, sovereignty over natural resources, codes of conduct, and guidelines and recommendations of international organizations (Boyle, 1999).

With establishment of the United Nations (UN) in 1945 after World War II, international governance advanced greatly. The UN Charter authorized it to achieve harmony of actions among nations and to resolve common problems through diplomacy and international mechanisms. Initially constrained

Box 3.4 Some definitions in international law

Agreement: a compact entered into by two or more nations or heads of nations; in the wide sense, any act of coming into conformity; in the narrow sense, an accord between states, but less formal than a treaty; may or may not be obligatory; includes convention, treaty, protocol, accord, act, declaration, pact, provision, etc.

Convention: agreement concluded among states on matters of vital importance; often used *in lieu* of treaty, but usually restricted to agreements sponsored by an international organization; intended to be legally binding, but requires ratification.

Declaration: a document whose signatories express their agreement with a set of objectives and principles; may not be legally binding, but carries moral weight.

International law: the body of legal rules and norms that regulates activities carried out by agreement among nations; intended to be legally binding, but requires ratification.

Protocol: agreement that completes, supplements, amends, elucidates, or qualifies a treaty or convention; has the same legal force as the initial document.

Ratification: final confirmation of a treaty, convention, or other document by a nation's competent body (legislature or head of state), thereby becoming legally binding and securing that country's commitment to it; there is no prescribed length of time for ratification.

Regime: arrangements that contain agreed-upon strategies, principles, norms, rules, decision-making procedures, and programs that govern interactions of participants in specific areas, such as fishing, navigation, trade, and scientific research.

Resolution: text adopted by a deliberative body or an international organization; may or may not be binding.

Treaty (from Latin *tractere*, to "treat"): an agreement entered into by two or more nations or heads of nations; intended to be legally binding; requires ratification.

Sources: Fox (1992); Gamboa (1973); Gleick (2000); University of Virginia School of Law (online)

Table 3.4 UN environmental agencies with international marine programs.

UN agency (year formed)	Mandate	Examples of commissions or relevant programs
FAO Food and Agriculture Organization, an autonomous agency within UN system (1945)	To improve nutrition, food production and distribution; alleviate hunger-malnutrition; food standards; long-term strategy for conservation and management of natural resources.	Intergovernmental Oceanographic Commission (IOC); Fisheries Department involved in Convention on Biological Diversity and UNCLOS; programs on environmental quality.
UNESCO UN Educational, Scientific, and Cultural Organization (1946)	To advance universal respect for justice, rule of law, human rights, fundamental freedoms of all peoples; emphasizes interdisciplinary approach; promotes understanding; encourages scientific research and training.	Man & Biosphere Program (MAB); World Heritage sites; Coastal Regions and Small Islands Initiative; promotes international ocean science: works closely with International Council of Scientific Unions (ICSU) and Scientific Committee on Ocean Research (SCOR).
IMCO International Maritime Organization (IMO 1958); preceded by Inter-Governmental Maritime Consultative Organization1947)	To develop policies for international shipping; to facilitate technical cooperation; concern for marine environment, maritime safety, efficiency of navigation; prevention and control of marine pollution from ships.	Administers London Convention, and subsequent conventions, e.g., MARPOL and protocols for pollution; develops guidelines for ballast-water control of exotic species introductions; measures to prevent accidents; maritime legislation and its implementation.
UNEP UN Environmental Program (1972)	To coordinate environmental agreements and activities within United Nations system; to aid nations develop and adopt environmental policies, strategies, and actions.	Helps with formation of environmental treaties and agreements; funds and guides environmental strategies and action plans; coordinates regional seas programs and shared environmental problems of multinational, multi-cultural nations that border those seas; aids environmental negotiations, conventions (e.g., for biodiversity, climate change).

by national sovereignty, the UN was increasingly called upon to address growing transnational issues in international development, pollution, resource exploitation, and others, and to achieve preeminence through its specialized, problem-resolving agencies (Table 3.4) as mandated by their constitutions to undertake global, environment-related policy-making. Each agency has a policy-making body representing the views of member states, and expertise in research and management to which issues may be referred. The policy-making architecture and related capacity of these specialized UN agencies provide a global approach for addressing complex maritime/marine issues (Hinds, 2003). In particular, routine discharge of ballast water and sewage is regulated under international pollution standards for the high seas (Table 3.5). The UN, however, unlike national authority, has no counterpart to a national legislature, lacks regulatory power, and requires its agencies to seek consensus among nations.

Environmental activity expanded exponentially after World War II (Fig. 3.3). As numerous environmental organizations became established, discussions among them lead to international treaties and intergovernmental organizations collectively called the "world environmental regime," broadly defined as "a partially integrated collection of world-level organizations, understandings, and assumptions that specify the relationship of human society to nature" (Meyer *et al.*, 1997). International regimes are usually directed toward specific topics such as fishing, scientific research, navigation, trade, biodiversity, and their consequences; e.g., the *Convention on Biological Diversity, Agenda 21,* and various protocols. Most rest on one or more constitutive documents that are not necessarily legally binding. Regimes are particularly necessary for resolving transboundary environmental issues that arise when activities within the jurisdiction of one nation have consequences affecting other nations. When a fish population migrates between national jurisdictions, fishing in one area can affect fishing in another area, resulting in disputes over the level of take, even when precise species distributions may not be known. A general trend is to reallocate jurisdiction under voluntary agreements, from national to transnational to supranational authorities, usually lacking power as a "soft-law" mechanism. The final version of a compromised protocol is often a diluted solution that lacks civil-society commitment (Brown, 2011).

International ocean policy is especially complex. It combines individual interests of sovereign nations within an ocean commons matrix of traditional practices, where freedom of the seas brings economic benefits in sea trade, international preeminence in sea power, national security, and global stability (Kraska, 2008). Throughout history, all nations held legitimate right to the high seas, with freedom to navigate, to use resources, and to fish as a common use right. Through the 15th century's maritime mobility, small nations gained vast

Table 3.5 Major international conventions on vessel pollution. Many follow protocols on specific aspects. Dates indicate signing of agreements; ratification takes almost a decade.

Date	Instrument	Intent
1963	*Treaty Banning Nuclear Weapons Tests in the Atmosphere, in Outer Space and Under Water*	To prevent nuclear pollution, globally.
1969	*International Convention Relating to Intervention on the High Seas in Cases of Oil Pollution Casualties*	To prevent or mitigate oil pollution by accidents involving ships outside territorial waters. A protocol extending to other hazardous substances (chemicals) entered into force in 1983.
1969	*International Convention on Civil Liability for Oil Pollution Damage*	To ensure adequate compensation from damage of oil pollution. Placed liability compensation on ship owners releasing or discharging oil.
1971	*International Convention for the Establishment of an International Fund for Compensation for Oil Pollution Damage*	To provide further compensation to oil pollution victims. Placed the burden of compensation on ship owner, with time limits on amount payable. Funded by oil importer contributions.
1972	*London Convention. Protocol bans radioactive wastes and incineration at sea*	Control all sources of pollution to marine environment by dumping of wastes.
1973/1978	*International Convention for the Prevention of Pollution from Ships (MARPOL) and its 1978 Protocol (MSARPOL) supercede the International Convention for the Prevention of Pollution of the Sea by Oil (1974)*	Addressed pollution by oil, noxious liquid substances, harmful substances carried in packaged forms, sewage, and garbage. Widely regarded as the most important instrument of its type. Almost all other agreements since this time have depended on the principles therein. Areas identified as "special areas" receive higher protection.
1974	*Paris Convention for prevention of marine pollution from land-based sources*	Further restrictions on dumping of wastes at sea.
2001	*International Convention on the Control of Harmful Anti-fouling Systems on Ships*	Prohibits use of harmful organotins in anti-fouling paints used on ships; establishes a mechanism to prevent potential future use of other harmful substances in anti-fouling systems.

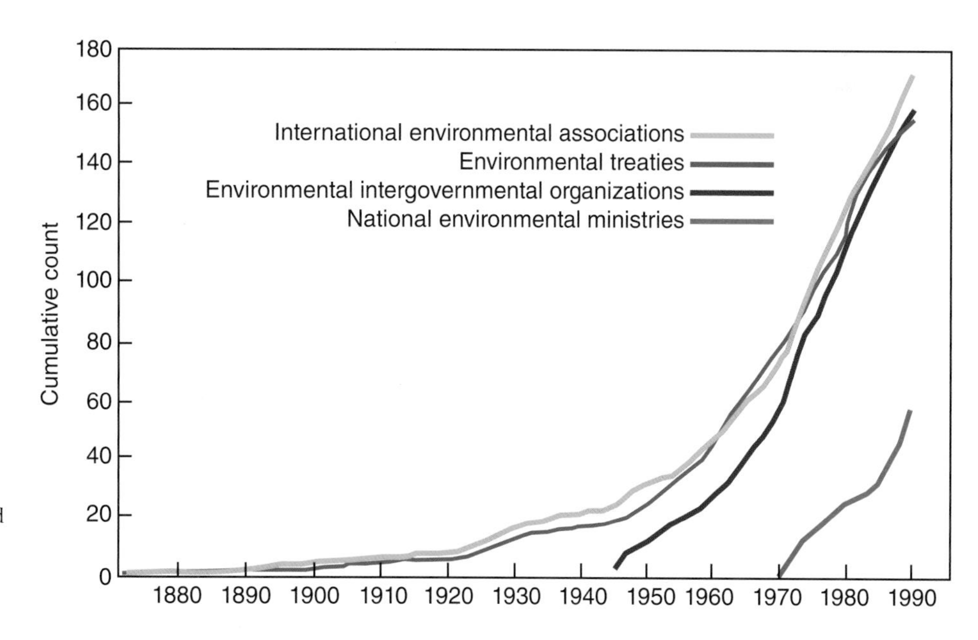

Fig. 3.3 International environmental activities accumulated with greater intensity after World War II, with an extraordinary expansion of international non-governmental and governmental organizations, and treaties. From Meyer *et al.* (1997), with permission.

empires and power, and relied upon such freedoms for safety and prosperity and for conducting international trade. As freedom of the seas waxed and waned, tensions between governments grew in the exercise of government authority over the sea (Kraska, 2011). In the late 1600s, nations endorsed piracy as a profession, pitting one ruler against another on the open seas, and privateers were rewarded for capturing goods of other countries. With increasing activity and recognition of the ocean's importance, particularly after World War II, development, management, and protection increased and

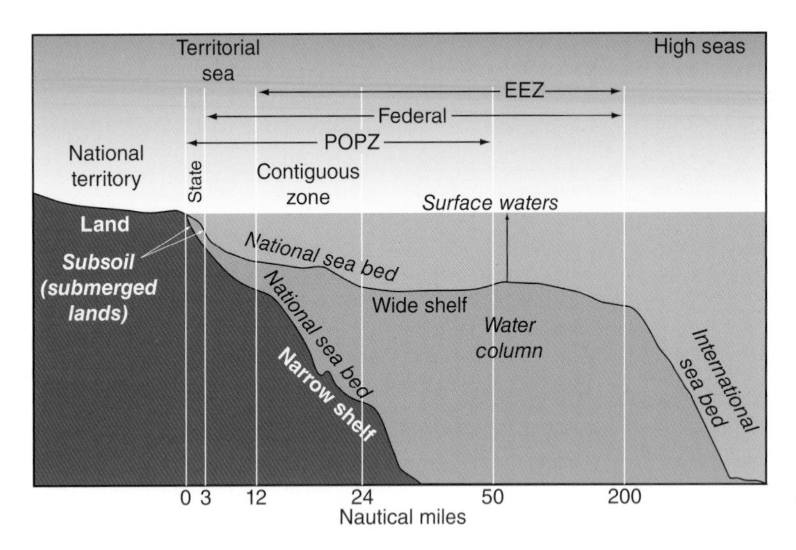

Fig. 3.4 Coastal and marine jurisdictional boundaries recognized under international law subdivide ocean space horizontally and vertically, with different regulations for surface water, the water column, seabed, and subsoils (i.e., submerged coastal lands, usually under state or provincial control). Ocean space allocated to national sovereignty includes a contiguous zone that extends another 12 nmi and where coastal nations can exercise control over customs, immigration, and fiscal or sanitary (pollution) matters. Nations exert control over living and non-living resources within the Exclusive Economic Zone (EEZ), beyond which are the high seas (open ocean; commons, international) under UNCLOS, but where certain restrictions may apply. A Prohibited Oil Pollution Zone (POPZ) is declared by some nations under the *International Convention for the Prevention of Pollution from Ships* (MARPOL). Compare with ecological subdivisions, Fig. 4.7.

jurisdictional boundaries became established over coastal areas and marine resource uses (Fig. 3.4). Fisheries, marine-mammal conservation, shipping, oil and gas, and mining issues required legal distinctions and established authority over use of designated areas and resources. Furthermore, new uses of the oceans for international trade, national security, oil, gas, energy development, and transport made oceans and coastal seas contested arenas.

3.5.2.1 International frameworks for conservation

The UN Conference on the Human Environment in 1972 was a major world event for environmental issues. It brought global recognition to human-environment interactions in a framework for environmental action, recognizing that "Man has the fundamental right to freedom, equality and adequate conditions of life, in an environment of a quality that permits a life of dignity and well-being." It provided Principles and Recommendations for nations to act, with communications about environmental issues (e.g., air pollution), training, and working relationships among agencies on such issues as clean water and population growth. It also recognized that when resources and environment are involved, a sovereign nation should not engage in activities that negatively affect the sovereignty of others.

In 1984, the UN General Assembly followed up by establishing the World Commission on Environment and Development (WCED)—the *Brundtland Commission*. Its 1987 report, *Our Common Future* (WCED, 1987), proposed *sustainable development*, a controversial term that divided nations: developed nations of the global *north*, many with already depleted resources, promoted resource sustainability, while the lesser-developed global *south* sought economic development of natural resources to benefit rising human populations.

In 1992 the UN convened the largest group of world leaders ever held at the United Nations Conference on Environment and Development (UNCED)—the "*Rio Earth Summit*"—in Rio de Janeiro to meet on environmental concerns. One hundred seventeen heads of state, concerned scientists, conservation organizations, and thousands of delegates and participants representing 178 nations placed humans at the *center of concerns for sustainable development* (Principle 1 of the *Rio Declaration*). This summit focused on developing international cooperation for shared environmental concerns, including conservation of large ecosystems that require large-scale transboundary management, as for example regional water resources. The Rio Conference initiated two important, legally binding conventions: The *Framework Convention on Climate Change* (the *Climate Change Convention*) that led to the adoption of the Kyoto Protocol in 1997, and the *Convention on Biological Diversity* (CBD) that urged nations to develop national strategies for the conservation and sustainable use of biological diversity. A comprehensive plan for action, *Agenda 21*, outlined strategies that set forth rights and obligations of nations to be carried out at all levels of organization; Chapter 17 addressed ocean protection (Table 3.6). Among its major themes, "Marine and Coastal Biodiversity" was specifically addressed in the *Jakarta Mandate on the Conservation and Sustainable Use of Marine and Coastal Biological Diversity* adopted in 1995 (Table 3.7).

The 2002 World Summit on Sustainable Development held in Johannesburg (South Africa) strongly reaffirmed UNCED and *Agenda 21*. Political leaders agreed to strive for *a significant reduction in the current rate of loss of biodiversity by 2010*, challenging the ecological and conservation community to detail the rates of biodiversity change (Dobson, 2005). Currently, available evidence indicates that biodiversity loss has not slowed, but rather is increasing (Secretariat of the CBD, 2013). In 2000, the United Nations Secretary-General, Kofi Annan, in his report to the UN General Assembly, called for an assessment of the consequences of ecosystem change for human well-being and for the scientific basis for action needed to enhance the conservation and sustainable use of those systems and their contribution to human well-being. Consequently, the *Millennium Ecosystem Assessment*, based on government requests and synthesis by 1360 experts, found that the most important drivers of change relate to habitat and climate, invasive alien species, pollution, human population, over-exploitation, technology, and lifestyle, with fishing being most

Table 3.6 Agenda 21 of the *Rio Declaration on Environment and Development* provided 27 principles to guide nations through goals and objectives, including a *Program of Action for Sustainable Development*. Section 2, Chapter 17, is on ocean protection; other chapters are identified by title only. Data from UNEP (2013a) Agenda 21. www.unep.org/.

Rio Declaration on Environment and Development
Principle 1: "Human beings are at the center of concerns for sustainable development."

Agenda 21: Program of Action for Sustainable Development
Sec. I. Social & Economic Dimensions: cooperation to accelerate sustainable development; combating poverty; changing consumption patterns; demographic dynamics and sustainability; protecting and promoting human health; promoting sustainable human settlement development; and integrating environment and development into decision-making.
Sec. 2 Conservation & Management of Resources for Development: atmosphere; land resources; deforestation; desertification and drought; mountains; agriculture and rural development biological diversity; biotechnology; freshwater; toxic chemicals; hazardous wastes; solid wastes and sewage; and radioactive wastes.
Chapter 17: Protection of oceans and all kinds of seas, including enclosed and semi-enclosed seas, and coastal areas and the protection, rational use and development of their living resources.
(A) Integrated management and sustainable development of coastal area, including exclusive economic zones
(B) Marine environmental protection
(C) Sustainable use and conservation of marine living resources of the high seas
(D) Sustainable use and conservation of marine living resources under national jurisdiction
(E) Addressing critical uncertainties for the management of the marine environment and climate change
(F) Strengthening international, including regional, cooperation and coordination
(G) Sustainable development of small islands
Sec. 3 Strengthening the Role of Major Groups: women; children and youth; indigenous people; non-governmental organizations; local authorities; workers and trade unions; business and industry; scientific and technological community; and farmers.
Sec. 4 Means of Implementation: financial resources; transfer of environmentally sound technology; science for sustainable development; education, public awareness, and training; capacity-building in developing countries; institutional arrangements; legal instruments and mechanisms; and information for decision-making.

Table 3.7 The *Jakarta Mandate on the Conservation and Sustainable Use of Marine and Coastal Biological Diversity* (1995) is a global consensus on the importance of marine and coastal biological diversity and is part of the Ministerial Statement at the Convention on Biological Diversity to implement conservation and sustainable use of marine and coastal biological diversity. Six key thematic issues were identified, being addressed through a multiyear program of work and advocating the ecosystem approach. Data from the Secretariat of the Convention on Biological Diversity (2013). Jakarta Mandate. UNEP (2013a), Agenda 21 online http://www.unep.org/.

Thematic issues	Operational objectives
1. Integrated marine and coastal area management (IMCAM)	• Review existing instruments • Promote development and implementation at the local, national, and regional levels • Develop guidelines and indicators for ecosystem evaluation and assessment
2. Marine and coastal living resources	• Promote ecosystem approaches to sustainable use of marine and coastal living resources • Make available to parties information on marine and coastal genetic resources
3. Marine and coastal protected areas	• Facilitate research and monitoring activities on value and effects of marine and coastal protected areas, or similarly restricted areas, on sustainable use of marine and coastal living resources • Develop criteria for establishment, management of marine and coastal protected areas
4. Mariculture	• Assess consequences of mariculture for marine and coastal biological diversity and promote techniques to minimize adverse impacts
5. Alien species and genotypes	• Achieve better understanding of the causes and the impacts of introductions of alien species and genotypes • Identify gaps in existing or proposed legal instruments, guidelines, and procedures and collect information on national and international actions • Establish an "incident list" of introductions
6. Ecosystem approach	• Precautionary; science-based; experts; involve local and indigenous communities; three levels ofimplementation (national, regional, global)

important in marine ecosystems (Millennium Ecosystem Assessment, 2005). These factors also cause loss in ecosystem function and services.

3.5.2.2 International maritime law

The *United Nations Convention on the Law of the Sea* (UNCLOS) exemplifies the most advanced and binding international law-making. Its framework was initiated with conferences held in 1930, 1958, and 1960 (UNCLOS, 2012). In 1965, 32 nations claimed 12 nmi territorial seas, increasing to 67 by late 1970s. In 1973, a negotiated framework, signed by 159 nations in 1982 under UNCLOS III and ratified in 1994, illustrated the increasing importance of oceans and the slow process of negotiation. Several industrialized countries objected to provisions for seabed mining (and others) and did not sign the treaty, most notably the United States, Great Britain, and Germany. UNCLOS III represents the most ambitious, most historic, and most far-reaching of international agreements to that time, providing the legal framework for all activities in the oceans and seas with a set of rules for use (UN General Assembly, 2009). Nevertheless, the U.S. claimed 200 nmi marine EEZs by Presidential Proclamation in 1983, followed by 27 countries claiming territories greater than 12 nmi by the beginning of the 21st century; 14 extended claims to 200 nmi, a phenomenon described as "creeping national jurisdiction." The greatest transfer of resources in recorded history thus occurred when UNCLOS III gave nations sovereign rights over all resources, living and non-living through extension of their EEZs seaward to 200 nmi. Sovereignty now divides most ocean space into segments that force nations to agree inter alia on management regimes for shared resources (Fig. 3.4). EEZs for island nations provide jurisdictional extensions that are often many times the size of the nation's land area. UNCLOS III governs the high seas beyond areas of national jurisdiction through consensus and agreement among nations, with the intent that conservation and sustainable use of marine biodiversity should be consistent with the legal framework of UNCLOS.

As a "constitution of the oceans" (MARIBUS, 2012), UNCLOS is among the most notable of maritime agreements (Table 3.8).

Table 3.8 Major events that influenced international maritime law and facilitated "creeping" offshore jurisdictions. Compiled from Archer *et al.* (1994); Wilder (1998); *Encyclopædia Britannica* (1999–2000) online.

Date	Event	International Agreement
450 AD	Codification of Roman Law. Evolved from law of ancient Rome (735 BC–5th century AD)	Legal system forming western law. Established Law of Procedure and absolute ownership, unlike Germanic systems and English law
529–535	Freedom of the Seas	First legal document of the sea
1493	*Mar clausum* by Papal bull *Inter Cetaera*	Gave Spain exclusive rights to land and sea west of Azore Islands
1588	England defeats *Spanish Armada*	Saved England from invasion and Dutch Republic from extinction. Delivered heavy blow to Spain
1625	*Mare Liberum* by Hugo Grotius	Freedom of the Seas; seas are international territory; defense for Holland's Dutch East India Company
1702	*De Dominio Maris*. Von Bynkershoek's cannon-shot rule	Codified coastal states rights to adjoining sea within range of shore-based artillery, about 3 nmi
1793	President Jefferson claims 3 nmi territorial sea for U.S.	Ripened into globally accepted standard over which a nation could assert ownership of the seas.
1938	U.S. exploration of Gulf of Mexico outside territorial sea	Oceans beyond 3 nmi fall under international law as common property of all nations. Drilling technology and discovery of oil in Gulf of Mexico spurred U.S. to extend its jurisdiction
1945	Truman Proclamations: U.S. establishes offshore control	Other nations assert claims. Constricts freedom of navigation
1958	First UN Conference on Law of the Sea (UNCLOS I); not ratified	Produced four separate conventions: Territorial Seas; Fisheries; Continental Shelf; High Seas. Failed ratification over Rights of Innocent Passage
1960	UNCLOS II: not ratified.	Nations claim 12 nmi territorial sea
1976	US Fisheries Conservation and Management Act	Expanded American fisheries jurisdiction from12 to 200 miles, eroding political power of distant-water fishing fleets and increasing coastal fishing
1982	UNCLOS III: not ratified	High seas resources become *mare nostrum* (our seas). 200 nmi EEZ established
1983	Reagan Presidential Proclamation 5030	Declared 200 mile EEZ for U.S., in line with central provisions of UNCLOS III
1994	UNCLOS III: ratified	All coastal nations claim 200 nmi EEZs
1999	Clinton Presidential Proclamation	Extends U.S. contiguous zone from 12 to 24 nmi offshore, for enforcement of environmental, customs, and immigration laws

It gave nations opportunities to evolve their own management strategies without compelling them to do so. It established overarching rules governing all uses of the world's oceans and seas and their resources, containing provisions for enforcing international pollution standards, fisheries soft laws, and binding dispute-settlement procedures. Conflict resolution is placed under the aegis of the signatories themselves, making UNCLOS a unique instrument in international law, with far-reaching implications.

UNLOS III in 1982 established the Commission on the Limits of the Continental Shelf (UNCLOS, 2012) to facilitate implementation of LOS to the outer limits of all nations' territorial sea and continental shelf beyond 200 nmi, constituting a last major redrawing of the world map. With a deadline of May 2009 established for submitting claims, the Commission received 48 national claims for vast extensions of maritime territories that began an undersea land-grab that made boundary demarcation a contentious diplomatic issue. Following the deadline, Russia placed its flag under the North Pole to claim the Arctic, which other Arctic states opposed, e.g., Canada, the United States, Norway, and Denmark. And in the South China Sea, a dispute among nations over the Spratly Archipelago that has plagued the region for decades, even when the Islands lacked economic importance prior to 1982, reflects a unique history of geopolitics when oil was discovered. Under UNCLOS III, islands that can sustain humans or an economic life are entitled to a 200 nmi limit, and title to the contested Spratlys could determine ownership of significant oil and gas resources. As this dispute is unresolved, the eventual outcome rests with the nations involved. To help resolve issues of energy extraction, the autonomous International Seabed Authority established by UNCLOS has developed rules, regulations, and procedures relating to deep seabed mining ("Mining Code").

3.5.2.3 International fisheries

UNCLOS addresses conservation and sustainable use of fishes and marine biodiversity beyond areas of national jurisdiction. Of particular relevance to fisheries is Part V for the EEZ and Part VII for the High Seas. UNCLOS takes a precautionary, ecosystem approach to management due to uncertainty of ocean ecosystems beyond areas of national jurisdiction (i.e., deep-sea ecosystems) and the vulnerability, resilience, and functioning of associated biota. The UN FAO is the only intergovernmental organization worldwide formally mandated by its constitution to undertake fisheries and aquaculture data collection, compilation, analysis, and diffusion of information.

The United Nations General Assembly supports sustainable fisheries on the high seas (A/RES/66/68). Nations and Regional Fishery Management Organizations (RFMOs) manage fisheries to prevent significant adverse impacts on areas identified as vulnerable marine ecosystems (VMEs), a concept described in FAO's *International Guidelines for the Management of Deep-sea Fisheries in the High Seas* (March 2007; Auster *et al.*, 2011). "Vulnerability" is the likelihood that a population, community, habitat, and ecosystem characteristics will experience substantial alteration due to short-term or chronic disturbance (FAO, 2009). Regional implementation, however, is problematic because of a lack of specificity and ecological uncertainties (Auster *et al.*, 2011).

The international, national, and local dimensions of fisheries include legally binding rules that become injected into national policies, legislation, and international treaties. Fisheries governance involves both hard law, such as national laws and international treaties, and soft law that lacks legally binding obligations, such as the FAO Code of Conduct for Responsible Fisheries and FAO International Plans of Action (Lugten, 2006). Since the 1990s, there has been a shift from hard to soft law in fisheries management (Allison, 2001). Soft-law international instruments are carefully negotiated and drafted with a basic understanding of good-faith commitment and a desire to influence the development of state practices. But while such soft-law instruments have substantially increased in number since the 1990s, fisheries have continued to decline, as too many boats seek too few fish.

Strong economic incentives are driving illegal and unreported fishing. Current worldwide illegal and unreported fishing losses total $10 to $23.5 billion annually and remove 11 to 26 million tons of fish (Fig. 2.8), which contributes to over-exploitation of stocks and hinders the recovery of populations (MRAG, 2005; Agnew *et al.*, 2009). At most risk are developing countries whose generally poor fisheries management and lack of control cause loss of major economic benefits, along with high costs of environmental degradation.

Fisheries governance is also hindered by the very uneven distribution of fish, which has been compensated by trade since time immemorial. Trade plays an important role in fishermen's livelihoods, even at the level of "subsistence" fisheries. In recent decades, international fish trade has increased rapidly, facilitated by widespread use of refrigeration, improved transportation, and communications. However, voluntary guideline obligations to curb overfishing are vague (Hewitt *et al.*, 2009) and nations often lack the capacity to undertake resource assessments, to develop management systems, and to effectively monitor user activities.

3.5.2.4 Regional mechanisms

Regional programs have the advantage of matching the geographic scale of many marine resource and environmental issues to large-scale ecosystem boundary conditions, such as for regional seas. Regional initiatives offer comprehensive institutional frameworks for international cooperation and as links to national governance.

The UNEP Regional Seas Programme has played a catalytic role in developing and implementing regional-seas programs since the 1970s (Table 3.9). UNEP facilitates information exchange, response options, and national ocean-management strategies. Regional governments adopt conventions, protocols, and action plans to address transboundary resource problems, pollution, management, institution building, protected areas, and finance.

One of the first and most successful regional agreements was the *Antarctic Treaty* (Table 3.2), which grew out of a scientific program—the International Geophysical Year (1957). This Treaty ensures that Antarctica is used for peaceful purposes, with international cooperation in science. An important innovation was that it suspended national sovereignty. The Treaty's jurisdiction encompasses a natural oceanographic

Table 3.9 UNEP Regional Seas conventions. Most have many aspects in common; protocols have been added to address priority concerns, of which samples are listed. Dates are for adoption of conventions; many are not yet in force. Several other regional seas programs exist, but many have not yet achieved conventions. Compiled from UNEP (2013) http://www.unep.org/regionalseas/programmes/default.asp.

Regional Sea	Instrument/administrator	Some major objectives
Baltic Sea	*Convention on the Protection of the Marine Environment of the Baltic Sea Area* (Helsinki Convention, 1974)	Pollution; protection of biodiversity; alien species; monitoring program; protected areas; integrated watershed management.
Northeast Atlantic	*Convention for the Protection of the Marine Environment* (Oslo and Paris conventions, 1974; revised as OSPAR, 1992)	Formulated regional consensus for cooperative actions on resources management and biodiversity.
Mediterranean Sea	*Convention for the Protection of the Marine Environment and the Coastal Region of the Mediterranean* (Barcelona Convention, 1976)	"Blue Plan" for long-term regional, coastal management, emphasizing pollution control and including a protected-area network.
Arabian Gulf	*Kuwait Regional Convention for Co-operation on the Protection of the Marine Environment from Pollution* (Kuwait Convention, 1978)	"Kuwait Action Plan" for combating pollution and for transboundary movements and disposal of hazardous waste.
Eastern Africa	*The Convention for the Protection, Management and Development of the Marine and Coastal Environment of the Eastern African Region* (Nairobi Convention, 1985)	Framework strategy for comprehensive approach to coastal area development; major concern for wild fauna and flora; pollution in cases of emergency.
West and Central Africa	*Convention for Co-operation in the Protection and Development of the Marine and Coastal Environment of the West and Central African Region* (Abidjan Convention, 1981)	Comprehensive strategy for conservation and development; pollution in cases of emergency.
Red Sea and Gulf of Aden	*Regional Convention for the Conservation of the Red Sea and Gulf of Aden Environment* (Jeddah Convention, 1982)	Comprehensive strategy for conservation; principal concern is for pollution by oil and other harmful substances.
Wider Caribbean	*Convention for the Protection and Development of the Marine Environment of the Wider Caribbean Region* (Cartagena Convention, 1983)	Action Plan for Caribbean Environment Program; specially protected areas for wildlife; oil spills; reduction and control of land-based sources of pollution.
South Pacific	*Convention for the Protection of Natural Resources and Environment of the South Pacific Region* (Noumea Convention, 1986).	Multilateral cooperation on protection of natural resources, dumping and pollution.

boundary, the oceanic Antarctic convergence, which encompasses unique biodiversity and significant fishing interests—notably krill, *Euphausia superba*, a basic food for many whales, seals, and penguins. The jurisdiction of the *Antarctic Treaty* was extended in 1980 through the *Convention for the Conservation of Antarctic Marine Living Resources* (CCAMLR), which endorsed ecosystem management. This Treaty has become a model for international cooperation, especially in scientific research; its *Agreed Measures for the Conservation of Antarctic Fauna and Flora* was an important consequence.

The Wider Caribbean is bounded by over 30 island and continental nations speaking English, French, Spanish, and Dutch which have ratified the 1983 *Convention for Protection and Development of the Marine Environment of the Wider Caribbean Region* containing protocols for regional cooperation in pollution control and protected areas and wildlife. The UNESCO-sponsored program Caribbean Coastal Marine Productivity (CARICOMP-1), a cooperative research network of over 25 marine laboratories, synoptically monitored using standardized methods the trends in structure and functionality of coral reefs, seagrasses and mangroves for over 20 years. The data were freely available online through a data management center at the University of the West Indies in Jamaica. CARICOMP-1 is coming to an end in 2013 with the publication of summary papers. A new cooperative network of laboratories, CARICOMP-2, will provide regional scientific input to developing regional programs in ecosystem-based management and regional governance, particularly the Caribbean Large Marine Ecosystem Project (http://www.clmeproject.org/).

FAO has also inaugurated a series of regional commissions and councils. The Intergovernmental Oceanographic Commission (IOC) studies phenomena such as El Niño and its major effects on climate, marine diseases, wildlife, and fisheries. A contentious case concerns the Atlantic bluefin tuna (*Thunnus thynnus*), arguably the world's most valuable fish, managed by the International Council for Conservation of Atlantic Tuna (ICCAT). Between 1970 and 2000, the western-Atlantic population declined to approximately 20% of its spawning stock biomass. ICCAT set quotas based on the assumption that eastern and western populations are distinct. Recent evidence (Block *et al.*, 2001) indicates that these populations intermix and that the species may not recover until and unless fishing is restricted throughout its entire range.

Despite many successes, decades of international programs have proven weak (Soares, 1998). This is due to gaps between approval of intergovernmental and international conventions, agreements, resolutions, and recommendations and their timely implementation (Hinds, 2003), as well as being hindered by the conflicting goals between resource exploitation and conservation.

3.6 MANAGEMENT CONCEPTS

Management is a goal-oriented mechanism for addressing specific policy objectives through strategic action plans under designated authorities (e.g., agencies). Management is a dynamic tool, responding adaptively to new information in a field that too often lacks sufficient information and resources and for which consensus among diverse constituencies must be reached. Marine management concepts are rapidly evolving from sector-based management towards ecosystem-based approaches for solutions to long-term resource issues.

3.6.1 Fisheries management

Key principles of traditional fisheries management are the regulation of exploitation and the management of fish stocks. The objective is to regulate fishing activities—when, where, and how to fish—so that fishing becomes sustainable and fish populations remain abundant and healthy. Science plays a key role in policy-directed management, evolving toward ecosystem-based fisheries management (Box 3.5).

The concept of Large Marine Ecosystems (LMEs) is an attempt to define fishery-based ecosystems for areas covering

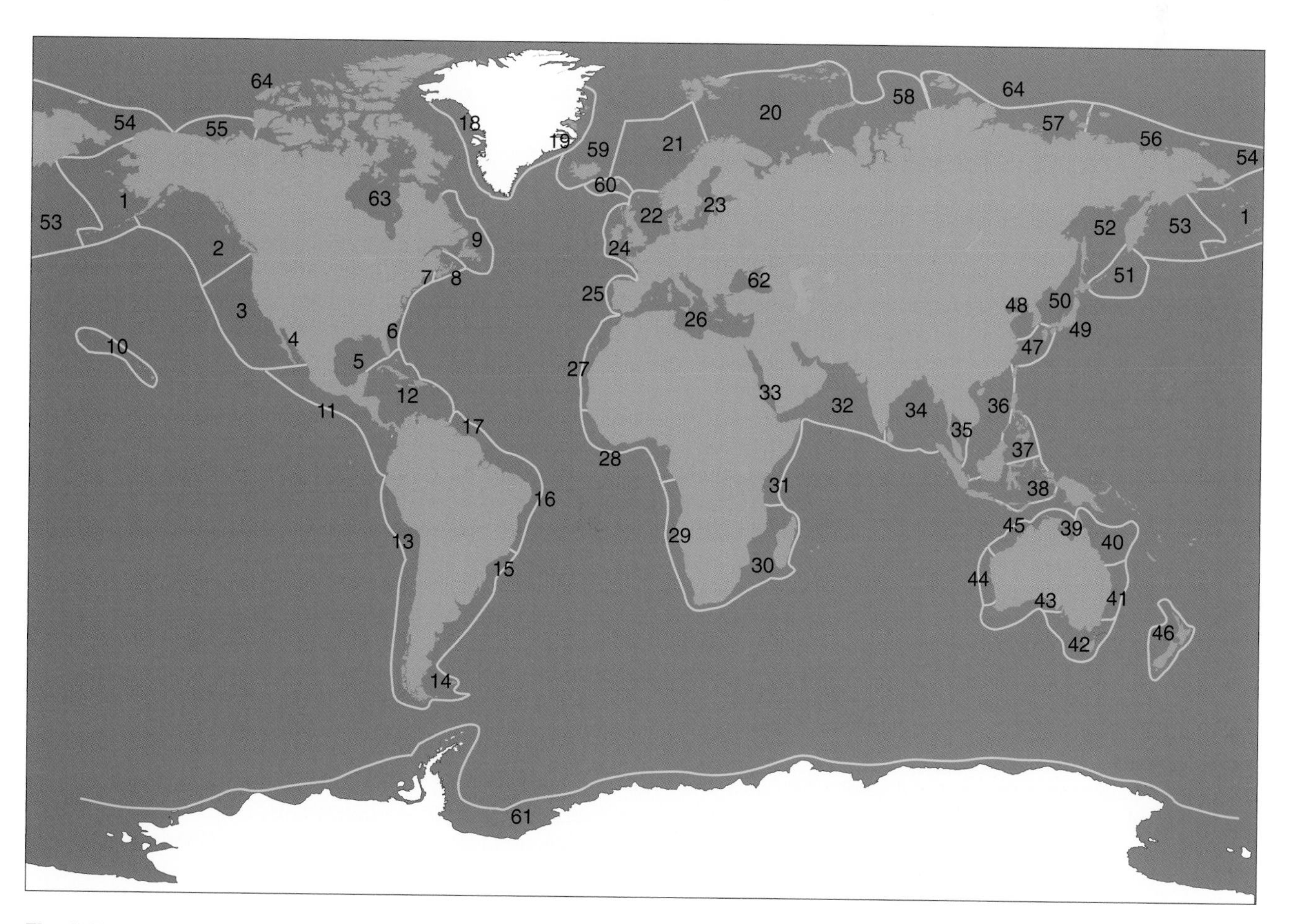

Fig. 3.5 Large Marine Ecosystems, now 64 in number. This concept incorporates ecosystem-based management, including concern for productivity, fish and fisheries, pollution, ecosystem health, socioeconomics, and governance, and is through a smaller number of international projects. See text for further explanation. From Sherman K, Aquarone MC, Adams S (2007). *Global Applications of the Large Marine Ecosystem Concept 2007–2010*. NOAA Technical Memorandum NMFS-NE-208. NOAA National Marine Fisheries Service, Northeast Fisheries Science Center, Woods Hole, Massachusetts.

200,000 km^2 or more of ocean space and characterized by distinct hydrography, productivity, and trophic interactions (Fig. 3.5; Sherman *et al.*, 2009). The 64 designated LMEs contribute approximately $12 trillion annually in ecosystem services to the global economy. Thus, LMEs are as much economic as they are environmental assets. Within many LMEs, however, overfishing is most severe, marine pollution is concentrated, and eutrophication and anoxia are increasing. Since 1995, the Global Environment Facility (GEF, Section 3.7.2) has provided substantial funding to support country-driven projects that introduce ecosystem-based assessment and management practices for the recovery and sustainability of LME goods and services (NOAA, online).

3.6.2 Coastal management

Coastal management incorporates complex interactions of laws, programs, and efforts to evaluate trade-offs and make decisions about how to use, conserve, and value resources and opportunities within the coastal realm (Frontispiece, Ch. 4). Coastal areas are important economic zones and are valued for their ecosystem functions and services. Finding the right balance between conservation and use engages all levels of government, research institutions, private citizens, industry participants, and non-governmental organizations (considered as "stakeholders") for establishing priorities and zoning, resolving user conflicts, and gaining partnerships among multiple levels of government and the public.

The coastal management planning process is distinguished by consensus building, generally conceived as a set of public goals or policies, a framework of procedures for carrying out those policies, and a set of organizations or agencies to implement procedures. A significant challenge is adopting land-sea interactions within an ecological unit (Ch. 4) for management and resolving conflicts among users.

Integrated coastal zone management (ICZM) is a conceptual extension of CZM, strongly influenced by UNCED as set out in *Agenda 21, Chapter 17*. By shifting toward predominantly bottom-up approaches among stakeholders, and by increased emphasis on long-term intergenerational sustainability, ICZM attempts to overcome single-sector management, fragmented

Box 3.5 Fisheries science: acquiring knowledge to support policy and management

Edward D. Houde
University of Maryland, Solomons, Maryland, USA

What is fisheries science?

Fisheries science broadly seeks to understand processes that control the dynamics and well-being of exploited fish and invertebrate populations and to predict their responses to fishing mortality. As such, fisheries science is quantitative ecology that investigates issues related to life history, population dynamics, habitats, and trophodynamics. It is a mix of fundamental and applied science conducted in support of policies enacted to assure sustainable fisheries. Policies generally are the legislative or executive mandates requiring responsible stewardship of fisheries and other marine resources. Fisheries science also serves resource managers who use scientific knowledge of fish population biology, habitats, predators, and prey to devise appropriate regulations for meeting policy goals. Scientific knowledge and advice often are delivered to managers in the form of numerical models to guide designation of catch targets and thresholds that guard against overfishing. Managers usually sit on commissions or in agencies with regional jurisdiction. In the United States, this translates into eight regional Fisheries Management Councils that strive to conduct their business (rules and regulations) based on the "best science available," thus carrying out the sustainable fishery policy required by the *Magnuson-Stevens Fishery Conservation and Management Act*.

In the past 50 years, the scope of fisheries science has broadened from principally addressing questions on population dynamics and demographics to inclusion of broader ecological research on effects of the environment, consequences of heavy fishing on predator-prey interactions, climate change, and effects of contaminants, pollutants, and disease. As requirements for ecosystem-based fisheries management have evolved, this "new" fishery science is increasingly conducted by research teams with broad interdisciplinary expertise.

Science to serve management

Two kinds of science, referenced here as Modes 1 and 2, serve fisheries management. Both are important for acquiring knowledge that managers use in addressing widespread overfishing in marine ecosystems. Mode 1 rests firmly on traditional science goals; it must be objective, legitimate, credible, and transparent. These classic criteria describe the science of discovery. Mode 1 fishery science is fundamental inquiry on the biology, ecology, and dynamics of fish populations or stocks (segments of a population) and usually is conducted in research laboratories and institutions by individuals or small research groups. Graduate student thesis research and other in-depth investigation into fish biology, ecology, and demographics fall into this mode. Mode 1 science can be slow and deliberate, must stand up to peer review and, by itself, usually is not sufficient to serve the impatient needs of fisheries management and policy. Consequently, fishery science has become increasingly reliant on Mode 2 science. Mode 2 science generally is sharply focused on answering management questions in a timely manner. It can consist of fundamental investigations or modeling research, or be a synthesis of scientific knowledge. It is usually conducted by teams and depends on consensus building by collective expertise of technical committees or advisory groups. Mode 2 fishery science frequently is delivered as a product of "Stock Assessment Workshops" hosted by management agencies or commissions. It may be sponsored by government, industry, NGOs, and sometimes special interests, and is conducted at regional levels (for example, blue crab in Chesapeake Bay) or national-international levels (for example, bluefin tuna in the Atlantic Ocean). Mode 2 science under the best circumstances is objective and transparent; it may or may not be subjected to rigorous peer review. Expert groups conduct assessments and often synthesize vast amounts of information on fish biology, population dynamics, habitats, hydrography, interacting species, and socio-economics for management agencies, either directly or via prestigious sponsoring institutions (e.g., National Research Council). In broad context, Mode 2 fishery science is the approach by which global and regional issues are being addressed, for example, climate change and its effect on productivity of ecosystems, and interactions with fisheries.

A large share of conservation and fishery science now consists of research on degraded ecosystems, depleted fish populations, and on effects of fishing on ecosystems. As policies evolve and we adopt precautionary, ecosystem-based approaches in fisheries management, the need to conserve ecosystem services, as well as manage for high, sustainable fishery yields, is becoming increasingly prominent. Traditional management institutions may require diversification and restructuring to manage fisheries in the broader context of ecosystem-based management. These needs are highly evident in stressed coastal ecosystems and estuaries such as Chesapeake Bay. As such, a combination of fundamental, in-depth science on the ecology of organisms and habitats (Mode 1) and synthetic science to advise management (Mode 2) provide complementary pathways to: (i) rebuild fisheries—for example, blue crab; (ii) to maintain restored fisheries—for example, striped bass; and, hopefully, the knowledge to (iii) restore collapsed fisheries—for example, oysters and American shad. Historically, the emphasis of fisheries science was provision of knowledge and advice to support single-species management. Recently, the emphasis has shifted and needs for knowledge on multi-species interactions, essential fish habitat, effect of fishing on the ecosystem and on untargeted organisms (the bycatch), and conservation of ecosystem services have become dominant themes for modern fisheries science that supports ecosystem-based management.

jurisdictions, and hierarchical command-and-control procedures. Thus, ICZM aims to achieve sustainable resource use among various economic and environmental sectors, but is challenged by insufficient attention to integration of economic, social, and political forces within ecological boundaries. Nevertheless, ICZM remains highly attractive as a unifying concept, providing for collaboration, communication, coordination, and information exchange among multiple disciplines and various sectors.

3.6.3 Marine Protected Area management

MPAs are managed for environmental and biodiversity protection, and for scenic or socio-economic values. They encompass many different management types, from strict protection to multiple use, carried out by public and private organizations, and within distinct management regimes (UNEP, 2009). Many do not restrict fishing, and some do not restrict oil and gas extraction. The largest MPAs (Section 3.3) are managed under a range of different management concepts. Australia's Great Barrier Reef Marine Park (GBRMP), established in 1975 for coral-reef conservation and marine spatial zoning, permits human activities that include fisheries and tourism while seeking high-level protection in specific areas. Its management agency, the Marine Park Authority, was established through an Act of Australia's Parliament. The largely unspoiled U.S. Papahānaumokuākea Marine National Monument that protects coral reefs, atolls, shoals, and islands is managed as a permanent "no-take" marine reserve through Presidential Proclamation 8031. Britain's Indian Ocean Chagos Marine Reserve is managed as a Strict Nature Reserve, and in the Pacific Ocean Kiribati's Phoenix Islands Protected Area Trust restricts fishing, being managed through cooperative partnerships financed through an endowment. These large remote areas of the ocean invite tourism and ecotourism opportunities, enhancing economic opportunities for island people.

3.6.4 Biosphere reserves

Biosphere reserves are internationally recognized under UNESCO's Man and the Biosphere Programme. They intend to integrate conservation with human activities, being a first attempt to institute a concept of hierarchical spatial planning by means of recognizing core, buffer, and transition zones. Today, the global network includes 580 sites in 114 countries (UNESCO, 2011); only a few are coastal-marine (Ch. 11). Biosphere reserves include three interconnected functions: conservation, development, and logistics. Appropriate zoning schemes combine core protected areas with surrounding buffer and transitional zones and are thereby particularly suitable for marine application. Local stakeholders, often with highly innovative and participatory governance systems, foster sustainable development. Biosphere reserves also foster dialogue for conflict resolution over natural resources by integrating cultural and biological diversity and traditional knowledge in management. They also can demonstrate sound sustainable practices and policies based on research and monitoring and act as sites of excellence for education and training.

Biosphere reserves have no force in international law, but can become legally official through national authority. As such, they can build and promote a global network of places designed to mesh human activity with biological and scenic assets according to community mechanisms.

3.6.5 Restoration management

Restoration management is increasingly becoming a major conservation priority, borne from the recognition of pervasive species depletions and ecosystem degradation. The goal of restoration is not necessarily to restore ecosystems to a pristine condition, due to the shifting baseline phenomenon (Dayton *et al.*, 1998, Jackson *et al.*, 2011) whereby identification of original, pristine conditions cannot be known. Rather, the goal is to renew degraded, damaged, or destroyed ecosystems through active human intervention, to prevent further degradation, and to achieve sustainable ecosystem states. Restoration of an area's natural resources, habitats, and services to some sustainable, resilient state is viewed as essential.

An active and growing area of marine restoration ecology involves government-mandated restoration of natural resources injured by human-use incidents, such as oil and chemical spills, pollutant releases, or physical destruction of habitat (Peterson and Kneib, 2003). U.S. federal laws, notably the *Comprehensive Environmental Response, Cleanup, and Liability Act* (CERCLA) of 1980, and the *Oil Pollution Act* (OPA) of 1990, dictate that restoration actions be undertaken to provide equivalent compensation for losses or injuries to natural resources held in public trust and to the ecological services that those resources would have provided (Burlington, 1999). Biodiversity restoration requires many management tools, which include sustainable fisheries management, pollution control, maintenance of essential habitats, and the creation of marine reserves. In this way, society invests in the productivity and reliability of the ecological goods and services that coastal ecosystems and oceans provide to humanity (Worm *et al.*, 2006).

3.6.6 Ecosystem-based management

Ecosystem-based management (EBM) is being widely considered as an effort to conserve species, maintain biodiversity, and to place human uses in an environmental context (McLeod and Leslie, 2009; Ch.13). It strives for integration of all management concepts mentioned above by focusing on protection, restoration, and management of functioning ecosystems within a spatially designated area as conceived in marine spatial planning (MSP; Ch. 13). It adopts ecosystem and precautionary principles as means for addressing the ecological impact of fisheries, environmental degradation, and other human-caused effects, while maintaining ecological integrity and vital economic interests that benefits society, e.g., healthy seafood, clean beaches, ocean benefits, and reducing the consequences of expanding anoxic and hypoxic zones.

EBM is made explicit in the *Marine Mammal Protection Act*, the *Magnuson-Stevens Fishery Conservation and Management Reauthorization Act*, and CCAMLR. Scientists and managers see challenges as to how to maintain the resiliency of systems that provide critical ecosystem services while also overcoming undesirable phenomena (Levin and Lubchenco, 2008). EBM assumes that uncertainty requires precaution and that new information requires adaptive management, which strongly suggest that EBM is experimental, requiring a strong element of research and monitoring. Thus, achieving positive results involves collaboration among biological, environmental, social sciences, as well as public understanding for promoting government policy.

3.7 AGENTS FOR CONSERVATION

Social movements, scientific evidence, financial institutions, and conservation groups that spur action have the capacity to sustain conservation efforts by altering public opinion, mobilizing voters, and/or creating new, non-legal norms of behavior and changed values. These can change the direction of society and alter the abuse of resources.

3.7.1 Environmental non-government organizations (NGOs)

People power has achieved national and international importance through non-governmental organizations (NGOs). NGOs are exceedingly diverse in their interests and methods, but together form effective communication channels among policy, politics, science, and the public. Tens of thousands of NGOs exist worldwide, only about half of which in developing nations are older than a decade or two, but until recently coastal-marine programs were relatively neglected. Most NGOs are non-profit and depend on voluntary efforts and contributions, with their survival often depending on the courage and persistence of a few dedicated individuals. Only the largest and most powerful have scientific expertise; others may develop ties with universities and government research organizations. NGO interests are overwhelmingly directed towards crisis situations, such as charismatic endangered species, depleted fisheries, habitat protection, and environmental pollution, with programs often lacking comprehensive strategies. Notable exceptions concern strategies for "hot spots" of species richness and biodiversity protection via public or private protection. NGOs are constrained to raise most of their funds by marketing issues that are attractive to the public. Programs for coral reefs, wetlands, sea turtles, marine mammals, and most recently "no-take" fisheries reserves have gradually intensified since the 1990s through the efforts of NGOs.

The first NGO of global significance was the International Union for the Conservation of Nature and Natural Resources (IUCN, the World Conservation Union) in 1948. IUCN has since grown into a large organization with worldwide influence and strong connections with the UN and national governments. Other national and international NGOs emerged, especially from the 1960s onwards and mostly in the developed world. Many then tended towards emotional "animal welfare" issues, but since that time their scientific credibility has grown significantly. Each NGO employs a variety of tactics to promote species protection and/or environmental conservation. Lawsuits represent a sample of NGO actions that are effectively used to promote environmental protection and to influence environmental policy-making and enforcement. Interest in the marine environment has also increased substantially, principally through using charismatic species—polar bears, whales, seals, tuna, sea turtles, corals etc.—as metaphors for climate change, over-exploitation, oil spills, and the like.

Many NGOs work closely with UN agencies to play important roles in numerous environmental conventions. Collaborative partnerships are continually formed to advance conservation in innovative ways. UNEP, FAO, and NGOs often work together on fisheries management to blend fisheries and conservation in MPAs. Throughout the world, community groups are organizing around coastal watersheds for regional spatial planning and management. Separately, or occasionally en masse, NGOs lobby governments, publicize information of strategic importance, and influence international conferences. They can be major actors in negotiation, as reflected at the Rio Conference in 1992, where more than 1400 NGOs were accredited to participate in discussions leading to conventions on biodiversity and climate change. Coalitions of NGOs continue to be represented at many international meetings, such as those of the *International Whaling Commission*, *London Dumping Convention*, and *Convention on International Trade in Endangered Species*. NGOs have also pressured international institutions to enlarge their environmental activities. Thus, NGOs have become a powerful force, influencing the direction of environmental and development policies through advocacy and "on-the-ground" action.

3.7.2 Development and financial assistance organizations

Conservation actions require considerable financial support. Funds come from development banks, national agencies, private foundations, and other sources (Table 3.10). The World Bank is a significant funding source, a partner in environmental programs, and a primary funder for projects to support the *Biodiversity Convention* and the *Stockholm Convention on Persistent Organic Pollutants* (POPs), among others. The International Monetary Fund (IMF), established in 1944 by the UN, enables countries (and their citizens) to act with one another to ensure the stability of the international monetary system, which is

Table 3.10 International banks and funds. Examples. The World Bank and the UN have almost the same membership. International funding is competitively available through international, intergovernmental organizations, such as the Global Environmental Facility. Coastal-marine resources receive small portion of total funds available, proportional to public interest. From www.imf.org; web.worldbank.org; www.undp.org; www.iadb.org/.

Name	Formation–goal	Functions
World Bank (WB, 1944)	Integrates nations into wider world economy; promotes long-term economic growth to reduce poverty in developing countries. Program on Global Sustainable Fisheries Management and Biodiversity Conservation in Areas Beyond National Jurisdiction	Largest single source of development lending; exerts policy leadership; trustee for Global Environment Facility (GEF) Trust Fund, an independent international financial entity created (1991) by UNEP, UN Development Program, and World Bank to help developing countries deal with environmental concerns
International Monetary Fund (IMF, 1944)	Monitors world currencies; helps maintain orderly system of payments between countries; lends money to members with serious imbalance of payments	Major influence on development policies of developing countries; monitors transactions in international trade and investment
UN Development Program (UNDP, 1965)	Provides developing nations with policy advice on a range of issues pertaining to poverty, institutional capacity, and globalization	Assists nations and territories; *Capacity 21* was launched at UNCED (1992) to assist nations implement *Agenda 21*; as of 2001, *Capacity 21* supported 21 efforts in 75 nations
Development Banks	Provide financial support and professional advice for economic and social development in developing countries	*Four Regional Development Banks*: African; Asian; European Bank for Reconstruction and Development; Inter-American Development Bank Group; The World Bank Group *Multilateral Financial Institutions*: European Commission (EC) and European Investment Bank (EIB); International Fund for Agricultural Development (IFAD); Islamic Development Bank (IDB); The Nordic Development Fund (NDF) and The Nordic Investment Bank (NIB); OPEC Fund for International Development (OPEC Fund)

essential for promoting sustainable economic growth, increasing living standards, and reducing poverty (IMF.org, online). Most public and private development and assistance organizations work closely with UN agencies (Table 3.4). Their main objectives are to assist developing nations in policy development, national strategies, infrastructure, and specific conservation projects. Private philanthropic organizations, particularly in the U.S., also fund environmental projects worldwide; e.g., the *Census of Marine Life* (Ch. 5).

Multinational development banks have often supported economic development projects with deleterious environmental consequences. But pressures from governments, NGOs, and the public have gradually influenced them to be more concerned with conservation. Hence, most multinational banks have adopted policies for sustainable development, with specific goals for biodiversity, fisheries, ocean law, shipping, pollution, global climate change, regional seas, freshwater, and related issues. The GEF is a cooperative effort of UNEP, the United Nations Development Programme (UNDP), and the World Bank, and is influential in supporting biodiversity conservation in developing nations. The GEF, participating countries, and other donors have provided significant support for LME projects. Unlike development banks, the World Trade Organization deals with global rules of trade among nations. Its main function is to ensure that trade flows as smoothly, predictably, and freely as possible.

A variety of assistance organizations are also promoting economic instruments that increasingly are being accepted as means to change human behavior. Economic incentives are being applied to protect forests and fisheries and to establish and manage protected areas. Many economic incentives address the "externality" costs of resource depletion and pollution such that producers, transporters, and consumers face full social and environmental costs of pollution and resource extraction. For example, the "polluter pays principle" reflects a shift in the burden of proof. Assistance organizations do not generally support scientific research per se, but may support assessment and monitoring programs that influence management. The PEW Charitable Trust, for example, is an independent non-profit that serves the public interest by providing information, advancing policy solutions, and supporting civic action, and has been especially effective by focusing on problems of climate change and large-scale protection of the global marine environment. Its *Global Ocean Legacy* program has supported creation of Large Ocean Reserves. The Sloan Foundation was the main supporter of the *Census of Marine Life* (Box 5.1).

3.8 CONCLUSION

Coastal and marine environmental change is forcing society to confront a paradox: conserve and restore diminishing marine assets as demands for ocean use intensify. Addressing this paradox requires a better understanding of marine science (Chs. 4, 5) and the challenges that each case study presents (Chs. 6–12). The need is to connect conservation mechanisms to issues at appropriate scales, with an understanding of ecosystem performance and resiliency, social justice, and equity. Challenges to national sovereignty, good governance, and ocean protection are emerging in the 21st century from the formation of gigantic economic trading blocs and mega-corporations, from rapid communications along the information highway, and from privatization. Solutions require innovative thinking in order to achieve sustainable use, to diminish or halt the loss of biodiversity, and to promote environmental sustainability.

REFERENCES

Adams AB, ed. (1962) *First World Conference on National Parks*. U.S. Dept. Interior, Washington, D.C.

Agnew DJ, Pearce J, Pramod G, Peatman T, Watson R, Beddington JR, Pitcher TJ (2009) Estimating the worldwide extent of illegal fishing. *PLoS ONE* **4**, e4570.

Allison EH (2001) Big law, small catches: Global ocean governance and the fisheries crisis. *Journal of International Development* **13**, 933–950.

Anderson RS (1995) The Lacey Act: America's premier weapon in the fight against unlawful wildlife trafficking. *Public Land Law Review* **16**, 27–60.

Archer JH, Connors DL, Laurence K, Columbia SC, Bowen R (1994) *The Public Trust Doctrine and the Management of America's Coasts*. University Massachusetts Press, Amherst.

Armstrong JM, Ryner PC (1980) *Ocean Management: Seeking a New Perspective*. The Traverse Group, Inc., under Contract AO-A01-78-00-1307 from U.S. Department of Commerce Office of Policy, U.S. Government Printing Office, Washington, D.C., Stock Number 003-000-00557-7.

Auster PJ, Gjerde K, Heupel E, Watling L, Grehan A, Rogers AD (2011) Definition and detection of vulnerable marine ecosystems on the high seas: problems with the "move-on" rule. *ICES Journal of Marine Science* **68**, 254–264.

Bean MJ, Rowland MJ (1997) *The Evolution of National Wildlife Law*. Environmental Defense Fund and World Wildlife Fund—US, Washington, D.C.

Berzin AA (2008) The truth about Soviet whaling. In *The truth about Soviet whaling: a memoir* [translated by Ivashchenko YV] (eds Ivashchenko YV, Clapham PJ, Brownell Jr. RL). *Marine Fisheries Review* **70**, 1–59.

Björklund M (1974) Achievements in marine conservation, I. Marine parks. *Environmental Conservation* **1**, 205–223.

Block BA, Dewar H, Blackwell SB, Williams TD, Prince ED, Farwell CJ, Boustany A, Teo SLH, Seitz A, Walli A, Fudge D (2001) Migratory movements, depth preferences, and thermal biology of Atlantic bluefin tuna. *Science* **293**, 1310.

Blumm MC, Ritchie L (2005) The pioneer spirit and the public trust: the American rule of capture and state ownership of wildlife. *Environmental Law* **50**, 101–147.

Bodansky D (1998) International environmental law in United States Courts. *IEL in US Courts* **7**, 57–62.

Borgese EM, Ginsbury N, Morgan JR, eds (1994) *Ocean Yearbook* 11. The University of Chicago Press, Chicago.

Boyle AE (1999) Some reflections on the relationship of treaties and soft law. *The International and Comparative Law Quarterly* **48**, 901–913.

Brooks KB, ed. (2009) *Before Earth Day: the Origins of American Environmental Law, 1945–1970*. University Press of Kansas, Lawrence.

Brown A (2011) Biodiversity. In *Global Environmental Politics* (ed. Kütting G). Routledge, Oxon., UK, 151–161.

Burdick WL (2004) *The Principles of Roman law and Their Relation to Modern Law*. The Lawbook Exchange, Ltd., Clark, New Jersey.

Burlington LB (1999) Ten year historical perspective of the NOAA damage assessment and restoration program. *Spill Science & Technology Bulletin* **5**, 109–116.

Caddy JF, Seijo JC (2005) This is more difficult than we thought! The responsibility of scientists, managers and stakeholders to mitigate the unsustainability of marine fisheries. *Philosophical Transactions of the Royal Society B* **360**, 59–75.

Cahn M (2002) Linking science to decision making in environmental policy: bridging the disciplinary. Forthcoming, in *Policymaking*. SUNY Press. www.csun.edu/~cahn/rulemaking.html

Carden K (2006) Bridging the divide: the role of science in species conservation law. *Harvard Environmental Law Review* **30**, 165–259.

CARICOMP (2001). The Caribbean Coastal Marine Productivity Program. *Bulletin of Marine Science* **69**, 819–829.

CEQ (2010) *Final recommendations of the interagency ocean policy task force, July 19, 2010*. The White House Council on Environmental Quality, Washington, D.C., 1–77.

Clapham P, Baker CS (2008) Modern whaling. In *Encyclopedia of Marine Mammals* (eds Perrin WF, Würsig B, Thewissen JGM). Academic Press, San Diego, 1239–1243.

Clapham P, Ivashchenko Y (2009) A whale of a deception. *Marine Fisheries Review* **71**, 44–52.

Cobb C, Halstead T, Rowe J (1995) If the GDP is up, why is America Down? *Atlantic Monthly* October, 59–78.

Coglianese C (2001) Social movements, law, and society: the institutionalization of the environmental movement. *University of Pennsylvania Law Review* **150**, 85–118.

Copeland CW (2008) The Federal rulemaking process: an overview. *Congressional Research Service*, Report RL32240. wikileaks.org/wiki/CRS-RL32240 Feb 2, 2009.

COP (2010) Tenth meeting of the Conference of the Parties to Convention on Biological Diversity in Nagoya, Aichi Prefecture, Japan.

Costanza R, Andrade F, Antunes P, van den Belt M, Boesch D, Boersma D, *et al.* (1999) Ecological economics and sustainable governance of the oceans. *Ecological Economics* **31**, 171–187.

Daly HE, Cobb Jr. JB (1994) *For the Common Good: Redirecting the Economy toward Community, the Environment and the Sustainable Future*. Beacon Press, Boston, Massachusetts.

Dayton PK, Tegner MJ, Edwards PB, Riser KL (1998) Sliding baselines, ghosts, and reduced expectations in kelp forest communities. *Ecological Applications* **8**, 309–322.

Dobson A (2005) Monitoring global rates of biodiversity change: challenges that arise in meeting the Convention on Biological Diversity (CBD) 2010 goals. *Philosophical Transactions of the Royal Society B* **360**, 229–241.

Doukakis P, Parsons ECM, Burns WCG, Salomon AK, Hines E, Cigliano JA (2009) Gaining traction: retreading the wheels of marine conservation. *Conservation Biology* **23**, 841–846.

Dutton I, Hotta K (1995) Introduction of coastal management. In *Coastal Management in the Asia-Pacific: Issues and Approaches* (eds Hotta K, Dutton IM). Japan International Marine Science and Technology Federation, Tokyo, 3–18.

FAO (2009) Report of the technical consultation on international guidelines for the management of deep-sea fisheries in the high seas, Rome, 4–8 February and 25–29 August 2008. *FAO Fisheries and Aquaculture Report* **881**, 1–86.

FAO (2011) A world overview of species of interest to fisheries. In *FAO Fisheries and Aquaculture Department* [online]. Rome. Updated. [Cited 14 June 2011]. www.fao.org/fishery/topic/2017/en

Fiorino DJ (1995) *Making Environmental Policy*. University of California Press, Berkeley.

Fox JR (1992) *Dictionary of International and Comparative Law*. Oceana Publications, Inc., Dobbs Ferry, New York.

Gamboa MJ (1973) *A Dictionary of International Law and Diplomacy*. Central Lawbook Publishing Company, Inc., Quezon City, Philippines, and Oceana Publications, Inc., Dobbs Ferry, New York.

Gleick PH (2000) *The World's Water 2000–2001. The Biennial Report on Freshwater Resources*. Island Press, Washington, D.C.

Harada M (1995) Minamata disease: methylmercury poisoning in Japan caused by Environmental Pollution. *Critical Reviews in Toxicology* **25**, 1–24.

Hendriks IE, Duarte CM, Heip CHR (2006) Editorial: biodiversity research still grounded. *Science* **312**, 1715.

Hewitt CL, Everett RA, Parker N (2009) Examples of current international, regional and national regulatory frameworks for preventing and managing marine bioinvasions. In *Biological Invasions in Marine Ecosystems*, Chapter 19 (eds Rilov G, Crooks JA). *Ecological Studies* **204**, Springer-Verlag, Berlin, Heidelberg, 335–352.

Hinds L (2003) Oceans governance and the implementation gap. *Marine Policy* **27**, 349–356.

Hinrichsen D (1998) *Coastal Waters of the World*. Island Press, Washington, D.C.

ICOSRM (2008) *Federal Ocean and Coastal Activities for CY 2006 and 2007*. Report to the U.S. Congress. Report prepared by the Interagency Committee on Ocean Science and Resource Management Integration. www.ocean.ceq.gov

IUCN (1980) *World Conservation Strategy*. International Union for the Conservation of Nature and Natural Resources, United Nations Environment Programme, World Wildlife Fund, Gland, Switzerland.

IUCN (2005) *The Durban Action Plan*. Revised version, March 2004. IUCN, Gland, Switzerland, 219–266.

IUCN (2010) *Red list of IUCN Threatened Species*. Version 2010.4. www.iucnredlist.org. Downloaded on 14 April 2011.

IUCN, UNEP, WWF (1991) *Caring for the Earth: a Strategy for Sustainable Living*. IUCN, Gland, Switzerland.

Ivashchenko YV, Clapham PJ, Brownell Jr. RL (2011) Soviet illegal whaling: the Devil and the details. *Marine Fisheries Review* **73**, 1–19.

Ivashchenko YV, Clapham PJ (2012) Soviet catches of bowhead (*Balaena mysticetus*) and right whales (*Eubalaena japonica*) in the North Pacific and Okhotsk Sea. *Endangered Species Research* **18**, 201–217.

Ivashchenko YV, Brownell Jr. RL, Clapham PJ (2013) Soviet whaling in the North Pacific: revised catch totals. *Journal of Cetacean Research and Management* (in press).

Jabour J (2008) Successful conservation—then what? The de-listing of *Arctocephalus* fur seal species in Antarctica. *The Journal of International Wildlife Law and Policy* **11**, 1–29.

Jackson JBC, Alexander KE, Sala E (2011) *Shifting baselines: the past and the future of ocean fisheries*. Island Press, Washington, D.C.

JOCI (2008) *Changing Oceans, Changing World*. Joint Ocean Commission Initiative. Washington, D.C.

Kamieniecki S (2006) *Corporate America and Environmental Policy*. Stanford University Press, Stanford, California, 1–327.

Keller AC, Gerber LR (2004) Monitoring the endangered species act: revisiting the eastern north Pacific gray whale. *Endangered Species UPDATE* **21**, 87–92.

Kelleher G, Bleakley C, Wells S (1995) *Priority Areas for a Global Representative System of Marine Protected Areas*. Great Barrier Reef Marine Park Authority, World Bank Environment Department, IUCN, Washington, D.C., Four volumes.

Kraska J (2008) The law of the sea convention: a national security success—global strategic mobility through the rule of law. *The George Washington International Law Review* **39**, 543–552.

Kraska J (2011) *Maritime Power and the Law of the Sea. Expeditionary Operations in World Politics*. Oxford University Press, New York, 1–464.

Levin SA, Lubchenco J (2008) Resilience, robustness, and marine ecosystem-based management. *BioScience* **58**, 27–32.

Lubchenco J, Palumbi SR, Gaines SD, Andelman S (2003) Plugging a hole in the ocean: the emerging science of marine reserves. *Ecological Applications, Supplement: The Science of Marine Reserves* **13**, S3–S7.

Lugten GL (2006) Soft law with hidden teeth: the case for a FAO international plan of action on sea turtles. *Journal of International Wildlife Law and Policy* **9**, 155–173.

MARIBUS (2012) Chapter 10, Law of the Sea. World Oceans Review. MARIBUS, Hamburg, Germany. Online. www.maribus.com

McCay BJ (1998) *Oyster wars and the public trust*. University of Arizona Press, Tucson, 1–246.

McCay BJ (2008) The littoral and the liminal: challenges to the management of the coastal and marine commons. *Mast* **7**, 7–28

McLeod K, Leslie H (2009) *Ecosystem-Based Management for the Oceans*. Island Press, Washington, D.C.

Mergler D, Anderson HA, Chan LHM, Mahaffey KR, Murray M, Sakamoto M, Stern AH (2007) Methylmercury exposure and health effects in humans: a worldwide concern. *Ambio* **36**, 3–11.

Meyer JW, Frank DJ, Hironaka A, Schofer E, Tuma NB (1997) The structure of a World Environment Regime, 1870–1990. *International Organization* **51**, 623–651.

Millennium Ecosystem Assessment (2005) *Ecosystems and Human Well-being: Biodiversity Synthesis*. World Resources Institute, Washington, D.C.

Moore JA (1993) *Science as a Way of Knowing*. Harvard University Press, Cambridge, Massachusetts.

Morison SE (1958) *Strategy and Compromise*. Little, Brown, & Co., Boston and Toronto.

MRAG (2005) *Review of Impacts of Illegal, Unreported and Unregulated Fishing on Developing Countries Synthesis Report*. Marine Resources Assessment Group Ltd., London, UK. www.illegal-fishing.info/

Mueller TL (1994) Federal endangered species act. In *Guide to the Federal & California Endangered Species Laws*, Chapter II. January, 6–76.

NOAA (2006) *North Pacific fur seal treaty of 1911*. Online. celebrating200years.noaa.gov/events/fursealtreaty/welcome.html

Nowlan L (2001) *Arctic legal regime for environmental protection*. IUCN Environmental Policy and Law Paper No. 44. The World Conservation Union, Gland, Switzerland.

NRC (2002) *The Drama of the Commons* (eds Ostrom E, Dietz T, Dolšak N, Stern PC, Stovich S, Weber EU). National Research Committee on the Human Dimensions of Global Change, Division of Behavioral and Social Sciences and Education, Washington, D.C., National Academy Press.

Office of Oil and Gas (2005) Overview of U.S. Legislation and Regulations Affecting Offshore Natural Gas and Oil Activity. Energy Information Administration, Department of Energy, Washington, D.C., 1–20.

OPTF (2010) *Final Recommendations of the Interagency Ocean Policy Task Force*. The White House Council on Environmental Quality, Washington, D.C. Online.

Oreskes N, Conway EM (2010) *Merchants of Doubt*. Bloomsbury Press, New York.

Osherenko G (2006) New discourses on ocean governance: understanding property rights and the public trust. *Journal of Environmental Law and Litigation* **21**, 317–381.

Peterson CH, Kneib RT (2003) Restoration scaling in the marine environment. *Marine Ecological Progress Series* **264**, 173–175.

Prideaux M (2003) *Conserving cetaceans: the convention on migratory species and its relevant agreements for cetacean conservation*. Whale and Dolphin Conservation Society, Munich, Germany.

Ray GC, Potter Jr. FM (2011) Historical perspectives. The making of the marine mammal protection act. *Aquatic Mammals* **37**, 520–552.

Ruhl JB, Salzman J (2006) Ecosystem services and the public trust doctrine: working change from within. *Southeastern Environmental Law Journal* **5**, 223–239.

Sachs J, Baillie JEM, Sutherland JW, Armsworth PR, *et al.* (2009) Biodiversity conservation and the millennium development goals. *Science* **325**, 1502–1503.

Salzman J, Thompson Jr. BH (2010) *Environmental Law and Policy*, Third Edition. Foundation Press, New York.

Sax JL (1970) The public trust doctrine in natural resource law: effective judicial intervention. *Michigan Law Review* **68**, 471–566.

Schevill WE, ed. (1974) *The Whale Problem: Status Report*. Harvard University Press, Massachusetts.

Secretariat of the Convention on Biological Diversity (2013) The Jakarta Mandate, Montréal, Canada. Online. www.biodib.org

Sherman K, Aquarone MC, Adams S (2007) *Global Applications of the Large Marine Ecosystem Concept 2007–2010. NOAA Technical Memorandum* NMFS-NE-208. NOAA National Marine Fisheries Service, Northeast Fisheries Science Center, Woods Hole, Massachusetts, 1–71.

Sherman K, Aquarone MC, Adams S, eds (2009) *Sustaining the World's Large Marine Ecosystems*. IUCN, Gland, Switzerland, viii+142.

Smith RJ, Muir RDJ, Walpole MJ, Balmford A, Leader-Williams N (2003) Governance and the loss of biodiversity. *Nature* **426**, 67–70.

Smith ZA (2004) *The Environmental Policy Paradox*, Fourth Edition. Pearson Prentice Hall, New Jersey.

Soares M, ed. (1998) *The Oceans: Our Future*. Official Report of the World Commission on the Oceans, Cambridge University Press, Cambridge, 147–152.

Toropova C, Meliane I, Laffoley D, Matthews E, Spalding M, eds (2010) *Global Ocean Protection: Present Status and Future Possibilities*. Brest, France; Agence des aires marines protégées, Gland, Switzerland, Washington, D.C. and New York, USA; IUCN WCPA, Cambridge, UK; UNEP-WCMC, Arlington, USA; TNC, Tokyo, Japan; UNU, New York, USA; World Conservation Strategy, 1–96.

UNCLOS (2012) The United Nations Convention on the Law of the Sea (A historical perspective). Online. www.un.org/Depts/los/convention_agreements/convention_historical_perspective.htm#Historical%20 Perspective

UNEP (2005) *Register of international treaties and other agreements in the field of the environment*. Division of Policy Development and Law UNEP/Env.Law/2005/3. United Nations Environment Programme, Nairobi.

UNEP (2009) *A New On-line System to View and Study the World's Marine Protected Areas*. The United Nations Environment Programme World Conservation Monitoring Centre, Cambridge, UK. www.wdpa-marine.org

UNEP (2013a) *Agenda 21*. United Nations Sustainable Development. Online. www.unep.org/

UNEP (2013b) Regional seas program. Online. www.unep.org/regionalseas/programmes/default.asp

UNESCO (2011) *Main characteristics of Biosphere Reserves*. Online. www.unesco.org

UN General Assembly (2009) *Oceans and the Law of the Sea*. Report of the Secretary-General Sixty-fourth session, Agenda item 76. A/64/66Add.2, 1–69.

Wade P, Kennedy A, LeDuc R, Barlow J, Carretta J, Shelden K, Perryman W, *et al.* (2011) The world's smallest whale population? *Biology Letters* **7**, 83–85.

Wagner FH (2001) Freeing agency research from policy pressures: a need and an approach. *BioScience* **51**, 445–450.

Walpole M, Almond REA, Besançon C, Butchart SHM, *et al.* (2009) Tracking progress toward the 2010 biodiversity target and beyond. *Science* **325**, 1502–1503.

Wallace BP, DiMatteo AD, Bolten AB, Chaloupka MY, Hutchinson BJ, *et al.* (2011) Global conservation priorities for marine turtles. *PLoS ONE* **6**, e24510.

WCED (1987) *Our common future*. The World Commission on Environment and Development (The Brundtland Report). Oxford University Press, Oxford, UK, 1–400.

Weaver TJ (2010) *Rebuild In Depth: Oceans*. World Economic Forum, January.

Wilder RJ (1998) *Listening to the sea*. University of Pittsburgh Press, Pittsburgh, Pennsylvania.

Woody T (2011) Wildlife at risk face long line at U.S. agency. *New York Times* April 24.

WRI, IUCN, UNEP (1992) *Global Biodiversity Strategy: Guidelines for Action to Save, Study, and Use Earth's Biotic Wealth Sustainably and Equitably*. World Resources Institute, Washington, D.C.

WEF (2011) *Global Risks 2011*, Sixth Edition. World Economic Forum (eds Van der Elst K, N), CH-1223 Cologny/Geneva, Switzerland. www.weforum.org

Worm B, Barbier EB, Beaumont N, Duffy JE, Folke C, *et al.* (2006) Impacts of biodiversity loss on ocean ecosystems. *Science* **314**, 787–790.

Yablokov AV (1994) Validity of Soviet whaling data. *Nature* **367**, 108.

Yang T, Percival RV (2009) The emergence of global environmental law. *Ecology Law Quarterly* **36**, 615–659.

CHAPTER 4

MARINE SYSTEMS: THE BASE FOR CONSERVATION

We feel clearly that we are only now beginning to acquire reliable material for welding together the sum total of all that is known into a whole; but on the other hand, it has become next to impossible for a single mind to fully command only a small specialized portion of it.

E. Schrödinger (1944)

4.1 A SYSTEMS APPROACH

This chapter presents an overview of the marine environment as a system of global significance. It highlights major attributes in which biological diversity plays a critical role important to ecosystem structure and function in a constantly changing environment. The global ocean covers more than 70% of the Earth and holds 97% of Earth's water, with complex linkages to land, air, astronomical forces, and to chemical, biological, and hydrologic cycles. Uncertainty is pervasive, especially about ecological stability resulting from energy and fisheries exploitation of already-disturbed ocean systems, exacerbated further by the vast variability of the ocean system itself. In the wake of climate change and with poor understanding of the potential hazards such disruptions might bring, a systems perspective is essential to guide conservation efforts.

4.2 DYNAMIC PLANETARY FORCES

The global ocean is a vast, continuously moving body of water driven by planetary forces connected to the Moon and Sun. Planet Earth is habitable due to a hospitable climate and a capacity to capture and recycle water, energy, and nutrient chemicals, globally balanced in levels of organization modified over geologic time. Thermodynamics, gravity, motion, and chemistry in the many hierarchical dimensions of time and space drive its behavior. As the ocean plays a major role in this whole-Earth system, modern technologies—satellites, submersibles, and monitoring systems—are revealing its mysteries.

4.2.1 The global ocean and climate

The global ocean absorbs >97% of solar radiation, which powers the global circulation pattern (Bigg *et al.*, 2003). Its thermal capacity is a thousand times greater than the atmosphere, making the ocean a major heat reservoir for the planet (Riebesell *et al.*, 2009). Its interactions with continents and the deep sea result in the redistribution of large quantities of heat, water, gases, particles, and momentum. The ocean expands in volume when heated and contracts when cooled, causing sea levels to rise and fall on scales related to global climate change. And through complex processes and feedbacks in recycling of energy and vital elements (e.g., carbon, nitrogen, others), the protean ocean maintains a delicate balance with climate at all scales of interactions.

The nature of the ocean and its role in climate regulation relates to its major component, water (H_2O). Water is a polar chemical compound composed of two hydrogen atoms and one oxygen atom, with a molecular size of less than a nanometer and instantaneous reactions in less than nanoseconds. The water molecule has unique properties that give the ocean attributes of thermal capacity, surface tension, viscosity, elasticity, and solvency. Under specific conditions, water is transformed into gas or ice, and can hold gas molecules (carbon dioxide, CO_2; oxygen, O_2; methane, CH_3; etc.) in solution to be exchanged with the atmosphere. Due to its vibrating hydrogen bonds, water absorbs heat and can release large amounts of heat without changing much in temperature. The massive extent of ocean water thus plays important roles in the global water cycle and climate, with carbon dioxide connecting climate to ocean acidity, ocean warming, and climate change.

4.2.2 Solar radiation and energy transfers

Solar radiation received by the oceans is influenced by Earth's tilt and its axis of rotation. This interaction creates uneven distributions of energy and seasonal, latitudinal, and thermodynamic differences in the global land-ocean-atmosphere system. The only places that receive perpendicular radiation from the Sun at some time of the year are between the Tropics of Cancer and Capricorn, located 23°30′ latitude north and south of the equator, respectively. These low latitudes gain more solar radiation than is lost, unlike high latitudes that lose more energy than they gain. The transfer of excess heat from low to high latitudes maintains the energy balance between the atmosphere and the oceans, producing seasonal surpluses of heat at lower latitudes and deficits of heat at

Marine Conservation: Science, Policy, and Management, First Edition. G. Carleton Ray and Jerry McCormick-Ray.

higher latitudes. Any alteration of this energy-transfer system may affect climate. This differential heat absorption and re-radiation from the ocean, in combination with landmass configuration, Earth's rotation, and gravitational forces, highlights the ocean's importance in the global, cybernetic system.

The Sun's electromagnetic energy is also a driving force for physical and bio-energetic processes. The Sun emits a radiation spectrum of ultraviolet (30–400 nm), visible "light" (400–700 nm), and infrared (700–3000 nm) that the ocean differentially absorbs according to its water depth and transparency. Fifty percent of solar radiation is absorbed within the top half-meter of ocean surface (Soloviev and Lukas, 2006). Light penetrates only a few tens to hundreds of meters depth (the "photic zone", 300 m maximum) where the photic zone receives only 1% or less of the surface value, limiting photosynthesis, animal vision, and photoperiodic responses of plants and animals (Fig. 4.1). In this thin surface layer where 90% of sea life occurs, plants capture radiant energy, utilize carbon dioxide, release oxygen, and package elements (carbon, nitrogen, phosphorus, sulfur, etc.) into complex living matter.

4.2.3 Earth's rotation, gravity, and fluid motions

Circulation patterns of the world's oceans are products of complex interactions driven by Earth's rotation, winds, and the configuration of ocean basins. Due to Earth's rotation on its axis, surface fluids (i.e., air, water) shift from a straight line of flow to a direction that is approximately perpendicular to the original direction of flow, a phenomenon known as the Coriolis effect. This force has a strength that is proportional to the speed of Earth's rotation, which differs with latitude, tending to deflect moving currents and objects (e.g., plankton, fish) to the right in the Northern Hemisphere and to the left in the Southern Hemisphere. For example, a particle at 60° north latitude moving northward and not attached to the Earth initially moves eastward at about 1500 km h^{-1}, but at 30° latitude moves only 800 km h^{-1}. Coriolis force is an important factor in forming cyclonic weather systems and in affecting long-distance migration paths of species such as marine mammals, sea turtles, and sea birds. Within five degrees of the equator, the Coriolis force is weak and hurricanes do not form.

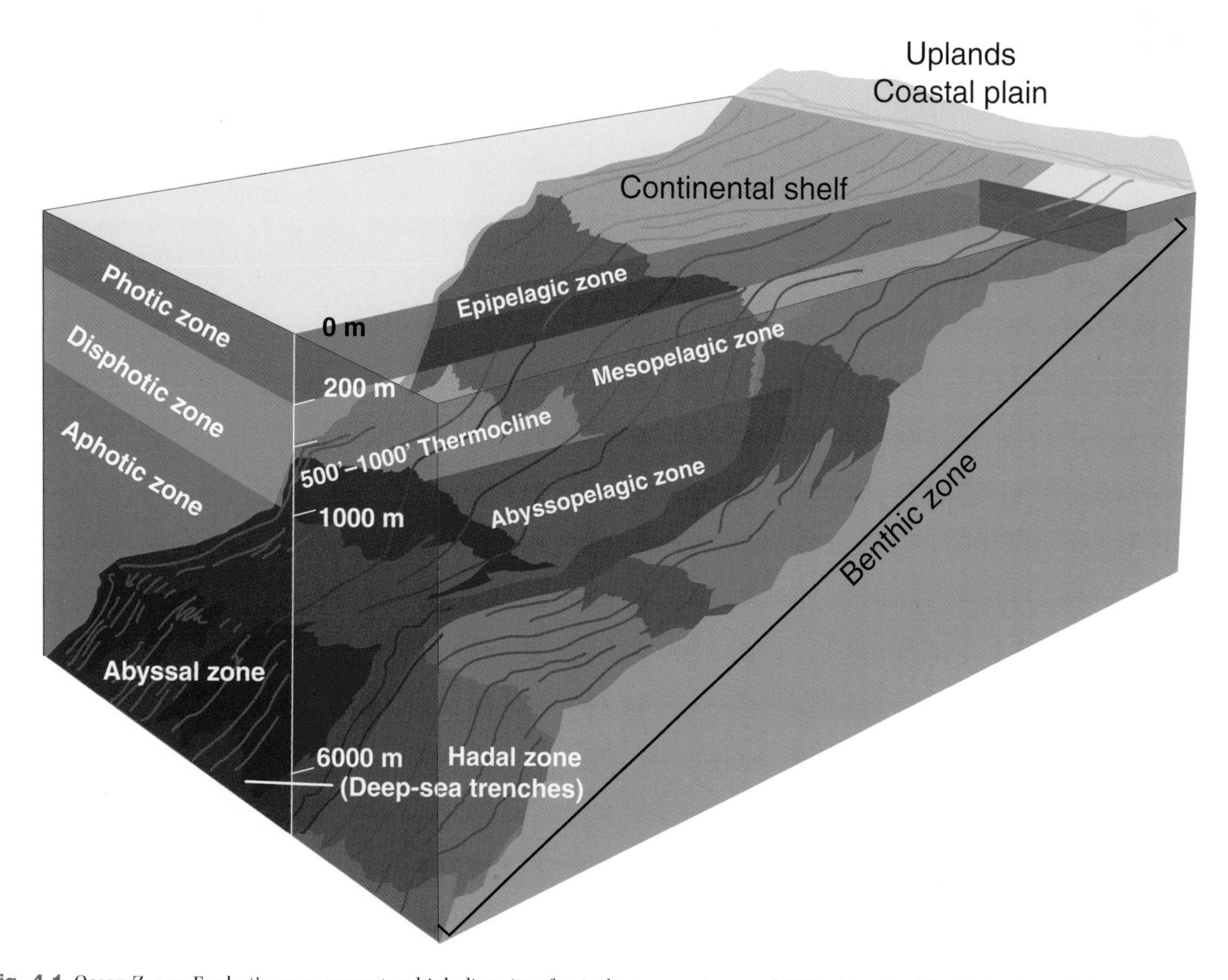

Fig. 4.1 Ocean Zones: Euphotic zone supports a high diversity of epipelagic organisms and 90% of marine life. Disphotic, twilight zone supports mesopelagic species. Aphotic zone lacks sunlight, where relatively few species live in low abundance. Below the photic zone, the bathyal, abyssal, and hadal are deep-sea regions. Benthic zones (brown) grade from shelf at the shore to continental slope and rise (together the "margin") and then to the deep-sea bed. See text for explanation.

Earth's rotation also powers ocean movements, further affected by wind and tidal forces, with horizontal and vertical circulation patterns modified by variations in water density, gravity, wind friction, and continental boundaries.

Another phenomenon known as the Ekman spiral acts to change current direction with depth. When wind sets surface water into a direction of motion, this direction down to a depth of about 100 m is deflected (theoretically) 45° to the right in the Northern Hemisphere and to the left in the Southern Hemisphere. Hence, deeper layers are progressively deflected to the right (or left) of the overlying layer's movement. When movements of all ocean layers are combined, net deflection is approximately 90°. At large scales, an Ekman circulation pattern is observed as ocean gyres that cover large oceanic basins (e.g., Pacific Gyre) with a center of lighter, less dense water becoming elevated to form a "hill," where gravity forces water to flow outward and downward, a phenomenon known as geostrophic flow. Ocean gyres are associated with large surface currents, the ocean "rivers" that affect regional climates around the world and serve as transport for many marine organisms. Major oceanic current systems that are in balance with Earth's rotational and gravitational forces are significant ocean features, and play significant roles in regional climate.

In all ocean basins (Table 4.1) on both sides of the equator, strong western and eastern boundary currents flow for thousands of kilometers. In general, western boundary currents flow from the equator to deliver warm water toward the poles, and eastern boundary currents flow from high latitudes toward the equator and deliver cold water to the tropics. And as warm air from the tropics moves toward the poles at faster rates than the Earth spins (the Coriolis effect), winds are created that affect ocean currents. Most importantly, westerly winds (40–50° latitudes) blow from west to east to force equatorial water eastward. Trade winds (20° latitudes), on the other hand, blow east to west. The North and South Equatorial Currents in lower latitudes flow west and the Equatorial Counter Current flows east. In El Niño years, equatorial currents in the Pacific Ocean intensify. Such ocean current systems not only affect climate, but also can be important transport systems for migratory marine mammals, sea turtles, oceanic fishes, and larvae of many species.

In the Atlantic Ocean, the best-known western boundary current is the Gulf Stream. This warm surface current transports warm water from the Gulf of Mexico and tropical Atlantic to the northern, colder Atlantic Ocean towards Europe. The northward transport of heat by the Gulf Stream moderates Europe's northern climate. At speeds of 97 km day^{-1}, this powerful current moves 100 times as much water as all the rivers on Earth (USGS, online). The Gulf Stream is modified by continental boundaries and follows the edge of the coastal-ocean boundary. It helps form the clockwise-flowing mid-Atlantic ocean gyre. The Gulf Stream provides a transport mechanism and habitat for many forms of life, its initial course being set by the North American continental slope, where warm- and cold-core rings spin off. In the Pacific, the Kuroshio Current, the world's second largest current, flows across the Pacific at speeds up to 121 km day^{-1}, and is approximately 1000 m deep. It forms south of Japan as the western boundary current of the North Pacific Gyre.

Table 4.1 Major oceans of the world and their significance.

Ocean	Areal extent 10^6 $mile^2$ (km^2)	Significance
Pacific Ocean	64 (165)	Global weather phenomena El Niño/La Niña; fisheries (60% of 1996 world's total fish catch); ocean-atmosphere interactions, climate control, global carbon fluxes; covers 1/3 earth surface; highest mountain on Earth (Mauna Kea, 33,476 ft/10,203 m) and deepest trench (Mariana Trench, 36,198 ft/11,033 m deep)
Atlantic Ocean	30 (77)*	Covers 20% Earth surface; receives ~4x more riverine inflow from land than either Pacific or Indian Oceans; initiates thermocline circulation (conveyor belt) in transport of heat and salt on a planetary scale; world's most heavily trafficked sea routes. Supports major fishing, dredging of aragonite sands (Bahamas), crude oil/natural gas production (Caribbean Sea, Gulf of Mexico, North Sea)
Indian Ocean	26 (68)*	Third largest ocean. The most expressed monsoon system; contains major sea routes (oil and petroleum products); oil and gas fields (40% world's offshore oil production), fish, shrimp, sand, gravel aggregates, placer deposits, polymetallic nodules
Southern Ocean	8 (20)*	Fourth largest ocean, surrounds Antarctica; freezes in winter; potential large/giant oil and gas fields; manganese nodules, possibly placer deposits, sand & gravel, freshwater storage in icebergs; supports major fisheries; marine mammals
Arctic Ocean	5 (14)	Contains widest continental shelf; winter ice cover (thinning); ecosystem slow to change and recover from disruptions or damage; receives large watershed inputs; contains sand and gravel aggregates, placer deposits, polymetallic nodules, oil and gas fields. Supports major fisheries, marine mammals (seals, walruses, whales)

*The International Hydrographic Organization in Spring (2000) delimited a fifth world ocean by removing the Southern Ocean from the Atlantic Ocean (from Longhurst, 1998b).

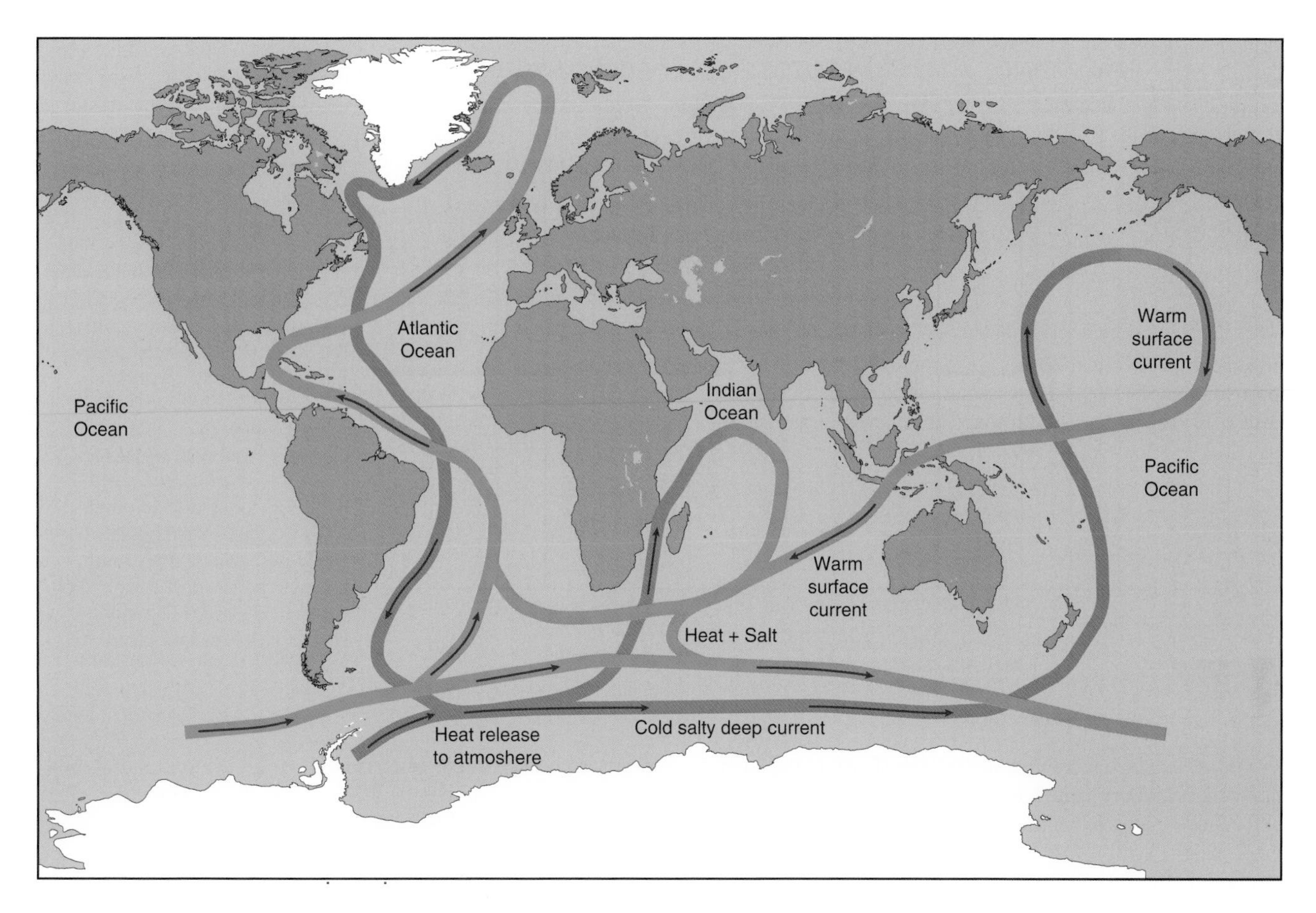

Fig. 4.2 Ocean conveyor belt: a simplified, conceptual model of the redistributionof global water masses. High-salinity North Atlantic water cools and sinks into cold, deep, high-salinity North Atlantic Deep Water current moving south into the Southern, Indian, and Pacific oceans, where upwelling moves into shallow and warm waters, then in a return current to the North Atlantic. Part of this global ocean circulation includes the buoyancy-driven thermohaline circulation, which transports heat from the tropics to the northern North Atlantic and causes a northward heat flux through the Atlantic. Based on data from Broecker (1991); IPCC (2001); Richardson (2008).

On a planetary scale, ocean surface circulation connects with the deep thermohaline (deep mass) circulation in a global pattern of water movement, captured in the scientific concept as the "global oceanic conveyor belt" (Broecker, 1991, 1997; Fig. 4.2). This circulation transports heat, energy, and solutes through processes that can be thought to begin in the North Atlantic with the North Atlantic Current and formation of North Atlantic Deep Water (NADW), initiated through evaporation and sinking of cold, salty surface water. This deep water flows southward to form the Antarctic Circumpolar Current, then northward along the ocean bottom into the Pacific and Indian Oceans, gradually warming and mixing with the overlying surface water. If North Atlantic surface-water salinity somehow drops too low to allow for the formation of deep-ocean water masses, the system can weaken or shut down entirely, as apparently happened during the Little Ice Age (about 1400 to 1850 AD). Broecker (2006) and colleagues suggest that the ocean's overturning was responsible for the rapid climate fluctuations experienced during Earth's last glacial period. Because of the ocean's crucial role in the Earth's climate system and its massive uptake, transport, and storage of heat and carbon dioxide (CO_2), scientists are intensively debating the conveyor-belt concept (Lozier, 2010, 2012), particularly in the context of climate change.

4.2.4 Major geologic movements

The Earth is made up of moving layers. The Earth's upper mantle and crust (the lithosphere) were built from its hot molten core. The hard lithosphere is broken into dynamic, continental, tectonic plates that move and cause earthquakes, volcanic activity, and continental drift. Over hundreds of millions of years, the Earth's land surface has been rearranged in patterns very different than today, as plates have drifted apart or together in a series of supercontinents during the ancient Archean Eon, forming Rodinia, and subsequently Pangaea surrounded by Panthalassa (ocean) about 400 million years ago. The Pangean plates began to separate about 135 million years ago into two landmasses: Laurasia to the north and Gondwanaland to the south, intercepted by the Tethys Ocean. As the plates continued to drift apart, the continents and oceans took their present positions (Blakey, 2008).

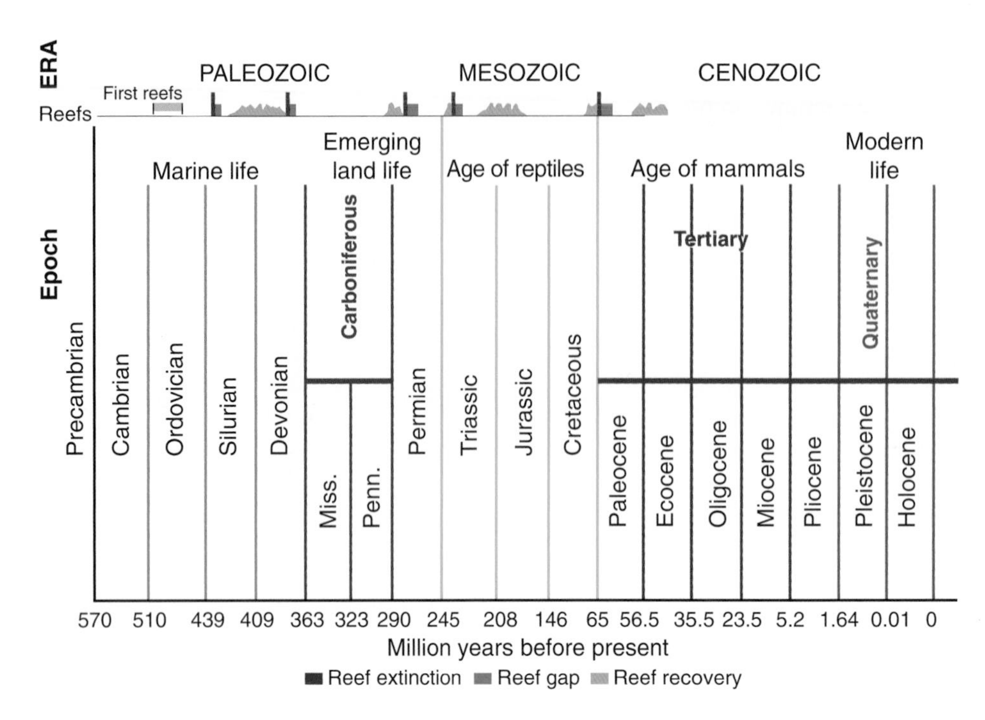

Fig. 4.3 Earth and its life have changed dramatically during 570 million years of geologic time, evolving through major eras (Paleozoic to Cenozoic) and epochs (Precambrian to the Holocene), with coral reef extinctions and recoveries into the present, diverse modern age. Reefs were first observed in the Ordovician and underwent five major extinction events (vertical dark purple bars) that left the Earth without living coral reefs for at least four million years (reef gaps = red, smaller rectangles), and recovery into different, prolific reef growth (pink pattern). Modern reefs are not shown. Based on data from USGS (2010) and Veron, JE (2008).

This geological history provides the basis for tracking Earth's time and evolution. Time scales called geological eras, periods, and epochs (Fig. 4.3) highlight changes in life forms up to the modern era. Within the 300 million-year Paleozoic Era and formation of Pangaea, life moved onto land, and most invertebrate and vertebrate groups and vascular plants evolved. This period of no less than two ice ages and diversifying marine life ended with the Permian/Triassic extinction 250 million years ago, the greatest mass extinction event in geologic history. In total, five mass extinctions have occurred during Earth's history, followed by new and diversified marine and terrestrial life. The Mesozoic Era is marked by the formation of the present land and ocean configuration—the Age of Reptiles—when reptiles flourished and birds and small mammals evolved only to suffer a second great extinction event at the end of the Cretaceous Period. Biotic evolution then resulted in the Age of Mammals, which continues today. Five extinction events affected coral reefs as much as any other ecosystem, with recovery requiring at least four million years during "reef gaps" before living reefs returned (Veron, 2008).

4.3 MAJOR OCEAN STRUCTURES AND CONDITIONS

A conceptual view of the dynamic ocean system is facilitated by identification of its major structural features. Boundaries in the water column that affect species distributions include pressure, light, temperature, oxygen, pH, and salinity, each of which change with depth, season, and circumstances. Combinations of these factors and a shared common history create water masses, i.e., identifiable bodies of water with physical properties distinct from the surrounding water, most reliably determined by temperature and salinity. Water-column features and ocean patterns are influenced by continental landmasses that contribute lithospheric and continental shelf boundary conditions.

4.3.1 Physical structuring

Water masses, currents, eddies, gyres, and upwelling areas form prominent coherent features of the water column, with peripheries marked by steep transition zones known as fronts. Frontal systems (tidal, shelf-edge, oceanic) are separated by physical discontinuities of current speed, temperature, salinity, and density, with sharp boundaries and abrupt ecological changes (Longhurst, 1998a). Vertical structure is created by sharp changes in temperature (thermoclines), oxygen, and salinity/density (pycnoclines) that can stratify the water column into different biological zones. Thus, when water movement is slowed at a boundary, as between water masses, sea bottom, land and air, and encountering hard objects, a steep transition over relatively short distances creates a spatial zone of discontinuity, which can demarcate distinct but sometimes "leaky" seascapes and biogeographic patterns (Ch. 5).

Physical forces that drive circulation patterns and create eddies and fronts at mesoscales (1–500 km) form large-scale coherent oceanographic, chemical, and biological patterns, and also form patches down to millimeters in size. Physical forces that create eddies and fronts may play a role in the generation of phytoplankton and zooplankton patchiness at all scales (Martin, 2003). Physical oceanographic processes such as turbulent advection, upwelling, convergence, and vertical mixing can drive various biological responses such as growth, grazing, and behavior, and strongly regulate and are correlated with planktonic spatial patterns. At frontal zones, marine productivity and biomass increase to sometimes exceptionally

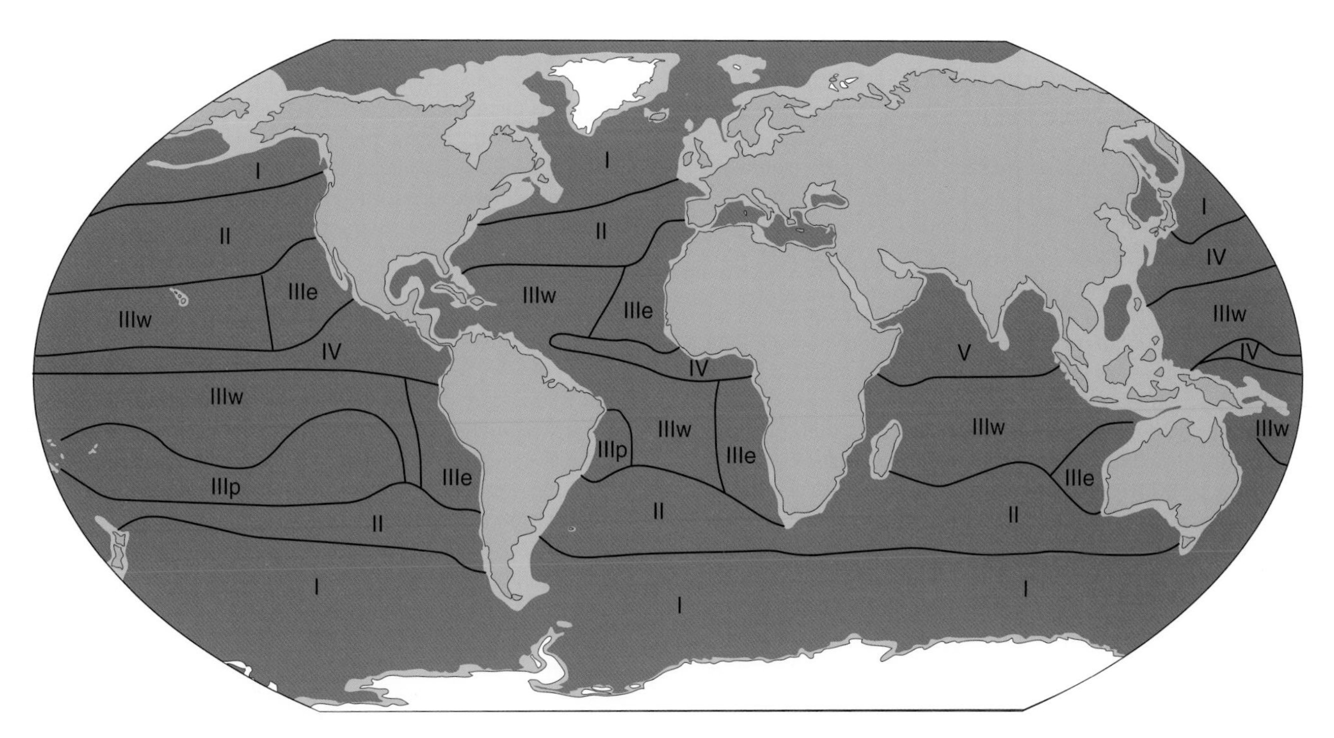

Fig. 4.4 Continental shelf (light blue) extends offshore around the land. Physio-oceanographic features distinguishocean provinces, classified as: (I) variable eastward; (II) weak and variable; (III) trade wind; (IV) strong westward and equatorward; and (V) monsoons with seasonal reversals. Category III is further subdivided into: (IIIe) strong equatorward; (IIIw) westward; and (IIIp) strong poleward. Based on data from Hayden *et al.* (1984); Holligan and Reiners (1992).

high levels that attract top predators, where organisms aggregate on a variety of spatial scales, depending on their size, evolutionary inheritance, life history, and physiological capacity (Bost *et al.*, 2009). In areas of surface-water discontinuities where strong and dynamic interfaces separate eddies and fronts, many large predators (tuna, birds, turtles, and cetaceans) come to feed (Kai *et al.*, 2009). Furthermore, eddies and their attendant fronts entrap fish eggs and larvae to form important pelagic habitats for fisheries production (Govoni *et al.*, 2010).

The relatively shallow continental shelf that surrounds nearly all landmasses is an extension of the continental geologic crust that creates a benthic boundary condition between land and sea (Fig. 4.4). This shelf begins at the littoral zone and extends to outer edge depths of approximately 200 m, over which "neritic water" is distinguished from ocean pelagic deep water off the shelf (Huthnance, 1995), although both are commonly referred to as "pelagic." Similarly, marine sediment that covers most of the shelf is distinct in composition from that under the open ocean beyond the shelf. Depth, motion, and seafloor stability establish conditions that influence the water column's heterogeneous structure.

At the edge of the continental shelf, identified as the continental/ocean margin (slope and rise; Fig. 4.1), the deepening seafloor abruptly descends from 200 to approximately 4000 m. Here, the ocean-shelf exchange and stratification are interrupted by water masses of different characteristics (notably temperature and salinity), and special internal ocean processes are created (Huthnance, 1995). The relatively narrow continental margin that covers only about 11% of the global seabed contains sharp environmental gradients and tectonic activity (Buhl-Mortensen *et al.*, 2010; Levin *et al.*, 2010). Over the abruptly deepening continental slope, ocean processes interact with complicated topography marked by submarine canyons that exceed any canyon on land. The depth changes from approximately 2000 to 5000 m then becomes gradual, distinguishing the margin from the deep-sea floor at the continental rise (Menot *et al.*, 2010). The continental margin, a most heterogeneous environment characterized by multiple water masses, distinct hydrographic characteristics, and stratified water, separates the neritic and ocean systems.

The dynamic continental margins are highly productive, and can be unstable (Helly and Levin, 2004). Some receive massive river inflows with exceptional inputs of floodwater, macrophytic detritus, suspended organic matter, and debris of terrestrial origin, and are marked by coastal longshore transport systems. Some margins are naturally hypoxic and low in biodiversity. In other margins, upwelling brings nutrients to the surface that stimulates productivity and brings large marine predators such as elephant seals (Ch. 12) and sperm whales (*Physeter macrocephalus*) into these major feeding areas. Tectonic activity can jolt, and subduction can squeeze, triggering downward-flowing turbidity flows that carve into the substrate and force reduced fluids out of the system that would otherwise fuel chemosynthetic (seep) ecosystems. These events can be episodic, and violent cascades can lead to erosion and material deposition, which impose variable temporal and spatial constraints on living fauna. In some margins at mid-bathyal depths, naturally hypoxic waters can smother the seabed.

The open ocean, with an average depth of approximately 4000 m, is distinguished by its deep bathyal, abyssal, and hadal zones (Fig. 4.1). The approximately 4°C bathyal zone in 1000 to 3000 m depths falls below the photic zone, and overlaps with the upper abyssal zone in 2000 to 6000 m depths. Abyssal water, which originates at the air-sea interface of polar regions, spreads over the 300 million km^2 abyssal plain 1 km below the surface, an area that represents 83% of ocean and 60% of Earth's surface. This is the world's largest habitat and last remaining wilderness, a cold, lightless, high-pressure environment rich in rare species (Grassle and Maciolek, 1992; Van Dover, 2000), but of low abundance, a high level of endemism, and patchy biotic distributions (Vinogradova, 1997). Mobile epibenthic megafauna at 4100 m depth can exhibit inter-annual changes in abundance from one to three orders of magnitude that reveal time scales relevant to the biota that live there (Ruhl, 2007). In the abyssal benthos, long dead carbonates of protozoan origin forms massive areas of benthic mud and siliceous ooze (Van Dover, 2000). Bodies of dead whales form islands of production (Ch. 5). The deepest ocean zone is the hadal, in water depths greater than 6000 m, water temperatures of 1.0–2.5°C, and dominated by ocean trenches ventilated by deep currents. The deepest trench is the Marianas Trench at almost 11,030 m (11 km). The hadal environment is under tremendous hydrostatic pressure, and species are often restricted to local areas; 95% occur only in a single trench or groups of adjacent trenches (Vinogradova, 1997).

Submarine mountain ranges of high relief occur in the mid-ocean. They result from planetary-scale processes of ocean crust formation at spreading centers where ocean plates diverge and where new oceanic crust is formed. When plates separate, global heat is lost, and ridges with fissures or cracks along their crests are formed. In these mid-ocean ridge systems, active volcanism accounts for a significant number of Earth's total volcanism (www.mbari.org). Hydrothermal vent systems are a conspicuous feature of Earth's oceanic crust. Spatial and temporal scales of venting are influenced by the rate of new crust formation and the amount of tectonic activity. Unique endemic faunal vent assemblages were first discovered in 1977 (Van Dover, 2000). Numerous others have since been found that host assemblages of giant clams and mussels, tubeworms, eyeless shrimp, and bacteria that all depend on sulfur as the primary energy source, rather than oxygen gained through photosynthesis. As vent biota are intimately linked to the geologic and chemical environment, they depend on obtaining reduced inorganic chemicals from chemosynthesis, a process known since 1887 to occur in surface marine sediments described by the Russian scientist, Sergei Winogradski, as the "black layer" (Kiel and Tyler, 2010).

Seamounts are prominent, widespread underwater topographic features ("mountains") that rise from the sea bottom, but do not break the surface. On abyssal plains and continental slopes, seamounts can be interspersed with topographic features that may extend >1000 m vertically upward from the seafloor (McClain *et al.*, 2009). Estimates of their numbers vary, but up to 200,000 may exist (Clark *et al.*, 2010). Seamounts host rich communities of fish (Williams *et al.*, 2010) and algae (Littler *et al.*, 2010). Isolating mechanisms on some seamounts create highly endemic faunas (Richer de Forges *et al.*, 2000); some lack endemic species yet contribute structures different from the seafloor and potentially provide larvae to recolonize suboptimal, non-seamount habitats (McClain *et al.*, 2009).

Emergent features in any depth of water are islands or island chains classified as continental, marginal, or oceanic. Islands exhibit continental features proportional to their size, the largest being microcosms of continents. Continental and marginal islands are often formed by sedimentary processes and have structural links to continents. Others such as Fiji and the Azores are oceanic islands surrounded by deep oceanic water, with many formed by volcanism, e.g., the Hawaii chain. Due to their isolation, oceanic islands often exhibit unique biological conditions with dynamic barriers and steep gradients to the surrounding ocean. Although islands cover only about 3% of the world's surface, they support a disproportional amount of biodiversity, especially endemics. Biota on oceanic islands may be as different as are the islands themselves, hosting 12 m sunflowers, 250 kg tortoises, marine iguanas, and many others that occur nowhere else. Islands known as atolls, as Charles Darwin correctly hypothesized, are submerged ocean mountains crowned with a ring of actively growing coral reefs; most occur in the Indian and Pacific oceans; only a few are Caribbean, notably in the western Caribbean and The Bahamas. The Saba Bank Atoll in the Caribbean Netherlands is one of the three largest among 400–500 atolls that exist worldwide.

Many islands totally lack capacity for surface or ground water storage, and the biota are adapted to episodic rain events and desert conditions surrounded by a salty sea. Estuarine conditions and anadromous/catadromous species are often lacking. Islands are vulnerable habitats, exhibiting the world's highest extinction rates. About half of the 724 recorded animal extinctions in the last 400 years were island species (CBD Secretariat, 2010, online).

4.3.2 Chemical structuring

Chemical reactions involving temperature, dissolved sodium, chloride, carbon dioxide, carbonate, crystalline calcium carbonate, etc., give the ocean its saltiness, with a salinity of approximately 36 (°/oo) and a mildly basic pH of approximately 7.8–8.2. Water temperature and salinity greatly affect water density, and interact with light to characterize ocean systems.

Living structures depend on three elements that make up 99% of organic molecules: carbon, hydrogen, and oxygen. Nitrogen forms a crucial organic molecule in the blueprint of a cell—its nucleic acid. The development of an oxygenated atmosphere in ancient times advanced life beyond single-celled organisms, which with improving oxygen concentrations is broadly linked to ecological diversification and biological complexity (Knoll *et al.*, 2006). When oxygen was lacking more than two billion years ago, an important constituent of Earth's early atmosphere was hydrogen gas, likely released by microbes (chemoautotrophs; Falkowski, 2012). When oxygen-producing cyanobacteria, a major primary producer in primordial seas, arose to be the source of oxygen for the planet, this most significant event changed the history of life and the evolution of life itself (Canfield, 2005). As oxygen levels in the atmosphere rose, shallow oceans became mildly oxygenated, but deep oceans remained anoxic (Holland, 2006). When atmospheric oxygen reached modern levels during the early Phanerozoic Eon (544 ma to present), visible life emerged and diversified. Shallow oceans became oxygenated and deep-ocean oxygen fluctuated considerably on geologically short time scales (Holland, 2006). This allowed a high diversity and abundance of marine life to evolve, until approximately 250 million years ago at the end of the Permian as much as 82% of all genera and perhaps 80–95% of all marine species went extinct. On land, vertebrates, plants, and insects also underwent mass extinctions presumably due to climatic effects, including acid rain, global warming, and volcanic eruptions that involve chemical change (Erwin *et al.*, 2002). Knoll *et al.* (1996) proposed that the deep anoxic ocean overturned and high concentrations of carbon dioxide came to the surface. Carbon dioxide in excess is soluble in seawater and increases acidity, which may have affected calcifying taxa, replacing them with highly productive non-calcifying taxa.

Today, oxygen and nitrogen are the most abundant gases dissolved in seawater. These with carbon, phosphorous, and others, play key roles in forming biological structures and complex life forms through ecological transformation. Oxygen minimum zones are a permanent, natural feature of the ocean, with over one million km^2 of permanently hypoxic shelf and bathyal sea floor in existence today, over half (59%) of which occurs in the northern Indian Ocean with dissolved oxygen levels of $<0.5\,ml\,l^{-1}$ (Helly and Levin, 2004). Due to warming oceans, nutrient enrichment, high atmospheric concentrations of carbon dioxide, and enormous changes in ocean chemistry are occurring, inducing ocean acidification and causing rapid declines in mid-water oxygen concentrations (Brewer and Peltzer, 2009). Coastal "dead zones," i.e., areas deficient in oxygen (hypoxic) or depleted of oxygen (anoxic; Box 2.5) restrict development of complex life forms.

4.4 PLANETARY CYCLES

The Earth's planetary system appears to be unique in the universe; a system maintained by the cycling of matter, water, and living resources (Kleidon *et al.*, 2010).

4.4.1 Water cycle

Water is a naturally circulating resource that is constantly recharged (Oki and Kanae, 2006). Global movement of water on, in, and above the Earth (the water cycle) is driven by convection and atmospheric motion that solar radiation and evaporative processes create. Water connects atmospheric, terrestrial, and ocean processes in a cycle that has continued for millions of years, as water changes from liquid to vapor or ice and back again. Ocean currents and evaporation move water into clouds that fall back to Earth as rain, fog, hail, snow, or sleet. Water moves through forests, plants, and soils, draining from land in surface runoff or stored in lakes, aquifers, reservoirs, and groundwater (Fig. 4.5). Freshwater from land sources eventually moves to the ocean. Globally, the high variability and availability of water make it a vulnerable natural resource, even more critical under scenarios of climate change.

The global water system is central to Earth dynamics. It integrates and regulates biogeochemical and biogeophysical processes that maintain terrestrial and aquatic ecosystem integrity (Crossland *et al.*, 2005). It controls terrestrial ecosystem dynamics through interactions with biota and climate, ecosystem feedbacks, and watershed connectivities, all of which modulate hydrologic processes and rates of flow (D'Odorico *et al.*, 2010). Climate controls globally variable moisture, precipitation, temperature, water storage, and water availability. Thick snow falls at certain times and places in colder climates, compacting into ice to form glaciers and ice sheets that when melted provide water sources. In warmer regions or periods of drought, water availability affects terrestrial and coastal ecosystems, especially estuaries (Ch. 6).

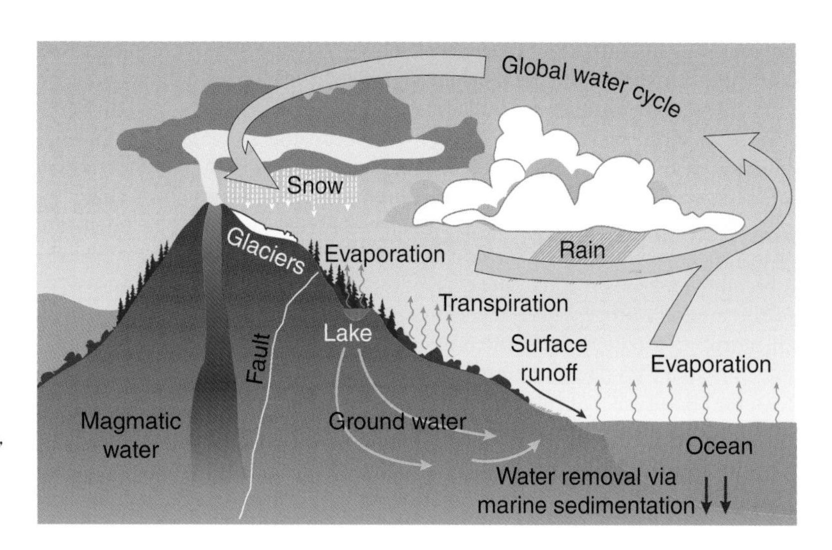

Fig. 4.5 The global water cycle is driven by the sun as water is recycled through land, ocean, air, forests, plants, and soils. The sun heats water, water evaporates, falls as snow and rain, is stored in glaciers, lakes, aquifers, reservoirs, groundwater and the ocean, and drains from land as surface stream flow and groundwater seepage back to the sea.

4.4.2 Biogeochemical cycle

At Earth's surface, matter and energy are exchanged and reused through extremely complex biogeochemical cycles (Fig. 4.6a; Hedges, 1992). Interacting processes operate on timescales of microseconds to eons, in domains as small as a living cell to domains encompassing the entire atmosphere-ocean system. In marine systems, nutrient chemicals (nitrogen, phosphorus) can become limiting unless captured and recycled by organisms and returned to the system through upwelling, bioturbation, river flow, etc. Elemental carbon, in close association with nitrogen and phosphorus, moves through living matter, where biochemistry and global biogeochemical cycling are linked in characteristic stoichiometric ratios: e.g., C:N:P Redfield ratio (Redfield, 1958), a measurement tool indicating that ocean water is in chemical equilibrium. As essential constituents of living systems, organisms regulate the rates of recycling of these elements and where and which chemical form accumulates. Thus, living organisms play key roles in biogeochemical reactions and global recycling pathways, facilitated by atmospheric motion and transformations of elemental chemicals (Schlesinger, 1997).

Carbon cycling is of particular importance (Fig. 4.6b). The global ocean is the largest reservoir of organic (biogenic) carbon on Earth's surface (Hedges, 1992), where colloids in surface waters form a globally significant fraction of dissolved organic carbon (Kepkay, 2000). Marine organisms mediate carbon flow in a series of processes referred to as the "biotic pump," consuming and cycling dissolved and particulate organic matter. Marine photosynthetic activity captures more than half of the global carbon (Falkowski *et al.*, 2000; Nellemann *et al.*, 2009), with biotic growth and respiration; microzooplankton grazing also plays important roles (Calbet and Landry, 2004). Among other small organisms, ubiquitous unicellular cyanobacteria produce organic carbon, heterotrophic bacteria metabolize carbon, and viruses are important in recycling (Wilhelm and Suttle, 1999). Carbon dioxide gas (CO_2) released to surface waters through respiration and decay is balanced, in part, by the calcification and growth of organisms.

Large amounts of calcium carbonate fall out of surface waters from sinking dead marine plant and animal exoskeletons (e.g., microscopic protozoans, foraminiferans, coccolithophores). This sinking of biogenic particles drives respiration in the water column and helps maintain a strong vertical ocean gradient of inorganic carbon. Under various circumstances, a reverse biological pump moves carbon back into the water column, such as when deep-sea fishes spawn on the bottom and their eggs rise up to the thermocline to hatch, and larvae become consumers in surface waters. If carbon is not released or transported out of deeper waters (e.g., bioturbation, Ch. 5), it becomes trapped on and in the ocean floor.

The marine nitrogen cycle is perhaps the most complex of all. Nitrogen is a major component of the atmosphere, and its role in primary production is critical, being fixed by biota and bacteria into useable nitrate (NO_3) and ammonium ion forms (NH_4), or released as inert nitrogen gas (N_2) to the atmosphere through denitrification. Nitrogen circulates through biota, the atmosphere, and back through biota, thereby exerting a significant influence on cycles of many other elements, in particular carbon and phosphorus. Phosphorus also plays a major role in coastal systems and global oceanic primary production, being a vital macronutrient in living systems. Most chemical forms of nitrogen in the ocean are bioavailable, but dissolved nitrogen gas (N_2), the most abundant form of nitrogen, is generally not. Nitrification and denitrification are critical processes linked to the global carbon cycle and climate (Canfield *et al.*, 2010; Gruber and Galloway, 2008).

The sulfur cycle is of special interest because the ocean is Earth's main sulfur sink (Liss *et al.*, 1994). Sulfate is the second most abundant anion in seawater, with oxygen playing a key role in the oxidative part of the sedimentary sulfur cycle. In ocean surface waters, decomposition of phytoplankton releases a volatile gas, dimethyl sulfide (DMS), that is significant in cloud formation and climate, and contributes to atmospheric sulfur dioxide (SO_2). DMS is also linked to coral spawning, fish abundance, and to squid aggregations over coral reefs (DeBose and Nevitt, 2008).

Humans have affected virtually every major biogeochemical cycle (Falkowski *et al.*, 2000), and the way that nutrients cycle and place constraints on the rates of biological production and land/sea ecosystem structures. Understanding the complex interactions and relationships among nutrient cycles and other biogeochemical and climatological processes requires a systems approach (Falkowski *et al.*, 2000). Atmospheric carbon dioxide (CO_2) exchanges rapidly with the oceans and results from both biotic and human activities. The ocean stores an estimated 93% of all of Earth's CO_2 (Nellemann *et al.*, 2009) as well as a third of anthropogenic CO_2 from fossil fuel and deforestation emissions (Siegenthaler and Sarmiento, 1993). The marine carbon cycle controls CO_2 partitioning between oceans and atmosphere, with ocean productivity playing an important part. Coastal sediments under high carbon loading and high oxygen demand may become anaerobic and sulfate reducing (Middelburg and Levin, 2009). The carbon cycle is influenced by the rate of atmospheric change of CO_2 and depends on biogeochemical and climatological processes, and human interactions, which with other long-lived greenhouse gases (e.g., methane, CH^4; nitrous oxide, N_2O), trap heat to cause global warming.

4.5 MAJOR PLANETARY INTERFACES

The interfaces between land and sea, benthos and water column, and air and sea are dynamic boundaries characterized by steep, transitional gradients and inter-exchanges. These interfaces result in directional fluxes of nutrients, energy, and materials important to species and ecological processes in adjacent systems. The pulses of energy and materials across these interfaces drive land and seascape formation, further modified by tectonics, climate, biota, and sea-level change.

4.5.1 Land-sea interface

Rivers and ocean tides mingle at shallow-water interfaces. There, terrestrial and oceanic forces overlap spatially, and

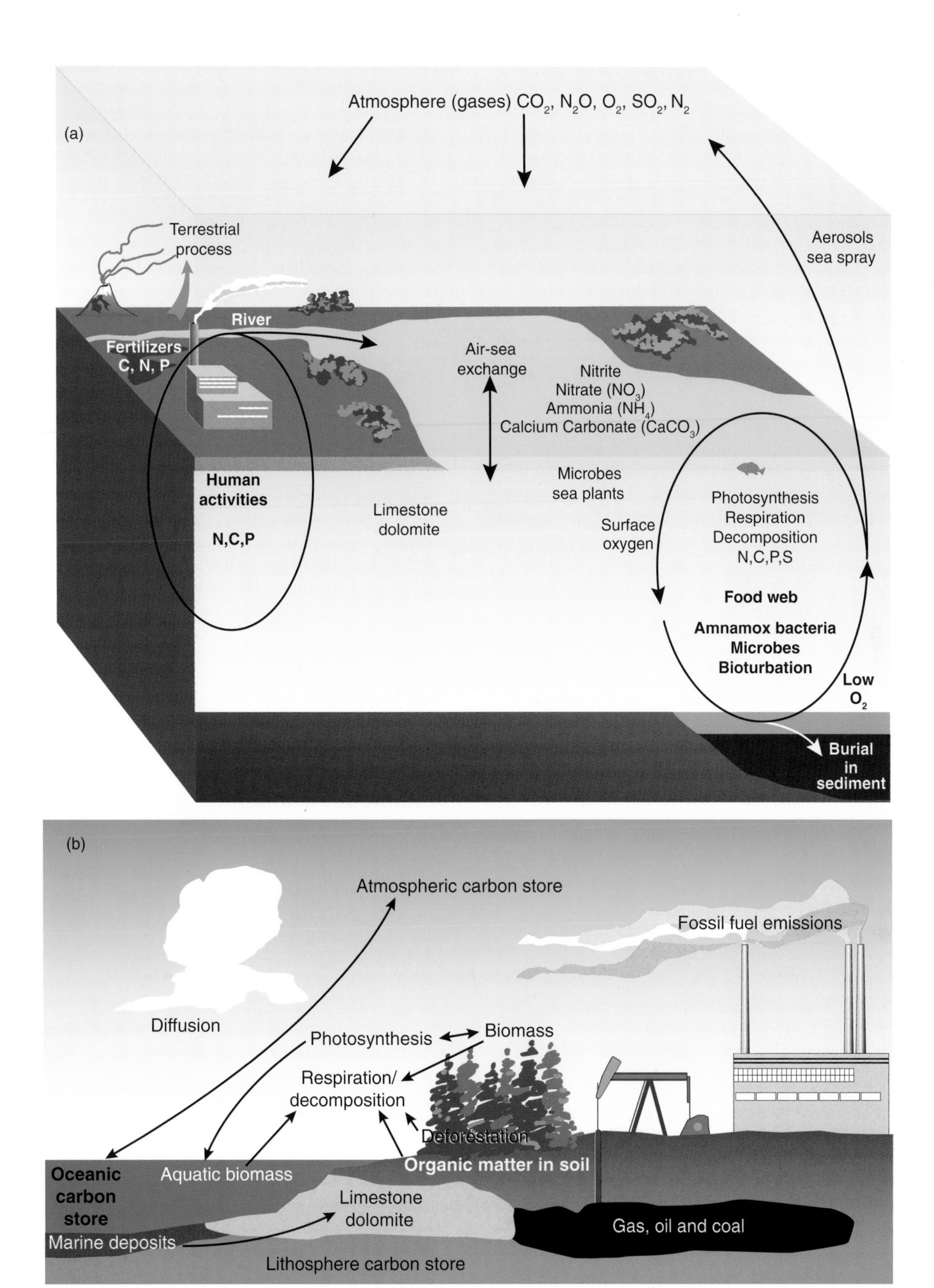

Fig. 4.6 Biogeochemical cycles are extremely complex. (a) Nitrogen (N_2O, N_2, NO_3, NH_4), carbon (CO_2, $CaCO_3$), phosphorus (P), and sulfur (S) move through various compartments on land, in air, in the water, to become transformed through biological, geological, and physical processes. Dimethyl sulfide (DMS) is produced by bacteria in phytoplankton and released to the atmosphere. Phosphorus accumulates in both organic and inorganic sediments, with no biological process generating an important gas flux to the atmosphere in the phosphorus cycle, as sea-salt aerosols can. (b) Carbon is stored in elemental rocks of carbonate minerals (limestone, shale), in thick coal beds, petroleum reservoirs, and in the atmosphere as carbon dioxide. In seawater, carbon exists in several forms (dissolved organic and inorganic, and particulate organic forms of living and dead matter), and large amounts of carbon dioxide can be absorbed from the atmosphere. Carbon in its dissolved inorganic form in seawater is a major, active reservoir. Carbon dioxide gas exchange with the atmosphere occurs through a gradient affected by winds and other environmental factors.

alternations of salt and freshwater, desiccation and drowning, intense heat and cold, and unique adaptations for survival are required to withstand variable tidal and wind energies that can move, crush, and smother. Tides powered by Earth, Moon, and Sun move energy and ocean water toward coastlines, further modified by changing weather, winds, seafloor topography, local water depths, and coastline configuration. Coalescing rivers and underground flows deliver freshwater and terrestrial materials through watersheds, where rooted vegetation controls surface runoff, stabilizes substrates, and cleans and transpires water in modified energy gradients to the ocean (Crossland *et al.*, 2005). On land, hydraulic gradients result in groundwater seepage near shore that may contribute to flows further out on the shelf through confined aquifers delivering chemicals and nutrients into coastal sediments (Burnett *et al.*, 2003). New materials far from original sources, driven by terrestrial and marine forces, create a nearshore, shallow-water, benthic system exposed to varying motion and constant change.

At this interface, the intermingling of different chemicals influences coastal water quality. Watersheds deliver freshwater (rain, rivers) with rich supplies of positively charged calcium, carbonate, and bicarbonate ions that interact with seawater delivery of abundant, negatively charged, dissolved chloride and sodium ions to affect coastal water density, electrolytes, and nutrients. When freshwater meets seawater, colloids are formed and deposited on the benthos, especially in estuaries, to play important roles in biogeochemical cycling of trace elements (Guo and Santschi, 1997). And the degree of fresh or saline water affects distributions and biogeography of coastal species (Ch. 6), while loading of carbon and nutrient chemicals affect the structure and dynamics of the coastal environment.

4.5.2 Benthic-pelagic interface

The seabed is the most extensive habitat on the planet, excepting the ocean itself. It provides habitat for living structures that are among the most prominent and most productive ecosystems (seagrass beds, biogenic reefs, algal flats, clam beds, etc.). The largest faunal assemblage in areal coverage of any biome on Earth resides on and in the marine benthos (Snelgrove, 1998), where life histories of many benthic organisms involve hatching, cysting, and larval releases to the water column. Between the benthos and the overlying water body, the seabed involves a two-way exchange, or flux, of matter that contribute nutrients, particulate matter, and microorganisms to the pelagic system (i.e., plankton), with profound influences on the dynamics of populations and communities (Raffaelli *et al.*, 2003). The soft mud/sandy bottom and harder gravel/rocky bottoms play important roles in biogeochemical cycling, hydrodynamic modulation, and pelagic/benthic coupling and exchanges. On or above this interface or within the mixed sediment layer, detritus, debris, and fecal pellets are consumed, degraded, exported, or trapped. Macroorganisms are key drivers of biogeochemical fluxes between water and sediment by aerating the benthos through burrows, bioturbation (Ch. 5), and active and passive bio-irrigation. Microorganisms degrade plant products into small organic remnants and release denitrifying gases (N_2O). Under low light energy and an external organic carbon source, microorganisms (chemoautotrophs) reproduce and release hydrogen gas (Falkowski, 2012). Hence, a flux of matter and chemicals crosses this interface through processes that involve complex exchanges governed by hydrology, biological production, and organic and dissolved material fluxes through processes of decomposition and regeneration.

4.5.3 Air-sea interface

At the marine-atmosphere interface, globally significant exchanges of heat, momentum, and water vapor take place (Sikora and Ufermann, 2005). The ocean's heat capacity within its top 2–3 m is the same as the atmosphere above it (Soloviev and Lukas, 2006). The atmospheric layer is a variably thick medium through which exchanges with land and ocean result in large quantities of heat and momentum, energy, gases, particles, and materials (Box 4.1). At this interface, wind energy is converted into surface waves (Wang and Huang, 2004), kinetic energy is dissipated (Terray *et al.*,

Box 4.1 Dust-to-dust: wind-blown material

Michael Garstang
[Retired, Department of Environmental Sciences, University of Virginia], Charlottesville, Virginia, USA
Amber J. Soja
NASA National Institute of Aerospace, Hampton, Virginia, USA

Upwards of a billion tons of terrigenous material is lifted into the atmosphere and distributed around the globe every year (Fig. B4.1.1). Most of this material originates over the global deserts. Almost half of the above amount leaves the west coast of North Africa, originating over the Sahara and Sahel. Vast quantities of dust pour off the Gobi Desert into the Pacific Ocean. Saharan dust has been found over the Near East, in the Alps, in Finland, and in Scotland and pervasively in the Caribbean, the Amazon Basin, the coastal southern United States, and Mexico.

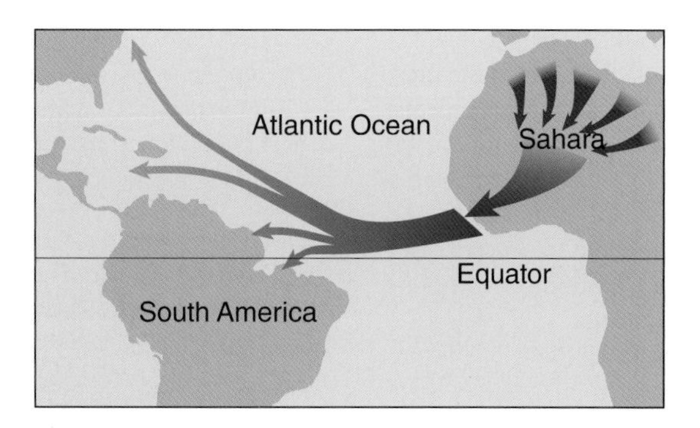

Fig. B4.1.1 Aerial transport of dust from continent to continent and ocean. Courtesy of M. Garstang.

Sources of the airborne particles or aerosols are numerous. A large fraction is soil dust, but organic matter from dry vegetation, products from biomass burning, and input from human activity, including cooking fires, agriculture, and industry, all contribute to the total airborne load. Particle sizes range from 1–200 mm for smoke from intense fires, to submicron sizes, which include pollen and spores. Silicon is often the dominant element, with abundant iron originating from the iron-rich red lateritic (oxisol) soils of the subtropical deserts. As many as 22 elements have been found in African dust captured in the middle of the Amazon rainforest. Phosphates, potassium, and nitrogen are present in amounts which when deposited reach kilograms per hectare per year.

The large particles and the greater part of the mass of aerosols are deposited in the coastal waters of the source region, and constitute a significant input of nutrients to these waters. Primary productivity in the coastal waters is dependent upon a suite of abundant (e.g., iron, carbon, nitrogen, and phosphorus) and trace nutrients (e.g., silicon, copper, and zinc), all present in the airborne load. Iron limits productivity in some oceanic regions and is abundant in the airborne soil dust. Evidence from the Pleistocene period suggests that increases of export production, which may have contributed to lower CO_2 concentration in the glacial atmosphere, were accompanied by a greater supply of iron from wind-blown aerosols. While upwelling is conventionally believed to provide the nutrients required for phytoplankton production, aeolian events may play a far greater role than has been previously recognized.

Patterns of phytoplankton blooms detected from the Coastal Zone Color Scanner carried on the Nimbus 7 satellite show expansion and contraction of blooms off the west coast of Africa which are spatially and temporally synchronous with dust outbreaks. As dust is transported by the atmosphere away from source regions, the total airborne load and deposition rates decrease. Both, however, remain significant. Annual transport through a hypothetical wall erected from the ocean surface to 4 km altitude and extending from 10 to 25° N latitude along 60° W longitude is estimated at 25–37 MT per year. This dust load enters the Caribbean Sea and continues onward across the Florida Peninsula and into the Gulf of Mexico. Similar transports extend northwestwards over the Sargasso Sea and southwestwards into the Amazon Basin. Estimates based upon calculations made over land in the nearly closed system of the Okavango Delta in northern Botswana show that dust from the atmosphere can on an annual basis contribute between 6 and 60% of the nutrient load. Previous studies had assumed all of the nutrients in this delta to be waterborne. Similar conditions may exist over coastal waters remote from the major sources of airborne material. Waters such as the Sargasso Sea, known to be largely oligotrophic, may receive significant nutrient supplies from atmospheric deposition. The same transport and deposition processes that deliver airborne nutrients to coastal waters can import trace elements from industry and agriculture (e.g., pesticides, fungicides, other organics). Such deposition has been suggested as a possible cause of coral die-off in the Caribbean. Advances in remote sensing from satellites now allow daily monitoring of dust over the oceans as well as detailed descriptions of long-range transport.

Sources: Garstang *et al.* (1998); Kaufman *et al.* (2005); Liu *et al.* (2008); Mahowald *et al.* (2005); Prospero (1999); Swap *et al.* (1992)

1996), and turbulent flux and breaking waves contribute 85% of the atmospheric water vapor and accentuate releases of important gases. Chemical processes involving marine aerosols may significantly impact tropospheric oxidation processes, sulfur cycling, radiation balance, climate, and ocean surface fertilization (Keene *et al.*, 1998). Wave dynamics, gas transfer, nutrient and pollutant mixing, and plankton photosynthetic efficiency are major factors in air-sea exchanges.

A host of important gases that influence climate are exchanged, absorbed, and emitted at this interface. The ocean is both a sink and source for trace gases, and surface waters are supersaturated with gases, especially sulfur gases, e.g., hydrogen sulfide and dimethyl sulfide (DMS; Section 4.4.2). Any volatile organic compounds such as iodine and nitrates can react with chloride atoms released from sea-salt aerosols (Keene and Jacob, 1996; McFiggans *et al.*, 2002). Iodine in high concentrations in the troposphere contributes to ozone destruction (Carpenter, 2003), and air-sea fluxes in the coastal ocean may be higher than in the open ocean (Carpenter *et al.*, 2009). Atmospheric aerosols in significant amounts are important in regulating the composition of atmospheric greenhouse gases that contribute to ocean warming and acidification when mixed into the ocean interior.

4.6 THE DYNAMIC COASTAL REALM

The land/sea, benthic/pelagic, and air/surface interact most intensely in the broad domain of the coastal realm (Frontispiece). This region encompasses the approximately 200 m land elevation to the approximately 200 m marine depth (Pernetta and Milliman, 1995; Crossland *et al.*, 2005; Ducklow and McCallister, 2004; Longhurst, 1998a) in accord with geologic rises and falls of sea level. Ketchum (1972) defined this region as a "coastal zone," i.e., a "band of variable width" that functionally "is the broad interface between land and sea where production, consumption, and exchange processes occur at high rates of intensity." This region exhibits features and functions very different from land or ocean alone, qualifying it as a third major subdivision of Earth—a "realm," used here to signify the highest biogeographic category (Fig. 4.7; Pielou, 1979), ranking it equally with land and open-ocean realms. This geochemically and biologically active region plays a dominant role in biogeochemical cycles (Gattuso *et al.*, 1998), production, biodiversity, and services disproportionate to its global size (Table 4.2), with a resiliency that persists through seasonal and annual cycles, perturbations, and alterations of sea-level change. Human activities and conservation efforts need to account for the coastal realm's unique set of conditions.

4.6.1 Sculpting coastal land and seascapes

Interacting forces on different time and space scales sculpt the coastal realm into hierarchical patterns. Geologic and climatic events shape and move, adding much complexity to a strongly inhomogeneous and dynamic system. Understanding large-scale coastal behavior requires understanding short-term events and long-term processes (Schwarzer *et al.*, 2003), where

Table 4.2 Coastal realm attributes of global significance. Based on data from Holligan and Reiners (1992); Pernetta and Milliman (1995); Gattuso *et al.* (1998); Beck *et al.* (2003); Borges and Gypens 2010.

Significance	Attribute
Spatial coverage	18–20% of Earth surface; 8% of ocean surface; <0.5% of ocean volume
Biogeochemistry process	~50% of global denitrification; 80% of global organic matter burial; 90% of global sedimentary mineralization; 75–90% of global sink of suspended river load, associated elements, and pollutants; ~20% of surface pelagic oceanic calcium carbonate stock; >50% of present-day global carbonate deposition
Production	New primary production rates significantly higher than open oceans; ~1/4 global primary production supply; ~14% of global ocean production; ~90% of world fish catch; Major nursery for fish and shellfish production
Human services	40% of the world population; 2/3 of world's major (>1.6 million people) cities; Concentrated global trade; Major oil and mineral reserves; Major sport fisheries supported; Major military centers for national defense

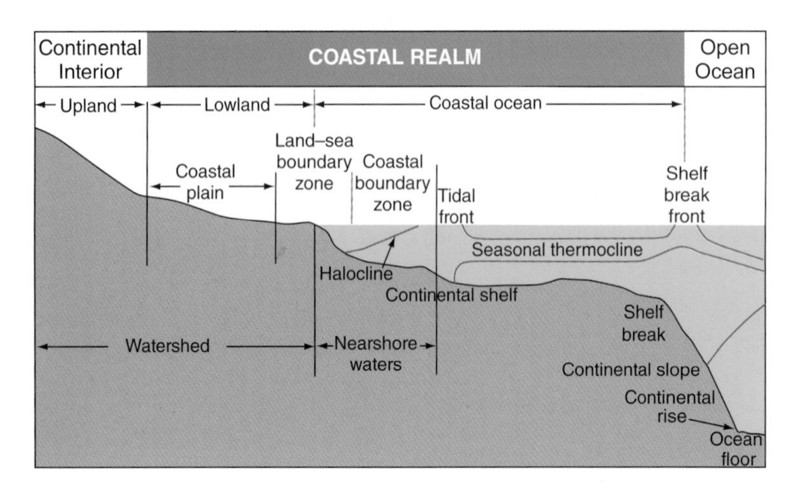

Fig. 4.7 Diagrammatic representation of coastal realm structure. Major boundaries occur in close proximity (land-sea, air-sea, benthos-water column) and strongly influence biotic distributions. Based on data from Holligan and Reiners (1992); Pernetta and Milliman (1995).

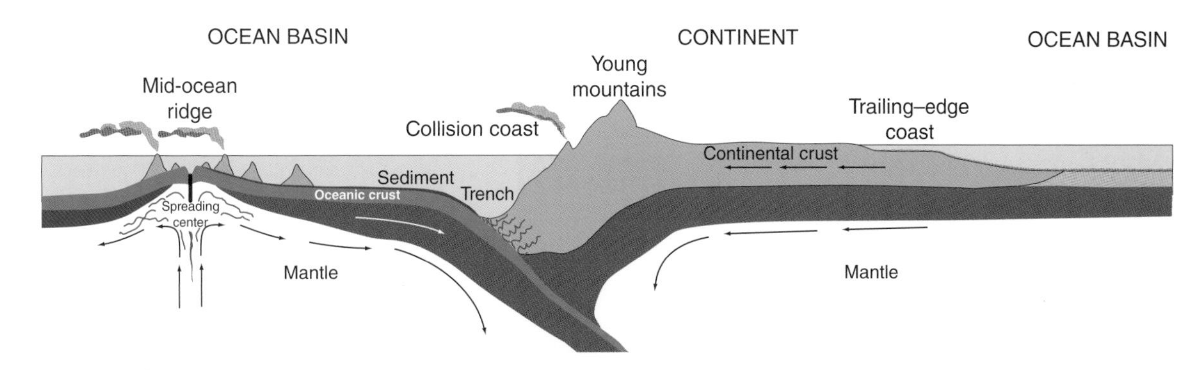

Fig. 4.8 Coasts and tectonic plates. A cross-section across South America and extending into the Atlantic illustrates a trailing-edge east coast. A spreading center at the eastern Pacific Rise through the Peru-Chile trench creates conditions for a collision coast. From Inman DL, Nordstrom CE (1971) Tectonic and morphologic classification of coasts. *Journal of Geology* **79**, 1–21 © 1971 The University of Chicago. With permission from University Chicago Press.

Fig. 4.9 Mountains and sea interact to create clouds and coastal weather phenomena resulting from land/water contrasts. Thermally driven effects result from contrasts in heating, modulated by contrasts in surface friction between land and water, and interactions between larger-scale meteorological systems (hurricanes, typhoons) that pass over water and coastlines. These produce distinct smaller-scale systems, and orographic effects driven by steep coastal terrain induces strong winds and longshore flows. This photograph from Betty's Bay, South Africa, shows cloud formation due to the orographic effect of onshore winds rising against the mountain. Photograph © Ray & McCormick-Ray.

antecedent conditions over geologic time have established large-scale coastal conditions, further modified by short-term oceanographic, climatic, and watershed events.

Plate tectonics create a great variety of coastal types important to biodiversity and conservation. Collision coasts, characterized by active volcanic and earthquake zones, occur where continental plates collide and one of the plates is subducted. The terrestrial relief is relatively straight, with mountain ranges and a steep narrow continental shelf (Fig. 4.8). On trailing-edge coasts, plates spread apart and the continental shelf is generally wide, giving rise to barrier islands and large regionally variable estuaries and lagoons. Neo-trailing-edge coasts are recently formed coasts that exhibit rifting (Red Sea, Sea of Cortez). Afro-trailing-edge coasts occur on both coasts of southern Africa (Atlantic and Indian Ocean), and Amero-trailing-edge coasts occur on North and South America east coasts. Coasts of marginal seas have typically curved coastlines with back-arc basins (seas within island arcs; e.g., Aleutians and Kurile Islands). Such coasts are frequently modified by large rivers and deltas, provide more protection from the open ocean than other coasts, and are most biologically diverse.

Coastal conditions create climates important to coastal diversity. Elevated coastal surfaces and uneven landscapes complicate meteorological transformations, generating winds that create sea-surface waves, atmospherically induced coastal-ocean currents, upwelling, fog, haze, stratus clouds, and orographic effects (Fig. 4.9). Changes in heat flux can introduce instabilities, cloud cover, and unique coastal phenomena (land and sea breezes, thunderstorms, atmospheric and oceanic fronts). The convergence of marine air over coastlines can also result in strong convection, heavy precipitation, and runoff that together increase erosion and pollution loads and disperse sediments. Coastal meteorology is thus unique due to exchanges between the atmosphere and heterogeneous coastal surfaces, complicated further by interactions with the

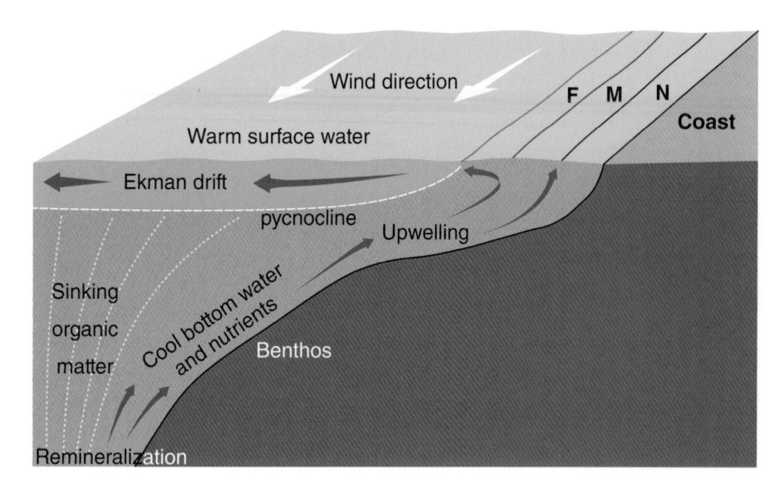

Fig. 4.10 Coastal upwelling. An oceanographic process usually driven by longshore winds that push warmer, nutrient-poor open surface waterawayfrom the coast, drawing nutrient-rich water to sunlit surface waters, enhancing photosynthesis. Water mass types: [N] nearshore waters of high primary productivity dominated by small cell-size phytoplankton and few zooplankton organisms and fishes; [M] mid-zone waters with abundant large phytoplankton, large zooplankton, very abundant small filter-feeding planktivorous fishes, and numerous sea birds; [F] frontal zone waters contain few plankton and planktivorous fishes, but plentiful carnivores (fishes, sea birds, marine mammals). Based on data from Bakun (1996); Gross and Gross (1996); Mann and Lazier (1991).

oceans (NRC, 1992). And coastal climate at global scales dramatically affects CO_2 exchange, photosynthesis, and biological productivity, played out at smaller scales through rapid and significant atmospheric events such as storms, wind events, and interactions caused by pressure gradients and topography.

Winds also interact with geomorphological and ocean processes to shape coastal landforms, sea-ice fields, coastal waters, and intertidal areas. Maximum wind speeds come from hurricane activity in areas where sea-surface temperatures usually exceed 26°C and with intense pressure gradients, causing massive flooding and high waves that wash over extensive land/sea areas (Trenberth, 2005). Storms and tectonic events deliver long-wavelength ocean surges (tsunamis) that cause severe beach, cliff, and dune erosion, much flooding, and human mortality. Winds also interact with ocean Coriolis force and Ekman transport (Section 4.2.3) to upwell nutrient-rich bottom water into sunlit surface water above thermal stratification (Fig. 4.10), stimulating high phytoplankton productivity, intense biological activity, and high fisheries production that often occurs close to shore (Mann and Lazier, 1991) and mostly on the west coasts of continents, e.g., U.S., Peru, and southwest Africa.

Sea-level changes sculpt the coasts, being intimately associated with global climate cycles of cooling, warming, and geologic adjustments on regional scales. Warming the ocean expands its volume and increases storm activity, storm surges, and strong waves that move sediment and erode coasts. Ocean thermal expansion raises sea levels, which involves a combination of global sea-level change (eustasy) and gravitational equilibrium of Earth's crust (isostasy). Through isostasy, Earth's crust maintains equilibrium relative to a fixed point where shelf subsidence is compensated by an uplift of coastal land, typically on rugged coastlines with narrow shelves. These vertical and horizontal displacements facilitate new coastal geometries and topographies. For example, ice sheets can influence isostatic changes; the weight of Antarctic ice sheets has depressed the entire continent hundreds of meters, and in the present interglacial period, sea level has varied across the coastal plain from a maximum upper height at 50–100 m land elevation to a lower limit of approximately 100–200 m at the outer edge of the continental shelf (Fig. 4.11).

New technological advancements in satellite altimetry and geodetic leveling have increased the accuracy of global sea-level measurements (Nicholls and Cazenave, 2010), providing new evidence of higher 20th century levels than in the 19th century. However, the rise and fall of sea level due to isostasy and eustasy varies with location and may not always be apparent. If the land surface subsides as ocean volume increases, the rate of submergence will be greater than induced by changes in ocean volume alone. Furthermore, sediment flux to coasts from increased human activities affects coastal-ocean, flood-plain, and delta-plain functions (Syvitski et *al.*, 2005; Syvitski and Kettner, 2012). Massive withdrawal of water, gas, and oil can cause land to subside, and massive amounts of sediment from sediment-laden rivers deliver added weight onto the continental shelf (Milliman and Haq, 1996).

4.6.2 Land-ocean interactions

Watersheds sculpt the coastal realm into many permutations of land and seascape patterns. Watersheds can be simple, compound, or complex, depending on whether their drainage system originates near shore, or upland and across a coastal plain, in either case to empty into coastal waters through five land/sea drainage compartments (Fig. 4.12). Watersheds interact with coastal receiving-basins in a freshwater gradient and delivery of sediment. Seaward, in water depths generally less than 20 m (the subtidal shoreface entrainment volume) and with much mixing, surface water interacts with waves and tidal mixing to entrain a volume of oscillating water with sediment. Seaward in shelf depths of 20–50 m, water temperatures are generally homogeneous with depth in winter but may be stratified (often nearly two-layered) in summer (Mooers, 1976). At mid-shelf depths of 50–150 m, the water column is generally vertically stratified, with frequent exposures to coastal jets and surface fronts. At the outer shelf-break in 150–250 m depths, surface and bottom fronts are common, and shelf and oceanic waters interact most intensively. Farther

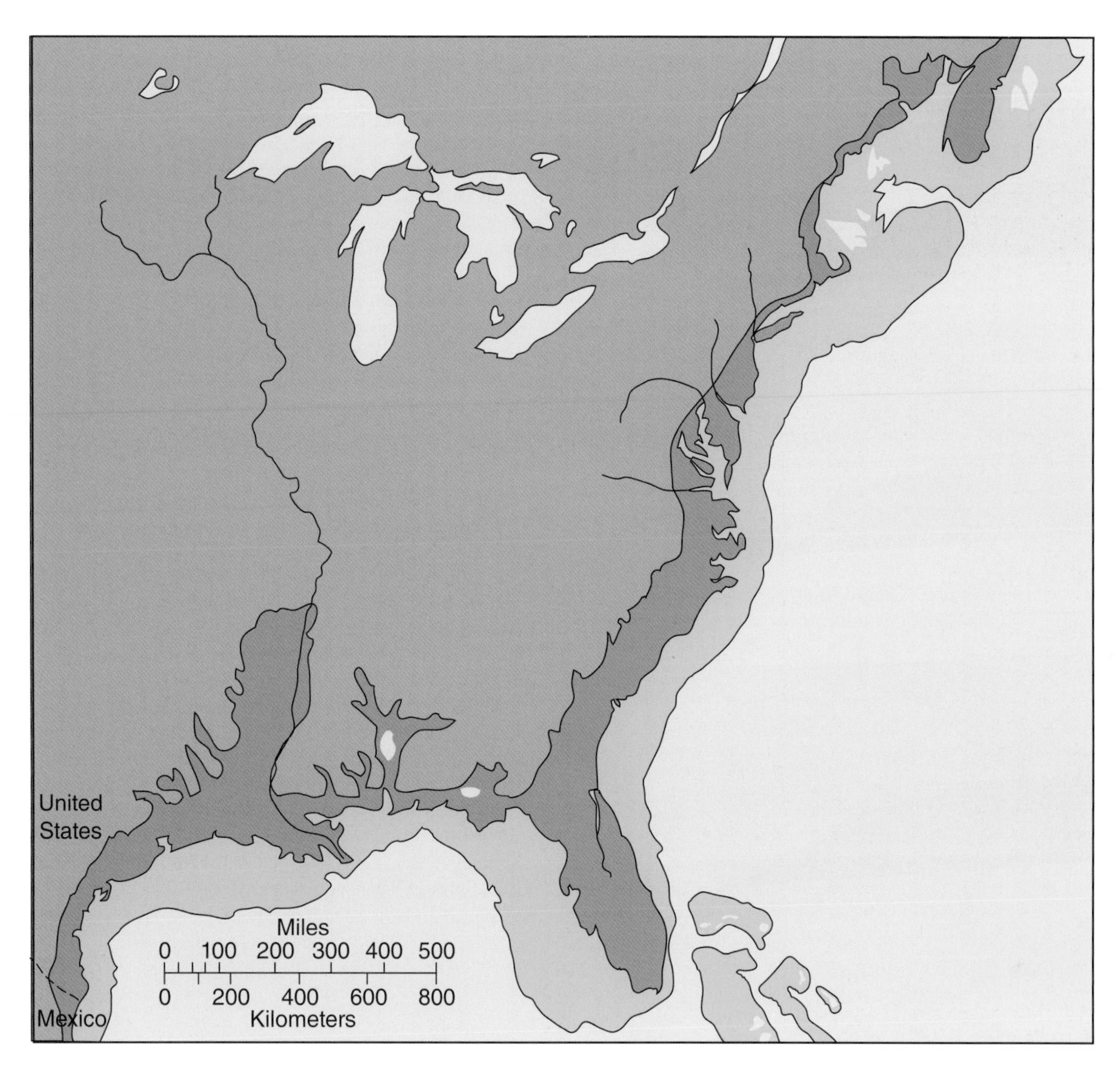

Fig. 4.11 Sea-level rise and fall helps define the area of the coastal realm. Brown color represents historic maximum sea-level rise as uplands today; dark green defines present-day coastal plains; aqua blue represents the extent of land during minimum sea level that occurred during the latest Pleistocene ice age. Based on data from Emery (1969).

offshore over the continental slope, oceanic water masses and boundary currents usually dominate.

4.6.3 Geomorphologic patterns

Shorelines derived from river-driven forces may protrude into coastal waters as deltas (Fig. 4.13) and headlands, or recede into protected embayments exposed to freshwater-tidal interactions as in estuaries or bays. Others forces create steep, rugged, high-energy shores and rocky outcrops exposed to intense sea, sun, and air, or into low, flat, lower-energy beaches. All are exposed in varying degrees to hydrological and meteorological forces that pulse and flux across their boundaries.

Offshore, deltas and headlands strongly influence coastal circulation, hydrodynamics, and sedimentation patterns. Deltas provide habitat for shellfish, birds, and juvenile fish, and also support extensive agriculture, fisheries, and, occasionally, rich deposits of coal, oil, and natural gas. The Mississippi River Delta, for example, receives large amounts of sediment, nutrients, and debris every year from its extensive, complex watershed that covers 31 states and more than 3.2 million km^2 of the United States. This drainage, highly modified by human activities, moves terrestrial products and sediment into shallow estuarine areas, forming a delta that protrudes into shallow Gulf of Mexico waters. Intensive channeling and diking of the watershed and hurricane activity have changed this delivery system, contributing to delta erosion and extensive loss of habitat for juvenile fishes, crabs, oysters, and waterbirds.

Embayed coasts broadly include estuaries, bays, and lagoons and support extraordinary production and ecological complexity. Estuaries are geologically ephemeral and depend on

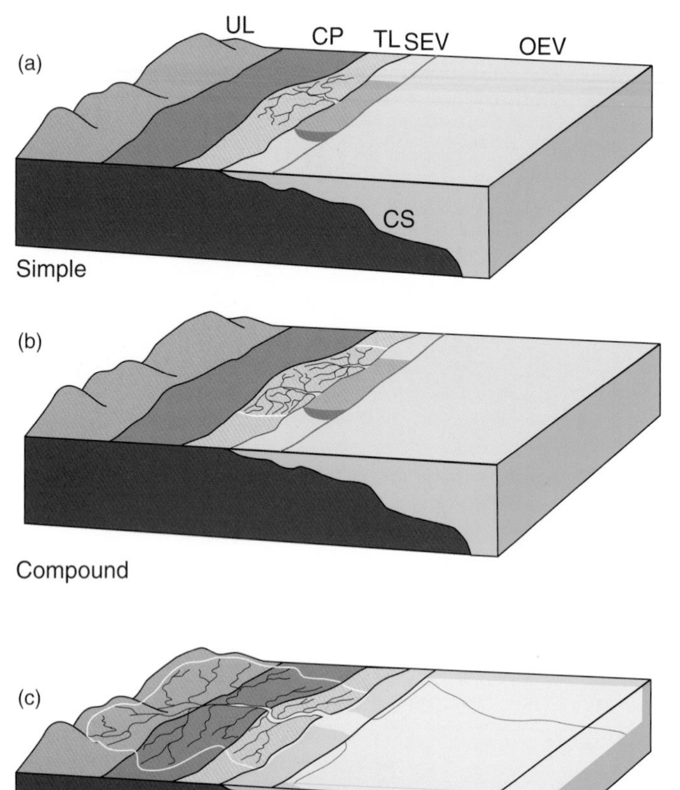

Fig. 4.12 Watershed types defined broadly by comparison of their reach across five divisions: [UL] uplands; [CP] coastal plain; [TL] tidelands; [SEV] shoreface entrainment volume; [OEV] offshore entrainment volume; [CS] indicates continental shelf. (a) Simple systems: terrestrial-marine exchanges through only one watershed unit; e.g., tidelands [TL] that drain into the shoreface entrainment volume [SEV]. (b) Compound systems reach across more than one watershed, e.g., multiple streams or estuaries drain into a common shoreface volume within a longshore reach of coast. (c) Complex systems involve two or more units; e.g., a large drainage that includes all three terrestrial subdivisions [UL, CP, TL] has sufficient flow to bypass the shoreface volume to empty directly into the offshore entrainment volume [OEV]. Many other permutations of this five-part scheme are possible. From Ray GC, Hayden BP (1992) Coastal zone ecotones. In *Landscape Boundaries: Consequences for Biotic Diversity and Ecological Flows* (ed. di Castri F, Hansen AJ). Springer-Verlag, New York, pp. 403–420. With kind permission from Springer Science+Business Media B.V.

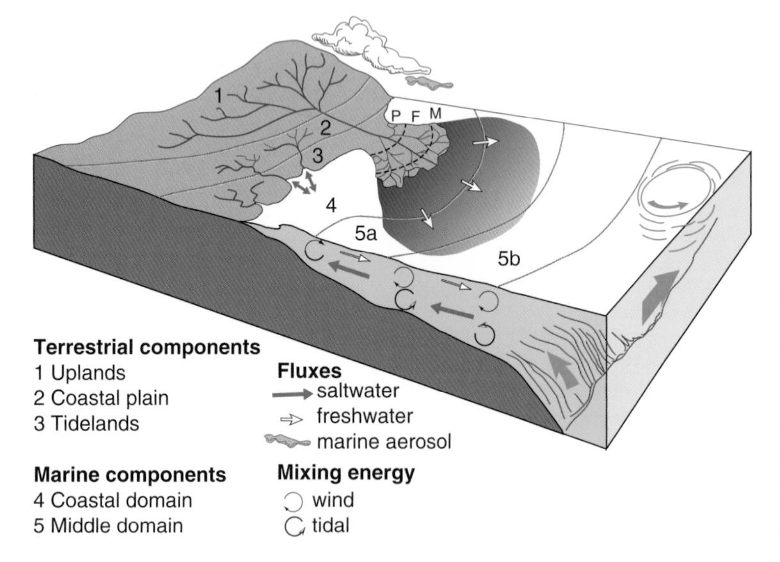

Fig. 4.13 Coastal watershed subdivisions (1–3; Fig. 4.12) interact with the offshore entrainment volume (5a,b). Drainage from subdivision 1 forms a delta that is influenced by circulation patterns of tides and winds, and consists of a delta plain [P], delta front [F], marine prodelta [M], with a freshwater plume that extends offshore, indicated by the dark-blue shaded area. Ray and McCormick-Ray (1989). Coastal and marine biosphere reserves. In: *Proceedings of the Symposium on Biosphere Reserves*. The 4th World Wilderness (eds Gregg, Jr, WP, Krugman SL, Wood Jr. JD) U.S. Dept Interior, Atlanta, Georgia, pp. 68–78.

freshwater input, being semi-enclosed water bodies that were built by the latest major episodes of sea-level fall and rise (Ch. 6). They have free connections to the open ocean in which varying amounts of seawater are diluted by freshwater. Rivers deliver dissolved inorganic chemicals of positively charged freshwater ions, especially cations of calcium, magnesium, sodium, ammonium, and hydrogen, that interact with ocean electrolytes of sodium, potassium, and chloride salts. This interaction forms distinct transitional regions between fresh and saline water that vary with location and season. In summer, warm, buoyant freshwater flows seaward over incoming, relatively cold seawater, between which is an internal

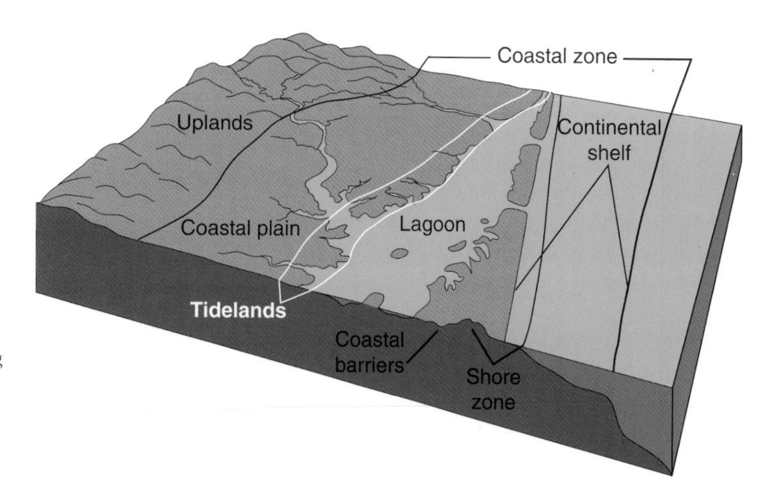

Fig. 4.14 Lagoons interface with land and sea and are protected by a coastal barrier (land, coral reef), with variable influences from watershed drainages. From Figure 1: "Coastal barriers within the broader context of the coastal zone," G. Carleton Ray and William P. Gregg Jr., "Establishing Biosphere Reserves for Coastal Barrier Ecosystems," in *BioScience*, vol. 41, no. 5, May 1991. © 1991 by the American Institute of Biological Sciences. Published by the University of California Press.

boundary. Interactions with ocean tides create estuarine gradients, from fresh and mildly brackish to fully saline zones. Salt content (salinity based on ppt, parts per thousand, or PSU, Practical Salinity Unit) is a useful indicator for subdividing estuarine zones: limnetic (freshwater 0.5 ppt; oligohaline 0.5–5.0 ppt; mesohaline 5–18 ppt; polyhaline 18–30 ppt; euhaline >30 ppt). Salinity demarcates living conditions based on osmotic regulation that determines distributions of higher taxa (Telesh and Khlebovich, 2010), separating biota into spatially different types of communities that are among the most productive on Earth.

Other embayed coasts have less freshwater influence. Bays and gulfs can be large enough to be considered marginal seas: Bengal Bay (Indian Ocean) and Biscay Bay (eastern Atlantic) are similar in scale to the gulfs of California (eastern Pacific) and Guinea (eastern Atlantic). Sounds combine bay and estuary features: e.g., Long Island Sound (western Atlantic). And lagoons (Fig. 4.14) are important and widely distributed coastal features, often associated with estuaries, but usually have lesser freshwater inputs, smaller tides, and react differently to hydrological and meteorological forces. Also commonly associated with estuaries and lagoons are coastal marshes that maintain and improve water quality by trapping sediment and reducing silt delivered by rivers, otherwise destined to fill channels and settle on shellfish and seagrass beds, coral reefs, and tidal flats. At the low end of a wind-and-wave energy spectrum, are tidal mud and sand flats; they occur on coasts with low-energy tides and unconsolidated shores, especially along shallow, estuarine-channel banks where sediment sorting, tidal movements, and benthic chemistry create horizontal and vertical zonation patterns. Clam beds flourish on these biologically rich habitats, which also host mixtures of algae and vascular plants in or anchored to shallow sand and mud bottoms, and on which small organisms may grow abundantly on fronds and leaves. Some tidal flats serve as sources of dissolved salts and nutrients and as local heat reservoirs.

Open coasts are dominated by waves of intense energy, generated mostly by tides and winds, but also by earthquakes, fallen objects, mudslides, and other disturbances. Oscillating waves encountering the bottom become asymmetrical and break into a surf zone of high turbulence and powerful longshore currents. Tremendous force is applied to shore and sediment interfaces by waves 12 m-high, which may reach velocities of approximately 16 ms^{-1} and accelerations reaching 1000 ms^{-2}, or about 100 times the acceleration of gravity. A wave 3 m-high at its crest line can transmit 100 kW of energy per meter, moving sediments in water depths up to about 20 m depth (Inman and Brush, 1973). Coastal and topographic structures such as coral and oyster reefs are important structures that modify wave energy and behavior.

Open coasts are among the most dynamic of coastal environments, often being formed of unconsolidated sediment. High-energy beaches represent a net balance between wave energy and sediment supply, a balance that can be changed by storms, hurricanes, tsunamis, earthquakes, and human activities. Waves deliver sand that becomes deposited behind beaches into sand dunes, which provide habitat for unique biological communities adapted to cope with poor soils, drought, heat, and desiccation. Dunes, stabilized by grasses and other vegetation, buffer shores against erosive waves and wind and help regulate the water table. Dunes provide shelter for many species, with adjacent beaches filtering large volumes of seawater, recycling nutrients, and supporting coastal fisheries. A single beach can contain several hundred invertebrate species, each uniquely adapted to this highly dynamic, three-dimensional environment. Beaches attract a diversity of reproducing organisms, including nesting sea birds, sea turtles, and fishes (Schlacher *et al.*, 2007; Defeo *et al.*, 2009; Dugan and Hubbard, 2010). Some oystercatchers (*Haematopus* spp.) and the endangered North American piping plover (*Charadrius melodus*) nest only on exposed beaches.

Rocky coasts are rugged, high-energy environments, being exposed to the full energy of currents, waves, and wind, but support many sessile invertebrates and macroalgae able to withstand those forces. The sea palm (an alga, *Postelsia palmaeformis*) attaches to rocky cliffs and reefs and thrives under strong wave action in the temperate eastern Pacific. Intertidal rocks contain unique fauna exposed to breaking waves, high-velocity water, and salt spray projectiles. In temperate latitudes, rocky shores display distinct biotic zonation patterns: an upper zone of littorinid snails and lichens, a middle zone of barnacles and mussels, a lower zone of algae, and a subtidal zone where kelp forests (i.e., a diversity of macroalgae and flourishing kelp gardens) can dominate. The high physical disturbances in subtidal zones decrease with depth. Research on zonation, competition, and predation on rocky coasts is a scientific pillar of community ecology.

Islands provide habitat for a diversity of unique species. Continental islands (Section 4.3.1) share a relatively low proportion of terrestrial biodiversity, but often support endemic species. Island size is presumed to be an important factor in species diversity; some data suggest that as island area increases so do numbers of species. The "island biogeography" hypothesis has been developed for terrestrial species, but is not well understood for coastal-marine environments. Rather, habitat variety surrounding islands, as well as species invasions initiated by humans, may have greater influence on species richness, being independent of island size.

Sea ice is a prominent feature of both Arctic and Antarctic polar regions. Many portions are seasonally ephemeral, more so as global climates warm (Ch. 7). Sea-ice area increases from summer to winter by about 35% in the Arctic and 85% in the Southern Ocean. Lower seasonal Arctic variability in sea-ice cover is due to its being surrounded by land, unlike the Southern Ocean that surrounds the Antarctic continent, and thus is lower latitude. Sea ice forms an important habitat in both regions; diatoms accumulate on its underside where fish, crustaceans, and other species find refuge. Sea ice also forms substrate for abundant marine birds and mammals (penguins, seals, walruses) to rest and breed (Ch. 7). Although sea ice hosts much biota, it can be inimical to benthic and shore life when currents and wind move it shoreward to scour the benthos deep into subtidal areas, inhibiting algae growth and sessile organisms. Abutment of sea ice on shores, however, is advantageous to some forms of life such as Arctic foxes (*Alopex lagopus*) that venture onto sea ice, but do not swim. When sea ice is relatively contiguous with shores, it expands foraging opportunities far onto sea ice for both foxes and polar bears (*Ursus maritimus*). Polar bears hunt for seals on sea ice and may swim and walk hundreds of miles on ice in pursuit of prey; the foxes follow, seeking scraps of polar bear kills.

Superimposed on coastal realm seascape structures are familiar biogenic structures: marshlands, seagrass and macroalgal beds, and shellfish and coral-reef habitats, among others. These sessile, shallow-water benthic living seascapes generally depend on currents to deliver food and nutrients and to disperse wastes and reproductive products. Through a sum of interactions over time, geologic and biogenic structures influence their hydrological surroundings by mobilizing or depositing soft sediment, increasing spatial heterogeneity, providing and amplifying options for colonization, and increasing biodiversity. Although biogenic seascapes occupy only a small fraction of the total area of marine and estuarine systems, their importance is out of scale with their size. For example, autotrophic seascapes (wetlands, marshes, mangroves, algal flats, seagrass beds), which form between low and high water levels, modify alternating periods of tidal action. Those near shore depend on hydrologic and sedimentary dynamics, where minimal wave action consolidates soft sediment clays and silts and accumulates fine to very fine sand (Perillo *et al.*, 2009). Corals, mollusks, coralline algae, sapporellid and vermetid worms, and other "biological engineers" (Ch. 5) often cement to one another to form extensive reef systems (e.g., oysters, Ch. 6). These living seascapes directly or indirectly control nutrients and the availability of sediment, modulate current flows, and modify, maintain, or create habitat for other species. Biogenic seascapes thus add dimensional complexity as they concentrate intensive biotic activity into high production for internal use and for export.

4.7 THE COASTAL REALM: AN ECOSYSTEM OF GLOBAL IMPORTANCE

Ecosystem ecologists concerned with land and seascape transformations focus on large-scale biophysical processes, accumulation of matter, and energy flow. A focus on interactions among linked processes at multiple scales draws attention to the coastal realm as a high-level, complex ecosystem of global significance, a system filled with surprises, where species are most often hidden from easy view.

Dynamic forces in the coastal realm operate on many scales to dissipate energy, create structures, and sustain ecosystems. Physical environments establish the context in which organisms capture and transform kinetic and radiant energies into ecological efficiencies. Biotic capacity draws on inherited blueprints acquired over evolutionary time that give rise to high production, persistence, and resiliency. Seascape structures that persist under conditions of intense dynamic forces exhibit high ecological biomass, productivity, and diversity, thereby providing entry points into marine ecosystems and spatial units for conservation. Biophysical seascapes and their scale-related interconnections (processes, nutrient networks, natural history, and metapopulation linkages) establish a spatial context for measuring biomass, chemical fluxes, and qualities of resiliency, stability, persistence, and change. While temporal variability may modify seascape patterns, many characteristic biotic communities maintain their identities. But will coastal functions persist at characteristic levels if biota or seascape patterns change, or will costly human interventions be required?

4.7.1 Attributes of the coastal realm

Coastal realm attributes make apparent that this realm is best understood as a system. Through linked environments, species, and processes, this realm supports distinctive biota, persistent biophysical structures, exceptionally high primary productivity, and exceptional secondary animal biomass production. It is also the major source of the world's commercially valued shellfish and fisheries, and a carbonate factory for biologically diverse reefs and tropical beaches. Its massive capacity to maintain water quality and high production is maintained through a series of energy gradients from headwaters to the ocean. Through periodic and aperiodic pulses and fluxes, biological opportunities are enhanced, energy needs are subsidized, reproductive success is facilitated, and coastal functions are restored under conditions of change. Two other major ecological attributes stand out.

4.7.1.1 Dimensional complexity

Coastal realm environments require a diversity of biological and ecological mechanisms to efficiently capture energy, transform nutrients, and adjust to change. A fundamental, self-evident property of all living things is geometric (exponential) growth that operates under a specific set of limited resources; i.e., all populations of living organisms grow geometrically when unaffected by their environments (Berryman, 2003). Coastal realm structures add ecological dimensions that increase opportunities for growth, population expansion, and survival. Frontal systems, eddies, and benthic topography are seascape structures that are quasi-stationary, seasonally persistent, or prominent year-around features on which migratory species and major fisheries depend (Belkin *et al.*, 2009). Living coral (Ch. 8) and oyster reefs (Ch. 6) create hard bottom structures that buffer energies while providing dependable shelter and habitat for a diversity of species to grow, feed, and reproduce. Such "hot spots" of high biological activity add dimensional complexity and expand opportunities for feeding and shelter, resulting in high production in an otherwise less-productive environment. Reefs also protect lagoons for seagrass beds to flourish and to support high biomass of fishes and invertebrates, as well as endangered dugongs, manatees, and green turtles. Reefs, marshes, and seagrass beds are linked at larger scales through feeding networks and reproductive pathways (Ch. 8). Nevertheless, reef diversity and complexity carry high maintenance costs, such that little energy remains available for export, making these environments particularly vulnerable to disturbance, e.g., fishing (Fig. 4.15).

In temperate regions, macroalgal communities such as kelp beds are extremely productive. The largest and among the most productive alga is the giant kelp (*Macrocystis pyrifera*), which can reach more than 60 m in length and grows along cool-temperate rocky coasts. Water motion plays a key role in top-down, consumer-driven food webs (Gerard, 1984; Halpern *et al.*, 2006). Kelp support very large standing stocks of associated species (Heck *et al.*, 1995), including marine mammals, fishes, crabs, sea urchins, mollusks, other algae, and epibiota that collectively make them among the most diverse and productive ecosystems of the world (Mann, 1973; Steneck *et al.*, 2002; Gattuso *et al.*, 1998).

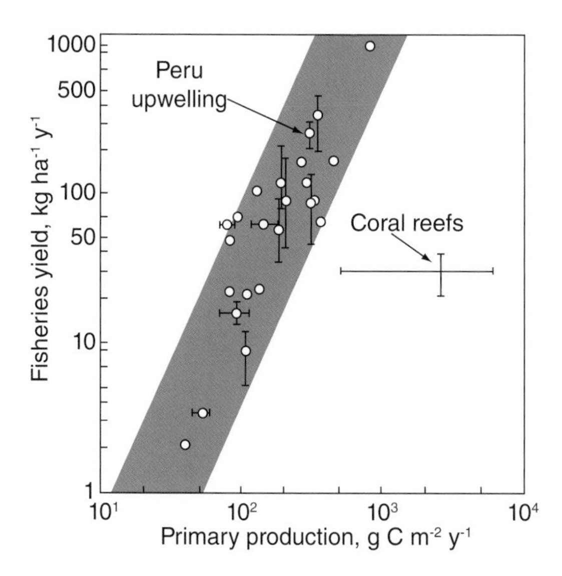

Fig. 4.15 Comparison of fisheries yield and gross primary production for Peru upwelling vs.coral reefs. Coral reefs are high primary producers but energy production does not support high fisheries yield. Their low net productivity makes them especially vulnerable to fishing. From Birkeland C, ed. (1997). *Life and death of coral reefs*. Chapman and Hall, New York. © Chapman and Hall 1997. With kind permission from Springer Science+Business Media B.V.

Coastal marshes and tropical mangroves add other dimensions. These seascapes filter, recycle, and modulate coastal energies, being interconnected by hydrology and species at particular stages of their life cycles (Ch. 5). Collectively and individually, these structures support juveniles of commercial fishes and sustain large populations of migratory waterfowl. Their products (abiotic materials, larvae, etc.) are hydrodynamically delivered to offshore habitats, a scale-dependent vector that links topographic heterogeneity to community structure (Guichard and Bourget, 1998).

4.7.1.2 Spatial heterogeneity

Spatial heterogeneity among coastal types, as between high-energy and low-energy coastal systems, creates major differences in biotic composition and in processing and exchanging materials. Scales of coastal heterogeneity and hydrologic interactions, driven by short-term fluctuations and events (e.g., storms, precipitation), produce many interacting variables that influence biological richness and ecological performance. Spatially varied coastal realm structures have nested within them smaller-scale seascape patches of high to low production, depending on their different species assemblages. Through life-history options to disperse, colonize, feed, and reproduce, many organisms move between land and sea, benthic and pelagic environments, and rarely stay within any one habitat throughout their lives. Short-term, spatially variable abundance, diversity, and function add robust dimensions to ecological processes that may be lost when only long-term averages across coastal types are measured.

4.7.2 Ecosystem properties

The coastal realm, when studied as a whole system, displays a distinctive character not apparent from its individual components alone. When observed as an interconnected whole, autotrophic production and habitat heterogeneity yield greater relevance to fisheries, algal blooms, metapopulations, and transient or migrating species moving among compartments. Ecosystem attributes result from properties of the ecosystem—energy, biochemistry, and evolution—harnessed, transformed, and modified through networks of interactions that connect organisms with the environment as a functional whole.

4.7.2.1 Energy

Organisms capture physical energy from their environment, package it, transform it, and cycle it. Transformation of one kind of energy into another involves thermodynamic laws, natural history, and food webs (Ch. 5). How energy is captured, stored, transformed, and used is a property of the ecosystem.

Thermodynamic laws reveal energy in a closed system. The first thermodynamic law generally states that energy may be transformed, but is never created or destroyed (conservation of energy). Heat and work are two forms of energy, and when work is performed, a significant portion of energy is lost as heat. The second law involves direction, which is more relevant for ecosystems. It generally states that for isolated systems that do not exchange energy or mass with their surroundings, transformation always involves some irreversible loss of energy or degradation; i.e., energy moves irreversibly toward increasing disorder, i.e., entropy (Prigogine, 1980).

Central to all species' life histories and survival is energy acquisition, storage, and transfer. Living cells capture and process energy in very organized compartments that contribute to an organism's growth, metabolism, and reproduction. Primary producers (plants, phytoplankton) capture solar energy, generally expressed as grams of carbon fixed per meter squared per year (gm^{-2} yr^{-1}). Gross production is the total amount of organic matter produced in an area over a given time, and net production is the amount of organic matter produced in excess of an organism's physiological needs (growth, maintenance, reproduction, etc.) and that is available to the ecosystem. Biomass is the total weight of living material in a specified area. This capture and transfer of bounded, useful energy (called *exergy*; Kay, 2000) provides high-quality fuel for the ecosystem that is transferred through food webs (Jørgensen and Fath, 2006). Organisms and ecosystems have evolved essential thermodynamic attributes to maintain a high state of order with low entropy (Jørgensen, 2002).

Organisms capture and move energy through irreversible processes, with substantial loss to the ecosystem. Energy for life comes from cellular process involving the oxidation of carbon-hydrogen bonds (Falkowski, 2012). Energy conversion and transfer involves expending energy in biological metabolism, movement, growth, and reproduction, with energy loss through respiration and generation of heat, and in transfers between biological compartments and food webs. More precisely, photosynthetic capture of solar energy has a thermodynamic efficiency of approximately 3–5% that becomes available for maintenance of plant physiological processes. Transfers from plant to predator are about 10 to 20% efficient, with considerable variability among herbivores, carnivores, cold-blooded (ectothermic) invertebrates, fishes, and some warm-blooded (endothermic) fishes, birds, and mammals. Yet high productivity and availability of resources has yielded high biomass, observed as large populations of marine mammals, flocks of seabirds, and abundant fishery resources. Sustaining this large biomass depends on high turnover rates of high-quality primary producers and small consumers at the base of the food web, as well as high turnover rates of energy among top consumers.

Marine food chains tend to be longer than terrestrial ones, hence more complex. For example, a top land predator such as a wolf exhibits three trophic levels (vegetation to herbivore to wolf), but tuna involve four or five trophic levels (phytoplankton to zooplankton to small fish and/or larger fish to tuna). Thus, less total annual production is apportioned to top marine predators than to terrestrial predators, which is partially compensated by higher turnover rates for marine primary producers (phytoplankton) than for land plants. Phytoplankton may have low biomass at any one time, which when compounded over seasonal periods results in high production. Furthermore, energy demands of most large terrestrial predators, mostly endothermic birds and mammals, are greater than for most top marine predators, which are mostly ectothermic invertebrates and fishes. The exceptions are marine birds and mammals, all of which have high energy demands; these energy demands are compensated by a number of morphological and behavioral adaptations, for example, adopting short food chains—sea cows (Sirenia) are herbivores and baleen whales eat zooplankton. Killer whales (*Orcinus orca*) are an exception; their position at the very top of marine food webs forces them to be few in number and relatively low in biomass.

The coastal realm represents <0.5% of ocean volume (Chen and Borges, 2009), yet its overall net primary productivity is generally greater than for either the open ocean or land (Valiela, 1984). Coastal ecosystems host autotrophic producers that process and release a large production of detritus (Duarte and Cebrián, 1996). These primary producers have high turnover rates, processing high inputs from watersheds, airsheds, and coastal upwelling, thereby performing a major role in the carbon cycle (Duarte *et al.*, 2005). They play critical roles through rate-setting processes, with turnover rates that can be measured in hours as in the case of bacteria and up to a day for some phytoplankton, in comparison with one to two times a year or even decades for most terrestrial plants. Most large marine plants such as kelps grow new leaves at exceptionally high rates as measured annually, in contrast to terrestrial woody plants that devote much of their biomass to hard tissue as support against gravity, meaning that production is low relative to biomass. Tidal marshes, mangroves, and seagrasses not only transform sunlight through photosynthesis into primary production, but also modulate tidal and hydrologic energies in varying degrees, providing protective habitat for breeding, feeding, and resting for fish and wildlife. While the open ocean yields more than 60% of total marine production due to its overwhelming area and volume (Longhurst *et al.*, 1995; Table 4.3), it lacks the high biomass concentration and turnover rates of primary producers that together fuel coastal-realm food webs.

4.7.2.2 Biogeochemicals

The coastal realm is the most geochemically and biologically active area of the biosphere (Gattuso *et al.*, 1998), receiving massive inputs of terrestrial organic matter, nutrients, sediment, and detritus that organisms process and the system exchanges with the open ocean (Chen and Borges, 2009). Chemicals are properties of ecosystems, notably carbon, nutrient chemicals (nitrogen, phosphorus), sulfur, and iron most critical to biological systems. These elements are captured and redistributed through bioactive compartments, being cycled through processes that enable microbes, plants, and animals to create physical structures and regulate nutrient fluxes

Table 4.3 Estimated total net primary production (NPP) in billion tons of carbon per year for estimated area of global ocean and different coastal sectors, and percentage. Based on data from Duarte and Cebrián (1996); Jennings *et al.* (2001).

MARINE PRIMARY PRODUCERS	Total NPP (10^9t C y^{-1})	Area Covered (10^6km^2)	Area (% of total)	NPP (% of total)
Ocean: Phytoplankton	43.00	332.00	88.46	81.10
Coastal: Phytoplankton	4.50	27.00	7.19	8.49
Macroalgae	2.55	6.80	1.81	4.81
Mangroves	1.10	1.10	0.29	2.07
Coral reef algae	0.60	0.60	0.16	1.13
Seagrasses	0.49	0.60	0.16	0.92
Marsh plants	0.44	0.40	0.1	0.83
Microphytobenthos	0.34	6.80	1.81	0.64
Total	**53.00**			

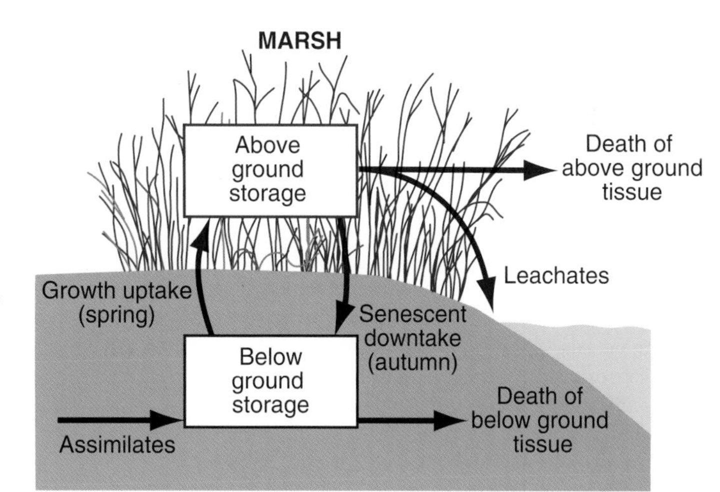

Fig. 4.16 Saltmarsh-grass systems filter pollutants, buffer wave action, support fisheries, contribute detritus, and more. *Spartina alterniflora* re-oxygenates anaerobic marsh soils by enhancing bacterial and nitrogen-producing activity. In winter, storage is in roots; in summer growing season, most storage is above ground. When young and growing vigorously, saltgrass is a net carbon importer, but with maturation becomes a major exporter, providing an energy subsidy to surrounding systems.

(Berhe *et al.*, 2005). Biologically important processes in wetlands, estuaries, seas, coastal waters, and sediments are linked to increased greenhouse gas concentrations of methane (CH^4) and nitrous oxide (N_2O) in the atmosphere (Bigg *et al.*, 2003). The net uptake of CO_2 in the coastal realm represents about 20% of the world's ocean uptake of anthropogenic CO_2 (Thomas *et al.*, 2004. Although much uncertainty remains about the transformation, transport, and cycling of carbon and nitrogen (Thomas *et al.*, 2004; Cai, 2011), the interchange of biogeochemicals through biotic reservoirs focuses attention on coastal species and ecosystems.

The coastal realm supports abundant marine species that translocate chemicals across regions. Migratory species transport significant amounts between land and sea, as for example, migratory fish, mammals, and birds that move between coastal-ocean and estuarine environments and deposit substantial amounts of ocean nutrients to coastal systems annually. Anadromous fish on return from the sea (e.g., shad, salmons) spawn in freshwaters and, through their waste products or dead bodies, enhance stream productivity important to their developing young (Ch. 6). Birds and marine mammals transfer organic matter from sea to land, an especially significant function in high-latitude systems; e.g., bears eat spawning salmon and deposit waste products in forests. Individual species in coastal habitats rework, import, and export considerable amounts of materials at local scales (Fig. 4.16), and their aggregate in mass at landscape scales contributes to ecosystem performance. Organisms construct hard carbonate reefs by capturing abundant supplies of calcium carbonate through a process constrained by eutrophication, anoxia, and watershed activities that affect carbonate chemistry (Feely *et al.*, 2004; Borges and Gypens, 2010). Seagrass beds represent only 0.1 to 0.2% of the global ocean, but contribute above- and below-ground biomass at high rates of production that makes this plant community among the most productive on Earth (Duarte, 2002; Duarte and Chiscano, 1999), thus contributing significantly to nutrient cycling and food-web structure (Orth *et al.*, 2006). Degradation plays a key role; as autotrophic production is processed, leaves and roots are degraded, and nutrients in the form of detritus are buried or made available for transport to deep, offshore waters (Suchanek *et al.*, 1985).

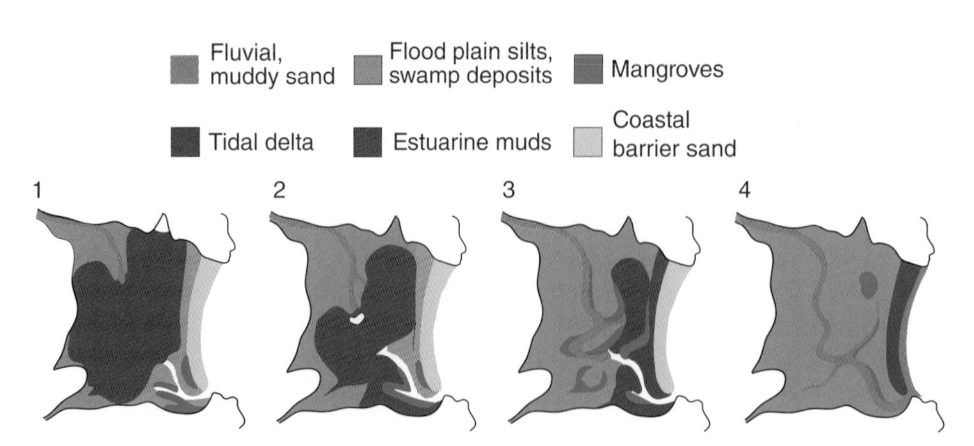

Fig. 4.17 Evolutionary change for two Australian estuaries: (a) drowned river valley; (b) barrier estuary. As a river continues to flow in and a barrier closes off the tidal exchange, water covers estuarine muds and fluvial muddy sand. The deeper-water portions underlain by estuarine muds become smaller as water drains into channels, and mangrove re-locate. This process can be altered or reversed by storm events. From Roy (1984). New South Wales estuaries: their origin and evolution. In *Coastal Geomorphology in Australia* (ed. Thom BG). Academic Press, Sydney, pp. 99–121.

A major concern for biogeochemical processes is the exponential spread of dead zones in coastal oceans (Diaz and Rosenberg, 2008; Box 2.5). The rapid spread of ocean deoxygenation since the 1960s has major consequences (Middelburg and Levin, 2009); with rising temperatures, ocean acidification will result in substantial changes affecting marine ecosystems and biogeochemical cycles (Gruber, 2011).

4.7.2.3 Evolution

An important property of ecosystems is evolution, i.e., change over time. Biota respond and adapt to dynamic coastal ecosystems through inheritance and behavior to better obtain nutrients, materials, and energy, and to survive over generations of time. The outcome of these actions and interactions over large and small scales can enhance a particular ecosystem's growth, expansion, maturity, or under less optimum conditions, move the ecosystem into senescence and decay.

Large-scale coastal, geomorphologic evolution involves a number of factors in which antecedent conditions set the stage for a sequence of change (Stephenson and Branderb, 2003). Macroscale forces of sea-level rise, glaciation, tectonics, storms, winds, and climatic events, interacting with changing chemistry of the atmosphere, impose historical constraints and conditions for future change. Shorelines cross continental shelves in sequences of sea-level advance and retreat. Sediment is mobilized and redeposited through feedbacks among topography, biology, and fluid dynamics to initiate new landforms. Lagoons and marshes transgress across the landscape, where large portions of estuaries sequentially become mud flats through interactions and feedbacks among physical and biotic processes (Fig. 4.17). Living structures keep pace by adapting, moving, adjusting, and systematically evolving through self-generation and self-organization into steady-state patterns (Koppel *et al.*, 2008), operating within larger and smaller, external and internal pulsing mechanisms (Odum *et al.*, 1995), thereby to balance against forces of disequilibrium. To enhance survival options, organisms may adapt to tidal pulses, for example by utilizing the energy of oscillating water flow for transport, feeding, and larval dispersion. If life-history strategies fail to keep pace with change, species perish. If change is too sharp, it can be catastrophic, as exhibited in geologic history with five episodes of mass extinction (Fig. 4.3; Knoll *et al.*, 1996; Veron, 2008).

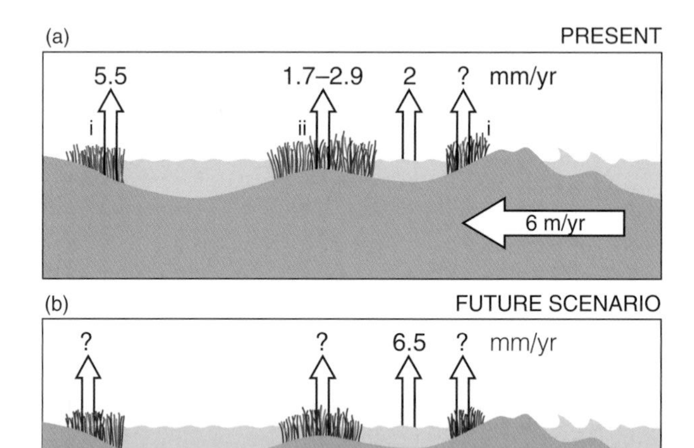

Fig. 4.18 Sea-level rise presently measured (a) and predicted (b) at the Virginia Coast Reserve Long-Term Ecological Research Site indicates differences in rates of change at three locations of this barrier island/lagoon. Horizontal arrows indicate landward erosion rate of the barrier island. Upward arrows indicate rates in change of marsh elevation and sea level. (a) Under present sea-level rise, mid-lagoon marshes will likely disappear (ii) while mainland fringing marshes will persist, because sea-level rise of 4 mm yr^{-1} is more than 1.7–2.9 for mid-lagoon marshes, but less than 5.5 for mainland fringing marshes (i). (b) Under future scenario (2100 AD), marsh elevation must increase for marshes to keep pace and persist at sea-level rise rates of 6.5 mm yr^{-1} (IPCC, 2007). Courtesy of Karen McGathery and Patricia Wiberg.

How organisms move, colonize, and stabilize coastal ecosystems is difficult to predict because of diverse evolutionary responses to environmental changes among taxa. For example, under different forecasted sea-level-rise scenarios that affect lagoon and marsh development, marsh growth rates can differ significantly (Fig. 4.18). Organisms acquire energy and nutrients as they organize over time into spatial patterns that evolve into quasi-equilibrium states. Of considerable influence is biological capacity for high gross turnover rates, expansion in numbers (individuals, species), and increase in living matter (biomass). But as all natural systems are constrained by energy, mass, momentum, and other properties, interference with system structure may force the ecosystem to move toward maximum disorder, with a potential to change into a disorganized steady state (Kleidon *et al.*, 2010). While prediction is difficult for a host of species, many marine species that disperse as juveniles will be affected by increased stormy weather as predicted by climate change; some that undertake long-distance migration near the ocean surface (e.g., sea turtles, whales, etc.) may reach unfavorable locations (Monzón-Argüello *et al.*, 2012). Opportunistic plant species with capacities for widespread seed dispersal and high levels of energy utilization and fecundity may colonize new areas, only to compete with native biota responding to changing conditions. And the loss of certain "key" species (Ch. 5) can result in consequences difficult to predict.

4.8 THE ECOSYSTEM CONCEPT

The ecosystem has long been recognized as a fundamental unit of nature. It is a special kind of system in steady-state quasi-equilibrium, recognized by its structure and identified by emergent organization and measurable quantities of production, complexity, and resiliency (Box 4.2). Composed of species, chemicals, materials, and energy exchanges that interact with the surrounding environment, ecosystems are driven by processes that create both "order from order" and "order from disorder" (Schrödinger, 1944; Kay, 2000). The appearance of high-level order from low-level chaos is explained by theoretical non-equilibrium thermodynamics (Prigogine, 1980; Kay 2000). As "open" and locally produced steady-state systems, ecosystems are far from equilibrium and act within a larger system of increasing disorder under conditions of the second law of thermodynamics. This law establishes the universal rule that spontaneous change is always accompanied by degradation into a more dispersed and chaotic state (Jørgensen, 2002).

Henry Louis Le Châtelier's (1850–1936) basic principle about a system in equilibrium is that when a stress or disturbance is brought to bear, the system tends to counteract the disturbance to achieve a new equilibrium state. The interaction of two systems not in mutual equilibrium tends to drive them to a final common equilibrium. In biological systems, "homeostasis" is the tendency of the system to maintain internal stability when a disturbance is applied, which is supported by growing evidence of the stabilizing effect that species play in maintaining steady state (Ernest and Brown, 2001). Homeostasis in ecosystems depends upon feedbacks among components and properties that serve to achieve a dynamic equilibrium between order and disorder, growth and dissipation. That is, the ecosystem is presumed to exist in a reasonably stable yet dynamic state or condition, maintained through coupled interactions arising from antecedent conditions, captures of energy, and counter forces against disequilibrium dictated by the second law of thermodynamics.

Systems theory in ecology arose notably during the 1950s and 1960s with the hope that ecology might turn into an exact science that potentially could predict and be guided by a set of uniform theoretical foundations (Voigt, 2011). Ecosystem theory is concerned primarily with organismal communities, their abiotic environment and interactions with one another. With organisms in mind, ecosystem science has been approached in at least three different ways: the organism as the focus of the ecosystem; the ecosystem as a set of processes involving the roles of organisms in the transfer and alteration of matter and energy; and the ecosystem as a geographic area of sufficiently similar topography, climate, and biota (Blew, 1996).

Ecosystems process continuous fluxes of energy and materials across their boundaries (Fig. 4.19). Ecosystems are regulated by physical exchanges and species' transformations of materials and energy delivered in pulsed dispersals and migrations. Within ecosystems, biological communities display timed oscillations in cycles of predator and prey abundance and form aggregations to feed or reproduce. Functional attributes of ecosystems under dynamic, presumably multi-stable conditions may undergo a variety of trajectories in both time and space, as observed by Orians (1975): (i) constancy, lacking change in some system parameter (e.g., species number); (ii) inertia, resisting external perturbations; (iii) elasticity, the rate at which the system returns to its former state following a perturbation; (iv) amplitude, the area over which the system is stable; (v) cyclic stability, the system cycles about some central point or zone; and (vi) trajectory stability, the system moves toward some final endpoint or zone despite differences in starting points, or the system converges to a particular state from a variety of starting positions. If biological and physical systems are in harmony, ecosystem performance can be amplified; if not, it can degrade. System adjustment involves delays, lags in recovery, and resiliency, where thermodynamics, complexity, positive and negative feedbacks, and thresholds are important variables (DeAngelis *et al.*, 1986). In some cases, feedbacks from natural histories of species enhance conditions for their own survival: e.g., self-stereotaxis of oysters and anadromous fishes that deliver nutrients to natal streams (Ch. 6).

Feedback dynamics is key to maintaining ecological function. System feedback is a process whereby when change occurs in one quantity, a second quantity in turn changes the first. Positive feedbacks increase and amplify change in the first quantity; negative feedbacks reduce it, resulting in attenuation, time delays, oscillatory instability, etc. When critical thresholds are passed, homeostatic mechanisms no longer operate and amplification can drive ecosystems towards new regimes or equilibrium states, in shifts that may be smooth, abrupt, or discontinuous (Fig. 4.20; Lees *et al.*, 2006); however, ecosystem response is not a fixed property (Scheffer and Carpenter, 2003). Kay (1991) queried whether change occurs along the original developmental pathway or a new one; i.e., if the system organizes or disorganizes to return to an original state, or whether the ecosystem flips into some new, catastrophic state unacceptable to humans.

Combinations of hierarchy, structure, energy, and self-organization can result in an emergent condition (Nielsen and Müller, 2000; Jørgensen, 2002), that is, a higher level of organization markedly different from lower levels (Allen and

Box 4.2 What is an ecosystem?

A "system" is composed of parts with inter-dependent connections and feedbacks that form a complex whole and maintains an internal steady state, or homeostasis, despite a changing external environment. A system can be thought of as a construct of the human mind to explain a phenomenon, and its delineation can be somewhat arbitrary.

Weinberg (1975, 2001) conceived three types of systems (Fig. B4.2.1). System 1 is a *small-number system*, characterized by "organized simplicity," as for example in mechanical systems (automobiles) whose parts follow basic physical laws, e.g., Newton's laws. Such laws explain interactions and predict outcomes by means of mathematical equations. System 3 at the other extreme is a *large-number system*, characterized by "unorganized complexity" of a large number of identical components that are random in behavior and yield to overall statistical averaging, for example, random motions of molecules of a perfect gas, on the order of 10^{23} molecules. The individual motions of the molecules are virtually unknowable, as they collide and rebound unpredictably. Their overall motions, however, at a given temperature follow the predictive power of gas laws.

Ecosystems do not fit either of these categories. Rather, they are System 2 *medium-number systems* characterized by structural and functional interactions among an intermediate number of components. Weinberg's term for these is "organized complexity." The components express a wide spectrum of process rates and spatial characteristics. Ecosystems are not easily amenable to either mechanical or statistical solutions. Some degree of abstraction is required to study ecosystems, but they are not merely abstractions. Rather, they are best understood as organized units of nature. A key to understanding them is to make their space- and rate-dependent organization and linkages explicit. That is, ecosystems cannot be understood on any single spatio-temporal scale, and no single type of observation can be extrapolated to define the nature of the underlying processes.

Ecosystems are in constant change, whereby they may be said to lose identity, exhibit trajectories, or exist in a state of dynamic equilibrium. The systems that exist at any one time have been selected from all systems of the past, being the best "survivors." System survival thus refers to the length of time a system exists and its evolutionary history. Thus, an ecosystem's identity becomes synonymous with its resiliency and viability.

Sources: O'Neill *et al*. (1986); Weinberg (1975, 2001)

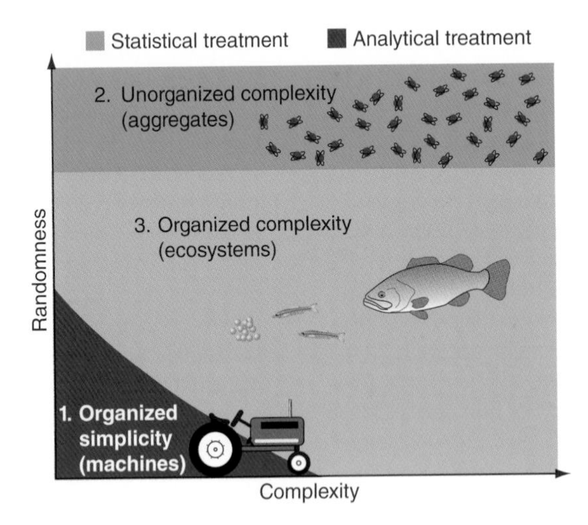

Fig. B4.2.1 Three types of systems differ in degree of complexity and organization. (1) Organized simplicity, e.g. machines: small-number systems contain so few components that analytical treatment applies. (2) Unorganized complexity, e.g., gas molecules large-number systems contain so many identical, randomly interacting components that their statistical properties appear to be deterministic, and statistical treatment can be applied. (3) Organized complexity, e.g. ecosystems: medium-number systems that contain an intermediate number of components that interact non-randomly, and therefore cannot be analyzed by traditional methods and often appear to be stochastic, a situation that is encapsulated into the "law of medium numbers," e.g., Murphy's law, *Anything that can happen will happen*. Based on data from Weinberg (1975, 2001).

Starr, 1982). When major and multiple pressures pass an ecosystem's critical point, the system may switch to an alternate stable state—a process called a "regime shift." Some shifts may be driven by external forcing (climate change, alien invasions, cultural eutrophication, overfishing, etc.) or internal perturbations, or be triggered by a disproportionately small time-lagged force in a process referred to as "hysteresis." When a critical threshold is passed, substantially stronger driving forces may be required for recovery to an initial steady state. At critical thresholds, the forward reaction may reverse to a lower critical threshold (Scheffer *et al.*, 2009), through failures in ecosystem structure, or surprise, or heightened complexity that makes the system vulnerable to "collapse" (DeAngelis *et al.*, 1986). For a world experiencing dramatic anthropogenic changes in environmental conditions and severe perturbations, changes to alternative stable states may have serious, unexpected consequences (Schröder *et al.*, 2005).

The different ways that ecosystems respond to change, and survive, indicates that they can be viewed as "complex adaptive systems" (Brown, 1995; Levin, 1998). Species and biotic communities tend to persist with a resiliency that is rarely predictable and an outcome that is rarely if ever stable. Environmental conditions constrain or enhance a system's natural capacity to expand, but through localized interactions and selection, biotic processes emerge into higher levels of organization that maintain ecological stability. In a hierarchy of interactions, constraints imposed on biotic capacity come from lower and higher levels; i.e., lower- and higher-level components interact to add new levels of organization and increasing levels of complexity that are vulnerable to change, and even collapse (Fig. 4.21). Positive feedbacks reinforce change in the direction of the deviation, with the potential to destabilize the system (Milsum, 1968; DeAngelis *et al.*, 1986); negative feedbacks do the opposite.

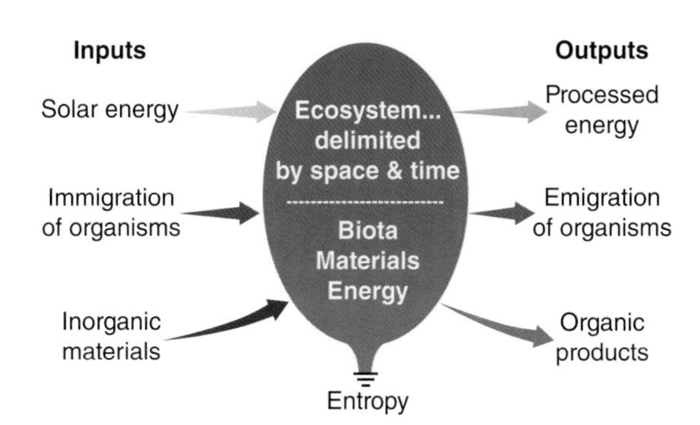

Fig. 4.19 A conceptual delineation of a spatially explicit ecosystem in time and space. The ecosystem is open to inputs, thermodynamically at non-equilibrium, and subject to degradation as a result of increased entropy. Output quality depends on ecosystem efficiency. Boundaries may be natural (watershed contours, coastal-ocean fronts, interface of water-benthos) or arbitrary (whatever spatial feature is of special interest, e.g., species range, trophic, etc.). Based on data from Odum (1989).

In a normal, quasi-stable equilibrium state under optimum conditions, ecosystems move over time toward ever-greater complexity. Complex marine ecosystems may maintain stability through alternative steady states, consistent stable states, or dynamic regimes (Knowlton, 1992; Scheffer and Carpenter, 2003; Folke *et al.*, 2004; Petraitis and Dudgeon, 2004; Schröder *et al.*, 2005; Daskalov *et al.*, 2007). How long an ecosystem may endure in a quasi-equilibrium state is a measure of its *persistence*, and how fast it returns to that state is a measure of its *resiliency*. If a deviation occurs in one direction, negative feedbacks force the system in the opposite direction to maintain system *identity*. To maintain functional resiliency, some ecosystems may switch easily between alternative states (Holling, 1973; Walker and Myers, 2004; Ives and Carpenter, 2007; Walker and Salt, 2012). Complex dynamic systems (ecosystems, financial markets, climate), however, can reach tipping points and abruptly shift from one equilibrium state to another.

4.9 ECOSYSTEM BASE FOR CONSERVATION

Although the "ecosystem" concept is fundamental to conservation and management, its concept needs clarification (Jax, 2005, 2007), especially for highly variable and dynamic coastal and marine environments. That is, as the observer delineates an ecosystem of interest, he/she imposes a perceptual bias, a filter through which the system is viewed (Levin, 1992).

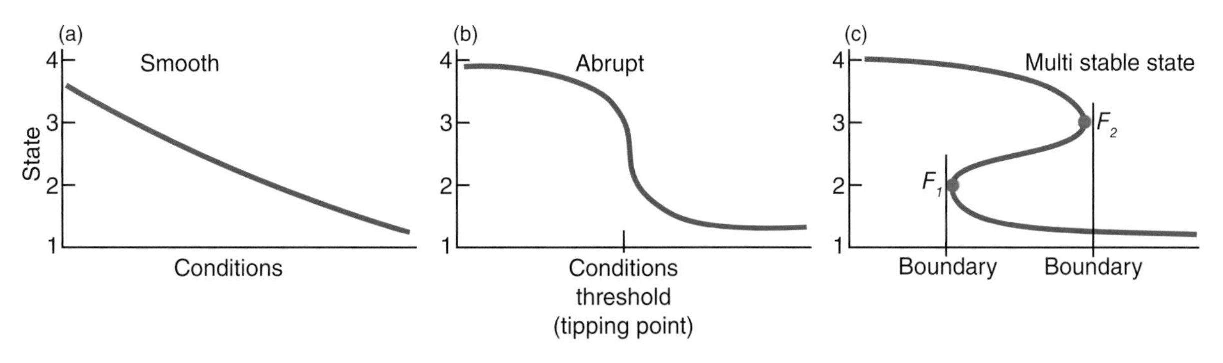

Fig. 4.20 Ecosystem responses to change in external conditions. (a) Smooth transition; (b) profound transition, when conditions approach a critical level; (c) switch to alternate stable state (F_1, F_2) over a range of conditions (hysteresis). Response is not a fixed property of a system, although some systems tend to respond in a more non-linear way than others. From Scheffer M, Carpenter SR (2003) Review: catastrophic regime shifts in ecosystems: linking theory to observation. *TRENDS in Ecology and Evolution* **18**, 648–656.

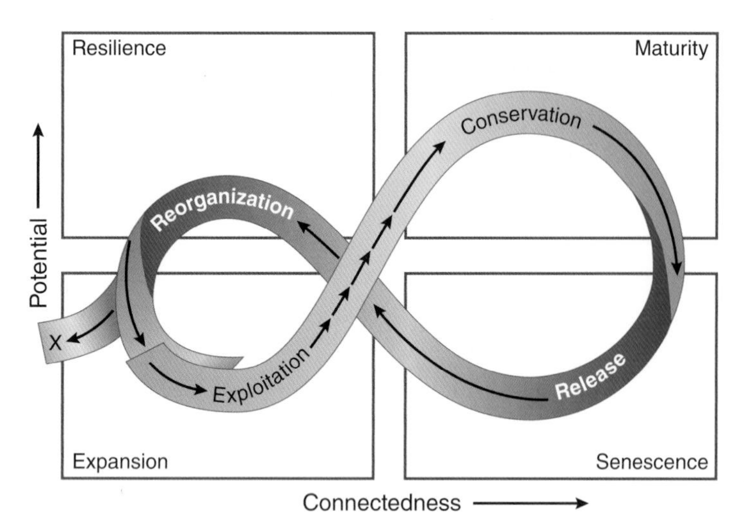

Fig. 4.21 Panarchy concept of an ecosystem (Holling and Gunderson, 2002). Ecosystems are complex adaptive systems that cycle through four functional stages: expansion, maturity, senescence, and resilience. The ecosystem loses and gains potential, i.e., inherent in accumulated resources, biomass (exergy) where nutrients can leak away (x). The ecosystem moves from low to high connectedness as aggregated elements gain advantages among controlling variables. Initially, a growing and expanding ecosystem rapidly changes towards maturity, with conservation of aggregated elements that control or mediate influences of external variability. The ecosystem loses potential and connectedness. Low connectedness is associated with loosely connected, diffuse elements dominated by outward relations and affected by outside variables with accumulating assets (exploitation). Arrows indicate flow speed in the cycle: short, closely spaced arrows indicate slow changes. From Holling CS (2001) Understanding the complexity of economic, ecological, and social systems. *Ecosystems* **4**, 390–405. With kind permission from Springer Science+Business Media B.V.

Functional ecosystems are spatial components of the marine and coastal environment, interacting holistically and viewed at different hierarchical scales. The open-ocean and coastal realms as ecosystems become apparent through recognition of their supporting processes and inter-relationships that provide services for humans and other organisms, through the roles organisms play within them and their subsystems, and through processes that sustain smaller ecological units. Scientific questions thus relate to: how open-ocean and coastal realm systems and their subsystems function hierarchically as ecological units, how functional attributes are sustained, how specific parts contribute to persistence and stability of attributes, and at what scale individual components fulfill particular kinds of functions—ecological production, biomass, biodiversity, resiliency, and services important to people, i.e., food, clean water, oxygen, etc. In particular, recognizing the coastal realm as a complex adaptive ecosystem exposed to most of the issues described in Chapter 1, and that plays significant roles in global and local processes, establishes the context and urgency for ecosystem-based approaches at appropriate scales to meet marine conservation needs.

REFERENCES

Allen TFH, Starr TB (1982) *Hierarchy: Perspectives for ecological complexity*. University of Chicago Press, Chicago.

Bakun A (1996) *Patterns in the ocean: ocean processes and marine population dynamics*. California Sea Grant College System, National Oceanographic and Atmospheric Administration, La Jolla.

Beck MW, Heck Jr. KL, Able KW, Childers DL, Eggleston DB, *et al.* (2003) The role of nearshore ecosystems as fish and shellfish nurseries. *Issues in Ecology* **11**, 1–12.

Belkin IM, Cornillon PC, Sherman K (2009) Fronts in large marine ecosystems. *Progress in Oceanography* **81**, 223–236.

Berhe AA, Carpenter E, Codispoti L, Izac A-M, Lemoalle J, Luizao F, *et al.* (2005) Nutrient cycling. In *Ecosystems and Human Well-Being*, Chapter 12, **vol. I**, Current State and Trends. Millennium Ecosystem Assessment, Island Press, Washington, D.C., 331–353.

Berryman AA (2003) On principles, laws and theory in population ecology. *Oikos* **103**, 695–701.

Belkin M, Cornillona PC, Sherman K (2009) Fronts in large marine ecosystems. *Progress In Oceanography* **81**, 223–236.

Bigg GR, Jickells TD, Liss PS, Osborn TJ (2003) Review: the role of the oceans in climate. *International Journal of Climatology* **23**, 1127–1159.

Birkeland C, ed. (1997) *Life and death of coral reefs*. Chapman and Hall, New York.

Blakey RC (2008) Gondwana paleogrography from assembly to breakup—a 500 m.y. odyssey. In *Resolving the late Paleozoic ice age in time and space* (eds Fielding CR, Frank TD, Isbell JL). Geological Society of America, Special Paper 441, 1–28.

Blew RD (1996) On the definition of ecosystem. *Bulletin of the Ecological Society of America* **77**, 171–173.

Borges AV, Gypens N (2010) Carbonate chemistry in the coastal zone responds more strongly to eutrophication than to ocean acidification. *Limnology and Oceanography* **55**, 346–353.

Bost CA, Cotté C, Bailleul F, Cherel Y, Charrassin JB, Guinet C, *et al.* (2009) The importance of oceanographic fronts to marine birds and mammals of the southern oceans. *Journal of Marine Systems* **78**, 363–376.

Brewer PG, Peltzer PT (2009) Limits to marine life. *Science* **324**, 347–348.

Broecker WS (1991) The giant ocean conveyor. *Oceanography* **4**, 79–89.

Broecker WS (1997) Thermohaline circulation, the Achilles heel of our climate system: will man-made CO_2 upset the current balance? *Science* **278**, 1582–1588.

Broecker WS (2006) Abrupt climate change revisited. *Global and Planetary Change* **54**, 211–215.

Brown JH (1995) *Macroecology*. University of Chicago Press, Chicago, Illinois.

Buhl-Mortensen L, Vanreuse A, Gooday AJ, Levin LA, Priede IG, Buhl-Mortensen P, *et al.* (2010) Biological structures as a source of habitat heterogeneity and biodiversity on the deep ocean margins. *Marine Ecology* **31**, 21–50.

Burnett WC, Bokuniewicz H, Huettel M, Moore WS, Taniguchi M (2003) Groundwater and pore water inputs to the coastal zone. *Biogeochemistry* **66**, 3–33.

Cai W-J (2011) Estuarine and coastal ocean carbon paradox: CO_2 sinks or sites of terrestrial carbon incineration? *Annual Review of Marine Science* **3**,123–145.

Calbet A, Landry MR (2004) Phytoplankton growth, microzooplankton grazing, and carbon cycling in marine systems. *Limnology and Oceanography* **49**, 51–57.

Canfield DE (2005) The early history of atmospheric oxygen: homage to Robert M. Garrels. *Annual Review of Earth and Planetary Sciences* **33**, 1–36.

Canfield DE, Alexander N, Glazer AN, Falkowski PG (2010) The evolution and future of earth's nitrogen cycle. *Science* **330**, 192–196.

Carpenter LJ (2003) Iodine in the marine boundary layer. *Chemical Reviews* **103**, 4953–4962.

Carpenter LJ, Jones CE, Dunk RM, Hornsby KE, Woeltjen J (2009) Air-sea fluxes of biogenic bromine from the tropical and north Atlantic Ocean. *Atmospheric Chemistry and Physics Discussions* **9**, 1805–1816.

CBD (online) Island biodiversity—what's the problem. Convention on Biological Diversity Secretariat, www.cbd.int.

Chen CTA, Borges AV (2009) Reconciling opposing views on carbon cycling in the coastal ocean: continental shelves as sinks and near-shore ecosystems as sources of atmospheric CO_2. *Deep-Sea Research II* **56**, 578–590.

Clark MR, Rowden AA, Schlacher T, Williams A, Consalvey M, Stocks KI, *et al.* (2010) The ecology of seamounts: structure, function, and human impacts. *Annual Review of Marine Science* **2**, 253–278.

Crossland CJ, Kremer HH, Lindeboom HJ, Crossland JIM, LeTissier MDA (2005) *Coastal fluxes in the Anthropocene*. Springer, Berlin, 231pp.

Daskalov GM, Grishin AN, Rodionov S, Mihneva V (2007) Trophic cascades triggered by overfishing reveal possible mechanisms of ecosystem regime shifts. *Proceedings of the National Academy of Sciences of the United States of America* **104**, 10518–10523. www.pnas.org

DeAngelis DL, Post WM, Travis CT (1986) Positive feedback in natural systems. *Biomathematics* **15**. Springer-Verlag, Berlin, 290pp.

DeBose JL and Nevitt GA (2008) Dimethylsulfoniopropionate is linked to coral spawning, fish abundance and squid aggregations over a coral reef. *Proceedings of the 11th International Coral Reef Symposium*. Ft. Lauderdale, Florida, 7–11 July, 275–279.

Defeo O, McLachlan A, Schoeman DS, Schlacher TA, Dugan J, Jones A, *et al.* (2009) Threats to sandy beach ecosystems: a review. *Estuarine, Coastal and Shelf Science* **81**, 1–12.

Diaz RJ, Rosenberg R (2008) Spreading dead zones and consequences for marine ecosystems. *Science* **321**, 926–929.

D'Odorico P, Laio F, Porporato A, Ridolfi L, Rinaldo A, Rodriguez-Iturbe I (2010) Ecohydrology of terrestrial ecosystems. *BioScience* **60**, 898–907.

Duarte CM (2002) The future of seagrass meadows. *Environmental Conservation* **29**, 192–206.

Duarte CM, Cebrián J (1996) The fate of marine autotrophic production. *Limnology and Oceanography* **41**, 1758–1766.

Duarte CM, Chiscano CL (1999) Seagrass biomass and production: a reassessment. *Aquatic Botany* **65**, 159–174.

Duarte CM, Middelburg JJ, Caraco N (2005) Major role of marine vegetation on the oceanic carbon cycle. *Biogeosciences* **2**, 1–8.

Ducklow H, McCallister SL (2004) The biogeochemistry of carbon dioxide in the coastal ocean. In *The Sea* (eds Robinson AR, McCarthy, Rothschild BJ). The President and Fellows of Harvard College.

Dugan JE, Hubbard DM (2010) Loss of coastal strand habitat in southern California: the role of beach grooming. *Estuaries and Coasts* **33**, 67–77.

Emery KO (1969) The continental shelves. *Scientific American* **221**, 107–122.

Ernest SKM, Brown JM (2001) Homeostasis and compensation: the role of species and resources in ecosystem stability. *Ecology* **82**, 2118–2132.

Erwin DJ, Bowring SA, Yugan J (2002) End-Permian mass extinctions: a review. In *Catastrophic Events and Mass Extinctions: Impacts and Beyond* (eds Koeberl C, MacLeod KG). Geological Society of America Special Paper 356, Boulder, Colorado, 363–383.

Falkowski PG (2012) The global carbon cycle: biological processes. In *Fundamentals of Geobiology*, First Edition (eds Knoll AH, Canfield DE, Konhauser KO). Blackwell Publishing Ltd, Oxford, UK, 5–19.

Falkowski PG, Scholes J, Boyle E, Canadell J, Canfield D, Elser J, *et al.* (2000) The global carbon cycle: a test of our knowledge of earth as a system. *Science* **290**, 291–296.

Feely RA, Sabine CL, Lee K, Berelson W, Kleypas J, Fabry VJ, Millero FJ (2004) Impact of anthropogenic CO_2 on the $CaCO_3$ system in the oceans. *Science* **305**, 362–366.

Folke C, Carpenter S, Walker B, Scheffer M, Elmqvist T, Gunderson L, Holling CS (2004) Regime shifts, resilience, and biodiversity in ecosystem management. *Annual Review of Ecology and Systematics* **35**, 557–581.

Garstang M, Ellery WN, McCarthy TS, Scholes MC, Scholes RJ, Swap RJ, Tyson PD (1998) The contribution of aerosol- and water-borne nutrients to the functioning of the Okavango Delta ecosystem, Botswana. *South African Journal of Science* **94**, 223–229.

Gattuso J-P, Frankignoulle M, Wollast R (1998) Carbon and carbonate metabolism in coastal aquatic ecosystems. *Annual Review of Ecology and Systematics* **29**, 405–434.

Gerard VA (1984) The light environment in a giant kelp forest: influence of *Macrocystis pyrifera* on spatial and temporal variability. *Marine Biology* **84**, 189–195.

Govoni JJ, Hare JA, Davenport ED, Chen MH, Marancik KE (2010) Mesoscale, cyclonic eddies as larval fish habitat along the southeast United States shelf: a Lagrangian description of the zooplankton community. *ICES Journal of Marine Science* **67**, 403–411.

Grassle JF, Maciolek NJ (1992) Deep-Sea species richness: regional and local diversity estimates from quantitative bottom samples. *The American Naturalist* **139**, 313–341.

Gross MG, Gross E (1996) *Oceanography: A View of Earth*. Prentice Hall, Upper Saddle River, New Jersey, USA.

Gruber N (2011) Warming up, turning sour, losing breath: ocean biogeochemistry under global change. *Philosophical Transactions of the Royal Society A* **369**, 1980–1996.

Gruber N, Galloway JN (2008) An earth-system perspective of the global nitrogen cycle. *Nature* **451**, 293–296.

Guichard F, Bourget E (1998) Topographic heterogeneity, hydrodynamics, and benthic community structure: a scale-dependent cascade. *Marine Ecology Progress Series* **171**, 59–70.

Guo L, Santschi PH (1997) Composition and cycling of colloids in marine environments. *Review of Geophysics* **35**, 17–40.

Halpern BS, Cottenie K, Broitman BR (2006) Strong top-down control in southern California kelp forest ecosystems. *Science* **312**, 1230–1232.

Hayden BP, Ray GC, Dolan R (1984) Classification of coastal and marine environments. *Environmental Conservation* **11**, 199–207.

Heck KL, Able AW, Roman CT, Fahay MP (1995) Composition, abundance, biomass, and production of macrofauna in a New England estuary: comparisons among eelgrass meadows and other nursery habitats. *Estuaries* **18**, 379–389.

Hedges JI (1992) Global biogeochemical cycles: progress and problems. *Marine Chemistry* **39**, 67–93.

Helly JJ, Levin LA (2004) Global distribution of naturally occurring marine hypoxia on continental margins. *Deep-Sea Research I* **51**, 1159–1168.

Holland HD (2006) The oxygenation of the atmosphere and oceans. *Philosophical Transactions of the Royal Society B* **361**, 903–915.

Holligan PM, Reiners WA (1992) Predicting the responses of the coastal zone to global change. *Advances in Ecological Research* **22**, 211–255.

Holling CS (1973) Resilience and stability of ecological systems. *Annual Review of Ecology and Systematics* **4**, 1–23.

Holling CS (2001) Understanding the complexity of economic, ecological, and social systems. *Ecosystems* **4**, 390–405.

Holling CS, Gunderson LH (2002) Resilience and adaptive cycles. In *Panarchy* (eds Gunderson LH, Holling CS). Island Press, Washington, D.C., Chapter 22, 25–26.

Huthnance JM (1995) Circulation, exchange and water masses at the ocean margin: the role of physical processes at the shelf edge. *Progress in Oceanography* **35**, 353–431.

Inman DL, Brush BM (1973) The coastal challenge. *Science* **181**, 20–32.

Inman DL, Nordstrom CE (1971) Tectonic and morphologic classification of coasts. *Journal of Geology* **79**, 1–21.

IPCC (2001) *Climate Change: The Scientific Basis.* Contribution of Working Group I to the Third Assessment Report of the Intergovernmental Panel on Climate Change (eds Houghton JT, Ding Y, Griggs DJ, Noguer M, Linden PJ, Dai X, *et al.*). Cambridge University Press, Cambridge, UK and NY, USA, 881pp.

IPCC (2007) *Climate Change 2007: Synthesis Report.* An Assessment of the Intergovernmental Panel on Climate Change. Adopted section by section at IPCC Plenary XXVII, Valencia, Spain, 12–17, November 2007.

Ives AR, Carpenter ST (2007) Stability and diversity of ecosystems. *Science* **317**, 58–62.

Jax K (2005) Function and “functioning” in ecology: what does it mean? *Oikos* **111**, 641–648.

Jax K (2007) Can we define ecosystems? On the confusion between definition and description of ecological concepts. *Acta Biotheoretica* **55**, 341–355.

Jennings S, Kaiser MJ, Reynolds JD (2001) *Marine fisheries ecology*. Blackwell Science Ltd, Oxford, UK, 417pp.

Jørgensen SE (2002) *Integration of ecosystem theories: a pattern*, Third Edition. Kluwer Academic Publishers, Dordrecht, The Netherlands.

Jørgensen SE, Fath BD (2006) Examination of ecological networks. *Ecological Modelling* **196**, 283–288.

Kai ET, Rossi V, Sudre J, Weimerskirch H, Lopez C, Hernandez-Garcia E, *et al.* (2009) Top marine predators track Lagrangian coherent structures. *Proceedings of the National Academy of Sciences of the United States of America* **106**, 8245–8250.

Kaufman YJ, Koren I, Remer LA, Tanré D, Ginoux P, Fan S (2005) Dust transport and deposition observed from the Terra-Moderate Resolution Imaging Spectroradiometer (MODIS) spacecraft over the Atlantic Ocean. *Journal of Geophysical Research* **110**, D10S12.

Kay JJ (1991) A nonequilibrium thermodynamic framework for discussing ecosystem integrity. *Environmental Management* **4**, 483–495.

Kay JJ (2000) Ecosystems as self-organizing holarchic open systems: narratives and the second law of thermodynamics. In *Handbook of Ecosystem Theories and Management* (eds Jørgensen SE, Müller F). Lewis Publication, CRC Press, 135–160.

Keene WC, Jacob DJ (1996) New directions reactive chlorine: a potential sink for dimethylsulfide and hydrocarbons in the marine boundary layer. *Atmospheric Environment* **30**, i–iii.

Keene WC, Sander R, Pszenny AAP, Vogt R, Crutzen PJ, Galloway JN (1998) Aerosol pH in the marine boundary layer: a review and model evaluation. *Journal of Aerosol Science* **29**, 339–356.

Kepkay PE (2000) *Colloids and the ocean carbon cycle. The Handbook of Environmental Chemistry*, Chapter 2 (ed. Wangersky P), **vol. 5**, Part D Marine Chemistry. Springer-Verlag, Berlin, Heidelberg.

Ketchum BH, ed. (1972) *The Water's Edge: Critical Problems of the Coastal Zone*. The Massachusetts Institute of Technology Press, Cambridge, Massachusetts.

Kiel S, Tyler PA (2010) Chemosynthetically-driven ecosystems in the deep sea. In *The Vent and Seep Biota. Topics in Geobiology* (ed. Kiel S). **33**, 1–14.

Kleidon A, Malhi Y, Cox PM (2010). Maximum entropy production in environmental and ecological systems. *Philosophical Transactions of the Royal Society B* **365**, 1297–1302.

Knoll AH, Bambach RK, Canfield DE, Grotzinger JP (1996) Comparative earth history and late Permian mass extinction. *Science* **273**, 252–225.

Knoll AH, Javaux EJ, Hewitt D, Cohen P (2006) Eukaryotic organisms in Proterozoic oceans. *Philosophical Transactions of The Royal Society* **361**, 1023–1038.

Knowlton N (1992) Thresholds and multiple stable states in coral-reef community dynamics. *American Zoologist* **32**, 674–682.

Koppel J van de, Gascoigne JC, Theraulaz G, Rietkerk M, Mooij WM, Herman PMJ (2008) Experimental evidence for spatial self-organization and its emergent effects in mussel bed ecosystems. *Science* **32**, 739–742.

Lees K, Pitois S, Scott C, Frid C, Mackinson S (2006) Characterizing regime shifts in the marine environment. *Fish and Fisheries* **7**, 104–127.

Levin LA, Sibuet M, Gooday AJ, Smith CR, Vanreusel A (2010) The roles of habitat heterogeneity in generating and maintaining biodiversity on continental margins: an introduction. *Marine Ecology* **31**, 1–5.

Levin SA (1998) Ecosystems and the biosphere as complex adaptive systems. *Ecosystems* **1**, 431–436.

Levin SA (1992) The Problem of Pattern and Scale in Ecology: The Robert H. MacArthur Award Lecture. *Ecology* **73**, 1943–1967.

Liss PS, Malin G, Turner SM, Holligan PM (1994) Dimethyl sulphide and *Phaeocystis*: a review. *Journal of Marine Systems* **5**, 41–53.

Littler MM, Littler DS, Brooks BL (2010) Marine macroalgal diversity assessment of Saba Bank, Netherlands Antilles. *PLoS ONE* **5**, e10677.

Liu Z, Omar A, Hair J, Kittaka C, Hu Y, Powell K, C Trepte, *et al.* (2008) CALIPSO lidar observations of the optical properties of Saharan dust: a case study of long-range transport. *Journal of Geophysical Research* **113**, D07207.

Longhurst A (1998a) *Ecological geography of the sea*. Academic Press, San Diego, 398pp.

Longhurst A (1998b) Oceans and seas. In *World of Earth Science* (eds Lerner KL, Lerner BW). Gale Cengage, 2003. eNotes.com. 2006. 29 Mar, 2010.

Longhurst A, Sathyendranath S, Platt T, Caverhill C (1995) An estimate of global primary production in the ocean from satellite radiometer data. *Journal of Plankton Research* **17**, 1245–1271.

Lozier MS (2010) Review: deconstructing the conveyor belt. *Science* **328**, 1507–1511.

Lozier MS (2012) Overturning in the North Atlantic. *Annual Review of Marine Science* **4**, 291–315.

Mahowald NM, Baker AR, Bergametti G, Brooks N, Duce RA, TD Jickells, *et al.* (2005) Atmospheric global dust cycle and iron inputs to the ocean. *Global Biogeochemical Cycles* **19**, GB4025, 15pp.

Mann KH (1973) Seaweeds: their productivity and strategy for growth. *Science* **182**, 975–981.

Mann KH, Lazier JRN (1991) *Dynamics of marine ecosystems*. Blackwell Scientific Publications, Boston, 466pp.

Martin AP (2003) Phytoplankton patchiness: the role of lateral stirring and mixing. *Progress in Oceanography* **57**, 125–174.

McClain CR, Lundsten L, Ream M, Barry J, DeVogelaere A (2009) Endemicity, biogeography, composition, and community structure on a northeast Pacific seamount. *PLoS ONE* **4**, e4141.

McFiggans G, Cox RA, Mössinger JC, Allan BJ, Plane JMC (2002) Active chlorine release from marine aerosols: roles for reactive iodine and nitrogen species. *Journal of Geophysical Research* **107** (D15), 4271.

Menot L, Sibuet M, Carney RS, Levin LA, *et al.* (2010) New perceptions of continental margin biodiversity. Chapter 5, In *Life in the World's Oceans* (ed. McIntyre AD). Wiley-Blackwell, Oxford UK, 79–101.

Middelburg JJ, Levin LA (2009) Coastal hypoxia and sediment biogeochemistry. *Biogeosciences Discuss* **6**, 3655–3706. www.biogeosciencesdiscuss.net/6/3655/2009/

Milliman JD, Haq BU, eds (1996) *Sea-level Rise and Coastal Subsidence: Causes, Consequences, and Strategies*. Kluwer Academic Publishers, The Netherlands.

Milsum JH (1968) Mathematical introduction to general system dynamics. In *Positive Feedback* (ed. Milsum JH). Pergamon Press, New York, 23–65.

Monzón-Argüello C, Dell'Amico F, Marco A, López-Jurado LF, *et al.* (2012) Lost at sea: genetic, oceanographic and meteorological evidence for storm-forced dispersal. *INTERFACE Journal of the Royal Society*, online before print February 8, 2012.

Mooers CNK (1976) Overview of the physical dynamics of the continental margin. *Hydrological Sciences Bulletin* **21**, 467–471.

Nicholls RJ, Cazenave A (2010) Sea-level rise and its impact on coastal zones. *Science* **328**, 1517–1520.

Nielsen SN, Müller F (2000) Emergent properties of ecosystems. In *Handbook of Ecosystem Theories and Management* (eds Jørgensen SE, Müller F). Lewis Publications, CRC Press, 195–216.

NRC (1992) *Coastal Meteorology: a Review of the State of the Science*. Panel on Coastal Meteorology, National Research Council. National Academy Press, Washington, D.C.

Nellemann C, Corcoran E, Duarte C, Valdes L, DeYoung C, Fonseca L, Grimsditch G, eds (2009). Blue Carbon. A rapid response assessment. United National Environment Programme, GRID-Arendal. www.grida.no

Odum EP (1989) *Ecology and Our Endangered Life-Support Systems*. Sinauer Assoc., Inc., Sunderland, Massachusetts.

Odum WE, Odum EP, Odum HT (1995) Nature's pulsing paradigm. *Estuaries* **18**, 547–555.

Oki T, Kanae S (2006) Review: global hydrological cycles and world water resources. *Science* **313**, 1068–1072.

O'Neill RV, DeAngelis DL, Waide JB, Allen TFH (1986) *A Hierarchical Concept of Ecosystems*. Monographs in Population Biology 23. Princeton University Press, Princeton, New Jersey.

Orians GH (1975) Diversity, stability and maturity in natural ecosystems. In *Unifying Concepts in Ecology* (eds van Dobben WH, Lowe-McConnell RH). W. Junk BV Publishers, The Hague, 139–150.

Orth RJ, Carruthers TJB, Dennison WC, Duarte CM, Fourqurean JW, Heck Jr. KL, Hughes AR, *et al.* (2006) A global crisis for seagrass ecosystems. *BioScience* **56**, 987–996.

Perillo GME, Wolanski E, Cahoon DR, Brinson MM, eds (2009) *Coastal Wetlands: an Integrated Ecosystem Approach*. Elsevier, Amsterdam, The Netherlands.

Pernetta JC, Milliman JD (1995) *Land-ocean Interactions in the Coastal Zone—Implementation Plan*. IGBP Global Change Report. International Geosphere—Biosphere Programme, Stockholm, No. 33.

Petraitis PS, Dudgeon SR (2004) Detection of alternative stable states in marine communities. *Journal of Experimental Marine Biology and Ecology* **300**, 343–371.

Pielou EC (1979) *Biogeography*. John Wiley and Sons, New York.

Prigogine I (1980) *From being to becoming*. W. H. Freeman and Company, New York, 272pp.

Prospero PM (1999) Long-term measurements of the transport of African mineral dust to the southeastern United States: implications for regional air quality. *Journal Geophysical Research* **104**, 15917–15927.

Raffaelli D, Bell E, Weithoff G, Matsumotoc A, Cruz-Mottad JJ, P Kershawe, *et al.* (2003) The ups and downs of benthic ecology: considerations of scale, heterogeneity and surveillance for benthic–pelagic coupling. *Journal of Experimental Marine Biology and Ecology* **285–286**, 191–203.

Ray GC, Gregg Jr. WP (1991) Establishing biosphere reserves for coastal barrier ecosystems. *BioScience* **41**, 301–309.

Ray GC, Hayden BP (1992) Coastal zone ecotones. In *Landscape Boundaries: Consequences for Biotic Diversity and Ecological Flows* (ed. di Castri F, Hansen AJ). Springer-Verlag, New York, 403–420.

Ray GC, McCormick-Ray MG (1989) Coastal and marine biosphere reserves. In *Proceedings of the Symposium on Biosphere Reserves*. The 4th World Wilderness (eds Gregg, Jr. WP, Krugman SL, Wood Jr. JD). U.S. Dept. Interior, Atlanta, Georgia, 68–78.

Redfield AC (1958) The biological control of chemical factors in the environment. *American Scientist* **46**, 205–221.

Richardson PL (2008) On the history of meridional overturning circulation schematic diagrams. *Progress in Oceanography* **76**, 466–486.

Richer de Forges BR, Koslow JA, Poore GCB (2000) Diversity and endemism of the benthic seamount fauna in the southwest Pacific. *Nature* **405**, 944–947.

Riebesell U, Körtzinger A, Oschlies A (2009) Sensitivities of marine carbon fluxes to ocean change. *Proceedings of the National Academy of Sciences USA* **10**, 20602–20609.

Roy PS (1984) New South Wales estuaries: their origin and evolution. In *Coastal Geomorphology in Australia* (ed. Thom BG). Academic Press, Sydney, 99–121.

Ruhl HA (2007) Abundance and size distribution dynamics of abyssal epibenthic megafauna in the northeast Pacific. *Ecology* **88**, 1250–1262.

Scheffer M, Carpenter SR (2003) Review: catastrophic regime shifts in ecosystems: linking theory to observation. *TRENDS in Ecology and Evolution* **18**, 648–656.

Scheffer M, Bascompte J, Brock WA, Brovkin V, Carpenter SR, Dakos V, *et al.* (2009) Early-warning signals for critical transitions. *Nature* **461**, 53–59.

Schlacher TA, Dugan J, Schoeman DS, Lastra M, Jones A, Scapini F, *et al.* (2007) Sandy beaches at the brink. *Diversity and Distributions* **13**, 556–560.

Schlesinger WH (1997) *Biogeochemistry: an Analysis of Global Change*. Academic Press, 588pp.

Schröder A, Persson L, De Roos AM (2005) Direct experimental evidence for alternative stable states: a review. *Oikos* **110**, 3–9.

Schrödinger E (1944) *What is life?* Cambridge University Press, Cambridge, UK.

Schwarzer K, Diesing M, Larson M, Niedermeyer R-O, Schumacher W, Furmanczyk K (2003) Coastline evolution at different time scales—examples from the Pomeranian Bight, southern Baltic Sea. *Marine Geology* **194**, 79–101.

Siegenthaler U, Sarmiento JL (1993) Atmospheric carbon dioxide and the oceans. *Nature* **265**, 119–125.

Sikora TD, Ufermann S (2005) Marine atmosphere boundary layer cellular convection and longitudinal roll vortices. In *Synthetic Aperture Radar Marine User's Manual* (eds Jackson JR, Apel JR). Office of Research and Applications, NOAA, Washington, D.C., Chapter 14, 321–330.

Snelgrove PVR (1998) The biodiversity of macrofaunal organisms in marine sediments. *Bioscience and Conservation* **7**, 1123–1132.

Soloviev A, Lukas R (2006) *The Near-Surface Layer of the Ocean*. Springer, The Netherlands, 574pp.

Steneck RS, Graham MH, Bourque BJ, Corbett D, Erlandson JM, Estes JA, Tegner MJ (2002) Kelp forest ecosystems: biodiversity, stability, resilience and future. *Environmental Conservation* **29**, 436–459.

Stephenson WJ, Branderb RW (2003) Coastal geomorphology into the twenty-first century. *Progress in Physical Geography* **27**, 607–623.

Suchanek T, Williams SL, Ogden JC, Hubbard, DK, Gill IP (1985) Utilization of shallow-water seagrass detritus by Caribbean deep-sea macrofauns: *del* $_{13}C$ evidence. *Deep Sea Research* **32**, 201–214.

Swap R, Garstang M, Greco S, Talbot R, Kallberg P (1992) Saharan dust in the Amazon Basin. *Tellus* **44B**, 133–149.

Syvitski JPM, Kettner A (2012) Sediment flux and the Anthropocene. *Philosophical Transactions of the Royal Society A* **369**, 957–975.

Syvitski JPM, Vörösmarty CJ, Kettner AJ, Green P (2005) Impact of Humans on the Flux of Terrestrial Sediment to the Global Coastal Ocean. *Science* **308**, 376.

Telesh IV, Khlebovich VV (2010) Principal processes within the estuarine salinity gradient: a review. *Marine Pollution Bulletin* **61**, 149–155.

Terray EA, Donelan MA, Agraval YC, Drennan WM, Kahma KK, Williams III AJ, *et al.* (1996) Estimates of kinetic energy dissipation under breaking waves. *Journal of Physical Oceanography* **26**, 792–807.

Thomas H, Bozec Y, Elkalay K, de Baar HJW (2004) Enhanced open ocean storage of CO_2 from shelf sea pumping. *Science* **304**, 1005–1008.

Trenberth K (2005) Uncertainty in hurricanes and global warming. *Science* **308**, 1753–1754.

USGS (1970) *The national atlas of the United States of America*. U.S. Dept. Interior Geological Survey, Washington, D.C.

USGS (2010) *Divisions of geologic time—major chronostratigraphic and geochronologic units*. Fact Sheet 2010–3059, 2pp.

Van Dover CL (2000) *The Ecology of Deep-Sea Hydrothermal Vents*. Princeton University Press. Princeton, New Jersey.

Veron JE (2008) Mass extinctions and ocean acidification: biological constraints on geological dilemmas. *Coral Reefs* **27**, 459–472.

Valiela L (1984) *Marine Ecological Processes*. Springer-Verlag, New York, 546pp.

Vinogradova NG (1997) Zoogeography of the abyssal and hadal zones. *Advances in Marine Biology* **32**, 325–387.

Voigt A (2011) The rise of system theory in ecology. In *Ecology Revisited* (eds Schwarz A, Jax K). Springer, Dordrecht, Part 5, Chapter 15, 183–194.

Vinogradova NG (1997) Zoogeography of the abyssal and hadal zones. *Advances in Marine Biology* **32**, 32–387.

Walker B, Myers JA (2004) Synthesis: thresholds in ecological and social–ecological systems: a developing database. *Ecology and Society* **9**, 3. Online. www.ecologyandsociety.org/vol9/iss2/art3

Walker B, Salt D (2012) *Resilience Practice*. Island Press, Washington DC.

Wang W, Huang RX (2004) Wind energy input to the surface waves. *Journal of Physical Oceanography* **34**, 1276–1280.

Weinberg GM (1975, 2001) *An Introduction to General Systems Thinking*. John Wiley & Sons, New York. Reprinted, *Silver Anniversary*, Dorset House Publication Company, NY.

Wilhelm SW, Suttle CA (1999) Viruses and nutrient cycles in the sea. *BioScience* **49**, 781–788.

Williams JT, Carpenter KE, Van Tassell JL, Hoetjes P, Toller W, *et al.* (2010) Biodiversity assessment of the fishes of Saba Bank Atoll, Netherlands Antilles. *PLoS ONE* **5**: e10676.

CHAPTER 5

NATURAL HISTORY OF MARINE ORGANISMS

What, is the jay more precious than the lark
Because its feathers are more beautiful?
Or is the adder better than the eel
Because his painted skin contents the eye?

Petruchio in Shakespeare, W. *Taming of the Shrew* (Act IV, Scene iii)

5.1 WHAT IS NATURAL HISTORY?

Petruchio makes an important point, that appearances are not necessarily indicators of ecological value or priorities for conservation. Humans naturally relate other species to themselves, as "large" or "small," "fast" or "slow," "friendly" or "fierce," and living "long" or "short" lives. Rather, studies of natural history allows understanding of species from *their own perspectives*—to get a "feel" for how species adapt and function in their natural environments (Wilson, 2012). That is, conservation, evolution, and interpretations of species behavior are rooted in natural history, which is all about observation (Box 1.1). Or as George Bartholomew (1986) put it: "Indifference to a phenomenon's natural context can result in a paralyzing mismatch between the problem and the questions put to it."

Natural-history studies encompass both life-history strategies and evolutionary adaptations, directed by the most powerful of scientific theories about life—evolution. It involves studies of living forms as they make a living in their own environments, and how they adapt and respond to complex and changing environments. But because human senses are limited in the marine world, technical modalities—scuba, electronic tags, satellites, hydrophones, ships, submersibles, etc.—are required to extend and expand marine natural-history studies underwater.

Natural history is approached today from several different perspectives. For ecologists, community and ecosystem pattern and process are paramount. For conservation biologists, the focus is on ways species and their communities function. For managers, maintenance of particular species and their habitats is a mandate. For social scientists, biodiversity can be seen as a social "good." For businessmen, species and their habitats represent economic value, biotechnology, trade, or tourism—or, conversely, an impediment to development. For politicians, biodiversity can represent conflicts between nature preservation and social or economic values. But for all, better knowledge of species and how they function within their ecosystems has the potential to alter notions about how species are able to thrive and provide environmental services in a rapidly changing world. If biodiversity loss and ecosystem degradation are to be reversed, knowledge of natural history becomes ever more important, to help explain species fitness, persistence, resistance, resilience, and vulnerabilities to disturbance. It is first based on careful annotation in the field that documents and describes the "what" and "where" of organisms, followed by the "how" and "why," and subsequently by hypotheses for testing.

Here, we briefly summarize aspects of the extraordinary diversity of marine species' life histories, biological patterns in space and time, and the nexus of species natural history and ecosystem functions. We do not attempt to explain all of extant natural-history theory. Rather we emphasize, through examples, matters especially pertinent to conservation, with the objective of engaging readers to look further, to experience marine life directly, and to assist in developing new and better understanding to achieve conservation goals.

5.2 DARWINIAN EVOLUTION

Theodosius Dobzhansky (1973) famously noted that "nothing in biology makes sense except in the light of evolution." Charles Darwin based his book, *On the Origin of Species* (1859), on natural history and concluded that species change by processes of natural selection. At the same time, Alfred Russell Wallace came to the same conclusion. From these and subsequent field studies, the "theory of evolution" is now supported by a large body of fact drawn from paleontology, morphology, genetics, physiology, behavior, and systematics, and explains how life has evolved to occupy all of Earth's ecosystems.

Through decades-long periods of observation, Darwin established five major theories (Mayr, 1997): (i) that organisms steadily evolve over time—the theory of evolution as such; (ii) that different kinds of organisms descended from a common ancestor—the theory of common descent; (iii) that species multiply over time—the theory of speciation; (iv) that evolution takes place through gradual change of populations—the theory of gradualism; and (v) that the mechanism of evolution is competition among vast numbers of unique individuals for resources, which leads to differences in survival and

Marine Conservation: Science, Policy, and Management, First Edition. G. Carleton Ray and Jerry McCormick-Ray.

reproduction—the theory of natural selection. The first four theories were rapidly adopted at the time; the fifth was controversial, largely because of lack of a mechanism, namely genetics, proposed shortly after *Origin of Species* by Gregor Mendel in 1866.

Scientists now consider evolution as a process by which organisms have evolved into modern forms through such mechanisms as competition, cooperation, genetics, mutation, migration, natural selection, and biological expansion, among others. Evolutionary trends include both programmed intrinsic factors and complex, open-ended, contingent, and non-programmed, extrinsic biological and physical factors. Presently, biologists almost universally accept natural selection, and research proceeds more fully to explain it. Natural history remains the key underlying attribute.

5.3 DIVERSITY OF MARINE LIFE

"Biodiversity" is a recently coined contraction of "biology" and "diversity," and is widely used to describe the variety of life. The *Convention on Biodiversity* (Ch. 3) defines biodiversity as: the "variability among living organisms from all sources including, *inter alia*, terrestrial, marine and other aquatic ecosystems and the ecological complexes of which they are part; this includes diversity within species, between species and of ecosystems." This definition suggests that descriptive science matters (Grimaldi and Engel, 2007).

Darwin's *Origin of Species* was successful because it provided vast descriptive evidence from the study of life. However, only a minority of species has names, and for only a tiny fraction of those is natural history reasonably well known. Essential questions are: What is a species? Why have so many species evolved? How might species change their ways, disappear, or evolve into new forms in a rapidly changing world? In what ways have species accommodated to environmental heterogeneity and to the dynamics of marine systems? And by what means does biodiversity yield so many benefits to humans?

5.3.1 Taxonomic diversity

Taxonomy is a branch of science concerned with evolutionary relationships and classification, and is a rapidly evolving field of study. As May (1990) put it: "Without taxonomy to give shape to the bricks, and systematics to tell us how to put them together, the house of biological science is a meaningless jumble." Taxonomy examines species' genetic and morphological characteristics and develops a classification that reflects evolutionary history. The most widely accepted taxonomy uses Charles Darwin's criteria of inheritance (genealogy) and similarity (morphology) to describe phylogeny (relationships), as defined by its membership in a hierarchical classification of taxa: i.e., species, genus, family, order, class, phylum, kingdom, and domain. The most fundamental measure of diversity lies at the phylum level, whereby marine systems exhibit about twice the diversity for animals as land or freshwater (Fig. 5.1).

In 1866, Ernst Haeckel suggested that the tree of life had only two main branches—plants and animals. Scientists later found that some single-celled species, e.g., flagellates, exhibit

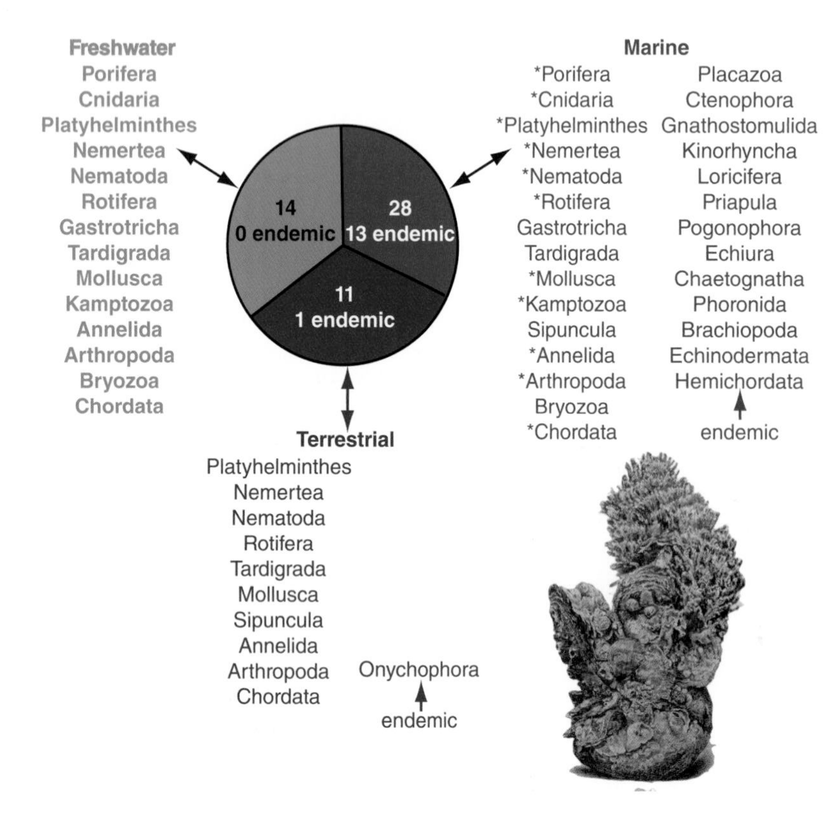

Fig. 5.1 Phyletic diversity is richest in marine environments and poorest terrestrially. Eleven phyla (*) have symbiotic species, and an additional four phyla (not listed: Orthonectida, Dicyemida, Nematomorpha, and Acanthocephala) are exclusively symbiotic and often parasitic. *Note*: Some phyla are listed elsewhere under different names, and different classifications exist; as many as 37 animal phyla have been proposed (May, 1988). *Inset*: Oyster patch. More than a dozen phyla may have occurred on this clump of eastern oysters (*Crassostrea virginica*)—more than for all terrestrial environments together. Based on data from May (1988).

both plant and animal characteristics. Today, five kingdoms of life (excepting viruses) are recognized: Bacteria, Protozoa, Fungi, Plantae, and Animalia, with a sixth recently proposed, Chromista (algae) (Cavalier-Smith, 2004). All six kingdoms contain both unicellular and macroscopically visible organisms. "Microbes" include bacteria that are ancient prokaryotic cells, i.e., they lack cell nuclei; this group includes many more phyla than do macroorganisms. Bacteria probably diversified by symbiogenesis, whereby two or more species combine genetic material. All other kingdoms are eukaryotic (have cells with nuclei).

One of the first questions biologists ask is: How many species exist (May, 1988)? Total known eukaryotic species number approximately 1.8–1.9 million (May and Harvey, 2009); credible estimates for total numbers are 5–10 million, with defensible estimates of 3–100 million. The portion that are marine is unknown, but it seems almost certain that named marine species might one day outnumber terrestrial ones; this is predictable because rates of species discovery are astounding and because three-dimensional marine systems are two orders of magnitude greater in size than the land, and ecologically as diverse as well. Surprisingly, two decades ago deep-sea communities of the continental slope and rise were discovered, and may be among the most diverse of all; samples of infauna at 1500–2500 m depths off the U.S. east coast contained 798 species from 71 families, and 14 phyla (Grassle and Maciolek, 1992), possibly only the "tip of the iceberg" of what still remains to become known to science.

The *Census of Marine Life* constitutes the most ambitious accounting of marine life to date (Box 5.1; Snelgrove, 2010). In 1982, discovery of a tiny, benthic-dwelling invertebrate, *Nanaloricus mysticus*, added a new phylum, Loricifera, to the catalog of marine life. The recent discovery of abundant deep-sea annelids (Acrocirridae) with unique bioluminescence "bombs" emphasizes our limited knowledge of deep-sea communities (Osborn *et al.*, 2009). Microbes offer a new view of biodiversity. The Global Ocean Sampling project has yielded "a tidal wave of microbial DNA" (Bohannon, 2007), for which the "boundary between living and dead is fuzzy" (Pearson, 2008); the number of microorganisms defies imagination—on the order of a million cells cm^{-3}, the oceanic total being equivalent to the total biomass of terrestrial plant life. Most abundant of all in the world's oceans are viruses (Danovaro *et al.*, 2008). Rates of discovery of vertebrates are not to be neglected. Nelson (1984) estimated that 21,700 species of fishes might exist, of which 60% are marine and coastal; Nelson (2006) estimated 32,000 species, a 47% increase in less than three decades! A major question is: why do so many species coexist? For coral-reef fishes, the answer seems to lie in post-settlement processes, although a variety of factors maintain high species diversity from system to system (Box 5.2).

However, the living world is richer, more diverse, and more dynamic than taxonomy alone suggests. For example, diversity is diminished when all life-history stages of a species are lumped under one name. Most marine species—especially invertebrates and some fishes—exist in several distinct forms

Box 5.1 The *Census of Marine Life*

Paul Snelgrove
Memorial University of Newfoundland, St. John's, Newfoundland, Canada

J. Frederick Grassle
Rutgers University, New Brunswick, New Jersey, USA

Who lives in the sea, where do they live, and how abundant are they? As basic as these questions seem, the major knowledge gaps they represent formed the basis for the *Census of Marine Life*. In 1996, frustrated by inaction on widespread concern of accelerating human impacts and potential extinction of marine species before they were even recognized by science, U.S. scientist Fred Grassle approached Jesse Ausubel, a program officer with the Sloan foundation. Their initial discussion eventually led to the launch of the global *Census of Marine Life* in 2000, a ten-year program that would engage more than 2700 researchers from 80 nations around the world to raise over $650 million for marine biodiversity research. The program emphasized communication and collaboration, dividing the work among 14 different field projects and an additional three focused on data and integration. It spanned from the shoreline to the abyss, from microbes to whales, and everything in between, as it considered oceans past, present, and future.

The timing proved fortuitous. Recent advances in genetic and ocean observation technologies meant that Census researchers could identify different species, even microbes, unambiguously. The use of genetic barcoding using specific genes and genetic sequences to differentiate species resolved many taxonomic challenges. The possibility of up to a billion different kinds of microbes in the oceans fundamentally changed our past view of the microbial world that "everything is everywhere." Simultaneously, remotely operated vehicles using new high-definition cameras illuminated the deepest and least-studied ocean habitats, facilitating the discovery of new hydrothermal vents and seeps, revealing patchier and more complex habitats than previously recognized. In coastal zones, Census researchers engaged school children as "citizen scientists" to help collect data from intertidal habitats around the world using standardized methodologies. Some researchers found new ways to look backward in time using ships' logs and

(*Continued*)

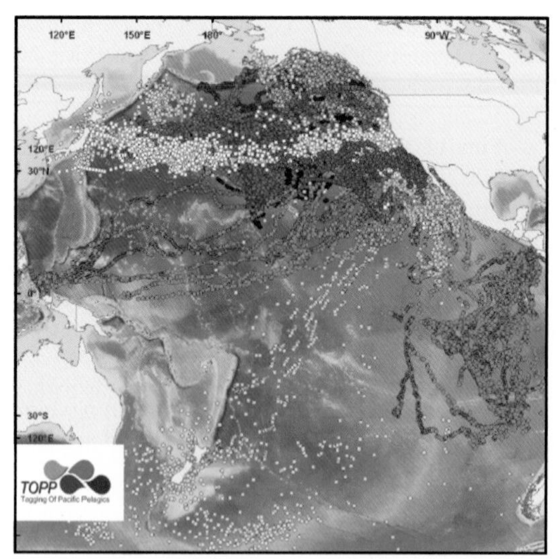

120°E
150°E
180°
90°W
TOPP
Tagging Of Pacific Pelagics
● Black footed Albatross
● California Sea Lion
○ Humpback Whale
● Laysan Albatross
● Northern Elephant Seal
○ Sooty Shearwater
● Albacore
● Blue Shark
○ Humboldt Squid
● Leatherback Turtle
● Loggerhead Turtle
● Mako Shark
○ Mola
○ Pacific BluefinTuna
● Salmon Shark
● Thresher Shark
○ White Shark
○ Yellowfin Tuna

Fig. B5.1.1 Migration pathways of species tagged by the Tagging of Pacific Predators project. From Block *et al.* (2010). Reproduced with permission of John Wiley & Sons.

Data SIO, NOAA, U.S. Navy, NGA, GEBCO
© 2012 Cnes/Spot Image
Google earth

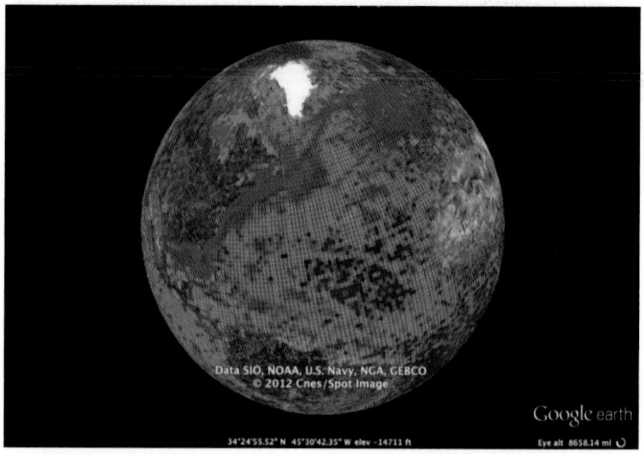

Fig. B5.1.2 OBIS data records for the global ocean as visualized in Google Earth. Red denotes comparatively well-sampled areas whereas blue denotes poorly sampled areas. This image emphasizes surface waters, noting that sampling effort typically declines rapidly with depth. From Vanden Berghe (2010). Reproduced with permission of John Wiley & Sons.

monastery records to reconstruct oceans past, and others found new ways to incorporate biodiversity knowledge into more holistic ocean management.

Census scientists found ocean life more diverse, more dynamic, more impacted by humans, and, in some cases, more resilient than previously realized. New species were discovered in just about every habitat sampled from the intertidal to the deep sea, from the ocean's surface to the seafloor, and in taxa spanning from microbes to fishes. Dramatic and previously unrecognized declines in multiple species targeted by fisheries indicated 65–98% declines in major taxa of marine mammals, fishes, and seabirds, contrasted with rediscovery of a few species thought long extinct. Census scientists used new satellite tags to identify major migration routes used by marine mammals, birds, turtles, and large fishes in the open ocean (Fig. B5.1.1), and locations where they congregate. Some of the migrations span thousands of kilometers across ocean basins through international waters, and tags augmented with small environmental sensors allowed ocean life to act as oceanographic "ships," collecting novel data from remote ocean regions that research vessels rarely visit. Identifying these "blue highways" has already helped in conservation efforts for threatened species such as leatherback turtles.

Using the *Ocean Biogeographic Information System* (OBIS; www.iobis.org), the Census' major database, scientists constructed global views of ocean life and demonstrated the power of data sharing, a central tenet of the program. OBIS contains over 30 million records of life in the ocean, and helped to identify global biodiversity hotspots for well-known taxa, showing tropical peaks for coastal taxa such as fishes and temperate peaks for oceanic groups such as squids. OBIS also identified knowledge gaps, clearly showing that the deep ocean, the polar regions, and many tropical regions remain effectively unexplored (Fig. B5.1.2). Statistical extrapolations identified taxonomic gaps, showing that 91% of marine species remain undiscovered, not including the vastly under-sampled microbes. Indeed, 4–5 new marine species are still described every day, and that number is constrained by limited numbers of taxonomic specialists and large numbers of unknown species. Whereas science has already discovered some 75–90% of marine fishes, we know only a small fraction of other groups such as nematodes and even less about microbes. Clearly, much remains to be discovered even after the Census.

Sources: Block *et al.* (2010); Lotze and Worm (2009); McIntyre (2010); Mora *et al.* (2011); Ray and Grassle (1991); Snelgrove (2010); www.coml.org

Box 5.2 How do so many kinds of coral-reef fishes coexist?

Mark A. Hixon
University of Hawai'i at Manoa, Honolulu, Hawai'i, USA

Coral-reef fishes compose the most speciose assemblages of vertebrates on Earth. The variety of shapes, sizes, colors, behavior, and ecology exhibited by reef fishes is truly amazing. Reef fishes are dominated by about 30 families, mostly the perciform chaetodontoids (butterflyfish and angelfish families), labroids (damselfish, wrasse, and parrotfish families), gobioids (gobies), and acanthuroids (surgeonfishes). Worldwide, about 8,000 species of marine fishes inhabit coral reefs at some stage of their life cycle. Hundreds of species may coexist on the same reef at one time or another.

A key question for the conservation of coral-reef fishes is: How do so many species coexist? This question is important because conservation requires the identification and protection of natural mechanisms that maintain high species diversity. It is best answered at the level of the ecological guild, which is defined as a group of species that use the same general suite of resources (food, space, etc.) in the same general habitat, such as butterflyfishes that feed on coral polyps inhabiting a reef slope. The central issue is that, as population sizes of species within a community grow to levels where resources are in short supply, one or a few species within each guild should outcompete other species, thereby reducing local species diversity. What prevents such competitive exclusions?

Four hypotheses provide clues to the question of coexistence of reef fishes (Fig. B5.2.1). Present information both corroborates and refutes each hypothesis at different reefs, suggesting that all four hypotheses may be valid at some time and place.

A review of the bipartite life cycle of reef fishes is necessary before examining these hypotheses. Many reef fishes (exceptions are gobies, blennies, pipefishes, and a few others) are broadcast spawners, whose gametes and larvae undergo pelagic dispersal, with varying degrees of local retention. Typically, after about a month, late-stage larvae settle in reef or near-reef habitats. Recruitment is the measure of settlement, estimated by counts of newly settled fish. The accuracy by which recruitment actually measures settlement is a major issue in distinguishing among these hypotheses.

The niche diversification and competitive lottery hypotheses both assume that competition is strong among juveniles and adults on the reef, so that coexistence of species is maintained despite the risk of competitive exclusion. The basic idea for the former (sometimes called the "competition hypothesis") is that high overlap in resource use within a guild, combined with competition between the constituent species, selects for lower overlap or diversification of niches. This scenario results in resource partitioning, whereby species within a guild that overlap greatly in diet tend to forage in slightly different microhabitats; alternatively, species that forage in the same location may have slightly different diets. However, a description of resource partitioning provides only a pattern, not the process that caused that pattern.

Some guilds seem to coexist despite an apparent absence of resource partitioning. For example, territorial, herbivorous damselfishes are highly aggressive toward each other, and if all suitable habitat space is occupied by territories, how do such species coexist without niche diversification? The competitive lottery hypothesis (sometimes

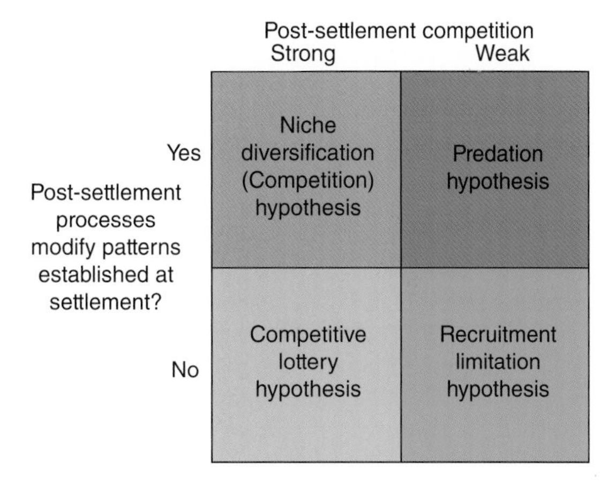

Fig. B5.2.1 Four hypotheses explaining the coexistence of many species of coral-reef fishes. From Jones GP (1991). Postrecruitment processes in the ecology of coral reef fish populations: a multifactorial perspective. In *The Ecology of Fishes on Coral Reefs* (ed. Sale PF), pp. 294–328. Academic Press, San Diego, CA.

(Continued)

called the equal chance hypothesis) offers a relatively complex explanation, based on several restrictive assumptions. First, there has to be a strong prior residency effect, whereby a fish that finds a place to live on the reef can successfully defend its territory against all comers. Second, late-stage larvae of all species have to be available to settle in any space that opens on the reef, be it by the death of a territorial fish or by the creation of new habitat space by storms or other disturbances. These larvae are analogous to lottery tickets, in that whichever individual finds the open space first is the winner of that space. Under these conditions, it is proposed that no single species can gain the upper hand in the competition for living space, despite the lack of resource partitioning. In reality, the rate of competitive exclusion may only be slowed rather than prevented, since no two species are truly equal, by definition.

The remaining two hypotheses both assume that competitive exclusion of species is not an issue because some factor keeps population sizes below levels where resources become limiting. Some fish populations have low larval settlement rates, so that living space is not as limiting as the former hypotheses assume. The recruitment limitation hypothesis proposes that low larval supply prevents juvenile and adult populations from reaching levels where substantial competition occurs, in which case post-settlement mortality is density-independent—that is, occurs at a constant proportional rate. Unfortunately, the definition of recruitment limitation has changed through time, so that recruitment is sometimes measured up to months past settlement, and early post-settlement processes are thus ignored. In fact, shortly after settlement, many reef fishes undergo density-dependent mortality in which case mortality rate increases with local population size.

Finally, as an alternative to recruitment limitation, the predation hypothesis suggests that competitive exclusion is prevented by predation rather than low larval supply. In fact, both density-dependent and density-independent predation on newly settled reef fishes, which are typically less than 2 cm long, is usually severe. Many different species of generalized reef fishes and macroinvertebrates—mostly species not normally considered piscivorous—have been found to consume new settlers. There is mounting observational and experimental evidence that such intense predation keeps populations of many reef fishes in check, precludes competitive exclusions, and thereby maintains high local species diversity.

The picture that emerges from the past several decades of research on coral-reef fishes is that a variety of factors maintain high species diversity, and that the relative importance of these factors varies from system to system. This situation indicates the truth of John Muir's admonition that "when we try to pick out anything by itself, we find it hitched to everything else in the universe." Such complexity suggests that the conservation of coral-reef fishes can be best accomplished by preserving entire systems from direct human impact in fully protected marine reserves.

Sources: Jones (1991); Polunin and Roberts (1996); Sale (1991, 2002)

during one lifetime (Fig. 5.2); each form displays unique traits in food preference, habitat, etc. (as is also true for most terrestrial invertebrates and fishes). Such life-cycle diversity indicates unique ways that marine life has evolved, suggesting a correspondingly diverse number of roles that each kind of organism plays within its community. Size is yet another measure of diversity, broadly represented in the ocean, where species are classified as: microfauna, 0.001–0.1 mm in length or breadth (bacteria are even smaller); benthic meiofauna, <0.5 mm; benthic macrofauna, >0.5 mm; and megafauna, very large organisms. Each size class fulfills different functions (see also Ch. 9, Section 9.3.2.2).

Biodiversity is also a characteristic of each environment. The benthos is higher in biodiversity than pelagic systems, and coastal areas are higher than the open ocean (Gray, 1997). A host of different species lives within the sediment (infauna), on sediment (epifauna), near the bottom (demersal), or on benthic structures. Near coasts, coral reefs have high biodiversity. However, equating different climatic zones to biodiversity can be misleading; the tropics have high coral and fish species diversity, but temperate and polar regions hold the most diversity of pinnipeds and penguins. Thus, each environment has a biodiversity "signature," both in kind (what species live there) and number (abundance).

5.3.2 Measuring biodiversity

A central problem for biodiversity conservation is assessment. Species assessment depends on indices, for which statistical generalizations are subject to considerable uncertainty. These include: (i) *richness*, the numbers of species within a sampling unit; (ii) *abundance*, the numbers of individuals within that unit; and (iii) *evenness*, a measure of the relative abundances of all species present. These three measures may be applied across three geographical scales: (a) *alpha* diversity for within-area diversity; (b) *beta* diversity for between-area diversity; and (c) *gamma* diversity for the relative numbers of species within and among large geographical regions (Roff and Zacharias, 2011). A serious drawback for applying these measures is sampling error; that is, diversity indices rarely include all or even most taxa within a specified unit, especially for marine systems, which are difficult to sample. New methods have recently become available that allow diversity indices to be drawn from data consisting simply of species lists, arising from unknown or uncontrolled sampling effort (Ch. 9, Section 9.3.2).

Two criteria commonly used in conservation are species richness and endemism, used to identify "hot spots" of diversity as candidates for protected areas. In this case, "richness"

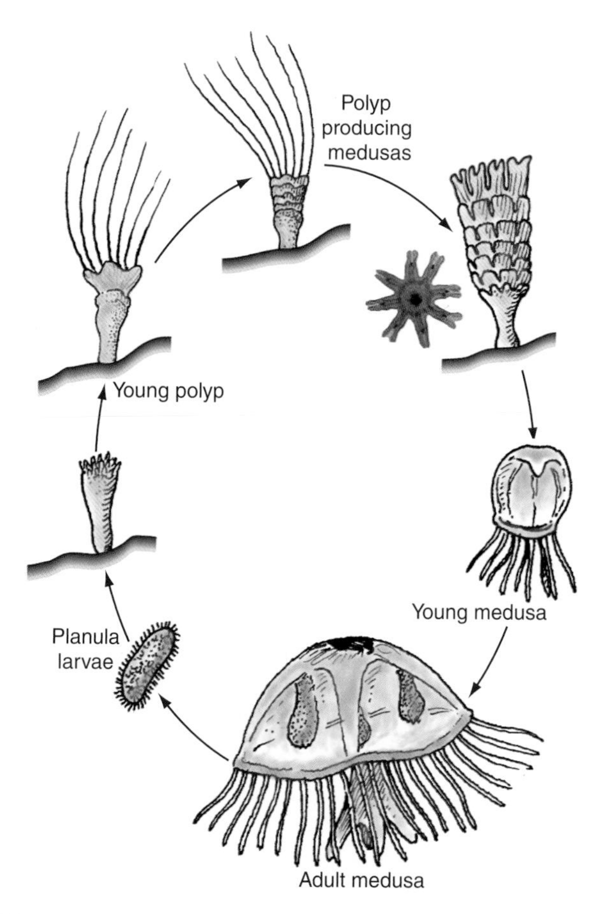

Fig. 5.2 Many coelenterates (Cnidaria: jellyfish, corals, etc.) have complex life cycles (Buchsbaum *et al.*, 1987; Moore, 2001). The moon jelly (*Aurelia*) is taxonomically a single species, but exhibits several different life-cycle stages, each with its own lifestyle. Its adult, sexual medusa, and planula larva stages are pelagic; its polyp stage is sessile and asexually reproduces medusas.

becomes a presumed ecological asset. Endemism is used to express a species' confinement to a certain area, that is a minimum range where a species only occurs and nowhere else; this is useful at small scales, but because every species is endemic at some scale, endemism can be arbitrary. For example, the "living fossil" coelacanth fish (*Latimeria chalumnae*) apparently occurs only in isolated, relatively deep, ledgy areas of the Indian Ocean. In contrast, sea turtles and giant whales are endemic to whole ocean basins. Local endemics are often of very high conservation priority under the assumption that these species are most vulnerable to disturbances. However, wide-ranging species can be equally vulnerable when any one life-history function is restricted to a small area or is generally under threat; e.g., breeding areas and nurseries of wide-ranging fishes, sea turtles, albatrosses, and marine mammals. Rapoport's rule has been proposed to deal with the scale issue; it states that the mean size of species' ranges decreases toward lower latitudes and that fewer, relatively abundant species exist within large home ranges at high latitudes. However, this rule seems to be a phenomenon only of relatively high latitudes (Rohde, 1996). Furthermore, endemicity and richness are often poorly correlated among different taxa, e.g., for coral reefs (Hughes *et al.*, 2002). Therefore, application of richness and endemism for conservation of "hot spots" seems to be limited to specific cases in which those criteria are constrained by species' natural histories.

5.3.3 Diversity of form and function

Marine species are taxonomically most diverse, but many have similar body forms, thereby underestimating diversity. A species' form, physiology, and anatomical features are indicators of its inheritance and function. When related species exhibit similar traits, "parallelism" is observed; e.g., similar claws of different species of crustacea, or streamlined body shapes in many groups of fishes. But when several unrelated organisms exhibit similar forms (anemones, bryozoans), evolutionary convergence is said to occur; for example, filter feeding is similar for gills of mollusks and fishes, as well as baleen of whales. Also, fishes and whales of completely different phyla depend on flexible but ridged body form. As Thompson (1961) said of speedy cetaceans: "... in the whale or dolphin *stiffness* must be ensured in order to enable the muscles to act against the resistance of water in the act of swimming." Thus, form is a good indicator of function, but may or may not be a good indicator of taxonomic relationships.

Body form reflects natural-history interactions with the environment. Colonial species (e.g., protozoans, corals, sponges, tunicates, etc.) challenge the basic concept of the individual (Hamilton *et al.*, 1987); e.g., are colonial species "individuals"? Colonial animals are composed of many genetically identical individuals (clones) that act together as a single organism. Conversely, in some colonial sponges and coelenterates, different individual colony members perform different functions, which when assembled together constitute an individual organism. Other colonial species take different forms under different environmental conditions. A single genotype interacting with the environment may exhibit variable observable characteristics (phenotypic plasticity) impossible to evaluate when species are poorly defined morphologically (Tautz *et al.*, 2003). Individuals that may be morphologically similar, but are genetically distinct, have been identified as "sibling species" (e.g., the common edible mussel, *Mytilus edulis*, and *Montastraea* corals). If all such artifacts of taxonomy were to become known, species diversity would dramatically increase (Knowlton, 1993), but ecological function would not. To help solve this problem, species identification is being increasingly based on genetics rather than body form (morphology). DNA evidence is rapidly improving understanding of basic biogeographic patterns.

Size is another important factor when considering ecological function. Size limits the numbers of individuals and numbers of species in any environment (May, 1988). Marine organisms have a much greater size/weight range than terrestrial ones, extending approximately 21 orders of magnitude from the smallest bacteria at 10^{-13} g to blue whales at 10^{8} g. The phenomenon described as Cope's rule is the tendency for organisms in evolving lineages to increase in size over time. Size can benefit survival, mating success, and fecundity, but larger size can also lead to an increase in development time and greater vulnerability to environmental crises (Hone and Benton, 2005). Larger-sized individuals also take up more space in limited environments. Unlike terrestrial plants, primary producers (autotrophs) in the sea are predominantly small (phytoplankton), with the exception of seagrasses and giant kelps. Smaller marine plants and animals are better able to assimilate dissolved nutrients directly through external tissues due to high surface-to-volume ratios. Large kelp attached to deep substrates gain light advantages when they grow towards the surface. Marine consumers (heterotrophs) generally consume smaller prey than themselves. A few species defy this "big fish eat little fish" rule; e.g., large sharks, barracudas, bluefish, and orcas tear apart prey larger than themselves. On the other hand, the largest marine species (whales and whale sharks) eat tiny zooplankton in vast quantities; conversely, numerous small amphipods can devour large whale carcasses.

The expected confluence of form and function can lead to non-intuitive conclusions unless placed in context. One tiny autotrophic dinoflagellate can switch feeding mode to become a fish predator (*Pfiesteria piscicida*; Burkholder *et al.*, 2001). Some predatory fishes disguise form in order not to reveal function (Fig. 5.3a). Others use disguise so as not to become prey themselves (Fig. 5.3b). And, walruses (*Odobenus rosmarus*) do not dig for benthic food with their tusks, as had been formerly assumed (Ch. 7).

Fig. 5.3 Camouflage hides species' identity, exemplified by: (a) spotted scorpionfish, *Scorpaena plumieri*. Look in the center for the eye between the two blue *Chromis*. (b, c) Color phases of peacock flounder, *Bothus lunatus*. The pale phase was adopted almost immediately after swimming from the coral head and landing on sand. Look for the light blue outlines. Photographs © Ray & McCormick-Ray.

5.4 LIFE HISTORY

Life history expresses how species make a living in their natural environments. Marine species fall into five broad categories:
• *pelagic*: species that swim freely in the water column, sometimes for considerable distances;
• *demersal*: free-swimming species that live near the bottom;
• *symbiotic*: species that live in close association with other species; i.e., parasites, commensals, and mutualists;
• *structural*: sessile "engineers" (foundation species) that build biogenic structures, such as reefs, algal forests, and seagrass beds;
• *interstitial*: motile species that inhabit living structures and interstices or surfaces of abiotic rocky and sedimentary environments; i.e., cryptic species and infauna.

Many species' life histories often represent a mix of these types. For example, larvae of most invertebrates and fishes are pelagic, although adults may be sessile or parasitic. Nevertheless, all must conform to life in water.

5.4.1 Life in a fluid ocean environment

Life in the ocean environment is very different from terrestrial life. Change is constant and surprises are inevitable. To meet life-history requirements under varying changes of temperature, salinity, density, oxygen, and intense motion, an individual thrives or tolerates, makes physiological adjustments, or retreats. Sea life also contends with water's molecular arrangement of two hydrogen atoms and one oxygen atom (H_2O) that gives water its unique properties (Skinner, 2010).

Temperature is a major controlling factor in sea life distribution, abundance, growth, and reproduction. Ocean temperatures vary from approximately −1.9 to 40°C, whereas air temperature ranges over >80°C. In air, oxygen is relatively abundant, with approximately 9 moles O^2 m^{-3} at 10°C; in seawater, maximum oxygen concentration at 10°C is approximately 0.35 moles m^{-3}, which at 30°C falls to around 0.2. Therefore, cold polar water is relatively oxygen-rich, whereas many tropical organisms live at or near their upper thermal and lower oxygen limits. Most organisms require oxygen and give off carbon dioxide as they respire. Invertebrates and fishes are "cold-blooded" (ectotherms, poikilotherms), deriving their oxygen from water. Some fast-swimming fishes (tunas, marlins, and swordfishes) meet oxygen demands by having special tissues that allow them to achieve higher body temperatures than the surrounding water. High oxygen demands of "warm-blooded" species (endotherms, also termed homeotherms: birds, mammals) preclude them from living without access to air. Sulfur-reducing bacteria can live without oxygen, obtaining energy from chemosynthesis under hypoxic or anoxic conditions.

Light plays a major role in the distribution of marine organisms. Most species live in the euphotic zone where light is sufficient for photosynthesis (Fig. 4.1). Just below the surface within this euphotic zone, light intensity is at least halved, and then rapidly decreases with depth. The clearest water transmits only about 1% of ambient light to depths of 200 m, where only the blue end of the spectrum penetrates. Turbidity, density, turbulence, back-scattering due to particles, and the Sun's angle relative to the surface also strongly affect light penetration. In the deep sea below the aphotic zone and thermocline, cold temperatures, lack of light, and high hydrostatic pressures have profound effects on the types of organisms that can live there.

Water plays major roles in physiology and behavior. Solutes move easily across cell membranes. Sodium ions in sea salt affect an organism's internal water content, requiring osmotic and ionic regulation by the organism to maintain homeostatic balance between extra- and intracellular fluids; some organisms regulate (osmoregulators) and others do not (osmoconformists). Intertidal mollusks may respond to unfavorable salinities by closing their shells. Some estuarine-marine reptiles, birds, and marine mammals have salt glands with secretory capacity consistent with seawater. Diamondback terrapins (*Malaclemys terrapin*), being both terrestrial and fully estuarine reptiles, have salt glands intermediate in secretory capacity. And sea turtles and many seabirds possess mechanisms to eliminate excess salt.

Water keeps objects of lesser density afloat. Buoyancy is the upward force that makes objects float—icebergs, ships, human swimmers; i.e., Archimedes' principle states that a force equal to the weight of the displaced fluid buoys up a body immersed in the fluid. But if the object's weight is heavier than the weight of water displaced, it sinks. Most plants and animals are heavier than the water displaced and tend to sink, thus requiring specialized mechanisms to control buoyancy. Oil droplets or long external spines prevent or reduce planktonic organisms from sinking. Algae have gas floats to hold their photosynthetic blades near the surface. As fish move through various depths of water, they require physiological adjustments because as water pressure increases with depth, the volume of gas decreases (Boyle's law). Most fishes have a membrane-enclosed, gas-filled swim bladder within the body cavity, which serves as a hydrostatic organ (and, in some species, for sound detection and production) without expending much energy.

Because water is relatively more opaque than air and limits vision, marine animals require special sensory mechanisms for food detection, finding mates, and for other purposes. Most invertebrates communicate and locate food by means of chemical receptors. Sharks sense pressure waves to aid prey detection. And because sound travels about 10 times faster and much farther in water than air, marine organisms have evolved unique mechanisms for communication over considerable distances. Some seals, walruses (Ch. 7, Section 7.4.1.1), and a few whales use stereotyped songs for courtship purposes. Porpoises and dolphins use sound production and reception for echolocation through a high-frequency, bio-sonar system, similar to bats.

Size, shapes, and behavioral adaptations aid organisms in their interactions with complex hydrodynamic forces, which operate under physical laws of fluid motion and conservation of mass and energy, and where streamlined, fusiform body forms facilitate locomotion (Fig. 5.4). Because water movement is slowed by contact with an object, a velocity gradient field is generated in a boundary zone. Approaching a hard surface, water encounters a "no-slip condition" that varies with substrate type and water flow character (Denny, 1988).

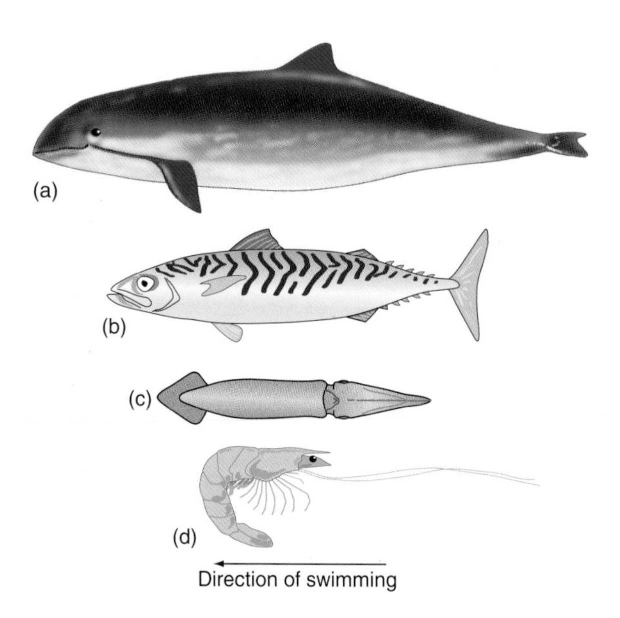

Fig. 5.4 Many swimming animals have evolved body shapes to reduce drag (Vogel, 1983). Most swimmers, such as (a) porpoise and (b) pelagic fish move forward, while (c) squid and (d) shrimps swim backward. The usual method of locomotion of the shrimp is by crawling forward on the substrate.

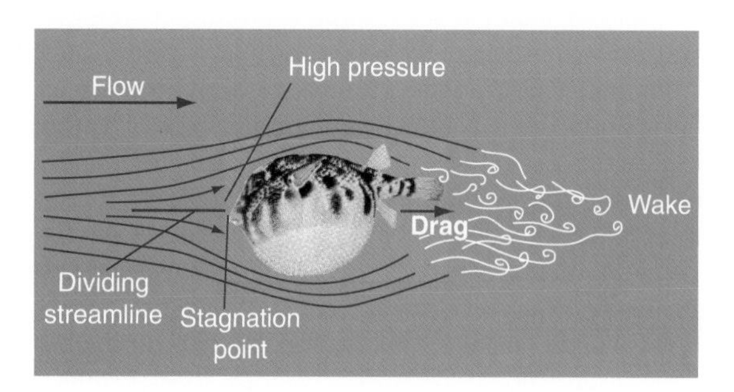

Fig. 5.5 Fluid-dynamic drag results when an object interrupts flow (Munson *et al.*, 1994; Vogel, 1994). On a round object (e.g., puffer fish), frictional drag results from relatively high pressure on its upstream face and from relatively low pressure in a turbulent downstream wake. This pressure difference tends to push the object (fish) downstream in the direction of flow. The hydromechanics of the tail reduce the effects of drag.

Because of the nature of water molecules and water properties of viscosity, density, and incompressibility, flowing water tends to form parallel (laminar) streamlines that are sensitive to change, especially when speed changes. With increasing velocity, streamlines separate upon encountering an object, compress over the object, and finally break down into a turbulent wake. Drag results from relatively high pressure on the object's upstream face and relatively low pressure in a turbulent downstream wake, and is measured as removal of momentum from the moving fluid. This pressure difference tends to push the object downstream in the direction of flow (Fig. 5.5). For

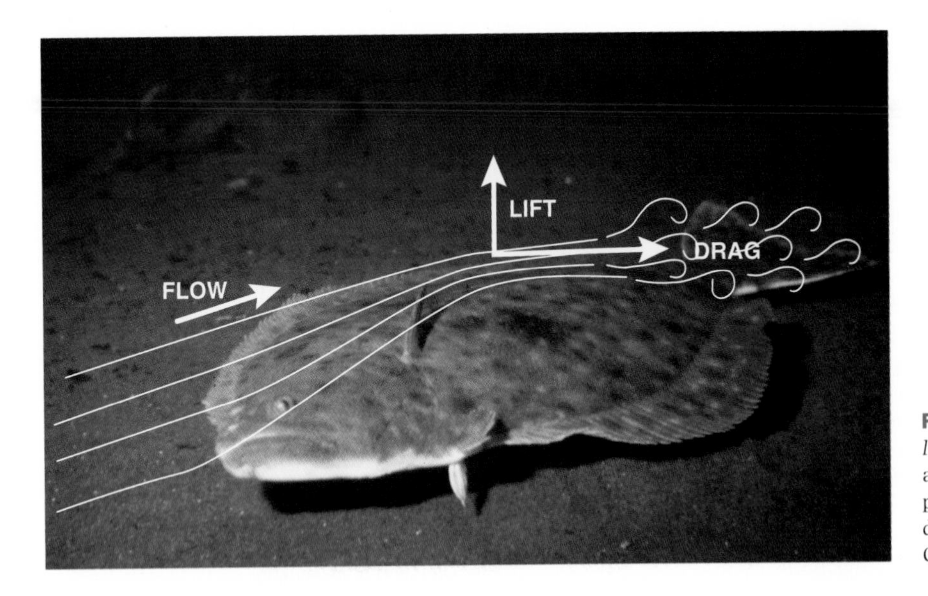

Fig. 5.6 A southern flounder (*Paralichthys lethostigma*) with different curvature above and below experiences lift, the Bernoulli principle, that acts at right angles to the direction of flow. Photograph © M. A. de Camp.

curved organisms, such as limpets attached to the bottom and flatfishes, compressed streamlines accelerate velocity, increase pressure, and create lift—Bernoulli's principle (Fig. 5.6). And sessile organisms, which totally depend on water movements for delivery of food particles, dispersal of larvae, removal of wastes, and to oxygenate the water also affect, and are affected by water patterns. Currents, turbulence, and the forces of lift, drag, and pressure act on reef communities, which alter water flow patterns as they grow (Fig. 5.7). When any of water's forces become too great, benthic organisms topple, sediment is moved, banks erode, and objects in the water column are thrown into turbulent motion. But through active and passive mechanisms, organisms can enhance hydrodynamic performance (Fish and Lauder, 2006); the reduction of turbulence by the skin of dolphins (Ridgway and Carder, 1993) remains an active field of study. The conservation of mass and the principle of continuity explain what happens to water forced through a narrowing, confined passage (Fig. 5.8); the rate of inflow must equal the rate of outflow, forcing an increase in velocity that allows for that same volume to pass and to produce turbulence. These physics of flow aid such species as sponges in filter feeding (Fig. 5.9), but have consequences when rivers are "engineered." London Bridge once rested on wide piers that doubled current speeds downstream and required construction of "cut-waters." This made the situation worse, increasing risk for small boats attempting to "shoot the bridge" (Vogel, 1983).

Experimental studies show that when flow velocity is low, streamlines form in parallel patterns (laminar flow), but as velocity increases, streamlines breakdown and become turbulent (Fig. 5.10). Flow velocity and the ability of organisms to make a living are affected by water density, temperature, and viscosity ("stickiness"). Density causes seawater to be approximately 50 times more viscous than air, to affect an organism's swimming and filtering. Temperature affects density and viscosity; seawater is about 25% more viscous at 0°C than at

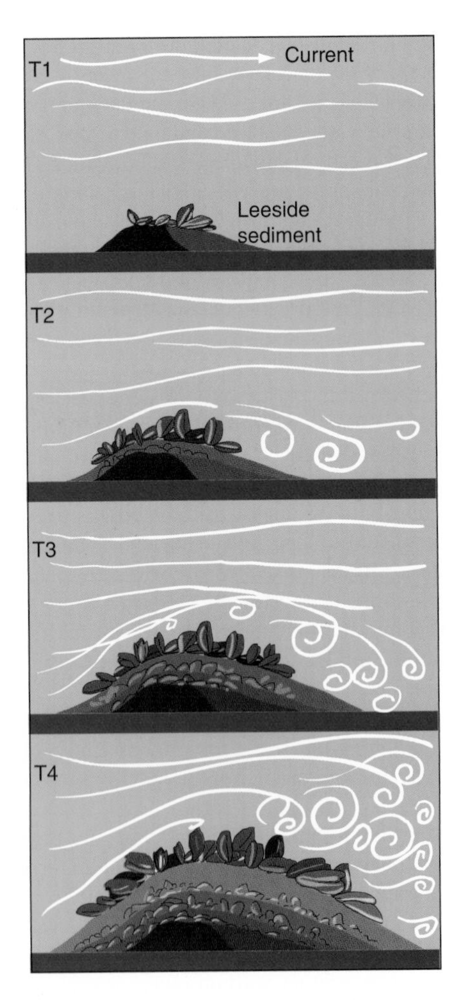

Fig. 5.7 Currents interact with an oyster bed that grows and expands. As a bed grows over time (T1–T4), turbulence increases in the water column. Based on data from Vogel (1983).

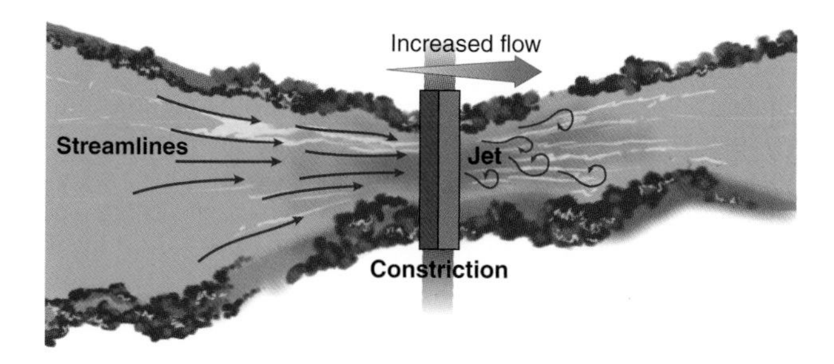

Fig. 5.8 The principle of continuity is illustrated by an analogy of water flow through a constricted area, where streamlines of fluid converge and flow volume does not change. As water enters the restriction, flow rate increases, a deeper channel is formed, and the emerging jet of water becomes turbulent on the exit side (Denny, 1988).

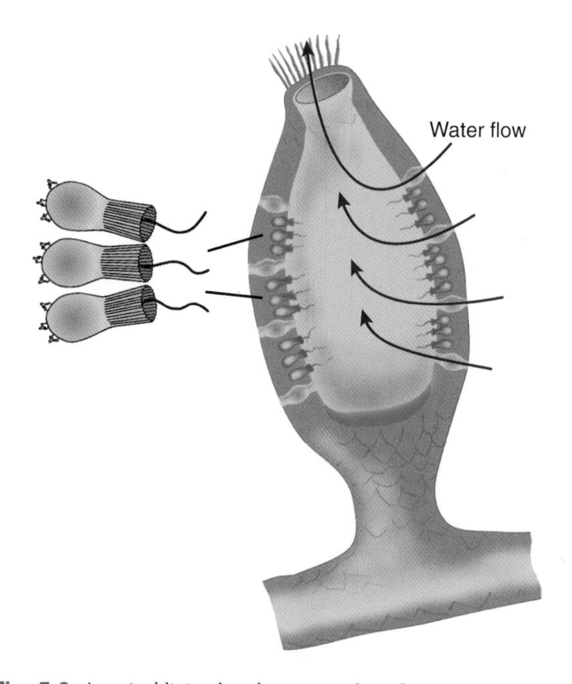

Fig. 5.9 A typical living benthic sponge benefits from the principle of continuity by means of many small (incurrent) and one or a few large (excurrent) openings on its surface and an interior canal system. Small flagellated cells (exaggerated in illustration) pump a volume of water through the larger chamber every few seconds, equal to the volume of the entire sponge, nourishing sponge cells and microorganisms (Diaz and Ward, 1997).

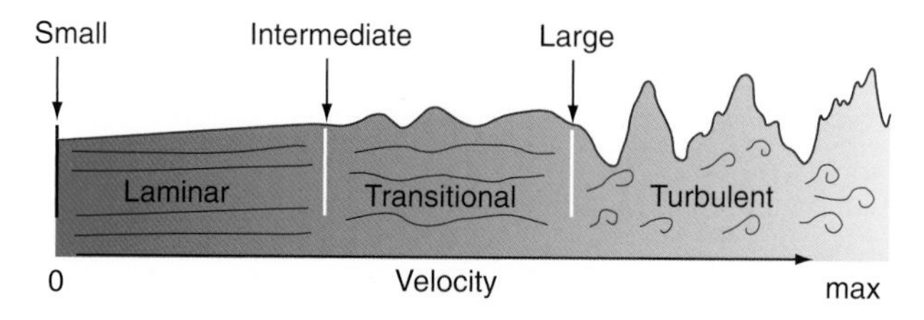

Fig. 5.10 Streamlines in an unbounded fluid follow smooth paths at low velocity, but as velocity increases, flow becomes transitional then highly irregular (turbulent) (Denny and Wethey, 2001; Munson *et al.*, 1994; Vogel, 1994).

20°C. Viscosity complicates prey capture, filter feeding, locomotion, and survival. Small organisms <~10 m are strongly affected by water's viscosity, while larger organisms generally can overcome its constraints. A small organism's motility and inertia (its natural resistance to change) are very small relative to a larger predator.

In tide pools and wave-swept shores, abrupt changes are a fact of life—one reason why intertidal areas are relatively low in species numbers, although production can be very high (Leigh *et al.*, 1987). Water's force is much greater than wind at equal speeds due to water's density, momentum, and viscosity. However, winds can attain much greater velocities. At tidal shores, water velocity is commonly only 1.6 ms^{-1}, but splashing wave velocity can reach more than 10.5 to 25 ms^{-1} when associated with breaking waves (Denny, 1989; Denny and Blanchette, 2000; Denny *et al.*, 2003). Considerable force results, and to resist it, organisms have evolved mechanisms for firm attachment; for example, algae and barnacles cement to rocky shores, and kelp anchor by holdfasts to the subtidal floor. Organisms take refuge from high velocities, turbulence, lift, and drag by seeking such protective shelters as rock crevices, other organisms, and reefs. Species inhabiting intertidal, wave-swept shores are subject not only to forces of waves and currents, but also to alterations in submergence, exposure to air, extreme heat and cold, desiccation and drowning, salt and freshwater, moving objects, and collisions.

5.4.2 Life-history strategies

The sea is a turbulent bouillabaisse of life, with each species' survival depending on continual adjustments and accommodation to new sets of environmental conditions. Because of the properties of water, marine organisms have evolved life-history strategies very different from terrestrial species, with peculiarities that influence many aspects of their evolution and survival (Strathmann, 1990).

An organism is a basic living unit; its life history represents an *adaptive complex* that has been evolved to meet environmental conditions. Each species' life history represents a series of selective compromises that result in each individual's best chance of passing its genes on to the next generation—i.e., its fitness. For species such as marine mammals and seabirds with direct development, life histories are comparatively simple, and the young and adults both lead similar lives, often in social groups. However, most aquatic invertebrates and fishes navigate through several life-history stages during their lifespans, requiring a series of different habitats. Survival mechanisms can be biochemical, physiological, morphological, and/or behavioral. In any case, an individual can make physiological adjustments to a stressor (acclimation), adjust gradually over a longer term (acclimatization), and/or evolve during many generations (genetic changes) in response to environmental change. Behavioral responses to stimuli vary from automatic, innate, and directional (taxes), to unidirectional (kineses). A wide variety of sensory perceptions and communications make use of tactile, visual, auditory, density/pressure, and chemical signals, each of which may be genetically programmed or learned. Combinations of behaviors, directed by sensory perceptions, are key to life histories.

5.4.2.1 Reproduction, dispersal, and recruitment

Marine reproductive strategies vary enormously, and each species has evolved to maximize its fitness. Microorganisms reproduce rapidly: bacteria divide to become clones of the original; viruses enter host cells to replicate and to be released to infect new cells; and unicellular protozoa and algae can reproduce both sexually and asexually. Marine flowering plants (seagrasses) have evolved reproductive strategies that typically involve alternate generations (sporophyte and gametophyte), or vegetative asexual reproduction in which clones or fragments continue to grow and reproduce themselves. Most invertebrates and bony fishes produce small larvae that undergo several juvenile stages before transforming into adults. Some marine invertebrates (e.g., sponges, corals) undergo both sexual and asexual reproduction. Palolo worms (*Eunice viridis*) gather to breed during periods of the full moon, as do groupers (Ch. 8, Section 8.3.1.2). In a few corals, blue-light-responsive cryptochromes provide a clue to the moon's intensity, triggering spawning (Levy *et al.*, 2007). The Caribbean coral *Porites astreoides* responds to the combined effects of several variables that cause half of the colony to become hermaphroditic and the other half female (Chornesky and Peters, 1987). Fishes have evolved a suite of reproductive strategies (Box 5.3); some build nests and care for young, and

Box 5.3 Fish life-history strategies, selection, and population regulation

Kirk O. Winemiller
Texas A&M University, Department of Wildlife and Fisheries Sciences, College Station, Texas, USA

A remarkable diversity of reproductive strategies is observed among the fishes. For example, certain live-bearing sharks and the coelacanth (*Latimeria chalumnae*) produce one or a few large offspring at a time, whereas the ocean sunfish (*Mola mola*) releases over 600 million tiny pelagic eggs in a single spawning bout. Female bay anchovy (*Anchoa mitchilli*) mature at 2–3 months and shed small batches of eggs every 1–4 days, whereas female white sturgeon (*Acipenser transmontanus*) mature at 15–30 years and 1.5–2.0 m length with only a small percentage of females spawning (100,000–4.7 million eggs per clutch) during a given year. Life-history strategies result from trade-offs among attributes that have either direct or indirect effects on reproduction and fitness. Comparative studies have yielded a robust pattern of fish life-history syndromes, with three primary life-history strategies defining the endpoints of a triangular continuum (Fig. B5.3.1). One endpoint, the *periodic strategy*, defines species that have delayed maturation at intermediate or large sizes, produce large numbers of small eggs, and tend to have short reproductive seasons and rapid larval and first-year growth rates. Another endpoint, the *opportunistic strategy*, characterizes species that mature rapidly at small sizes, produce relatively small numbers of eggs, and have long reproductive periods with multiple spawning bouts. A third endpoint, the *equilibrium strategy*, defines species that produce relatively small cohorts of large eggs or neonates, often in association with a long reproductive season, and have well-developed parental care.

Periodic strategists receive two benefits from delayed maturation and large adult body size: capacity to produce large numbers of eggs, and enhanced adult survival. Periodic fishes often have synchronous spawning that coincides either with migration into favorable habitats or with favorable periods within the temporal cycle of the environment. These fishes cope with large-scale spatial heterogeneity by producing great numbers of tiny offspring, at least some of which thrive once favorable locations are encountered. Normally, early larval survival is very low in the marine environment. For the few fortunate larvae that encounter areas of high resource density, growth is rapid. As a result of upwellings, gyres, convergence zones, and other oceanographic features, physical parameters (salinity, temperature), primary production, and zooplankton densities are unevenly distributed in the open ocean. Massive cohorts of small pelagic eggs facilitate dispersal of marine fishes during early life stages. Mortality due to settlement in hostile habitats (advection) is balanced over the long term by survival benefits derived from the recruitment of a certain fraction of larval cohorts into suitable regions or habitats. At higher latitudes, environmental variation is strongly seasonal. Periodic fishes exploit this seasonal variation by releasing large numbers of progeny during periods favorable for their growth and survival.

The opportunistic strategy yields a high intrinsic rate of population increase (*r*) and is associated with high population turnover. Although the size of egg cohorts tends to be small in these small fishes, reproductive effort is actually high because they reproduce early and often. In extreme cases, serial spawning results in an annual biomass of spawned eggs that greatly exceeds female body mass. Small fishes with early maturation, high reproductive effort, and high intrinsic rates of increase are efficient colonizers. These populations can quickly compensate for high adult mortality. The opportunistic strategy is observed in anchovies (Engraulidae), silversides (Atherinidae), and killifishes (Cyprinodontidae)—small species often found in dynamic habitats or faced with high predation risk. Experiments

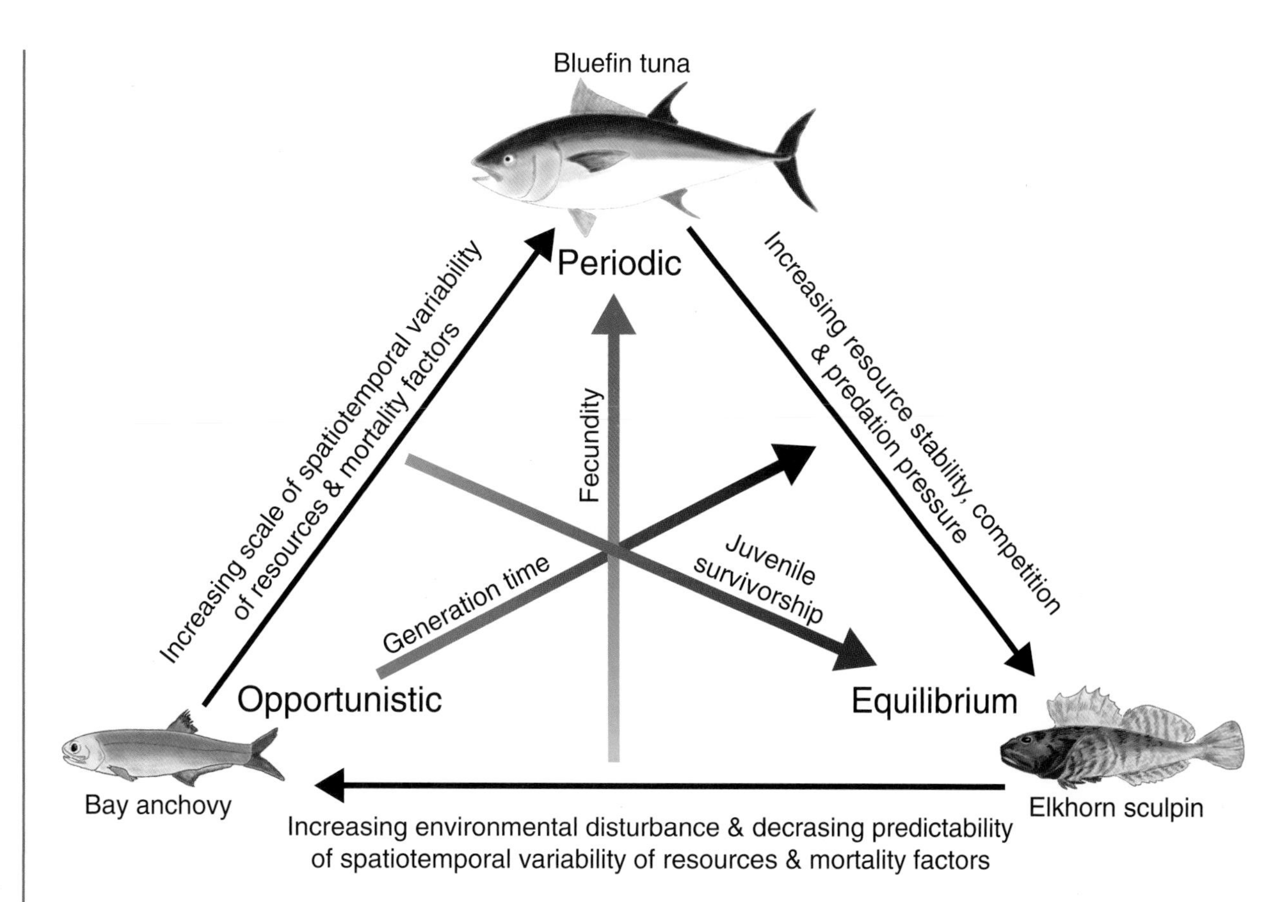

Fig. B5.3.1 Fish life-history strategies reflect tradeoffs between age at maturation, fecundity, and early life stage survival. Different patterns of environmental variation favor different trait combinations. From Winemiller (1995). Fish ecology. In *Encyclopedia of Environmental Biology*, Vol 2 (ed. Nierenberg WA), pp. 49–65. Academic Press, San Diego, CA.

with small fish species, such as guppies (*Poecilia reticulata*) and silversides (*Menidia menidia*), have demonstrated rapid evolution toward more extreme opportunistic characteristics in response to greater predation mortality. Size-selective fisheries have been found to select for earlier maturation at smaller sizes.

Equilibrium fishes have large eggs and parental care that results in larger, more advanced juveniles at the onset of independent life. Marine ariid catfishes (oral brooding of a few large eggs) and sharks, rays, and other live-bearing fishes with long gestation periods and large neonates provide extreme examples of the equilibrium strategy. Parental care seems to be more common in tropical nearshore and reef fishes (e.g., pipefishes, sea horses, eelpouts, sculpins, and some gobies) compared with tropical pelagic and temperate marine fishes.

Of course, intermediate life-history strategies occur within the triangular gradient of life histories. For example, live-bearing is usually associated with few young, but rockfishes (Scorpaenidae) and other cool-water live-bearing fishes often produce large batches of small eggs. Also, divergent life-history strategies frequently coexist in the same habitats. Species' ecological niches influence the manner in which environmental variation yields natural selection.

This triangular life-history continuum has a quantitative foundation. Fitness can be estimated by r, the intrinsic rate of natural increase of a population or genotype. The intrinsic rate of increase can be approximated as

$$r \sim \ln(R_0)/T$$

where R_0 is the net replacement rate, T is the mean generation time, and

$$R_0 = \sum l_x m_x$$

In this equation l_x is age-specific survivorship and m_x is age-specific fecundity, resulting in

$$r \sim \ln\left(\sum l_x m_x\right)/T$$

(Continued)

Therefore, population growth rate depends directly upon fecundity, survivorship, and timing of reproduction, and a change in any one parameter affects the values of the others. For example, there generally is a tradeoff between clutch size and offspring size. Averaged over many generations, the three parameters must balance, or the population eventually will grow exponentially or decline to extinction. Three endpoint strategies result from trade-offs among age of maturation (positively correlated with mean generation time), fecundity, and survivorship. The periodic strategy corresponds to high values on the fecundity and age at maturity axes (the latter a correlate of population turnover rate) and low values on the juvenile survivorship axis. The opportunistic strategy of high r (via rapid maturation) corresponds to low values on all three axes. The equilibrium strategy corresponds to low values on the fecundity axis and high values on the age at maturity and juvenile survivorship axes.

Large body size in periodic strategists enhances adult survivorship during suboptimal conditions and permits storage of energy and biomass for future reproduction. The possibility of perennial reproduction represents a bet-hedging tactic whereby, sooner or later, reproduction coincides with favorable conditions that facilitate strong recruitment. Spawning tends to be periodic and synchronous, so that discrete annual cohorts may dominate a population for many years. Correlations between parental stock densities and densities of recruits have been shown to be negligible in populations of periodic-type fishes. Recruitment frequently depends on climatic conditions that influence water movement, egg/larval retention zones, and productivity, and on other environmental factors that determine early growth and survival. For periodic fishes, the variance in larval survivorship that serves as input for population projections lies well beyond our current measurement precision and accuracy. Even under pristine conditions, the fate of most larvae is an early death. Therefore, it follows that some minimum level of spawning must occur during each spawning period if strong cohorts are to develop during the unpredictable exceptional years. Management of periodic strategists requires maintenance of some minimum adult stock density so that periodic favorable conditions can be exploited, as well as protection of spawning and nursery habitats. Because recruitment is determined largely by unpredictable interannual environmental variation, this minimum density will be impossible to determine with any degree of precision.

Theoretical studies have shown that reducing mean generation time is the most effective strategy for maximizing the intrinsic rate of increase in a density-independent setting. Many opportunistic fishes are found in shallow marginal habitats, the kinds of environments that experience the largest and most unpredictable changes on small spatial and temporal scales. Tidal dynamics change water depth in shallow habitats such as tide pools and salt marshes. In the absence of intense predation and resource limitation, opportunistic-type populations quickly rebound from localized disturbances, and these populations ought to show large variation in abundance, with infrequent strong density-dependence. Because they tend to be small and often occur in marginal habitats, opportunistic fishes usually are not exploited at a large scale. Some important commercial fishes, like menhaden (*Brevoortia patronus*), are intermediate between opportunistic and periodic strategists.

The equilibrium strategy should be favored in density-dependent settings, and this may be why it is more common among sedentary reef fishes than among estuarine and pelagic fishes. Compared with opportunistic and periodic strategists, equilibrium strategists tend to show moderate fluctuations in population density, and should conform better to stock-recruit models. Because equilibrium strategists produce relatively few offspring, early survivorship must be relatively high in order to maintain a stable population. Relatively few equilibrium fishes are commercially exploited on a large scale. Shark fisheries are increasing globally and seem to be unsustainable at current levels of exploitation. Management of equilibrium fishes should stress habitat integrity and healthy adult stocks that can produce surplus yields that may be harvested at modest levels and replaced via natural compensatory mechanisms.

Sources: Conover *et al*. (2005); Winemiller (2005)

others bear only a few, very large offspring (most sharks). "Re-entrant" marine vertebrates (reptiles, birds, and mammals with terrestrial ancestors) have evolved a remarkable suite of reproductive adaptations for breeding on land or sea ice, while feeding at sea, as for example sea lions (Fig. 5.11). The variety seems endless.

Species populations are replenished and maintained through such life-history mechanisms as dispersal and recruitment. Dispersal means movement away from a birthplace; recruitment means return of young to an appropriate habitat, often the natal one. Factors that control recruitment operate on many time-space scales and often are so variable and unpredictable that they seem random. Larvae that develop directly from eggs into juveniles and disperse in the water column on their own, do so at considerable risk. If dispersal leads individuals to unsuitable environments, recruitment may be unsuccessful, as in the case of sessile species such as barnacles that disperse larvae by means of currents. In contrast, diatoms and microbes appear not to be limited by dispersal, being dependent on local to regional selection processes (Cermeño

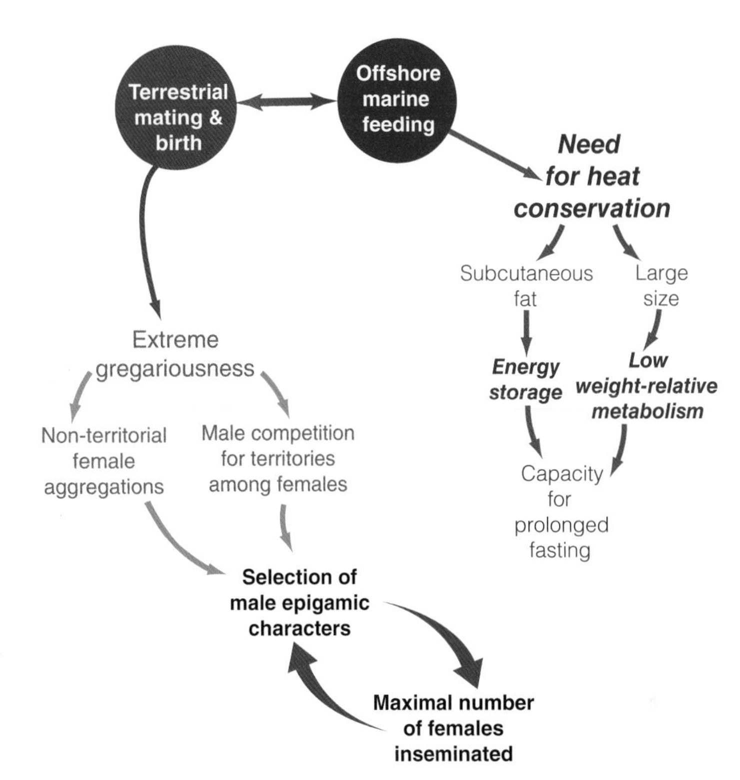

Fig. 5.11 Sea-lion life history combines energetic demands from marine feeding (light green) and reproductive and behavioral physiology (light red) on land (Bartholomew, 1970). At sea, heat conservation (dark green) benefits from large body size and subcutaneous fat. On land, extreme gregariousness and male competition enhances reproductive success (fitness, dark red).

and Falkowski, 2009). Plankton that form spores may suddenly appear in the water column and just as rapidly disappear. In cases where environments are ephemeral, such as shallow marshes, creeks, and nearshore ponds, species may produce eggs stored in a resting state in sediments, i.e., an "egg bank" that functions as a time-dispersal system and allows rapid recolonization when appropriate environmental conditions reappear. This situation can be risky when generations do not overlap; a single year of failed reproduction can eliminate the population from the habitat.

Larvae of species that are dispersed into the water column in massive numbers are particularly vulnerable (Strathmann, 1990). Planktonic distribution by currents may carry eggs and larvae over long distances to unfavorable, or to favorable habitats (Roberts, 1997). The recently discovered giant tube worm (*Riftia pachyptila*) inhabits widely separated hydrothermal vents that largely occur on oceanic ridges; its larval life has been shown experimentally to average 38 days, allowing dispersal of 100 km under average current rates, in which dispersal along ridges appears to be limited mostly by periodic reversals of current flows (Marsh *et al.*, 2001); in this case, dispersal to isolated sea mounts would depend on off-ridge transport and/or faster current flows. However, such passive dispersal over long distances is highly inefficient due to diffusion and mortality of larvae along the way (Cowen *et al.*, 2000). An option is "self-recruitment" in which larvae return to the same area as the population that produced them (Box 5.2; Ch. 8, Section 8.3.1.2; Jones *et al.*, 1999; Swearer *et al.*, 1999). On a much larger scale, Atlantic bluefin tuna (*Thunnus thynnus*) return to Mediterranean and Gulf of Mexico reproductive areas annually, and their populations mix to some extent (Rooker *et al.*, 2008).

In all cases, dispersal and recruitment are functions of the ecology and natural history of the species involved. For the vast majority of species, including those of commercial importance, the precise factors controlling recruitment are speculative or unknown. Boicourt's (1982) observation remains relevant, that larval transport "stands at the state of the art in both physical and biological fields."

5.4.2.2 Migration

Migration, contrary to dispersal, is periodic travel by an organism or species population that is temporally and geographically predictable and consistent. Organisms have many reasons to migrate, for example for finding mates or for seeking appropriate breeding or feeding areas. Reasons for migration may seem obvious, but environmental cues are poorly understood. Red knots (*Caldaria cantus*, a shorebird) migrate annually from southern South America to Arctic breeding grounds and back, covering 30,000 km (Niles *et al.*, 2009); their major springtime stop is in Delaware Bay to feed on eggs of horseshoe crabs (*Limulus polyphemus*) to gain energy to complete the trip. Bar-tailed godwits (*Limosa lapponica*) employ wind-selected migration, assisted by extreme fat loads, to pass 11,000 km, one-way, through at least six latitudinal zones; along the way, they select vertical and lateral winds and compensate for wind drift to adjust flight speed (Gill *et al.*, 2005). Marine mammals such as baleen whales travel hundreds, even thousands, of miles between rich, plankton-filled, high-latitude seas to feed, then to warmer, less-productive waters to bear young and nurse. European eels (*Anguilla anguilla*) undertake a 5000 km migration from Europe to the Sargasso Sea; miniaturized satellite tags showed that eels swim initially against prevailing currents, and undertake vertical diel migrations between 200 and 1000 m depths, to enter favorable currents and possibly to avoid predation (Aarestrup *et al.*, 2009). Similarly, related Japanese eels (*A. japonica*) enter a particular current that transports them to East Asia (Tsukamoto, 2006). These examples of highly migratory species suggest an endogenous map sense, aided by a visually guided magnetic compass, now known to exist in several species (Wiltschko and Wiltschko, 2005; Alerstam, 2006). But some migrations remain local. Some crustaceans undertake "marches" to breed, such as lobsters at sea and several species of land crabs that move across land to coastal waters to breed. And, California grunions (*Leuresthes tenuis*, a small fish) migrate shoreward to spawn in beach sands during the highest lunar tides; the eggs hatch in a month and larvae develop before being carried away by the next highest lunar tide—a remarkable example of phenological synchrony (Section 5.4.2.5).

5.4.2.3 *r*-K selection

Some species are slow-growing, slow to colonize, and spend much time caring for their few offspring. Others are prolific and rapidly colonize new areas. These two contrasting lifestyles are referred to as K and *r*, or as equilibrium and opportunistic species, respectively. K and *r* describe adaptive complexes among species, involving differing sets of interactions and feedbacks (Table 5.1). K-selected species' populations are density dependent and appear to fluctuate around environmental carrying capacity; *r*-selected species are not so constrained, and are more likely to be invasive. Density-dependent regulation, resource availability, and environmental fluctuations are integral to current demographic theory and important for population regulation (Reznick *et al.*, 2002).

The *r*-K adaptive complex implies that populations are regulated by an *adaptive suite* of intrinsic and extrinsic factors; e.g., genetic drift, natural rates of increase, capacity to disperse, resource limitations, environmental change, etc. (Ricklefs, 1973). On the contrary, K- and *r*-species have often been differentiated on the basis of absolute body size, on the assumption that large species have relatively low reproductive rates and long life spans. However, large size is only one part of the adaptive complex, and can be misleading if comparisons are made among taxa with widely differing habitats and life histories. Insects are often assumed to be *r*-selected, as they are small, generally have short life spans, and reproduce rapidly; nevertheless, reproductive rates vary enormously. Conversely, marine mammals have been widely perceived as K-selected, due to their large size and low reproductive rates, which also vary. Marine mammals exhibit order-of-magnitude differences in size and population doubling times, and can be either obligate or facultative consumers. Thus, marine mammals exhibit their own range of *r*-K adaptations inter alia (Fig. 5.12). This does not suggest that marine mammals may be exploited as if they were *r*-selected fish (Estes, 1979).

Whether a species is K- or *r*-selected strongly influences how it may be conserved. Communities dominated by K-selected species would, theoretically, be more stable and resilient than communities dominated by opportunists. K-selected species, being strong competitors, optimize fitness; conversely, *r*-selected species more readily overexploit resources. However, caution is needed to avoid the internal inconsistency of simplistic interpretation (Wilbur *et al.*, 1974). The green sea turtle has high fecundity consistent with *r*, but its long life, late maturity, and large size (Ch. 8, Section 8.3.2.4) would seem K-selected. Furthermore, some species are regulated not by food supply, but by predation. Thus, interpretation of the K-*r* concept requires understanding of the entire adaptive suite of life-history

Table 5.1 Characteristics of the *r*-K Continuum. Based on data from Pianka (1970).

Correlates	*r*-Selection	K-Selection
Climate	Variable and/or unpredictable, uncertain	Fairly constant and/or predictable, more certain
Mortality	Often catastrophic, nondirected, density-independent	More directed, density-dependent
Survivorship	Juvenile survivorship low compared to adults	Juvenile survivorship high compared to adults
Population size	Nonequilibrium, variable in time; usually below carrying capacity, fills ecological vacuums, recolonization required	Equilibrium, fairly constant in time; at or near carrying capacity, recolonization not usually necessary
Intra- and interspecific competition	Variable, often lax	Usually keen
Selection favors	Rapid development, high *r*, early reproduction, smaller body size, semelparity (individuals reproduce only once)	Slower development, lower resource threshold, delayed reproduction, larger body size, iteroparity (individuals reproduce repeatedly)
Life span	Usually short	Longer
Leads to	Productivity	Efficiency

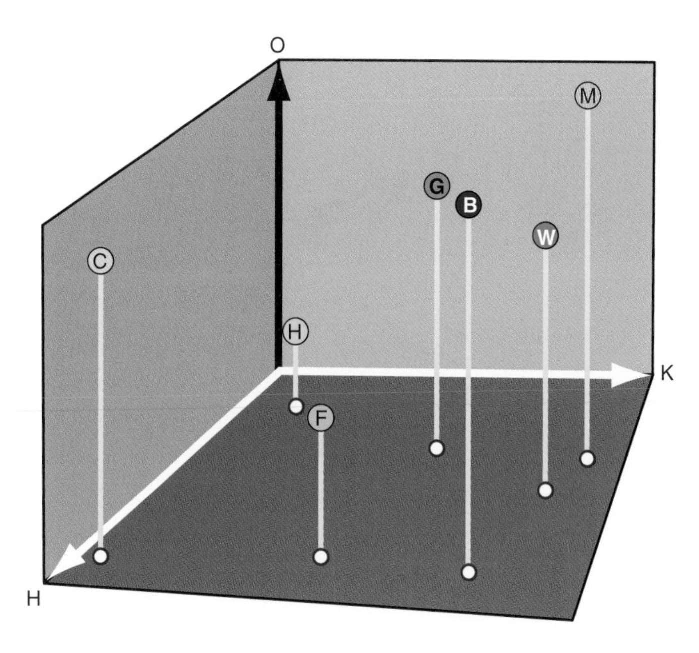

Fig. 5.12 Relative *r*–K positions of some marine mammal species in theoretical space (Ray, 1981). Three relative non-parametric axes for comparison are: [O] food obligateness; [K] maturation, body size and/or length of life; [H] environmental/habitat maturity, stability, and predictability. Within the box: (H) harbor seals (*Phoca vitulina*) are small, live in variable environments, have facultative food habits, mature at a relatively early age, bear young once a year, and are relatively *r*-selected; (B) blue whales (*Balaenoptera musculus*) are very large, live in relatively predictable environments, are food restricted, mature late, are long-lived, and bear young only every 4–6 years and are relatively K-selected; (C) crabeater seals (*Lobodon carcinophagus*) of the Southern Ocean; (W) walrus (*Odobenus rosmarus*) of Arctic shelves; (F) northern fur seal (*Callorhinus ursinus*) of the temperate North Pacific; (G) gray whale (*Eschrichtius robustus*) of the temperate to subarctic North Pacific; (M) bowhead whale (*Balaena mysticetus*) of the coastal Arctic.

attributes involved, as well as ecological dimensions and causal effects pertaining to evolution of life histories.

5.4.2.4 Food, feeding, and food webs

All species seek to optimize their rate of reproduction and food intake. Food particles in the sea are patchily distributed and oceanographic processes are important for the concentration, transport, and availability of food. Marine species have evolved a large variety of mechanisms to obtain nourishment and optimize energy extraction. Microorganisms, many invertebrate parasites, and several free-living invertebrates engulf (phagocytize) whole particles, sometimes larger than themselves, or absorb dissolved organic matter directly through cell surfaces. Herbivores generally consume plants piece by piece; e.g., sea urchins that graze on macroalgae, periwinkle snails that consume salt marsh grasses, and manatees that eat submerged aquatic vegetation. Sessile, benthic species exhibit a wide variety of morphological forms for increasing food encounters. Filter feeding that obtains nourishment from the water column is an especially common and efficient marine feeding mechanism for extracting microalgae, food particles, and small organisms from the water. For many invertebrates, filter feeding involves cilia that produce feeding currents for pulling particles into the mouth, where food morsels are identified and sorted and undesired particles discarded in pseudofeces (e.g., oysters, Ch. 6). Small copepods' slow swimming speed and use of feeding currents suggest feeding adapted to turbulent diffusion. Many corals obtain nourishment through mutualistic associations with photosynthetic zooxanthellae, but remain capable of predation.

Predators engulf prey by using a variety of food-gathering mechanisms—claws and jaws, cilia, predatory ambush, and specialized mechanisms that trap and entangle prey (e.g., mucus nets). To acquire food, the predator must first find it (sensory perception), then employ mechanisms that override prey defenses (feeding strategy), and also possess appropriate mechanisms to extract materials and energy (anatomy). Tunas and other pelagic predators are speedy enough to capture fast-swimming, schooling fishes. Sharks, some teleost fishes, and orcas (*Orcinus orca*) are unique in their ability to tear flesh from prey larger than themselves, and like wolves, form social packs for hunting. Humpback whales (*Megaptera novaeangliae*) in groups concentrate zooplankton and small fishes by exhaling air to form bubble nets.

Different life-history stages of a single species, from larval to juvenile to adult, often consume foods that vary widely in size and species. For many fishes, pelagic larvae consume plankton, juveniles switch to small invertebrates, and adults may be top predators. Oceanic-stage (juvenile) green turtles (*Chelonia mydas*) eat sea jellies and other small floating organisms (Parker *et al.*, 2008) then switch to seagrasses as adults. Arctic ringed seals (*Phoca hispida*) may consume zooplankton, benthic fauna, or fishes, depending on sea-ice conditions, season, and water depth. Orcas consume fishes, seals, and even other whales, depending on season, location, and availability of prey. The filter-feeding baleen blue whale (*Balaenoptera musculus*) specializes on eating small krill (*Euphausia superba*) in Antarctic (Southern Ocean) waters, but eats a wider variety of food in the Northern Hemisphere. In all cases, food and feeding connect organisms to a community of species and with their total environment. When environments change, so may food and feeding opportunities. In the case of wandering albatrosses (*Diomedea exulans*), improved breeding success and significant adult weight gain has increased due to greater intensity and poleward movement of westerly winds in the Southern Ocean, and benefits from the increased duration of their foraging trips (Weimerskirch *et al.*, 2012).

At the other end of the predator-prey spectrum, whale falls—dead whales that fall to the bottom of the sea—appear

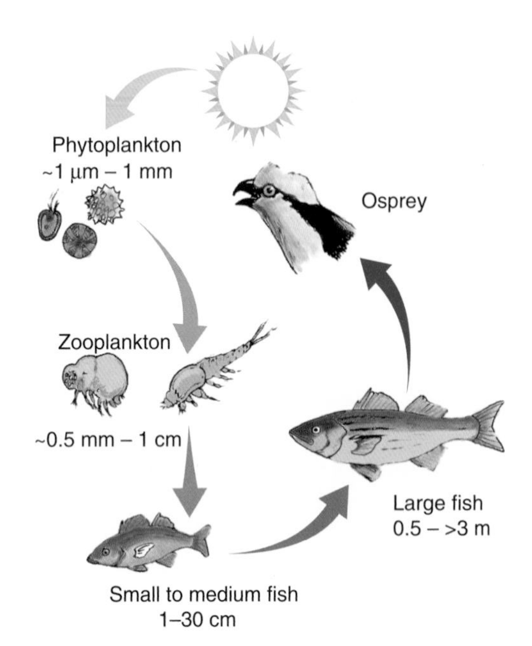

Fig. 5.13 A five-level food chain. Energy is captured from solar radiation to phytoplankton, with a transfer efficiency of about 3%. Energy is transmitted up the chain with transfer efficiencies from approximately 10 to 20% (generally greater for ectotherms than for endotherms). Intricate food webs and high overturn rates of primary producers and zooplankton increase efficiency, making predators more abundant than food chain energetics might predict.

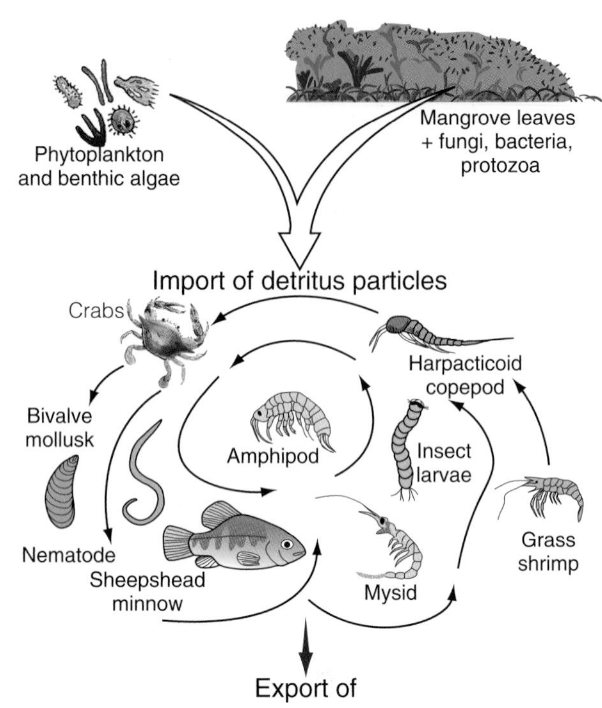

Fig. 5.14 Detritus particles derived from plants are cycled through food webs when detritivores and deposit-feeders consume particles and expel them as feces. Feces, particles, detritus, and organic nutrients are exported to small and large predators outside this system. From Odum and Heald (1975). The detritus-based food web of an estuarine mangrove community. In *Estuarine Research* (ed. Cronin LE). Academic Press, New York, pp. 265–287.

to be specifically suited for small deep-benthic consumers (Rouse *et al.*, 2004; Haag, 2005). Species of *Osedax*—Latin for "bone-devourer"—worms were discovered in January 2002 feeding on the bones of a dead gray whale; females are about the length of an index finger, and one female can carry >100 microscopic males. *Osedax* has no eyes, no mouth, and no stomach, but tunnels into whalebones by means of its "root system" and turns them into "Swiss cheese" by activities of a unique symbiotic bacterium, thereby increasing the rate of removal of the carcass. The abundance of *Osedax* suggests that the frequency of whale falls has historically been great, and that whales might play a substantial role in attracting worms, mollusks, and other species to recycle organic material. Communities that depend on whale falls are estimated to survive for up to a century on the fats and sulfides made available by whale carcasses. How larvae find new whale falls is unknown.

Food chains were once conceived as linear, i.e., from producers through a series of consumers, to decomposers, thereby passing energy through successive trophic levels (Fig. 5.13). However, linear trophic structure does not accurately reflect the extremely complex energy exchanges of ecosystems. Rather, food webs exhibit a reticulate structure of feedbacks and connectivities, serving as surrogates for community structure (Cohen, 1989). Food-web studies indicate that each species interacts with a limited number of others, theoretically placing an upper limit on species numbers. The transfer of energy through food webs is relevant to both the persistence of species and to community resistance to environmental change and disturbance (deRuiter *et al.*, 2005). Food webs are flexible and dynamic, and energy transfer in the form of detritus is exported out across ecosystem boundaries (Fig. 5.14). Thus, the basic concept of food webs can be extended to a broader framework of ecological networks that are more inclusive of different components of ecosystem biomass flow, taking into consideration different kinds of species interactions that are not strictly trophic, such as meta-community connections (Brose *et al.*, 2004; Dunne, 2006). In this sense, food webs may serve as indicators of ecosystem sustainability and resilience.

A special case of feeding behavior is "central-place foraging," in which species forage from a special location and return to that location with food for themselves and/or their young. For example, sea birds and seals leave rookeries or haulouts to feed at sea and return to the same location to feed young or to rest; walruses do the same from moving ice (Ch. 7). This behavior involves trade-offs between distance traveled, abundance and size of prey, and duration of foraging trips that must be accommodated in order to optimize energy (Pyke *et al.*, 1977; Elliott *et al.*, 2009).

5.4.2.5 Phenology

Life in the seas is adapted to daily (circadian), regular beats of light and tidal rhythms, and also to monthly, semi-monthly, and seasonal cycles for feeding, mating, and reproduction. Phenology refers to the duration, timing, and coordination of species' physiology and behavior in accord with environmental conditions. For example, successful feeding and reproduction often must be timed to seasonal warmth or water conditions so that offspring have access to adequate nutrition, or for a species to reproduce successfully. In such cases, environmental cues—e.g., tides, moon phases, and seasonal temperatures, among others—ensure coordination of events with physiological needs (oysters and fishes, Ch. 6; pinnipeds, Ch. 7; Nassau groupers, Ch. 8). In particular, marine predators often congregate into limited areas to feed; some are timed to plankton that rise upwardly at night, and others to small schooling fishes that collect in shallow, slow, counter-rotating vortices at the ocean surface (Langmuir circulation that results from interactions of wind speed and wave movement; Thorpe, 2004). Hatching of northern shrimp (*Pandalus borealis*) is timed to match food availability, but is delayed in females of the more southerly population that migrate offshore into colder waters so that their young can feed on the productive spring bloom (Koeller *et al.*, 2009). And, baleen whales migrate long distances from warm temperate or tropical waters to feed in higher latitudes on the seasonally high zooplankton production of colder seas.

Matches and mismatches in phenological timing can enhance or threaten a species' survival. An exquisite example of phenology concerns seals that bear pups on sea ice (Ch. 7). Newborn pups are vulnerable to cold and quickly need to acquire a thick insulating coat of subcutaneous fat ("blubber") during brief nursing periods—a race against time that depends on weight gain before their ice habitat melts; if ice melts too soon, pups are forced to enter the cold water at considerable energetic expense. A potential mismatch concerns Antarctic krill (*Euphausia superba*) that form expansive "superswarms" in pelagic environments at various life-history stages (Nicol, 2006). Should environmental conditions change, so might krill behavior and distributions, with potential threats to food sources for dependent penguins, seals, and whales.

Phenological phenomena are vulnerable to short-term environmental variability and to long-term environmental change. Biological clocks (endogenous timing) are important in the timing of life-history events, roughly coinciding with a 24-hour cycle. But how endogenous timing and environmental cues interact is most often unclear. If phenology at any scale between a species' behavior and/or its social or ecological requirements are mismatched, the results could be catastrophic, a condition that climate change can induce.

5.5 BIOLOGICAL ASSOCIATIONS

No species is totally isolated. Some species are territorial and live solitary lives, except during reproduction when they pair to breed. Others form groups, large or small, seasonally or permanently. In all cases, cooperation and competition are prevalent, involving interactions between individuals and groups within and between species. Populations, communities (biocoenoses), and ecosystems are biological associations that form ecological units; these associations have increasingly become the focus of conservation strategies and are paramount for understanding ecological complexity (Jax, 2006).

Many fish and invertebrates "school", a behavior that reflects harmony of action and movement in a truly egalitarian state in which members are alike in both influence and importance (Shaw, 1978). Formation of schools requires initial conditions of critical population density and a few leaders whose actions others follow and that propagates over long distances (Makris *et al.*, 2009). Small fishes (herring, sardines, anchovies, etc.) and large fishes (tuna, mackerel, etc.) form massive schools; under schools of tuna, dolphins form schools—a school upon a school. Many invertebrates also school (zooplankton, squids, crabs, etc.). Corals cannot be said to "school," but their clonal members can be as numerous in a single colony as a large school of fish.

5.5.1 Associations among species: Symbiosis

Symbiosis describes living together between individuals of two species and takes three basic forms: commensalism (one partner benefits), mutualism (both species benefit), and parasitism (one species benefits at the expense of the other). These associations are generally long-term rather than transitory, and one or both partners usually have structural and/or behavioral adaptations that foster the association. The evolutionary advantage of symbiosis is creation of biological novelty and innovation through permanent or long-term interactions, as made evident by speciation, pathogenesis (the evolution of disease), and the "evolutionary arms race" (Paracer and Ahmadjian, 2000). Symbiosis thus highlights the recognition that species often depends on other species.

Symbioses are common among marine species. The relationship can be either facultative (optional) or obligate (necessary). Many symbioses have energetic, trophic, or reproductive underpinnings. Several species of fish and shrimp gain food by relieving larger fishes of burdensome parasites at "cleaning stations." Mutualism also occurs between photosynthetic dinoflagellates (zooxanthellae) and many hermatypic (reef-building) corals; zooxanthellae lend color to corals as they gain protection from predation, and corals benefit from nutrients derived from zooxanthellae. Many sponges form associations with blue-green algae, and some anemones are colored green by chlorophyte symbionts. Very few mollusks host algal symbionts; an exception is the giant clam (*Tridacna gigas*). Some species form behavioral or physiological dependency on other species (e.g., remora fish that depend on particular shark species, specialized barnacles on whales, etc.), and some others may gain a free ride on other species (phoresy), such as algae on sea turtles. A remarkable three-way symbiosis among seagrasses, a bivalve (clam), and its sulfide-reducing bacteria may be key to the long-term success of seagrasses (van der Heide *et al.*, 2012); natural accumulation of organic matter should result in toxic sediment sulfide levels, but the endosymbiotic bacteria in lucinid clams reduce sulfide stress for seagrasses, allowing them to flourish. Thus, conservation of the clams, in this example, is important for seagrass conservation.

Some species modulate the environment to benefit other species—a temporally displaced form of mutualism. A "death assemblage," for example, created by dead dominant species can also create a "life assemblage" through post-mortem

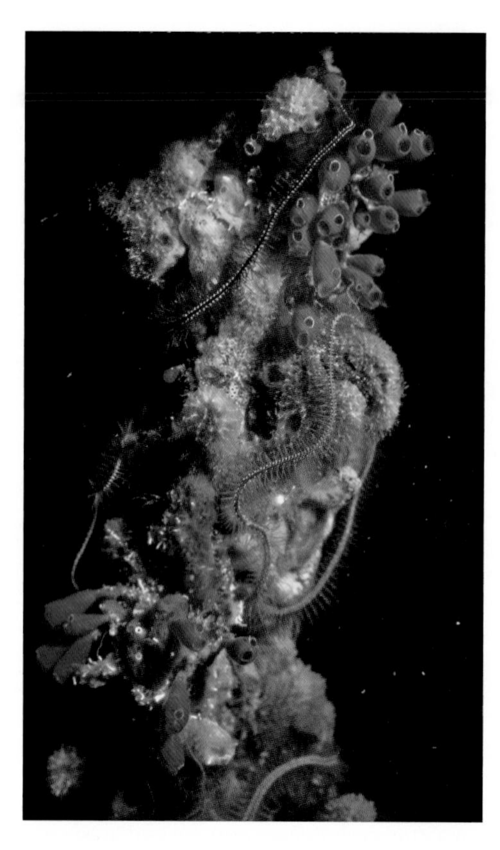

Fig. 5.15 A possible, symbiotic, community of tunicates, brittle stars, and others on red mangrove prop roots in Belize. Photograph © Ray & McCormick-Ray.

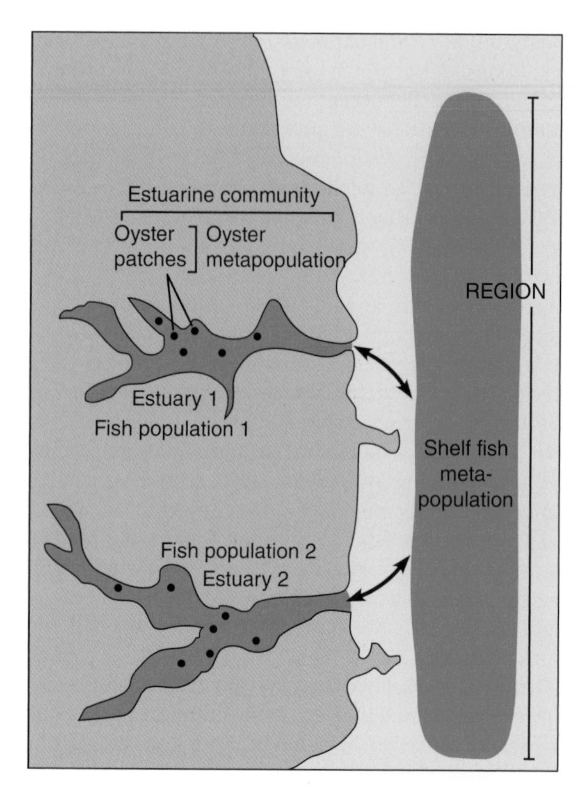

Fig. 5.16 Many temperate estuaries include metapopulations of oysters and fishes. Oyster beds exemplify intra-estuarine metapopulations (NRC, 1995). Anadromous fishes exemplify extra-estuarine metapopulations, in which costal-ocean species require estuaries at some portion of their life histories, especially fishes that assemble as adults and subadults over the shelf and depend on estuaries for breeding or as nurseries for their young. In this case, fisheries management requires consideration of multiple estuarine populations, whereas oysters may be managed in single estuaries.

feedback (Kidwell *et al.*, 2001); i.e., deep-sea species that depend on dead bodies of whales (Section 5.4.2.4). And animal parts such as skeletons and dead shells provide substrates that can enhance a community of benthic species (oyster shell, Ch. 6). Additionally, whole biotic communities can form possible symbiotic associations; epiphytic algae attached to seagrasses, and many other species that play host to a variety of organisms. A striking example of a possible symbiotic, community association occurs on prop roots of red mangroves (Fig. 5.15). Such associations are recurrent, but reasons are unclear.

5.5.2 Associations within species: Populations and metapopulations

A biological population is a group of individuals within a single species that produces viable offspring. The most basic property of a population is the number of reproductive individuals it contains. Under most circumstances, populations tend to preserve a characteristic level of abundance, being described colloquially as "abundant," "common," or "rare." Population dynamics depends on demographic factors that include social structure, life-history variation, dispersal, colonization options, and minimum population size. Mathematical population-dynamics analysis (MSY, maximum sustainable yield) is a tool used, with mixed results, to project population viability when species are exploited, for example for commercial fisheries. Much variation exists in population structures that complicate understanding and projecting future population size. For example, inbreeding that can occur in small populations can greatly reduce average individual fitness due to loss of genetic variability; in large populations, environmental carrying capacity can affect population growth. Therefore, population dynamics involves more than a statistical procedure and to be meaningful must include details of the species' ecology, behavior, and habitat.

A species population is considered closed, open, and/or interconnected depending on dispersal and recruitment patterns. Patchy recruitment of individuals of a species and regular, long-distance connectivity between reproductive source populations and recruiting populations depend on "open" populations, while those species that depend on direct development or "self-recruitment" from a common larval pool within relatively short distances depend on "closed" populations (Steneck and Wilson, 2010). Some populations are connected in a metapopulation—a population of populations, each component of which occurs in a suitable area (Fig. 5.16). For example, most larvae-dispersed species have more than one interbreeding population (lobsters, many herrings). Metapopulation dynamics connects patches through dispersal, with important implications for species' survival; i.e., if one local population goes extinct, larvae from other occupied patches may recolonize the site—the "rescue effect." Related to metapopulation is the concept of contingency in which populations are maintained through alternative pathways; e.g., Hudson River striped bass (*Morone saxatilis*) have three different pathways for reproduction (Secor *et al.*, 2001).

5.5.3 Multiple species associations: Guilds, communities, and niches

Guilds focus attention on groups of species that exploit the same class of environmental resources in a similar way (Simberloff and Dayan, 1991). Chesapeake Bay hosts eight abundant species of filter-feeding Clupeid fishes (herrings) that have evolved opportunistic, filter-feeding lifestyles that complement one another. When one species becomes depleted, others may increase in abundance and thereby conserve the ecological function of plankton-eating (planktivory; Ray, 2005). Other guilds are associations among unrelated species. For example, seabirds and marine mammals in a shared feeding area may be associated together in at least four ways, each of which preserves top-predator biomass (Skov *et al.*, 1995): (i) many seabirds that concentrate near their breeding colonies to feed; (ii) seabirds and porpoises that feed together in shelf waters; (iii) several small- to medium-sized cetaceans and seabirds that concentrate mainly along shelf edges and remote banks and ridges; and (iv) large cetaceans and seabirds that feed on small, oceanic crustaceans. Guilds that relate to several species performing similar roles to maintain ecological function can be conceptualized to simplify the diversity of ecological roles in a system.

Species can also form functionally interconnected communities that are spatially and temporally defined by prominent species and/or associations (Ricklefs, 1973; Allen and Hoekstra, 1992). Communities are commonly named for a dominant structure or physical feature (salt marsh, mudflat, sea ice, pelagic, benthic, etc.), and are composed of species that compete and/or cooperate for food, space, or other resources in complex webs of interactions. Communities are not merely assemblages of species. Rather, community structure and dynamics represent the summed natural histories of all organisms present; but communities are not equivalent to ecosystems, as the physical environment is not specifically included. The concept of "niche" defines each species' role in its community. Hutchinson (1965) visualized a species' "niche" as a hypothetical ecological hypervolume, or way of life, in contrast to habitat, which is the physical space where the species lives. Theoretically, no two species can coexist in the same niche space without risking competitive exclusion. A full portrait of an organism's niche includes requirements for space, food, physical conditions, and appropriate conditions for mating and other behaviors.

Species within communities are interconnected by exchanges of energy and materials, immigration and emigration, successional change, biogeochemical transformations, and physical change. Understanding communities requires study of structure and function in time and space, and depends heavily on knowledge of species' natural histories together with environmental dynamics. Measurements of production, exploitation, and assimilation efficiencies allow reconstruction of energy flows that bind species together. As natural selection affects and depends on interactions among species within the community as a whole, co-evolution among species is often the glue that bonds the community together in interdependent associations.

Community relationships develop over time, during which some species are incorporated and others are excluded. Community complexity is a product of the diversity of all interacting organisms, where competition, cooperation, and predation are dominant factors that shape community structure and give rise to ecosystem attributes. Competition restricts sharing, propagates antagonistic relationships, and promotes low connectance among species. In contrast, cooperation tends to promote connectedness, but how natural selection can lead to cooperation seems non-intuitive; i.e., do cooperators pay costs and defectors pay none? Several scenarios have been proposed (Nowak, 2006), among which are kin and group selection. Kin selection favors reproductive success (fitness) for related individuals, i.e. depends on altruistic cooperation among relatives to ensure the survival of their shared heritage. Group selection involves cooperation (reciprocity) among non-relatives, in which there are benefits for the group as a whole; i.e., group cooperators receive more benefits than defectors, beyond what might be predicted by simple genetic relatedness. Wilson (2012) proposes that group selection leads to "eusociality," the most advanced form of social behavior, shared by social insects and humans.

How communities function, survive, and prosper is best understood in the context of interconnected biophysical units, i.e., of ecosystems. Much is being gained through the study of ecological networks, their architecture, and how species extinction or persistence is influenced by this architecture (Bascompte, 2010). When observed at larger scales, mutualistic networks presumably would promote highly connected communities of interacting groups of species. Such networks may, however, be vulnerable to species loss, structural modifications, and food-web and trophic cascades that alter ecosystem processes (Thébault *et al.*, 2007). In such cases, community composition can change with location and scale of observation, important parameters for understanding ecosystem structure, function, and stability (Worm and Duffy, 2003; Worm *et al.*, 2006). The number of species present (richness), the particular species present (composition), their relative abundances (evenness), their interactions (non-additive effects), and their temporal and spatial variation all contribute to ecosystem structure and resiliency (Chapin *et al.*, 2000).

All of these ecological units—ecological networks, communities, and areas of relative species richness, evenness, and endemism, etc.—share problems of boundary definition (Jax, 2006). Boundaries have been identified spatially as biogeographic units (watershed, habitat type, oceanic, coastal, benthic, pelagic, etc.) or functionally as networks of relatedness (food webs, biogeochemical pathways, etc.). The difficulty in boundary delineation lies in the way species and ecosystems function, because several different, functionally related units may exist and intersect within the same spatial volume, and each species may depend on several spatially or temporally distinct ecosystems during its life cycle. To complicate matters further, ecosystem boundaries change hierarchically in space and over time. Boundary delineation thus serves as a tool for comprehension and communication, but may or may not differentiate scale-related similarities and differences among ecosystems.

5.6 BIOGEOGRAPHIC PATTERNS IN SPACE AND TIME

Marine biotic patterns of the present day have resulted from millions of years of history. The outcome has been a biosphere with roots in the past, observed as spatial patterns best represented by species associations (assemblages and communities), rather than by presence and absence of dominant species or percent endemism (Adey and Steneck, 2001). Mapping these associations helps to relate biotic relationships to environmental parameters and promotes understanding of how interactions among species and environments have come about. As such, biogeography facilitates understanding of evolutionary and adaptive processes among suites of communities, habitats, and ecosystems in order to provide a basis for conserving nature's variety. Just as taxonomy helps scientists sort out evolutionary relationships, biogeography may be viewed as a taxonomy of environments that hierarchically defines ecosystems.

Spalding *et al.* (2007) present a global, hierarchical classification that focuses on coastal and shelf areas and recognizes 12 realms, 62 provinces, and 232 ecoregions. The finest-scale units are ecoregions, defined as areas of relatively consistent species composition that are distinct from adjacent systems. The intent is to identify ecologically representative areas to facilitate protection of the full range of biodiversity—genes, species, higher taxa, communities, evolutionary patterns, and ecological processes—in a spatial context, e.g. as potential MPAs. The boundaries are approximate, recognizing that boundaries shift and overlap continuously with weather patterns, seasons, and fluctuations in environmental conditions, and will become unstable over time, eventually requiring changes in conservation strategies and tactics.

A major challenge is that terrestrial, freshwater, and marine systems have been treated separately and differently. Terrestrial provinces have been classified and studied largely on the basis of vegetation under the assumption that plants reflect climatic regimes. Watersheds cut across these provinces and have been mapped separately. Coastal and marine provinces have been variously defined by physiography and physical characteristics of water masses (Hayden *et al.*, 1984), or by planktonic distributions (Longhurst, 1998). Conversely, benthic classification is based on physical (geological) features (mud, sand, boulders, etc.). Such biogeographies, each useful in its own right, would benefit from an integrated biogeographic approach that recognizes land-sea, air-sea, and benthos-water column interfaces interdependently (Ch. 4, Section 4.5).

5.6.1 Biotic spatial patterns

What makes coastal-marine systems biogeographically distinct is their biota. At a global scale, the highest diversity of fishes, marine mammals, seabirds, sea plants, and sea turtles live in the coastal realm (Ch. 4) and nowhere else. Fishes constitute more than half of all known vertebrates worldwide. Nelson (1984) recorded that of 21,723 described species of fishes, 8411 (approximately 39%) live in freshwater, 10,200 (approximately 47%) occur from estuaries to the outer continental shelf, and 2700 (approximately 12%) are oceanic. Marine mammals are similar; 123 modern species have been described, two are known only from dead specimens (beaked whales, family Ziphiidae) and three are extinct as a result of human exploitation. The natural histories of the remaining 118 species are well enough known that a reasonable assessment of habitats is possible; 6 (5.1%) are freshwater, 5 (4.2%) are terrestrial/oceanic, only 9 (7.6%) are fully oceanic, and 98 (approximately 83%) spend significant portions of their lives, or live solely, in the coastal realm. Other groups of species would be expected to follow similar patterns. That is, biota, not surprisingly, reflect major subdivisions of coastal and marine realms. At the mesoscale, the coastal realm is recognizable at three hierarchical scales according to physical and biological characteristics (Fig. 5.17); cross-shelf fish distributions help verify this biogeographic pattern (Fig. 5.18).

5.6.2 Transitional patterns: Gradients

Ecological boundaries are delineated spatially and physically, as for watersheds and habitats, and/or biologically, as for productivity. Such boundaries, however, can be indistinct or even ephemeral, and vary according to the variable and scale chosen. Transition zones (ecotones) represent gradients between systems and are characterized by unique time and space scales and by strengths of interactions between adjacent systems. Striking transitions occur between systems, such as at the terrestrial-aquatic interface, between biologic structures, or at community boundaries (e.g., seagrass beds, reefs, etc.), due to gradients in hydrology, food availability, larval transport, and/or oxygen content. At least three classes of environmental gradients affect species richness: (i) consumable resource gradients, such as nutrients (for plants, light is a resource); (ii) regulator gradients (temperature, salinity) such as rates of physiological processes; and (iii) gradients that may have no direct influence on organisms, but are correlated with both resources and regulators (latitude, water depth). Species are most often assumed to be adapted to optimum physiological conditions, but usually live along a regulator gradient—for temperature, neither too hot nor too cold, or for salinity, neither too fresh nor too salty. Nearshore, differing salt tolerances among species result in zoned distributions, as in "mangroves" where Caribbean red mangroves (*Rhizophora* sp.) occur in deeper water than black mangroves (*Avicennia* sp.). Other species often do best at high levels of a resource gradient, e.g., a preferred nutrient level or food source. But if two or more species are similarly adapted, competition may mask gradient effects.

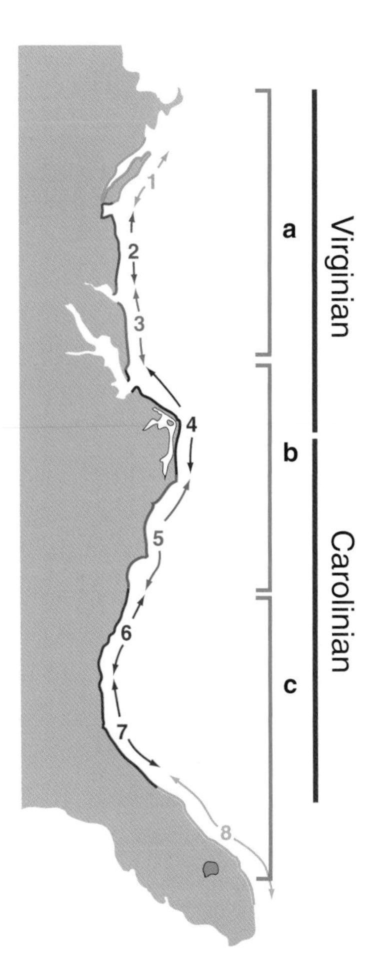

Fig. 5.17 A hierarchical bio-geo-physiographic classification of the U.S. east coast coastal realm (Hayden and Dolan, 1979; Hayden *et al.*, 1984). Biogeographic provinces, Virginian (I) and Carolinian (II), of the eastern United States are derived from coastal species assemblages. Physiographic regions (a, b, c) and smaller subregions (1–8) represent a classifications of island-lagoon marsh subsystems that co-vary with offshore bathymetry. The defining characteristics for these biogeographic regions are geology, seasonal variations in ocean surface currents, and companion seasonal variations in atmospheric winds and currents.

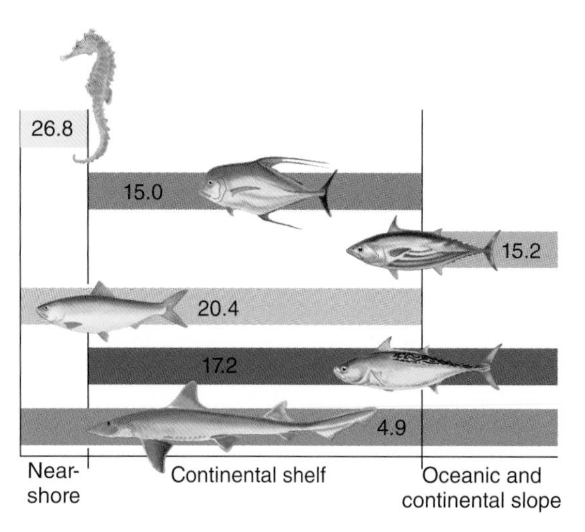

Fig. 5.18 Six overlapping, cross-shelf distribution patterns (represented by shaded areas) of the coastal realm are evident for 553 species of fishes of the U.S. east coast's Carolinian and Virginian biogeographic provinces (Ray, 1991). The numbers are rounded percentages of the numbers of species in each category. Nearshore is defined as water depths of <20 m. The continental shelf and oceanic waters are separated at the shelf edge. Species typical of these patterns are, from top to bottom: lined seahorse, *Hippocampus erectus*; African pompano, *Alectis ciliaris*; skipjack tuna, *Euthynnus pelamis*; Atlantic menhaden, *Brevoortia tyrannus*; little tunny, *Euthynnus alletteratus*; and smooth dogfish, *Mustelus canis*.

At large scales, transitional gradients require special consideration. At global scales, latitude gradients in species richness are recognized; early explorers observed the tropics as teeming with life, whereas temperate zones held fewer kinds of animals and plants, against which high-latitude seas seemed barren. Such observations depend upon which taxon or which type of ecosystem is examined. For example, fish species diversity generally decreases toward higher latitudes, but species of some groups such as cods (Gadidae) increase at higher latitudes. Others are endemic to high latitudes; icefishes (Nototheniidae) are restricted almost entirely to the Southern Ocean. For marine mammals, sea cows (Sirenians) are restricted to the tropics, but pinnipeds (walruses, seals, and sea lions) are most varied and abundant in higher latitudes. Species richness may relate also to longitudinal or radial gradients. Fish and coral diversity in tropical Indo-Pacific reefs decrease toward higher latitudes and also eastward from a center in Indonesia toward Hawaii and westward toward Africa (Hoeksema, 2007). Waters surrounding islands exhibit radial gradients in which diversity decreases outward in all directions. And for pelagic species, oceanic gradients occur vertically at thermoclines and pycnoclines (Ch. 4). Gradients can be abrupt or gradual depending on the underlying geophysical structure.

5.7 BIOTIC FUNCTIONAL DIVERSITY

Functional diversity defined by Steele (1991) is "the variety of different responses to environmental change, especially the diverse space and time scales with which organisms react to each other and to the environment." At the benthic-water interface, a diversity of species and their metabolic and bioturbation activities serve to process carbon, nitrogen, and sulfur through different routes in the system. At the land/sea interface, various seascape structures remove and redeposit sediment to buffer shores against erosive and climatic events. Within the water column, a diversity of communities influences the direction and transport of solutes and particles, reworks and recycles organic material, and provides sinks for carbon dioxide. Microbes are responsible for about half of Earth's primary productivity, and their diversity is very great (Rees, 2005). Viruses, the most abundant organisms in the world's oceans, are major causes of mortality, drivers of global geochemical cycles, and reservoirs of the greatest genetic diversity on Earth (Suttle, 2005). With regard to phytoplankton blooms, their production of marine aerosols can modulate the properties of marine clouds with significant influences on Earth's radiation budget—effects comparable to those over highly polluted areas (Meskhidze and Naenes, 2006).

Species diversity and functional diversity appear to be related (Micheli and Halpern, 2005). While species richness exerts predominant ecological effects on ecosystems, species evenness (Section 5.3.2; Ch. 9) may be even more important (Wittebolle *et al.*, 2009). That is, theoretically, the probability is that at high evenness, species' resistance to perturbation is high; but when evenness is low, i.e., dominated by one or a few species, resistance to perturbation will be maintained only if the dominant species are tolerant. As ecosystems become more dominated by fewer, domesticated and/or invasive species, changes in evenness and losses of functional diversity become especially worrisome; i.e., ecosystems potentially become less resistant to climate, sea level, and shoreline change. Functional diversity presents options to accommodate change, and appears to maintain ecological resiliency.

5.7.1 Food-web and trophic functions

Feeding often has a major influence on marine ecosystems (Section 5.4.2.4), and feeding relationships in marine food webs are far more complicated than indicated by the ancient proverb, "big fish eat little fish." The functional roles played by feeding animals can have "top-down" effects on marine ecosystem pattern and process, especially as predators transfer energy in food webs and throughout the ecosystem or, conversely, primary production can amplify consumer biomass through a "bottom-up" effect. Coral reefs provide examples. Randall (1974) was one of the first to observe that fishes that graze on corals can "adversely affect" coral reefs: "The continual rain of sediment on reefs from scarid [parrotfishes] and acanthurid [surgeonfishes] fishes as well as the deposition of the crushed remains of many hard-shelled invertebrates [by other fishes] would seem to have a greater impact on coral reefs than the fishes which feed directly on coral." Meyer and Schultz (1985) noted that grunts that feed by night on seagrass beds produce excretory products (nitrogen and phosphorus) when they rest by day on reefs, thereby enhancing reef growth. Since then, many studies have indicated that reef fishes are important sources of both disturbance and nutrition on coral reefs and that the removal of those fishes by humans can have significant effects on reef condition.

Observations of such local effects have led to concepts of trophic cascades, in which strong connectivities among species influence whole communities and ecosystems. Two sorts of cascades have been proposed. Species-level cascades are pertinent for relatively "simple" systems containing few dominant species, wherein species have strong effects on one another. In salt marsh systems, crabs and terrapins limit herbivorous snails that otherwise would overgraze saltgrasses (Fig. 5.19). Community-level cascades consist of changes among species or processes that substantially alter ecological function throughout an entire system. But when large numbers of species share multiple trophic relationships, predictions about how cascades operate are often impossible.

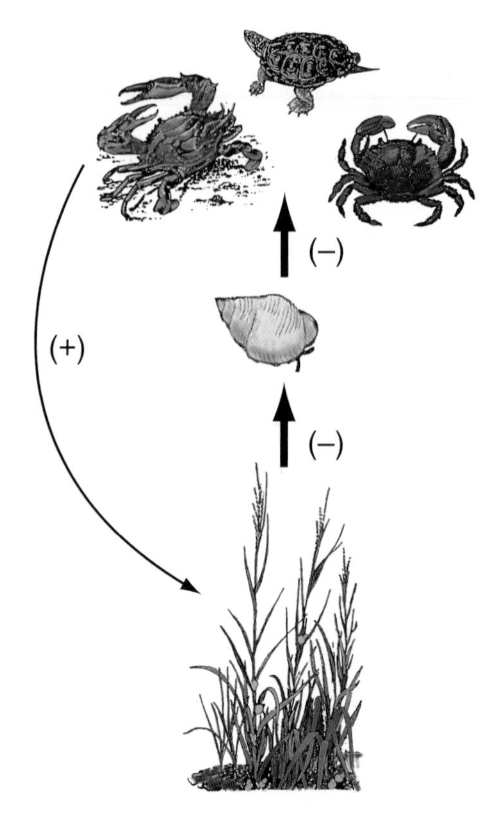

Fig. 5.19 Marsh-inhabiting predators—blue crabs (*Callinectes sapidus*), diamondback terrapins (*Malaclemys terrapin*), and mud crabs (*Penopeus* spp.)—exert top-down controls on the abundance of western Atlantic salt marshes by controlling densities of herbivorous snails (marsh periwinkle, *Littoraria irrorata*), facilitating luxuriant saltgrass growth. In the absence of predation, snail densities increase, resulting in runaway consumption of the marsh canopy and ultimately transforming the marsh into a barren mudflat. From Silliman and Bertness (2002). Diagram courtesy of Jane K. Neron, with permission from B. R. Silliman.

Trophic interactions also strongly influence the magnitude and direction of linkages to biodiversity, productivity, and ecosystem stability (Worm and Duffy, 2003). Changes in predator abundance, as from overfishing, can have far-reaching consequences (Duffy, 2002). For instance, removal of large apex predators (tunas, sharks, billfishes, sea basses, etc.) may cause functional downgrading of food webs, opportunities for invasive species to flourish, and alteration of biogeochemical cycles (Estes *et al.*, 2011). Depletion of predatory sharks has allowed cownosed rays (*Rhinoptera bonasus*) to increase and feed on valued scallops, which caused the shutdown of a century-long fishery (Myers *et al.*, 2007). Similarly, high-trophic-level feeders such as salmon sharks (*Lamna ditropis*) removed by fisheries bycatch had unexpected effects through "mesopredator release" in which smaller predators become so abundant that they affected the commercially valued pollock fishery (*Theragra chalcogramma*; Wright, 2010). In other cases, loss of forage fishes can affect food webs. A surprising series of events in the Benguela Current system of northwestern South Africa resulted from overfishing sardines (*Sardinops sagax*), which allowed jellyfish to dominate the ecosystem and bearded gobies (*Sufflogobius bibarbatus*) to thrive, whose behavior brought benthic dead-end ecosystem products back into the system to make it more productive than otherwise (Utne-Palm *et al.*, 2010).

The smallest organisms in the ocean form a "microbial food web," which plays a major role in production processes. These <5 μm organisms (picoplankton) transfer much more newly synthesized carbon through food webs than previously thought (Richardson and Jackson, 2007). Bacteria and bacterivorous microflagellates assimilate dissolved carbon produced by phytoplankton in the water column and channel it back to the grazing community; other microbes and viral lysis play key roles in nutrient cycling and in controlling dense cell populations (Bratbak *et al.*, 1994). In deep-sea ecosystems, viruses create an essential source of organic detritus in a "viral shunt," which allows the ecosystem to cope with severe resource limitation (Danovaro *et al.*, 2008). Beneath 1000 m, all bacterial heterotrophic production becomes organic detritus.

5.7.2 Disturbance

"Disturbance" is in some respects a misleading term because it seems to infer negative effects, whereas it is often beneficial. In many cases, disturbance is a normal characteristic of systems. Physical and biological disturbances can disrupt spatial structures and interfere with temporal oscillations, species' behavior, and community interactions, but also may increase biotic activity and production. Biotic communities normally flourish under optimal conditions, but when disturbance occurs, communities must reorganize. Absence of disturbance may help explain why deep-water environments appear more uniform and relatively stable than elsewhere.

The rate, frequency, and magnitude of disturbance—such as from storms, dredging, pollution, or arrival of new (exotic or invasive) species—are important controls on marine ecosystem development. Disturbance can be harmonic and regular (waves, tides) or disruptive and irregular (tidal waves, storms). It can be destructive or provide new opportunities. A common form of disturbance is benthic "bioturbation," a process described by Charles Darwin in one of his last books (1881), *The Formation of Vegetable Mould, through the Action of Worms, with Observations on their Habits* (online). The disturbance of sediment structure caused by biotic activity exerts major influences, including sediment aeration and irrigation, distribution of organisms, nutrient dispersal, and community structure and function (Fig. 5.20). Many deposit-feeders mix and recycle sediment within the oxidized surface; so-called "conveyor-belt species" pump reduced sediment to oxidized surfaces. By finding food in sediments, the movements of large and small organisms affect the availability of microbial resources in sediments, release trapped gases and nutrients, and, through variation in flows of materials and modification of fluid dynamics, provide biogeochemical heterogeneity (Gutiérrez *et al.*, 2003; Gutiérrez and Jones, 2006). In the Bering Sea, large biota such as benthic feeding whales and walruses can cause massive displacement of sediment and biota (Fig. 5.21; Ch. 7, Section 7.5), affecting benthic geomorphology, biodiversity, and biogeochemistry on large, regional scales. In these cases, bioturbation may be considered "normal" to system behavior, influencing species diversity and benthic heterogeneity at several scales, with both positive and negative effects.

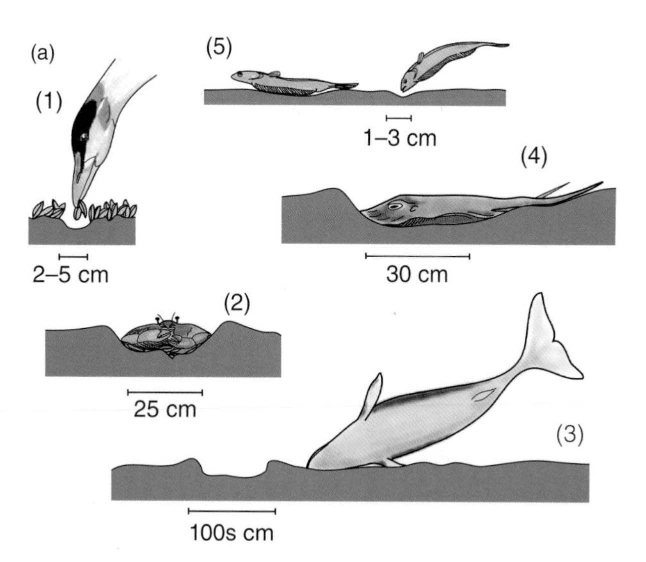

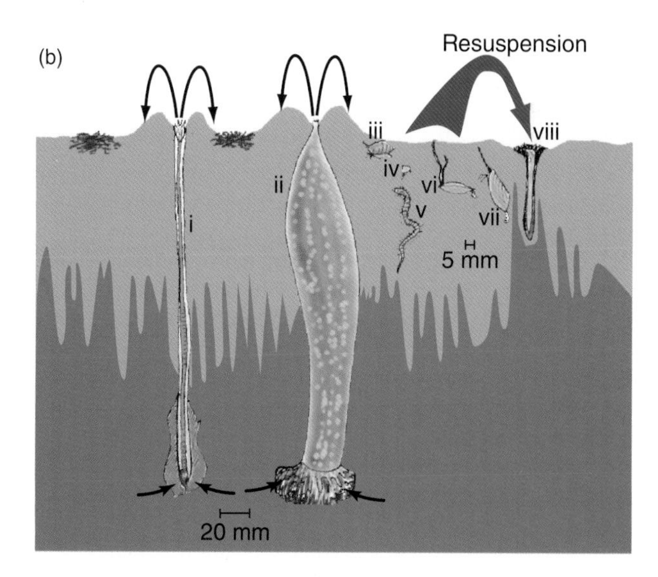

Fig. 5.20 Bioturbation from animal mobility and feeding displaces sediment particles at different scales and influences species diversity and feeding-type diversity. (a) Benthic patchiness generated by bottom-feeding disturbances: (1) eider duck; (2) crab; (3) gray whale; (4) ray; (5) flatfish. Hall *et al.* 1994. reproduced with permission of John Wiley & Sons. (b) Many deposit-feeders mix and recycle sediment. Arrows indicate routes of sediment ingestion and feces egestion by relatively large sediment organisms: (i) burrowing polychaete worm; (ii) sea cucumber; (iii) snail; (iv) clam; (v) errant polychaete worm; (vi) clam; (vii) clam; (viii) anemone. Conveyor-belt species (i, ii) overturn sediment and pump reduced sediment to the oxidized surface. Other species (iii–vii) cycle particles within the oxidized surface. Rhoads, DC (1974). Organism-sediment relations on the muddy sea floor. *Oceanography and Marine Biology Annual Review* **12**, 263–300.

Fig. 5.21 A walrus rooting in the Bering Sea benthos for infaunal food, with significant effects on transport of sediment, release of nutrients, and changes in benthic communities. See Ch. 7 for explanation. Photograph © Goren Ehlme, with permission.

Wave forces in nearshore environments can disturb substrates and topple structures. Waves frequently displace small rocks in shallow, subtidal areas, causing fewer species to be supported than otherwise. Waves generated by storms in tropical seas strike coral reefs and overturn large corals, destroying old beds, creating new habitat, forcing redistribution of living structures, and potentially increasing biodiversity. Such physical disturbances are best seen as fundamental structuring aspects of those ecosystems, to which organisms have adapted. Research findings are ambivalent about whether or not species and environments are closely coupled in physically accommodated environments, or whether there is a direct relationship between species richness and environmental or community stability. Disturbance may be a controlling factor. Many studies indicate that competition and predation tend to stabilize communities, so long as physical conditions are reasonably constant over longer time periods, i.e., the stability-time hypothesis (Sanders, 1968). On the other hand, the intermediate disturbance hypothesis posits that various levels of disturbance act to subdue the effects of competition, allowing dominant, subdominant, and opportunistic species to coexist. For a rocky intertidal community, where disturbance plays a major role, competition for primary space results in clear dominance and species hierarchies (Dayton, 1971). Recovery time from disturbance varies according to the size and location of disturbance. Small-scale disturbances have shorter recovery times than extensive disturbances, such as that of hypoxia or anoxia (Box 2.5). Conversely, fishing effects can be massive and recovery time slow (Hall *et al*., 1994).

5.7.3 Structurally and functionally modifying species

Some species by their presence affect the structure of ecosystems and exert large positive effects on local population densities and sizes, species boundaries at multiple scales, and species composition and diversity (Bruno and Bertness, 2001). Some species directly or indirectly control the availability of resources to other organisms, providing a foundation on which other species depend, or function in the ecosystem to affect others. "Keystone" and "foundation" species call attention to controlling linkages across levels of ecological organization.

It is not always simple to determine which species are "key" or "foundation." Some clues may be found in: (i) the durability of species' impacts; (ii) the number and types of resource flows and structures that are modulated; and (iii) the number of other species affected. Oysters, corals, saltgrasses, and mangroves can structurally dominate their communities while also altering hydrology, nutrient cycles, and sediment stability. Observations of territorial damselfishes (Pomacentridae) provide examples of how defense of algal grazing territory can enhance the diversity of reef algae by decreasing overall grazing by fishes and urchins—an example of a local "keystone" effect by means of intermediate disturbance (Williams, 1980; Hixon and Brostoff, 1983).

The term "keystone" was originally intended to describe species that prevent resource monopolization by consuming prey species, inhibiting competition, and preventing competitive exclusion (Paine, 1974). A broader definition of "keystone" relates to trophic cascades. For example, sea otters (*Enhydra lutris*) prey on sea urchins in the North Pacific, which inhibits overgrazing of macroalgae; in the otter's absence, urchins proliferate and overgraze kelp to create "urchin barrens" (Estes and Palmisano, 1974). On the other hand, foundation species—so-called "ecosystem engineers"—create structures, as in the case of oysters that build reefs (Ch. 6), thereby initiating systemic state changes. Keystone and foundation species can have enormous implications for conservation, as they often are critical for maintaining ecosystem function (Crain and Bertness, 2006; Wright and Jones, 2006).

An intriguing example of biotic modification of the oceans on regional scales concerns stable "Redfield ratios" in phytoplankton and seawater of carbon, nitrogen, and phosphorus

(C:N:P = 106:16:1, Redfield, 1934). In oceanic environments, interactions between organisms and their environment are reciprocal; that is, "the environment not only determines the conditions under which life exists, but the organisms influence the conditions prevailing in their environment" (Redfield, 1958). The Redfield ratio is important for estimating carbon and nutrient fluxes in determining which nutrients are limiting in primary production processes. Excess nitrogen from watersheds into the coastal can alter the ratio, leading to phytoplankton blooms and hypoxia. N:P:SI (silicon) ratios have also been perturbed, leading to changes in coastal plankton (Jickells, 1998), with potential cascading effects. Trends in river inputs, in combination with losses of intertidal habitats important for denitrification, will likely alter coastal phytoplankton, with unknown consequences for food webs.

5.8 "SEASCAPE" AS AN ORGANIZING PRINCIPLE

Taylor *et al.* (2002) note that: ". . .taking a landscape perspective in which the linkages between species and ecosystems play out in space offers an unprecedented opportunity to enhance the linkages between these traditionally separate subdisciplines in ecology." Strengthening knowledge of the ties between species and ecosystems has long been recognized as an essential topic for conservation. "Seascape ecology" is the emerging marine equivalent of landscape ecology as an interdisciplinary field that studies spatial structure, function, and change (Liu and Taylor, 2002). Seascapes represent directly observable habitats at mesoscales. Biogenic structures at the seascape-scale are of particular interest for conservation. They include coral reefs and shellfish beds, seagrass beds, mangrove forests, and dense aggregations of a variety of species on the benthos among intertwined tube-building polychaete worms. The longevity of a biogenic structure may be centuries as for coral reefs, or relatively short and ephemeral but spatially reoccurring, as for bivalve or polychaete reefs (Callaway *et al.*, 2010). Analysis of such seascapes through spatially explicit combinations of natural history, spatial statistics, and modeling can facilitate understanding of population and metapopulation dynamics, *r* vs. K strategies, effects of species invasions, and relationships between ecosystem function and biodiversity.

Seascape ecology is concerned with how habitats are located relative to one another, the causes of spatial patterns, and the ways in which spatial configurations affect ecological processes, such as flows of matter, nutrients, energy, and organisms. As such, seascape ecology can provide practical guidelines for conservation (Wu and Hobbs, 2007). For example, fisheries management models (e.g., population dynamics) that have traditionally treated habitats as homogeneous, closed systems without considering temporal and spatial variation at multiple scales, would benefit by reflecting heterogeneous, patchy habitat structures as inclusive in Large Marine Ecosystems; e.g., to recognize such patterns to better maintain the hierarchical composition, structure, functional organization, biodiversity, and processes characteristic of ecosystems (Wiens *et al.*, 2002). Seascape ecology thus represents a paradigm shift by treating species within habitats in time and space. In this context, terrestrial principles (Dale *et al.*, 2002) are transferable to the oceans:

- individual species and networks of interacting species affect the structure and functioning of ecological systems;
- each site has a specific set of organisms and abiotic conditions that uniquely determines ecological processes;
- the presence, size, and patterns of habitat patches and abiotic conditions on the [sea]scape affect ecological systems;
- disturbances are ubiquitous in nature and often are an integral part of ecological systems; and
- ecological processes change with time.

Such an approach has been applied, at least in part, to oyster reefs, for protected-area planning, and sea-ice/marine-mammal relationships (Chs. 6, 7, 9, 10, 12). Seascape application forms an important interdisciplinary bridge within ecology, with considerable potential to strengthen the ties between species and ecosystems and between population ecology and ecosystem ecology (Turner and Cardille, 2007).

Historically, metapopulation models have considered colonization and extinction for all patches to be equal, and have not incorporated the effects of matrix quality, heterogeneity, and spatial structure that have important effects on model predictions (Fahrig, 2007). Levins (1969) was first to propose the study of population processes in heterogeneous environments, in which the number of occupied patches increases with increasing colonization rate and decreasing extinction rate; derivative models have been termed "patch occupancy," "presence-absence," and "extinction-colonization," all of which assume that habitat patches are either occupied or not, and the only processes operating are colonization and population extinction. These models have led to the widely held assumption that more dispersive species should fare better in fragmented, human-modified environments. Landscape population models lead to the opposite conclusion by explicitly including the environmental matrix (Fahrig, 2002; Fig. 7.27). Due to unsuitable quality of the environmental matrix between habitat patches, some individuals of a species may experience increased mortality when emigrating from patch to patch. Thus, at higher emigration rates, population persistence may decrease due to added dispersal mortality, i.e., loss of habitat increases the proportion of the population that spends time in the matrix, where reproduction is not possible and where mortality rate is higher than in breeding habitat. This leaves open the question: when does the heterogeneity of the matrix itself, independent of its average quality, affect population persistence? This question clearly needs to be addressed when species or their predators have different affinities for different matrix cover types.

Understanding the patterns, causes, and consequences of spatial heterogeneity for species' persistence is a research frontier. The Atlantic sturgeon (*Acipenser oxyrinchus*) provides an example (Niklitschek and Secor, 2005). The sturgeon's low tolerance for temperatures >28°C is a critical threshold for the first two summers of life. During 1993–2002, most down-estuary refuges were unsuitable due to hypoxia and high salinities beyond the regulatory capabilities of young fish. Surface summer temperatures were, therefore, spatially restricted and ranged from 0–35% of the modeled surface area. The value and size of nursery habitat was highly sensitive to climate oscillations and anthropogenic interventions affecting temperature or dissolved oxygen. The study predicted that achieving dissolved oxygen criteria could increase habitat by 13% for an average year, while increasing water temperature by only 1°C would reduce suitable habitat by 65%.

Seascape ecology faces complex theoretical and logistic challenges, especially for marine species, due to significant gaps in natural-history understanding. Information is sparse on emigration, colonization rates, and critical limits, which severely limits the ability to model species' habitat preferences. This situation strongly suggests that population-dynamics models must be altered to include spatially explicit pattern and process. Increased knowledge of natural history is key, particularly as applied to fisheries and establishment of Marine Protected Areas (Ch. 13).

5.9 NATURAL HISTORY: THE BASIS FOR CONSERVATION

This chapter provides a brief overview about how marine species make a living. Each species represents a *complex adaptive suite* of evolutionary inheritance, behavior, physiology, and morphology. This means that various aspects of natural history must be viewed together, not separately, and in an evolutionary context. The biotic and ecological complexity that has developed during the past millions of years has depended, in large part, on the anomalies of geography and history, mass extinction events (Jablonski, 2001), and the evolution of heterogeneous spatial and temporal patterns. The close relationship between a species and its habitat suggests that species-environment relationships may be co-evolutionary, and so may be the close relationships among biodiversity, ecosystem identity, and resilience (Ch. 13).

Population numbers for all species fluctuate to some extent from year to year. Some species fluctuate wildly (*r*-selected), while others (K-selected) adjust to major disturbances through physiological, behavioral, and reproductive adaptations for their persistence, resistance, and resilience. At any one time, a species may be present in some patches and not others, persisting by means of source–sink interactions and drawing on diverse habitat options. A population's fitness, measured as the net production of offspring that survive to breed, or as a product of competitive abilities, is also affected by a host of environmental and demographic factors. Small populations may be at high risk because of inbreeding, random genetic drift, and loss of genetic variability, whereas large populations are expected to persist for longer times. What is to be emphasized is that dispersive species may fare worse in a fragmented environment than has previously been projected by metapopulation models (Fahrig, 2007). Understanding these relationships depends on analysis of the nexus between ecosystems and species natural histories.

In the context of human-caused perturbations (Ch. 2), wide gaps in knowledge about species' ranges and habitats mean that predictions about future consequences will, in most instances, be difficult. Even less is known about how species respond to specific kinds of disturbances and environmental change, how fast new species are invading new locations and changing community structures, and how species can recover from severe declines. Recent extinction rates are estimated to be orders of magnitude faster than the estimated historical average, and could increase many-fold due to emerging human factors, including climate change. This places current extinction at the tipping-point of a sixth great extinction event (Fig. 4.3), differing from the previous Big Five in that it is associated with human activities. Scores of rare and endangered species may already be among the "living dead," especially in the still largely unknown ocean (Box 5.1). For marine systems, species may disappear, unrecognized, with the expected result that the natural functions of ecosystems will be increasingly compromised. Of particular consequence is the decimation of large apex predators, which highlights "the urgent need for interdisciplinary research to forecast the effects of trophic downgrading on process, function, and resilience in global ecosystems" (Estes *et al.*, 2011). In this context, natural history requires a renaissance.

REFERENCES

Aarestrup A, Økland F, Hansen MM, Righton D, Gargan P, *et al.* (2009) Oceanic spawning migration of the European eel (*Anguilla anguilla*). *Science* **325**, 1660.

Adey WH, Steneck RS (2001) Thermogeography over time creates biogeographic regions: a temperature/space/time-integrated model and an abundance-weighted test for benthic marine algae. *Journal of Phycology* **37**, 677–698.

Alerstam T (2006) Conflicting evidence about long-distance animal navigation. *Science* **313**, 791–794.

Allen TFH, Hoekstra TW (1992) *Toward A Unified Ecology*. Columbia University Press, New York.

Bartholomew GA (1986) The role of natural history in contemporary biology. *BioScience* **36**, 324–329.

Bartholomew GA (1970) A model for the evolution of pinniped polygyny. *Evolution* **24**, 546–559.

Bascompte J (2010) Structure and dynamics of ecological networks. *Science* **329**, 765–766.

Block BA, Costa DP, Bograd SJ (2010) A view of the ocean from Pacific predators. In *Life in the world's oceans: diversity, distribution and abundance* (ed. McIntyre AD). Wiley-Blackwell, Oxford, 291–311.

Bohannon J (2007) Ocean study yields a tidal wave of microbial DNA. *Science* **315**, 1466–1467.

Boicourt WC (1982) Estuarine larval retention mechanisms on two scales. In *Estuarine Comparisons* (ed. Kennedy VS). Academic Press, New York, 445–457.

Bratbak G, Thingstad T, Heldal M (1994) Viruses and the microbial loop. *Microbial Ecology* **28**, 209–221.

Brose U, Ostling A, Harrison K, Martinez ND (2004) Unified spatial scaling of species and their trophic interactions. *Nature* **428**, 167–171.

Buchsbaum R, Buchsbaum M, Pearse J, Pearse V (1987) *Animals Without Backbones*. University of Chicago Press, Chicago, Illinois.

Bruno JF, Bertness MD (2001) Habitat modification and facilitation in benthic marine communities. In *Marine Community Ecology* (eds Bertness MD, Gaines SD, Hay ME). Sinauer Associates, Inc., Massachusetts, 201–218.

Burkholder JM, Glasgow HB, Deamer-Mella N (2001) Overview and present status of the toxic complex (Dinophyceae). *Phycologia* **40**, 186–214.

Callaway R, Desroy N, Dubois SF, Fournier J, Frost M, Godet L, Hendrick VJ, Rabaut M (2010) Ephemeral bio-engineers or reef-building polychaetes: how stable are aggregations of the tube worm *Lanice conchilega* (Pallas, 1766)? *Integrative and Comparative Biology* **50**, 237–250.

Cavalier-Smith T (2004) Only six kingdoms of life. *Proceedings of the Royal Society of London* **271**, 1251–1262.

Cermeño P, Falkowski PG (2009) Controls on diatom biogeography in the ocean. *Science* **325**, 1539–1541.

Chapin FS III, Zavaleta ES, Eviner VT, Naylor RL, Vitousek PM, Reynolds HL, *et al.* (2000) Consequences of changing biodiversity. *Nature* **405**, 234–242.

Chornesky EA, Peters EC (1987) Sexual reproduction and colony growth in the scleractinian coral *Porites Astreoides*. *Biological Bulletin* **172**, 161–177.

Cohen JE (1989) Food webs and community structure. In *Perspectives in Ecological Theory*, Chapter 13 (eds Roughgarden J, May RM, Levin SA). Princeton University Press, Princeton New Jersey, 181–202.

Conover DO, Arnott SA, Walsh MR, Munch SB (2005) Darwinian fishery science: lessons from the Atlantic silverside (*Menidia menidia*). *Canadian Journal of Fisheries and Aquatic Sciences* **62**,730–737.

Cowen RK, Lwiza KMM, Sponaugle S, Paris CB, Olsen DB (2000) Connectivity of marine populations: open or closed? *Science* **287**, 857–859.

Crain CM, Bertness M (2006) Ecosystem engineering across environmental gradients: implications for conservation and management. *BioScience* **56**, 211–218.

Dale VH, Fortes DT, Ashwood T (2002) A landscape-transition matrix approach for land management. In *Integrating Landscape Ecology into Natural Resource Management* (eds Liu J, Taylor WW). Cambridge University Press, Cambridge, UK, 265–293.

Danovaro R, Dell'Ano A, Corinaldesi C, Magagni M, Noble R, Tamburini C, *et al.* (2008) Major viral impact on the functioning of deep-sea ecosystems. *Nature* **454**, 1084–1087.

Dayton PK (1971) Competition, disturbance, and community organization: the provision and subsequent utilization of space in a rocky intertidal community. *Ecological Monographs* **41**, 351–389.

Denny MW (1988) *Biology and the Mechanics of the Wave-Swept Environment*. Princeton University Press, Princeton.

Denny MW (1989) A limpet shell shape that reduces drag: laboratory demonstration of a hydrodynamic mechanism and an exploration of its effectiveness in nature. *Canadian Journal of Zoology* **67**, 2098–2106

Denny MW, Blanchette CA (2000) Hydrodynamics, shell shape, behavior and survivorship in the owl limpet *Lottia gigantea*. *The Journal of Experimental Biology* **203**, 2623–2639.

Denny MW, Miller LP, Stokes MD, Hunt LJH, Helmuth BST (2003) Extreme water velocities: topographical amplification of wave-induced flow in the surf zone of rocky shores. *Limnology and Oceanography* **48**, 1–8.

Denny M, Wethey D (2001) Physical processes that generate patterns in marine communities. In *Marine Community Ecology* (eds Bertness MD, Gaines SD, Hay ME). Sinauer Associates, Inc. Sunderland, MA, 3–37.

deRuiter PC, Wolters V, Moore JC, Winemiller KO (2005) Food web ecology: playing Jenga and beyond. *Science* **309**, 68.

Diaz MC, Ward BB (1997) Sponge-mediated nitrification on tropical benthic communities. *Marine Ecology Progress Series* **156**, 97–107.

Dobzhansky T (1973) Nothing in biology makes sense except in the light of evolution. *American Biology Teacher* **35**, 125–129.

Duffy JE (2002) Biodiversity and ecosystem function: the consumer connection. *Oikos* **99**, 201–219.

Dunne JA (2006) The network structure of food webs. In *Ecological Networks: Linking Structure to Dynamics in Food Webs* (eds Pascual M, Dunne JA). Oxford University Press, Oxford, UK, 27–86.

Elliott KH, Woo KJ, Gaston AJ, Benvenuti S, Dall'Antonia L, Davoren GK (2009) Central-place foraging in an Arctic seabird provides evidence for Storer-Ashmole's halo. *The Auk* **126**, 613–625.

Estes JA (1979) Exploitation of marine mammals: *r*-selection of *K*-strategists? *Journal of the Fisheries Research Board of Canada* **36**, 1009–1017.

Estes JA, Palmisano JF (1974) Sea otters: their role in structuring nearshore communities. *Science* **185**, 1058–1060.

Estes JA, Terborgh J, Brashares JS, Power ME, Berger J, Bond WJ, *et al.* (2011) Trophic downgrading of Planet Earth. *Science* **333**, 301–306.

Fahrig L (2002) Effect of habitat fragmentation on the extinction threshold: a synthesis. *Ecological Applications* **12**, 346–353.

Fahrig L (2007) Landscape heterogeneity and metapopulation dynamics. In *Key topics in Landscape* Ecology (eds Wu J, Hobbs RJ). Cambridge University Press, Cambridge, UK, 78–91.

Fish FE, Lauder GV (2006) Passive and active flow control by swimming fishes and mammals. *Annual Review of Fluid Mechanics* **38**, 193–224.

Gill Jr. RE, Piersma T, Hufford G, Servranckx R, Riegen A (2005) Crossing the ultimate ecological barrier: evidence for an 11,000-km-long flight from Alaska to New Zealand and eastern Australia by bar-tailed godwits. *The Condor* **107**, 1–20.

Grassle JF, Maciolek NJ (1992) Deep-sea species richness: regional and local diversity estimates from quantitative bottom samples. *The American Naturalist* **139**, 313–341.

Gray JS (1997) Marine biodiversity: patterns, threats and conservation needs. *Biodiversity and Conservation* **6**, 153–175.

Grimaldi DA, Engel MS (2007) Why descriptive science still matters. *BioScience* **57**, 646–647.

Gutiérrez JL, Jones CG, Strayer DL, Iribarne OO (2003) Mollusks as ecosystem engineers: the role of shell production in aquatic habitats. *Oikos* **101**, 79–90.

Gutiérrez JL, Jones CG (2006) Physical ecosystem engineers as agents of biogeochemical heterogeneity. *BioScience* **56**, 227–236.

Haag A (2005) Whale fall. *Nature* **433**, 566–567.

Hall SJ, Raffaelli D, Thrush SF (1994) Patchiness and disturbance in shallow water benthic assemblages. In *Aquatic Ecology, Scale, Pattern, and Process* (eds Giller PS, Hildrew AG, Raffaelli DG). Blackwell Science, Ltd., Oxford, 333–375.

Hamilton NRS, Schmid B, Harper JL (1987) Life-history concepts and the population biology of clonal organisms. *Proceedings of the Royal Society of London. Series B, Biological Sciences* **232**, 35–57.

Hayden BP, Dolan R (1979) Barrier islands, lagoons, and marshes. *Journal of Sedimentary Petrology* **49**, 1061–1072.

Hayden BP, Ray GC, Dolan R (1984) Classification of coastal and marine environments. *Environmental Conservation* **11**, 199–207.

Hixon MA, Brostoff WN (1983) Damselfish as keystone species in reverse: intermediate disturbance and diversity of reef algae. *Science* **220**, 511–513.

Hoeksema BW (2007) Delineation of the Indo-Malayan centre of maximum marine biodiversity: the coral triangle. In *Biogeography, Time, and Place: Distributions, Barriers, and Islands*, Chapter 5 (ed Renema W). Springer-Verlag, Berlin-Heidelberg, Germany, 117–178.

Hone DWE, Benton MJ (2005) The evolution of large size: how does Cope's Rule work? *TRENDS in Ecology and Evolution* **20**, 4–6.

Hughes TP, Bellwood DR, Connolly SR (2002) Biodiversity hotspots, centres of endemicity, and the conservation of coral reefs. *Ecology Letters* **5**, 775–784.

Hutchinson GE (1965) *The ecological theater and the evolutionary play*. Yale University Press, New Haven, Connecticut.

Jablonski D (2001) Lessons from the past: evolutionary impacts of mass extinctions. *Proceedings of the National Academy of Sciences* **98**(10), 5393–5398.

Jax K (2006) Ecological units: definitions and application. *The Quarterly Review of Biology* **81**, 237–258.

Jickells TD (1998) Nutrient Biogeochemistry of the Coastal Zone. *Science* **281**, 217–222.

Jones GP (1991) Postrecruitment processes in the ecology of coral reef fish populations: a multifactorial perspective. In *The Ecology of Fishes on Coral Reefs* (ed Sale PF). Academic Press, San Diego, CA, 294–328.

Jones GP, Millcich MJ, Lunow C (1999) Self-recruitment in a coral reef fish population. *Nature* **402**, 802–804.

Kidwell SA (2001) Preservation of species abundance in marine death assemblages. *Science* **294**, 1091–1094.

Kidwell SM, Jablonski D (1983) Taphonomic feedback: ecological consequences of shell accumulation. In *Biotic Interactions in Recent and Fossil Benthic Communities*, Topics in Geobiology (eds Tevesz MJS, McCall PL). Plenum Press, NY, 195–248.

Knowlton N (1993) Sibling species in the sea. *Annual Review of Ecology and Systematics* **24**, 189–216.

Koeller P, Fuentes-Yaco C, Platt T, Sathyendranath S, Richards A, Ouellet P, *et al.* (2009) Basin-scale coherence in phenology of shrimps and phytoplankton in the North Atlantic Ocean. *Science* **324**, 791–793.

Leigh Jr EG, Paine RT Quinn JF, Suchanek TH (1987) Wave energy and intertidal productivity. *Proceedings of the National Academy of Sciences* **84**, 1314–1318.

Levins R (1969) Some demographic and genetic consequences of environmental heterogeneity for biological control. *Bulletin of the Ecological Society of America* **15**(3), 237–240.

Levy O, Appelbaum L, Leggat W, Gothlif Y, Hayward DC, Miller DJ, Hoegh-Guldberg O (2007) Light-responsive cryptochromes from a simple multicellular animal, the coral *Acropora millepora*. *Science* **318**, 467–470.

Liu J, Taylor WW (2002) *Integrating Landscape Ecology into Natural Resource Management*. Cambridge University Press, Cambridge, UK.

Longhurst A (1998) *Ecological Geography of the Sea*. Academic Press, San Diego, California.

Lotze HK, Worm B (2009) Historical baselines for large marine animals. *Trends in Ecology and Evolution* **24**, 254–262.

Makris NC, Ratilal P, Jagannathan S, Gong Z, Andrews M, Bertsatos I, *et al.* (2009) Critical population density triggers rapid formation of vast oceanic fish shoals. *Science* **323**, 1734–1737.

Marsh AG, Mullineaux LS, Young CM, Manahan DT (2001) Larval dispersal potential of the tubeworm *Riftia pachyptila* at deep-sea hydrothermal vents. *Nature* **411**, 77–80.

May RM (1988) How many species are there on Earth? *Science* **241**,1441–1449.

May RM (1990) Taxonomy as destiny. *Nature* **347**, 129–130.

May RM, Harvey PH (2009) Species uncertainties. *Science* **323**, 687.

Mayr E (1997) *This is Biology: the Science of the Living World.* Belknap Press of Harvard University Press, Cambridge, Massachusetts.

McIntyre AD, ed. (2010) *Life in the world's oceans: diversity, distribution and abundance* (ed. McIntyre AD). Wiley-Blackwell, Oxford, 384pp.

Meskhidze N, Naenes A (2006) Phytoplankton and cloudiness in the Southern Ocean. *Science* **314**, 1419–1423.

Meyer JL, Schultz ET (1985) Migrating haemulid fishes as a source of nutrients and organic matter on coral reefs. *Limnology and Oceanography* **30**, 146–156.

Micheli F, Halpern BS (2005) Low functional redundancy in coastal marine assemblages. *Ecology Letters* **8**, 391–400.

Moore J (2001) *An Introduction to the Invertebrates*. Cambridge University Press Cambridge.

Mora C, Tittensor DP, Adl S, Simpson AGB, Worm B (2011) How many species are there on earth and in the ocean? *PLoS Biol* **9**, e1001127.

Munson BR, Young DF, Okiishi TH (1994) *Fundamentals of Fluid Mechanics*, Second edition. John Wiley & Sons, Inc., New York.

Myers RA, Baum JK, Shepherd TD, Powers SP, Petersen CH (2007) Cascading effects of the loss of apex predatory sharks from a coastal ocean. *Science* **315**, 1846–1850.

Nelson JS (1984) *Fishes of the World*, Second Edition. John Wiley and Sons, New York.

Nelson JS (2006) *Fishes of the World*, Fourth Edition. Wiley Interscience, New York.

Nicol S (2006) Krill, currents, and sea ice: *Euphausia superba* and its changing environment. *BioScience* **56**, 111–120.

Niles LJ, Bart J, Sitters HP, *et al.* (2009) Effects of horseshoe crab harvest in Delaware Bay on red knots: are harvest restrictions working? *BioScience* **59**, 153–164.

Niklitschek EJ, Secor DH (2005) Modeling spatial and temporal variation of suitable nursery habitats for Atlantic sturgeon in the Chesapeake Bay. *Estuarine, Coastal and Shelf Science* **64**(1), 135–148.

Nowak MA (2006) Five rules for the evolution of cooperation. *Science* **314**, 1560–1563.

NRC (1995) *Understanding Marine Biodiversity*. Committee on Biological Diversity in Marine Systems, National Research Council, Washington, D.C.

Odum WE, Heald EJ (1975) The detritus-based food web of an estuarine mangrove community. In *Estuarine Research* (ed. Cronin LE). Academic Press, New York, 265–287.

Osborn KJ, Haddock SHD, Pleijel F, Madin LP, Rouse GW (2009) Deep-sea, swimming worms with luminescent "bombs". *Science* **325**, 964.

Paine RT (1974) Intertidal community structure: experimental studies on the relationship between a dominant competitor and its principal predator. *Oecologia* **15**, 93–120.

Paracer S, Ahmadjian V (2000) *Symbiosis: An Introduction To Biological Associations*. Oxford University Press, Oxford, UK.

Parker DM, Balaxs GH (2008) Diet of the oceanic green turtle, *Chelonia mydas*, in the North Pacific. In *Compilers Proceedings of the Twenty-Fifth Annual Symposium on Green Turtle Biology and Conservation* (eds Kalb HJ, Rohde A, Gayheart K, Shanker K). NOAA Technical Memorandum NMFS-SEFSC-582, 94–95.

Pearson A (2008) Who lives on the sea floor? *Nature* **454**, 952–953.

Pianka ER (1970) r and K Selection. *American Naturalist* **104**, 592–597.

Polunin NVC, Roberts CM, eds (1996) *Reef Fisheries*. Chapman & Hall, London.

Pyke GH, Pulliam HR, Charnov EL (1977) Optimal foraging: a selective review of theory and tests. *Quarterly Review of Biology* **52**, 137–154.

Randall JE (1974) The effect of fishes on coral reefs. *Proceedings of the Second International Coral Reef Symposium* **1**, Great Barrier Reef Committee, Brisbane, Australia.

Ray GC (1981) The role of large organisms. In *Analysis of Marine Ecosystems* (ed. Longhurst AR). Academic Press, New York and London, 397–413.

Ray GC (1991) Coastal-zone biodiversity patterns. *BioScience* **41**, 490–498.

Ray GC (2005) Connectivities of estuarine fishes to the coastal realm. *Estuarine Coastal and Shelf Science* **64**, 18–32.

Ray GC, Grassle JF (1991) Marine biological diversity. *Bioscience* **41**, 453–461.

Redfield AC (1934) On the proportions of organic derivations in sea water and their relation to the composition of plankton. In *James*

Johnstone Memorial Volume (ed. RJ Daniel). University Press of Liverpool, 177–192.

Redfield AC (1958) The biological control of chemical factors in the environment. *American Scientist* **46**, 205–221.

Rees J (2005) Bio-oceanography. *Nature (Insight)* **437**, 335.

Reznick D, Bryant M, Bashey, F (2002) *r*- and *K*-Selection revisited: the role of population regulation in life-history evolution. *Ecology* **83**, 1509–1520.

Rhoads DC (1974) Organism-sediment relations on the muddy sea floor. *Oceanography and Marine Biology Annual Review* **12**, 263–300.

Richardson TL, Jackson GA (2007) Small phytoplankton and carbon export from the surface ocean. *Science* **315**, 838–840.

Ricklefs RE (1973) *Ecology*. Chiron Press, Newton, Massachusetts.

Ridgway SH, Carder DA (1993) Features of dolphin skin with potential hydrological performance. *IEEE Engineering in Medicine and Biology* **12**, 83–88.

Roff J, Zacharias M (2011) *Marine Conservation Ecology*. Earthscan, London, UK.

Roberts CM (1997) Connectivity and management of Caribbean coral reefs. *Science* **278**, 1454–1457.

Rohde K (1996) Rapoport's Rule is a local phenomenon and cannot explain latitudinal gradients in species diversity. *Biodiversity Letters* **3**, 10–13.

Rooker JR, Secor DH, De Metrio G, Schloesser R, Block BA (2008) Natal homing and connectivity in Atlantic Bluefin tuna populations. *Science* **322**, 742–744.

Rouse GW, Goffredi SK, Vrijenhoek RC (2004) *Osedax*: bone-eating marine worms with dwarf males. *Science* **305**, 668–671.

Sale PF, ed. (1991) *The Ecology of Fishes on Coral Reefs*. Academic Press, San Diego, California.

Sale PF, ed. (2002) *Advances in the Ecology of Fishes on Coral Reefs*, Second edition. Academic Press, San Diego, California.

Sanders HL (1968) Marine benthic diversity: a comparative study. *American Naturalist* **102**, 243–282.

Secor DH, Rooker JR, Zlokovitz E, Zdanowicz VS (2001) Identification of riverine, estuarine, and coastal contingents of Hudson River striped bass based upon otolith elemental fingerprints. *Marine Ecology Progress Series* **211**, 245–253.

Shaw E (1978) Schooling fishes: the school, a truly egalitarian form of organization in which all members of the group are alike in influence, offers substantial benefits to its participants. *American Scientist* **66**, 166–175.

Silliman BR, Bertness MD (2002) A trophic cascade regulates salt marsh primary production. *Proceedings of the National Academy of Sciences* **99**, 10500–10505.

Simberloff D, Dayan T (1991) The guild concept and the structure of ecological communities. *Annual Review Ecology and Systematics* **22**, 115–143.

Skinner JL (2010) Following the motions of water molecules in aqueous solutions. *Science* **328**, 985–986

Skov H, Durinck J, Danielsen F, Bloch D (l995) Co-occurrence of cetaceans and seabirds in the northeast Atlantic. *Journal of Biogeography* **22**, 71–88.

Snelgrove PVR (2010) *The Census of Marine Life: Making Ocean Life Count*. Cambridge University Press, 286pp.

Spalding MD, Fox HE, Allen GR, *et al.* (2007) Marine ecoregions of the world: a bioregionalization of coastal and shelf areas. *BioScience* **57**, 575–583.

Steele JH (1991) Marine functional diversity. *Bioscience* **41**, 470–474.

Steneck RS, Wilson JA (2010). A fisheries play in an ecosystem theater: challenges of managing ecological and social drivers of marine fisheries at multiple spatial scales. *Bulletin of Marine Science* **86**, 387–411.

Strathmann RR (1990) Why life histories evolve differently in the sea. *American Zoologist* **30**(1), 197–207.

Suttle CA (2005) Viruses in the sea. *Nature* **356**, 356–361.

Swearer SE, Caselle JE, Lea DW, Warner RR (1999) Larval retention and recruitment in an island population of a coral-reef fish. *Nature* **402**, 799–802.

Tautz D, Arctander P, Minelli A, Thomas RH, Vogler AP (2003) A plea for DNA taxonomy. *TRENDS in Ecology and Evolution* **18**, 70–74.

Taylor WW, Hayes DB, Ferreri CP, Lynch KD, Newman KR, Roseman EF (2002) Integrating landscape ecology into fisheries management: a rationale and practical considerations. In *Integrating Landscape Ecology into Natural Resource Management* (eds Liu J, Taylor WW). Cambridge University Press, Cambridge, UK, 366–389.

Thébault E, Huber V, Loreau M (2007) Cascading extinctions and ecosystem functioning: contrasting effects of diversity depending on food web structure. *Oikos* **116**,163–173.

Thompson D (1961) *On Growth and Form*, Abridged edition (ed. Bonner JT). Cambridge University Press, Cambridge, UK.

Thorpe SA (2004) Langmuir circulation. *Annual Review of Fluid Mechanics* **36**, 55–79.

Tsukamoto K (2006) Spawning of eels near a seamount. *Nature* **439**, 929.

Turner MG, Cardille JA (2007) Spatial heterogeneity an ecosystem processes. In *Key topics in Landscape Ecology* (eds Wu J, Hobbs RJ). Cambridge University Press, Cambridge, UK, 62–77.

Utne-Palm AC, Salvanes AGV, Currie B, Kaartvedt S, Nilsson GE, *et al.* (2010) Trophic structure and community stability in an overfished system. *Science* **329**, 333–336.

Vanden Berghe E, Stocks KI, Grassle JF (2010) Data integration: The Ocean Biogeographic Information System. In *Life in the world's oceans: diversity, distribution, and abundance* (ed. McIntyre AD). Wiley-Blackwell, Oxford, UK, 333–353.

van der Heide T, Govers LL, Fouw J, Olff H, van der Geest M, van Katwijk MM, *et al.* (2012) A three-stage symbiosis forms the foundation of seagrass ecosystems. *Science* **336**, 1432–1434.

Vogel S (1983) *Life in Moving fluids*. Princeton University Press, Princeton, New Jersey.

Vogel S (1994) *Life in Moving fluids*, Second edition. Princeton University Press, Princeton, New Jersey.

Weimerskirch H, Louzao M, de Grissac S, DeLord K (2012) Changes in wind pattern alter albatross distribution and life-history traits. *Science* **335**, 211–214.

Wiens JS, VanHorne B, Noon BR (2002) Integrating landscape structure into natural resource management. In *Integrating Landscape Ecology into Natural Resource Management* (eds Liu J, Taylor WW). Cambridge University Press, Cambridge, UK, 23–67.

Wilbur HM, Tinkle DW, Collins JP (1974) Environmental certainty, trophic level, and resource availability in life history evolution. *American Naturalist* **108**, 805–817.

Williams AH (1980) The threespot damselfish: a non-carnivorous keystone species. *The American Naturalist* **116**, 138–142.

Wilson EO (2012) *The Social Conquest of Earth*. Liveright Publishing Corporation, New York, London.

Wiltschko W, Wiltschko R (2005) Magnetic orientation in birds and other animals. *Journal of Comparative Physiology A: Neuroethology, Sensory, Neural, and Behavioral Physiology* **191**, 675–693.

Winemiller KO (1995) Fish ecology. In *Encyclopedia of Environmental Biology*, **vol. 2** (ed. Nierenberg WA). Academic Press, San Diego, 49–65.

Winemiller KO (2005) Life history strategies, population regulation, and their implications for fisheries management. *Canadian Journal of Fisheries and Aquatic Sciences* **62**, 872–885.

Winslow F (1882) Methods and results. Report of the oyster beds of the James River, Va. and of Tangier and Pocomoke sounds, Maryland and Virginia. Government Printing Office, Washington, D.C.

Wittebolle L, Marzorati M, Clement L, Balloi A, Daffonchio D, Heylen K, *et al.* (2009) Initial community evenness favours functionality under selective stress. *Nature* **458**, 623–626.

Worm B, Duffy JE (2003) Biodiversity, productivity and stability in real food webs. *Trends in Ecology and Evolution* **18**, 628–632.

Worm B, Barbier EB, Beaumont N, Duffy JE, Folke C, Halpern BS, Jackson JBC, Lotze HK, *et al.* (2006) Impacts of biodiversity loss on ocean ecosystem services. *Science* **314**, 787–790.

Wright B (2010) Predators could help save pollock. *Science* **327**, 642.

Wright JR, Jones CG (2006). The concept of organisms as ecosystem engineers ten years on: progress, limitations, and challenges. *BioScience* **56**, 203–209.

Wu J, Hobbs RJ (2007). *Key topics in Landscape Ecology*. Cambridge University Press, Cambridge, UK.

Wu J (2007) Scale and scaling: a cross-disciplinary perspective. In *Key topics in Landscape Ecology*. Cambridge University Press, Cambridge, UK, 115–142.

CHAPTER 6

CHESAPEAKE BAY: ESTUARINE RESTORATION WITH AN ENVIRONMENTAL DEBT

As man has uprooted the greatest forests, so can he also annihilate the richest oyster beds . . . The preservation of oyster beds is as much a question of statesmanship as the preservation of forests.

Karl Möbius (1883)

6.1 THE GREAT SHELLFISH BAY

Estuarine systems worldwide are being developed, modified, and degraded. These naturally adaptive and productive systems that have sustained human uses since civilization began are critical to coastal economies. Yet over-exploitation, habitat transformation, and pollution are functionally compromising their expected performance. The ecologist Joel Hedgpeth (1977) warned that the most significant factor contributing to degradation of estuaries is failure to treat them as natural systems rather than as conveniences to serve many conflicting purposes. As these vital ecosystems are being transformed in the Anthropocene era (Ch. 1), the balance between sustaining their health and economic benefits for a growing human population requires innovative, protective, and restorative solutions that consider ecosystem-based management (McLeod and Leslie, 2009).

The Chesapeake Bay is the largest U.S. estuary and among the largest, most complex and most productive in the world. The indigenous Algonquin native people called it "Chesepiooc," the great shellfish bay. This estuarine system has provided socio-economic benefits, not only in fisheries production but also as a center for international trade, military bases, and employment and enjoyment for an expanding human population. Over decades of intensifying use that is trending the Bay into an altered state of degradation, human intervention has become mandatory. This case study highlights the scientific and socio-political nature of this dynamic ecosystem that historically supported high abundances of estuarine species. While most people agree that the Bay needs restoration, an ecological debt and its costly burden pit public and private sectors to a duel of legal and public persuasion.

6.2 ECOLOGICAL LINKAGES TO NATURAL WEALTH

When the explorer Giovanni da Verrazano sailed through the Virginia Capes in 1524, he observed a verdant, forested landscape. He encountered extensive marshes, abundant natural resources, and oyster reefs that obstructed ship movements. The landscape held large trees, 40% higher than today, with prosperous stands of American chestnut trees (near-extinction today). A 1000-year-old white oak measured 13 ft in diameter 16 ft above the base and 10 ft in diameter 31 ft from its base (Hiltz, 2005), and one cypress tree measured 18 ft in diameter (Blankenship, 1999). Mature forests attracted large mammals, wildfowl, seasonally abundant waterfowl, and passenger pigeons (now extinct) that blackened the sky (MacCleery, 1994). Extensive natural marshes looked like fields of hay. George Percy wrote in 1607: ". . . upon this plot of ground [Lynnhaven Bay, near Norfolk Virginia], we got a good store of mussels and oysters, which lay on the ground as thick as stone . . ." The early explorer Captain John Smith described an abundance of fish and diverse species that sustained his ailing party in the first colonial settlement at Jamestown in 1608. Father White in Maryland's first settlement in the 1660s described the Potomac River as ". . . the sweetest and greatest river I have seene so that the Thames is but a little finger to it. There are noe marshes or swampes about it, but solid firm ground, with great variety of woode, not choaked up with under shrubs, but commonly so farre distant from each other as a coach and fower horses may traveale without molestation . . . All is high woods except where the Indians have cleared for corne. It abounds with delicate springs which are our best drinke." The English colonists moved into the Bay region and established a social system, encouraging an expanding human population to modify the region with increasing intensity (Fig. 6.1), accelerating estuarine change.

6.2.1 A complex ecosystem

The Chesapeake Bay is a complex, coastal-plain, estuarine system of the mid-Atlantic region (Fig. 6.2). It contributes

Marine Conservation: Science, Policy, and Management, First Edition. G. Carleton Ray and Jerry McCormick-Ray.

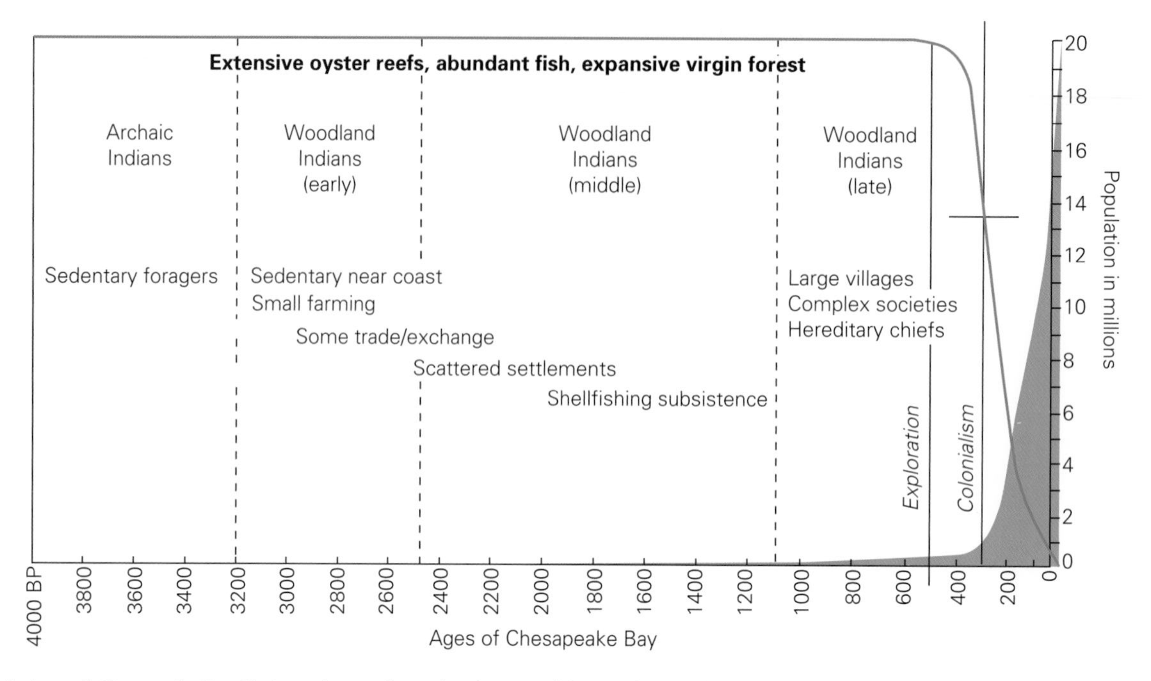

Fig. 6.1 Ages of Chesapeake Bay. Native culture subsisted on oysters, fishes, and native forest resources for thousands of years. European culture quickened the slow pace of change that depleted estuarine species. X-axis shows years before present (0 represents year 2000) followed by rapid human population increase that is expected to continue. Based on data from Custer (1986).

ecological products to fisheries and conservation-targeted wildlife—dolphins, whales, fishes, sea turtles, and migratory birds—and plays an important ecological role of regional importance in the wider mid-Atlantic ecosystem. As a partially mixed estuary (Pritchard, 1956), it depends upon variable tidal fluxes and wind/riverine pulses to drive its circulation (Goodrich and Blumberg, 1991), to support estuary-dependent species, and to maintain regional biodiversity and ecosystem health.

The Bay's characteristic health and resiliency depend on internal and external pulses in complex ecological networks, with timing of events that are linked spatially and temporally across the Bay. The Bay is an essential breeding and feeding habitat for many coastal invertebrates, fishes, birds, and juveniles of many species that depend on the Bay as critical "nursery" habitat to feed, shelter, and support their young before dispersing to new locations. Physiological timing and access to appropriate habitats are critical for their survival, abundance, and recruitment back into populations and communities widely distributed across the Bay and into coastal waters. Some species create landscape and seascape structures (oyster reefs, marshes, seagrass beds) that function as habitats for other species, and these structures adapt, persist, or change with fluctuating conditions.

Bay health and resiliency depend on its 166,000 km^2 watershed system that delivers terrestrial products into the semi-enclosed 6500 km^2 Bay basin; the 25:1 ratio of watershed drainage to Bay size is the highest for any semi-enclosed coastal water body in the world (Fig. 6.3). Surface drainage originates in high elevations, transgresses into streams through forests, grasslands, and hilly piedmont terrain, through many state and local jurisdictions, farmlands, and various other land uses, and down abrupt falls into eight major tributary systems: Susquehanna, Patapsco, Patuxent, Potomac, Rappahannock, York, James, and Choptank (Fig. 6.2; Table 6.1). Downstream from fall lines, where cities (e.g., Richmond, Virginia) historically developed, freshwater tributaries deliver terrestrial materials (fluvial products) across a relatively flat, variably wide (24–145 km) sedimentary coastal plain and through tidal marshes where riverine water grades into saline water before entering the Bay. When tributary surface waters are included with Bay surface area, the area almost doubles, highlighting the extent of tributary water. With total surface freshwater runoff and groundwater contributing more than 90% of freshwater to the Bay, delivery of freshwater to the Bay varies greatly among tributaries, in complex mixtures of estuarine and ocean water. The mixing is made more complex by seasonal change, climatic events of drought, storms, and hurricanes, and by Bay geomorphology.

Bay tributaries deliver variable amounts of freshwater, north to south and east to west, which create a skewed, dynamic entry of freshwater into the Bay's variable estuarine system. The Bay's largest sub-watershed in the north, the Susquehanna River, crosses three state jurisdictions and contributes about half of the Bay's total freshwater supply as it drains through various industrial activities that include hydraulic-fracture (fracking) mining for natural gas and extensive agriculture. Its flow is constrained by the Bay's largest hydroelectric dam, with many others along its path. At its mouth, extensive freshwater marshes filter fluvial materials and pollutants before exiting into the northern Bay. The Potomac River is the second largest watershed, crosses four

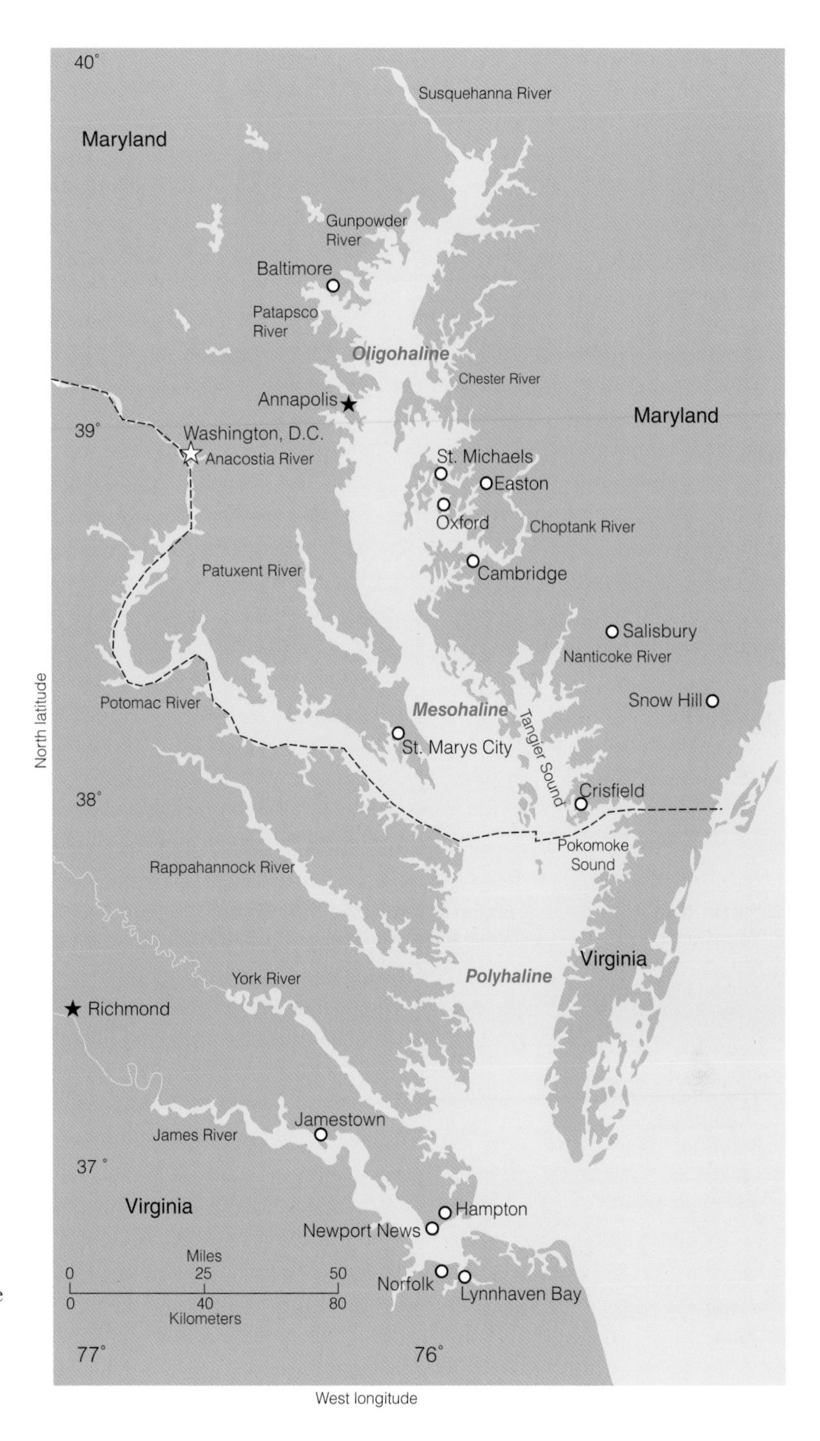

Fig. 6.2 Chesapeake Bay place-names, major tributaries, and salinity regions. Stars locate capitals: state (black) and federal (white); circles locate historic and urban centers; dashed line indicates Maryland/Virginia state boundary. Major salinity provinces for the Bay are indicated as oligohaline (<10 psu), mesohaline (10–20 psu), and polyhaline (>20 psu) (based on data from Harding and Perry (1997), which are subject to change with season and storm events).

state jurisdictions, delivers about 17% of the Bay's total freshwater into its middle portion, and encompasses major population centers, including the rapidly expanding U.S. Capital at Washington, D.C. The James River is the third-largest watershed, located entirely within Virginia's jurisdiction, and encompasses one of the largest U.S. ports at Norfolk, which is a major military center. The James contributes about 14% of the Bay freshwater into the southwest portion. The drier eastern barrier peninsula ("Eastern Shore") across the Bay where salt marshes and agricultural land are extensive, contributes less than 4% of the total, and falls under Maryland and Virginia jurisdictions, and a small portion of Delaware's.

6.2.1.1 Physical structure

The 320 km-long, funnel-shaped Bay basin is relatively shallow. Water depths average 8.4 m, decreasing to 6.5 m if depths of major tributaries are included. A 30–50 m deep and 2–4 km-wide paleo-channel (Colman *et al.*, 1990) that is important to ship traffic and benthic-water flow, is flanked by

Fig. 6.3 Chesapeake Bay watershed (blue) and inclusive states, highlighting freshwater drainage into the Bay.

Table 6.1 Major tributaries (see map location Fig. 6.3). Total freshwater input to Bay averages 2300 m^3s^{-1}.

Tributary-drainage basin (Bay location)	Drainage basin (square miles) % freshwater to Bay	State jurisdiction(s)	Watershed land use	Assessment of overall condition Ecosystem health indicators
Susquehanna (upper Bay)	27,500 50–60%	NY; Penn; Md.	Pop. 4 million For. 66% Agr. 29% Dev. 2%	Moderate
NW Maryland (northwest)	1,670	Md.	Pop. 2.1 million For. 36%; Agr. 34; Dev. 17%	Moderately good
Patuxent (west)	957 2%	Md.	Pop. 600,000 For. 42%; Agr. 34; Dev. 11%	Poor
Potomac (west)	14,679 17%	W. Va.; DC; Md; Va	Pop. 5.2 million For. 57% Agr. 32% Dev. 5%	Poor
Rappahannock (west)	2,845 <4%	Va	Pop. 241,000 For. 57%; Agr. 31; Dev. 2%	Poor
York (west)	3,270 3rd smallest %	Va	Pop. 373,000 For 61%; Agr. 21; Dev. 2%	Very poor
James (southwest)	10,432 ~14%	Va	Pop. 2.5 million For 71%; Agr. 17; Dev. 5%	Moderate
Eastern Shore (largest river is Choptank R.)	5,048 (Choptank = 795) <1%	Md; Va; Del.	Pop. 467, 265 Agr. 39% For. 25% Dev. 2%	Choptank = poor Nanticoke = moderate Fishing Bay = poor Wicomico = poor

Pop. = population; For. = forest cover; Dev. = developed; Agr. = agriculture; Va = Virginia; Md = Maryland, DC = Washington, D.C.; Penn = Pennsylvania; NY = New York. Source: EPA (1983); CBP (2001, 2011); Harding *et al.* (1999); Boicourt *et al.* (1999); Chesapeake Bay Watershed Assessments; Chesapeake Ecocheck.

shallow and variably wide (5–30 km) silty-mud shoals. Only a small proportion of the Bay exceeds 10 m depths, making it susceptible to intense interactions, where water pushed by winds interacts with bathymetry, irregular shorelines, and fluctuations of currents and tides that play important roles in water motion and mixing. The effects of Earth's rotation in the wide, southern portion of the Bay further complicate water movements and salinity (Coriolis effect, Ch. 4).

Water motions, tidal beats, and changing water masses complicate the pelagic nature of this system, with strong influences on marshes, reefs, and seagrasses. Tidal energy provides a major mechanical force and affects hydrologic structure. Tides deliver salt water from the Atlantic Ocean in predictable, semi-diurnal, oscillating beats, sending a progressive wave northward (Fig. 6.4) that influences water movements, shoreline erosion, sedimentation patterns, and current transport processes. Tidal range progressively decreases northward, from about 1 m at the Bay's entrance to fractions of a meter at mid-Bay, where tides are replaced by a partial standing wave. Thus, within a 24-hour period, the energy of an entire wavelength is contained in the Bay when a second wave enters. Dominant flow directions shift with changing tides, and tidal energies push water over shoals into tributaries and shallow creeks, where asymmetric tidal beats alter flow speeds and directions. Depth, water-mass properties, bottom roughness, and topographic structures further complicate water movements.

Estuarine water is chemically active. Rivers deliver positively charged ionic compounds (iron, nitrogen, dissolved organic matter, etc.) that interact with negative ions (sodium, chlorides, etc.) of ocean water. This ionic interaction causes a major change in ionic strength that results in flocculation and colloid formation. Colloids are bio-reactive, and during estuarine mixing, flocculation of colloidal iron and organic matter form particulate iron that sinks to the benthos (Cotrim da Cunha *et al.*, 2007). Also, trace metals are incorporated into colloidal organic matter from respiration, photo-oxidation, and aggregation, including mercury (Guo and Santschi, 1997; Kepkay, 2000; Cotrim da Cunha *et al.*, 2007). Colloidal reactivity and ability to aggregate and concentrate trace metals on the benthos play important roles in the carbon cycle and food webs. Colloidal organic carbon is a globally significant fraction of dissolved organic carbon in the surface ocean (Kepkay, 2000).

In this estuarine environment, salinity, temperature, and water movement interact to form major pelagic structural features, creating water-mass characteristics marked by sharp, dynamic boundaries that affect the Bay's ecology. Buoyant, warm, surface freshwater from rivers flows seaward over opposite-flowing, cold, dense, saline water flowing landward. In the region between them, a sharp density transition creates a boundary known as the pycnocline (Fig. 6.5), a major structural and transport feature. At the pycnocline, current velocity drops, pH and ionic composition are altered, chemicals change,

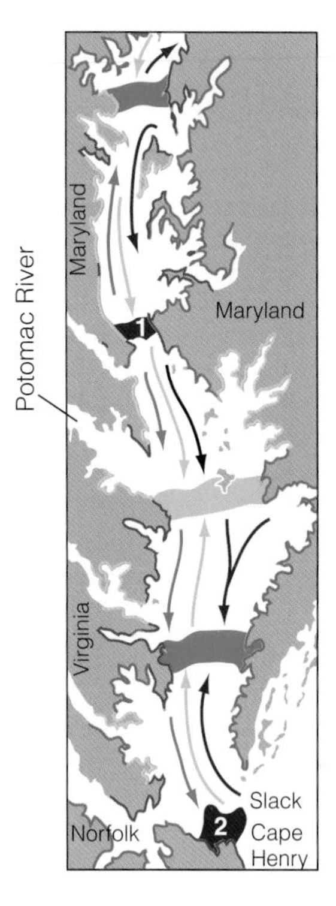

Fig. 6.4 A progressive tide enters the Bay mouth at time 0 and moves northward to mid-Bay (1) when another wave enters (2). Blue bands indicate slack tide every six hours, color-matched to dominant current flows (arrows). A complete semi-diurnal tide wave is thus contained in the Bay at all times, interacting with variable winds, complex shoreline structures, bottom topography, currents, Earth rotation, isostasy, and watershed inflows. These complex flows of motion all contribute to varying shoreline erosion patterns (purple = severe; red = high; yellow = moderate; green = slight). Based on data from Stevenson & Kearney (1996).

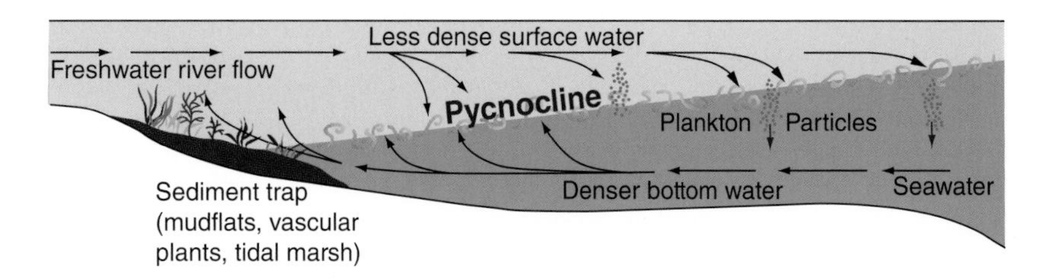

Fig. 6.5 Conceptual schema of a pycnocline, a water-column stratification where there is a sharp salinity change over a short vertical distance. Intense mixing occurs as buoyant surface freshwater flows seaward over opposite, landward-flowing, dense, bottom seawater. This stratification interacts with a tidally averaged residual flow, whereby particles (plankton, larvae) can be carried horizontally over some distance, or retained and deposited at slack tide. Based on data from Correll (1978).

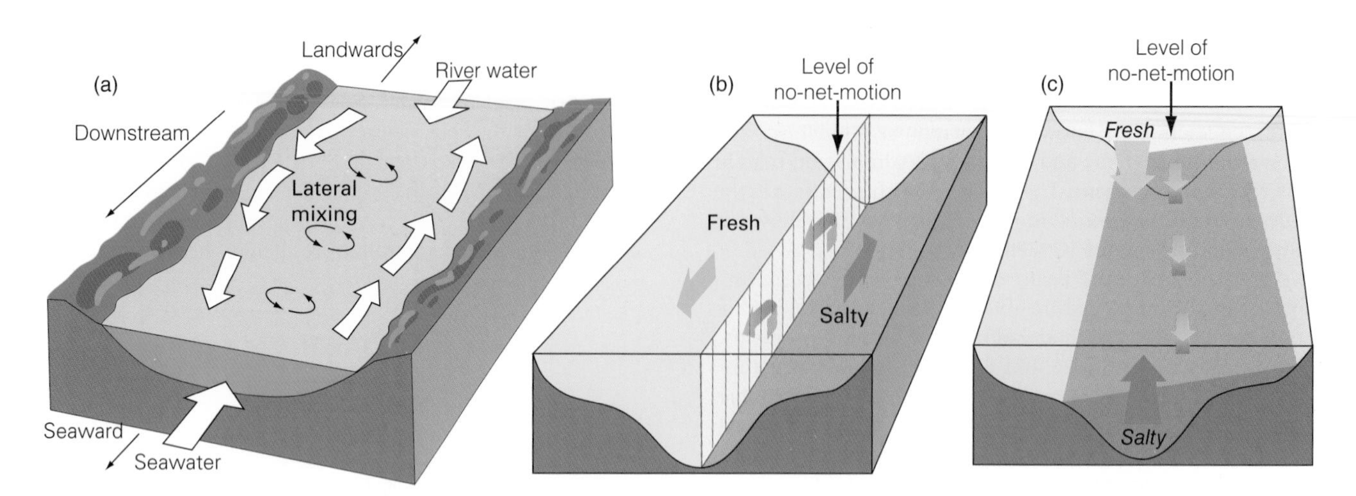

Fig. 6.6 Conceptual illustration of different estuarine circulation patterns, which can shift with season and freshwater inflow. (a) Net, non-tidal, residual circulation with a lateral mixing pattern in a well-mixed, vertically homogeneous estuary. Bearmann GG, ed. (1985) Waves, Tides and Shallow Water Processes. The Open University. With permission from Elsevier. (b) Under well-mixed conditions, a vertical two-way circulation pattern can occur with vertical no-net-motion (winter), and change under conditions of a moderately stratified, inhomogeneous estuary (summer); and (c) in summer where a mid-water, horizontal circulation pattern has a slight horizontal-tilt due to the Earth's rotation, as studied in the James River, Va. b, c from Nichols (1972). Sediments of the James River estuary. In *Environmental Framework of Coastal Plain Estuaries*. Memoir 133. The Geological Society of America, Colorado, pp. 169–212.

flocculation and colloids (gels, sols, emulsions) occur, and nutrients are absorbed or released through differential exchanges. The asymmetric ebb-and-flood tidal motions and freshwater inflow set up a two-layered, internal estuarine circulation system, also known as a residual estuarine current, that lacks net motion and functions like a conveyor belt in transport processes. As water and salt in the pycnocline are moved from lower to upper levels, a vertical-motion velocity becomes entrained and particles are transported horizontally. Entrained particles (colloids, nutrients, plankton, larvae, sediment) may be transported long distances or deposited in areas of no-net-motion, i.e., hydrological "null points," where turbidity is high and all but the fine-clay fraction of mineral particles are deposited over rather short distances. Freshwater inflow and changes in salinity, temperature, turbulence, and mixing all affect the pycnocline, and cold winter months break it up. This physical stratification effectively creates an efficient, selective trap or transport for river-borne nutrients, pollutants, and organic matter, and a mechanism for larval transport. A well-mixed, vertically homogeneous estuary (Fig. 6.6a,b) forms a net, non-tidal residual circulation pattern in contact with air, that becomes a moderately stratified inhomogeneous estuary formed in summer (Fig. 6.6c). Under conditions of high nutrient concentrations in warm summer months and below the pycnocline, dissolved oxygen can become depleted.

The Bay is a partially mixed estuary, where salinity, oxygen, and nutrients are strongly influenced by river runoff (Goodrich and Blumberg, 1991; Li *et al.*, 2005). Salinity, a measure of saltiness (Ch. 4), is critical to understanding species distributions and abundances and estuarine biogeography. Salinity tracks spatial salt gradients in Bay and tributary systems, from areas of low salt content in brackish water to full strength seawater. In brackish water, hydrochemistry is accelerated, metals and phosphorus are mostly removed, chemical and biological reactions are rapid, and biotic diversity is low. Ionic transformation is most active in areas up to 6 psu (ppt), variable up to 10, and relatively inactive at higher salinities. Due to differences in basin geometry, watershed area, inflow quantity and quality, and proximity to tidal influences, a variety of salinity gradients exist among tributaries, each being subject to differences in drought, storm, human activities, and seasonal changes. Salt content effectively creates not only areas of active chemical transformations, but also physiological barriers to sensitive species.

6.2.1.2 Biogeographic pattern

Biotic distributions are affected by the Bay's physically complex, heterogeneous environment. Physiological tolerances deter ocean species from entering freshwater systems and freshwater species from entering ocean water, unlike estuarine species such as oysters, blue crabs, some fish, and many year-round resident species that tolerate variable degrees of salinity and temperature. Freshwater species remain in salinities less than 5 psu, and fully marine species occur in greater than 30 psu. Species unable to tolerate wide fluctuations in salinity (stenohaline) stay within their range of physiological tolerance: e.g., freshwater species are at highest diversity at heads of tributaries, and saltwater species are at highest diversity at saltier mouths of tributaries. Lowest biodiversity occurs in physiologically stressful brackish water. Thus, estuarine biotic communities are spatially complex, sustaining resilient biota adapted to natural change. Many species move in or out of the Bay and into the watershed, making definition of an estuarine species difficult (Ch. 5; Ray, 2005). In the Bay ecosystem, biota assemble into communities and create habitats in biogeographic patterns, where repeating, pulsed oscillations between the physical environment and internal community pulsing together may relate to ecosystem performance (Odum *et al.*, 1995).

One such repeating pulsing pattern is observed in the seasonal abundance and distribution of Bay offspring. Most estuarine species are highly fecund, beginning life at very small sizes and undergoing at least three distinct life stages (larva, juvenile, adult). Each stage of development requires a different diet, habitat, and location, where oscillating tidal currents and water motion determine the concentration and delivery of food and oxygen, waste removal, transport, and dispersal to appropriate habitats. Estuarine-associated species use a diversity of habitats for reproduction, growth, and recruitment, their life-history stages being connected across the region as metapopulations with contingent memberships (Rooker and Secor, 2005; Secor, 1999). Different species capture food, being adapted to varying degrees of tolerance to wind, tidal currents, circulation patterns, water density, and mixing. Timing is critical for energy conservation and for optimum reproduction; adults release their reproductive products (sperm, eggs, larvae) into the water column for dispersal under optimal environmental conditions and availability of food. Each species may encounter steep environmental discontinuities over short distances that it must overcome through survival mechanisms and utilization of ecological subsidies linked to estuarine dynamics.

While some species are year-round Bay residents, others move in or out to find preferred areas to feed, reproduce, mature, or hide; sessile species committed to a benthic location risk survival when environmental conditions change. Oscillating environmental conditions expose these species to tidal, seasonal, and climatic fluxes. For example, on a small local scale in shallow water at slack tide, stationary biota are exposed to varying amplitudes (cm s^{-1}) of salinity, current velocity, and surface elevation, all of which change with time and location (Fig. 6.7). Some conditions often have lag effects (hysteresis) that affect biota hours, days, weeks, months, or even years later, depending on scale. Although the dynamic interplay of multi-scale interactions is difficult to observe, measure, or predict, estuarine animals have evolved mechanisms to survive, or they perish.

Year-round sessile molluskan species (oysters, clams) typically colonize along salinity and tidal gradients in self-organizing, spatially adapted biogeographic patterns governed by their natural histories and tolerance mechanisms (Fig. 6.8). In doing so, they create seascape habitats, as do tidal wetland communities that also self-organize in spatial patterns. In the York River, nine wetland types are structured along salinity gradients with exposure to tidal inundation (Box 6.1). When freshwater and environmental patterns change, species adjust, communities shift, and biogeographic patterns change.

6.2.2 Estuarine ecosystem efficiency

Biota coevolve with dynamic estuarine systems. Through dynamic interplays among physical processes and functional

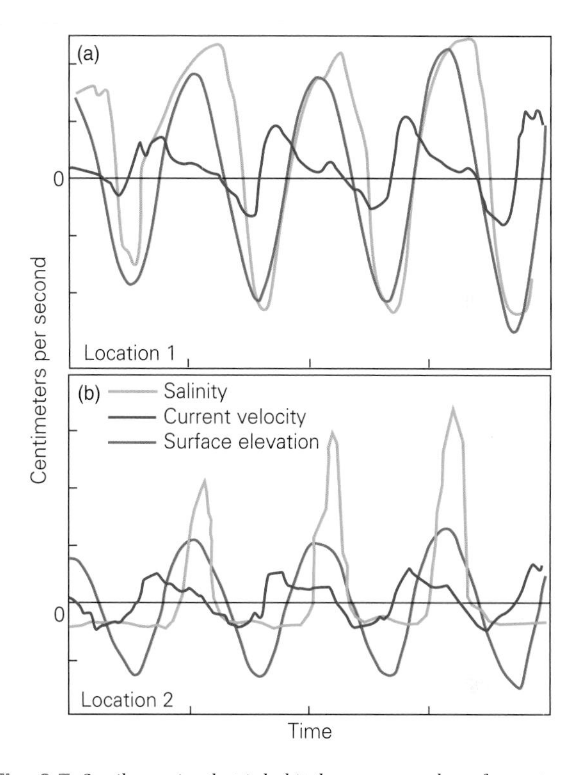

Fig. 6.7 Sessile species that inhabit the upper reaches of an estuary undergo exposures to extreme environmental variations at slack tide (0) due to lag times in salinity, current velocity, and surface elevation operating over several tidal cycles and that differ within and between sites (a, b). Based on data from Uncles *et al.* (1991).

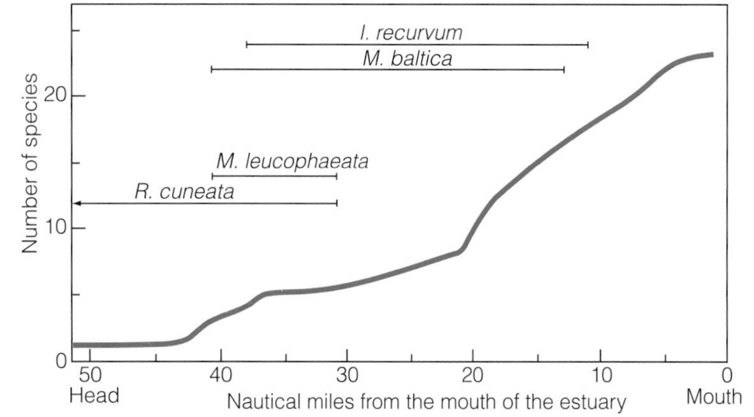

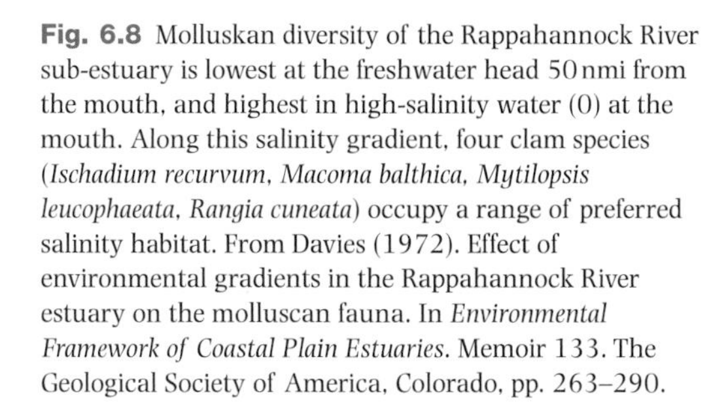

Fig. 6.8 Molluskan diversity of the Rappahannock River sub-estuary is lowest at the freshwater head 50 nmi from the mouth, and highest in high-salinity water (0) at the mouth. Along this salinity gradient, four clam species (*Ischadium recurvum, Macoma balthica, Mytilopsis leucophaeata, Rangia cuneata*) occupy a range of preferred salinity habitat. From Davies (1972). Effect of environmental gradients in the Rappahannock River estuary on the molluscan fauna. In *Environmental Framework of Coastal Plain Estuaries*. Memoir 133. The Geological Society of America, Colorado, pp. 263–290.

Box 6.1 Tidal marshes of the York River, Chesapeake Bay

Lori A. Sutter
Virginia Institute of Marine Science, College of William and Mary, Gloucester Point, Virginia, USA

James E. Perry
Department of Coastal and Ocean Policy, Virginia Institute of Marine Science, College of William and Mary, Gloucester Point, Virginia, USA

The York River tributary of the Chesapeake Bay has a large number of wetland communities distributed along gradients of salinity and tidal inundation. The plant communities found in these wetlands are determined by the location along these gradients, which can vary both spatially and temporally. The combined stresses of inundation and salt water, while limiting the types of biota that can survive in the marshes of the lower portion of the Bay, also provide for a diverse number of tidal wetland habitats.

Nine common vegetated marsh types have been described in the tidal freshwater, oligohaline, mesohaline, and polyhaline sections of the York River (Table B6.1.1). These are arranged in the York River landscape along a salinity gradient with the polyhaline marshes at the mouth and tidal freshwater marshes farther upstream from saltwater influence. In the lower portion of the river only a few vascular plants are able to tolerate the combined effects of tidal inundation and high salt content. In upstream reaches, water column salinity is low to non-existent. Without the stress of salinity, more species of vascular plants are able to survive. Here, tidal inundation can be the principal factor affecting community composition and function.

The tidal wetlands of the Chesapeake Bay perform a number of important ecological functions and anthropomorphic services that are highly valued by humans. The most important of these functions and services are primary

Table B6.1.1 Marsh communities of the York River.

Community type	Salinity range	Dominant species	Distribution
Salt marsh cordgrass (i.e., smooth cordgrass)	poly- to mesohaline	*Spartina alterniflora* (saltmarsh cordgrass) [usually monospecific stands]	mean sea level (MSL) to approximately mean high water (MHW)
Saltmeadow	meso- to polyhaline	*Spartina patens* (saltmeadow hay) Distichlis spicata	slightly increased elevation located landward of the salt marsh cordgrass community
Black needlerush	meso- to oligohaline	*Juncus roemerianus* (black needlerush) [usually monospecific stands]	landward or interspersed among the saltmeadow community
Saltbush		*Iva frutescens* (salt bush) *Baccharis halimifolia* (groundsel tree)	upward boundary of the tidal marsh
Big cordgrass	oligohaline	*Spartina cynosuroides* (big cordgrass)	slightly above MHW, but variable in range
Cattail	saline tidal reaches	*Typha angustifolia* (narrow-leaved cattail)	isolated stands in brackish marshes, often near the upland margin where there is freshwater seepage
Reed grass	all	*Phragmites australis* ssp. *australis*, *P. a.* ssp. *Americanus* (reed grass)	above MHW and almost always associated with disturbance
Salt panne	meso- to polyhaline (become hyper-saline)	*Salicornia virginica, S. europea* and *S. bigelovii* (saltworts) [sparsely vegetated]	shallow depressions, which often form within the interiors of large salt marsh cordgrass communities
Brackish marsh	meso- to oligohaline	no single species dominates: salt marsh cordgrass, saltmeadow hay, saltgrass, black needlerush, saltbushes, three-square bulrush (*Schoenoplectus americanus*), big cordgrass and cattail	distributed vertically from MSL, where salt marsh cordgrass dominates, to the upper limits of tidal inundation, where the saltbushes occur
Freshwater mixed marsh	fresh	no single species dominates: *Nuphar luteum* (yellow pond lily) *Peltandra virginica* (arrow arum) *Pontederia cordata* (pickerel weed) *Zizania aquatica* (wild rice)	under tidal influence factors affecting the dominance include season, elevation and salinity

Source: Based on data from Perry and Atkinson (2009)

production and detritus availability, wildlife and waterfowl support, shoreline erosion buffering, and water-quality control.

With sea-level rise (SLR), salinity and inundation period are expected to increase; these factors are likely to bring changes to the tidal marsh communities that line the river. The predominant paradigm is that salt marshes will either accrete to maintain their position or drown; salt marshes will replace brackish marshes and brackish marshes will replace freshwater marshes. Because the topography steepens upstream with little space to allow freshwater marsh development, even if the tide can move upriver, it is possible that these systems could disappear. These changes will have resulting impacts on the functions and services of the marshes that require further study to fully understand, but we can predict the following: First, if salt marshes drown, there would no longer be a shoreline buffer function to protect coastal landscapes. If these systems transition to mudflats or open water, we will lose sediment trapping functions, thereby reducing water quality, and potentially silting of shellfish beds, submerged aquatic vegetation, and navigation channels. Pollutant filtration once offered by these systems may also be lost.

The brackish marsh community is considered a microcosm of all the communities found in saline water and is ranked along with the saltmarsh cordgrass community as one of the highest-valued marsh areas in Virginia because of its productivity, diversity, and value for flooding, buffer, erosion, and water-quality control. Because of brackish marsh location in low to moderate salinity areas, many are known spawning and nursery grounds for finfish and crabs. They also are important as valuable foraging areas and habitats for a wide diversity of wildlife species. If these systems "march upstream," it is unclear whether or not small organisms will be able to keep pace with the upstream movement.

The freshwater mixed community has one of the highest annual productivities in tidal wetlands of this region. These marshes are valuable for wildlife/waterfowl as the plants produce a diversity of abundant seeds, roots, and tubers. Typically, tidal waters are important spawning and nursery grounds for many resident and anadromous fishes such as striped bass, shad, and river herrings. The marshes are also important as flood and erosion buffers and sediment filters; however, much of the aboveground vegetation dies back in the winter creating broad mudflats. Sediments are readily trapped during the growing season, enabling most of these areas to maintain themselves under SLR conditions. Salinity intrusions during years of drought may significantly change community structure within one year's time so that more salt resistant species may dominate. Furthermore, if there is no room for these communities to move upstream, the river system may lose all of the benefits they offer. The loss of high productivity may change the food that is pulsed into the river system, potentially affecting nursery functions both temporally and spatially. Finally, SLR may facilitate the ability of salt-tolerant native species (e.g., *Spartina alterniflora*) and invasive species (e.g., *Phragmites australis*) to outcompete native species that cannot tolerate increasing levels of salinity.

assemblages, biota capture energy, transform geochemistry, and evolve into ecological patterns that contribute to estuarine production. Biota interrupt the continuous flow and oscillating patterns of energy, nutrients, chemicals, and sediment to create dynamic spatial patterns. Where species colonize, hydrologic flows can quickly change, and when biota form assemblages and functional communities, they influence ecological processes relating to Bay health, resiliency, and production.

6.2.2.1 Transformation of energy into estuarine production

The modern Bay supports an estimated 2700 species, 10% percent of which are fish. The Bay is a nursery habitat for many invertebrates and fishes, all of which depend on the Bay's conversion of photosynthetic energy into biomass and high-grade protein through energy-efficient connections. Small plants (phytoplankton) in the water column constitute a large biomass that captures sunlight and produces dissolved organic substances important to sub-microplankton (planktobacteria and marine viruses <0.2 μm) that form a trophic pathway—"microbial loop"—in estuarine production. Bacteria (picoplankton) and very small plants (0.2–200 μm, nanoplankton, microplankton, e.g., diatoms) are important food for very small animals (e.g., small rotifers, protozoa, oyster larvae), larger invertebrates, and some filter-feeding fish. Small animals (mesozooplankton, e.g., copepods, cladocerans >200 μm), larvae of fish and shellfish, and some large fish and shellfish all graze on this large primary production. Thus phytoplankton, microzooplankton, and other mesozooplankton together form a critical link between the Sun's energy and higher trophic levels. Specifically, larvae of the commercially valued striped bass (*Morone saxatilis*) feed on 20 μm–0.2 mm meso/micro-zooplankton (copepods, cladocerans, various larvae) in tidal fresh marshes; as they mature in about two years they shift to larger prey in different locations. Commercially valued American shad (*Alosa sapidissima*) filter-feed on 0.2–20 mm mesozooplankton during their entire lives, and bay anchovy (*Anchoa mitchilli*) filter-feed on zooplankton, their eggs and larvae being important food for planktonic filter feeders.

The linkage of species to appropriate food sources results in a web of ecological connections throughout the Bay. Abundant larvae and juveniles of many species find shelter and food in shallow, nutrient-rich waters, where oyster beds, seagrasses, and marshes provide nursery habitats for those juveniles to feed, grow, mature, and become food for other species.

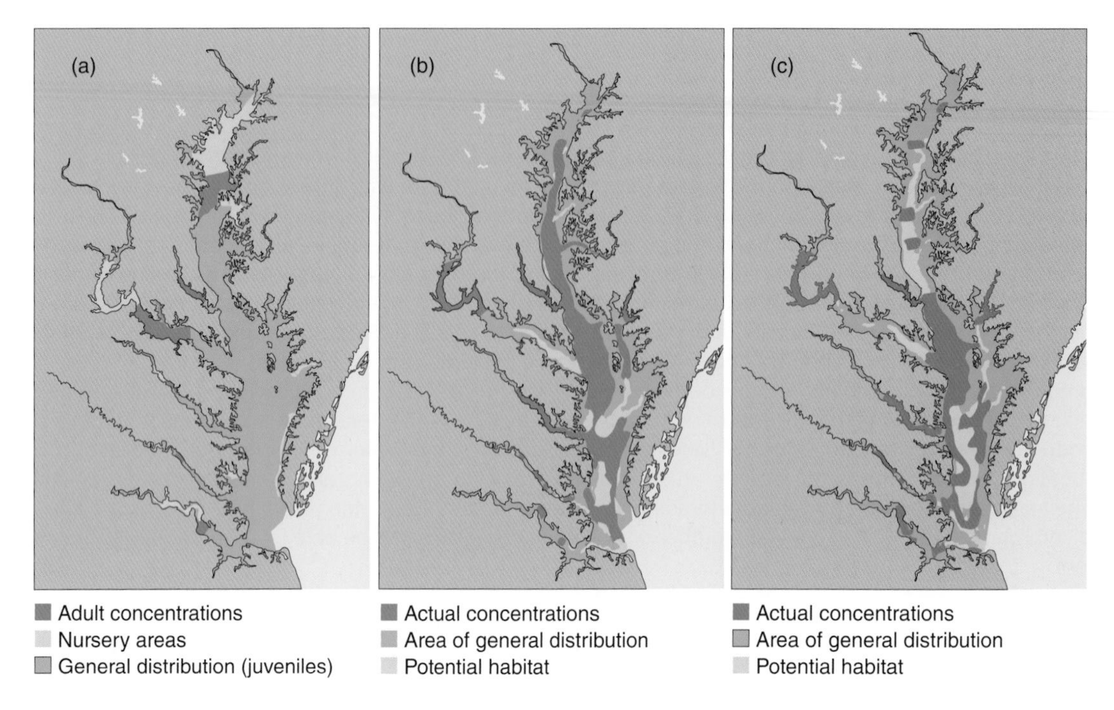

Fig. 6.9 The schooling, euryhaline fish, spot (*Leiostomus xanthurus*) spatially occupy the shallow sand/mud benthos of the Bay. Adults eat soft benthic invertebrates and animal detritus, their concentrations shifting location with season: (a) spring; (b) summer; (c) fall. Young reside in tidal creeks and shallow estuarine areas. Modified from Homer and Mihursky (1991).

Bottom-feeding species such as commercially valued spot (*Leiostomus xanthurus*) feed nearshore on benthic invertebrates and change location with season and age (Fig. 6.9). This species performs key roles in estuarine dynamics, e.g., by regulating benthic species densities, structuring distributions of microorganisms, and resuspending sediment through bioturbation. Important forage fish (spot, bay anchovy, and menhaden (*Brevoortia tyrannus*)) are food for commercially valued striped bass, bluefish (*Pomatomus saltatrix*), weakfish (*Cynoscion regalis*), and various flounders in a web of feeding associations. Thus, the Bay's shallow-water habitats attract abundant and diverse resident and migratory species of waterfowl, shorebirds, commercial fishes, sharks, ospreys, eagles, sea turtles, and marine mammals (e.g., bottlenose dolphins, *Tursiops truncatus*) in size-dependent, food-chain linkages that fuel and sustain the Bay ecosystem by means of energy-transfer efficiencies, community metabolism, and biomass production.

Hydrologic corridors provide transport subsidies that facilitate species movement to spatially extensive habitats. Tidal energies pulse water, nutrients, and chemicals through meandering tidal channels, increasing species survival options. Decaying marsh vegetation enters the detrital food web where crabs, fishes, and bivalves feed. Sessile species (clams, oysters) filter particles from the water, their presence acting to dissipate tidal energy, reduce tidal speeds, and stabilize soft bottom sediments. Regular beats of tides exchange water and materials through channels and marshes that sustain oyster beds, seagrass beds, and mudflat communities, and create landscape patterns behind which forests grow (Fig. 6.10). These tidal pulses also increase options for coexistence among species as they deliver eggs, gametes, larvae, and seedlings of resident and transient species to spatially and temporally segregated habitats.

Wide-ranging, coastal-ocean species are abundant in the Bay. Large, estuary-dependent anadromous fishes travel great distances in the coastal ocean to spawn in brackish and fresh waters. American shad (reaching 75 cm, 5.5 kg) migrate in coastal waters and enter estuaries to spawn, where juveniles feed and grow (Fig. 6.11). Striped bass (to 1.8 m, 57 kg) travel within 10 km of shore from the St. Lawrence River, Canada, to the St. Johns River, Florida, and enter estuaries and natal tributaries to spawn. Along the way both species encounter adverse, energetically expensive conditions (Standen, 2002). And their maturing young shift feeding locations, synchronously timed with appropriate temperature, salinity, and hydrology (see phenology, Ch. 5). Other estuarine-dependent fish such as menhaden (Fig. 6.12) and spot (see this section above; Fig. 6.9) also widely use the Bay, shifting location with life history and season.

Human activities in estuarine ecosystems alter energetic efficiencies. Fisheries extraction alone (trawling, dredging) significantly alters benthic structure and function. Removal of the largest individuals, with the added interference of dams, pollution, degraded water quality, and overabundant exotic species, causes inefficiencies in feeding, reproduction, and migration, which also disrupt ecological feedbacks and compromise the natural resilience of estuarine production.

Fig. 6.10 Communities in a tidal channel self-organize into a landscape pattern, evident at low tide from channel to oyster bed, and to mudflat, salt marsh, and higher forested land. Photograph © Ray & McCormick-Ray.

6.2.2.2 Biota capture and process chemicals through estuarine filtering

The Bay receives a broad mixture of chemicals from air, ocean, and land, brought into close proximity over relatively short distances and transferred through biogeochemical processes critical to ecosystem performance. Watersheds and airsheds deliver key nutrients (nitrogen, phosphorous, carbon) with minerals, contaminants, and sediments. Biota capture, exchange, concentrate, and transform these chemicals into habitats and food resources. Oxygen plays a critical role, being consumed through biotic and ecological metabolism in processes of growth and decay, with release of carbon dioxide and ammonium gases to the atmosphere.

Biota, habitats, and watershed/estuarine systems are large-scale filters in the Bay ecosystem (Kennedy, 1984; McGlathery *et al.*, 2007). At coastal interfaces where chemical transformation is regulated most intensely, the estuarine system traps, processes, and redistributes large amounts of matter, nutrients, pollutants, and sediment. Estuarine, non-tidal circulation, tidal mixing, and estuarine geometry effectively filter and deposit suspended sediment (Schubel and Carter, 1984). And biota filter, trap, process, and redistribute tons of chemicals, having specialized mechanisms to exploit chemicals, regulate exchanges, and discharge wastes, thus helping to restore essential habitats at various rates and scales of ecological organization (Fig. 6.13).

Organisms remove the most soluble forms of iron, calcium, and magnesium from the water, along with pollutants and toxicants bonded to particulates, dissolved in solution, or accumulated in food. Organism structure is composed of both minerals (calcium carbonate, calcium phosphate, silica) and biological materials. They release dissolved organic material (DOM) and gases (carbon dioxide, oxygen, methane, sulfides) through respiration and photosynthesis, and their dead bodies consume oxygen in decay processes. Rooted plants move chemicals above and below the surface, through aerobic and anaerobic substrates. Migratory species import and export tons of chemicals; e.g., post-spawning bodies of anadromous fishes contribute considerable amounts of ocean-derived metabolites, lipids, nitrogen, and other materials into their natal streams, thus nourishing shallow-water and marsh inhabitants (Garman and Macko, 1998). Defecation by millions of juvenile Atlantic menhaden, spot, and pinfish imports an estimated 1 kg of zinc, 56 kg of iron, 1 kg of manganese, and 0.2 kg of copper to a coastal-plain estuary in summer months (Cross *et al.*, 1975). Mollusks produce and deposit shells and feces to the benthos in such abundance that they can restructure the environment around them (Gutiérrez *et al.*, 2003; Gutiérrez and Jones, 2006). Oysters in particular can deposit tons of suspended matter annually, further enhanced by the attached species they support (Haven and Morales-Alamo, 1972). And ecological communities metabolize carbon differently, with strong seasonal pulsing in carbon release (Table 6.2). Ultimately, biota contribute substantially to estuarine metabolism through regulating the release of chemicals and determining the routes and locations of bioactive chemicals. Where species occur, how they behave, and where they are abundant often give clues to sources of biochemical processing otherwise difficult to detect.

Plants play key functional roles in maintaining coastal health (McGlathery *et al.*, 2007). They reduce floods, recharge groundwater, augment low nutrient flows, and influence downstream nutrients, debris, and sediment. They filter and transform chemicals into biomass, form spatially expansive forests, marshes, and seagrass beds that trap, utilize, stabilize, absorb, and process tons of chemicals otherwise washed directly into Bay waters. They create wetland habitat that covers an estimated 4% (approximately 1.5 million acres) of the Bay's coastal watershed, now much reduced from historic levels (NOAA, online). Species that tolerate horizontal and vertical tidal fluxes dominate fresh, brackish, and salt marsh communities. And salt-tolerant plants store large amounts of organic carbon in sediments, to thrive in hot summer temperatures under low freshwater supply and anoxic hypersaline sediments, where the deep-rooted salt marsh grass (*Spartina*

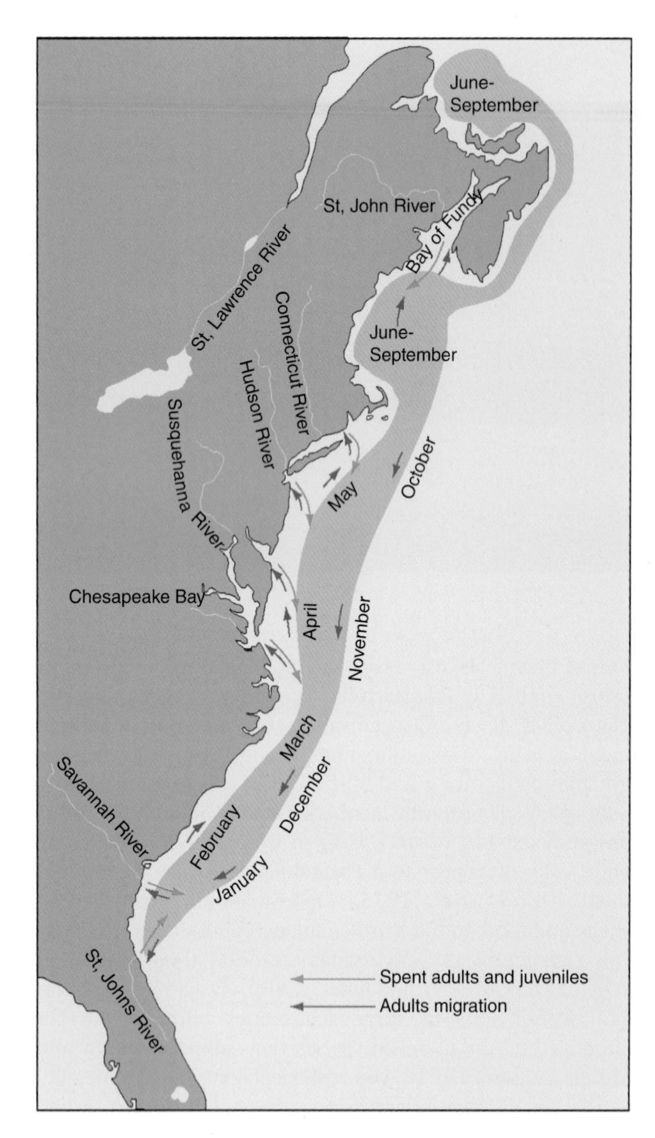

Fig. 6.11 Shad (*Alosa sapidissima*) move through the continental shelf in seasonally directed migratory movements. Along their migration route, adults enter estuaries in spring to spawn in their natal streams. Juveniles and some spawned adults exit estuaries in fall.

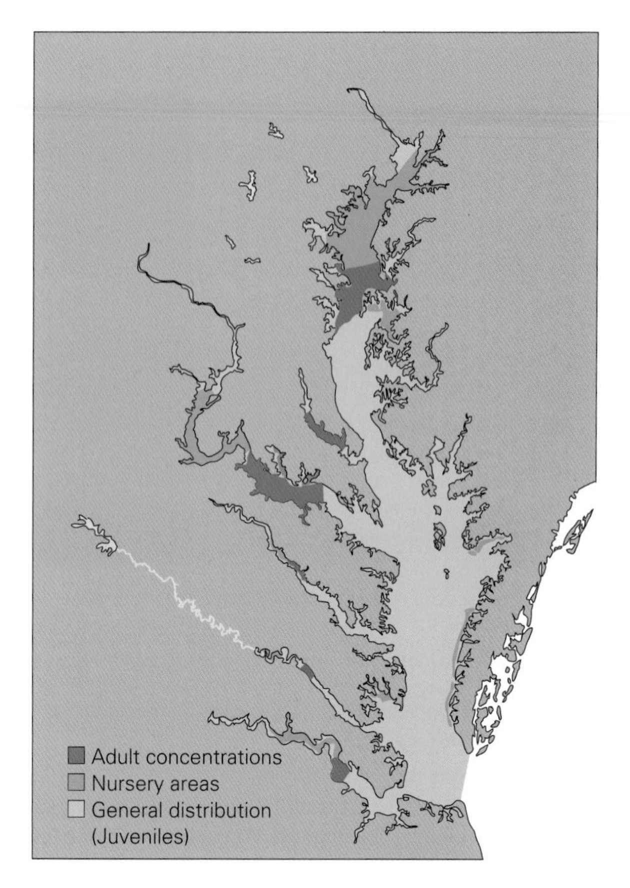

Fig. 6.12 Euryhaline menhaden (*Brevoortia tyrannus*) use parts of the entire Bay and tributaries at different life-history stages. They are a key prey for other commercial fishes, and constitute the largest fishery landings by volume of the Atlantic Coast, ranking second in the U.S. landings behind pollock in Alaska (NOAA, online). Redrawn from Funderburk *et al.* (1991).

alterniflora) stabilizes shorelines, removes toxic metals, and maintains habitats of the estuarine system.

On subtidal mudflats, seagrass, algae, and benthic microflora increase diffusion of phosphorus from sediment to water by trapping particulate phosphorus and releasing dissolved phosphate. They also consume nitrates, regenerate nutrients, accelerate nitrogen fixation, and release oxygen. Benthic algae remove phosphate from seawater, and eelgrass (*Zostera marina*) pumps phosphorus between bottom sediments and the water column. Eelgrass stores as much carbon as forests and supports fisheries by providing habitat—a much-diminished function due to global losses (Fourqurean *et al.*, 2012).

In the water column, phytoplankton consume nutrients. They grow in high concentrations where nutrient loads are high (eutrophic), lower under moderate loads (mesotrophic), and lowest under low nutrient conditions (oligotrophic). Saline water is normally nutrient-limited, where gradients in nutrient concentration generally decrease seaward from land. However, some tributaries are typically eutrophic and others typically vary from being eutrophic under exceptionally high riverine inputs to being mesotrophic during droughts. An early indication of eutrophy is increased abundance of phytoplankton, and phytoplankton in the Bay has significantly increased since the 1950s (Harding, 1994; Harding and Perry, 1997), being positively and strongly correlated with river flow (Boynton and Kemp, 2000) and human activities. River flow determines rates of flushing, but chemicals also move slowly through groundwater (Flewelling, 2009; Flewelling *et al.*, 2012), sometimes taking decades before reaching Bay water and causing lag effects (hysteresis).

Filter- and suspension-feeding species form an important functional guild (Ch. 5, Section 5.3.3) in the Bay. Such feeding is an important mechanism for reducing phytoplankton abundance and maintaining water quality and benthic production. Large concentrations of filter-feeding and foraging menhaden (*Brevoortia tyrannus*), for example, can potentially reduce phy-

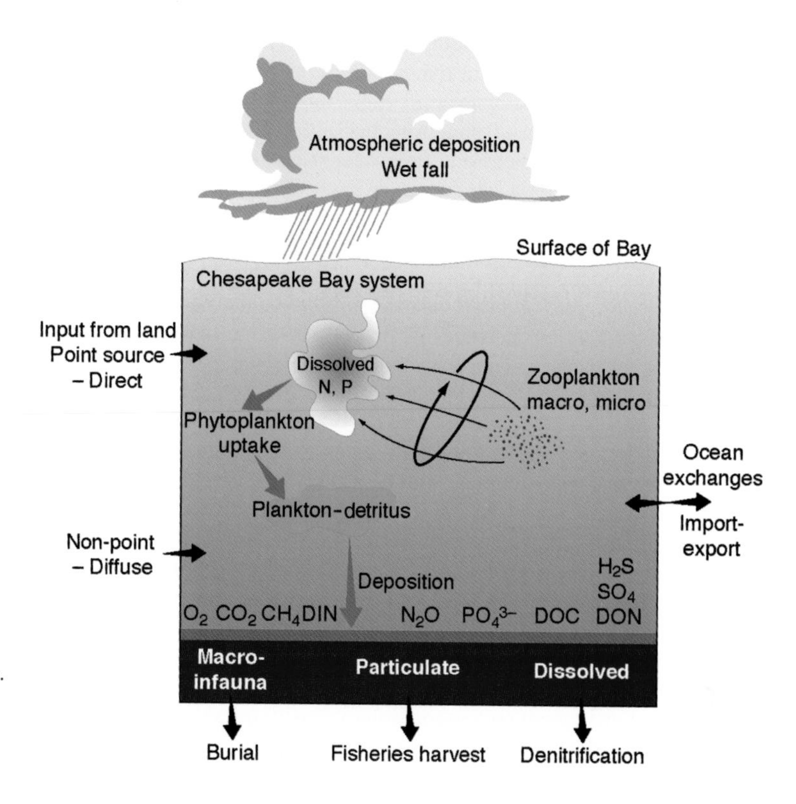

Fig. 6.13 A conceptual model of biogeochemical flow through the Bay. The Bay receives, exchanges, and processes chemicals entering from land, air, benthos, and the Atlantic Ocean. The biota and humans also contribute. DIN = dissolved inorganic nitrogen; DOC = dissolved organic carbon; DON = dissolved organic nitrogen. Data from Boynton *et al.* (1995).

Table 6.2 Seasonal carbon production in different communities. Mesohaline portion of Bay. Values express percentage of total annual production. Based on data from Baird and Ulanowicz (1989). The seasonal dynamics of the Chesapeake Bay ecosystem. *Ecological Monographs* **59**, 329–364.

Biological community	Seasonal % annual production				Annual production
	S	F	W	SP	C (g/m^2)
Primary producers	39	20	14	27	347
Water column invertebrates	55	11	6	28	209
Macro- and meiobenthos	48	16	4	32	33.7
Nekton	41	25	6	28	0.45
All communities	45	17	11	27	595.5
	% net production estimated				
Phytoplankton channeled through microbial loop	50	30	28	25	

S = summer, F = fall, W = winter, SP = spring.

toplankton biomass while simultaneously increasing primary productivity (Dalyander and Cerco, 2010). Historical populations of oysters in summer could filter Bay water within days (Newell 1988; Newell *et al.*, 2007), releasing metabolic products to the benthos at high rates (Haven and Morales-Alamo, 1972). Benthic filter/suspension-feeding animals (oysters, clams, mussels, polychaetes) link water-column production to benthic feeding crabs, demersal fish, and a diversity of other species. Today, menhaden and shellfish roles in the feeding guild are much reduced, being replaced by nuisance microbes and jellies, e.g., by *Mnemiopsis leidyi* that consumes large quantities of plankton, fish eggs, and larvae (Breitburg and Fulford, 2006).

Microorganisms are especially important in the Bay system for processing chemicals. They rapidly decompose a tremendous bulk of organic matter and, consequently, release greenhouse gases to the atmosphere. In the water column, dissolved carbon not taken up in the planktonic food web is processed by heterotrophic bacterio-plankton; bacteria on the benthos also degrade and transform organic matter and fecal pellets into dissolved, particulate matter. Aerobic bacteria (*Nitrosomonas, Nitrobacter*) mineralize nitrogen into soluble nitrogen compounds important for plant growth. And in the process of decay, aerobic bacteria convert ammonium into nitrate consumed by multi-cellular organisms and recycled. Under anaerobic conditions, however, bacteria consume and transform nitrogen, then release it as a gas (denitrification) to the atmosphere. Denitrifying bacteria thus remove critical nitrogen from the system, converting the biologically important ammonium ion into inert nitrogen gas through denitrification (Fig. 4.6). Furthermore, anaerobic bacteria in freshwater release methane gas, while marine bacteria release sulfide gas. Sulfur bacteria (e.g., *Desulfovibrio*) oxidize and reduce sulfur, the fourth most abundant element in seawater. Lethal hydrogen sulfide produced by sulfate-reducing bacteria contributes to anoxic bottom water, once rare in the Bay, but is today an annual event. Much remains to be known of mechanisms that underlie microbial participation in biogeochemical cycles, marine food webs, anoxia, and disease outbreaks, including how viruses affect carbon and nutrient flows and disease (NAS, 2001).

Benthic fauna are also involved in organic degradation, nutrient recycling, gas exchange, and denitrification, and are affected by conditions beneath the pycnocline (Section 6.2.1.1). Increasing organic matter in the water column can initially increase oyster growth and size, then cause decline (Kirby and Miller, 2005). By filter feeding on organic matter and phytoplankton, benthic animals (e.g., oysters, clams) control water-column nutrients and chemicals that become deposited in feces and shell and/or are mixed with sand to create a very hard, rough bottom that interacts with tides that stir and oxygenate the water. Reworked particles from the water column fall to the benthos and if deposited too slowly or too rapidly, can affect benthic fauna.

The rate of a biogeochemical process is an important measure of a chemical's distribution, transport speed, and impact on the system. A rate is measured as residence time, i.e., the average time that particles spend within an area or how fast they move through the system (Jay *et al.*, 2000). This rate influences larval recruitment and biogeochemical and ecological processes. Chemicals that enter the Bay can be retained about one year unless removed by organisms or sequestered in the sediment before exiting with the estuarine plume into coastal waters. Increased loads of nitrogen and phosphate enter the Bay in warm summer months when the water column is stratified, to stimulate phytoplankton production and decrease water clarity for seagrasses. When turbidity interferes with photosynthesis, plants die and oxygen is consumed. Degradation is initiated beneath the pycnocline, trapped in a positive feedback of benthic decay and anaerobic sediment. With increased stirring generated by seasonal winds, storms, precipitation and hydrology, the pycnocline can be de-stratified and the benthos re-oxygenated.

The mid-salinity, mesohaline region of the Bay is an especially important biogeochemical area. Although tributaries contain varying degrees of mesohaline water, the well-studied area north of the Potomac River is most important (Fig. 6.2). Here, the average residence time of water is about 42 days, Susquehanna River water dominates, brackish marshes are extensive, currents are slow, flocculation and sedimentation are high, turbidity is maximum, and particles settle to the bottom in high concentrations. Here, biological communities exhibit different seasonal and annual production values (Table 6.2). Historically abundant oysters have become replaced by the pervasive, annual formation of sub-pycnocline anoxia, with severe aesthetic and economic consequences.

6.2.2.3 Evolution of the Bay system

The Bay ecosystem has been shaped by small-scale sedimentary, hydrological, and biotic processes imbedded in larger-scale climatic, geological, and fluvial interactions. Before the Bay was formed about 15,000 years ago during the late Pleistocene ice-age epoch (Fig. 4.3), mean sea level was 100–150 m lower than today and the continental shelf was dry land (Fig. 4.11). The Susquehanna River and major tributaries flowed directly to the Atlantic Ocean, carving deep fissures that still remain. Pleistocene ice melt began about 10,000 years ago, flooding the wide, sedimentary coastal plain, mixing fresh with seawater, filling ancient river valleys, and isolating the coastal dunes of today's Eastern Shore peninsula, hence trapping estuarine water. Glacial melt slowed about 6000 years ago, and about 2000–3000 years ago sea level approached within about 3 m of its present level. Rivers delivered sediment that tidal currents redeposited, filling the Bay and causing shores to flood, which was further exacerbated by sea-level rise.

Today, the Bay's shoreline morphology continues to be modified by climatic events, isostasy, and biotic activities. Sedimentation and erosion are widespread dynamic features that occur at different rates around the Bay. The freshwater-dominated upper Bay exhibits high sedimentation rates, the middle Bay shows modest rates, and the marine-dominated lower Bay varies from modest to high rates. At local scales, tidal velocity reverses dominant flow directions that influence erosion, sedimentation, and the morphology of small tidal embayments. In shallower areas, dominant flood tides enhance near-bed transport landward while dominant ebb tides enhance such transport seaward through deeper areas of tidal channels. Shoreline change is further complicated by sinking land (subsidence), which occurs at different rates around the Bay due to varying tidal energies, erosional rates (Titus *et al.*, 2009), and ancient events.

Storms, weather systems, and climatic events initiate structural change in the system, and biota must adjust. Extreme weather fluctuations with alternating floods and droughts over extended periods of time can expose species to salinity shock, desiccation, or drowning. About 30 times a year, storms ("northeasters", hurricanes) typically deliver extensive precipitation that radically changes salinity. Acting over days and weeks, storms trigger a sequence of events: an initial response in about 7–26 days when the storm approaches, to storm passage when the estuary undergoes shock, rebound, and recovery (Nichols, 1994). Heavy rains can depress salinity for more than a month, and storms can open new inlets or permanently close others, thereby altering salinity and landforms. Floods move freshwater into normally saline areas, while droughts draw saline water into freshwater areas. Land- and seascape components react and adjust in time-averaged, system-wide responses with lag-times of variable duration. And as the system readjusts, it reorganizes.

The interactions of biota with sediment, erosion, deposition, and transport over time have built complex biogeomorphic patterns. Such patterns result from feedbacks among geomorphic, biologic, ecologic, and evolutionary processes. Through emerging studies of biogeomorphology, geomorphologists are attempting to integrate ecological and evolutionary concepts with landscape structures (Naylor *et al.*, 2002; Corenblit *et al.*, 2011). Biotic energy facilitates the construction of biogeomorphologic structures through ecological processes. Concepts about keystone species, ecosystem engineers, phenotypes, and niche (Ch. 5) become integrated into macroevolution, improving understanding of complex ecological patterns and their formations. Biogeomorphologic formations appear to result from biological feedbacks by means of dominant processes of bioconstruction, bioerosion, and bioprotection, where key species and communities play direct or indirect roles through utilization of available resources and interactions with other species that collectively modulate physical conditions, e.g., marsh vegetation (Reed, 2000) and oyster reefs (Section 6.3).

6.3 EASTERN OYSTER: QUINTESSENTIAL ESTUARINE SPECIES

The robust eastern oyster (*Crassostrea virginica*) is adapted to the high variability of estuarine conditions. Its high production, filtration, deposition, and behavior make it a quintessential estuarine species with important ecological performance in the ecosystem (Fig. 6.14). This benthic species meets variable estuarine conditions by cementing itself to substrates and being fecund: one female in one spawning season can discharge millions of eggs in summer months. This production occurs when the Bay is biologically most active, food sources are abundant, and risk of storms—but for occasional hurricanes—is reduced. Although the chance for a single egg to become fertilized and to mature into a reproducing adult is very small, timing, opportunity, and luck have allowed oysters to develop hard reef structures. The oyster is a winner in a dynamic system, as is well documented by its historical abundance.

Oysters start life in the water column at a very small size. Fertilized eggs 40–50 µm in diameter, the size of a silt particle, hatch and grow in the water column throughout a three-week period during which the pycnocline plays a role in their transport (Ruzecki and Hargis, 1989). They develop from larvae into 70–75 µm trochophores the size of a very-fine-sand particle, then into 250–400 µm veligers the size of a medium-sand particle (Fig. 6.15), grazing on micro-algal cells of <10 µm. They confront hydrologic and viscosity barriers, disease and numerous predators, and are dispersed by tidal currents. At the top of the pycnocline, slack tide may concentrate them with food particles. Mature veligers recruit to solid substrates on the benthos, especially to oyster shell (self-stereotaxis), where they attach irreversibly by electromagnetic adhesion to become "spat." Spat typically crowd together, growing rapidly and competitively, accreting vertically over successive generations (Fig. 6.16). Their growth and orientation are influenced by current-flow direction (stereotropism). As currents deliver food particles, hair-like feeding structures on their gills generate micro-currents that increase food encounters (Ward *et al.*, 1994). Under slow currents, high turbidity, and sedimentation, oysters can tolerate hypoxia for short periods, but they can smother under sustained low oxygen levels. Over a long time, dead shells and feces accumulate and mix into the sediment to form hard-bottom oyster beds along a benthic ridge in spatial patterns, as documented by Winslow (1882; Fig. 6.17). The rough bottom beds interact with current flows to agitate the water, alter flow patterns, and serve important roles in biogeochemical cycling (Fig. 6.18).

Oyster beds attract a diverse assemblage of crabs, invertebrates, fishes, and waterfowl. They provide habitat for benthic plants (algae, submerged aquatic vegetation) and diverse filter feeders (polychaetes, bivalves, etc.), and food for consumers (crabs, shorebirds, migratory waterfowl, people). Observed at multiple scales, oyster assemblages form "hot spots" of

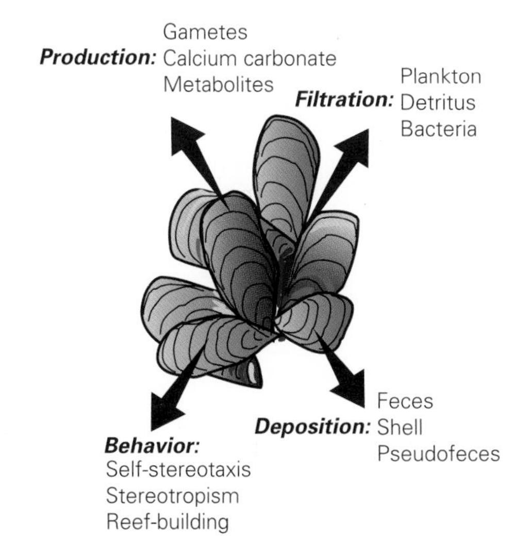

Fig. 6.14 The oyster is an ecosystem engineer and foundation species. Oysters filter water and remove organic matter, produce abundant gametes and metabolic products, deposit large quantities of feces and shell, and behave in response to other oysters and to current flow. Their historically high abundance in the Bay formed structural reefs that influenced hydrology, biogeochemical pathways, benthic communities, and estuarine function.

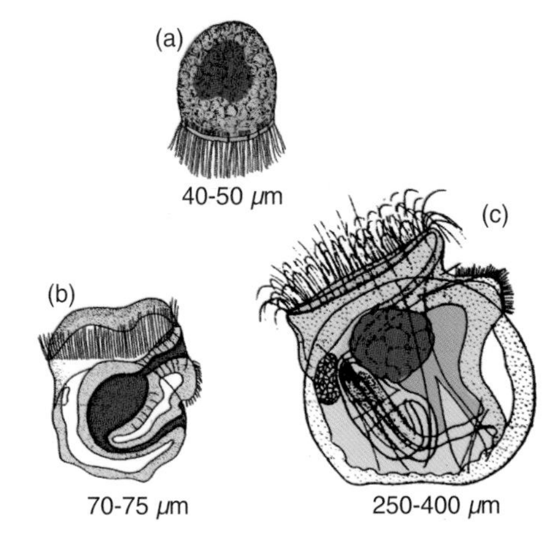

Fig. 6.15 Millions of oyster gametes are released and fertilized in the water column and develop into larvae, contributing to the microzooplankton population in three stages of maturation: (a) early larva; (b) trochophore; (c) veliger. From Galtsoff PS (1964) The American oyster Crassostrea virginica Gmelia. *Fishery Bulletin* **64**, 1–480.

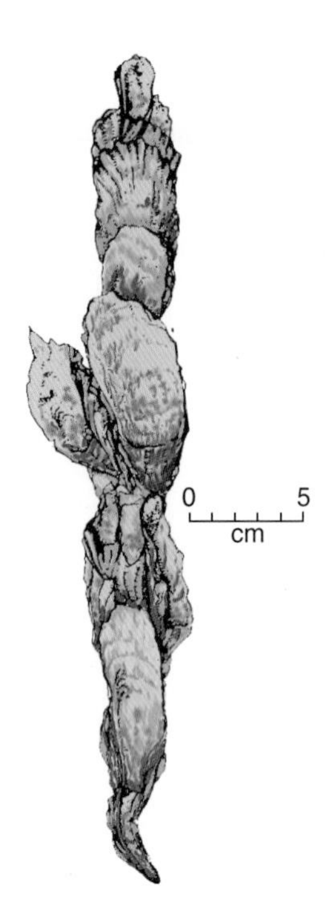

Fig. 6.16 Oysters cluster in vertical growth attachments over successive generations. From Galtsoff PS (1964) The American oyster Crassostrea virginica Gmelin. *Fishery Bulletin* **64**, 1–480.

functional biodiversity (Fig. 6.19). In addition to increasing species richness (Fig. 5.1), oyster beds increase the Bay's heterotrophic production, being among the highest producers of any benthic community (Bahr, 1974, 1976; Bahr and Lanier, 1981). This "biocenose," a community of species associations, is sensitive to change (Möbius, 1883).

A healthy, growing oyster reef co-adjusts with tidal-channel energy and shoreline processes, forming a coadapted, dynamic reef system within the Bay. Beds accrete at pace with sea-level rise (DeAlteris, 1988) to influence hydrology and topographic forms (Fig. 5.7). Beds concentrate a diversity of filter-feeding organisms whose combined interactions contribute to water-column clarity. Their feces and mucus deposited to the benthos bind into cohesive particles with sediment and dead shells to create the hard-bottom habitat patches that historically fringed Bay shores. These hard benthic structures at the interface between salt marshes, seagrass beds, and deep tidal channels, intercept tidal patterns and bottom currents, modulate hydrological regimes, increase turbulent mixing, and modify the adjacent environment.

Oyster-reef production and persistence in the Bay is well established. Oysters are recorded in the fossil record since the early Mesozoic, with shell basement underlying marshes in geological strata, and massive dead oyster shells forming two parallel bands at 70 and 50 m depths along most of the U.S. mid-Atlantic shelf (Emery, 1967). Historic records of oyster beds highlight their dominance in tidal channels and marshy lands that border the Bay (Fig. 6.20), potentially influencing marsh development by modulating the alteration of current flow energies and contributing to sedimentation (Fig. 6.21).

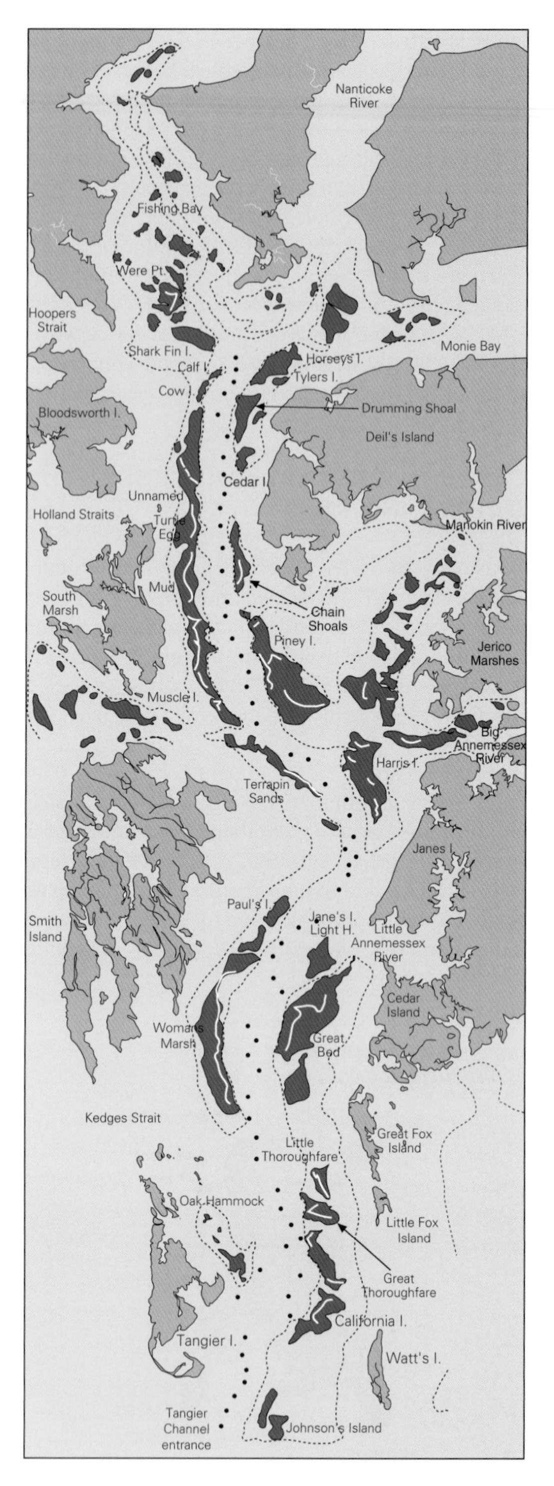

Fig. 6.17 Historical fringing oyster reef in Tangier Sound, eastern Chesapeake Bay. The oyster's historical abundance created a hard fringing reef in a spatial pattern bordering a meandering tidal channel (black dots indicate deep channel bottom) and intertidal marsh islands (brown). Hard-bottom beds (purple shaded areas) and a dense, solid ridge of concentrated oysters (white line on shaded bed) clearly demarcated the bed, outside of which oysters were sparse. From McCormick-Ray (1998). Oyster reefs in 1878 seascape pattern—Winslow revisited. *Estuaries and Coasts* **21**, 784–800 with kind permission from Springer Science+Business Media B.V. Redrawn from Winslow (1882).

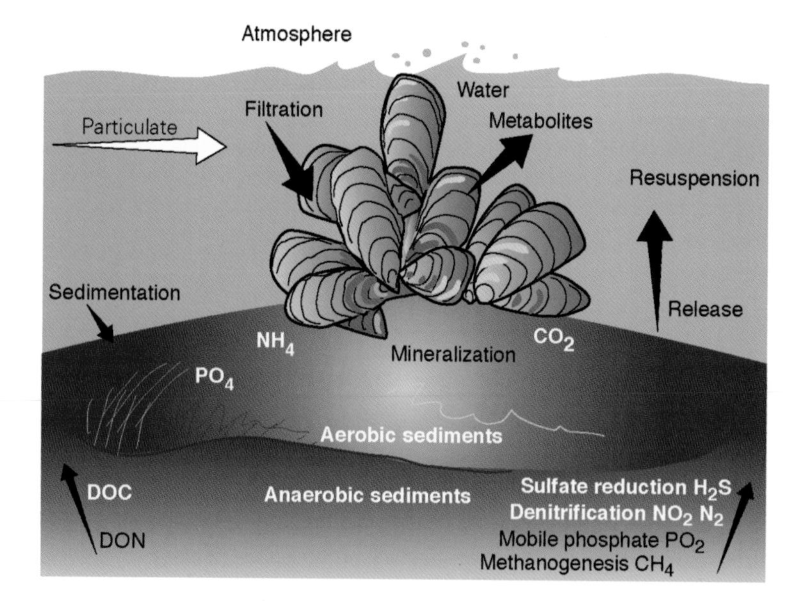

Fig. 6.18 Schema of biogeochemical processes of an oyster bed. Dense oyster populations remove large quantities of suspended particulate and dissolved organic material in the water column. They release metabolites, re-mineralize nutrients in forms useful to phytoplankton, and move carbon, nitrogen, and phosphorus at faster rates than pelagic food webs (Dame *et al.*, 1989). DOC = dissolved organic carbon; DON = dissolved organic nitrogen. From Dame (1996). *Ecology of Marine Bivalves*. CRC Press with permission from Taylor & Francis.

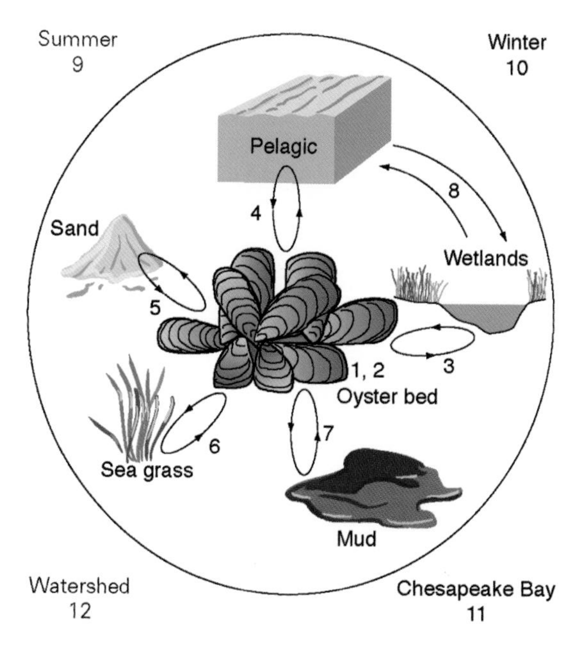

Fig. 6.19 The oysters' role in estuarine complexity/diversity is performed at multiple scales. (1) Oysters play individual roles as a source of food, substrate, water filtering, and benthic habitat. (2) Hard reef structure attracts diverse species. (3–7) Beds are linked to different habitats, to (8) increase optional choices for co-existence among resident and visiting species, which change seasonally: (9) in summer and (10) in winter. (11) They construct a fringing reef system along tidal channels, and (12) contribute Bay products beneficial to mid-Atlantic ocean species.

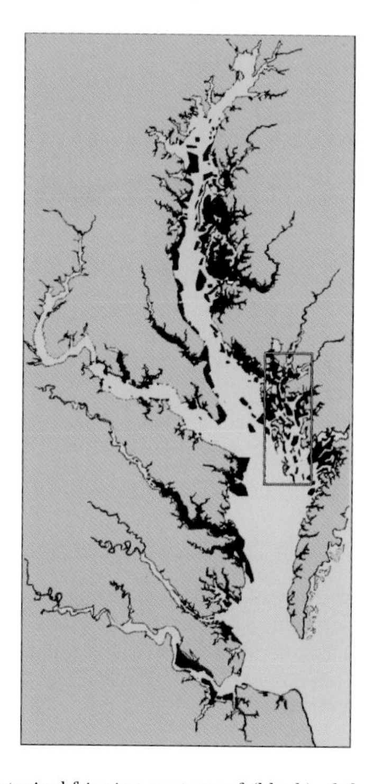

Fig. 6.20 Historical fringing oyster reef (black) of the Bay. Rectangle indicates Tangier Sound (Fig. 6.17). From Funderburk *et al.* (1991).

Thus, functionally, oyster ecological systems cascade hierarchically upward from local scales, to relationships of metapopulation seascapes important to valued fish and to regional attributes (Fig. 6.22, Fig. 5.18; McCormick-Ray, 2005; Ray, 2005; Secor and Rooker, 2005).

The historic record thus makes clear that this quintessential estuarine species is adapted to estuarine vagaries, and performs the role of a foundation species in the Chesapeake Bay (Ch. 5). Its resilience is a matter of record. In the historic Bay, oyster resilience was evident by decreased shell size when the human population that harvested them was expanding, and then increased size when the human population declined (Fig. 6.23). However, oyster beds are sensitive to change, especially excessive sediment that smothers the bed. The bed degenerates through a process of senescence, decay, and erosion, where dead parts (shell, skeleton, hard parts) are no longer deposited, soft silt covers the bottom, dependent species leave, and failed oyster recruitment halts recolonization. Other oyster recruits find new locations, where they colonize, mature, and develop healthy beds in a dynamic self-renewal of the seascape. Thus, larval connections to other oyster beds form into a reef system whose function may cascade up over time to interact with other species and habitats.

6.4 FROM RESOURCE ABUNDANCE TO ECOSYSTEM CHANGE

The oyster's economic value in the 1800s historically ranked them second only to whaling as the U.S.'s most profitable "fishery." This naturally produced wealth brought unrestrained expansion to the oyster business in an age of rapid industrialization and technological achievements. As long ago as 1892, Stevenson observed that for every region of the world where the oyster industry attained commercial importance, oysters passed through four stages: (i) natural reefs in a primitive condition furnishes the entire supply of oysters; (ii) reefs become somewhat depleted, produce small oysters many of which are transplanted to private grounds and are given indi-

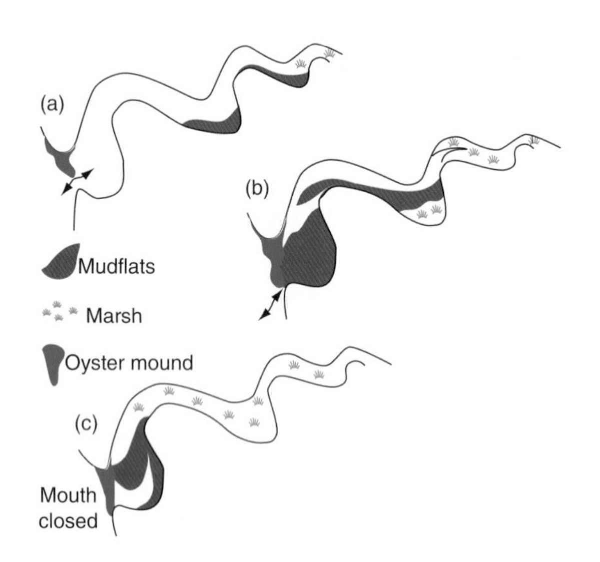

Fig. 6.21 Concept of oysters' role in marsh evolution time sequence: (a) oysters (blue patch) intercept current flow at mouth of tidal creek into marsh; (b) sediment is deposited and accumulates (brown area); (c) subsequently marshland develops. Modified from Hayes and Sexton (1989).

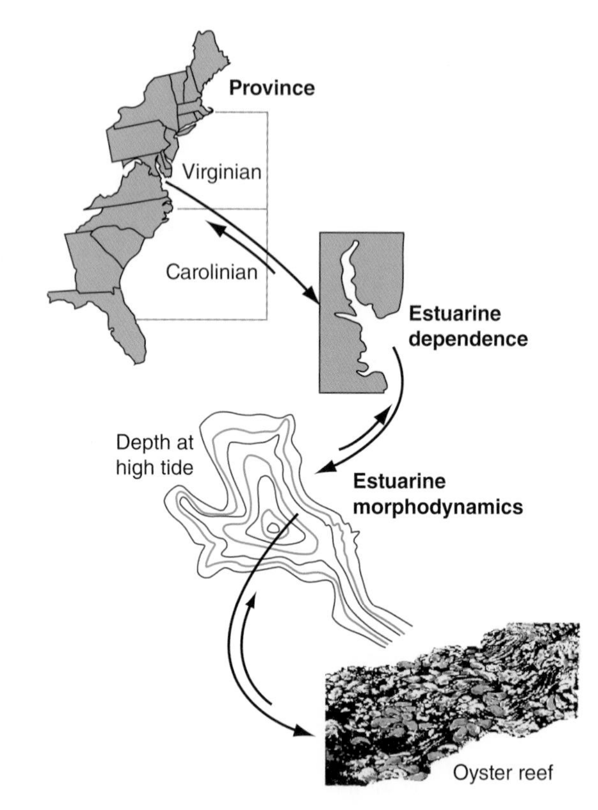

Fig. 6.22 Conceptual role of oysters in multi-scaled hierarchical feedbacks. The estuary-dependent oyster is an ecosystem engineer at local scales, influences function at estuarine scales, and serves as food and habitat for migratory species at regional scales. Role of oyster beds to oysters cumulative ecological influences on the Virginian–Mid-Atlantic Province. Ray GC, Hayden BP, McCormick-Ray MG, Smith TM (1997) Land-seascape diversity of the U.S. east coast coastal zone with particular reference to estuaries. In *Marine Biodiversity: Causes and Consequences* (eds Ormond RFG, Gage JD, Angel MV), pp. 337–371. With permission from Cambridge University Press.

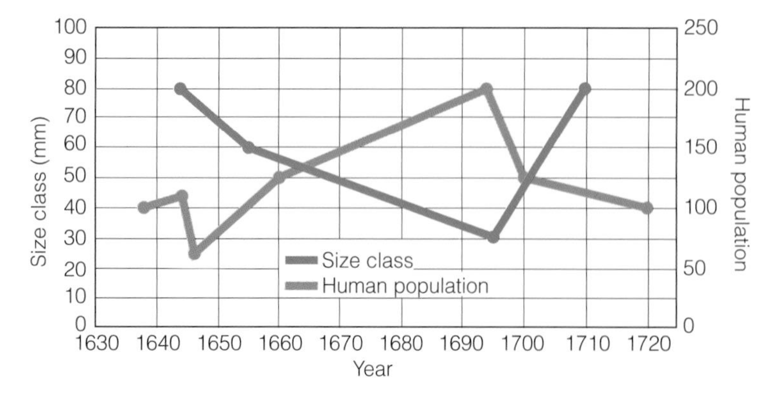

Fig. 6.23 Oyster size-class decreased then increased with changes in the human population, based on data at St. Mary's, Maryland, between 1630s to 1720. Republished with permission of the Washington Academy of Sciences. From Miller, H. (1986) Transforming a "Splendid and Delightsome Land": Colonists and Ecological Change in the 17th and 18th-Century Chesapeake. *Journal of the Washington Academy of Sciences* **76**, 173–187.

vidual protection to mature; (iii) public beds become depleted, available supply is very irregular and uncertain, and almost all oysters are small, and transplanted to private areas; and (iv) the industry depends almost entirely on areas under individual ownership or protection, and policymakers respond by designing different property rights regimes.

6.4.1 Natural-resource wealth in decline

The Chesapeake Bay ecosystem in the 1800s produced the richest populations of seafood in the world. Herrings (Family Clupeidae) of several species (including shad) and oysters were major commodities of importance to the national economy, as reported in the U.S. eighth census in 1860 (Ingersoll, 1881). As New England and European oysters became depleted, the Bay attracted entrepreneurs that made Baltimore a commercial hub for Bay resources. Formerly regarded a "poor man's food," the first Bay oyster house was opened in 1830 by C.H. Maltby from Connecticut, and the oyster industry grew rapidly by mid-century. Dredging introduced by New England oystermen was outlawed in 1820 because it destroyed oyster beds and outraged Bay oystermen. By 1858, oysters brought an estimated $20 million to the Bay. Virginia oysters were considered the finest, reflected in the species name, *Crassostrea virginica*.

The end of the American Civil War in 1865 brought thousands of unemployed people to seek a living by harvesting oysters in the Bay. An oyster captain could earn $2000 a year, when most Marylanders earned less than $500 and when Virginians were emerging from Civil War losses. Baltimore's advancing technology, new railroad system, and port facilities extended the Bay's oyster trade to affluent cities, western states, and European markets, making the Bay the world's oyster capital. Large, round oysters appealed to Baltimore's growing affluent class, and market value increased. As oystermen recognized that harvests exceeded recovery (Winslow, 1882b), growing commercial value ignited frenzy: "[T]he waters were thrown open to every one who would pay the military officials for a permit to oyster . . ." (Ingersoll, 1887). Feuds and wars erupted between neighbors and over boundaries. State legislators repealed the prohibition on oyster dredging, and the wars and declining oyster yields brought the U.S. Congress in 1871 to create the federal Commissioner of Fish and Fisheries to investigate the causes, survey the most valued beds, and establish state oyster commissioners. Maryland legislators appointed experts to the Maryland Oyster Commission, whose advice they did not follow (Kennedy and Breisch, 1983), and Virginia legislators appointed a three-member Virginia Fish Commission. Both states established ineffectual oyster police. In the 1890s, 900 dredgers harvested ever-fewer oysters that yielded thin profits to an inefficient industry. Similarly, the important shad fishery suffered a decline due to over-exploitation, habitat loss, wars between neighbors, and construction of dams obstructive to breeding habitat. Abundant blue crabs, menhaden, and striped bass also attracted commercial interest.

Bay oyster harvests that averaged more than 10 million bushels a year in the late 1800s had declined to 2–3 million bushels a year by 1900. The decline of valued oysters, shad, and herrings by 1900 (Cronin, 1986) brought public concern, and in the 1940s the General Assemblies of Maryland and Virginia agreed to investigate. The state governors appointed the Chesapeake-Potomac Study Commission to examine oyster and finfish conflicts over state boundaries, especially in the contentious Potomac River. Dr. Paul S. Galtsoff of the U.S. Fish and Wildlife Service reported that oyster harvests were only a fraction of those 50 years earlier, especially in the upper Bay and the Potomac and York rivers (Chesapeake-Potomac Study Commission, 1948). His 23-year oyster research in Virginia's York River confirmed that impacts of intensive dredging, overfishing, and pulp mill sulfate wastes had degraded the formerly high-quality York oysters (Galtsoff *et al.*, 1947). Other biologists reported the Bay's 1900 fish production had been depleted by half. The Commission then called for better state oyster management, formed a joint state authority to coordinate Bay-wide fisheries (especially shad and herring), and created an Interstate Potomac River Basin Commission to address Potomac conflicts. George Washington's *Compact of 1785* was replaced by the *Potomac River Compact of 1958* (amended in 1970). In 1950, the Atlantic States Marine Fisheries Commission (Public Law 539), representing the industry, fishery managers, and scientists, was created to manage fisheries within state jurisdictions (up to 3 mi offshore) for "better utilization of fisheries, shellfish, and anadromous fishes of the Atlantic seaboard . . ." For federal waters, the Mid-Atlantic Fishery Management Council was appointed to develop fishery plans and regulations. These divisions of fisheries authority proved contentious, as marine species at various stages of their lives cross arbitrary jurisdictional boundaries.

Resource-use conflicts are now increasingly settled in U.S. courts, where entrepreneurial and commercial groups have received a disproportionate share of wealth and power (Horwitz, 1977; Weinberg, 2003). A court settlement, as between poor oystermen arguing for common property rights and richer citizens seeking privatization of an industrialized resource such as oysters, becomes influenced by social relations and culture and has tended to favor utility claims with preference given to private property rights over public rights (McCay, 1998). This clash over public and private ownership resulted in different oyster-management practices between Maryland and Virginia (Rieser, 2006).

Thus, despite warnings of depletion, the oyster policies of both states have vacillated without firm commitment to conservation (Kennedy and Breisch, 1983). Wasteful practices and resource declines were evident when the hydraulic dredge was introduced and when anoxic water was first reported in Maryland in 1938 (Newcombe and Horne, 1938). Oyster declines were compounded by ecological change, synergistic effects from reef destruction, and over-harvesting. In the late 1950s, the decline dramatically accelerated with emergence of pandemic protozoan diseases, *Perkinsus marinus* (dermo) and *Haplosporidium nelsoni* (MSX). By the 1990s, the oysters' ecological role was declared irreversibly lost (Rothschild *et al.*, 1994; Hargis and Haven, 1999), which not only brought economic hardship to Bay watermen who depended on them, but indirectly affected ecological change. And by the end of the 20th century, less than 1% of the historic value of oysters and shad remained to be managed (Fig. 6.24).

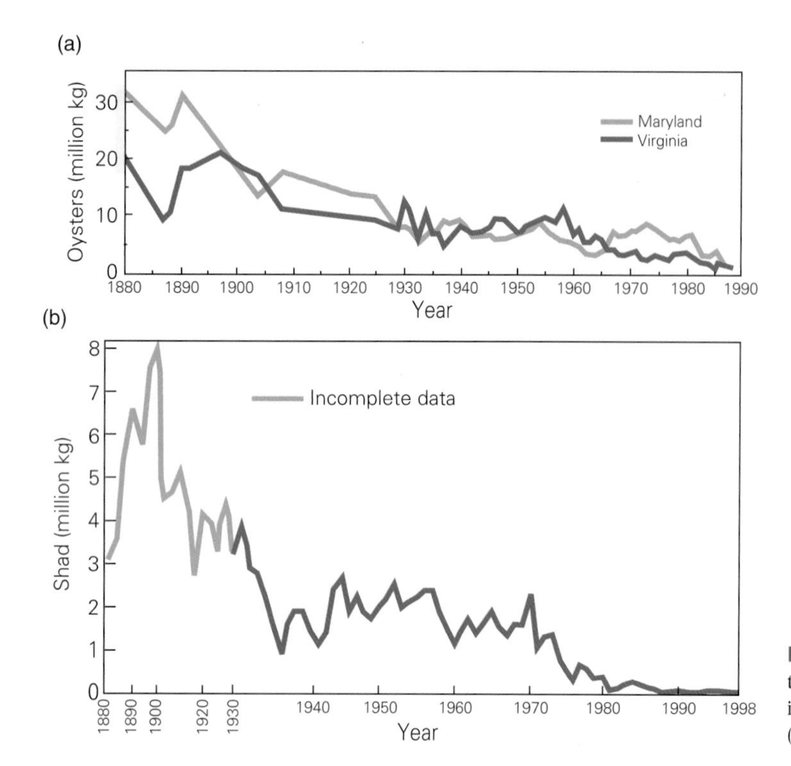

Fig. 6.24 Harvest trends show decline for over 110 years to present: (a) oysters; (b) shad (shaded light green line indicates incomplete data collection procedures). From CBP (1999).

6.4.2 Impact of watershed change

Settlements first sprouted up around the Bay in the 1600s under land-use policies inherited from England's Royal services, and continued after the American Revolution. Furs, timber, marsh "hay," and trees were extracted, beavers exploited to near extirpation, and virgin forests and wilderness were turned into the familiar farmland of English tradition, along with introduced farm animals and associated weeds and diseases. Plows uprooted native vegetation, leaving deep furrows in thin, deforested soils, exposing native organisms directly to heat, sun, rain, wind, and soils depleted of nutrients. An abundance of natural resources benefitted the expanding human population; forests provided fuel for heat and timber for shipbuilding, water provided power, transport, and waste disposal, and farmlands produced abundant food. Land use and water change is recorded in Bay sediment, where anoxic conditions and eutrophication have steadily increased since the first Europeans settled the land (Cooper and Brush, 1991).

The advancing agrarian society of traders, tobacco farmers, and planters came with long-term consequences. The wide Bay at its farthest navigable point on the Potomac River became transformed into Georgetown Harbor in 1751 and filled with silt by early 1800 (Gottschalk, 1945). By the mid-1800s, 40–50% of the watershed's forested landscape was under cropland. Sediment, trash, and pollutants poured into streams, harbors, and tributaries; Baltimore Harbor on the Patuxent tributary became the hub of industrial growth, and required dredging by 1831. The Army Corps of Engineers, created by the U.S. Congress in 1802, became charged with dredging and jetty construction in the Baltimore channel in 1824, initiating its long history in harbor maintenance. By the late 1800s, Baltimore Harbor was besieged with sewage, pathogens, contaminants, cholera, and tons of sediment that required long-term dredging. Removal of forests, expanding farmland and towns, and wetlands drained for public health concerns contributed tons of sediment and nutrients into streams, harbors, and the Bay (Brush, 2009). Sediment accumulating at the mouth of the Bay required continual dredging to facilitate important ship trade.

Human uses in the upper Bay and Potomac River since 1760 (especially since 1940) have increasingly diversified, and so have nutrient-enriching nitrogen sources (Jaworski *et al.*, 2007). Sediment core analysis since 1400 AD shows a correlation between increasing carbon deposition, planktonic diatoms, and cleared land (Fig. 6.25). As diatom diversity steadily decreased, anoxia also increased (Cooper and Brush, 1993), associated with land-use changes since 1900 to the present, with increasing human populations that converted rural land into urban land (Fig. 6.26). And since the 1970s, regrown forests on abandoned farmland have been declining at 100 acres per day, being replaced with cultivated agriculture and urbanized land. Urban sprawl has claimed 0.75 million Bay watershed acres in 30 years (Potomac Conservancy, 2010). Urbanized Washington, D.C., is connecting with urban Baltimore, Pennsylvania, and Virginia (Fig. 6.27). More automobiles and hard, impervious surfaces (Fig. 6.28) reduce slow rainwater percolation into groundwater and aquifers, and increase flooding runoff, air and water pollution, and acidification (Rice and Herman, 2012). Such development has increased freshwater demands for energy and utilities (Fig. 6.29). In the aggregate, pollutant loads from agriculture, auto-

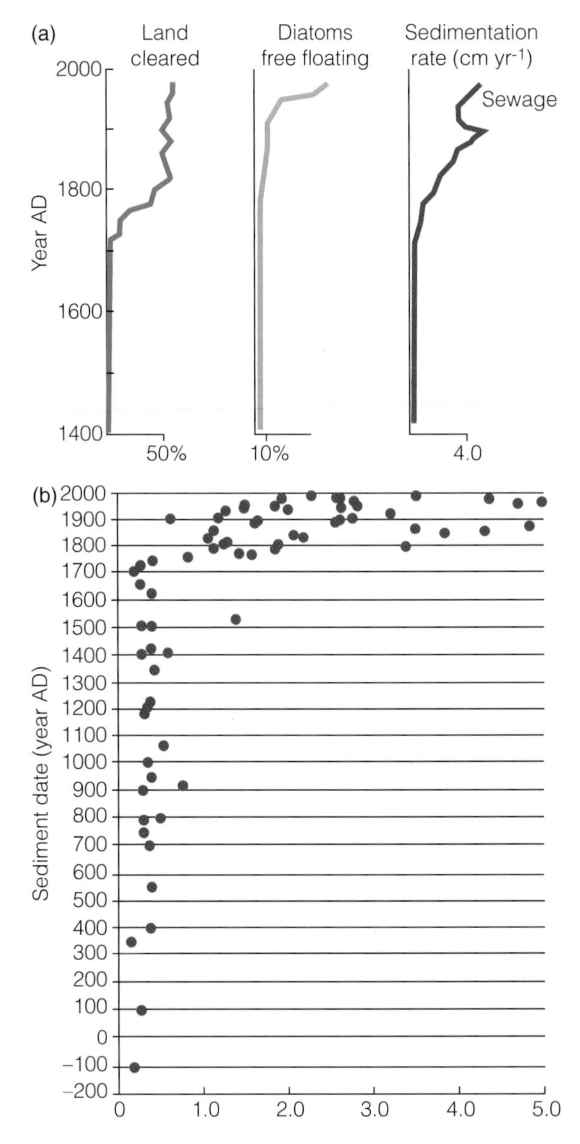

Fig. 6.25 Sediment core analysis show: (a) Stratigraphic profiles at Furnace Bay, Maryland, since 1400 show that increases in percent of cleared land are correlated with increases in free-floating diatoms and nutrient sedimentation rate. Used with permission of the Washington Academy of Sciences. From Brush, G. (1986) Geology and Paleoecology of Chesapeake Estuaries. *Journal of the Washington Academy of Sciences* **76**: 146–160. (b) Over 2500 years, the total organic carbon (TOC) (mg/cm) recorded in the sediment per year increased after 1800. From Cooper and Brush (1993). A 2500 year history of anoxia and eutrophication in Chesapeake Bay. *Estuaries and Coasts* **16**, 617–626, with kind permission from Springer Science+Business Media B.V.

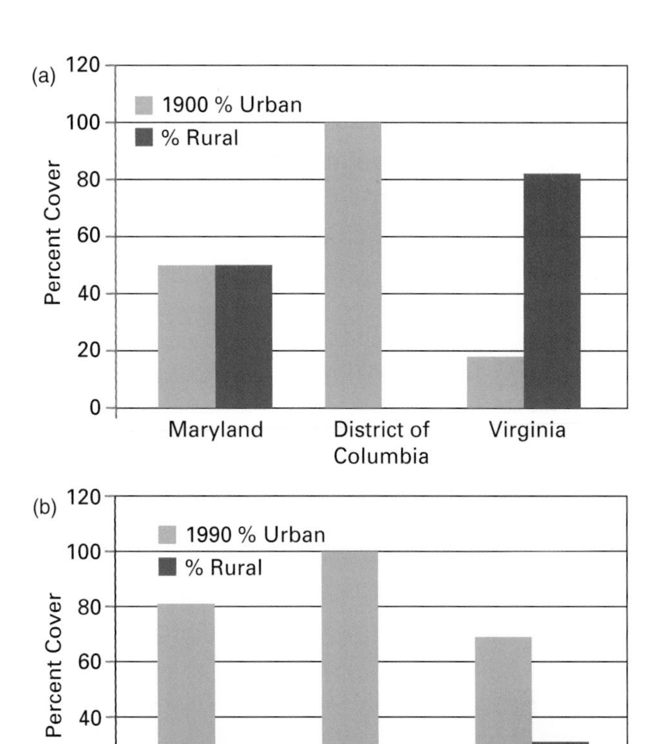

Fig. 6.26 Percent urban and rural land cover shown for 1900 (a) and 1990 (b) highlights increased urbanization and reduction in rural land. Source: U.S. Census Bureau (1995).

mobile exhaust, land uses, septic systems, wastewater treatment centers, and other sources are entering hundreds of streams and groundwater, loading nitrogen and phosphorus into the Bay (Table 6.3). The impact is documented in the mesohaline portion of the Bay, where sediment cores have recorded increased nitrate concentrations, anoxia, and associated deformed shells (Fig. 6.30; Brush, 2009; Zimmerman and Canuel, 2000).

6.4.3 Accumulating an ecosystem debt

Multi-scaled human interferences, such as toxic pollutants (Table 6.4), and ecological processes have become embedded in climatic, geologic, and system-wide evolutionary change. Nutrients and sediments added to Bay water are stimulating phytoplankton growth, creating anoxic bottom water, and changing biogeochemical exchanges. With nutrients and sediments pouring into the Bay, its volume, depth, and area should contract and flush freshwater seaward; however, channel dredging and freshwater/sediment blockage by dams and reservoirs may be contributing to the rise in Bay water levels, causing flooding that erodes shores (Pritchard and Schubel, 2001). And evidence shows that development along the forested southern shores of estuarine creeks and bays of Virginia's major rivers (e.g., James, York, Rappahannock) that were converted to agriculture, residential, and commercial and industrial uses, had increased erosion more than twice the rate of the opposite, undeveloped, northern wooded shores (Hardaway and Anderson, 1980). This erosion issue requires that shoreline management involve property owners, land-use planners, city, county, and state officials, resource managers, watermen, marina owners, and many others to reach agreement on appropriate action (Hardaway and Byrne, 1999).

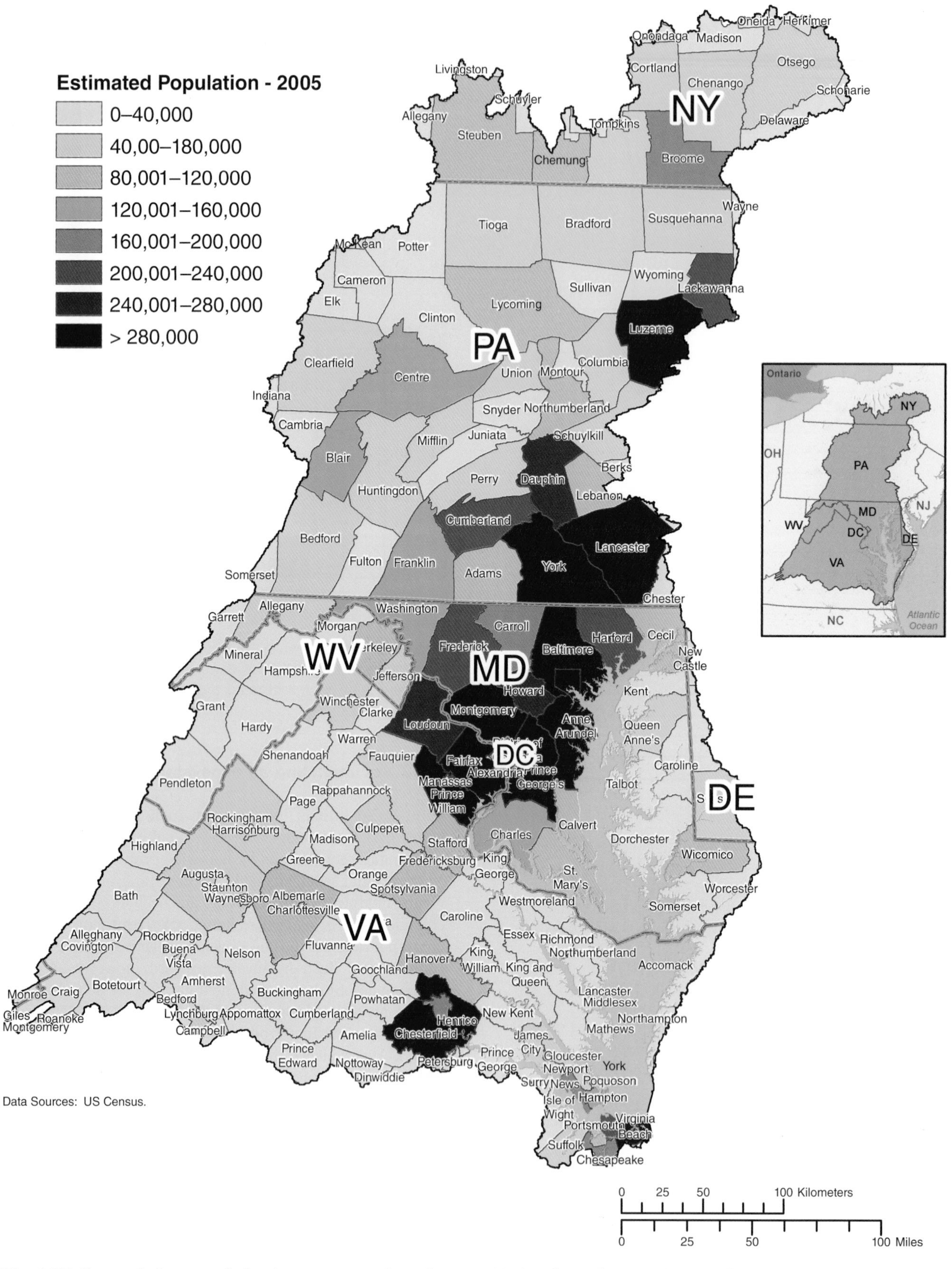

Fig. 6.27 Chesapeake Bay watershed and county estimated populations in 2005. Redrawn from CBP (2008) online.

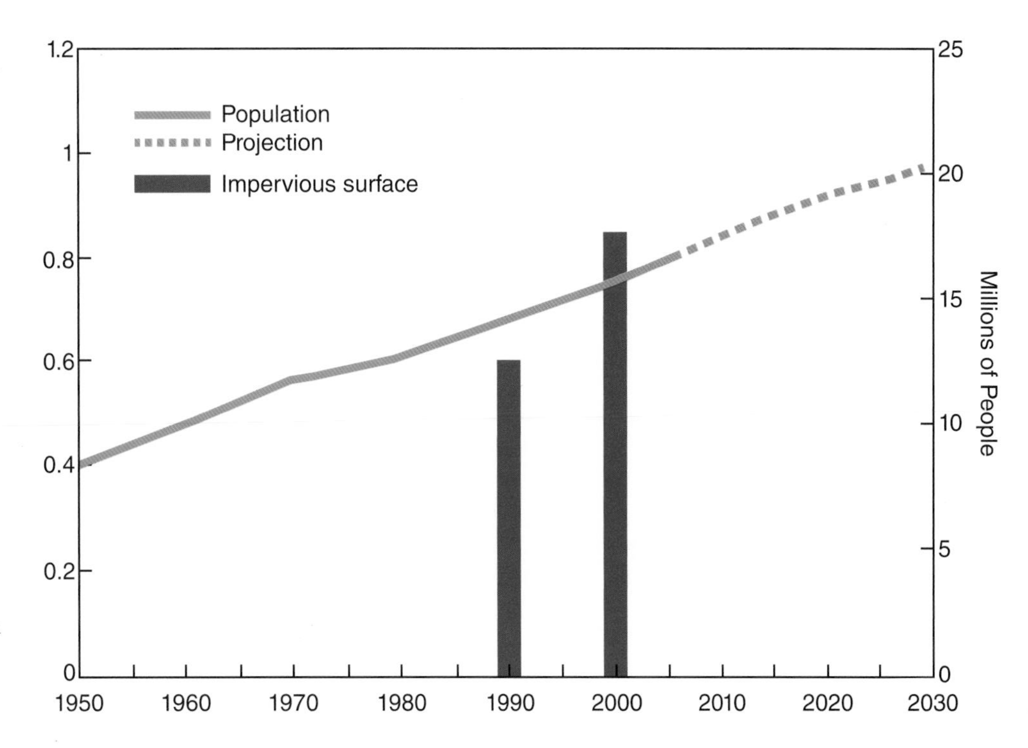

Fig. 6.28 Documented increases in human population and impervious surface, with projection, 1950–2030. Redrawn from CBP online.

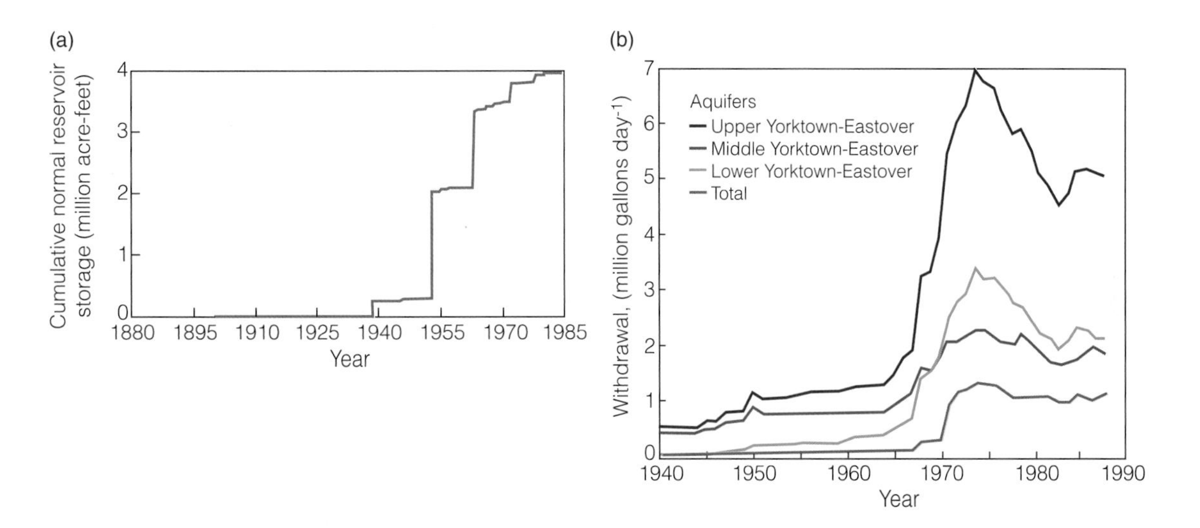

Fig. 6.29 Water use trends in Virginia: (a) reservoir water storage development, 1880–1985 (*National Water Summary*, 1987); (b) water withdrawal, 1940–90, Virginia's eastern shore. From Richardson (1992). *Hydrogeology and analysis of the ground-water-flow system of the eastern shore, Virginia*. Open-file report 91–480, US Geological Survey, Richmond, Virginia.

The environmental debt that is accumulating for the Bay due to multi-scaled human interferences in ecological processes is causing non-intuitive and unexpected ecological responses. That is, historical dependencies and estuarine dynamics together are operating in cycles of collapse and reorganization for which there are multiple possible outcomes (Ch. 4, Section 4.7). Changes in the Bay system resulting from downward trends of estuarine species such as oysters and shad (Fig. 6.24), the expanding and accelerating volume of midsummer hypoxic water (Hagy *et al.*, 2004), the decline of once extensive eelgrass beds (Orth *et al.*, 2010), high summer concentrations of stinging jellyfish (Purcell *et al.*, 1994), harmful algae blooms (Marshall *et al.*, 2005), and eutrophic water stimulating massive production of phytoplankton (Malone *et al.*, 1996; Boynton and Kemp, 2000), all contribute to the overall ecosystem debt (Table 6.5). Human activities, sea-level rise, and loss of island habitat have reduced some waterfowl populations while others have flourished (Box 6.2). Reduction of filter-feeding shellfish (clams, oysters) and their shell production (Gutiérrez *et al.*, 2003), and reduction of fish (especially juvenile menhaden) together with deleterious land-use practices and increased water turbidity are significantly

Table 6.3 Nitrogen and phosphorus (kg km^{-2} yr^{-1}) input to upper Bay, and Patuxent and Potomac sub-estuaries. Residential and commercial-industrial values are medians for the United States Environmental Protection Agency. Agricultural and forest values are also medians. From Magnien *et al.* (1992). External nutrient sources, internal nutrient pools and phytoplankton production in Chesapeake Bay. *Estuaries and Coasts* **15**(4), 497–516.

Land use category	Nitrogen load	Phosphorous
Residential		
Low density	543	99
Medium density	963	148
High density	1,803	247
Commercial-industrial	1,630	222
Cropland	889	222
Pasture	494	74
Feedlot, waste storage	289,731	25,935
Forest	247	24.7

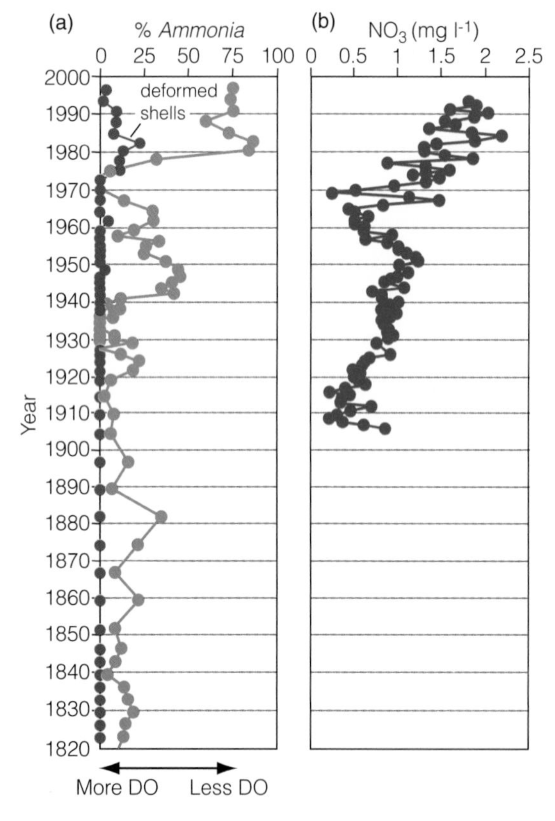

Fig. 6.30 Sediment cores revealed (a) increasing eutrophic and anoxic conditions since the 1800s, with increased benthic foraminiferan *Ammonia parkinsoniana* (blue dots), a pollution indicator, that increased more than 75% in the sediment between 1820 and 2000, associated with decreased dissolved oxygen (DO) and foraminiferan shell abnormalities (red dots) that appeared in the mid-1970s; (b) increases in nitrate (NO_3 in mg l^{-1}) over same time. Redrawn from Karlsen *et al.* (2000). Historical trends in Chesapeake Bay dissolved oxygen based on benthic foraminifera from sediment cores. *Estuaries and Coasts* **23**(4), 488–508, with kind permission from Springer Science+Business B.V.

Table 6.4 Examples and sources of contaminants in Bay. Based on data from CBP (1992); US Environmental Protection Agency for the CBP; US Government Printing 312-014/40145, Washington DC, April; CBP (1999). Kline (1998).

Contaminant	Source
Copper	Rivers, especially upper Bay. Diffuse in lower Bay (atmosphere; anti-fouling paint on boat hulls).
Mercury	Atmosphere: coal burning, incineration. Batteries, urban storm water.
Tributyltin	Anti-fouling paint for boats; industrial fungicide; high concentrations near recreational boating.
Atrazine	Runoff from farm fields. Agricultural herbicide. Uncertain severity.
PCBs	Sediments (buried), recycled between air and water. Widely spread throughout Bay. Now banned from most uses.
PAHs	Urban and industrial areas; urban storm water runoff, atmospheric deposition, oil spills, recreation and industrial boating.
Chlordecone (Kepone®)	Accidental spill by plant in James River. Used as stomach poison in baits. Extremely persistent, lipophilic.
Dioxins	Domestic, industrial processes (incineration of plastics, industrial processes, combustion of fossil fuels, pulp mills).
Phthalates	Butylbenzyl-phthalate, a plasticizer in vinyl floor tiles, adhesives, and synthetic leather; di-n-butylphthalate, a plasticizer in food packaging, PCV, some elastomers, and insect repellent.
Phenol	Coal cooking plants, chemical plants, gas works, oil refineries, pesticide plants, and wood preservative and dye manufacturing industries. Produced by municipal wastes, coal liquid, and discharges from paper, pulp and jute mills, and paint manufacture.
DDT	Very high local levels from ocean dumping. Persistent. Banned in U.S. 25 years ago.

changing the Bay, increasing the debt and uncertainty. Most critically, excess nutrients (eutrophication) caused by humans (Ch. 2) have restructured the Bay's ecosystem (Kemp *et al.*, 2009), impairing its ability to absorb and degrade organic materials that contribute to chronic hypoxia (Box 2.5). Sustained hypoxia affects higher forms of life by impairing reproduction, growth, and immune responses: first the fishes, shellfish, and crabs disappear, then a "dead" zone develops that is difficult to reverse (Breitburg *et al.*, 2009). A range of nonlinear positive feedbacks tends to reinforce and accelerate the eutrophication process (Kemp *et al.*, 2005). Without ecosystem-based intervention to reduce these impacts, continued losses in functional species, disruption of energy-efficient

Table 6.5 An ecosystem debt for Chesapeake Bay results from cumulative exposures to continuous inputs, extractions, and physical alterations that relate to cumulative environmental stressors that impact the ecosystem. See text for sources.

Source ***Inputs, extractions, physical alterations***	**Environmental stressors**	**Ecosystem impact**
Nutrient pollution	Overabundant algae; dinoflagellate blooms; brown-red tides; continuous over-loading	Alter biogeochemical cycling, water column, benthos; hypoxic/anoxic benthos; toxic algae/shellfish poisoning; fish kills
Invasive species	Nuisance species: stinging sea nettles (*Chrysaora quinquecirrha*); comb jelly (*Mnemiopsis leidyi*); invasive common reed (*Phragmites australis*); nutria (*Myocastor coypus*); flathead catfish (*Pylodictis olivaris*)	Displace native species; reduces biological diversity alternative species composition and ecosystem linkages
Toxic chemicals	Biological exposure to more than 1000 toxics. Legacy toxins (DDT, PCBs, PAHs, chlordane, Kepone®, arsenic, mercury, dieldrin, cadmium, zinc, malathion)	Bioaccumulation; biomagnification; sub-lethal toxicity; public health; unsafe seafood consumption
Air pollution	Anthropogenic inputs of CO_2, N_2O, SO^2; air-water exchange	Increases water acidity affecting calcium carbonate organisms
Urbanized bay	Noise, traffic, collisions; artificial light; mobilized sediment, deposition; harden shoreline; increase pathogens, toxins	Increased turbidity, endangered species; disrupts life-history events/phenologic timing; habitat loss, biogeographic change; disease epidemics
Watershed change	Urbanized watershed; altered freshwater-tributary oscillatory patterns; altered landscape; increased exposures to pollutants/sediment new toxics/ pharmaceuticals	Exposures to endocrine disruptor decrease water quality, quantity, hydrological cycling; introduced pathogenic bacteria/viruses; toxic chemicals
Fisheries reductions	Legacy depletions/continuous extraction; illegal fishing; harvesting impact, selective removal; bycatch removal; aquaculture; degraded water quality, habitat change	Food web impact (removes oldest, largest, most adapted); altered seascape patterns/ benthic structure
Shoreline change	Sea level rise; harden/restructure shores; altered flows of sediment, hydrology	Impacts wetland/seagrass habitat; fisheries; water clarity/quality; buffer loss
Degraded areas	Point-source degradation	Degrades ecosystem health; hypoxic/anoxic areas; loss of ecologic performance; contributes to diseases
Climate change	Severe weather; warming water; acidification	Biotic change; altered fisheries; shoreline loss

Box 6.2 Chesapeake Bay waterbirds, sea-level rise, and island restoration

R. Michael Erwin
[Retired, Department of Environmental Sciences, University of Virginia], Weaverville, North Carolina, USA

The Chesapeake Bay has a history of being one of the "crown jewels" for waterfowl and recreational hunting in North America. In addition to providing sport, waterfowl (ducks, geese, and swans) may also serve as biological indicators within the estuary. They, along with other waterbirds (a term that includes colonially nesting species gulls, terns, pelicans, cormorants, shorebirds such as sandpipers, plovers, American oystercatchers, raptors such as bald eagles (*Haliaeetus leucocephalus*) and osprey (*Pandion haliaetus*), and marsh-dwelling rails and bitterns) perch at the top of the Bay's trophic structure. If conditions deteriorate in fish and invertebrate communities in the Bay, many of these species suffer reduced productivity and eventually population declines.

An interesting paradox has developed within the Chesapeake with respect to waterbird populations and distributions. Some species have increased both in range and numbers especially in the past 20–30 years, while others seem to be in jeopardy. On a positive note, federal bans of organochloride pesticides that caused widespread egg-shell thinning in many bird species in the middle of the 20th century, with subsequent population crashes, have resulted in rebounds of a number of species. Bald eagles now number more than 1000 pairs, with a doubling time of only about eight years for the Bay population! Ospreys also have increased dramatically, especially in the major tributaries. The pelican family has also prospered as pesticides have declined; both the brown pelican (*Pelecanus occidentalis*) and double-crested cormorant (*Phalacrocorax auritus*) have expanded dramatically in the Bay since the

(*Continued*)

1980s. For reasons other than contaminant cleanup, some other species have also expanded their ranges and established populations in the Bay over the past 10–15 years. Notably this includes two nesting shorebirds, the American oystercatcher (*Haematopus palliatus*) and black-necked stilt (*Himantopus mexicanus*). Royal terns (*Sterna maxima*) also have colonized the Bay since the early 1980s and in some years have nested in large numbers on islands in Tangier Sound. Earlier studies suggest that predator disturbances on many of the ocean coast islands have caused shifts into the Bay for some of these species.

In recent years, however, major concerns have been directed at climate change and its many potential ramifications. For the Bay region, the most notable is sea-level rise and its potential effects on waterbird nesting habitat. Dramatic changes have long been recognized as marshes continue to be inundated by rising water on the Eastern Shore, and islands, even of substantial size, have disappeared (e.g., Sharp's Island, Maryland, where a hotel stood in the late 1800s), or have been sharply reduced in size (e.g., Poplar Island). Not all of this island loss, however, can be attributed to sea-level rise, as mainstem Bay islands receive no riverine sediments; in addition, erosion from commercial shipping and recreational boating can also exacerbate losses.

With respect to small island losses, however, a recent study conducted by the state of Virginia highlighted some of these island changes. In Tangier Sound, Virginia, 15 islands were analyzed for areal change using photographic records from 1994 and 2007 (Table B6.2.1). These same islands also had been surveyed for nesting waterbirds during this same time period. The resulting 21% decrease in island area coincided with a 50% decline in nesting wading birds (nine species of herons and egrets) and 64% loss of nesting ducks, primarily American black ducks (*Anas rubripes*). Common terns (*Sterna hirundo*), a threatened species in Maryland, have suffered the highest percentage loss. Although this pattern does not reflect cause and effect, it does seem symptomatic of the trend in nesting island habitat loss and declines in nesting populations of many key waterbird species. From future projected sea-level rise changes from recent climate change reports (see www.epa.gov/climatechange/effects/coastal/sap4-1.html), it appears that sea levels will exceed the capacity of the marshes to maintain pace in the mid-Atlantic region, especially if the ice sheets of Greenland and Antarctica are factored in. Without strong human intervention to stabilize or restore these islands, much nesting habitat for more than 25 species of wading birds, shorebirds, marsh birds, terns, gulls, and waterfowl is in jeopardy.

A Bay restoration "highlight" for the past decade has been Poplar Island, Talbot County, on Maryland's eastern shore. For the U.S. Army Corps of Engineers, Poplar Island is an 1140-acre "Beneficial Use of Dredged Material" island designated over the next 20 years to receive sediments from navigation channels in the mid-Bay. For the federal and state agencies charged with natural resource stewardship, it has been a golden opportunity to restore some of the area's waterbirds, wetland acreage, fish, and submerged aquatic vegetation that have declined as the island dwindled from about 1140 acres in the mid-1800s to fewer than five acres by the early 1990s. With half of the newly restored island designated to be uplands, and half wetlands, the island has seen impressive changes since construction began in 1999. Within a year or two of initial dike construction, both common and least terns (*Sterna hirundo* and *S. antillarum*, respectively) colonized the island, and by 2005, more than 800 pairs of common terns nested—the only colony for the species in the Maryland portion of the Bay! Also colonizing were snowy egrets (*Egretta thula*), ospreys, and large numbers of mallard ducks (*Anas platyrhynchos*) and double-crested cormorants. Tidal wetland cells have been planted with *Spartina* grasses and these wetlands have become magnets to feeding shorebirds and waterfowl.

The lessons learned from Poplar Island are being applied to the next generations of dredged-material islands, farther south at Barren Island and James Island in the Bay. Opportunities abound for federal, state, universities, and non-government organizations to work in partnership on these projects to restore island uplands, wetlands, and surrounding submerged aquatic vegetation and oyster reefs. While these large projects may help to sustain some populations of waterbirds, large islands also attract or support mammalian (foxes, raccoons) and avian (gulls) predators. Such has been the case at Poplar Island as well as Hart-Miller Island near Baltimore. Thus, there remains a critical need also to restore a number of smaller (under 20 acres) islands to allow waterbirds ample nest-site choices, and hence "place their eggs in many more baskets" than is currently possible.

Table B6.2.1 Changes in area of Tangier Sound (VA) islands (n = 15) and the size (pairs) of waterbird nesting populations during a 13-year period, 1994–2007. (Waterbird data from Center for Conservation Biology, College of William and Mary, and Virginia Game and Inland Fisheries Commission (latter also provided island change data.))

Parameter	1994	2007	Percent change
Island acreage	175	138	–21
Wading birds	1207	591	–51
Waterfowl	118	43	–64
Common terns	400	15	–91

Sources: Erwin *et al*. (2007a, 2007b, 2011); Watts *et al*. (2008)

transfers, system inefficiencies, and positive feedbacks move the Bay ecosystem in uncertain and unpredictable directions.

6.5 BAY RESTORATION: CHARTERING A COURSE

The Bay is not only an ecologically complex estuarine ecosystem, it is jurisdictionally and socially complex, challenging restoration efforts. The Bay ecosystem has transgressed from natural-resource abundance to depletion and one in need of restoration that requires ecosystem-based management (Ch. 13), unprecedented levels of knowledge and funding, coordinated governance, public involvement, and committed action at all levels of social and political organization.

A history of directives connects Bay health to historic water quality, involving legal decisions between federal and state authorities that began with the 1899 *Rivers and Harbors Act* to curb surface-water pollution, adjudicated by the U.S. courts. In the 1930s, Congress directed states to take measures to protect their water, and states approached industrial point-source pollution through weak administrative programs while ignoring the tough non-point pollution of farm runoff and human development (Salzman and Thompson, 2010). To address water quality in the 1960s under regulatory provisions of the 1899 *Rivers and Harbors Act*, Congress took a stronger position, and passed the 1965 *Water Quality Act* that directed states to curb water pollution (Downing *et al.*, 2003). But states lacked the political will to act and the necessary scientific data and technology on which to base water-quality standards.

Public outrage over environmental issues in the 1960s and 1970s forced a political shift in the nation's environmental policy (Ch. 3). Congress passed the *National Environmental Policy Act* (NEPA) in 1969 that mandated formation of the Environmental Protection Agency (EPA) to regulate and enforce water pollution measures. Congress passed the 1972 *Federal Water Pollution Control Act*, amended in 1977 as the *Clean Water Act* (CWA, Ch. 3), which included a citizen lawsuit provision. The Act also aimed to eliminate all pollutant discharges into the nation's water by 1985. Administered by the federal EPA, this law required states specifically to regulate point-source pollution, develop plans to control non-point pollution, and set water-quality standards for their waterways. Under a political climate of change in 1972, Congress directed EPA (P.L. 116) to undertake a comprehensive assessment of Chesapeake Bay water quality and its natural resources, and to identify appropriate management strategies for their protection (Senate Report 94-326). EPA compiled scientific data that were collected separately by Virginia and Maryland scientists into a holistic view of the Bay: *A Profile of Environmental Change* and *A framework for Action* (EPA, 1983a, b), which provided strong evidence of the loss of valued natural resources and the need for comprehensive action (Malone *et al.*, 1993).

These actions drew strong public awareness and support, and brought Maryland and Virginia into cooperative agreements. These states recognized their shared problems and formed the Chesapeake Bay Commission in 1980 to coordinate their Bay-related policies and commit to legislative solutions. The state of Pennsylvania soon joined to form a tri-state Commission. The highest authorities in the Bay region (Maryland, Virginia, District of Columbia, EPA, Bay Commission) then signed the 1983 *Chesapeake Bay Agreement*, a landmark Agreement to initiate a plan to negotiate a comprehensive and detailed working relationship directed by top government officials. This signing initiated a long-term plan to restore the Bay's oysters, seagrasses, fishes, marshes and water quality. Through increased environmental awareness and strong scientific studies, a culture of support sustained Bay restoration, but this effort proved far more complex than anyone had anticipated. Greatly increasing the complexity of management was the uneasy tension that erupted into political and legal debates over the roles of state and federal government, regulatory authority, and the private sector.

In order to address these contentious problems, high-ranking state and federal officials formed the policy-making Executive Council (EC) that developed a comprehensive plan, i.e., the *Chesapeake Bay Program* (CBP) administered by EPA as mandated under Section 117 of the federal *Clean Water Act*. The EC negotiated a detailed working relationship by directing the CBP to undertake voluntary agreements based on an organizational structure that linked scientific advisory boards, public education, industry, and citizen groups together. This became a multi-billion dollar effort that recognized the entire Bay watershed for restoration and management of the Bay's living resources, crossing numerous jurisdictions and programs to do so.

A second, 1987 Agreement reaffirmed the alliance among states, the federal EPA, District of Columbia, and the Bay Commission. State governors came to replace high-ranking officials on the Executive Council, agreeing that the watershed was the appropriate focus for Bay restoration. An important incentive was the recognition of the annual trillion-dollar revenue-generating capacity of the Bay, with benefits that extended to the entire mid-Atlantic region (Executive Order, 2009). This ambitious, enthusiastic, and visionary Agreement aimed to address water-quality issues through voluntary partnerships among stakeholders, with the primary goal that nutrient pollution would be reduced 40% by 2000. The Program advanced collaborative programs to monitor Bay water quality, reduce toxic inputs, establish a policy of "no-net-loss" for wetlands, and reduce watershed nutrients through regional strategies, undertaken through a Local Government Partnership Initiative. The upgrading of sewage treatment plants brought improved water quality to the tidal Potomac River, and the banning of phosphate from detergents in the late 1980s abruptly decreased nutrient input (D'Elia *et al.*, 2003; Kemp *et al.*, 2005). Action plans were initiated to reduce nutrients, toxics, and agricultural pollution, and to restore Bay seagrasses, tidal marshes, oysters, and anadromous fishes, the latter by providing fish passages in watershed streams, and more. However, pollution reduction under the *Clean Water Act* was not being curbed. Maryland, Virginia, and D.C. continued to list the Bay and several tidal tributary segments in 1996, 1998, and 2000 as "impaired." While the CBP focused on the Bay's natural resources, eutrophication was becoming a far more pervasive concern (Boesch *et al.*, 2001).

Nationwide, states' *Best Management Practices* were failing to achieve water-quality goals. Congress thus amended the *Clean*

Water Act in 1987 requiring federal permits to be issued for stormwater point-source discharges from construction, industrial, and municipal development. As scientific information was revealing the importance of estuaries as habitat for fisheries, Congress enacted the national *Estuaries and Clean Waters Act* of 2000 (S. 835). Chesapeake Bay became recognized as a national treasure, a model for the *National Estuary Program* to encourage estuarine habitat restoration. And to strengthen fisheries science in the Bay, Congress established a NOAA Bay office under the *NOAA Authorization Act* of 1992 (Public Law 102-567, reauthorized in 2002 P.L. 107-372) to move fisheries management toward multispecies approaches and ecosystem-based fisheries management (EBFM, Ch. 13). NOAA's cross-state federal authority came with incentives to protect and restore transboundary species that fell under separate state management programs with traditional, single-fisheries management programs. EBFM is most difficult to implement and poorly understood, requiring fisheries science itself to broadly evolve toward interdisciplinary approaches (Box 3.5).

Federal actions made clear that if local or state governments failed to meet CWA obligations, lawsuits by citizens and citizen groups could be brought through the federal court system. In 1998, the American Littoral Society and American Canoe Association filed a lawsuit against EPA over Virginia's failure to implement the CWA's Total Maximum Daily Load (TMDL) program for pollution control. TMDL is a management tool used for calculating a pollutant's maximum amount entering a water body and still meet water-quality standards, as required under Section 303(d) of the CWA. TMDL codifies a pollution budget and requires states to report to EPA those water bodies that fail to meet standards. The citizens' lawsuit settlement set the stage for establishing a federal TMDL for Bay pollutants that focused legal attention on water quality. Thus, as Virginia's waters revealed increased impairment and the 40% goal in reduction of nutrient pollution was not in sight, the Executive Council was reaching a third Agreement focused on the Bay.

The *Chesapeake Bay 2000 Agreement* reaffirmed top-level commitments of states and the federal government to Bay restoration. This Agreement specifically targeted restoration of fisheries, living resources, habitats and Bay water quality, with specified goals to be met by 2010. The first goal, "Restore, enhance and protect the finfish, shellfish and other living resources, their habitats and ecological relationships to sustain all fisheries and provide for a balanced ecosystem," and specifically to "achieve, at a minimum, a tenfold increase in native oysters in the Chesapeake Bay, based upon a 1994 baseline." The fourth goal, "water quality necessary to support the aquatic living resources of the Bay and its tributaries and to protect human health," fell within the federal *Clean Water Act* mandate. The CBP and its science and monitoring programs guided Executive Council decisions for voluntary commitments and public outreach to achieve the targeted goals by 2010, with sustained public awareness and participation that helped achieve intergovernmental cooperation across the watershed. But as the Program tracked its successes in water pollution control and toxic chemical dumping, in restoring thousands of wetland acres, and increasing striped bass numbers, oysters remained depleted and evidence of increasing toxic algal blooms, nuisance jellyfish, and anoxia made obvious that the Bay's overall condition was not improving.

A team of scientists addressed these discrepancies by merging several indicators of Bay health into a single score, which revealed that the Program's numerous measures to assess progress failed to account for the overall declining Bay condition. Such evidence brought the federal investigative arm of Congress, the Government Accountability Office (GAO), in 2005 to review and fault the Program for lack of a coordinated, comprehensive plan to cut pollution (GAO, 2005). Evidence made clear that the CBP would not achieve its goal of a "tenfold increase in oysters" by 2010, nor reduce nutrients and sediment to the Bay to levels that could restore living resources. Most critical, the Bay and tidal tributaries were still being listed as impaired under Section 303(d) of the CWA; i.e., states were not meeting nutrient reduction goals. The Program's high cost, controversial achievements, and poor evaluative efforts brought Congress to require improved direction to the Bay's overall failing condition. Hence, Congress specifically directed EPA to implement GAO recommendations in: *Chesapeake Bay Program: Improved Strategies are Needed to Better Assess, Report, and Manage Restoration Progress* (P.L.110-61; GAO-06-09). It also specified annual targets, activity reports, progress accounting with specified funding amounts and sources, and development of a Chesapeake Action Plan (CAP). EPA responded with the 2008 Report to Congress, expressing a detailed CAP to protect and restore the Bay: *Strengthening the Management, Coordination and Accountability of the Chesapeake Bay Program.*

Citizen lawsuits and state listing of waterways as "impaired" had redirected Bay restoration from a primary focus on natural resources to water-quality restoration under the regulatory authority of the CWA. This redirection forced EPA, and the states, to confront the difficult issue of declining water quality that resulted from increases in population, impervious surfaces, altered landscapes, and destruction of natural filtering mechanisms that were outpacing efforts in natural-resource restoration. Reducing nutrient and sediment loads from developed and developing lands under the federal stormwater program had proved difficult because the rate of new land development was faster than restoration efforts could achieve (EPA Inspector General, 2007). A lawsuit by the Chesapeake Bay Foundation, a private conservation organization, brought EPA to agree in 2009 to a legally binding settlement, enforcing mandatory pollution reductions and tougher regulation for farms, developments, and city storm systems under state authority.

The CWA is now the major instrument for Bay restoration. In 2010, EPA issued its *Chesapeake Bay Compliance and Enforcement Strategy* to guide uses of management tools toward pollution sources and non-compliance (EPA, 2010). It established the Chesapeake Bay TMDL in the form of a comprehensive "pollution diet," with "rigorous accountability measures to initiate sweeping actions to restore clean water in the Chesapeake Bay and the region's streams, creeks, and rivers" (Chesapeake Bay TMDL fact sheet). Pollution limits are now established for major river basins by jurisdiction, requiring states to survey their waters every two years, and to identify and list impaired waterways in a report to EPA. Any water body that fails to meet its water-quality standard (Bay and tributaries included) must have TMDL-based restoration plans, or a restoration plan based on EPA standards. Those plans are implemented through tighter discharge limits placed on point sources (e.g., sewage plants) and on Best Management Practices for non-point sources (runoff). The maximum allowable loads for nitrogen, phosphorous, and fine sediments are capped under Watershed Implementation Plans (WIPs) with TMDL accounting every two years. All local TMDLs for Virginia, Maryland, and Delaware are then included into an expanded, regional Chesapeake Bay TMDL for statistical consistency.

Accounting takes place under a Bay Watershed Model. This Model monitors the progress on capped limits, established at 184.9 million pounds of nitrogen, 12.5 million pounds of phosphorous, and 6.45 billion pounds of sediment each year. Nutrient allocations and sediment caps become linked in a super model with estuarine and airshed models. EPA then tracks and assesses progress, and if state or other jurisdictions do not meet commitments, specific federal mandatory action under the CWA can be applied. Among EPA's >36,000 watersheds nationwide, Chesapeake Bay's TMDL procedure is unprecedented in scope and scale. Complimenting these efforts, the Bay's Executive Committee reaffirmed its top-level commitments to restore water quality throughout the Bay watershed by 2025. Two-year milestones are showing positive results, but questions remain about the quality of the data (Blankenship, 2012).

President Obama boosted Bay restoration by issuing an Executive Order (EO 13508; May 12, 2009) that committed federal leadership to Bay ecosystem restoration. While maintaining existing programs among federal and state partners, this authorization gave renewed direction to ecosystem-based management. Although species are still managed under separate state management plans and by regional councils, the EO provides support for EBFM (CBP, online). However, funding the $420 million dollars needed to meet strategy goal implement in its 2012 Action Plan is proving difficult, due to political priorities of a divided Congress (FLCCB, 2012) facing financial deficits.

6.6 PEOPLE SHALL JUDGE

The Chesapeake Bay has moved from an ecosystem of historically high natural-resource abundance to a transformed, subsidized urban bay dependent upon human intervention. The central issue facing Bay restoration is water quality, which connects people's resource-use choices to Bay governance, federal and state management priorities, and ecosystem health. For ecological restoration, management requires boundary determination for ecosystems that shift, making management an inexact science. For ecosystem-based management (Ch. 13), scientific understanding of ecological dimensions, costly long-term programs, integration of federal, state, and local laws, applications of new management instruments, flexible and adaptive administrative programs, public participation, education, and incentives for cooperation are all required, while always keeping a central focus on the ultimate restoration goal: ecosystem resilience and resource sustainability.

Impaired rivers and Bay water are primarily due to excess nutrients (nitrogen, phosphorus) and sediment entering the watershed system. Bay partners are focusing on cost-effective approaches to reduce pollution, including upgrading wastewater facilities, implementing agricultural best practices, and funding local programs, but are challenged by the rate of new watershed development. In three years, EPA issued citations to Bay impairment that included 16 civil judicial settlements, 146 administrative orders, and one emergency order (EPA 2013, online) making clear their authority to enforce the law.

Fisheries management has benefitted from the Chesapeake Bay Program. Bay partners agree to restore oysters, seagrasses, marshes, and fishes, and to remove blockages on streams and rivers that prevent fish from migrating to spawning grounds. State moratoria and management programs have improved striped bass abundance, but at the expense of forage fishes (Walter *et al.*, 2003). Still at relatively low levels of abundance, native oysters have failed to qualify as threatened or endangered under the Federal *Endangered Species Act* because its long-term persistence is perceived as not at risk (EOBRT, 2007). Attempts to introduce an Asian-Pacific oyster (*Crassostrea ariakensis*) by the oyster industry failed against scientific scrutiny and public protest that made clear that its introduction was unacceptable. Native oysters are under intense management by states, NOAA, and the CBP to restore their reef structure, encourage aquaculture, curb disease, and promote water quality. Many are under private ownership. As fisheries (fin and shellfish) are important to both states and to the nation, Bay fisheries restoration is increasingly dependent on aquaculture, promoted under the 1980 *National Aquaculture Act* (P.L. 96-362) and water-quality improvement pursued by the Chesapeake Bay Program and EPA.

Numerous federal and state laws and programs are thus involved in the management, protection, and restoration of the Bay and its resources, for which public understanding and support are vital. These include: the Federal *Clean Air Act* and *Clean Water Act* as dominant tools to regulate pollution—Section 404 of the *Clean Water Act* limits how various environmentally sensitive lands including wetlands are to be used; the *Magnuson-Stevens Fishery Conservation and Management Act* for federal marine fisheries management; the *Coastal Zone Management Act* for state-federal coordination in coastal conservation; and the *Endangered Species Act* (ESA). And while the federal government provides important consistency roles, state and local government regulate, legislate, enforce, fund, and implement resource management subject to shifting political policies. Thus, while federalism plays a critical role in policymaking, states implement policy, elect governors and legislators, collect taxes, and establish budgetary priorities, and local government deals directly with citizens to serve local needs, generating revenue that development provides and taking measures to protect water resources. How land is developed and used, how much water is withdrawn from rivers and the ground, what natural-resource use restrictions apply or not, all depend on local support and decisions under the constitutional provisions of federal, state, and local governments.

Every Bay conservation initiative has had to confront strong opposition. In 2012, the American Farm Bureau and Home Builders Association issued a lawsuit against EPA in an attempt to redirect the Bay restoration process to local decisions. And congressmen from Virginia and Pennsylvania have introduced an amendment to the *Clean Water Act* that challenges EPA's authority to implement a TMDL, preferring state power and voluntary compliance to any Chief Executive, state, or D.C. request. Yet the key drivers determining Bay health and resiliency are collective choices by individuals, industry, court systems, and legislators.

The prediction among Bay scientists is this:

> "If sediment and nutrient loads continue at levels witnessed at the end of the 20th century, multiplied by a growing population and new development, water quality will worsen. Water clarity and oxygen levels will slide back toward conditions not seen since the 1980s . . . Escalating nutrient and sediment loads would result not only from a population expected to reach 19 million by the year 2030, but also from poor land use planning, with continued rapid loss of farm and forested lands, and only modest improvements in agricultural methods and wastewater treatment. These additional loads would largely defeat current efforts to restore underwater grasses, cause further loss of oxygen in the Bay's bottom waters, and undermine efforts to restore oysters due to worsening water quality" (STAC, 2002).

Also, scientists project that by year 2100, U.S. eastern regional water temperatures may increase by approximately 2° to 6°C above the historical average, CO_2 concentrations may increase 50 to 160%, relative sea level may rise 0.7 to 1.6 m, and both storm intensity and mean precipitation would increase in winter and spring (Pyke *et al.*, 2008; Boesch, 2008). In such a future, humans, biota, and the Bay system itself must make adjustments. Such changes would bring undesirable impacts to urban areas, coastlands, wetlands, commercial species, biodiversity, national parks, and protected areas (Jasinski and Claggett, 2009).

The Chesapeake Bay and its watershed form a complex, adaptive estuarine ecosystem. The Bay is a "common-pool resource" to which a large number of people have access and commonly overuse (Dietz, 2002), which has created the Bay's environmental debt from which recovery is slow and most difficult. The Chesapeake Bay Program's innovative, collaborative, expansive ecosystem-based approach encompasses a watershed where private property use is tied to economic and social well-being. Private property rights constitute a major issue for the Chesapeake Bay and are affecting the sustainability of the Bay's common-pool resources currently managed for the public by state and federal governments. Ecosystem-based management thus encompasses numerous political, social, and administrative levels of decision-making, together determining the outcome of Bay health and sustainability. Will democratic institutions and people's individual decisions sustain ecosystem-based approaches, or will a "tyranny of small decisions" (Odum, 1982) add up to an outcome that no one wants? The people shall judge.

REFERENCES

Bahr Jr. LM (1974) Aspects of the structure and function of the intertidal oyster reef community in Georgia. Ph.D. Thesis. Univ. Georgia, Athens.

Bahr Jr. LM (1976) Energetic aspects of the intertidal oyster reef community at Sapelo Island, Georgia (USA). *Ecology* **57**, 121–131.

Bahr LM, Lanier WP (1981) *The ecology of intertidal oyster reefs of the south Atlantic Coast: A community profile*. FWS/OBS-81/15. Biological Services Program, Dept. Interior, Washington, D.C.

Baird D, Ulanowicz RE (1989) The seasonal dynamics of the Chesapeake Bay ecosystem. *Ecological Monographs* **59**, 329–364.

Bearman GG, ed. (1989) *Waves, Tides and Shallow-Water Processes*. The Open University, Walton Hall, Milton Keynes, UK.

Blankenship K (1999) Logs of early explorers marveled at bay's vast forests. *Bay Journal* **8**. Online. patc.net/history/archive/virg_fst.html

Blankenship K (2012) States meeting most of the TMDL milestones. *Bay Journal* **22**(6), 1, 10–11.

Boesch DF, Brinsfield RB, Magnien RE (2001) Chesapeake Bay eutrophication: scientific understanding, ecosystem restoration, and challenges for agriculture. *Journal of Environ Quality* **30**, 303–320.

Boesch DF, ed. (2008) *Global warming and the free state: comprehensive assessment of climate change impacts in Maryland*. Report of the Scientific and Technical Working Group of the Maryland Commission on Climate Change. University of Maryland Center for Environmental Science, Cambridge, Maryland.

Boicourt WC, Kuzmic M, Hopkins TS (1999) The inland sea: circulation of Chesapeake Bay and the northern Adriatic. In *Ecosystems at the Land-Sea Margin, Drainage Basin to Coastal Sea. Coastal and Estuarine Studies* 55 (eds Malone TC, Malej A, Harding Jr. LW, Smodlaka N, Turner RE). American Geophysical Union, Washington D.C., 81–129.

Boynton WR, Garber JH, Summers R, Kemp WM (1995) Inputs, transformations and transport of nitrogen and phosphorous in Chesapeake Bay and selected tributaries. *Estuaries* **18**, 285–314.

Boynton WR, Kemp WM (2000) Influence of river flow and nutrient loads on selected ecosystem processes: a synthesis of Chesapeake Bay data. In *Estuarine Science: A Synthetic Approach to Research and Practice* (ed. Hobbie JE). Island Press, Washington, D.C., 269–298.

Breitburg DL, Fulford RS (2006) Oyster-Sea Nettle Interdependence and Altered Control Within the Chesapeake Bay Ecosystem. *Estuaries and Coasts* **29**(5), 776–784.

Breitburg DL, Hondorp DW, Davias LA, Diaz RJ (2009) Hypoxia, nitrogen, and fisheries: integrating effects across local and global landscapes. *Annual Review of Marine Science* **1**, 329–249.

Brush GS (1986) Geology and paleoecology of Chesapeake Bay: a long-term monitoring tool for management. *Journal of the Washington Academy of Sciences* **76**, 146–160.

Brush G (2009) Historical land use, nitrogen, and coastal eutrophication: a paleoecological perspective. *Estuaries and Coasts* **32**,18–28.

CBP (1992) *Chesapeake Bay basin comprehensive list of toxic substances*. U.S. Environmental Protection Agency for the Chesapeake Bay Program. U.S. Government Printing #312-014/40145, Washington D.C., April.

CBP (1999) *Environmental indicators: measuring our progress*. Chesapeake Bay Program, Annapolis, Maryland.

CPB (2001; 2011) Chesapeake Bay Program online. www.chesapeakebay.net/tsc.htm

Chesapeake-Potomac Study Commission (1948) *Report on fish and shellfish in the Chesapeake Bay and Potomac River with recommendations for their future management*. Daily Record Co. Press, Baltimore, MD, January 7.

Colman SM, Halka JP, Hobbs III CH, Mixon RB, Foster DS (1990) Ancient channels of the Susquehanna River beneath Chesapeake

Bay and the Delmarva Peninsula. *GSA Bulletin* **102**, 1268–1279, September.

Cooper SR, Brush GS (1991) Long-term history of Chesapeake Bay anoxia. *Science* **254**, 992–996.

Cooper SR, Brush GS (1993) A 2500 year history of anoxia and eutrophication in Chesapeake Bay. *Estuaries* **16**, 617–626.

Corenblit D, Baas ACW, Bornette G, Darrozes J, Delmotte S, Francis RA, *et al.* (2011) Feedbacks between geomorphology and biota controlling earth surface processes and landforms: a review of foundation concepts and current understandings. *Earth-Science Reviews* **106**, 307–331.

Correll DL (1978) Estuarine productivity. *BioScience* **28**, 646–650.

Cotrim da Cunha L, Buitenhuis ET, Le Quéré C, Giraud, X, Ludwig W (2007) Potential impact of changes in river nutrient supply on global ocean biogeochemistry. *Global Biogeochemical Cycles* **21**, GB4007.

Cronin LE (1986) Fisheries and resource stress in the 19th century. *Journal of the Washington Academy of Sciences* **76**, 188–198.

Cross FA, Willis JN, Hardy LH, Jones NY, Lewis JM (1975) Role of juvenile fish in cycling of Mn, Fe, Cu, and Zn in a coastal-plain estuary. In *Estuarine Research* I (ed. Cronin LE). Academic Press, Inc. New York, 45–63.

Custer JF (1986) Prehistoric use of the Chesapeake Estuary: a diachronic perspective. *Journal of the Washington Academy of Sciences* **76**, 161–172.

Dalyander PS, Cerco CF (2010) Integration of a fish bioenergetics model into a spatially explicit water quality model: application to menhaden in Chesapeake Bay. *Ecological Modelling* **221**, 1922–1933.

Dame RF (1996) *Ecology of Marine Bivalves*. CRC Press, Boca Raton.

Dame RF, Spurrier JD, Wolaver TG (1989) Carbon, nitrogen and phosphorus processing by an oyster reef. *Marine Ecology Progress Series* **54**, 249–256.

Davies TT (1972) Effect of environmental gradients in the Rappahannock River estuary on the molluscan fauna. In *Environmental Framework of Coastal Plain Estuaries*, Memoir 133 (ed. Nelson BW). The Geological Society of America, Boulder, Colorado, 263–290.

DeAlteris JT (1988) The geomorphic development of wreck shoal, a subtidal oyster reef of the James River, Virginia. *Estuaries* **11**, 240–249.

D'Elia CF, Boynton WR, Sanders JG (2003) A watershed perspective on nutrient enrichment, science and policy in the Patuxent River, Maryland: 1960–2000. *Estuaries* **26**,171–185.

Dietz T, Dolšak N, Ostrom E, Stern PC (2002) The drama of the commons. In *The Drama of the Commons* (eds Ostrom E, Dietz T, Dolšak N, *et al.*). National Research Council, National Academy Press, Washington, D.C., 3–35.

Downing DM, Winer C, Wood LD (2003) Navigating through Clean Water Act jurisdiction: a legal review. *Wetlands* **23**, 475–493.

Emery KO (1967) Estuaries and lagoons in relation to continental shelves. In *Estuaries* (ed. Lauff GH). Publication **83**, American Associations for the Advancement of Science, Washington D.C., 9–11.

EOBRT (2007) *Status review of the Eastern oyster* (*Crassostrea virginica*). Eastern Oyster Biological Review Team Report to the National Marine Fisheries Service, Northeast Regional Office. February 16, 105pp.

EPA (1983a) *Chesapeake Bay; a profile of environmental change*. EPA Region 3. United States Environmental Protection Agency, Philadelphia, Pennsylvania.

EPA (1983b) *Chesapeake Bay; a framework for action*. EPA Region 3. United States Environmental Protection Agency, Philadelphia, Pennsylvania.

EPA (2010) *Chesapeake Bay compliance and enforcement strategy*. Office of Enforcement and Compliance Assurance (OECA), U.S. Environmental Protection Agency, Washington D.C., May.

EPA (2013) Water enforcement for Chesapeake Bay. www.epa.gov/enforcement/water/progress-chesapeakebay.html#results.

EPA Inspector General (2007) *Development growth outpacing progress in watershed efforts to restore the Chesapeake Bay*. Report No. 2007-P-00031, Sept. 10. www.epa.gov/oig/reports/2007/20070910-2007-P-00031

Erwin RM, DF Brinker, Watts BD, Costanzo GR, Morton DD (2011) Islands at bay: rising seas, eroding islands, and waterbird habitat loss in Chesapeake Bay (USA). *Journal Coastal Conservation* **15**(1), 51–60.

Erwin RM, Miller J, Reese J (2007a) Poplar Island environmental restoration project: challenges in waterbird restoration on an island in Chesapeake Bay. *Ecological Restoration* **25**, 256–262.

Erwin RM, Watts BD, Haramis GM, Perry MC, Hobson KA, eds (2007b) Waterbirds of the Chesapeake Bay and vicinity: harbingers of change? *Waterbirds* **30** (Special Publication 1).

Executive Order (2009) Targeting resources to better protect the Chesapeake Bay is aim of U.S. Department of Agriculture's 202(b) Report. Online. executiveorder.chesapeakebay.net/post/Targeting-Resources-to-Better-Protect-the-Chesapeake-Bay-is-Aim-of-US-Department-of-Agriculture.aspx

FLCCB (2012) *Executive Order 13508 action plan*. FY2012, Developed by the Federal Leadership Committee for the Chesapeake Bay. U.S. Environmental Protection Agency, U.S. Department of Agriculture, U.S. Department of Commerce, *et al.* Washington, D.C.

Flewelling SA (2009) Nitrogen storage and removal in catchments on the eastern shore of Virginia. A Dissertation presented to the Graduate Faculty of the University of Virginia in Candidacy for the Degree of Doctor of Philosophy. Department of Environmental Sciences, University of Virginia.

Flewelling SA, Herman JS, Hornbereger GM, Mills AL (2012) Travel time controls the magnitude of nitrate discharge in groundwater bypassing the riparian zone to a stream on Virginia's coastal plain. *Hydrological Processes* **26**, 1242–1253.

Fourqurean JW, Duarte CM, Kennedy H, Marbà N, Holmer M, Mateo MA, *et al.* (2012) Seagrass ecosystems as a globally significant carbon stock. *Nature Geoscience*, published online, 20 May 2012.

Funderburk SL, Mihursky JA, Jordan SJ, Riley, D, eds (1991) *Habitat requirements for Chesapeake Bay living resources*, Second edition. Chesapeake Research Consortium, Inc., Solomons, Maryland.

Galtsoff PS (1964) The American oyster *Crassostrea virginica* Gmelin. *Fishery Bulletin* **64**, 1–480.

Galtsoff PS, Chipman Jr. WA, Engle JB, Calderwood HN (1947) Ecological and physiological studies of the effect of sulfate pulp mill wastes on oysters in the York River, Virginia. *Fishery Bulletin* **43**, Fishery Bulletin of the Fish and Wildlife Service **51**, U.S. Gov. Printing Office, Washington, D.C.

GAO (2005) *Improved strategies are needed to better assess, report, and manage restoration progress*. United States Government Accountability Office Report to Congressional Requesters, October, Washington, D.C.

Garman GC, Macko SA (1998) Contribution of marine-derived organic matter to an Atlantic coast, freshwater, tidal stream by anadromous clupeid fishes. *Journal of North American Benthological Society* **17**, 277–285.

Goodrich DM, Blumberg AF (1991) The fortnightly mean circulation of Chesapeake Bay. *Estuarine, Coastal and Shelf Science* **32**, 451–462.

Gottschalk LC (1945) Effects of soil erosion on navigation in upper Chesapeake Bay. *Geographical Review* **35**, 219–238.

Gross MG, Gross E (1996) *Oceanography: A View of Earth*. Prentice Hall, Upper Saddle River, New Jersey, USA.

Gutiérrez JL, Jones CG, Strayer DL, Iribarne OO (2003) Mollusks as ecosystem engineers: the role of shell production in aquatic habitats. *Oikos* **101**, 79–90.

Gutiérrez JL, Jones CG (2006) Physical ecosystem engineers as agents of biogeochemical heterogeneity. *BioScience* **56**, 227–236.

Guo L, Santschi PH (1997) Composition and cycling of colloids in marine environments. *Reviews of Geophysics* **35**, 17–40.

Hagy JD, Boynton WR, Keefe CW, Wood KV (2004) Hypoxia in Chesapeake Bay, 1950–2001: Long term change in relation to nutrient loading and river flow. *Estuaries* **27**(4), 634–658.

Hardaway CS, Anderson GL (1980) *Shoreline erosion in Virginia.* Sea Grant Program, Marine Advisory Service, Virginia Institute of Marine Science, Gloucester Point, VA, 25pp.

Hardaway Jr. CS, Byrne RJ (1999) *Shoreline management in Chesapeake Bay.* Virginia Institute of Marine Science College of William and Mary, Gloucester Point, Virginia.

Harding Jr. LW (1994) Long-term trends in the distribution of phytoplankton in Chesapeake Bay: roles of light, nutrients and streamflow. *Marine Ecology Progress Series* **104**, 267–291.

Harding Jr. LW, Degobbis D, Precali R. (1999) Production and fate of phytoplankton: annual cycles and interannual variability. In *Ecosystems at the Land-Sea Margin, Drainage Basin to Coastal Sea,* Coastal and Estuarine Studies 55 (eds Malone TC, Malej A, Harding Jr. LW, Smodlaka N, Turner RE). American Geophysical Union, Washington, D.C., 131–172.

Harding LW, Perry ES (1997) Long-term increase of phytoplankton biomass in Chesapeake Bay, 1950–1994. *Marine Ecology Progress Series* **157**, 39–52.

Hargis Jr. WJ, Haven DS (1999) Chesapeake oyster reefs, their importance, destruction and guidelines for restoring them. In *Oyster Reef Habitat Restoration: A synopsis and Synthesis of Approaches* (eds Luckenbach MW, Mann R, Wesson JA). Virginia Institute of Marine Science Press, Gloucester Point, VA, 329–358.

Haven DS, Morales-Alamo R (1972) Biodeposition as a factor in sedimentation of fine suspended solids in estuaries. In *Environmental Framework of Coastal Plan Estuaries* 18, *Geological Society of America,* 121–130.

Hayes MO, Sexton WJ (1989) IGC Field Trip T371: Modern clastic depositional environment, South Carolina. In *Coastal and Marine Geology of the United States, 28th International Geological Congress.* American Geophysical Union, Washington, D.C., T371, 1–70.

Hedgpeth JW (1977) Seven ways to obliteration: factors of estuarine degradation. In *Estuarine Pollution Control and Assessment, Proceedings of a Conference held at Pensacola, Florida, on February 11–13, 1975.* Environmental Protection Agency Report No. 440/1-77-007B, March 1977, vol. 2, 723–737, 23 ref., Office of Water Planning and Standard, Washington, D.C. Available from the National Technical Information Service, Springfield VA 22161.

Hicks SD (1964) Tidal wave characteristics of Chesapeake Bay. *Chesapeake Science* **5**, 103–113.

Hiltz A (2005) *Logging the virgin forests of West Virginia.* McClain Printing Company, 212 Main Street, Parsons, West Virginia. www.patc.us/history/archive/virg_fst.html

Homer ML, Mihursky JA (1991). Spot. In *Habitat Requirements for Chesapeake Bay Living Resources,* Second edition (eds Funderburk SL, Mihursky JA, Jordan SJ, Riley D). Chesapeake Research Consortium, Inc., Solomons, Maryland, 11, 1–19.

Horwitz MJ (1977) *The transformation of American Law, 1780–1860.* Harvard University Press, Cambridge, Massachusetts.

Ingersoll E (1881) The oyster industry. In *The Tenth Census of the United States.* The History and Present Condition of the Fishery Industries. Department of Interior, U.S. Government Printing Office, Washington, D.C., 156–175.

Ingersoll E (1887) The oyster industry. Part XX. The oyster, scallop, clam, mussel, and abalone industries. In *The Fisheries and Fishery Industries of the United States.* George Brown Goode, Section V, History and methods of the Fisheries II, Government Printing Office, Washington, D.C., 507–564.

Jasinski P, Claggett P (2009) *Responding to climate change in the Chesapeake Bay Watershed.* The Chesapeake Bay Watershed Climate Change Impacts (202d) Report. U.S. Department of Commerce and Department of the Interior's draft report under Section 202d of Executive Order 13508 (EO).

Jaworski NA, Romano B, Buchanan (2007) A treatise: the Potomac river basin and its estuary: landscape loadings and water quality trends 1895–2005. Unpublished. Online. www.umces.edu/president/Potomac/

Jay DA, Geyer WR, Montgomery DR (2000) An ecological perspective on estuarine classification. In *Estuarine Science* (ed. Hobbie JE). Island Press, Washington, D.C., 149–176.

Karlsen TM, Cronin SE, Ishman DA, Willard R, Kerhin R, Holmes CW, Marot M (2000) Historical trends in Chesapeake Bay dissolved oxygen based on benthic foraminifera from sediment cores. *Estuaries* **23**(4), 488–508.

Kemp WM, Boynton WR, Adolf JE, Boesch DF, Boicourt WC, Brush G, *et al.* (2005) Eutrophication of Chesapeake Bay: historical trends and ecological interactions. *Marine Ecological Progress Series* **303**, 1–2.

Kemp WM, Testa JM, Conley DJ, Gilbert D, Hagy JD (2009) Temporal responses of coastal hypoxia to nutrient loading and physical controls. *Biogeosciences* **6**, 2985–3008.

Kennedy VS, ed. (1984) *The estuary as a filter.* Academic Press, Orlando.

Kennedy VS, Breisch LL (1983) Sixteen decades of political management of the oyster fishery in Maryland's Chesapeake Bay. *Journal of Environmental Management* **164**, 153–171.

Kepkay PE (2000) Colloids and the ocean carbon cycle. In *The Handbook of Environmental Chemistry* 5, Part D Marine Chemistry (ed. Wangersky P). Springer-Verlag, Berlin, Heidelberg, 36–56.

Kirby MX, Miller HM (2005) Response of a benthic suspension feeder (*Crassostrea virginica* Gmelin) to three centuries of anthropogenic eutrophication in Chesapeake Bay. *Estuarine, Coastal and Shelf Science* **62**, 679–689.

Kline DL (1998) *Endocrine disruption in fish.* Kluwer Academic Publication, Boston.

Li M, Zhong L, Boicourt WC (2005) Simulations of Chesapeake Bay estuary: sensitivity to turbulence mixing parameterizations and comparison with observations. *Journal of Geophysical Research* **110**, C12004.

MacCleery DW (1994) *American forests: a history of resiliency and recovery, Third edition.* Forest History Society Issues Series, Forest History Society, Durham, North Carolina.

Magnien RE, Summers RM, Sellner KG (1992) External nutrient sources, internal nutrient pools, and phytoplankton production in Chesapeake Bay. *Estuaries and Coasts* **15**(4), 497–516.

Marshall HG, Egerton TA, Burchardt L, Cerbin S, Kokociński M (2005). Long term monitoring results of harmful algal populations in Chesapeake Bay and its major tributaries in Virginia, U.S.A. *Oceanological and Hydrobiological Studies* **XXXIV**, Supp. 3, University of Gdańsk, Institute of Oceanography, 35–41.

Malone TC, Boynton W, Horton T, Stevenson C (1993) Nutrient loading to surface waters: Chesapeake Bay case study. In *Keeping Pace with Science and Engineering.* National Academy Press, Washington, D.C

Malone TC, Conley, DJ, Fisher, TR, Gilbert, PM, Harding, LW, Sellner KG (1996) Scales of nutrient-limited phytoplankton productivity in Chesapeake Bay. *Estuaries* **19**, 371–385.

McCay BJ (1998) *Oyster Wars and the Public Trust: Property, Law and Ecology in New Jersey History.* University of Arizona Press, Tucson, AZ.

McCormick-Ray MG (1998) Oyster reefs in 1878 seascape pattern—Winslow revisited. *Estuaries* **21**, 784–800.

McCormick-Ray J (2005) Eastern Oysters. Historical oyster reef connections to the Chesapeake Bay—a framework for consideration. *Estuarine, Coastal and Shelf Science* **64**, 119–134.

McGlathery KJ, Sundbäck K, Anderson IC (2007) Eutrophication in shallow coastal bays and lagoons: the role of plants in the coastal filter. *Marine Ecology Progress Series* **348**, 1–18.

McLeod K, Leslie L, eds (2009) *Ecosystem-Based Management for the Oceans*. Island Press, Washington, 368pp.

Miller HM (1986) Transforming a "splendid and delightsome land": colonists and ecological change in the 17th and 18th century Chesapeake. *Journal of the Washington Academy of Sciences* **76**, 173–188.

Möbius K (1883) The oyster and oyster culture. In *Report of the Commissioner for 1880*, Part 8, Appendix H, pp. 683–751, U.S. Commission of Fish and Fisheries, Washington, D.C., 358 (Translation by Rice HJ). From the book, *Die Auster und die Austernwirthschaft*, 1877. Verlage von Wiegardt, Hempel and Parey, Berlin, 126pp.

NAS (2001) *Grand Challenges in Environmental Sciences*. Committee on Grand Challenges in Environmental Sciences, Oversight Commission for the Committee on Grand Challenges in Environmental Sciences. National Academy of Sciences, National Academy Press, Washington, D.C., 106pp.

Naylor LA, Viles HA, Carter NEA (2002) Biogeomorphology revisited: looking towards the future. *Geomorphology* **47**, 3–14.

Newcombe CL, Horne WA (1938) Oxygen-poor waters of the Chesapeake Bay. *Science* **88**, 80–81.

Newell RIE (1988) Ecological changes in Chesapeake Bay: are they the result of overharvesting the eastern oyster (*Crassostrea virginica*)? In *Understanding the estuary* (eds Lynch MP, Krome EC). Advances in Chesapeake Bay research, Chesapeake Research Consortium Publ. 129, Gloucester Point, VA, 536–546. Available at: www.vims.edu/GreyLit/crc129.pdf

Newell RIE Kemp WM, Hagy III JD, Cerco CF, Testa JM, Boynton WR (2007) Top-down control of phytoplankton by oysters in Chesapeake Bay, USA: Comment on Pomeroy *et al.* (2006). *Marine* Ecology Progress Series **341**, 293–298.

Nichols MM (1972) Sediments of the James River estuary. In *Environmental Framework of Coastal Plain Estuaries*, Memoir 133 (ed. Nelson BW). The Geological Society of America, Boulder, Colorado, 169–212.

Nichols MM (1994) Response of estuaries to storms in the Chesapeake Bay region. In *Changes in Fluxes in Estuaries: Implications From Science to Management* (eds Dyer KR, Orth RJ). Olsen & Olsen, Fredensborg, 67–70.

Odum WE (1982) Environmental degradation and the tyranny of small decisions. *BioScience* **32**, 728–729.

Odum WE, Odum EP, Odum HT (1995) Nature's pulsing paradigm. *Estuaries* **18**(4), 547–555.

Orth RJ, Marion SR, Moore KA, Wilcox DJ (2010) Eelgrass (*Zostera marina* L.) in the Chesapeake Bay Region of Mid-Atlantic Coast of the USA: challenges in conservation and restoration. *Estuaries and Coasts* **33**,139–150.

Perry JE, Atkinson RB (2009) York River tidal marshes. *Journal of Coastal Research SI* **57**, 43–52.

Potomac Conservancy (2010) State of the Nation's River 2010. Online. www.potomac.org/site/

Pritchard D (1956) The dynamics structure of a coastal plain estuary. *Journal of Marine Research* **15**, 33–42.

Pritchard DW, Schubel JR (2001) Human influences on the physical characteristics of the Chesapeake Bay. In *Discovering the Chesapeake Bay* (eds Curtin PD, Brush S, Fisher GW). The Johns Hopkins University Press, Baltimore, Maryland, 60–82.

Purcell JE, Nemazie DA, Dorsey SE, Houde ED, Gamble JC (1994) Predation mortality of bay anchovy *Anchoa mitchilli* eggs and larvae due to scyphomedusae and ctenophores in Chesapeake Bay. *Marine Ecology Progress Series* **114**, 47–58.

Pyke CR, Najjar RG, Adams MB, Breitburg D, Kemp M, Hershner C, *et al.* (2008) *Climate change and the Chesapeake Bay: state-of-the-science review and recommendations*. A Report from the Chesapeake Bay Program Science and Technical Advisory Committee (STAC), Annapolis, Maryland.

Ray GC (2005) Connectivities of estuarine fishes to the coastal realm. *Estuarine, Costal and Shelf Science* **64**, 18–32.

Ray GC, Hayden BP, McCormick-Ray MG, Smith TM (1997) Land-seascape diversity of the U.S. east coast coastal zone with particular reference to estuaries. In *Marine Biodiversity: Causes and Consequences* (eds Ormond RFG, Gage JD, Angel MV). Cambridge Univ. Press, Cambridge, UK, 337–371.

Reed DJ (2000) Coastal biogeomorphology. In *Estuarine science*, Chapter 13 (ed. Hobbie JE). Island Press, Washington, D.C., 347–361.

Rice KC, Herman JS (2012) Acidification of earth: an assessment across mechanisms and scales. *Applied Geochemistry* **27**(1), 1–14.

Richardson DL (1992) *Hydrogeology and analysis of the ground-water-flow system of the eastern shore, Virginia*. Open-file report 91-490, U.S. Geological Survey, Richmond, Virginia.

Rieser A (2006) Oysters, ecosystems, and persuasion. *Yale Journal of Law & the Humanities* **18** (Suppl. 49), 49–55.

Rooker JR, Secor DH, eds (2005) Connectivity in the life cycles of fishes and invertebrates that use estuaries. Special Issue, *Estuarine Coastal and Shelf Science* **64**(1), 1–148.

Rothschild BJ, Ault JS, Goulletquer P, Heral M (1994) The decline of Chesapeake Bay oyster population: a century of habitat destruction and overfishing. *Marine Ecology* **111**, 29–39.

Ruzecki EP, Hargis Jr. WJ (1989) Interaction between circulation of the estuary of the James River and transport of oyster larvae. In *Estuarine Circulation* (eds Neilson BJ, Brubaker J, Kuo A). The Humana Press, Clifton, New Jersey, 255–278.

Salzman J, Thompson Jr. BH (2010) *Environmental Law and Policy*, Third edition. Thomson Rueters/Foundation Press, New York City, NY.

Secor DH (1999) Specifying divergent migrations in the concept of stock: the contingent hypothesis. *Fisheries Research* **43**, 13–34.

Secor DH, Rooker JR (2005) Editorial: Connectivity in the life histories of fishes that use estuaries. *Estuarine, Coastal and Shelf Science* **64**, 1–3.

Schubel JR, Carter HH (1984) Estuary as a filter for fine-grained suspended sediment. In *The estuary as a filter* (ed. Kennedy VS). Academic Press, Orlando.

STAC (2002) *Chesapeake futures: choices for the 21st Century*. Scientific and Technical Advisory Committee, Chesapeake Bay Program, Annapolis, MD.

Standen EM, Hinch SG, Healey MC, Farrell AP (2002) Energetic costs of migration through the Fraser River Canyon, British Columbia, in adult pink (*Oncorhynchus gorbuscha*) and sockeye (*Oncorhynchus nerka*) salmon as assessed by EMG telemetry. *Canadian Journal of Fish and Aquatic Science* **59**, 1809–1818.

Stevenson CH (1892) The oyster industry of Maryland. *Bulletin of the United States Fish Commission* **12**, 203–297.

Stevenson JC, Kearney MS (1996) Shoreline dynamics on the windward and leeward shores of a large temperate estuary. In *Estuarine Shores* (eds Nordstrom KF, Roman CT). John Wiley & Sons, Chichester, 233–259.

Titus JG, Anderson KE, Cahoon DR, Gesch DB, Gill SK, Gutierrez BT, *et al.* (2009) *Synthesis and Assessment Product 4.1. The First State of the Carbon Cycle Report*. The North American Carbon Budget and Implications for the Global Carbon Cycle Coastal Sensitivity to Sea-Level Rise: A Focus on the Mid-Atlantic Region. Report by the U.S. Climate Change Science Program and the Subcommittee on Global Change Research. U.S. Climate Change Science Program Synthesis and Assessment Product 4.1. January.

Uncles RJ, Stephens JA, Barton ML (1991) The nature of near-bed currents in the upper reaches of an estuary. In *Estuaries and*

Coasts: Spatial and Temporal Intercomparisons, ECS 19 Symposium (ed. Elliott M, Ducrotoy JP). Olsen & Olsen, Fredensborg, Denmark, 43–47.

U.S. Census Bureau (1995) Urban and rural population: 1900 to 1990. Online. www.census.gov/population/www/censusdata/files/urpop0090.txt

Walter III JF, Overton JS, Ferry KH, Mather ME (2003) Atlantic coast feeding habits of striped bass: a synthesis supporting a coast-wide understanding of trophic biology. *Fisheries Management and Ecology* **10**, 349–360.

Ward JE, Newell RIE, Thompson RJ, Macdonald BA (1994) In viva studies of suspension-feeding processes in the Eastern oyster, *Crassostrea virginica* (Gmelin). *Biological Bulletin* **186**, 221–240.

Watts BD, Therres G, Byrd MA (2008) Recovery of the Chesapeake bald eagle nesting population. *Journal of Wildlife Management* **72**, 152–158.

Weinberg M (2003) *A Short History of American Capitalism*. A New History Press, Gloucester, MA, 325pp.

Winslow F (1882) *Methods and results. Report of the oyster beds of the James River, Va. and of Tangier and Pocomoke sounds, Maryland and Virginia*. U.S. Coast and Geological Survey, Appendix No. 11. U.S. Government Printing Office, Washington, D.C.

Winslow F (1882b) *Report of the oyster-beds of the James River, Virginia and of Tangier and Pocomoke sounds, Maryland and Virginia*. Coast-survey report for 1881. *Science* **2**, 440–443.

Zimmerman AR, Canuel EA (2000) A geochemical record of eutrophication and anoxia in Chesapeake Bay sediments: anthropogenic influence on organic matter composition. *Marine Chemistry* **69**, 117–137.

CHAPTER 7

BERING SEA SEALS AND WALRUSES: RESPONSES TO ENVIRONMENTAL CHANGE

G. Carleton Ray

Gary L. Hufford

[Retired, National Weather Service, National Oceanic and Atmospheric Administration], Eagle River, Alaska, USA

Thomas R. Loughlin

[Retired, NOAA National Marine Mammal Laboratory], Redmond, Washington, USA

Igor Krupnik

Arctic Studies Center, National Museum of Natural History, Smithsonian Institution, Washington, D.C., USA

The many threads of evidence, inquiry, and hypothesis . . . pattern themselves into a tapestry of history—the history of the landscape of Beringia, at one time dominated by sea, at another by a great plain; once covered with forest, then with treeless tundra; once populated by mammoth, horse, and bison, hunted by Paleolithic man; now the home of seal, walrus, and polar bear, sought by the world's most skillful sea-mammal hunters.

David M. Hopkins (1967)

7.1 A SHORT HISTORY OF DRAMATIC CHANGE

This chapter addresses the natural histories of Bering Sea pinnipeds (seals, fur seals, sea lions, and walruses), and their responses to human impact and rapid climate change. Aleut, Yupik, Iñupiat, and Chukchi peoples have exploited pinnipeds for millennia, and continue to do so, but in the modern era, climate change has become the dominant issue that affects Bering Sea natural resources and indigenous people.

The value and extent of the Bering Sea became known to the Western world only in the mid-18th century. Peter the Great, Tsar of Russia, appointed Captain Commander Vitus Bering in 1725 to explore unknown eastern shores beyond Russia and to claim resources and lands. On his second expedition Bering reached Alaska, accompanied by German physician-theologian Georg Wilhelm Steller. Steller documented many discoveries and several species that bear his name: Steller sea cow, Steller jay, Steller sea lion, and others. Bering's ship was wrecked on the Commander Islands, where he died in 1741. The news of rich resources spread rapidly among fur traders, and by 1743 the hunt was on, principally for sea otters (*Enhydra lutris*) and northern fur seals (*Callorhinus ursinus*). On Bering Island, Steller described the huge Steller sea cow (*Hydrodamalis gigas*; Fig. 2.2), which hunting crews wastefully exploited for food, leading to its extinction by 1768. A century later in 1867, the United States purchased Alaska from Russia—labeled "Seward's Folly," as the region was thought to hold little value—presenting an opportunity to expand U.S. territory. New England whalers, who had been active in the Bering and Chukchi Seas since the 1840s, rapidly increased their exploitation. And by the 1880s, bowhead whales (*Balaena mysticetus*) were so reduced that whalers turned to Pacific walruses (*Odobenus rosmarus divergens*), depleting them to less than a quarter of their estimated former population (Fay, 1982). Due to loss of these resources, indigenous peoples starved, exacerbated by alcoholism and diseases introduced by the whalers. By 1900, few whaling ships ventured to the Arctic for resources. By the end of the 19th century, the northern fur seal was critically depleted, and the sea otter was presumed extinct. The wasteful slaughter of

Marine Conservation: Science, Policy, and Management, First Edition. G. Carleton Ray and Jerry McCormick-Ray.

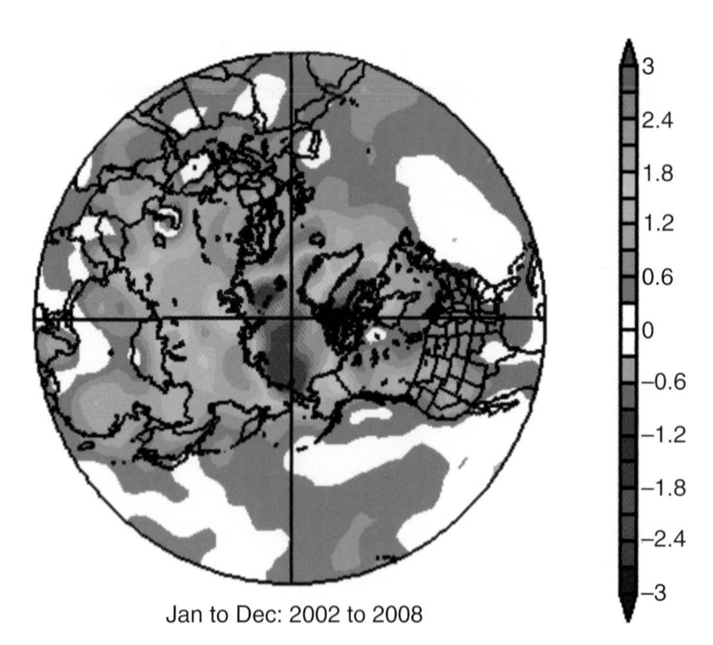

Fig. 7.1 Climate warming has been considerably greater in the Arctic than for any other large region on Earth. The bar on the right indicates increases in degrees Celsius from 2002–8. Image provided by the NOAA/ESRL Physical Sciences Division, Boulder Colorado: at www.esrl.noaa.gov/psd/

these valued species led to the international 1911 *Treaty for the Preservation and Protection of Fur Seals and Sea Otter* (*Fur Seal Treaty*, Ch. 3), initiating a remarkable recovery for both species.

Presently, climate change has introduced a high degree of uncertainty into Bering Sea conservation and management. The changing climate is affecting both marine ecosystems and indigenous peoples, and is occurring more rapidly in the Arctic than for any other large region on Earth (Fig. 7.1). Climate change raises questions about sea-ice diminishment, northward expansion of fisheries, intensified efforts to exploit oil and gas, legislative conflicts, the sustainability of healthy pinniped populations, and the ecological consequences of their changing status. From a socio-political point of view, conservation in the Bering Sea is undertaken against a backdrop of two contrasting, unequally positioned, cultural systems of noncommon antecedence: one founded on historical beliefs and subsistence needs, the other on industry, profits, and national-to-global economies. Fishing in the southern Bering Sea is presently the dominant commercial enterprise. By the 1980s, Alaska's seafood industry produced 40% of the world's salmon supply, and walleye pollock (*Theragra chalcogramma*) has now become the world's largest single-species fishery. Northward expansion of commercial fishing could carry substantial ecological concerns.

7.2 BIOPHYSICAL SETTING

The Bering Sea is characterized by a very high biomass of seabirds and marine mammals—warm-blooded and energy-intensive consumers—in marked contrast to the very high biological diversity of tropical seas (Ch. 8). Here, 29 marine mammals (seven small cetaceans, one more possible; six great whales, three more occasional; nine pinnipeds, one more occasional; sea otter; and polar bear) are full or part-time residents. Among these, pinnipeds are potential indicators of environmental change for reasons of their diversity of habitat choice, their widespread distributions, and their collectively very high consumption rates that demand continuing high ocean production.

The Bering Sea is separated from the North Pacific by the Aleutian Islands, covering 2.3 million km^2, and extending approximately 1500 km from the Aleutians to the 85 km-wide Bering Strait (Fig. 7.2). This sea is almost equally divided into two distinct ecological regimes: a southern deep-ocean basin heavily utilized by commercial fisheries, and a northern, relatively shallow shelf portion characterized by winter-spring sea ice, and not yet heavily fished. The southern basin is up to 3600 m deep and almost totally ice-free year-round. A steep continental margin, incised by seven of the largest submarine canyons in the world, is transitional between the shallow, gently sloping shelf and the deep basin. Volcanic intrusions—Pribilof Nunivak, St. Matthew, St. Lawrence, and the Diomede islands—are important breeding habitats for some of the largest populations of seabirds on Earth. Landward margins include mountainous shores, terrestrial lowlands, extensive wetlands, barrier islands, lagoons, and river deltas. Seasonal river flows influence the timing of river-ice entry and sediment delivery into the sea, as well as migration and spawning of the largest native salmon populations remaining today.

Physical factors heavily influence Bering Sea ecology. Ocean currents bring warm North Pacific waters into and through the Bering Sea (Fig. 7.3a) and directly influence oceanic productivity (Fig. 7.3b). The Bering Sea's climate is made complex by regional to global weather phenomena that determine species distributions. Atmospheric winds are seasonally dominated by the semi-permanent Aleutian Low (Fig. 7.4), which strongly influences storm tracks coming from the North Pacific. Usually, winter storms move eastward along the Aleutian Islands chain and into the Gulf of Alaska, with a

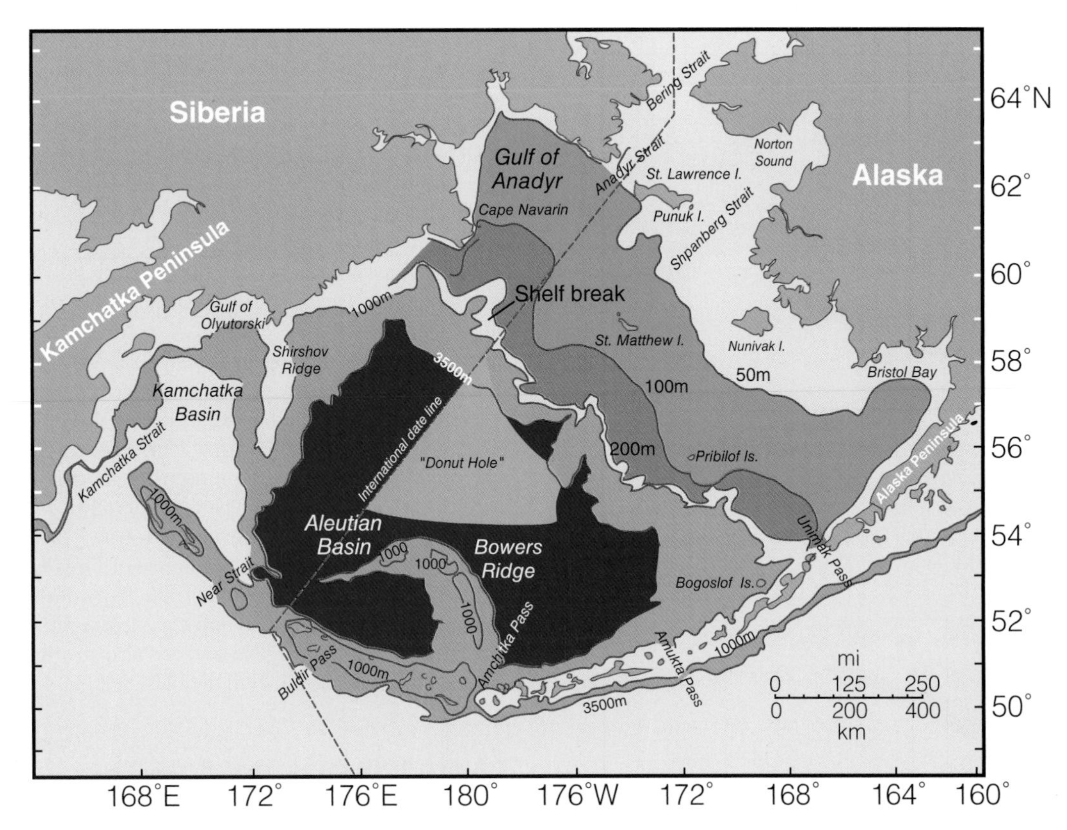

Fig. 7.2 Bering Sea geography and place names. The entire Sea is under U.S. and Russian jurisdiction, with the exception of the international "donut hole" outside U.S. and Russia EEZs. The dashed line indicates the separation of these two nations' jurisdictions at the International Date Line. Shades of blue indicate ecologically distinct domains where physical, chemical, and biotic factors influence reproduction, growth, and survival of marine organisms: light blue = Alaska coastal domain; medium blue = mid-shelf and Anadyr domains; darker blue = outer shelf domain; deep blue = Bering deep domain, surrounded by blue-gray shelf slope domain. From Loughlin and Ohtani (1999).

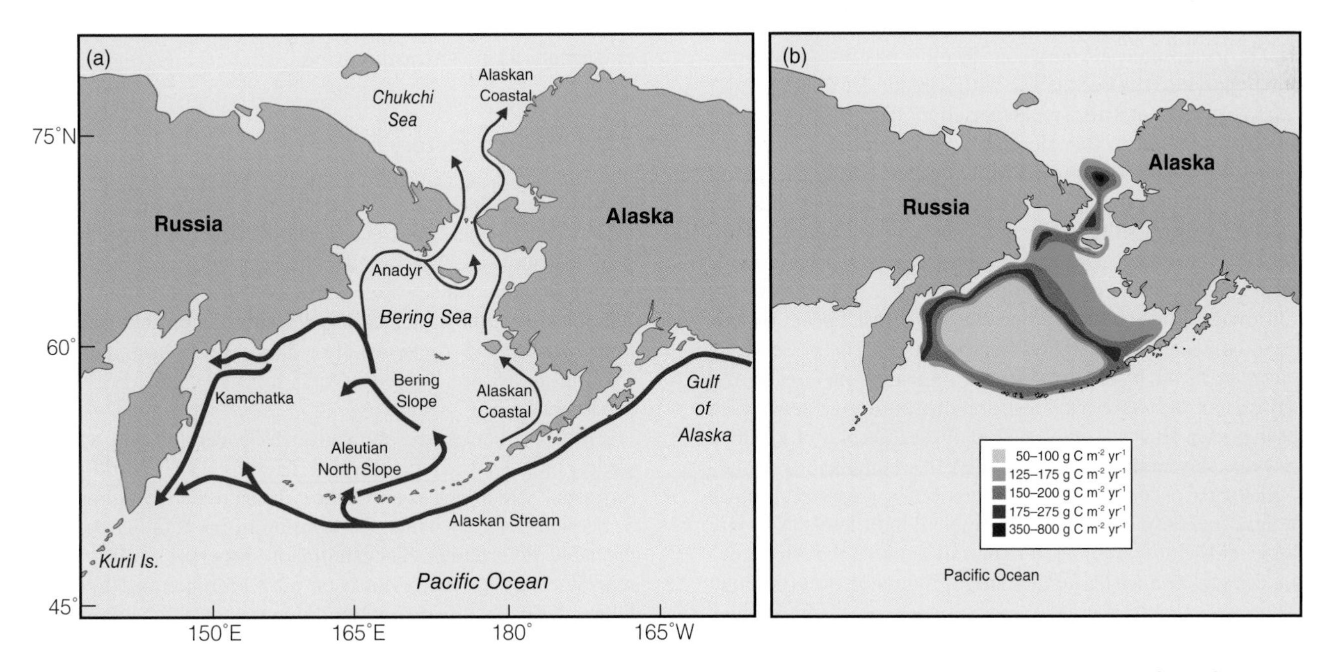

Fig. 7.3 (a) Bering Sea currents. Weak currents in the central and western Bering Sea increase the residence time of warm North Pacific waters, with significant effects on sea-ice melt. (b) Generalized pattern of primary production in the Bering Sea. The "green belt" is the region of high productivity at the shelf edge, with northwest and southwest branches. The northward extension reflects high productivity that is augmented by inputs from large rivers. From Springer *et al.* (1996). Reproduced with permission from John Wiley & Sons.

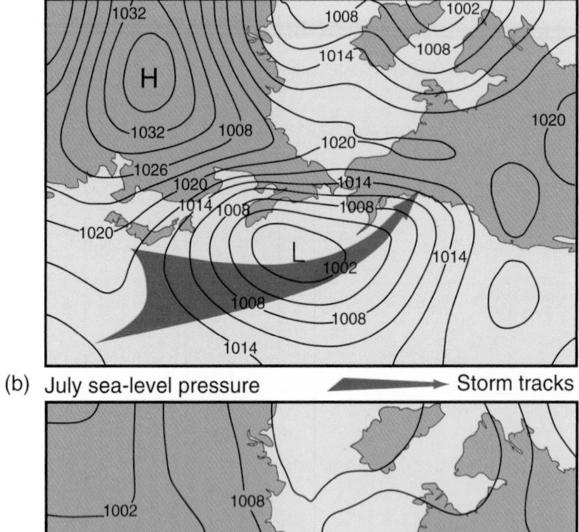

Fig. 7.4 The Aleutian Low (a) strong in winter and (b) weakening in summer. Arrows denote storm tracks. See text for explanation.

frequency of about four to five per month, producing cold outbreaks that are often followed by warm airflow from the North Pacific. In summer, storms generally move eastward and curve northward into the Bering Sea, with a frequency of only about two to three per month. Spatial variations of El Niño can cause an eastward shift in the Aleutian Low, resulting in increased air and seawater temperatures in the eastern half of the Bering Sea, but during La Niña the Aleutian Low appears to become less intense and shifts westward, resulting in cooling. Additionally, the Pacific Decadal Oscillation (PDO) affects shifts between cold and warm regimes: i.e., a cycle of warm waters and mild winters occurs when the northeast Pacific experiences cold ocean waters and harsh winters. The presence of warming North Pacific waters affects sea-ice melting in the central and western Bering Sea, and the southern extent of sea ice in spring. However, explanation of the interplay of El Niño/La Niña, the Aleutian Low, and the PDO remains unresolved.

Biological production is also influenced by seasonal changes in light intensity and duration. Ambient light levels from late fall to mid-winter are too low to permit significant phytoplankton growth, and sea ice further limits production over the shelf. Differences in temperature, salinity, currents, freshwater runoff, seasonal heating and cooling, sedimentation, sea-ice dynamics, wind stress, and horizontal and vertical mixing interact to subdivide the Bering Sea into domains of differing productivities relevant to biotic distributions (Fig. 7.2; Cooney, 1981; Loughlin *et al.*, 1999):

- *Alaska coastal domain*: shore to about 50 m depth; flow northward at 1–5 cms^{-1}; vertically well mixed and influenced by seasonal discharge of freshwater; summer surface temperatures to 14°C; winter water temperatures near freezing.
- *Mid-shelf domain*: approximately 50–100 m depth; little net circulation; surface layer wind-mixed; lower layer (>30 m) tidally mixed; surface temperatures 3–10°C in summer to near freezing in winter; lower layer 1–9°C.
- *Gulf of Anadyr domain*: currents generally strong; influenced by the Anadyr River; summer surface layer to 8–9°C; lower layer 0–2°C.
- *Outer shelf domain*: approximately 100–200 m depth; dominated by north-northwest current flow at 1–5 cms^{-1}; surface layer mixed by winds, temperatures to 9–10°C in summer; lower layer tidally mixed, winter temperatures 2–4°C.
- *Bering deep domain*: depth more than 200 m; greatly influenced by inflow from North Pacific; generally ice-free in winter; surface temperatures usually about 4–6°C.

In sum, the extraordinary abundance of birds and mammals is testimony to the high annual productivity of the Bering Sea. Factors affecting production and marine-mammal distribution are extraordinarily complex, and are also strongly affected by sea ice, which is examined in detail in Section 4.

7.3 MARINE MAMMALS OF THE SOUTHEASTERN BERING SEA

Three pinnipeds (Fig. 7.5) illustrate conservation issues of the southeastern Bering Sea: Steller sea lion (*Eumetopias jubatus*), northern fur seal (*Callorhinus ursinus*, family Otariidae, eared seals), and the harbor seal (*Phoca vitulina*, family Phocidae, "true" seals). Each species uses different portions of the sea to forage and each is affected by both natural and anthropogenic perturbations in contrasting ways.

7.3.1 Natural history

Factors that determine these three species' distributions include atmosphere and ocean climate, predator avoidance, prey distribution, reproductive strategy, and movement patterns among habitats. Avoidance of terrestrial predation might have been an important factor in determining present distributions since most rookeries (breeding areas) and haulouts (resting areas) are located at sites inaccessible to terrestrial predators.

All three species are opportunistic feeders, eating a wide variety of fishes and invertebrates. They are also "central-place foragers," which places energetic constraints on foraging (Ch. 5, Section 5.4.2.5). Thus, distribution of prey and energetics probably determines the extents of dispersal during non-reproductive seasons. Choice of prey is influenced by prey biomass, availability, water depth, degree of association with the bottom, reproductive behavior, degree of aggregation (e.g., solitary vs. schooling), and temporal and spatial distribution patterns. Benthic feeding is a common pinniped strategy, perhaps because the bottom limits prey alternatives for escape. Schooling behavior of prey optimizes energetic costs associ-

Fig. 7.5 (a) Northern fur seals at Pribilof Islands rookery. Photograph by K. Sweeney, NOAA National Marine Mammal Laboratory. (b) Steller sea lions at a rookery in the Aleutian Islands. Photograph by C. Fowler, NOAA National Marine Mammal Laboratory. (c) Pacific harbor seals at a haulout in Southeast Alaska. Photograph by D. Withrow, NOAA Alaska Fisheries Science Center.

ated with search and capture. All three species are the occasional prey of orcas (*Orcinus orca*) and sharks. The likelihood of shark attack is probably greater for Steller sea lions off the Washington, Oregon, and California coasts because of the greater abundance and diversity of sharks than in waters farther north.

7.3.1.1 Steller sea lion

This is the largest eared seal (Fig. 7.5a). Males can attain 1120 kg in weight and 3.25 m in length and are two to three times larger than females (Loughlin, 2009). Steller sea lions range across the North Pacific Ocean rim, from the Kuril Islands to California (Burkanov and Loughlin, 2005), and occupy numerous rookery (Fig. 7.6) and haulout sites. They are not known to migrate. Recent genetic studies of DNA, cell proteins, and morphology have indicated that six groups occur through their range, and that at least two subspecies exist, an eastern and a western one (Fig. 7.7; Phillips *et al.*, 2009).

Female sea lions may nurse their pups until they are four months to two years old, and are generally weaned just prior to the next breeding season. Individual sea lions may range widely. Individuals up to approximately four years of age tend to disperse farther than adults. As they approach breeding age, they have a propensity to stay in the general vicinity of breeding islands, and return to their island of birth to breed as adults. Principal prey of Steller sea lions includes a wide variety of fishes and invertebrates, and food preferences shift with positions along the coast (Sinclair and Zeppelin, 2002; Fig. 7.8). They tend to make relatively shallow foraging dives, with few dives recorded at greater than 250 m. Foraging-trip duration for females in summer is much less (about 24 hours) and covers less distance (average 17 km) than in winter (about 200 hours and 133 km, respectively). Yearling sea lions in winter exhibit foraging patterns intermediate between summer and winter females in trip distance (mean of 30 km), but shorter in duration (mean of 15 hours), with only an average of 1.9 hours per day spent diving (Loughlin *et al.*, 2003).

7.3.1.2 Northern fur seal

Mature males of this moderately sized pinniped attain up to 200–250 kg in weight and 1.9 m in length; males are two to three times larger than females (Fig. 7.5b; Gentry, 2009). During winter, the southern limit of their range extends across the Pacific Ocean from southern California to the Okhotsk Sea and Honshu Island, Japan, north of about 35° N latitude. In spring, most seals migrate north to breeding colonies in the Bering Sea. The largest colonies are on the Pribilof Islands and compose approximately 74% of the world population. Other breeding colonies are on the Commander Islands in the western Bering Sea and on Robben Island in the Okhotsk Sea and support approximately 15 and 9% of the population, respectively. Smaller breeding colonies reside on the Kuril Islands in the western North Pacific, Bogoslof Island in the central Aleutian Islands, and San Miguel Island off the southern California coast.

Northern fur seals are highly migratory and the most pelagic of pinnipeds. From November to March they remain at

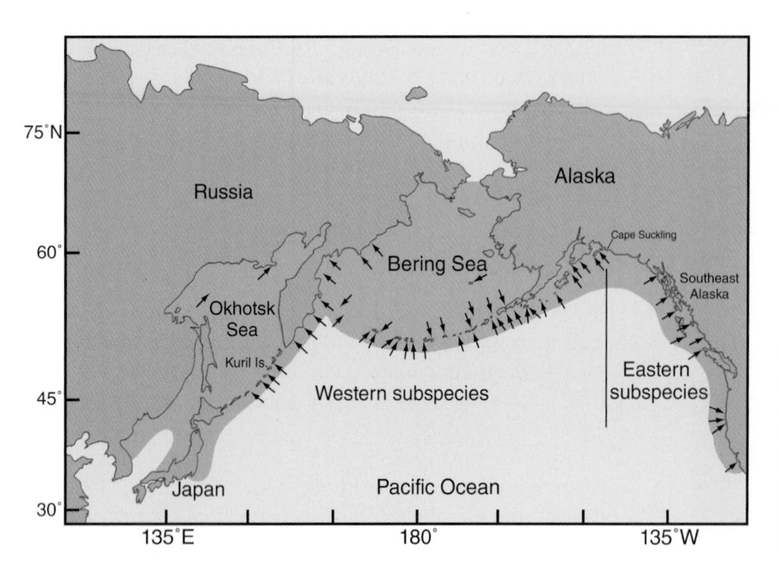

Fig. 7.6 Breeding range of Steller sea lions. The shaded area represents the approximate range of sea lions at sea. Arrows indicate breeding rookeries, some of which are too small to be shown on the map. Haulout areas are too numerous to be shown. The line separating the eastern and western subpopulations is at Cape Suckling, 144° W longitude. Phillips *et al.* (2009) has determined that the eastern and western "stocks" are subspecies.

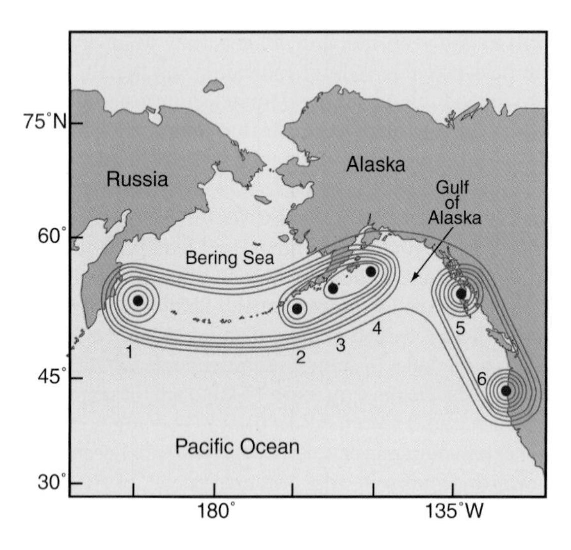

Fig. 7.7 Genetic identity among six grouped localities of Steller sea lions, based on similar mitochondrial DNA haplotypes. The groupings (solid circles) from west to east are: (1) Russia; (2) eastern Aleutian Islands; (3) western Gulf of Alaska; (4) central Gulf of Alaska; (5) southeastern Alaska; and (6) Oregon. Contour lines are drawn at 10% genetic identity and are joined at 30% identity. From Bickham *et al.* (1996).

sea. In March and April they gather along continental shelf breaks and begin to migrate to their respective breeding islands. They typically return to their natal sites to breed. Males come ashore and acquire breeding territories in late May and June. Most pups are born in July, nursed for about four months, and weaned in October or November. Northern fur seals prey primarily on schooling fish and gonatid squid. The species consumed vary with location and season (Call and Ream, 2012; Zeppelin and Ream, 2006). They feed primarily on fishes in continental shelf waters, but beyond the shelf appear to shift primarily to squid. Diving behavior is well studied and shows that females from the Pribilof Islands often dive to 200 m or more, for a maximum of 11 minutes (Gentry, 2009).

7.3.1.3 Pacific harbor seal

Both sexes of this relatively small seal weigh about 90–120 kg, but can weigh as much as 180 kg (Fig. 7.5c). Lengths range from 1.2–1.8 m; males tend to be slightly larger (Burns, 2009). In Alaska, they occur principally in the nearshore zone (Boveng *et al.*, 2003). They use hundreds of sites to rest or haul out along coastal and inland waters, including intertidal sand bars and mudflats in estuaries, intertidal rocks and reefs, sandy, cobble, and rocky beaches, islands, ice floes in fjords and inlets, log-booms, docks, and floats in marine areas. Group sizes typically range from small numbers of animals on some intertidal rocks to several thousand animals that occur seasonally in coastal estuaries. Harbor seals breed and feed in the same area throughout the year and are considered non-migratory. Depending on the region, they typically give birth on shore during a two-week period in spring and nurse their single pup for four to five weeks. After pups are weaned, they disperse widely in search of food. Breeding usually occurs in the water shortly after pups are weaned. Harbor seals commonly prey on many species of fish, squid, octopus, and small crustaceans, usually diving to less than 100 m for about two minutes. However, they are known to dive to >400 m depth and stay submerged >20 minutes (Eguchi and Harvey, 2005).

7.3.1.4 Ecological partitioning

Sea lions, fur seals, and harbor seals overlap in distributions, times spent in the Bering Sea, and items consumed (Fig. 7.9) but do so in different ways, thereby limiting competition for food (e.g., Call and Loughlin, 2005; Robson *et al.*, 2004). Fur seals reproduce on only one island group in the eastern Bering Sea where foraging locations for females often are >200 km from the islands. Time spent on land is only about two days for nursing, then back at sea for 4–10+ days foraging, amounting to approximately eight days a month, or approximately 35

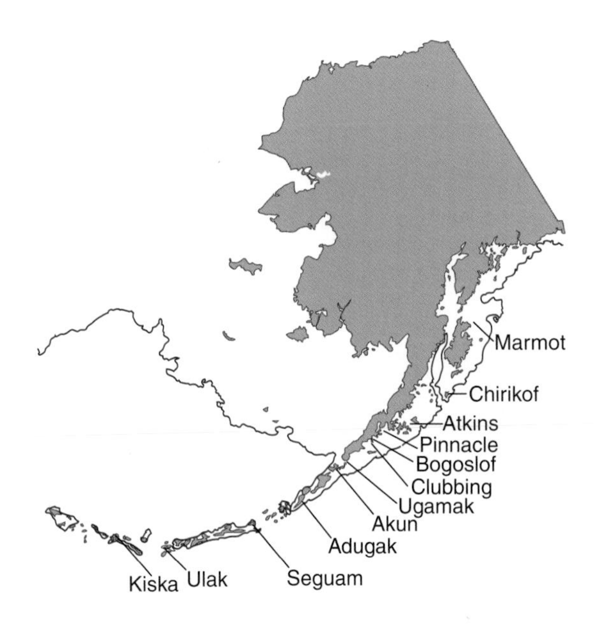

Percent frequency of prey occurrence
P=Pollock; Sa=Salmon; At=Arrowtooth flounder;
PC=Pacific cod; Sl=Sandlance; AM=Atka mackerel;
H=Herring; C=Cephalopods

Area	P	S	At	PC	Sl	AM	H	C
Marmot (64)	69	39	36	5				
Chirikof (74)	69	43	19	9				
Atkins (101)	86	46		6[1]				
Pinnacle (79)	67	71		9	33			
Bogoslof (74)	78	19[2]		1		30		
Clubbing (70)	87	33	26	16				
Ugamak (155)	51	48		3			33	
Akun (58)	36		33	19			55	
Adugak (73)	9[3]	24	73	5				
Seguam (117)		10		3		90		11
Ulak (105)	10		100	1				41
Kiska[4]				16-21		92-95		10-21

1 Sandlance also included
2 Cephalopods and deepsea smelts included
3 Cephalopods and herring included
4 Two sites

Fig. 7.8 Prey of the Steller sea lion. Percent frequency of occurrence of prey items collected from Steller sea-lion scats, June–August, 1990–8. Numbers in parenthesis, column 1, are sample sizes.

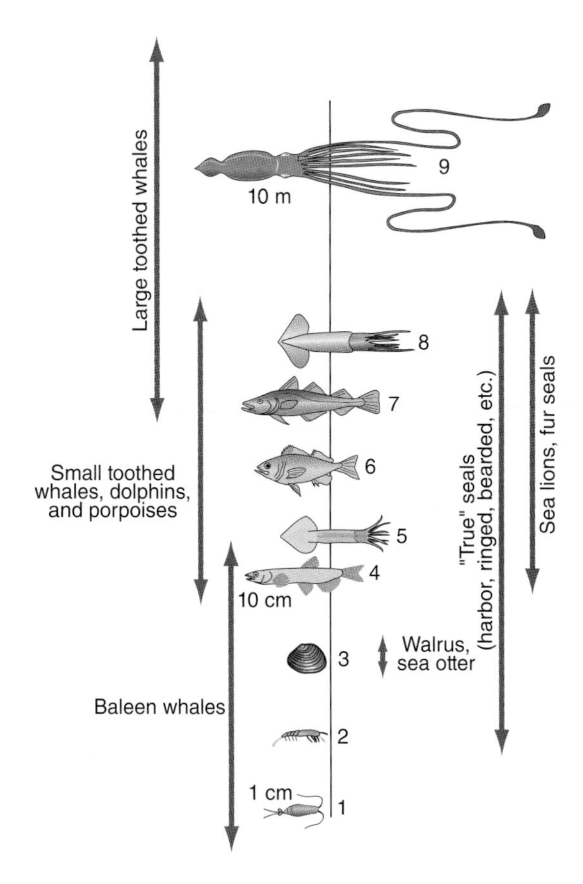

Fig. 7.9 Generalized size ranges of marine mammal prey in the Bering Sea and associations with prey partitioning. Prey size is shown to increase on a logarithmic scale from less than 1 cm to more than 10 m. (1) Calanoid copepods; (2) euphausids; (3) benthic invertebrates; (4) shallow- and mid-water fishes; (5) market squid; (6) benthic fishes; (7) demersal cods and their allies (principally juveniles); (8) benthic and mid-water squids; (9) giant squids and small marine mammals. From Loughlin *et al.* (1999).

days a year that the female spends on land. Adult reproductive males may be on land for more than two months, never leaving until the end of the breeding season, unless they lose their breeding territory to a competitor. Young animals may remain at sea for as long as two years.

Steller sea lions have shorter summer foraging trips than fur seals (<30 km), but extend those trips in winter when not held to shore by a pup. Rookeries are numerous and seem to be as far apart as the distance that females feed from the rookery, without overlapping with those from the nearest other rookery. This suggests that the distance between rookeries might be a result of competition among females. Sea lion pups have evolved to go without food for two days at most, compared to fur seal pups that may go days to weeks without milk. Harbor seals are on the opposite extreme, being more terrestrial, hauling out at each tidal cycle, and spending a large portion of their time resting or just languidly swimming near shore. Their rookeries are numerous and fairly close to one another (20–30 km apart). They do not venture far from shore to forage and tend to stay in the same general areas all year.

These three species also differ in their timing and duration in the Bering Sea. Fur seals are present there for only about six months (June to November). Sea lions are present all year, but are more dispersed in the non-breeding season, venturing farther offshore to feed and for longer periods than in summer and across hundreds of miles, rather than thousands of miles for fur seals. Harbor seals disperse only over small distances.

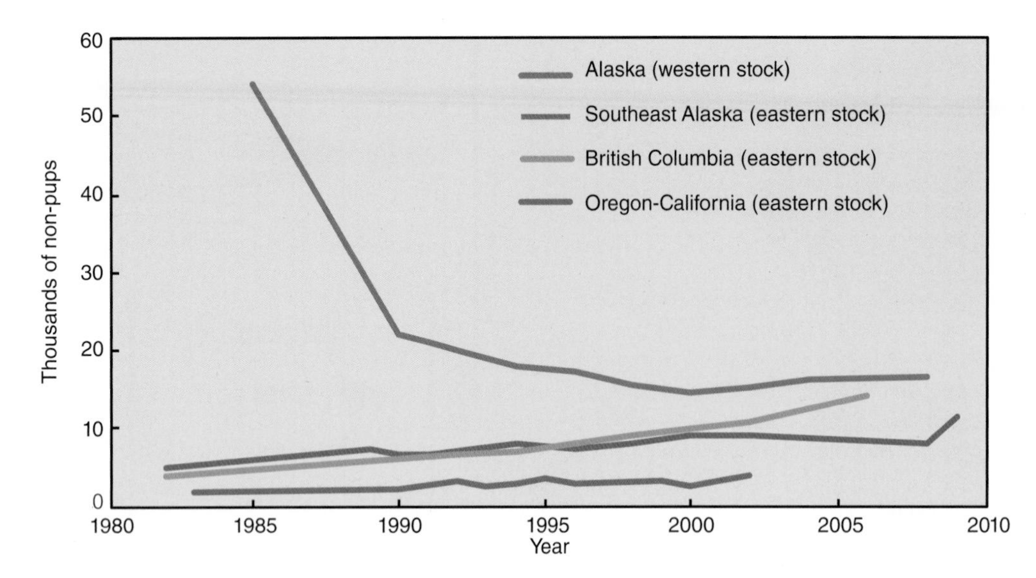

Fig. 7.10 Counts of adult and juvenile Steller sea lions at breeding rookeries and haulout sites in the United States from 1985 to 2010, depicting the dramatic decline of the western subspecies. In 1965, this subspecies was estimated at about 175,000. On the contrary, stocks of the eastern subpopulation have remained at low numbers, but have been relatively stable or increasing slightly. Graph courtesy of NOAA Marine Mammal Laboratory.

7.3.2 Status, trends, and implications of environmental change

All three southern Bering Sea pinnipeds can be counted reasonably accurately while on land, and assessments indicate that all have recently declined overall, although some population segments of each species may be stable or slightly increasing (Pitcher *et al.*, 2007; Simpkins *et al.*, 2003; Small *et al.*, 2003). Steller sea lion numbers in western Alaska are now only a small portion of those a few decades ago (Fig. 7.10). The western subspecies declined about 70% from the late 1950s to the 1990s in some areas; the rate of decline reached about 15% per year during 1985–89, but decreased in the 1990s to 5% annually. Between 2000 and 2004, this subspecies increased at approximately 3% per year—the only period of increase since trend information began to be collected. Results from a 2008–9 survey show that the population is now stable or declining slightly, with considerable regional variability. Therefore, this subspecies is still considered to be at risk of extinction within the next 100 years. In 2010, the National Marine Fisheries Service (NMFS) curtailed commercial fishing for Atka mackerel and Pacific cod in the central and western Aleutian Islands, important sea lion prey in those areas. This ruling is being challenged by the fishermen; the courts will likely be forced to intervene.

Causes for this decline are difficult to determine. Computer modeling and mark-recapture experiments suggest that the likely factor is decreased juvenile survival; lower reproductive success may also contribute. In some Alaskan areas where the diet includes numerous prey species, as in the eastern Aleutian Island area, sea lion numbers have been stable or increasing slightly, but in areas where sea lions primarily depend on one prey item, the population is declining. However, whether population trends are closely associated with diet diversity is equivocal. Possible effects of a declining prey base on Steller sea lions could include increased juvenile mortality, prolonged weaning periods, stunted pup growth, and increased effort to find and capture prey (Fig. 7.11). Other possible causes include disease,

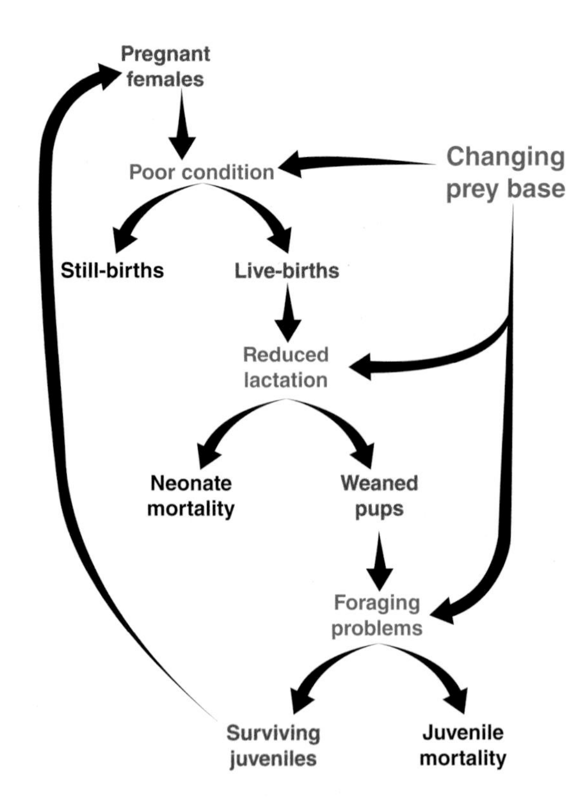

Fig. 7.11 Flow diagram suggesting the possible effects of a declining prey base on Steller sea lions in Alaska. Note that a changing prey base affects body condition, lactation, and foraging, all of which affect survival. From Loughlin and Merrick (1989).

pollution, effects of fisheries, and environmental change. Available evidence is insufficient to demonstrate that disease has played, or is playing, any significant role. The effects of fisheries and synergistic effects of environmental change are areas of intense research to determine causal effects.

Recent counts of eastern Pacific northern fur seals show that the population has declined by about 60% from a historical high of more than two million in the 1970s to about 655,500 in 2009. This decline is likely linked to both ecological and human causes. The Pribilof Islands portion of the population was designated as depleted pursuant to the *Marine Mammal Protection Act* on 17 June 1988 because the seals had declined >50% since the late 1950s; the population has continued to decline since this designation. There is no compelling evidence that northern fur seal carrying capacity has changed substantially since the late 1950s, and data on pup production are equivocal.

Genetic studies on Alaskan harbor seals suggest population subdivisions on a scale of 600–820 km between groups (Westlake and O'Corry-Crowe, 2002; O'Corry-Crowe *et al.*, 2003). Based on these studies NMFS has designated 12 management subdivisions that they term "management stocks." These 12 management stocks extend from the Pribilof Islands in the north, the Aleutian Islands in the west, and Clarence Strait (near British Columbia) in the south. The most current 2001–7 surveys indicate that numbers have declined in the Aleutian Islands and Bering Sea management units (about 9,000 seals total) while numbers elsewhere have increased slightly during the past two decades. The estimated number of harbor seals for all stocks is a little over 150,000, with over 90% of those in the Gulf of Alaska and Southeast Alaska management stocks.

In summary, none of these three pinniped populations is robust. Why these species are declining is unknown, but it is likely that many factors are working together. Eastern subpopulations of Steller sea lions in the central Gulf of Alaska and eastern Aleutian Islands may be slightly increasing, or stabilizing, but the decline in the western Aleutian Islands is severe. The fur seal story is complex in that they had been commercially exploited until the early 1980s, were subject to mortality in high seas salmon gill-net fisheries during the 1970s and 1980s, and for unknown reasons have declined in abundance at the Pribilof Islands. Very little attention has been given to harbor seals, but scientists and the public are noticing that there are not nearly as many as in the recent past.

These southern Bering Sea pinnipeds almost certainly have lived through many regime shifts in the two to three million years of their existence. What may be different about this most recent reduction in numbers of all three species is the coincident development of extensive fisheries targeting the same prey that sea lions, fur seals, and harbor seals depend on for food. Fisheries in the Bering Sea expanded enormously during the 1960s and 1970s. The existence of strong environmental influences, such as climate change, could also increase the sensitivity of sea lions, fur seals, and harbor seals to fisheries effects, or to changes in those ecosystems resulting from fisheries.

7.4 ICE-DEPENDENT PINNIPEDS OF THE NORTHERN BERING SEA

Unlike southern pinnipeds, five Beringian "pagophilic" (sea-ice dependent) pinnipeds depend on sea ice as habitat for reproduction, nursing, molt, and rest (Fig. 7.12): Pacific walrus (*Odobenus rosmarus divergens*), ribbon seal (*Histriophoca fasciata*), spotted seal (*Phoca largha*), bearded seal (*Erignathus barbatus*), and ringed seal (*Phoca hispida*). These species partition habitats during their winter-spring reproduction periods according to the character of the pack ice (Burns, 1970; Burns, 1981; Fay, 1974; Braham *et al.*, 1984; Lentfer, 1988; Ray and Hufford, 1989; Ray *et al.*, 2010). Reproduction in winter-spring depends on the synchronous timing, structure, and extent of sea ice. However, climate change is altering these sea-ice characteristics, making interpretation of sea-ice/habitat relationships difficult, particularly with respect to the phenological relationships of reproductive behavior.

7.4.1 Natural history

Natural history is key to understanding the importance of sea ice to all species of pagophilic pinnipeds for at least three reasons. First and foremost, sea ice provides reproductive habitat. Second, moving ice enlarges the ocean area over which these species feed. Third, all species are adapted to specific sea-ice "seascape" structure (floe associations, thickness, ridge formation, etc.). Critical to these is phenology (Ch. 5, Section 5.4.2.5); the time of the animals' birthing and mating must match sea ice formation and growth to be successful. That is, the duration of sea ice must be long enough to support walrus mothers and their calves during their northerly migration, and for pup seals to nurse and molt before taking to the water to feed.

7.4.1.1 Pacific walrus

Walruses (Fig. 7.12a,b) are circumpolar in distribution. Two subspecies are differentiated (Fay, 1982). The Atlantic walrus (*O. r. rosmarus*) ranges from the northwest Atlantic to the seas off central Siberia; the Pacific walrus is Beringian (shelf areas of the Bering, Chukchi, and East Siberian seas; Fig. 7.13). Pacific walruses historically occurred as far south as Unimak Pass, east to Southeast Alaska, and west to the tip of the Kamchatka Peninsula. Presently, they are confined to seasonal sea ice in winter and coastal haulouts (males in summer). Field observations indicate that they mainly occupy areas dominated by thick, angular, often-ridged, and moderately sized floes separated by intersecting leads (long openings) and lake-like openings called "polynyas" (Ray and Hufford, 1989; Ray *et al.*, 2010). New ice in leads and polynyas is tolerated as walruses can break ice up to 20 cm thick.

Walruses are among the most gregarious of mammals. Herds are composed of many groups that may collectively number in the thousands (Fig. 7.14). Typically, walruses are concentrated in two subpopulations in the north-central and southeastern Bering Sea from January through April to reproduce (Fig. 7.13). As sea ice disintegrates and retreats northward, walruses generally migrate with it, but should the ice reverse direction they may leave it periodically and swim north. By July, almost all females with newborn young, juveniles, and a few mature males occupy the marginal-ice zone of the eastern and western Chukchi Sea. Most mature males, however, move to coastal haulouts for the summer. From October through December, the entire population migrates

Fig. 7.12 Ice-dependent pinnipeds: (a) Two bull walruses; note the very thick, lumpy skin about the neck and shoulders. (b) Female walruses sheltering newborn calves. (c) Bearded seal. Photograph © E. Labunski. (d) Spotted seal female and pup. (e) Ringed seal, Photograph © E. Labunski (f) Female ribbon seal with pup. (g) Male ribbon seal. Other photographs © Ray & McCormick-Ray.

back to Bering Sea ice as it is forming. The sequence of ice formation (Section 7.4.2.1) is critical to this migratory pattern. Should sea ice vary in distribution, walruses would be expected to vary with it; therefore, due to varying ice conditions from year to year, their general distribution could potentially span most of the Bering Sea shelf area.

Male walruses reach 350 cm in length and 1700 kg in weight; females are a third smaller. Both sexes have formidable tusks, which help protect them from predation, principally by polar bears. Tusks also are used to assist hauling out on ice, as an anchor while resting (Fig. 7.15), and for sexual, dominance display, but not for digging for food, as has sometimes been assumed. Mating occurs in January through March in sea-ice environments. Mature males in the water engage in ritualized "song" displays that appear to establish male acoustic territories (Fig. 7.16). Walruses have among the lowest of mammalian reproductive rates. Delayed egg implantation and gestation occupy about 15 months; thus, the maximum reproductive rate is only one calf every two years per adult female at most (Fay, 1982). Maternal care is intensive; females provide body warmth to vulnerable calves (Fig. 7.12b) and closely guard them; calves may remain with their mothers for up to two years. Natural mortality of individuals more than one year old is very low, probably around 1% per year. Polar bears and orcas are their only natural predators at sea, but on the Alaska Peninsula brown bears are predators.

Walruses forage on and in sediments, in water depths rarely greater than 100 m. Their diet consists of a wide variety of benthic invertebrates. They appear to favor large, deeply buried clams (Fay *et al.*, 1984a); they detect food with their vibrissae and lips as they move forward along the bottom powered by their rear flippers (Fig. 7.15). Organisms in the sediment are rooted out, much in the manner of pigs rooting in soil. Biomass consumption indicates that walrus feeding may also have substantial ecological effects through bioturbation (Section 7.5), which may qualify them as a "key" foundation species (Ch. 5; Fay *et al.*, 1984b; Ray *et al.*, 2006).

7.4.1.2 Bearded seal

The bearded seal occurs widely in the Arctic and is the largest Arctic, ice-dependent seal (Kovacs, 2009; Fig. 7.12c). It occurs singly or in small groups in pack ice during winter and spring, but most frequently occurs in similar areas as do walruses (Ray

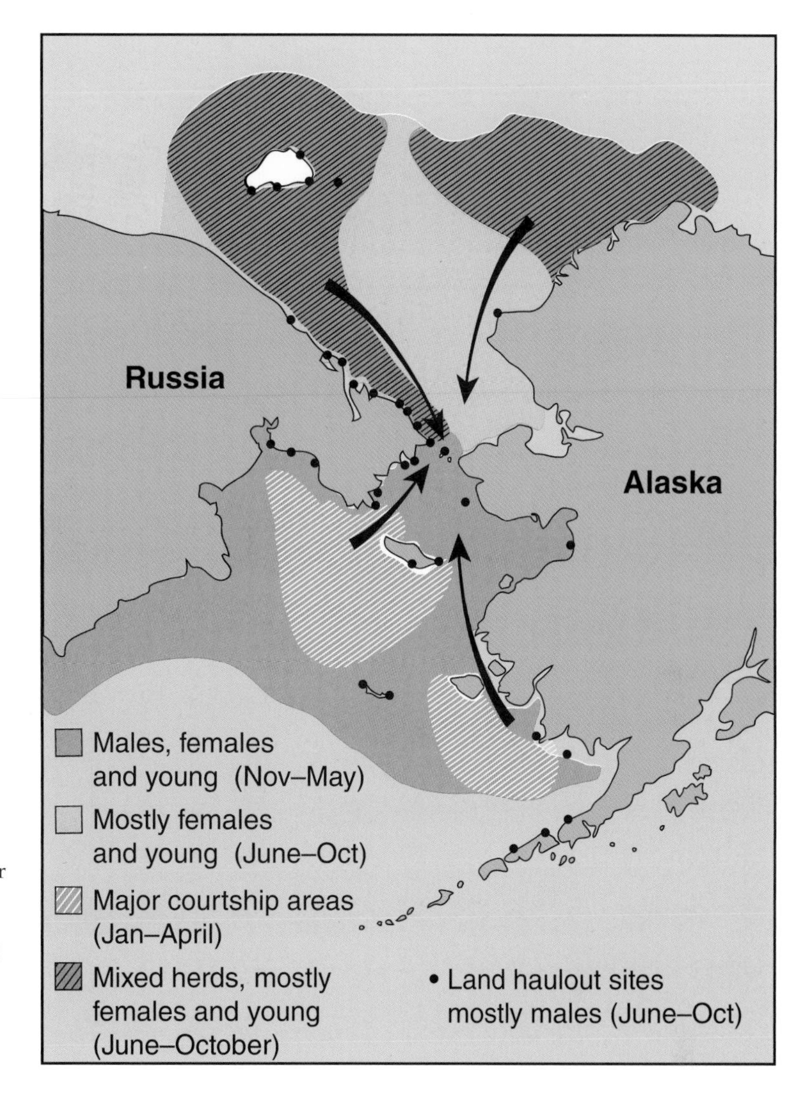

Fig. 7.13 Seasonal distribution of Pacific walruses prior to the 21st century. Two breeding concentrations occur in the Bering Sea in winter, west-central and southeast. The sexes then disperse differently in spring; females and young to the Chukchi Sea, males mostly to land haulout areas in the Bering and Chukchi seas. Diminishment of sea ice is changing these patterns. See text for explanation. From Fay (1982).

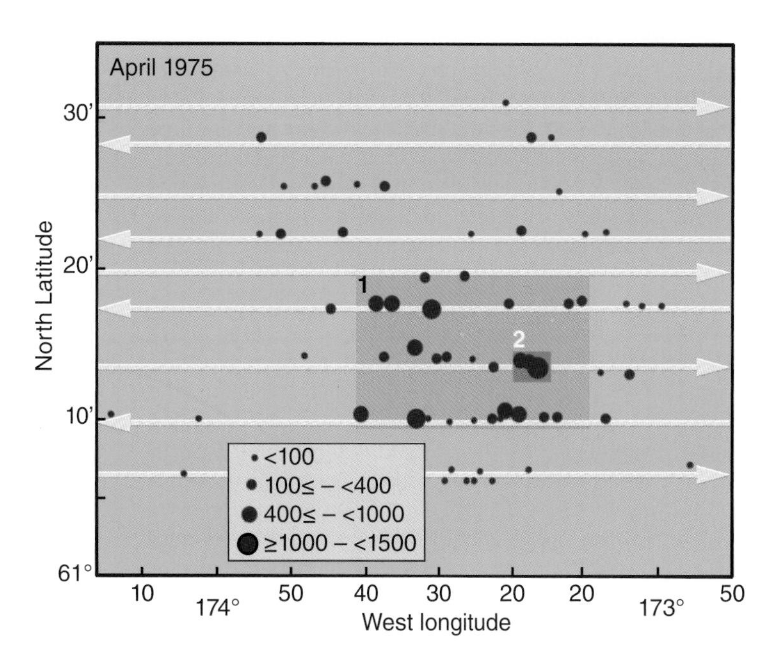

Fig. 7.14 On-ice walrus concentration in broken pack in the south-central Bering Sea, recorded April 1975 from NASA flights at 300–450 m altitude. Arrows indicate directions of flight lines. Box 1 encloses the bulk of the herd and Box 2 encloses two particularly large groups. Sizes of filled circles represent estimated numbers of animals visually observed. From Ray and Wartzok (1980).

Fig. 7.15 Depiction of walrus natural history based on observations of living animals at the New York Aquarium and from submersible diving operations in the Bering Sea in April 1972 off the icebreaker *Burton Island*. Individual walruses indicate swimming propelled by side-to-side swipes of the rear flippers, use of tusks for "anchoring" in sea ice, and a furrow due to walrus' benthic bioturbation. A small group of females (upper right, rear) is hauled out on sea ice before a sexually displaying male (not evident in painting). Painting by Robert E. Hynes/National Geographic Stock from Ray and Curtsinger (1979), with permission of the National Geographic Society Image Collection.

et al., 2010), and can break new or very thin ice (Burns, 1981). Males average 220 cm long and 250 kg; females are slightly larger. In winter-spring, males "sing" in reproductive arenas where ice floes surround leads and small polynyas (Ray *et al.*, 1969). In late March through April, pups are born without lanugo hair and are precocial, taking to water soon after birth. Nursing extends only 12–18 days, following which pups are abandoned. Precocial birth and short nursing time are, presumably, adaptations for avoiding polar bear predation. Bearded seals are benthic feeders that forage on crustaceans and mollusks in water depths up to about 130 m, tending to overlap with walruses in food consumed.

7.4.1.3 Spotted seal

The spotted seal is a coastal species except when breeding (Burns, 2009; Fig. 7.12d). It is superficially similar to the harbor seal, but is larger and generally paler with a spotted pattern. Males reach about 135 kg in weight and 170 cm in length; females are slightly smaller. Mating pairs form in March and stay together as a "family" (Fig. 7.23) through pup birth in March to early April and until the pup is weaned; they then mate. They tend to occupy floes of the outer, marginal portion of loose pack near the southernmost ice extent where floes are thin and dispersed and where polar bears rarely venture (Ray *et al.*, 2010). They often occur with ribbon seals but are generally more abundant easterly, i.e., from west of St. Matthew to outer Bristol Bay, Alaska. Pups often take shelter among pressure ridges in late March to mid-April, and rarely voluntarily enter the water while being nursed for three to four weeks. After weaning, pups remain alone on remnant sea ice until their white coat is shed and new hair is grown four to six weeks later. Their diet includes fishes, small crustaceans, and cephalopods. At all times of year, except during pupping, they may occur in concentrations of several dozen individuals.

7.4.1.4 Ringed seal

This smallest of Beringian seals is distinguished by its distinctive ringed pattern (Fig. 7.12e). Females attain 50–100 kg and 115–40 cm in length; males are slightly larger. Ringed seals occur throughout the Arctic and into adjacent icy seas (Hammill, 2009). They commonly occur in a heavy continuous sea ice (Ray *et al.*, 2010), being uniquely able to maintain breathing holes in ice of up to 2 m thick. Ringed seals are generally solitary, except when with a pup. Individuals are most concentrated in nearshore ice, but are most numerous over large areas of offshore, thick, semi-continuous sea ice (Fig. 7.23). They take advantage of large, flat, ice floes in spring and construct subnivean birth lairs in shorefast ice and pres-

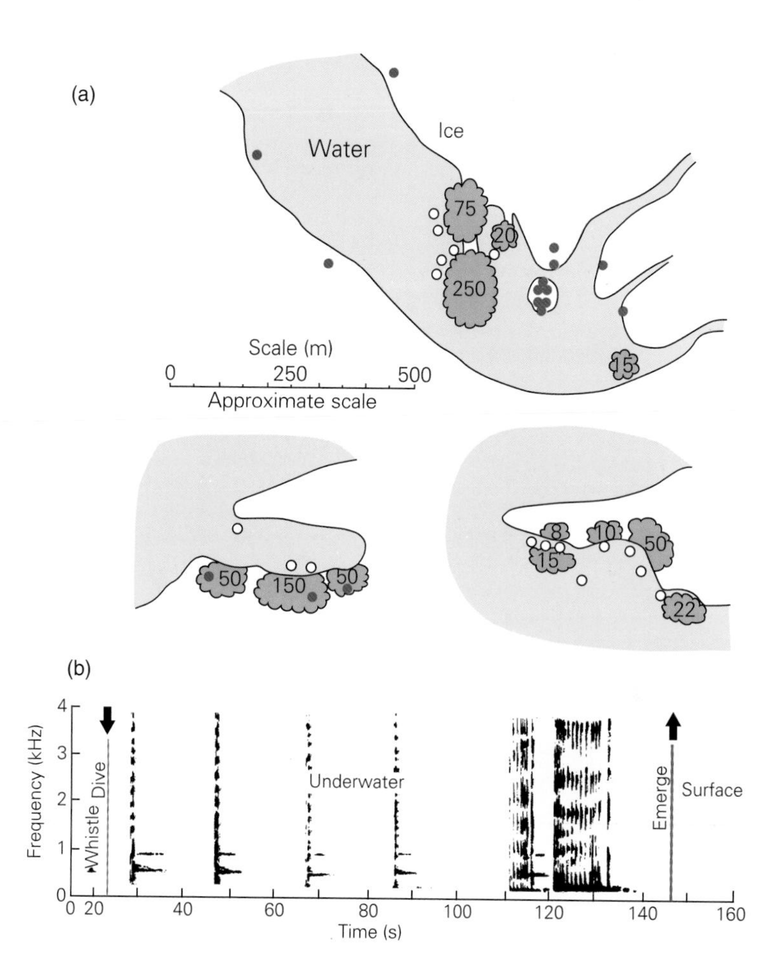

Fig. 7.16 Walrus lek and sonogram. (a) Pacific walrus reproduction patterns observed in the north-central Bering Sea in April 1972. Blue areas are water; white areas are sea ice. Wavy lines (brown) enclose female groups, with numbers of individuals indicated. Open circles represent bulls active in water; closed circles are inactive bulls on ice. (b) Sonogram of a male "song": a light whistle, emitted at the water's surface, is followed by a series of bell tones, emitted underwater, then a series of knocks, ending with a coda of rapid knocks, followed by surfacing. This sequence may be repeated for hours before a group of females on ice. From Fay *et al.* (1984b).

sure ridges to protect pups from polar bears. Pups are born in late March through April. Females nurse pups for up to seven weeks. Pups usually do not enter the water until they lose their lanugo hair in two to four weeks. Adults mate after weaning. Ringed seals have an extremely varied diet, including small fishes, crustaceans, and other invertebrates.

7.4.1.5 Ribbon seal

This most unmistakable and strikingly patterned seal is among the least known of all pinnipeds (Lowry and Boveng, 2009; Fig. 7.12f,g). Males reach 140kg in weight and 180cm in length; females are slightly smaller. Ribbon seals occupy loose pack, often with spotted seals, but are most abundant from a bit east of St. Matthew, westward into the Gulf of Anadyr and south along the Kamchatka coast. They are pelagic for most of the year, almost never coming onto land. In winter-spring, ribbon seals occupy the inner loose pack (Ray *et al.*, 2010) on floes of varying thickness, concentration, shape, and size; they appear to prefer fairly thick, often-ridged, snow-covered floes (Burns, 1981; Fedoseev, 2002). Their life history is similar to spotted seals during reproduction, but they do not form family groups. During March, ribbon seals gather in loose aggregations on sea ice to reproduce. They frequently occur on heavy floes of remnant ice (Box 7.1), where the marginal-ice seascape shows evidence of wave action and collisions. Pups are born from mid-March through early April. Nursing lasts for about three weeks and molting is complete by early July, at which time they become exclusively pelagic. Ribbon seals primarily consume fishes, small crustaceans, and cephalopods.

7.4.2 "Seascape" and habitat partitioning

Sea ice is most critical for pagophilic pinnipeds during their winter-spring reproductive periods. Habitat partitioning at that time has been strikingly illustrated by Braham *et al.* (1984; Fig. 7.17). Although those authors did not specifically refer to sea ice, the different patterns shown by walruses and seals strongly suggest that sea ice may be differentiated as habitat "seascapes" (Ray *et al.*, 2010) following principles of landscape ecology (Wu and Hobbs, 2007). Significantly, natural scaling properties of sea ice have also become better understood, and appear closely to match the interacting scales that determine pinniped/sea-ice habitat relationships (McNutt and Overland, 2003).

7.4.2.1 Sea-ice formation and habitat relationships

Arctic sea ice occurs across the Arctic as two major types: seasonal ice that forms each year and melts in summer, and multi-year ice that persists for more than one year. In winter,

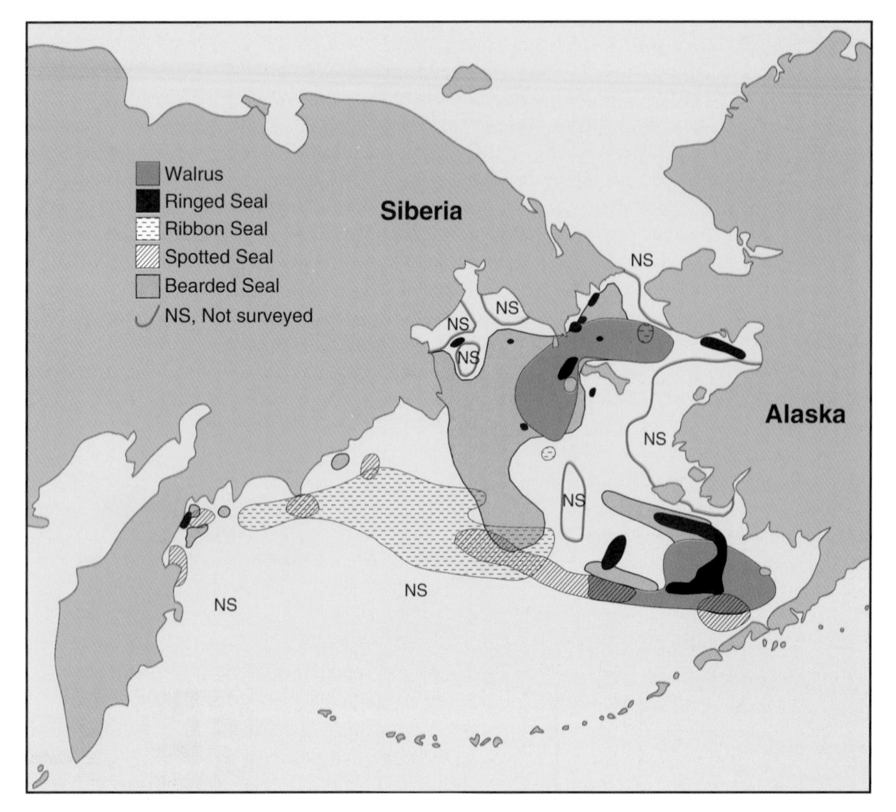

Fig. 7.17 Habitat partitioning among ice-dependent pinnipeds in winter. Each species is indicated to occupy discrete areas of the Bering Sea (assumed to correspond to seascape types): walruses in north west-central and southeastern areas; bearded seals over large areas of sea ice; spotted and ribbon seals in loose pack, with ribbon seals more westward; ringed seals patchily throughout in suitable ice conditions. NS = not surveyed. Modified from Braham *et al.* (1984).

seasonal sea ice normally covers about 75% of the Bering Sea shelf. Multi-year ice only occasionally occurs south of Bering Strait as intrusions from the Chukchi Sea. In summer, sea ice retreats into the Chukchi Sea, leaving the Bering Sea typically ice-free. This seasonal ice advance and retreat is the most extensive for any Arctic region.

Beringian sea ice is first formed along the coasts of the Chukchi and Bering seas in late October and extends farthest south by mid-March (the "climatological norm," see Fig. 7.24a,c,e). New ice is formed in a southward-moving "conveyor belt" over the shelf, where it meets warmer water and melts near the shelf-break (Pease, 1980). By April, dominant northeasterly winds decrease and atmospheric and oceanic temperatures begin to rise. Ocean surface currents, solar insulation, and winds then force the melting ice slowly northward through the Bering Strait. By the end of June only remnant ice remains in the northern portions of the Bering Sea.

Sea-ice formation and dynamics are best understood at multiple scales (McNutt and Overland, 2003; Fig. 7.18), each of which reflect sea-ice properties specific to marine-mammal natural histories. At local scales, individual floes coincide with species' floe preferences, characterized by such adjectives as "closed" vs. "open" pack, or "new," "ridged," or "heavy" ice. At intermediate (meso-) scales, sea ice behaves more as a plastic continuum, governed primarily by fracture mechanics and resulting in larger-scale "seascape" patterns equivalent to marine-mammal habitats; seascapes move more slowly, and can move as larger units according to how closely floes may be associated together. At the regional scale, sea ice responds to external atmospheric and ocean forcing on weekly and longer time scales, and is characterized by extent (how far north or south the ice extends), cover (how much of the region is covered), roughness (e.g., ridges), and mean thickness (mass). The regional scale coincides with the "range" over which each species may occur.

More precisely, at the seascape-scale, variable wind and current conditions result in six distinct seascape formations in winter-spring that have, until recently, been observed to be relatively consistent and predictable in timing and location (Fig. 7.19). "Broken pack" occurs in the central Bering Sea where thick, continuous ice is broken by oceanic swells that penetrate far into the pack and where leads and polynyas are frequent. "Loose pack" occurs at the southern extent of the pack and is particularly affected by atmospheric and oceanic conditions at the sea-ice margin. "Pack-ice-with-leads" occurs to the northwest and into Gulf of Anadyr, an area constrained by the land basin, and characterized by parallel leads. "Rounded pack" occurs where northward-moving currents confront southward-moving ice, resulting in very thick, heavily ridged, convergent floes. "Continuous pack" occurs near the Bering Strait region where the narrow strait concentrates floes and causes continuous stresses that form large pressure ridges. "Large polynyas" occur both within the pack and adjacent to land masses, according to wind conditions, which can rapidly change their extent. These six seascape types define potential winter-spring habitats within which pagophilic pinnipeds are patchily distributed in time and space, or conversely, unfavorable areas that they avoid. That walruses mostly occur in the same area as "broken pack" is indicated by historical records (Fay, 1982; Fig. 7.20). A principal component analysis of

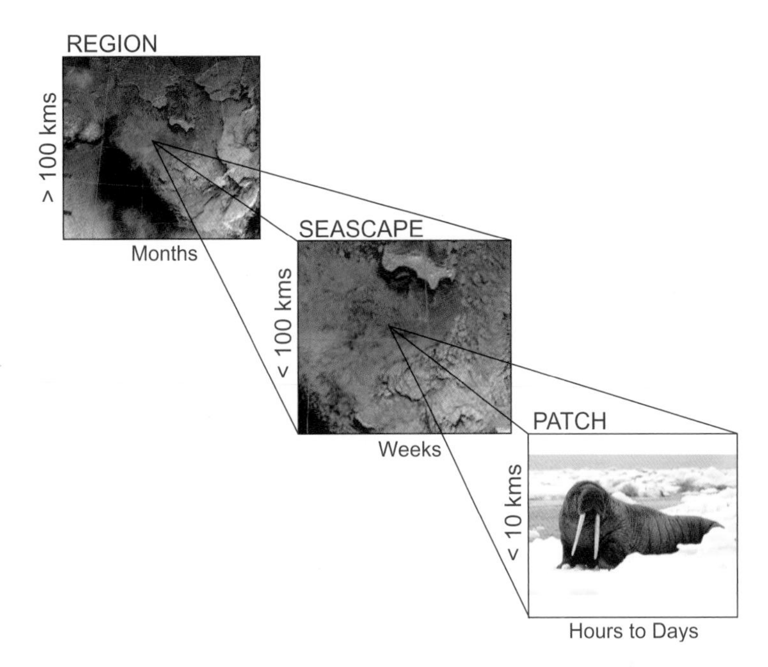

Fig. 7.18 A hierarchical relationship of regional sea-ice cover, sea-ice type, and local walrus natural history. The regional scale reflects the general range of the species; the seascape scale is specific to habitat options, i.e., sea-ice "seascapes" (Fig. 7.19); and the local scale is appropriate for species' floe-type preferences.

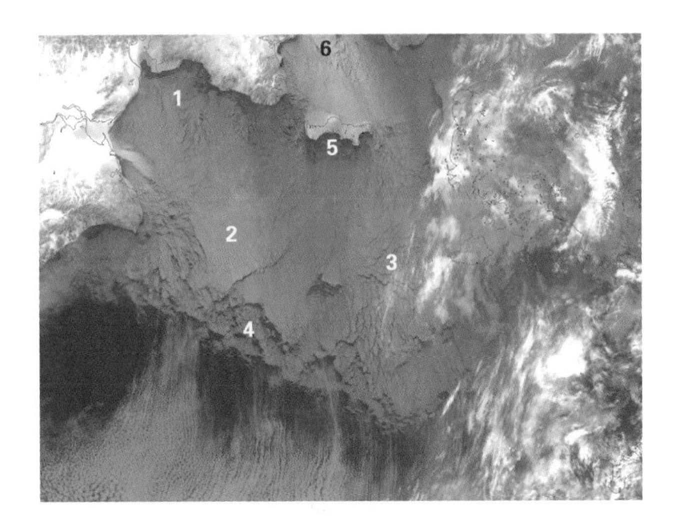

Fig. 7.19 Historical winter-spring seascape types exhibit distinct characteristics important to sea-ice dependent mammals in the Bering Sea. NOAA AVHRR (Advanced Very High Resolution Radiometer) infrared image of 30 March 1988: (1) broken pack; (2) loose pack; (3) pack ice with leads; (4) rounded pack; (5) continuous ice; (6) polynyas. See text for explanation. From Ray and Hufford (1989).

walrus winter distributions is consistent with this broken-pack association (Fig. 7.21). The extensive, north–south occurrence of rounded pack seems to divide St. Lawrence from Bristol Bay walruses (see Fig. 7.13). Observations of walruses on ice from icebreakers add additional support for these two subpopulations (Fig. 7.22). These findings agree with Native hunters' observation of "two waves" of walruses passing St. Lawrence Island during spring migration.

As for seals, repeated field observations confirm that spotted seals and ribbon seals consistently occupy loose pack. However, bearded and ringed seals seem not to be strongly associated with any particular seascape, but rather are more sensitive to local conditions within various seascapes. A combination of ship observations combined with regional satellite imagery enables "scaling up" from local field observations to the seascape-scale and allows testing of species/seascape relationships during critical reproductive periods (Fig. 7.23) and particularly under future scenarios of climate change.

7.4.2.2 Seascape trends and the "mixing bowl"

Changing climate (Fig. 7.1) is currently causing highly variable sea-ice conditions. Later onset of winter ice formation and earlier spring breakup have shortened the sea-ice season by 6–8 weeks (Walsh, 2008). Although sea-ice cover (the total area covered by sea ice) has diminished, regional sea-ice extent (how far north or south the sea ice occurs) seems not to be a significant factor in overall sea-ice cover. Variable wind conditions can create large polynyas due to southerly shifts of sea ice, but with little significant change of cover, i.e., total sea-ice habitat available (Fig. 7.24).

Sea-ice changes are further exacerbated by increased spring freeze-thaw episodes that create greater open-water areas, where winds can shift seascape patterns into a complicated "mixing bowl" of sea-ice types (Box 7.1). Thus, floes become disassociated and seascapes become less cohesive and consistent. Floes have been observed to accelerate one day and move little the next without a clear triggering mechanism. What is clear, however, is that floes can move more independently when floes are dispersed than when they are concentrated. This is because, when floes are closely packed, they are forced to move together, but when disassociated, their movements depend on their different amounts of above-water "sail" or submerged

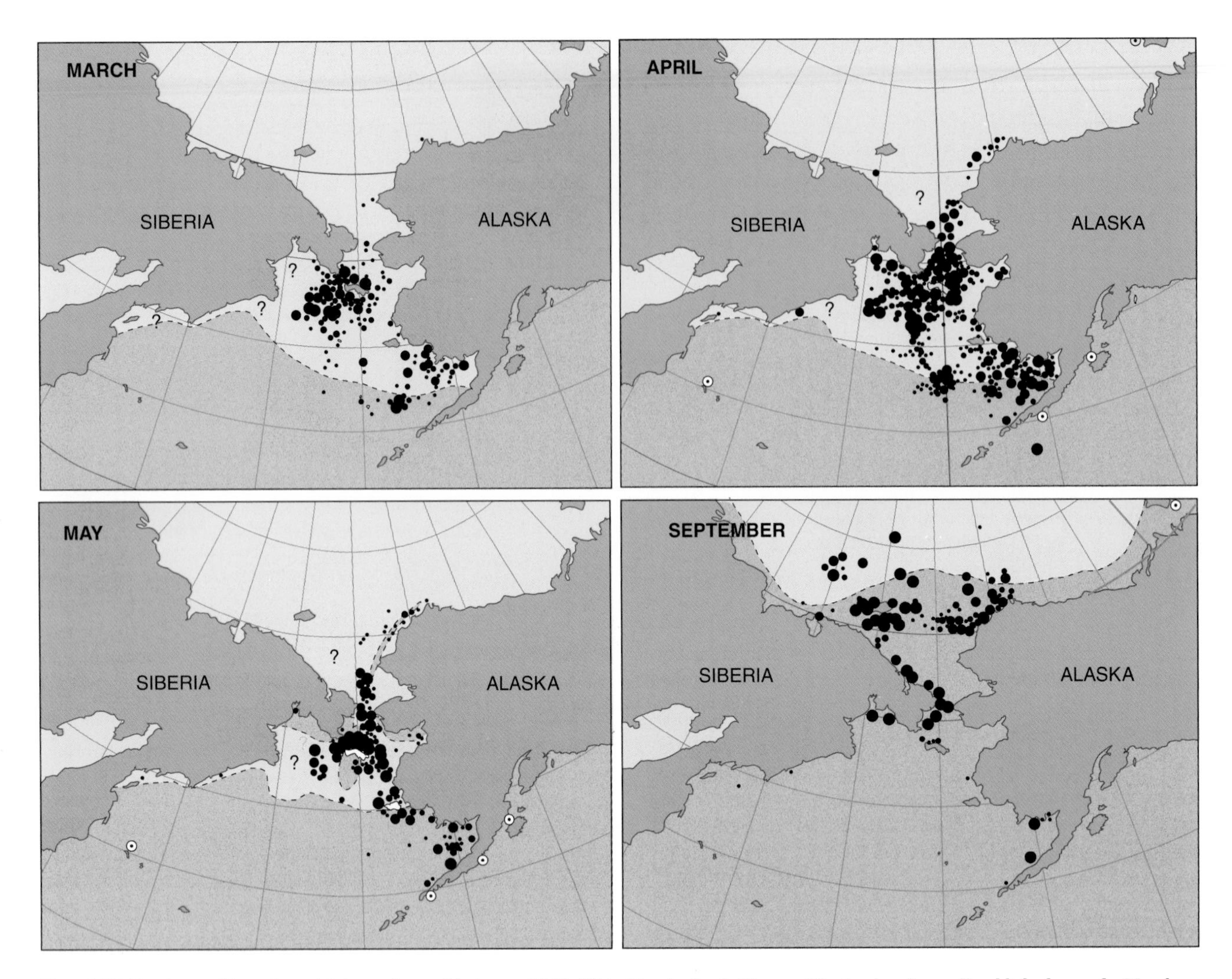

Fig. 7.20 Total monthly walrus sightings observed between 1930–78 in March, April, May, and September, from all published records. March and April distributions show that walruses favor the west-central broken pack (Fig. 7.19). The west-central subpopulation is separate from that of the southeast, possibly as a consequence of rounded pack as an unsuitable sea-ice type for walruses (Fig. 7.21); i.e. the rounded pack acts as a barrier between the two populations in winter. May distributions result from migration on northward-moving ice, and September distributions are for the Chukchi Sea. The numbers of sightings indicate survey effort and not population numbers. From Fay (1982).

"keel" that cause each floe to react independently to currents and/or winds. This mixing bowl effect on habitat is not predictable at present because it is very difficult to measure individual impacts of various stresses on sea-ice floes—wind, ocean currents, water and air temperature, ice pack internal interactions, shoreline boundaries, and bathymetry.

The consequences of the combined effects of the mixing bowl and rapid melt-out in spring on marine-mammal habitat are, first, that climate change is now resulting in a less well-structured seascape. The Bering Sea is slowly becoming ice-free earlier than the climatological norm of 1 July. Some floes are melting in place or melting before being advected very far into the southern Chukchi Sea. Thus, the consistent pattern of sea-ice types as observed in the past (Fig. 7.19) seems now less evident. Second, although the southerly extent of the ice seems not to influence the rapidity of melt-out, it may have a very significant effect on migration. Ice-dependent pinnipeds that rely on ice floes to "ride" into the Chukchi Sea must do so earlier in the spring and on more dispersed ice or be forced to undergo an energy-demanding swim.

7.5 DO LARGE MARINE MAMMALS MATTER?

The eight species of Bering Sea pinnipeds considered here clearly demonstrate both food and habitat partitioning, thus reducing competition for food and space while also maximizing fitness, i.e., their ability to perpetuate future generations. Reduction in their populations would be expected to have a cascading effect on lower levels of the food web and potentially higher entropy of the Bering Sea ecosystem as a whole.

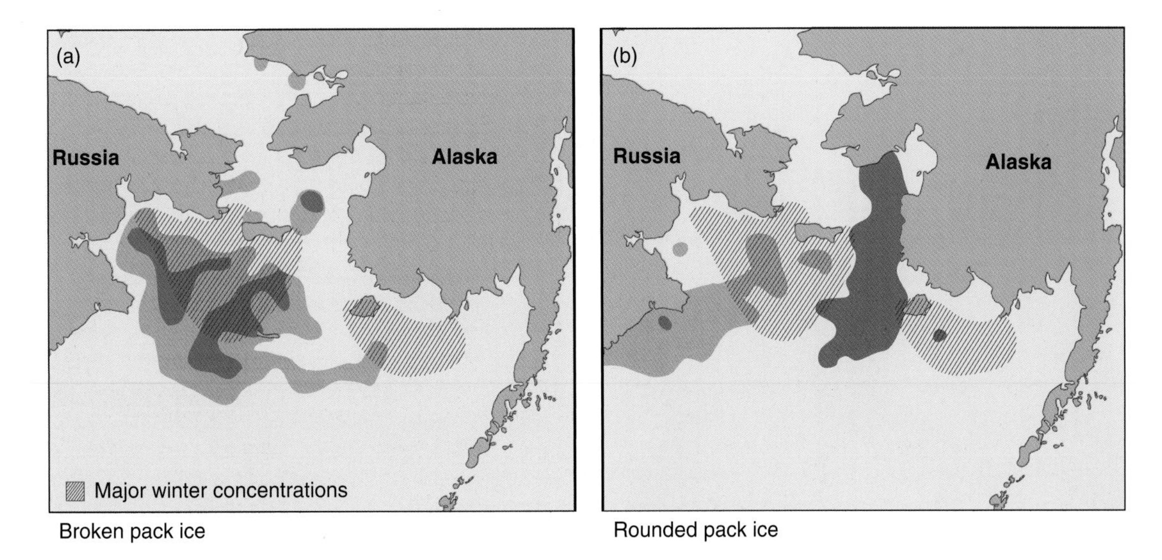

Fig. 7.21 Principal component analysis illustrates an association of walrus winter distributions (crosshatched, Fig. 7.13) with sea-ice types. The north-central Bering Sea subpopulation appears to co-occur with (a) broken pack, but not with (b) rounded pack. The Bristol Bay subpopulation appears not to be associated with either sea-ice type, but is probably due to the occurrence of suitable floes. Distributions of ice types result from statistical analyses of NOAA AVHRR imagery for March in 10 consecutive years, 1973–82; the darker-blue shading indicates greatest probability of occurrence. The rounded pack is suggestive of a strong barrier separating the two walrus subpopulations. From Ray and Hufford (1989).

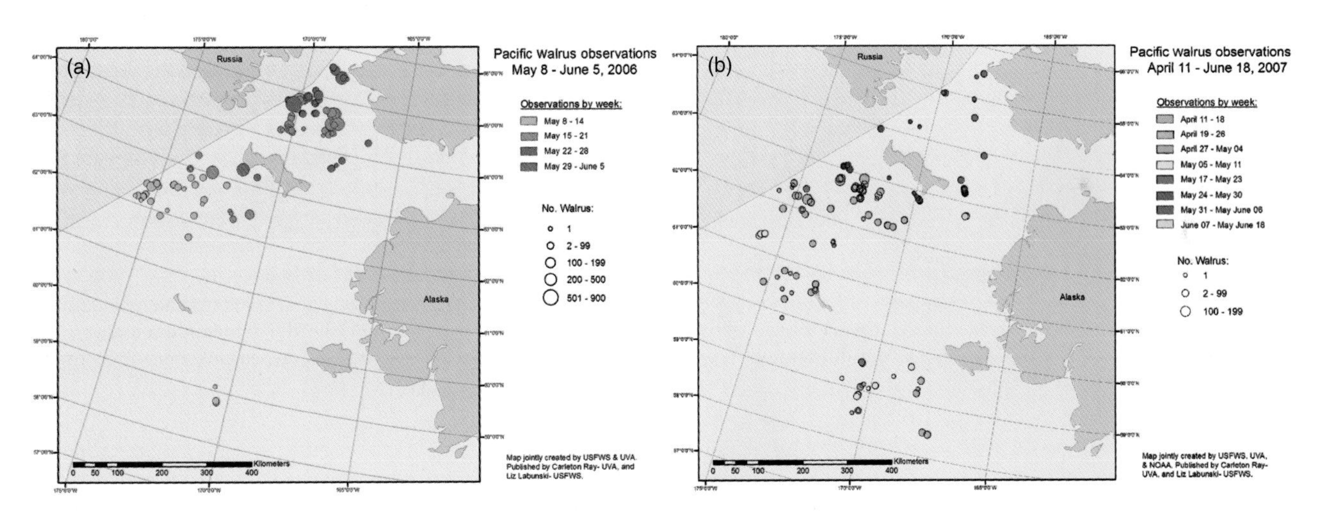

Fig. 7.22 Walrus observations from cruises of the icebreaker *Healy*. (a) 2006 spring migration and (b) 2007 late winter distribution after the reproductive season. Both years' distributions suggest two "waves" of walruses passing north, first from the west-central subpopulation, followed by lesser numbers from the southeast subpopulation. Observations are in accord with patterns from Fay (1982).

Although the southern and northern sections of the Bering Sea are fundamentally different ecologically, in common they share large populations of pinnipeds. Thus, the ecological consequences common to both regions would be expected to relate significantly to the depletion of these large consumers.

Pinniped consumption of Beringian biomass is extensive in species consumed (Fig. 7.9) and also massive in quantity. Fay (1982) calculated Pacific walrus total biomass consumption alone, assuming a population of 200,000 animals, to be approximately 8900 metric tons (mt) day^{-1}, or 3.25×10^6 mt per year. This represents a *net* rate of consumption, as walruses consume only soft parts of prey; if all organic matter was included, the *gross* "consumption" (i.e., amount of organic matter redistributed) reaches $9.5–12.6 \times 10^6$ mt a year. If annual consumption by all Beringian pinnipeds—walruses plus several hundred thousand bearded seals, spotted seals, ribbon seals, and ringed seals (NOAA, 2010)—were to be included, biomass removal would be enormous, substantially

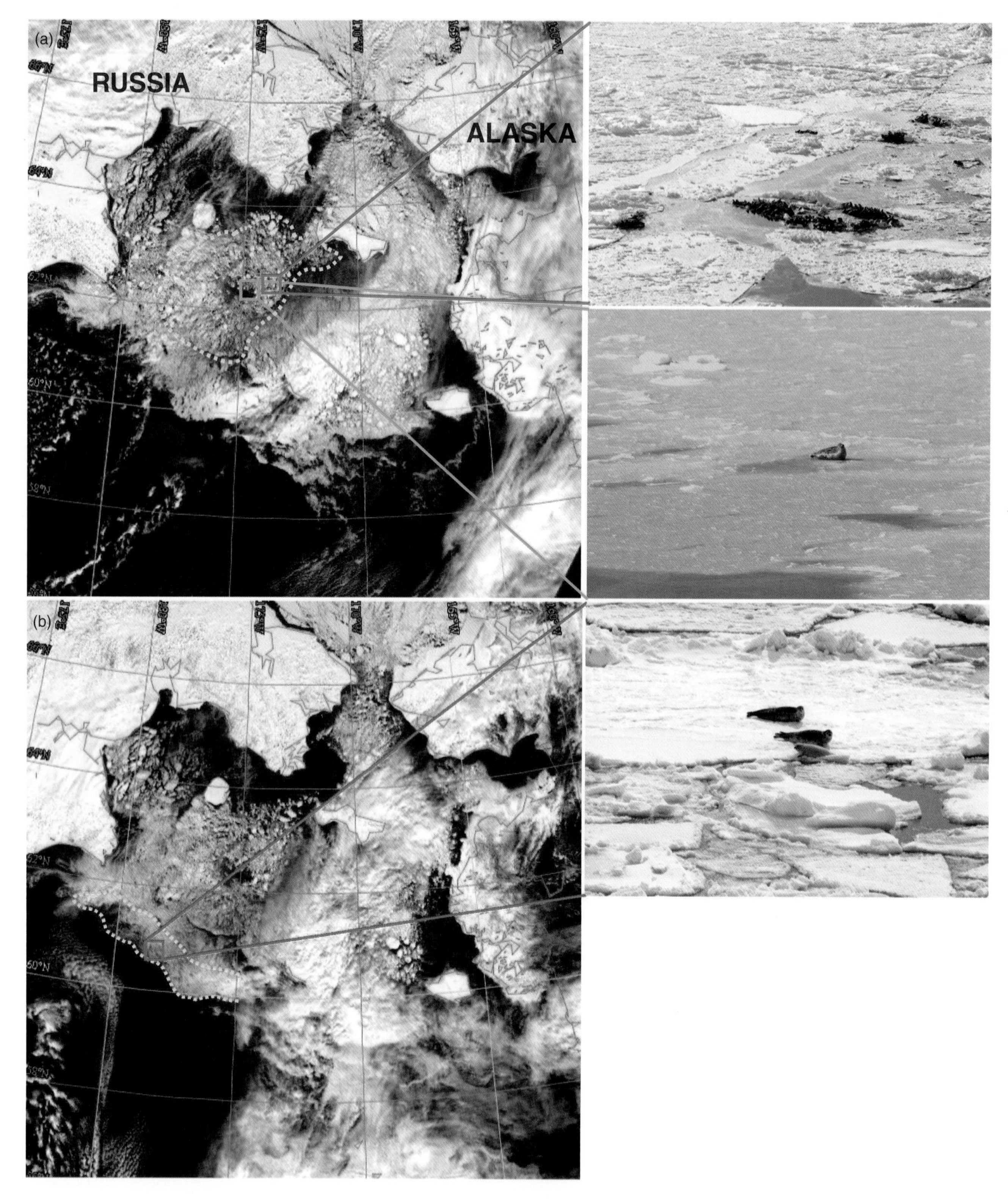

Fig. 7.23 Seascape and local scales of sea-ice types and pinnipeds. (a) Pacific walrus on ice floes observed within broken pack southwest of St. Lawrence Island, 20 April 2007. The sharp-edged character of individual floes and the intersecting leads separating floes are characteristic of broken pack. (b) Spotted seal "family" on outer fringes of loose pack, 15 March 2010. Areas enclosed by dashed yellow lines illustrate cover of broken pack (a) and loose pack (b), as determined by helicopter flights and interpretation of MODIS imagery. Environmental images from MODIS satellite imagery. From Ray *et al.* (2010). Reproduced with permission of John Wiley & Sons. Walrus and seal photographs © Ray & McCormick-Ray.

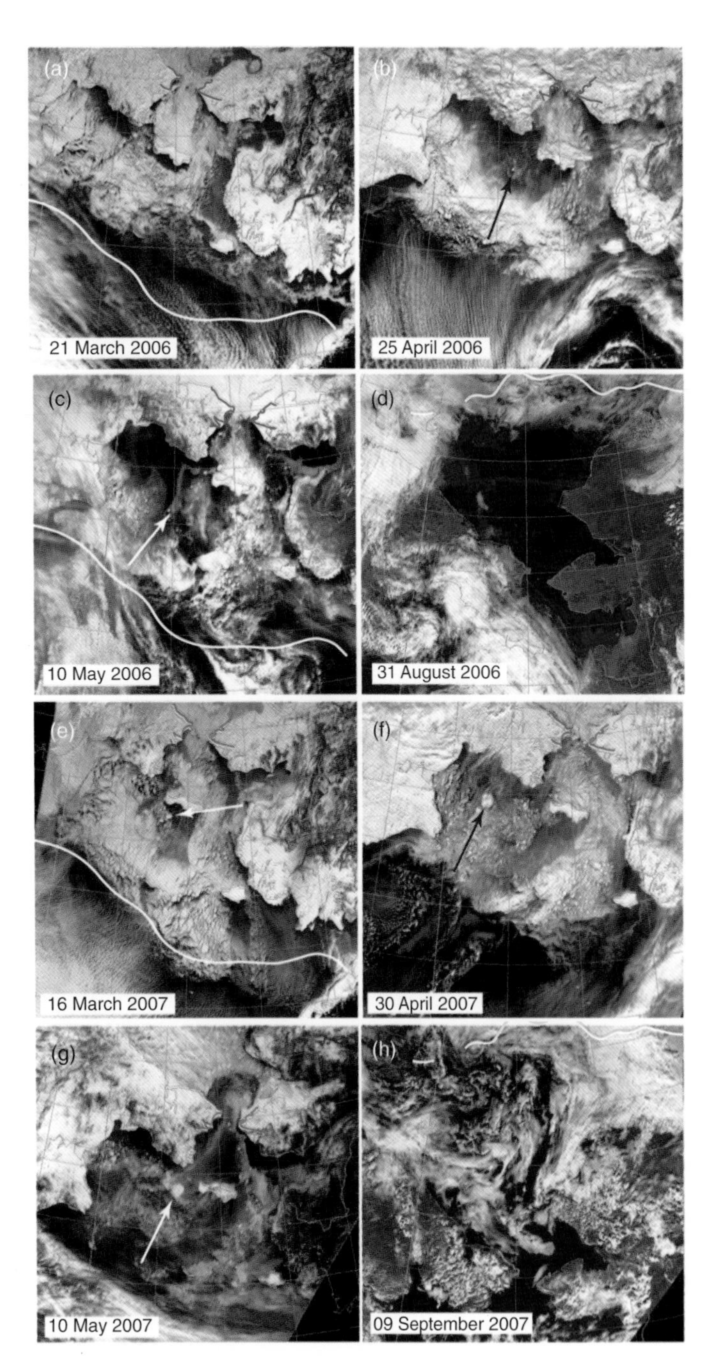

Fig. 7.24 Regional sea-ice cover from MODIS satellite imagery (resolution 250 m) for March, May, July, and late summer 2006 and 2007, showing change in northern Bering and Chukchi seas. White lines on (a), (c), and (e) represent the climatological norm for March at the time that ice extent is maximal. The sea-ice extents for 2006 and 2007 are very different. Arrows on (b) and (e) indicate areas of polynyas, which must be taken into account when estimating total ice cover. Again, 2006 and 2007 are strikingly different. Yellow lines on (d) and (h) indicate the 100 m depth contour in the Chukchi Sea during late summer, north of which walruses rarely feed; in both years, sea ice was near or north of that line, forcing walruses to occupy small areas of remnant pack (2006) or to retreat to land haulouts (2007). The arrow on (c) indicates the location of ice with ribbon seals that was tracked (Fig. B7.1.1, 2006, red line); arrows on (f, g) indicate tracked movement of a very large congealed floe (Fig. B7.1.1, 2007, red line).

exceeding all fisheries removals. Significantly, pinniped consumption constitutes recycling *within* the ecosystem, whereas fisheries represents biomass removal *from* the system; that is, the former would be expected to *increase productivity* due to increased turnover of resources, whereas fisheries depletes resources and thereby energy.

Furthermore, with respect to benthic consumption, walrus and bearded seal predation on in- and epifauna would be expected to exert top-down effects on benthic community composition and production (Ray *et al.*, 2006). In this respect, benthic community composition is of particular interest, as different species assemblages, whether resulting from predation or not, would contribute differentially to ecosystem performance (McCormick-Ray *et al.*, 2011). Biophysical effects of pinniped feeding are no less significant. For example, Johnson and Nelson (1984) estimated that feeding by walruses and gray whales—bioturbation—suspends approximately $120 \times 10^6\,m^3yr^{-1}$ of sediment in the north Bering and south-central Chukchi seas—twice the yearly sediment load of the Yukon River! Feeding bioturbation also has the potential to increase nutrient flux from the benthos to the water column by two orders of magnitude (Ray *et al.*, 2006), and thereby for productivity. Patchy patterns of benthic production have been observed by oceanographers and have been presumed to result

Box 7.1 The Bering Sea "mixing bowl"

Tracking individual floes from satellite imagery implies a "mixing-bowl effect." A combination of satellite imagery corresponded with ship (*USCGC Healy*) observations during 2006, 2007, 2008, and 2009. The first two years are illustrated here (Fig. B7.1.1). Thirteen floes tracked from 8 May to 11 June 2006 indicate general sea-ice movement. In April and early May, a major sea-ice melt left large expanses of open water. Consequently, 12 floes moved northward mainly in response to ocean circulation, with short perturbations in direction due to the winds. When winds were greater than 7.6 ms^{-1}, floes moved significant distances, especially through loose pack ice and open water. Floes in the central and western Bering Sea generally moved 9–50 km day^{-1}, which dispersed them over a much larger area. They accelerated with increased current speed as they moved toward and through the constricted Bering Strait. Two rounded floes (#9 and #12) in the eastern Bering Sea moved northward as they became entrained into the Alaska Coastal Current, their movements ranged from 16–74 km day^{-1} south of 64° N and greater than 92 km day^{-1} north of 64° N. A concentration of floes in remnant ice containing many ribbon seals was observed from the ship near the international dateline (Fig. 7.24c). The ice appeared to be a large concentration of shorefast and nearshore floes covered in places with sediment. Exceptionally clear weather and the size of the concentration allowed tracking for 33 days. High-resolution imagery showed the floe concentration to be >1 km in size. Prior imagery allowed backtracking to the origin off the mouth of the Anadyr River on 9 May in the Gulf of Anadyr. Storm winds superimposed on ocean currents on 12 May rapidly pushed the floe from near the Anadyr River southeast. The floe then lingered southeast of Cape Navarin for 14 days where the ocean current exerted sufficient stress on the floe to balance any wind stress. The result was that the floe became quasi-stationary and/or meandered near one location. Ocean currents then apparently pushed the floe northeastward slowly. Another wind event on 2 June rapidly pushed the floe farther northeast toward St. Lawrence Island, grounding it near the southwest tip of the Island where it melted in place. Overall the floe moved a total of 674 km. The track of this floe is an excellent example of the combined effects of winds and currents moving ice and associated seals.

In comparison, a major melt-out of sea ice occurred in the Bering Sea in early May 2007, despite its greater southerly extent (Fig. 7.24e). Strong northerly winds that typify winter conditions, and normally cease in early April, continued into May. The result was that nine floes were pushed southeastward by the winds, some south of the marginal ice. These floes melted, eliminating them as potential habitat by mid-May. A very large, unique, and unmistakable congealed floe was observed in satellite imagery on 10 April (Fig. 7.24f, g) and backtracked into the central Gulf of Anadyr to 14 March. This floe was pushed eastward by storm winds on 14 April toward the mouth of the Gulf of Anadyr. After a few days, the floe then drifted northeastward. Storm winds again pushed the floe east-southeast. The floe continued to slowly drift eastward, reaching St. Lawrence Island on 5 June where it broke up and melted in place. It is of interest that this floe track differed considerably from the large floe track in 2006.

The two years contrast sharply with respect to extent and cover of the pack. However, the sequences of events are similar. They also have in common a less concentrated pack than, for example, is shown historically in Fig. 7.19, leading to the observation that floe mixing is occurring and seascapes are becoming less organized.

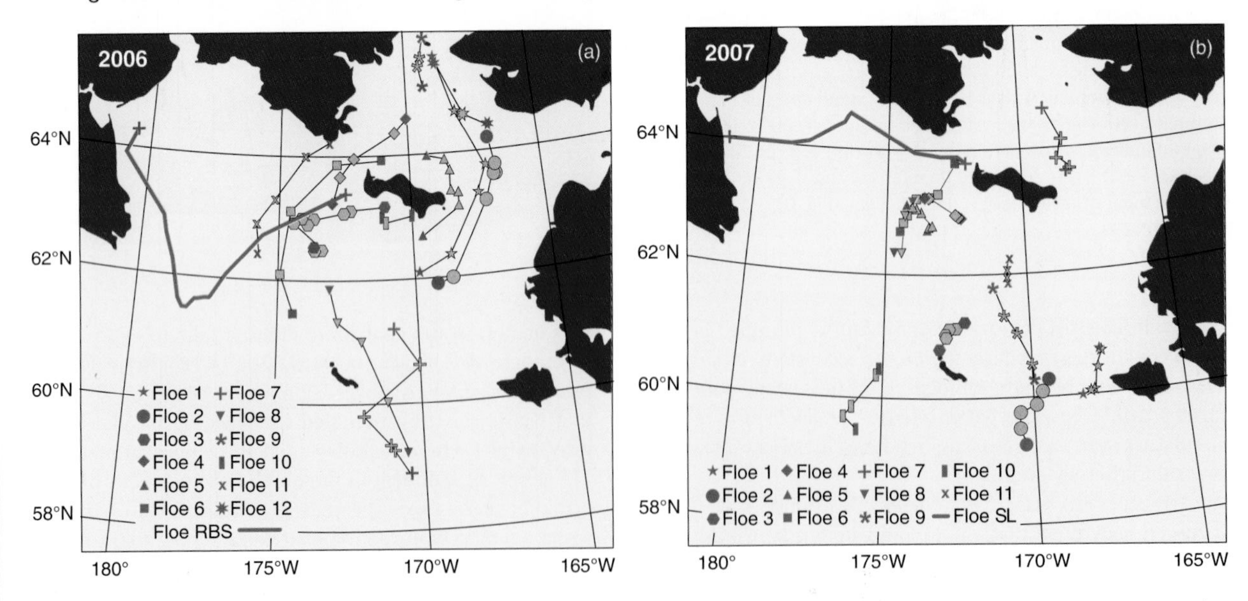

Fig. B7.1.1 Tracked floes, 2006–7. Floes are shown to variably change direction as forced by winds and currents, causing a mixing of floes and disruption of seascapes especially under conditions of sea-ice diminishment, as is presently occurring. Long red lines indicate large floe associations tracked for long distances (See also Fig. 7.24c, f, g). See text for explanation.

from primary production (e.g., Grebmeier *et al.*, 2006), but it seems possible that these patterns may also result from the feeding bioturbation by walruses, bearded seals, and gray whales, among others. The consequences of the combined effects of bioturbation have yet to be tested.

Spatio-temporal effects of pinniped feeding on production may originate in at least three ways. First, widely distributed species, such as those considered here, are likely be distributed as metapopulations (Ch. 5), as is known for Steller sea lions and fur seals, and suggested for walruses (Jay *et al.*, 2008). In this circumstance, spatio-temporal differences in feeding might be expressed. Second, all southern Bering Sea pinnipeds are "central-place foragers" (Ch. 5), that is, they leave rookeries or haulouts to feed, and return to the same general location afterwards. However, among the pagophiles, only the walrus may demonstrate this behavior. Field observations indicate that walruses remain within circumscribed areas of the broken pack when feeding, although they may not return to exactly the same floe following feeding bouts (Fig. 7.25; Jay *et al.*, 2010). In doing so, walruses would avoid the negative energetic trade-offs of central-place foraging (Ch. 5, Section 5.4.2.4), as their "central place" is in constant motion, allowing new food patches to be exploited. This situation changes for walruses on land, where local depletion of food resources would be likely. The 2007 sea-ice retreat north in summer (Fig. 7.24h) forced thousands of walruses to aggregate on terrestrial haulout sites on U.S. and Russian Chukchi Sea shores, a circumstance that has been repeated in subsequent years; in such cases, depletion of local resources would have been likely. Third, loss and structural changes of sea-ice habitat would likely result in walrus population depletion. Landscape population models suggest that loss and structural change of habitat, as would likely result from the "mixing bowl" effect, increase the proportion of time that the population spends in the portion of the habitat where reproduction may not be possible and/or where mortality is higher. This situation theoretically leads to a downward spiral toward regional extirpation or even extinction (Fahrig, 2007; Fig. 7.26). In this case, an uncertain future is predictable for walruses, and possibly for all Beringian ice-dependent pinnipeds. Ecological effects are to be expected.

The various effects of marine mammals on Bering Sea ecology are difficult to predict and may occur separately by species, or more likely to some degree, simultaneously. How marine mammals affect communities and production processes is fundamental for determining ecosystem change, and will also affect future management. However, present states of knowledge are not adequate for projections to be made.

7.6 THE CONFLICT ARENA

Resource conflicts are fundamentally different for the southern and northern portions of the Bering Sea. For the former, uncertainties about fisheries' effects on marine mammals presently dominate. For the latter, prospective oil and gas development, increased ship traffic, possibly including cruise liners, and northerly fisheries expansion resulting from diminishing sea ice pose significant challenges. In addition, the Bering and Chukchii seas are divided almost equally between the U.S. and Russia. For that reason, marine-mammal studies under the

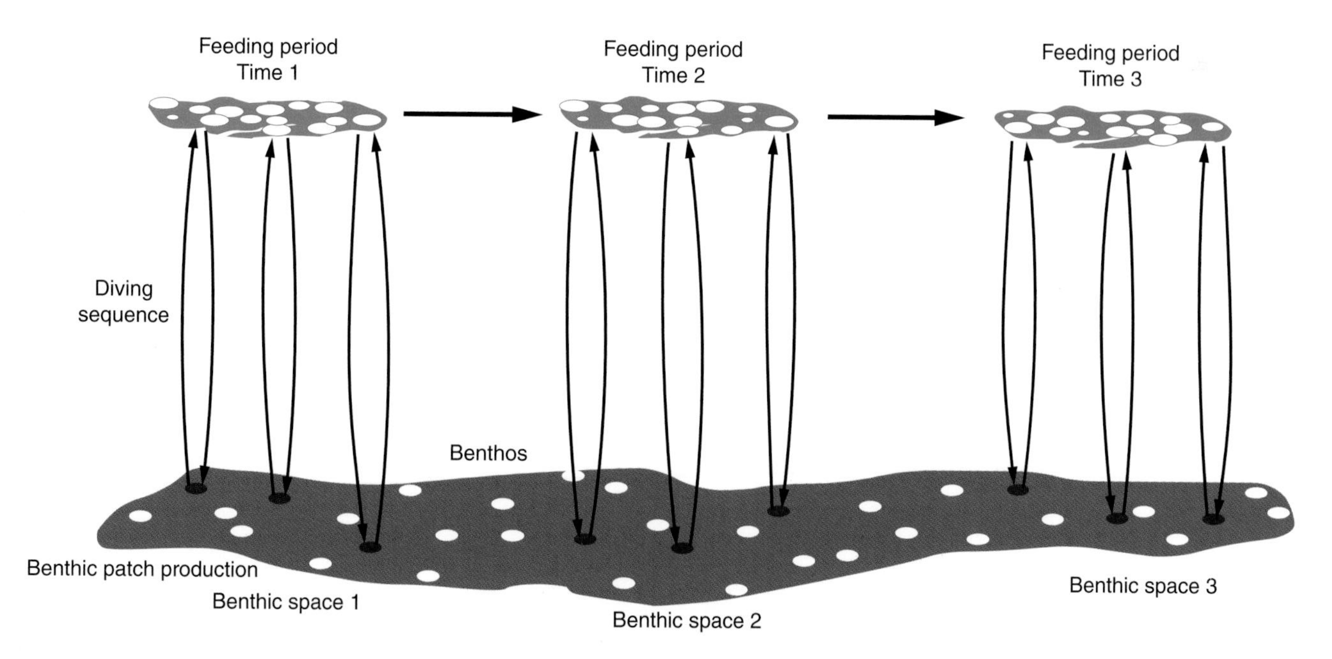

Fig. 7.25 Cartoon representing ice floe/benthic relationship hypothesized for walrus as due to "central place foraging." Hauled-out groups of walruses (white circles) rest on sea ice (blue "seascape" areas). Vertical arrows indicate walrus movements from their sea-ice floes to forage on the benthos then return to the same general area of the pack. The sea ice is moving with time (horizontal arrows). The continuous brown area below represents the benthos where patches of food are variably distributed. Black circles indicate areas where walruses have fed; open circles represent patches not fed upon. The result is a very patchy feeding distribution. See text for further explanation. From Ray *et al.* (2006). Pacific walrus: benthic bioturbator of Beringia. *Journal of Experimental Marine Biology and Ecology* **330**, 403–419.

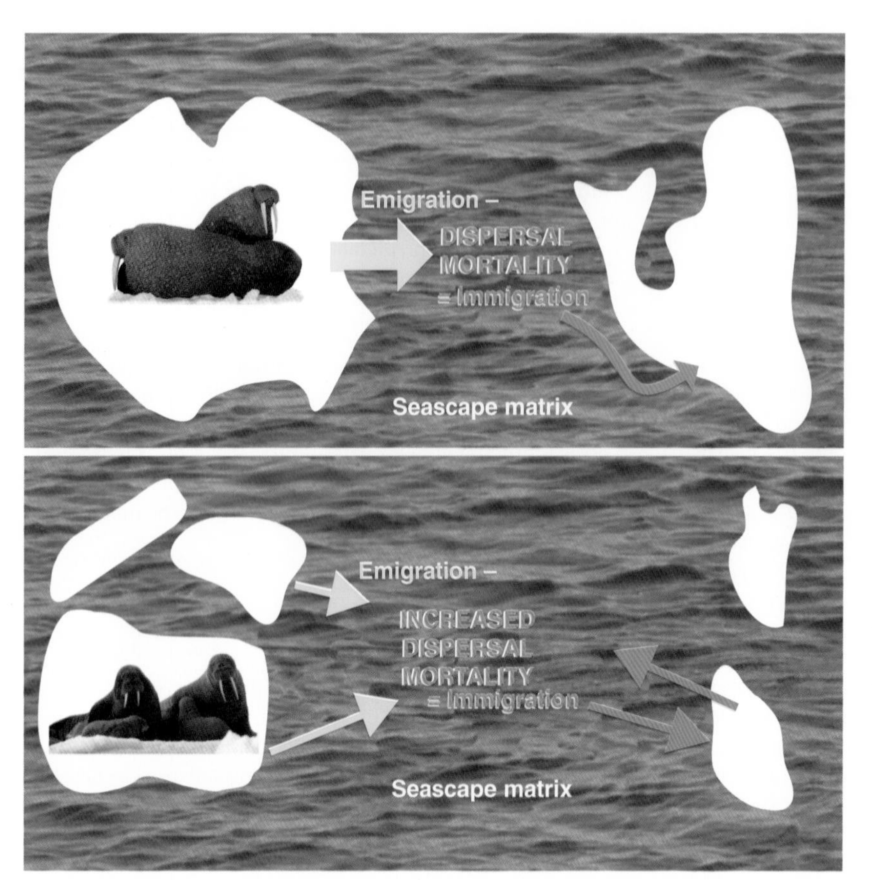

Fig. 7.26 Diminished sea ice resulting from the "mixing bowl" effect (see text) is hypothesized to increase walrus mortality (based on a concept from Fahrig, 2007). Two scenarios of sea-ice availability highlight the energetic costs for walrus as they disperse among floes. Upper: Under a relatively high proportion of available sea-ice habitat to water, relatively low mortality is expected. Lower: Under a relatively low proportion of available sea-ice habitat, dispersing walruses will expend more energy seeking appropriate sea-ice haulouts, thereby experiencing increased mortality.

U.S.-Russia Agreement on Cooperation in the Field of Protection of the Environment and Natural Resources have been essential and close cooperation continues (Box 7.2).

After World War II, high seas fisheries developed rapidly throughout the Bering Sea. Fisheries in Alaska were transformed by the passage of the *Magnuson Fishery Conservation and Management Act* (MFCMA, Ch. 3) and the *Marine Mammal Protection Act* (Ray and Potter, 2011). One of MFCMA's principal objectives was to restrict foreign fishing fleets so as to encourage development of the domestic fishing industry. As a result, the temporal and spatial distribution of the catch, especially of pollock, changed. Between 1963 and 1997 in both the Bering Sea/Aleutian Islands and Gulf of Alaska regions, pollock were fished increasingly in fall and winter in areas designated in 1993 as critical for Steller sea lions (Fig. 7.27). Commercial fisheries target several other important prey species eaten by sea lions, fur seals, and harbor seals and remove millions of metric tons of fish from the sea each year, potentially affecting their food supply. These pinnipeds are further affected directly by incidental catch in nets, entanglement in derelict debris (Fig. 2.14a), and shooting, or indirectly through competition for prey, disturbance, or disruption of prey schools. Incidental catches probably contributed to the early sea lion declines in the Aleutian Islands and western Gulf of Alaska, but are not presently considered to be an important component. However, the complexity of ecosystem interactions and limitations of data and models make it difficult to determine the extent to which pinnipeds are affected by fishing. For example, their primary prey (pollock) has continued to be abundant. Therefore, the declines seem not to be because of prey removal by fisheries, but may be due to the spatial or temporal availability of fish to predators. Biologists argue that fisheries cause "localized depletion" when trawlers take fish from small areas where dense fish schools occur in the same areas where sea lions and harbor seals feed (Fritz and Hinckley, 2005; Hennen, 2004).

The general conclusion is that no single factor, but a combination of factors, may be causing these declines. Therefore, what can be done? Actions to conserve sea lions include prohibitions on shooting, reductions on allowable incidental take in fisheries, placement of zones around rookeries to restrict trawling, designation of critical habitat, development of a Steller Sea Lion Recovery Plan, and other measures. NMFS does not want to impose regulations that might needlessly stifle the fishing industry, yet the government is required to protect and conserve the sea lion. There seems to be little doubt that sea lions and commercial fishing efforts concentrate on the same prey, yet data are not available to conclude that the fishing fleet is totally responsible for the decline. Nevertheless, fishing bears the brunt of responsibility since management of fishing is the parsimonious way to facilitate recovery. Furthermore the federal government has implemented numerous measures for the conservation of Steller sea lions, but none have been proposed for fur seals or harbor seals.

Box 7.2 Marine mammal studies under U.S.-Russia *Agreement on Cooperation in the Field of Protection of the Environment and Natural Resources*

Steven Kohl
Department of the Interior, Fish and Wildlife Service, Washington, D.C., USA

In 1972 the United States and Soviet Union signed an *Agreement on Cooperation in the Field of Protection of the Environment and Natural Resources* ("Environmental Agreement") to provide a framework under which the two nations could collaborate on issues of mutual interest. The Agreement was renegotiated in 1994 to replace the U.S.S.R. with the Russian Federation as signatory. Prior to 1972 there had been little joint marine mammal research or management activity between American and Russian scientists, with the exception of implementation of the 1911 *North Pacific Fur Seal Treaty* and occasional exchanges sponsored by the Academies of Sciences of the two nations. When the Environmental Agreement took effect, bilateral contacts increased considerably with the creation of a Marine Mammals Project under its auspices. A U.S.-Russia Working Group was set up to meet periodically in alternating countries to review the results of studies of shared cetacean and pinniped species and to adopt a program of joint activities for the following 18–24 months. The Working Group has continued uninterrupted to the present day; its 21st meeting was convened in Moscow in March 2010; the 22nd meeting was held in March 2013 in Seattle.

Federal and state government agencies, non-governmental organizations, major research institutions, and universities of both countries take part in the Marine Mammals Project. In the U.S. these include the National Marine Fisheries Service, Fish and Wildlife Service, U.S. Geological Survey Biological Resources Division, Alaska Department of Fish and Game, Alaska SeaLife Center, Monterey Bay Aquarium, and others. Among the Russian participants are the Russian Federal Fisheries and Oceans Research Institute (VNIRO) in Moscow, several branches of the Russian Pacific Federal Fisheries Research Center (TINRO), Academy of Sciences, Kamchatka Northeast Fisheries Agency (Sevvostrybvod), and the Federal Fleet Development and Research Institute (Giprorybflot) in St. Petersburg. Joint activities range from aerial surveys and shipboard studies to satellite tagging and shore-based work on haulouts.

In the years since 2005, cetacean studies have centered on gray, bowhead, beluga, and orca whales. Concern over western gray whales has resulted in several research cruises to monitor their feeding activities and reproductive success in the Sea of Okhotsk off Sakhalin Island. Deployment of satellite tags on large cetaceans is an annual U.S.-Russia effort, permitting studies of their movements, wintering areas, and degree of population discreteness. Intensive photography of gray and right whales has allowed American and Russian scientists to verify annual resightings with accuracy. Stepped-up collection of biopsy samples from beluga, gray, and orca whales has resulted in a corresponding increase in genetic research.

Pinnipeds continue to be the major focus of bilateral collaboration under the Marine Mammal Project. For walrus, a major current activity was the analysis and reporting of data collected during a 2006 comprehensive U.S.-Russian aerial and vessel-based survey of Pacific walrus throughout the Bering-Chukchi Seas region. Technological advances, including thermal scanning of walrus "hot spots" from aircraft and infrared photography, have been paired with more traditional survey methods such as visual observation to produce the first count of Pacific walruses since 1990. Recent joint studies of true seals (harbor, ribbon, spotted, ringed, bearded) have been carried out on their abundance, haulout spatial structure, feeding habits, genetics and mortality, with monitoring of Native subsistence harvests and ice conditions. For eared seals (fur seals, Steller sea lions), there has been intensive tagging and branding of newborn pups, analysis of telemetric data for survival rates and reproductive potential, and studies of diet composition, foraging behavior, and diseases.

Sea otters are a species of particular concern for American and Russian scientists. A reported 70% decline in sea otter populations in the northern Kuril Islands between 2004 and 2008 caused alarm, and mirrors a similar decrease in the Aleutian Islands that occurred at the end of the 1990s (Box 13.3). Joint studies seek to explain the reasons for such sharp drops in abundance, with diminished habitat carrying capacity leading to overexploitation of food resources suspected as a possible cause. At the same time, the Commander Islands and Kamchatka coastal populations of sea otters have been stable; Marine Mammal Project scientists are examining the comparative ecology of declining and stable populations.

All the species of marine mammals studied by American and Russian scientists are subject to increasing effects of climate change on their spatial and temporal distributions and alterations in the physical characteristics of their habitats. Joint work is underway to conduct Arctic marine ice-cover modeling at various times of the year and to determine key sea-ice habitat parameters affected by climate, through collection of remotely sensed and optical data, analysis of telemetric information provided by satellite downlinks, and examination of seasonal atmospheric circulation patterns. In the future, climate change and its ramifications will figure large in bilateral activities carried out under the Marine Mammal Project.

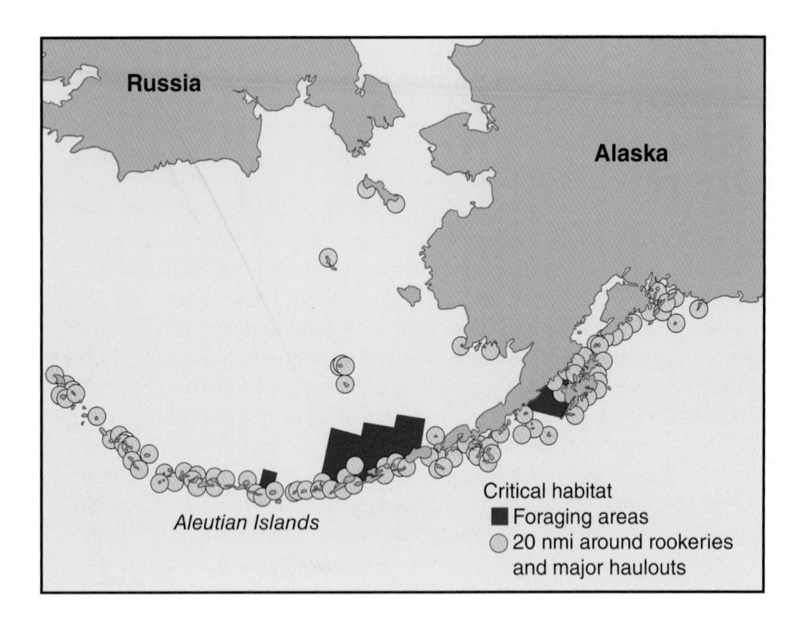

Fig. 7.27 Critical habitat for Steller sea lions in the Bering Sea, Aleutian Islands, and Gulf of Alaska, designated by the federal government in August 1993 as required by the *Endangered Species Act*. Critical habitat includes a 20 nmi radius around all rookeries and major haulouts, plus three aquatic foraging areas. From NOAA Office of Protected Resources.

In the north, detailed management plans have been developed only for walruses. From the passage of the MMPA in 1972 to the late 1980s, lack of management regulations and increasing demand for ivory for the cottage industry led to increased take, which was poorly recorded. In 1997, the U.S. Fish and Wildlife Service (FWS) and the Eskimo Walrus Commission established a cooperative agreement, including Russian counterparts, with two major components: a Marking, Tagging, and Reporting Program and a Walrus Harvest Monitoring Project. Together, these are intended to reduce waste, monitor subsistence take, collect biological information, and help control illegal take, trade, and transport. Presently, the total reported annual catch in 30 Native communities in Alaska and Russian Chukotka is estimated to be in the order of approximately 3000 animals; the presumed struck-and-lost rate is 40%, bringing the total take to approximately 4000–5000 animals per year. On the Russian side, subsistence hunting is currently regulated via agreements between Native communities and local district authorities that issue annual village catch quotas and collect harvest statistics. In Alaska, Native communities are not limited by federal law to the number of walruses that can be taken. Taking only the head for ivory is considered wasteful and is illegal. Current policy requires hunters to retrieve tusks (tagged for identification), heart, liver, "coak" (brisket), flippers, and "some red meat." The ivory, hides, and penis bone can be sold, but only if transformed into Native handicrafts or clothing. Non-Native catch is forbidden in both countries, and accidental kills, net entanglement, and boat collision losses are low. Shore haulout sites are strongly protected, but poaching and ship, air, and tourist disturbance are common.

A major problem concerns population assessment. Under the MMPA, an "optimum sustainable population" is an objective for all marine mammals. This is particularly difficult in the case of walruses and ice seals due to highly variable seasonal population shifts, inaccessibility, patchy distributions, time spent in water, and logistics in carrying out the assessment. In 2006, the FWS, in collaboration with Russian counterparts, conducted the most ambitious assessment to date, but with uncertain results: i.e., an estimated total of 129,000 animals, with a confidence range of 50,000–507,000 (Speckman *et al.*, 2010). Efforts are currently underway to assess ice seals. Problems of assessment leave management agencies in a difficult position under scenarios of climate change and loss of sea ice. Perhaps most importantly, very little research is being conducted on the ecological consequences of loss of marine mammals. Thus, the effects of rapid economic exploitation and sea-ice diminishment will remain highly speculative.

7.7 CULTURAL FACTORS: SUBSISTENCE HUNTING, TRADITIONAL KNOWLEDGE, AND COMMUNITY WELL-BEING

The pinnipeds of Beringia, particularly Pacific walrus, bearded seal, ringed seal, and, to a lesser extent, spotted seal, Steller sea lion, and fur seal, have been actively exploited as food, domestic, and commercial resources by aboriginal communities of the Bering and Chukchi Sea shores of Alaska and Siberia. Hence, the projected changes in pinniped distribution, abundance, and life cycle would certainly act as drivers of socioeconomic change. Understanding the nature of these changes is made urgent by growing evidence of shifts in climate, sea ice, and other physical-ecological parameters that may trigger dramatic restructuring of marine ecosystems of Beringia. Concerns have already been raised about negative impacts of these changes on the area's indigenous people, their economy, and well-being.

7.7.1 Historical factors

Archaeology offers solid evidence for an established human use of marine ecosystems of the coastal-shelf zone of the

North Pacific as early as 10,000 years ago. The earliest dated record within the Bering Sea proper from the Anangula Blade site in the eastern Aleutian Islands, dated 8750–8250 BP, suggests a marine economy with diversified use of resources. It is no accident that such an economy first emerged in the richer and more ecologically diverse southern portion of the Bering Sea. Along its southern margins on Kodiak Island and the Alaska Peninsula, several sites of about 6000 BP yield bones of harbor seal, porpoise, sea otter, Steller sea lion, waterfowl, albatrosses, salmon, cod, and halibut.

Indigenous marine hunting economies developed somewhat later in the northern Bering and southern Chukchi seas. The earliest evidence of indigenous marine hunting comes from Wrangel Island in the western Chukchi Sea and from the southern Chukchi Peninsula in the northern Bering Sea at about 3500–3800 BP, including midden pits with fractured bones of walrus, seals, and birds, artifacts made of walrus ivory, and engravings featuring scenes of walrus and whale hunting. These offer the earliest proof of human use of the Pacific walrus. In Alaska, a slightly later cultural complex of about 3300 years ago from Cape Krusenstern, southern Chukchi Sea, produced tools for maritime hunting as well as a litter of whale bones. All later coastal cultures of western and northwestern Alaska possessed sophisticated seal-hunting equipment and used walrus ivory for hunting tools, house implements, art, and decorated ritual objects, such as masks.

The first maritime hunters of Beringia were succeeded around 2200–1500 BP by people who were direct ancestors of the historic Inuit (Iñupiat and Yupik) and maritime Chukchi in Siberia. They lived in year-round coastal villages and possessed technology for effective year-round hunting of walrus, seals, and, later, whales—including large skin-covered boats. They also had dogs, used marine-mammal oil for heating and cooking, and stored large supplies of sea-mammal meat and blubber in underground ice cellars for use in wintertime. They built their villages at the best walrus- and seal-hunting locations—on cliffs, spits, and offshore islands in the midst of the sea, such as St. Lawrence, and Big and Little Diomede—for efficient open-water and sea-ice hunting.

The peak of the northern Bering Sea economy came with the development of whaling for bowhead whales from large skin boats, presumably around 1000 years ago. Whale hunting triggered population growth and the establishment of permanent villages at sites facing spring ice leads and polynyas along whale and walrus migration routes. The largest coastal villages housed several hundred residents with dozens of skin boat crews engaged in cooperative hunting.

The latest technological breakthrough came with European contacts. Russians introduced iron weapons, nets, and commercial hunting for fur seals and sea otters in the southern Bering Sea; they also exterminated the Steller sea cow (*Hydrodamalis gigas*) on the Commander Islands. The Americans after the 1840s brought firearms, wooden boats, whaling darting- and shoulder-guns, and later outboard motors. They also depleted northern Bering Sea stocks of bowhead and gray whales and Pacific walrus.

The newly introduced technologies made a dramatic impact on the efficiency of Native marine hunting, but Native tactics for hunting in ice leads, on ice floes, and in open water have not been altered profoundly. In many ways, present-day subsistence hunting in the northern Bering Sea remains a direct descendant of indigenous coastal cultures of the region. In the southern Bering Sea, local hunters largely shifted towards fur seals, other small seals, and sea otter. Also, abundant local fish, shellfish, and migrating bird resources continued to be actively exploited in both areas.

7.7.2 The walrus in Native subsistence economies

The Pacific walrus population has been exploited by indigenous hunters throughout its entire biological range in the Bering and southern Chukchi Sea for at least the past 2000 years. The area with the heaviest indigenous reliance on walrus is the junction of the Bering and the Chukchi seas, including the mainland shores of Northeast Asia and North America and the islands in between (St. Lawrence, Diomede, Sledge, and King Islands). Local Eskimo (Iñupiat and Yupik) and Chukchi may rightly be called the "walrus people." Traditionally, walruses provided 50–80% of their annual need for human and dog food, depending upon how well the hunting proceeded each year.

Aboriginal hunting equipment might seem "primitive" compared to modern rifles and motorboats. Nonetheless it was sophisticated and quite efficient (Fig. 7.28). Also, Native hunting, particularly of larger animals, tended to single out easier prey, such as juveniles, yearlings, pregnant females, and nursing cows because they were smaller, moved slowly, and were easier to catch. Furthermore, the number taken was not trivial and might have been at least several thousand animals per year. For example, Alaskan Inupiat hunters of Wales in the Bering Strait killed 322 walruses, 32 white whales, 80 bearded seals, and 4000–5000 other seals in 1890, for a village population of 539 (Thornton, 1931). On the Chukchi Peninsula, Siberia, the reported number of walruses killed by Native hunters in the early 1920s was about 2000–2500 animals per year, with at least a 30–50% loss rate of wounded and unrecovered animals. Native catch on the Alaskan side could have been a half of this, making a total of 3000–4000, or up to 5000 per year if killed and lost or wounded animals were included. In the late 1800s, Native catch was substantially higher, perhaps up to 8000–10,000, due to higher commercial demand for walrus ivory.

Those early data do not support a common assumption, namely, that Native subsistence hunting included some sort of "intuitive management" and was done strictly to cover for daily needs. Quite to the contrary, reported annual catches fluctuated dramatically from village to village, often by factors of two to three among years. The reasons for these oscillations were primarily natural. Local ice and weather conditions differed from year to year, thus affecting the position of advancing or retreating ice pack, and the availability of walrus within reach of hunters in small boats. More generally, resources available to Arctic subsistence hunters, either marine or terrestrial, are highly seasonal, and the seasons of "abundance" are usually brief. Also, resource availability is difficult to predict from year to year. These contingencies highlight the crucial significance of surplus catch and surplus food storage. Thus,

Fig. 7.28 Arctic Native people hunted walruses either from one-man kayaks or from much larger skin boats (umiaks), with a crew of 5–8 people. Kayaks were used (and still are being used today in some places) in southern Alaska, the central Canadian Arctic, and Greenland. Large skin boats propelled by sails and paddles were used by hunters in the Bering Strait area, on the Chukchi Peninsula, and in northern Alaska. The technology of hunting was similar in both cases. This figure illustrates (a) how a hunter in a kayak from Nunivak Island, Alaska, approached a walrus quietly on open water (or floating ice or coastal hauling grounds); (b) fastened a toggle-head harpoon with a skin float attached to the animal, then prepared to kill the animal with a heavier killing lance. From Fitzhugh and Kaplan (1982), courtesy of the National Museum of Natural History, Smithsonian Institution.

huge amounts of fresh meat and blubber were laid in storage or air-dried on open racks each year. The objective was to store more than enough food to last at least until the next season. In sum, walruses provided the most reliable local resource and offered the best return per hunting effort in terms of high-quality food that may be also stored in sufficient quantities. The economic contribution in terms of meat, blubber, hides for boat covers, and other products served as the major determinant of the size of human populations. When walrus hunting failed on St. Lawrence Island for two years (1878/9 and 1879/80), more than two-thirds of the island population died of starvation, hypothermia, and related accidents triggered by weakness and a desperate search for food.

Presently in both Alaska and Russian Chukotka, only Native people are allowed to hunt for walruses and seals (Fig. 7.29a) and to make full use of their products, including for commercial purposes such as ivory carving and souvenir production. Hunting is done with rifles from small motor-powered boats, but harpoons with floats and lances are still widely used. Walrus meat is actively consumed and stored for lean seasons (Fig. 7.29b), though it is rarely used for dog food, as dog teams are few in modern coastal villages, especially in Alaska, where people use snowmobiles and four-wheelers for traveling. Also, walrus skins are now hardly used for boat covers, except in a few communities, the reason being the adoption of faster aluminum boats. The economic value of ivory, walrus teeth, and skeletal bone for carving and souvenir production is extremely high, particularly on the Alaskan side, where it remains an important source of income.

7.7.3 Indigenous knowledge, co-management, and current environmental threats

Biologists working in the Bering Sea region have long experience in partnering with local hunters, studying the distribution, biology, and annual cycles of marine mammals. Nonetheless, traditional subsistence knowledge was largely treated as "anecdotal evidence," compared to systematic, natural-science research. Also, interests of local subsistence users were often ignored in biological assessments of marine mammal population health, and even more so in governmental efforts to protect depleted stocks and to establish legally binding management regimes.

This situation started to change in the 1970s, with the establishment of the first indigenous marine-mammal management organizations, such as the Alaskan Eskimo Whaling Commission (1976), Eskimo Walrus Commission (1978), Alaska Beluga Whale Committee (1988), Alaska Nanuuq (Polar bear) Commission (1994), Pribilof Islands Marine Mammal Commission (1998), and others. In 1992, a new umbrella organization, Alaska Indigenous People's Council for Marine Mammals (IPCoMM) was established, presently including 18 local marine-mammal commissions and regional groups. Also, a new term, "co-management," was brought into practice during the 1980s, assuming shared responsibility for the preservation, management, and scientific research on individual species, even local stocks. The actual level of partnership and data sharing varies for different marine-mammal species, being the strongest for the bowhead whale and the weakest for ice seals, with Pacific walrus somewhere inbetween.

Since the 1990s, indigenous management organizations, anthropologists, and marine biologists initiated a series of special programs for documentation of indigenous and traditional ecological knowledge (TEK) and subsistence hunting practices for many marine-mammal species. The main purpose was to document aboriginal management and conservation practices; an impressive amount of local ecological knowledge has been documented as well. Subsistence hunters helped identify certain morphological stocks at subspecies and/or metapopulation levels, specifically for bowhead whale and Pacific walrus, a feature that had not yet been recognized by marine biologists. Their knowledge was also instrumental in establishing local ranges and groupings for walrus, white whale, ice seals, and polar bear, and in documenting the specifics of seasonal migrations, reproductive cycles, and associations with sea ice. Substantial amounts of new publications have resulted from this cooperative work, both as gray literature and in academic publications (e.g., Bogoslovskaya, 2003; Bogoslovskaya *et al.*, 2007; Freeman *et al.*, 1995; Huntington *et al.*, 1999;

Fig. 7.29 (a) Animals killed by indigenous hunters are carefully examined during butchering for condition, health status, signs of parasites or sickness, and food content of the stomach. Often an experienced Elder (left foreground) is present to consult with young hunters. In this way, indigenous knowledge can contribute to a shared database with biologists. (b) Preparing *tuugtaq*, "meat balls" of walrus meat wrapped in skin, commonly of female walrus, that are sewn shut with the meat inside. This meat used to be stored in ice cellars, and was the most common winter food in many indigenous communities in the Northern Bering Sea-Bering Strait area. Photographs © G. Carleton Ray, Gambell, St. Lawrence I., Alaska.

Krupnik and Ray, 2007; Metcalf and Krupnik, 2003; Noongwook *et al.*, 2007; Oozeva *et al.*, 2004; Salomon *et al.*, 2011).

Certain indigenous organizations, like the Alaskan Eskimo Whaling Commission and the Nanuuq (Polar bear) Commission, accepted responsibility for enforcing community hunting quotas, conducting periodic game counts, imposing guidelines for non-wasteful catch, and training young hunters in effective traditional practices. The current trend, particularly in Alaska, is towards more collaboration, or at least consultation, among subsistence users and federal and state agencies in policy and management decisions concerning Beringian marine mammals.

The recently completed International Polar Year (IPY) 2007–8 featured several collaborative projects that combined efforts by scientists and indigenous residents in observation and knowledge documentation related to marine mammals and environmental change in Beringia and other polar regions (Krupnik *et al.*, 2011). Hunters, particularly in the communities adjacent to Bering Strait, report earlier spring migration of walruses and bowhead whales; more rapid retreat of the ice pack, which shortens the spring hunting season; and later fall advance of pack ice and associated marine species. As a result of IPY 2007–8, new collaborative observational efforts have been introduced, such as long-term ice monitoring in several communities (Krupnik *et al.*, 2010) and the Sea Ice for Walrus Outlook program since 2010 (SIWO online). The latter combines sea ice and walrus observations from scientists, local villages, satellites, and ships, and reports directly to walrus hunters in indigenous communities.

As a result of these recent developments, any future debates about the health of Beringian marine mammals and of their prospective listing as endangered or threatened species due to climate change and ice diminishment will most likely include substantial components of local biological and ecological knowledge, and observations by indigenous stakeholders.

7.8 ARE BERINGIAN PINNIPEDS AND THE BERING SEA ECOSYSTEM AT RISK?

The looming issue for the Bering Sea ecosystem, its species, and its Native people concerns emergent conditions and lag effects caused by climate change. Pinnipeds respond to environmental changes and human activities in ways that are poorly understood, and only uncertain hypotheses can be posed about their resiliency to rapid ecosystem shifts. Changes in abundances of large, slow-growing animals with low reproductive rates may lag behind projected environmental transitions, thereby making timely management measures difficult to implement. Therefore, current population assessments may reflect species' adjustments to past conditions, and trends may not become evident until some time after density-dependent responses are initiated. In addition, both the animal species and indigenous people of Beringia have been long subjected to recurring environmental shifts, often dramatic in time and scale, and thus have developed certain mechanisms for both short-term rapid response and long-term adjustment. Unfortunately, sufficient population or environmental baselines against which to measure change are insufficient for making future projections (Jackson *et al.*, 2011).

The result of changing conditions on Bering Sea pinnipeds is not necessarily extinction, but extirpation throughout much of their ranges, remnant populations for some of them, accompanied by expected ecosystem consequences. With respect to sea-ice-dependent pinnipeds, similarities between the hierarchical structure of sea ice and scaled habitat associations may be functional and evolutionary. Recent trends towards high inter- and intra-annual variability of sea-ice conditions at multiple scales and towards reduction and structural changes of seascapes may play reinforcing roles in shifting marine-mammal habitats and abundances. Consequently, it is highly likely that the variable conditions of the recent past will become more pronounced in the future, with implications for pinnipeds, other ice-dependent biota, the ecology of polar regions, and lifeways of many of its indigenous peoples.

In this context, a new, expanded research agenda is required, more directed toward ecosystem/habitat relationships and less toward numerical population assessments. Management agencies are forced to spend a majority of their resources on uncertain population enumeration, ironically a consequence of requirements of the the Marine Mammal Protection and Endangered Species acts' requirements for population assessment upon which management is presumed to be based. However, the major needs are for vastly improved knowledge of species' natural history, behavior, demographics, and ecological functions as these may be directly or indirectly related to scenarios of environmental change. Lacking that, the result may be to "miss the signal by focusing intently on what is all too commonly statistical noise" (Jackson *et al.*, 2011).

REFERENCES

Bickham JW, Patton JC, Loughlin TR (1996) High variability for control-region sequences in a marine mammal: Implications for conservation and biogeography of Steller sea lions (*Eumetopias jubatus*). *Journal of Mammalogy* **77**, 95–108.

Bogoslovskaya LS (2003) The Bowhead whale off Chukotka: Integration of Scientific and traditional knowledge. In *Indigenous Ways to the Present. Native Whaling in the Western Arctic* (ed. McCartney AP). Canadian Circumpolar Institute and University of Utah Press, Edmonton and Salt Lake City, 209–254.

Bogoslovskaya LS, Slugin I, Zagrebin I, Krupnik I (2007) *Osnovy morskogo zverobionogo promysla* [Subsistence sea-mammal hunting: Practical sourcebook]. Moscow-Anadyr, Russian Heritage Institute (in Russian).

Boveng PL, Bengtson JL, Withrow DE, Cesarone JC, Simpkins MA, Frost KJ, Burns JJ (2003) The abundance of harbor seals in the Gulf of Alaska. *Marine Mammal Science* **19**, 111–127.

Braham HW, Burns JJ, Fedoseev GA, Krogman BD. (1984) Habitat partitioning by ice-associated pinnipeds: distribution and density of seals and walruses in the Bering Sea, April 1976. In *NOAA Technical Report NMFS 12: Soviet-American Cooperative Research on Marine Mammals*, **Volume 1** – Pinnipeds (eds Fay FH, Fedoseev GA). U.S. Department Commerce, Washington, D.C., 25–47.

Burkanov VN, Loughlin TR (2005) Distribution and abundance of Steller sea lions, *Eumetopias jubatus*, on the Asian coast, 1720s–2005. *Marine Fisheries Review* **67**, 1–62.

Burns J (1970) Pagophilic pinnipeds. *Journal of Mammalogy* **51**, 445–454.

Burns JJ (2009) Harbor seal and spotted seal (*Phoca vitulina* and *P. largha*). In *Encyclopedia of Marine Mammals* (eds Perrin WF, Würsig B, Thewissen HGM). Academic Press, San Diego, CA, 533–542.

Burns J (1981) Ice as marine mammal habitat in the Bering Sea. In *The Eastern Bering Sea Shelf: Oceanography and Resources*, **vol. 2** (eds Hood DW, Calder JA). University of Washington Press, Seattle, 781–797.

Call KA, Loughlin TR (2005) An ecological classification of Alaskan Steller sea lion (*Eumetopias jubatus*) rookeries: A tool for conservation management. *Fisheries Oceanography* **14** (Supplement 1), 212–222.

Call KA, Ream RR (2012) Prey selection of subadult male northern fur seals (*Callorhinus ursinus*) and evidence of dietary niche overlap with adult females during the breeding season. *Marine Mammal Science* **28**, 1–15.

Cooney RT (1981) Bering Sea zooplankton and micronekton communities with emphasis on annual production. In *The Eastern Bering Sea Shelf: Oceanography and Resources* (eds Hood DW, Calder JA). NOAA Office of Marine Pollution Assessment, Washington University Press, Seattle, Washington, 947–974.

Eguchi T, Harvey JT (2005) Diving behavior of the Pacific harbor seal (*Phoca vitulina richardsi*) in Monterey Bay, California. *Marine Mammal Science* **21**, 283–295.

Fahrig L (2007), Landscape heterogeneity and metapopulation dynamics In *Key Topics in Landscape Ecology* (eds Wu J, Hobbs RJ). Cambridge University Press, Cambridge, UK, 78–91.

Fay F (1974) The role of ice in the ecology of marine mammals of the Bering Sea. In *Oceanography of the Bering Sea* (eds Hood D, Kelly J). University of Alaska Press, Fairbanks, 383–399.

Fay F (1982) Ecology and Biology of the Pacific Walrus *Odobenus rosmarus divergens Illiger*. *North American Fauna* (U.S. Depart. Interior, Fish and Wildlife Service) No. 74.

Fay FH, Bukhtiyarov YA, Stoker SW, Shults LM (1984a) Food of the Pacific walrus in winter and spring in the Bering Sea. In *NOAA Technical Report NMFS 12: Soviet-American Cooperative Research on Marine Mammals*, **Volume 1** – Pinnipeds (eds Fay FH, Fedoseev GA). Department Commerce, Washington, D.C., 81–88.

Fay F, Ray C, Kibal'chich A (1984b) Time and location of mating and associated behavior of the Pacific walrus, *Odobenus rosmarus divergens Illiger*. In *NOAA Technical Report NMFS 12: Soviet-American Cooperative Research on Marine Mammals*, **Volume 1** – Pinnipeds (eds Fay FH, Fedoseev GA). Department Commerce, Washington, D.C., 89–99.

Fedoseev G (2002) Ribbon seal, *Histriophoca fasciata*. In *Encyclopedia of Marine Mammals* (eds Perrin WF, Würsig B, Thewissen JGM). Academic Press, San Diego, 1027–1030.

Fitzhugh WW, Kaplan SA (1982) *Inua: Spirit World of the Bering Sea Eskimo*. Smithsonian Institution Press, Washington, D.C.

Freeman, MMR, Bogoslovskaya LS, Caulfield RA, Egede I, Krupnik I, Sevenson MG (1995). *Inuit, Whaling and Sustainability*. Altamira Press, Walnut Creek, California.

Fritz LW, Hinckley S (2005) A critical review of the regime shift – "junk food" – nutritional stress hypothesis for the decline of the western stock of Steller sea lion. *Marine Mammal Science* **21**, 476–518.

Gentry RL (2009) Northern fur seal (*Callorhinus ursinus*). In *Encyclopedia of Marine Mammals* (eds Perrin WF, Würsig B, Thewissen HGM). Academic Press, San Diego, CA, 788–791.

Grebmeier JM, Overland JE, Moore SE, Farley EV, Carmack EC, Cooper LW, Frey KE, Helle JH, McLaughlin FA, McNutt SL (2006) A major ecosystem shift in the northern Bering Sea. *Science* **311**, 1461–1464.

Hammill MO (2009) Ringed seal (*Pusa hispida*). In *Encyclopedia of Marine Mammals* (eds Perrin WF, Würsig B, Thewissen HGM). Academic Press, San Diego, CA, 972–975.

Hennen DR (2004) The Steller sea lion (*Eumetopias jubatus*) decline and the Gulf of Alaska/Bering Sea commercial fishery. Unpublished Ph.D. Dissertation, Montana State University, Bozeman, MT, 224pp.

Hopkins DM (1967) *The Bering Sea Land Bridge* (ed. Hopkins DM). Stanford University Press, Stanford, CA.

Huntington HP, and the Communities of Buckland, Elim, Koyuk, Shaktoolik, and Point Lay (1999) Traditional knowledge of the ecology of beluga whales (*Delphinapterus leucas*) in the eastern Chukchi and northern Bering Seas, Alaska. *Arctic* **52**, 49–61.

Jackson JBC, Alexander KE, Sala E (2011) *Shifting Baselines: the Past and the Future of Ocean Fisheries*. Island Press, Washington, D.C.

Jay CV, Outridge PM, Garlich-Miller JL (2008) Indication of two Pacific walrus stocks from whole tooth elemental analysis. *Polar Biology* **31**, 933–943.

Jay CV, Udevitz MS, Kwok R, Fischbach AS, Douglas DC (2010) Divergent movements of walrus and sea ice in the northern Bering Sea. *Marine Ecology Progress Series* **407**, 293–302.

Johnson KR, Nelson CH (1984) Side-scan sonar assessment of gray whale feeding in the Bering Sea. *Science* **225**, 1150–1152.

Kovacs KM (2009) Bearded seal (*Erignathus barbatus*). In *Encyclopedia of Marine Mammals* (eds Perrin WF, Würsig B, Thewissen HGM). Academic Press, San Diego, CA, 97–101.

Krupnik I, Allison I, Bell R, Cutler P, Hik D, López-Martínez J, Rachold V, Sarukhanian E, Summerhayes C, eds (2011) *Understanding Earth's Polar Challenges: International Polar Year 2007–2008*. Canadian Circumpolar Institute, Edmonton.

Krupnik I, Apangalook L, Apangalook P (2010) It's Cold, But Not Cold Enough: Observing Ice and Climate Change in Gambell, Alaska in IPY 2007–2008 and Beyond. In *SIKU: Knowing Our Ice. Documenting Inuit Sea Ice Knowledge and Use* (eds Krupnik I, Aporta C, Gearheard S, Laidler GJ, Kielsen LK, Holm L). Springer, Dordrecht, 81–114.

Krupnik I, Ray GC (2007) Pacific Walruses, indigenous hunters and climate change: bridging scientific and indigenous knowledge. *Deep-Sea Research II* **54**, 2946–2957.

Lentfer JW, ed. (1988) *Selected Marine Mammals of Alaska: Species Accounts with Research and Management Recommendations*. Marine Mammal Commission, Washington, D.C.

Loughlin TR (2009) Steller's sea lion. In *Encyclopedia of Marine Mammals* (eds Perrin WF, Würsig B, Thewissen HGM). Academic Press, San Diego, CA, 1107–1110.

Loughlin TR, Merrick RL (1989) Comparison of commercial harvest of walleye pollock and northern sea lion abundance in the Bering Sea and Gulf of Alaska. In *Proceedings of the International Symposium on the Biology and Management of Walleye Pollock*. University of Alaska, Fairbanks, Alaska, Alaska Sea Grant 89-1, 679–700.

Loughlin TR, Ohtani K (1999) Dynamics of the Bering Sea. University of Alaska Sea Grant.

Loughlin TR, Sterling JT, Merrick RL, Sease JL, York AE (2003) Diving behavior of immature Steller sea lions (*Eumetopias jubatus*). *Fishery Bulletin* **101**, 566–582.

Loughlin TR, Sukanova IN, Sinclair EH, Ferrero RC (1999) Summary of biology and ecosystem dynamics in the Bering Sea. In *Dynamics of the Bering Sea* (eds Loughlin TR, Ohtani K). University of Alaska Sea Grant Press, AK-SG-99-03, Fairbanks, Alasaka, 387–407.

Lowry L, Boveng P (2009) Ribbon Seal, *Histriophoca fasciata*. In *Encyclopedia of Marine Mammals* (eds Perrin WF, Thewissen JGM, Wursig B). MacMillan, New York, 945–948.

McCormick-Ray J, Warwick RM, Ray GC (2011) Benthic macrofaunal compositional variations in the northern Bering Sea. *Marine Biology* **158**, 1365–1376.

McNutt SL, Overland JE (2003) Spatial hierarchy in Arctic sea ice dynamics. *Tellus* **55A**, 181–191.

Metcalf V, Krupnik I, eds (2003) *Pacific walrus. Conserving our culture through traditional management*. Report of the Oral History Project under Cooperative Agreement #701813J506, Section 119. Eskimo Walrus Commission, Kawerak, Inc., Nome, Alaska.

NOAA (2010) Technical Memoranda, Status Reviews. Online.

Noongwook G, The Native Village of Savoonga, Huntington HP, George JC (2007) Traditional Knowledge of the Bowhead Whale (*Balaena* mysticetus) around St. Lawrence Island, Alaska. *Arctic* **60**, 47–54.

O'Corry-Crowe GM, Martien KK, Taylor BL (2003) The analysis of population genetic structure in Alaskan harbor seals, *Phoca vitulina*, as a framework for the identification of management stocks. *Southwest Fisheries Science Center Administrative Report LJ-03-08*. 54 pp.

Oozeva C, Noongwook C, Noongwook G, Alowa C, Krupnik I (2004) *Sikumengllu Eslamengllu Esghapalleghput. Watching Ice and Weather Our Way*. Washington, D.C., Arctic Studies Center.

Pease CH (1980) Eastern Bering Sea ice processes. *Monthly Weather Review* **108**, 2015–2023.

Phillips CD, Bickham JW, Patton JC, Gelatt TS (2009) Systematics of Steller sea lions (*Eumetopias jubatus*): Subspecies designation based on concordance of genetics and morphometrics. Occasional Papers, Museum of Texas Tech. University, Number 283, 1–15.

Pitcher KW, Olesiuk PF, Brown RF, Lowry MS, Jeffries SJ, Sease JL, Perryman WL, Stinchcomb CE, Lowry LF (2007) Status and trends in abundance and distribution of the eastern Steller sea lion (*Eumetopias jubatus*) population. *Fishery Bulletin* **107**, 102–115.

Ray C, Watkins WA, Burns JJ (1969) The underwater song of *Erignathus* (bearded seal). *Zoologica* **54**, 79–83 + plates I–III and phonograph record.

Ray GC, Curtsinger B (1979) Learning the ways of the walrus. *National Geographic* **156**, 564–580.

Ray C, Hufford G (1989) Relationships among Beringian marine mammals and sea ice. *Rapports et Proces Verbaux Reunion Conseil International Exploration de la Mer* **188**, 225–242.

Ray GC, McCormick-Ray J, Berg P, Epstein HE (2006) Pacific walrus: benthic bioturbator of Beringia. *Journal of Experimental Marine Biology and Ecology* **330**, 403–419.

Ray GC, Overland JE, Hufford GL (2010) Seascape as an organizing principle for evaluating walrus and seal sea-ice habitat in Beringia. *Geophysical Research Letters* **37**, L20504.

Ray GC, Potter Jr. FM (2011) The Making of the Marine Mammal Protection Act of 1972. *Aquatic Conservation* **37**, 520–552.

Ray GC, Wartzok D (1980) *Remote Sensing of Marine Mammals of Beringia*. Results of BESMEX: the Bering Sea Marine Mammal Experiment. Report under NASA Contract NAS2-9300, 1–77.

Robson BW, Goebel ME, Baker JD, Ream RR, Loughlin TR, Francis RC, Antonelis GA, Costa DP (2004) Separation of foraging habitat among breeding sites of a colonial marine predator, the northern fur seal (*Callorhinus ursinus*). *Canadian Journal of Zoology* **82**, 20–29.

Salomon A, Huntington HP, Tanape Sr. N (2011) *Imam Cimiucia: Our Changing Sea*. University of Alaska Press, Fairbanks.

Simpkins MA, Withrow DE, Cesarone JC, Boveng PL (2003) Stability in the proportion of harbor seals hauled out under locally ideal conditions. *Marine Mammal Science* **19**, 791–805.

Sinclair E, Zeppelin T (2002) Seasonal and spatial differences in the diet of western stock of Steller sea lions (*Eumetopias jubatus*). *Journal of Mammalogy* **83**, 973–990.

SIWO www.arcus.org/search/siwo

Small RJ, Pendleton GW, Pitcher KW (2003) Trends in abundance of Alaska harbor seals, 1983–2001. *Marine Mammal Science* **19**, 344–362.

Speckman SG, Chernook VI, Burn DM, Udevitz MS, Kochnev AA, Vasilev A, Jay CV (2010) Results and evaluation of a survey to estimate Pacific walrus population size. *Marine Mammal Science* **27**, 514–553.

Springer AM, McRoy CP, Flint MV (1996) The Bering Sea green belt: Shelf-edge processes and ecosystem production. *Fisheries Oceanography* **5**, 205–223.

Thornton HR (1931) *Among the Eskimos of Wales, Alaska, 1890–93*. The Johns Hopkins Press, Baltimore, Maryland.

Walsh JE (2008) Climate of the Arctic marine environment. *Ecological Applications* **18** Supplement, S3–22.

Westlake RL, O'Corry-Crowe G (2002) Macrogeographic structure and patterns of genetic diversity in harbor seals (*Phoca vitulina*) from Alaska to Japan. *Journal of Mammalogy* **83**, 1111–1126.

Wu L, Hobbs RJ, eds (2007) *Key Topics in Landscape Ecology*. Cambridge University Press, Cambridge, UK.

Zeppelin TK, Ream RR (2006) Foraging habitats based on the diet of female northern fur seals (*Callorhinus ursinus*) on the Pribilof Islands, *Alaskan Journal of Zoology* **270**, 565–576.

CHAPTER 8

THE BAHAMAS: CONSERVATION FOR A TROPICAL ISLAND NATION

The eastern United States possess very peculiar and interesting plants and animals, the vegetation becoming more luxuriant as we go south, but not altering in essential character . . . But if we now cross over the narrow strait, about fifty miles wide, which separates Florida from The Bahamas Islands, we find ourselves in a totally different country . . . such differences and resemblances cannot be due to existing conditions, but must depend upon laws and causes to which mere proximity of position offers no clue.

Alfred Russel Wallace *Island Life* (1880)

8.1 A NATION OF ISLANDS

For The Bahamas, "Sun, Sea, and Sand," as referred to by the Department of Tourism, attract the largest tourist economy in the Caribbean. Yet despite this advantage, The Bahamas faces a dilemma of maintaining a healthy resource base while seeking a sustainable economy. This challenge is not lost on The Bahamas' Government and its people, who have historically taken a lead among the world's small-island states in environmental legislation and international agreements. The Bahamas established the modern world's first land-and-sea park system under a single jurisdiction in 1958—The Bahamas National Trust—and the park system continues to grow. The Bahamas also was among the first nations to sign the 1992 *Convention on Biodiversity* and hosted the first Conference of the Parties to the Convention in Nassau in 1994.

Although the Bahamian economy is relatively strong among Caribbean nations, the nation remains vulnerable environmentally and economically. Many islands lack sufficient freshwater, coral reefs have declined, biodiversity is threatened by species invasions and climate change, and commercial development has degraded the nearshore reef-mangrove-seagrass complex—The Bahamas' most productive environment. Although solutions to issues that place environmental and economic sustainability at risk will not come easily, The Bahamas now finds itself in an enviable position relative to other Caribbean nations. It still possesses abundant resources and continues to implement measures for conservation of its resources, bolstered by its vision (BEST, 1997) of "a strong nation rooted in a healthy environment," implying maintenance of environmental integrity, a healthy resource base, and a sustainable economy.

This case study highlights major issues common to island nations worldwide, and illustrates that conservation of resources is fundamental for sustaining the economy and addressing the well-being of a growing population. In this context, this case study also addresses complimentary roles of government and NGOs, working together to implement national conservation policy.

8.2 BIOPHYSICAL AND SOCIAL SETTING

8.2.1 General characteristics and geology

The Bahamas, with the Turks and Caicos Islands, is the largest archipelago in the tropical Atlantic Ocean, equivalent in size to Australia's Great Barrier Reef (Fig. 8.1; Table 8.1). The land area is only 4% of its total area. Of the 11 largest islands, New Providence, where the capital, Nassau, is located, is the smallest; among the largest are Andros and Inagua, also with the least population density. The climate is seasonally subtropical (north) to tropical (south). Most of the land is less than 10 m above sea level; submerged banks are rarely more than 10 m deep, except where cut by tidal channels. Tides are semi-diurnal with an average daily range of approximately 0.8 m. Living organisms, e.g., mangroves, seagrasses, and coral reefs, situated mostly along island margins, form the most productive and biodiverse ecosystems. The Bahamas famed coral reefs constitute only about 1% of its total area.

The Bahamas Platform originated 200 million years ago with tectonic processes that separated North America from Africa (Dietz *et al.*, 1970). As continental drift continued, the Atlantic Ocean expanded, creating favorable conditions for coral-algal growth, calcareous-sand production, and sediment infilling, resulting in a series of large, thick, carbonate-capped platforms rimmed by marginal reefs, and separated by gullied slopes and canyons (Fig. 8.2). The Bahamas Platform is the oldest part of the Atlantic Ocean and was formed prior to the Caribbean islands. The Bahamas also has been, and remains, a vast carbonate factory (Broecker and Takahashi, 1966; Sealey, 1994). Sedimentary particles are formed when carbonates are

Marine Conservation: Science, Policy, and Management, First Edition. G. Carleton Ray and Jerry McCormick-Ray.

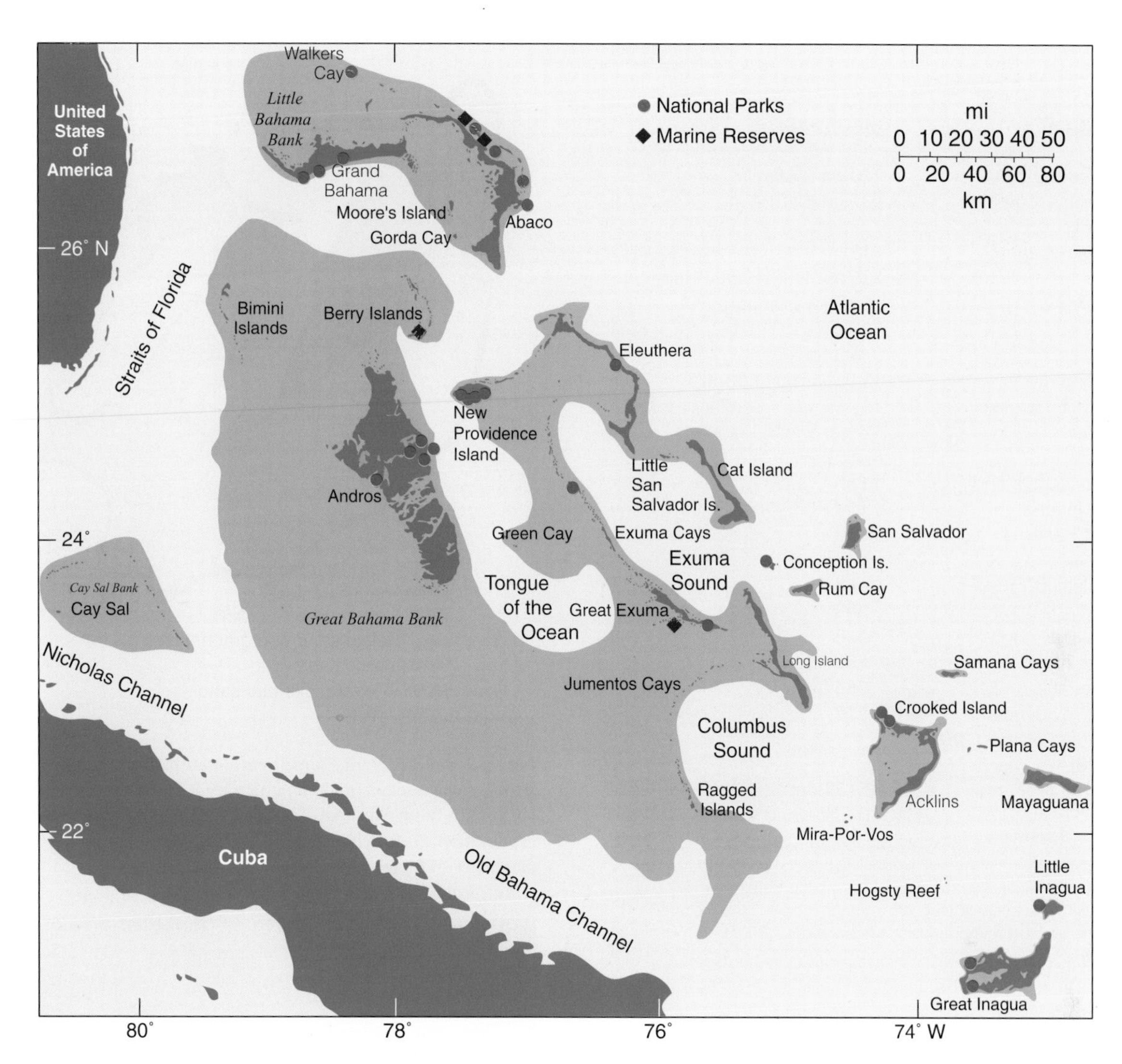

Fig. 8.1 Place names and protected areas of The Bahamas Archipelago; see also Table 8.1. National parks managed by The Bahamas National Trust are indicated by red dots, indicating location, but not size. Black diamonds indicate fishery reserves managed by the Department of Marine Resources. See text for explanation and consult The Bahamas National Trust (online) and the Department of Marine Resources (DMR, online) for details.

deposited around coral and shell fragments and animal fecal pellets. Skeletal sands (derived from corals and coralline algae), organic muds, clays, and silts are continuously redistributed by winds, currents, storms, hurricanes, and biotic activities. Sediment accumulates on beaches, forms dunes, and cements into rock (locally, "ironstone"). Presently, The Bahamas platform is an approximately 5 km-thick carbonate cap on the continental crust, which is presently thicker than the overlying ocean is deep. These deposits are among the largest on Earth at about 1.5 million km^3 in volume. The Great Bahama Bank is The Bahamas largest reef-fringed bank and lagoon, and thus may be considered to be among the world's largest atoll-like formations (Schlager and Ginsburg, 1981; Sealey, 1994).

During the past 150,000 years, much of The Bahamas' seascape has been intermittently drowned and exposed. Occasionally, sea levels dropped by >120 m from present levels. When sea levels were low, numerous caves were formed from solution of limestone. Cave roofs often fell, exposing caverns at the land surface, where water could collect in "blue holes" (Fig. 8.3), many of which are connected to the sea and exhibit tides. Dean's Blue Hole on Long Island is the deepest known (approximately 215 m). The longest underwater caves in The Bahamas are Dan's Cave on Abaco (>12,900 m), and the Lucayan Caverns, on Grand Bahama Island (approximately 9700 m).

8.2.2 Biogeography and biodiversity

The Straits of Florida, along with the Old Bahamas and Nicholas channels (Fig. 8.1), form effective ecological barriers for many organisms, distinguishing The Bahamas archipelago as a biogeographical ecoregion that differs from surrounding regions (Spalding *et al*., 2007). Its island groupings distinguish the Great Bahama Bank, Little Bahama Bank, and isolated islands such as San Salvador and form a basis for comparing species-environment associations. "Island biogeography"

Table 8.1 Characteristics of The Bahamas.

Parameter	Measurement
North–south extent*	720 km (448 mi)
Northwest–southeast extent†	937 km (528 mi
Greatest width	812 km (505 mi)
Total area‡	325,000 km^2 (124,000 mi^2)
Lands	35 major islands
	700 cays
	2,400 rocks
Total land area§	13,868 km^2 (4,500 mi^2)
Area for 11 largest islands§	12,455 km^2 = 89.8% of total
Andros	5,957 km^2
Abaco Islands and Cays	1,666
Great Inagua	1,544
Grand Bahama	1,373
Eleuthera	518
Long Island	448
Aklins	389
Cat Island	388
Exuma Island and Cays	291
Mayaguana	285
New Providence, including Nassau, the capital	210
Estimated reef area¶	~3,150 km^2 (1,260 mi^2)
	~1% of world total
	~15% of Caribbean

*Measured latitudinally. †Measured from Little Bahama's northwest bank to Great Inagua's southeast bank. ‡As included within the archipelagic boundary. §From Sealey (2001). ¶From Spalding *et al.* (2001).

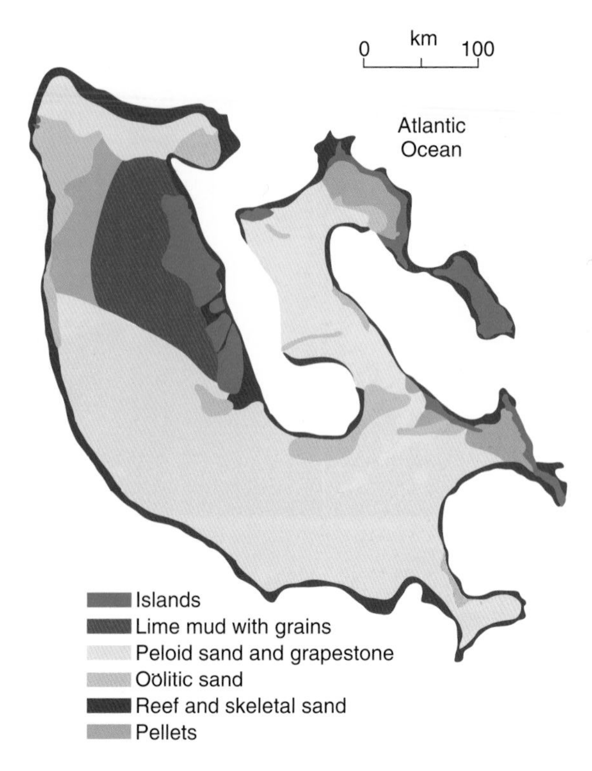

Fig. 8.2 Great Bahama Bank, a carbonate platform rimmed by reef structures of ancient origin and surrounded by U-shaped, flat-floored basins. The Bank is formed of several types of sediments. See text for explanation. From Schlager *et al.* (1981) Bahama carbonate platforms—The deep and the past, *Marine Geology* **44**, 1–24. With permission of Elsevier.

Fig. 8.3 (a) Nearshore blue hole, North Andros. (b) Uncle Bill's Blue Hole, on land, North Andros. Photographs © Ray & McCormick-Ray.

theory (MacArthur and Wilson, 2001) predicts that species and their communities would be expected to differ according to island sizes and spatial habitat arrangements, with implications for conservation. This complex research task remains as an opportunity to be conducted for The Bahamas.

Bahamian landscapes include pine forests, mixed hardwood coppice, agroecosystems, wetlands, sand beaches, and hypersaline ponds. Coastal-marine environments include tidal wetlands, mangrove forests, sand and mud flats, rocky shores, seagrass beds, and coral reefs; blue holes occur on land and in shallow water. Diverse land- and seascapes at different scales often occur in consistent patterns (Fig. 8.4a). The highest land on most islands is usually windward (eastward), where lagoons and reefs occur along coastal margins; low-lying wetlands and banks are generally leeward (westward). Northern islands support extensive hardwood and pine forests, the same islands where soils and climate are most suitable for agriculture. Southern islands are dry and windy, with salty aerosols, and support mainly stunted tree and scrub communities.

Terrestrial species diversity is low compared with larger Caribbean islands, due to the predominance of dry, soil-poor, karst topography (Fig. 8.4e). However, endemism is high (Fig. 8.5). An outstanding example is the peanut snail (*Cerion rubicundum*) whose different forms on individual islands provide an

Fig. 8.4 Landscape/seascape types at different scales. (a) The nearshore complex of environments at Staniard Creek and vicinity on the eastern side of North Andros: (i) sand- and mudflats; (ii) mangrove-lined creek; (iii) shallow-water lagoon where ledges and patch reefs occur; (iv) reef crest; (v) deep ocean; (b) intertidal red mangrove, *Rhizophora mangle*, recognized by its arching prop roots; (c) supraintertidal black mangrove, *Avicennia germinans*, showing diagnostic pneumatophores; (d) subtidal seagrass bed (turtle grass), *Thalassia testudinum*, habitat for queen conch, *Strombus gigas*; (e) Karst. Photographs © Ray & McCormick-Ray.

example of evolution in action (Gould, 1983; Fig. 8.5f). Smaller lizards are abundant (Fig. 8.5e); larger, relatively rare endemic rock iguanas (Figs 8.6c,d) occur in scrub and hardwood coppice. The endangered Bahama parrot (*Amazona leucocephala bahamensis*; Fig. 8.5b) presently occurs only on Abaco and Great Inagua.

Conversely, wetland and marine environments support high biodiversity. Mangroves grow best along creek margins, especially those with adequate water flow. Red mangrove (Fig. 8.4b) is the dominant pioneer plant in saltwater ponds and wetlands; black mangroves (Fig. 8.4c) are scattered over tidal flats, and white mangroves and buttonwood occur in drier environments. At the upper reaches of many larger wetland systems are salt marshes where crabs, insects, brine shrimp, and wading birds can be abundant. The Bahamas' national bird, the West Indian flamingo (*Phoenicopterus ruber*; Fig. 8.5a) inhabits selected, shallow, saline wetlands; Inagua's flamingo population of more than 50,000 is the largest in the Western Hemisphere. Seagrass and algal beds occur mostly in shallow creeks, lagoons, and sandy flats less than 10 m deep (Fig. 8.4d). These beds provide nursery areas for many commercial fish and are major feeding grounds for green sea turtles (*Chelonia mydas*, Fig. 8.15). The Bahamas hosts large stromatolites (Fig. 8.6), a rock-like microbial species association formerly known only from Western Australia. Stromatolites were discovered in the shallow waters of the Exuma Cays Land and Sea Park in the 1980s (Reid *et al.*, 1999).

The Bahamas is well known for its extensive coral reefs (Fig. 8.7a–e) and the very high diversity of fishes and invertebrates they support. Bahamian fish fauna alone consists of more than 600 known species, many of which are endemic. Fish communities exhibit complex, overlapping associations utilizing seagrass, coral reefs and mangrove habitats (Fig. 8.8). Important commercial species include members from several families, for example, groupers (Serranidae), snappers (Lutjanidae), jacks (Carangidae), and others. Small fishes such as damselfishes (Pomacentridae) are ubiquitous, and others such as butterflyfishes (Chaetodontidae) are mostly reef-dwellers. Grunts (Haemulidae) depend on shallow seagrass and sand habitats for nocturnal feeding, and structurally complex habitats (e.g., patch reefs, mangroves) for diurnal resting. Snappers inhabit both reefs and mangroves. Juveniles of many reef

Fig. 8.5 (a) The Bahamas' national bird, West Indian flamingo, *Phoenicopterus ruber*, males "marching" during breeding season; (b) endangered Bahama parrot, *Amazona leucocephalus bahamensis*, Inagua; (c) endemic rock iguana, *Cyclura cychlura*, Allen Cays, Exumas; (d) endemic rock iguana, *C. rileyi*, Exuma Cays Land and Sea Park; (e) abundant curly-tail lizard, *Leiocephalus carinatus*; (f) peanut shells, *Cerion rubicundum*, from North Andros (left) and Great Inagua (right); (g) golden hamlet, *Hypoplectrus unicolor*, golden color phase, at the place of its discovery in the Exuma Cays Land and Sea Park, 1958; (h) blackcap basslet, *Gramma melacara*, at the place of its discovery in the Exuma Cays Land and Sea Park, 1958; (i) Brotulid fish, *Lucifuga spelaeotes*, discovered in a blue hole on New Providence Island in 1968; this individual is from Andros. Photographs © Ray & McCormick-Ray.

Fig. 8.6 Stromatolites in Exuma Cays Land and Sea Park, most exceeding a meter in height. Photograph © Ray & McCormick-Ray.

Fig. 8.7 Diverse seascape features of coral-reef structures (a) Seascape pattern of patch reefs in the lagoon between land and the reef crest, North Andros; (b) coral/sponge reef with abundance of sponges in an area of high tidal velocity, <10 m depth; (c) vertical reef wall with large sheet coral, *Agaricia*, and sea whip, *Ellisella*, >30 m depth; (d, e) two historically abundant species much reduced by bleaching and warming are staghorn coral, (d) *Acropora cervicornis*, and (e) elkhorn coral, *A. palmata*. Photographs © Ray & McCormick-Ray.

species occur in creeks, mangroves, and seagrasses (e.g., Nassau grouper, Fig. 8.11).

Several new species discovered during the past half-century suggest that much remains to be known. The blackcap basslet (*Grammamelacara*, Fig. 8.5h) was first described from specimens collected in 1958 (Böhlke and Randall, 1963) during the Exuma Cays expedition (Section 8.4.1) and is now recognized as one of the commonest fishes of the reef wall. The golden hamlet (*Hypoplectrus* sp., Fig. 8.5g) was discovered nearby at the same time, and was thought to be a new color phase of its highly variable genus (Böhlke and Chaplin, 1968). A brotulid fish (*Lucifuga spelaeotes*, Fig. 8.5i) was discovered in the surface freshwater of a blue hole on densely populated New Providence Island in 1968 (Cohen and Robins, 1970); this fish appears to be widely distributed among Bahamian islands, but how it could have spread among islands separated by wide seawater expanses remains unknown.

In sum, the biodiversity of The Bahamas is a signature of its uniqueness among Caribbean islands and surrounding seas.

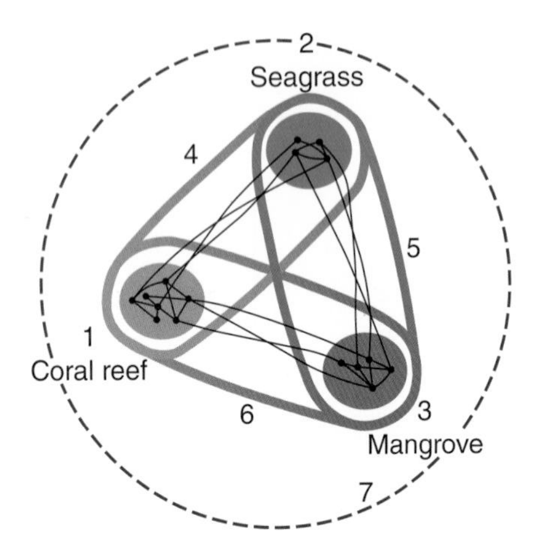

Fig. 8.8 Fishes exhibit a spatial, three-tiered hierarchy involving habitat partitioning in use of coral reefs, seagrasses, and mangroves. First, some species inhabit only one of the seven possible systems (1, 2, 3), e.g., certain coral-reef gobies and blennies occur in only one of the three habitats. Second, other species require two of the three habitats during their life histories (4, 5, 6), e.g., some grunts (family Haemulidae) and snappers (family Lutjanidae) that rest in daytime in coral reefs and feed during the night in seagrasses or mangroves. Third, some species are ubiquitous (7), e.g., barracudas. From Ray and McCormick-Ray (1992).

8.2.3 Historical social setting

About 6000 BP, Arawak Indians expanded from South America to Caribbean islands and developed their own culture, called "Taino" (Keegan, 1997; Wilson, 1997). They probably entered The Bahamas around 1500 BP, from Hispañiola, and survived on slash-and-burn agriculture and gathered a variety of terrestrial and marine life. Because most of their protein came from the sea, their villages were close to shore. Spanish records indicate a population of 40,000–60,000 at the time of contact. Little evidence remains today of their physical structures, but conch-shell middens provide evidence of resource use (Fig. 8.9).

In 1492, Columbus sailed throughout The Bahamas and named this place "Bajamar"—islands of a shallow sea—and called the Taino people "Lucayan," after "Lukku-Cairi," the name they gave themselves. Columbus found no material wealth in The Bahamas, but found gold in Hispañiola. The Spanish returned to The Bahamas to collect slaves to work the mines. By around 1510, the Spanish had systematically decimated most of the Taino through slavery, imported diseases, and massacres. Juan Ponce de Leon, who sailed in quest of the "Fountain of Youth," was last to encounter a native Bahamian, one old man, in 1513. Soon thereafter, the Taino ceased to exist as a people.

After extinction of the Taino, the islands remained uninhabited, being perceived as having little intrinsic value as well as a treacherous place for ships. Nevertheless, these islands and waterways held a strategic geopolitical position as the Spanish, French, English, and Dutch competed for empire and wealth in the New World. After defeat of the Spanish Armada in 1588, the English gained supremacy and King Charles I in

Fig. 8.9 A massive aggregation of queen conch shells, *Strombus gigas*, on Cay Sal beach and dune. Insert: embedded broken conch shells, one fragment dated at about 750 BP (C_{14}, Ray unpublished data) suggest that this area might have been a kitchen midden. Photograph © Ray & McCormick-Ray.

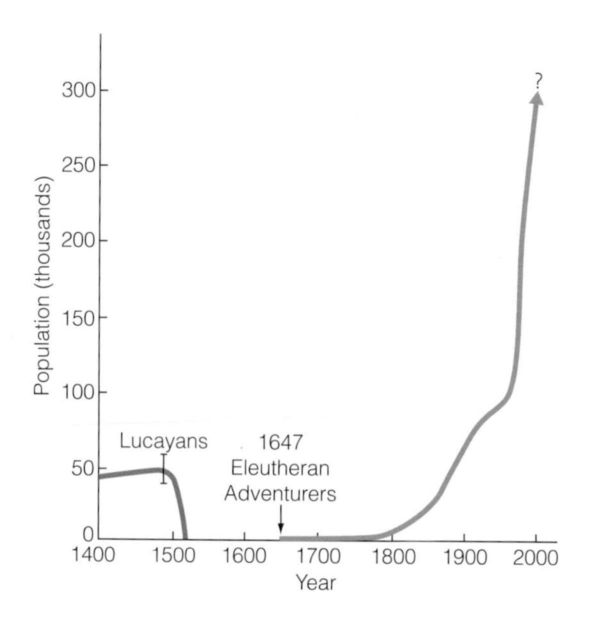

Fig. 8.10 Human population growth in The Bahamas. Pre-Columbian Lucayans are estimated to have numbered approximately 40,000–60,000, but were extinct by the early 16th century. The Bahamas remained unoccupied by humans until the mid-17th century, and the population did not again reach 50,000 until the late 19th century, after which the population grew rapidly. Data from Keegan (1997) and from various other historical and Bahamas Government sources.

1629 claimed possession of the Carolinas and The Bahamas. The islands became a British Crown Colony in 1717. Woodes Rogers, a former pirate, was appointed Royal Governor in 1718, bringing law and order. Settlers built plantations, mostly for cotton and sisal, and founded salt-producing industries. Slaves from Africa provided labor. The Bahamas thrived as the United States approached independence, and the population began to grow rapidly (Fig. 8.10). But as the Bahamian population grew, plantations failed to compete with U.S. cotton, and The Bahamas became economically depressed.

During the 1860s, the American Civil War returned prosperity to The Bahamas as it became a haven for Confederate blockade-runners. Shortly thereafter in 1870, sponging evolved into a major industry, significantly improving the economy. Nevertheless, during post-Civil War reconstruction, the economy declined again as the population rose to 53,000 by 1900. World War I again hit the economy hard, and The Bahamas became one of the poorest colonies in the British Empire. During the early 1920s, the economy grew again when the U.S. prohibited alcohol. Rumrunners thrived, creating an economic boon that also gave Bahamians opportunities to turn to tourism. The Great Depression again slowed growth until World War II when The Bahamas emerged as a major tourist and international financial center. In 1954, 100,000 visitors surpassed the colony's population of 85,000. The Bahamas achieved self-government in 1964, and on 10 July 1973, The Commonwealth of The Bahamas officially attained full independence as a member state of the British Commonwealth of Nations.

This history illustrates waves of external influences on The Bahamas' economy and its resources. Many Bahamians believe that The Bahamas' former history could repeat itself should overseas employment fall, foreign investment be reduced, hydrocarbon fuel costs soar, natural resources become further depleted, and especially if the fickle tourist industry were to falter.

8.3 CONSERVATION ISSUES

In pre-Columbian times, West Indian manatees (*Trichechus manatus*) and Caribbean monk seals (*Monachus tropicalis*) thrived in the Caribbean, and hordes of green turtles (*Chelonia mydas*) fed on extensive seagrass meadows. Now, the monk seal, last sighted in 1952 near Jamaica, is extinct, manatees have been extirpated over much of their range, and green turtles number in the tens of thousands rather than the millions. These examples reflect a history of change for Bahamian ecosystems. The following examples further illustrate a history of change and emergent issues that affect species, environments, and the well-being of the Bahamian people.

8.3.1 National icon: The Nassau grouper, *Epinephelus striatus*

8.3.1.1 The issue

The Nassau grouper (Fig. 8.11) is The Bahamas' most valuable food fish. Although it is relatively abundant in The Bahamas, overfishing has resulted in commercial extinction across much of the Wider Caribbean Region (i.e., including Bermuda, the Gulf of Mexico, the Caribbean proper, and the Antilles south to northern Brazil). Consequently, it is listed as "threatened" in the IUCN *Red List of Threatened Species*, but has not yet been listed under CITES (Ch. 3). As a top predator, the Nassau grouper plays an important role in the ecology of inshore and reef ecosystems. It also has high social value, being named after The Bahamas' capital, and is promoted as a fish of choice in restaurants. However, insufficient study has been dedicated to its ecological relationships and population dynamics, resulting in a tenuous scientific framework for conservation and management.

8.3.1.2 Natural history

The Nassau grouper is a tropical-subtropical species that ranges from Bermuda and South Carolina to the Gulf of Mexico, and throughout the Caribbean to Brazil. As for any "periodic" life-history strategist, juvenile survivorship appears to be most important for determining population size (Box 5.3). A series of life-history stages typifies Nassau grouper natural history (Fig. 8.12; Sadovy and Eklund, 1999); they mature at four to eight years of age and approximately 40–50 cm in length. The maximum-recorded age is 29 years, maximum length about 1 m, and maximum weight about 25 kg. Nassau groupers appear to maintain separate sexes throughout life, unlike many sea basses (family Serranidae) that change sex from female to male as they grow. At most times of year they are solitary, but in November through February they gather in large aggregations to spawn (Fig. 8.13). Past accounts have reported as many as 100,000 individuals at a single aggregation site (Smith, 1972). Almost all individuals are able to spawn

Fig. 8.11 Nassau grouper, *Epinephelus striatus*, iconic species of The Bahamas. Pictured is the typical banded color phase. For other color phases, see Fig. 8.12 and Fig. 8.13. Photograph © Ray & McCormick-Ray.

by age seven. Large fish can produce 5–6 million, 1 mm-diameter, eggs in a season.

Beginning a few weeks prior to spawning, aggregating groupers swim along the reef in groups of up to hundreds of individuals to reach spawning sites. Spawning occurs within 20 minutes before sunset during two to four day periods of the full moon (Sadovy and Eklund, 1999), which may provide a readily detectable cue for groupers to coordinate (Colin, 1992). Spawning groups of approximately 3–25 individuals, led by a dark-phase female, rush upwards to release eggs and milt (male fluid containing sperm) near the surface, then rapidly return to the substrate. Fertilized eggs hatch within a day after fertilization. Larvae two weeks old are able to swim. Larvae remain in the plankton for about six to seven weeks where they metamorphose into approximately 25 mm-long juveniles.

Aggregation behavior occurs in several species of fish. An aggregation is defined as "a group of conspecific fish gathered for purposes of spawning, with fish densities or numbers significantly higher than those found in the area of aggregation during the non-productive periods" (Domeier and Colin, 1997). It is not known how aggregation sites are selected, or why aggregation occurs. Reasons to aggregate to breed may be several; giving opportunities for males and females to meet, for sexual selection to occur (e.g., mate choice, sperm competition, etc.), optimizing fertilization, facilitating dispersal, and maxi-

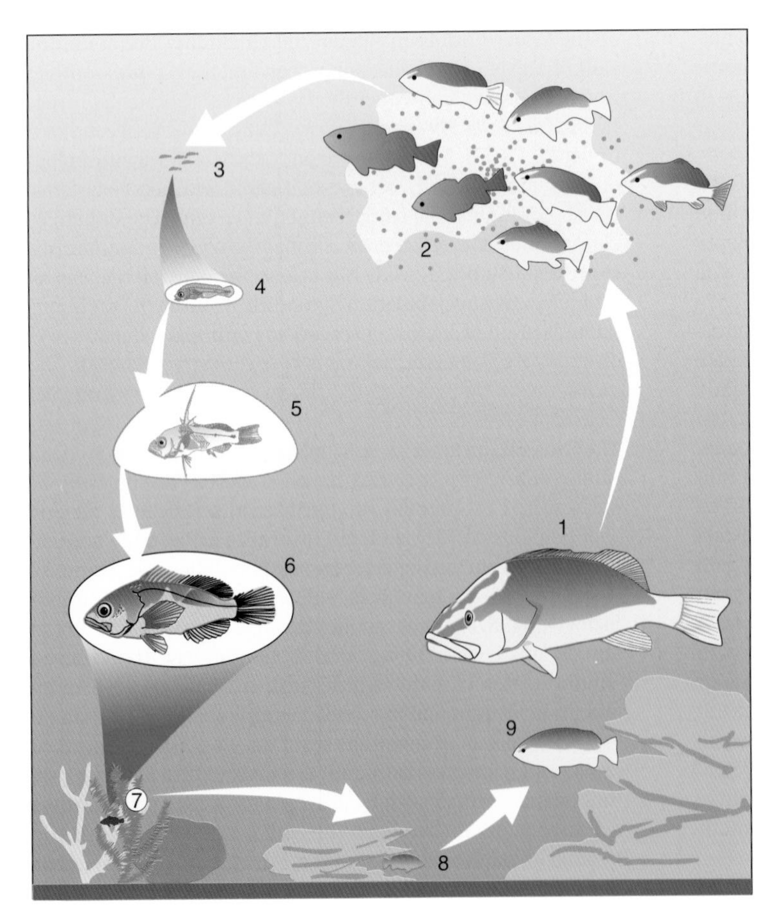

Fig. 8.12 Life history of the Nassau grouper (adapted from Sadovy and Eklund, 1999): (1) adult in bicolor phase; (2) spawning event, with dark-phase female in the lead and newly extruded eggs; (3) newly fertilized, pelagic eggs and hatching larvae; (4) ~3 mm pelagic, five-day-old preflexion larva; (5) 9 mm postflexion larva; (6) 25 mm transformed juvenile; (7) newly settled juvenile in algae; (8) half-grown juvenile in rocky ledge; (9) subadult in reef.

Fig. 8.13 Nassau grouper spawning aggregation, High Cay, near Andros Island, Bahamas, February, 1999. All four color phases are present, as named by Sadovy and Eklund (1999): (a) typical barred (see also Fig. 8.11), (b) bicolor, (c) white-belly, (d) dark. Photograph © Ray & McCormick-Ray.

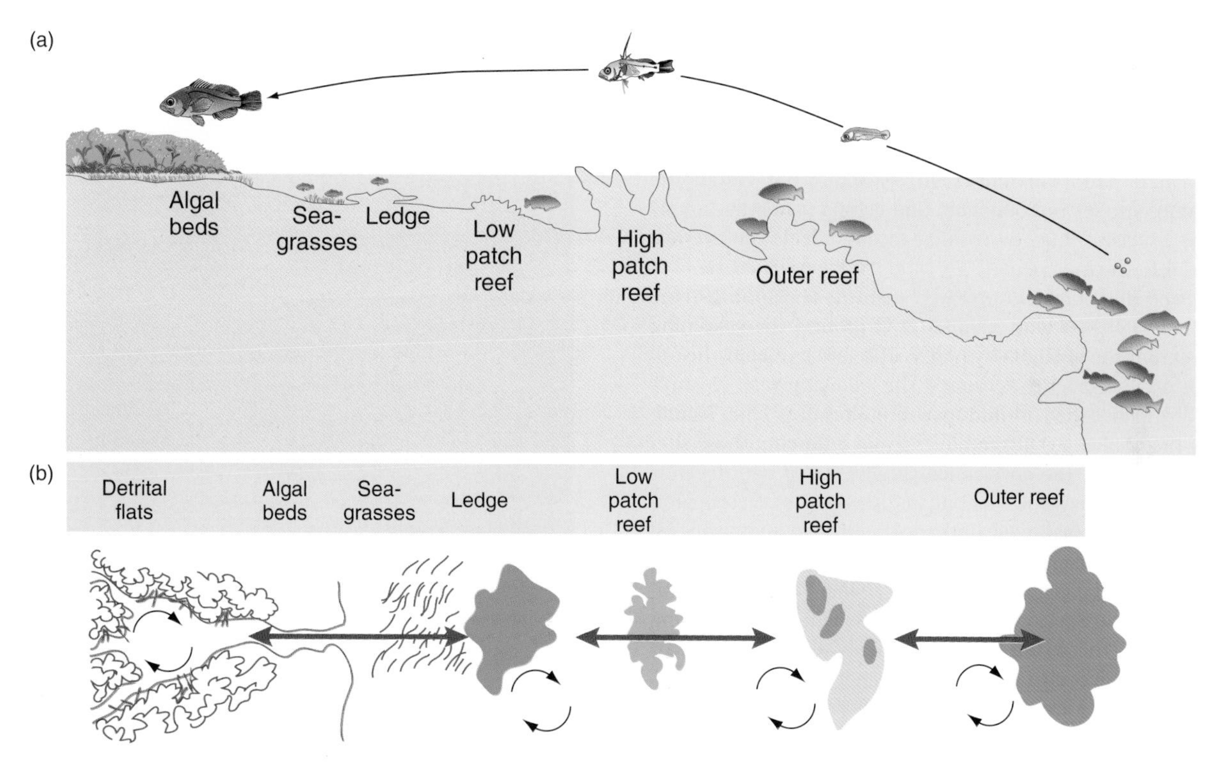

Fig. 8.14 Spatial inter-relationships among Nassau grouper habitats reflects life-history habitat shifts and hypothesized self-recruitment, as indicated by the arrows. (a) profile view and (b) plan view of habitat associations. Reproduction occurs at outer reefs, juveniles self-recruit to creeks and shallow-water environments, and individuals occur on sequences of ledges and patch reefs as they grow.

mizing chances for recruits to settle near their point of origin. Also, aggregation sites might be located to enhance local retention of eggs and weakly mobile larvae (Johannes, 1978; Colin, 1992; Aguilar-Perera and Aguilar-Davila, 1996; Colin *et al.*, 1997). It is likely that a combination of these factors has resulted in a particularly successful way of reproducing. Benefits are likely for adults and particularly larvae since a small reduction in mortality could have a large payback factor (Y. Sadovy, personal communication, 2012).

Another important feature of Nassau grouper life history is that they require multiple habitats through ontogeny (Fig. 8.14). All young life-history stages are highly subject to predation. Newly recruited juveniles often occur in macroalgal beds (*Laurencia*) in shallow waters (Eggleston, 1995; Dahlgren and Eggleston, 2000). If nursery habitats are degraded, juvenile survival may be poor. Following settlement, juveniles face trade-offs between predation and growth; that is, new recruits favor safe haven from predation, but as they grow, they move to shelters such as ledges and shallow-water patch reefs where predation may be more intense, but food is more abundant. Adults most commonly occur on coral reefs.

8.3.1.3 Conservation

The Nassau grouper is an example of a long-lived, large, late-to-mature species that naturally accumulates large standing stocks of old fish. Sadovy (1993) considers the Nassau grouper to be "unlucky" because its natural history is adapted for sustainability, but its tasty flesh makes it particularly vulnerable to overfishing. Species with low natural mortality and low growth rates tend to accumulate large standing stocks of older fish (Coleman *et al.*, 1996; Huntsman *et al.*, 1999). Thus, fishing is initially productive, as it "mines" large individuals, but once large fishes are depleted, population numbers and fishing success rapidly decline. The result for the Nassau grouper in the Caribbean as a whole is that spawning biomass has been reduced in only a few decades to only 1% of assumed pre-exploitation levels.

Fortunately, the practices that caused declines elsewhere have come under increasing control of The Bahamas Department of Marine Resources during the past decade and a half. Spearfishing on aggregations with compressors is now allowed only under permit, but is difficult to enforce. The minimum legal catch size (3 lb and 17 in) is somewhat below the minimum size for reproduction, and fishing generally targets the largest fish. Thus, overfishing mostly affects the average size of fish, the male/female sex ratio, and age structure.

Site-specific fishery closures were instituted beginning in the winter of 1998–9 with a small area around the spawning aggregation site at High Cay (Andros) before and after the full moons of November to February. The following year, closures were also established around spawning sites near Long Island. Given the challenges of monitoring and enforcing these site-specific closures, the government implemented countrywide closures in January 2004 for one month, with an accompanying sales ban of Nassau groupers during the closed season. All other groupers taken during the closure period were required to be landed with head and skin intact to enable species identification. This closure period has been increased incrementally, and at the time of this writing (2013) includes all of three months (December to February). However, this is a politically vulnerable situation. Unlike management of other commercially important species, decisions for the Nassau grouper are made annually by the office of the Prime Minister and his Cabinet prior to being incorporated into regulations of the Department of Fisheries. Insufficient knowledge on natural history and the distributions of populations continue to impede management.

8.3.2 Sea turtles in Bahamian waters

Alan B. Bolten
Department of Biology, Archie Carr Center for Sea Turtle Research, University of Florida, Gainesville, Florida, USA

Karen A. Bjorndal
Department of Biology, Archie Carr Center for Sea Turtle Research, University of Florida, Gainesville, Florida, USA

8.3.2.1 The issue

Bahamian waters are important foraging areas for three sea turtle species: green turtle (*Chelonia mydas*, Fig. 8.15a), hawksbill (*Eretmochelys imbricata*, Fig. 8.15b), and loggerhead (*Caretta caretta*, Fig. 8.15c). Leatherback (*Dermochelys coriacea*)

Fig. 8.15 (a) Green turtle, *Chelonia mydas*, showing the long pair of prefrontal scales that form its distinguishing characteristic. Photograph © Olga Stokes. (b) hawksbill, *Eretmochelys imbricata*, Little Cayman Island, and (c) loggerhead, *Caretta caretta*, Looe Key, Florida. Photographs © Ray & McCormick-Ray.

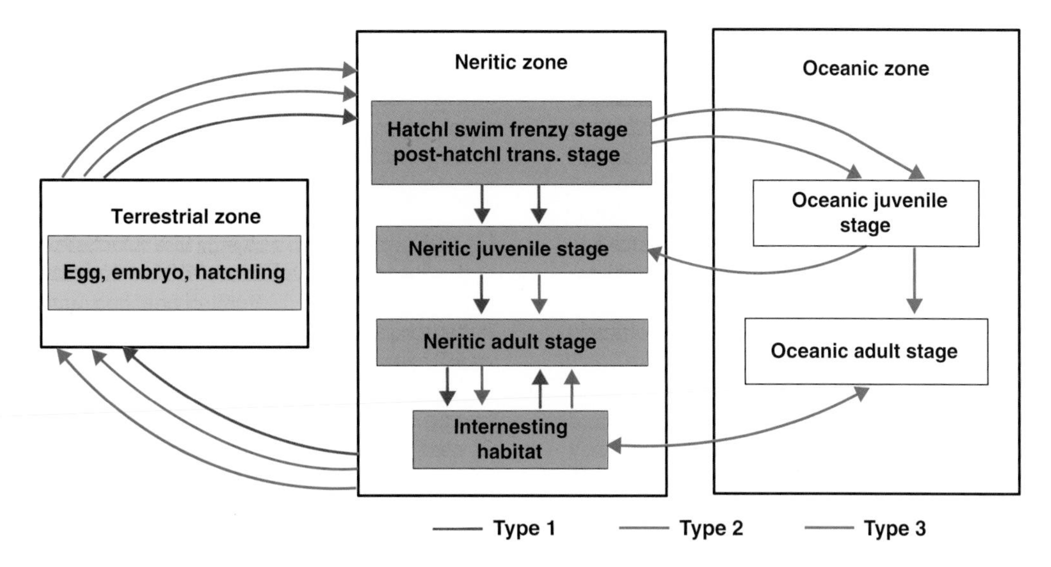

Fig. 8.16 Three sea turtle life-history patterns illustrating the different sequences of ecosystems inhabited during development and reproduction. From Bolten (2003) Variation in sea turtle life history patterns. In: *The Biology of Sea Turtles*, vol II (eds Lutz et al). CRC Press, pp. 243–257. With permission from Taylor & Francis.

and olive ridley (*Lepidochelys olivacea*) turtles also inhabit Bahamian waters in low numbers. The IUCN *Red List of Threatened Species* (IUCN, online) globally classifies the hawksbill and leatherback as "Critically Endangered," the green turtle and loggerhead as "Endangered," and the olive ridley as "Vulnerable."

Sea turtle populations in The Bahamas have suffered major population declines. Green turtles are at about 5% of carrying capacity in the Greater Caribbean (Jackson *et al.*, 2001). Nest production by the northwest Atlantic loggerhead population has declined approximately 43% from 1998 to 2006 (Witherington *et al.*, 2009), probably as a result of bycatch in commercial fisheries. The recent assessment of hawksbills indicates a decline in the Atlantic by about 80% during the last century (Mortimer and Donnelly, 2007). The Bahamas has an important role to play in the restoration and stewardship of all of these sea turtle populations, especially with regard to protection of feeding areas.

8.3.2.2 Natural history

Although there are only seven species of sea turtles worldwide, they occur in three major marine habitats: supralittoral, neritic, and oceanic and illustrate a diversity of life-history patterns (Bolten, 2003; Fig. 8.16). The flatback turtle (*Natator depressus*) of coastal waters of Australia is the only species that has a Type 1 life-history pattern. Loggerhead, green, and hawksbill turtles exhibit the Type 2 pattern, characterized by early juvenile development in the open ocean, and later juvenile development in the neritic zone. Leatherbacks and olive ridleys are Type 3, with complete development in the open ocean. Sea turtles also demonstrate dietary specializations from seagrasses (green turtle) to sponges (hawksbill) to a variety of invertebrates and fishes (loggerhead) to jelly organisms (leatherback) (Bjorndal, 1980). A major natural-history feature of all sea turtles is long-distance migration.

8.3.2.3 Habitat

The Bahamian Archipelago hosts only scattered nesting aggregations of green, hawksbill, and loggerhead sea turtles, but comprises some of the most productive foraging areas for sea turtles in the Wider Caribbean. The most important nesting area for loggerheads is Cay Sal Bank; after nesting they return to important foraging areas on the southern edge of the Grand Bahama Bank. Upwelling along this bank's edge results in nutrient-rich waters and, therefore, increased prey species, primarily crabs, other crustaceans, and hard-shelled mollusks. Limited loggerhead nesting also takes place on other islands, principally Abaco, Grand Bahama, Great Inagua, and San Salvador. Green turtles nest in low numbers on Abaco, Little Inagua, and Great Inagua. Hawksbill nesting is scattered at low densities throughout the archipelago.

Sea turtles play important roles in Bahamian ecosystems. The green turtle is the only herbivorous sea turtle and feeds almost exclusively on turtle grass (*Thalassia testudinum*; Bjorndal, 1980). As a direct result of grazing, cropped seagrass blades have a higher nutrient content than uncropped ones and thus become more nutritious for the turtles (Moran and Bjorndal, 2005, 2007). Because hawksbills have a specialized diet of sponges, which compete with corals for space, they help maintain reef biodiversity (León and Bjorndal, 2002). Loggerheads are important nutrient transporters from nutrient-rich foraging areas to nutrient-poor nesting beaches where nutrients left behind from sea turtle nests fertilize dune vegetation, thereby helping to stabilize beaches (Bouchard and Bjorndal, 2000).

8.3.2.4 The special case of the green turtle

Female green turtles reach sexual maturity at about 40 years of age. They nest every two or more years, and during a single season may nest as many as seven or more times at two-week intervals. On average, green turtle females deposit 115 eggs at each nesting. After 50 to 60 days, turtles of about 5 cm carapace length emerge from the nest, scramble down the beach, and enter the sea. High mortality occurs when birds, crabs, and fish prey on the hatchlings. Green turtle hatchlings are carnivorous in the open ocean, but where they go and for how long remains a mystery. After approximately three to five years, juveniles of 20–30 cm carapace length recruit to coastal waters, such as shallow mangrove creeks (Reich *et al.*, 2007).

One important question concerns how long juvenile green turtles stay in shallow creeks. Mark-recapture studies show that they grow slowly, taking about 12 years to reach 70 cm, when they leave shallow, coastal waters for deeper waters. A density-dependent relationship has been demonstrated for juvenile growth; when populations in creeks are high, growth is slower because of competition for food, but when populations decrease, growth rates increase (Bjorndal *et al.*, 2000). Recognizing this phenomenon is important because as conservation measures begin to have beneficial effects on the population, there may be a density-dependent effect that slows growth rates.

Another important question concerns migratory movements: where do juvenile green turtles in Bahamian waters come from? Use of mtDNA haplotype markers has shown that they are true ocean-goers. For example, juvenile green turtles that occur in Bahamian creeks have originated from nesting beaches in Tortuguero (Costa Rica), Yucatán (Mexico), Florida (USA), Aves Island (Venezuela), Matapica (Suriname), Atol das Rocas and Fernando de Noronha (Brazil), Ascension Island (UK), and Pailoa (Guinea Bissau), indicating connectivity between nesting beaches throughout the Atlantic system and foraging areas in The Bahamas (Lahanas *et al.*, 1998; Bjorndal and Bolten, 2008; Fig. 8.17a). Also, where do juveniles go when they leave Bahamian waters? Tagging studies have determined that juveniles range widely (Fig. 8.17b). Satellite telemetry has enabled tracking of movement patterns of large juvenile green turtles from shallow feeding areas in The Bahamas to feeding areas in other regions of the Caribbean (Fig. 8.17c).

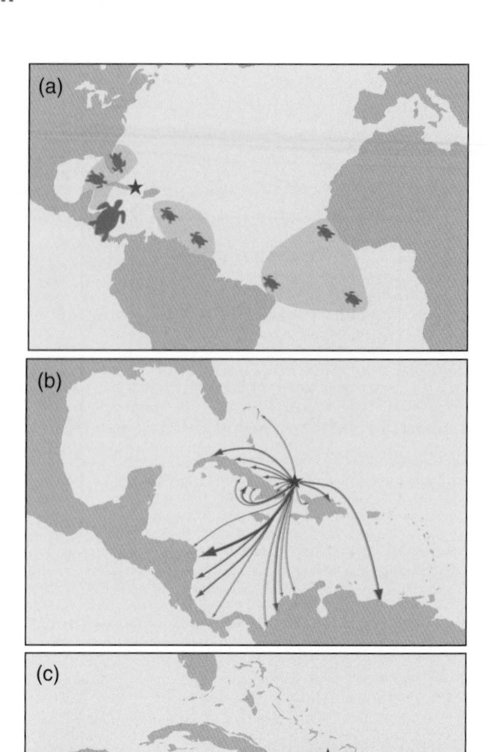

Fig. 8.17 (a) Mitochondrial DNA (mtDNA) sequences as genetic markers have demonstrated that green turtle rookeries throughout the Atlantic contribute to the population of juvenile green turtles at Great Inagua, The Bahamas (star); similar patterns are evident for other feeding grounds in The Bahamas. Sizes of turtles in the figure are proportional to the contributions of rookeries or rookery regions (dark-blue areas; Lahanas *et al.*, 1998; Bjorndal and Bolten, 2008). (b) Flipper tags with individual tag numbers and return address have determined destinations of green turtles after leaving foraging grounds at Great Inagua. In most cases, tags have been returned by fishers who have caught turtles for food. Widths of arrowheads are proportional to the number of tag returns (Bjorndal and Bolten, unpublished data). (c) Satellite telemetry has allowed tracking of movement patterns of three turtles from Great Inagua, The Bahamas, to other Caribbean feeding grounds. Movement patterns are consistent with locations from tag returns (Bjorndal and Bolten, unpublished data).

8.3.2.5 Conservation

IUCN's Marine Turtle Specialist Group has identified five major hazards for sea turtles: fisheries bycatch, direct take, coastal development, pollution and pathogens, and global warming. For The Bahamas "direct take" has historically been the most important factor. Therefore, on 1 September 2009, The Bahamas Ministry of Agriculture and Marine Resources issued the following press release after extensive consultations with the public:

> "The Bahamas Ministry of Agriculture and Marine Resources hereby announces that the Fisheries Regulations governing marine turtles have been amended to give full protection to all marine turtles found in Bahamian waters by prohibiting the harvesting, possession, purchase and sale of turtles, their parts and eggs. The new regulations also prohibit the molestation of marine turtle nests."

Therefore, the Government of The Bahamas has taken a bold step that could have an important impact on the conservation of Atlantic sea turtles. Sea turtles are a shared resource throughout the Wider Caribbean Region, which requires a coordinated, region-wide, management plan to ensure their survival. Negative effects on nesting aggregations (harvest of eggs or nesting females, or degradation of beach habitat) in one country could result in significant population changes in others. Therefore, protection by one country alone will not suffice, and over-exploitation by any one country may result in overall loss of sea turtle populations.

Regional management for sea turtles is recognized in the *Inter-American Convention for the Protection and Conservation of Sea Turtles* (IAC, online), which entered into force in May of 2001 and provides an intergovernmental legal framework for countries in the Americas to take actions benefiting sea turtle populations; The Bahamas has not yet ratified this treaty. However, CITES (Ch. 3) provides significant global protection for sea turtle populations by controlling international trade, and The Bahamas has ratified this convention. The Government of The Bahamas, in addition to regulatory measures, has established several Marine Protected Areas (MPAs) containing sea turtle habitats, including the Exuma Cays Land and Sea Park, Union Creek Reserve, and Conception Island. In the Union Creek Reserve, a 24-year study of annual survival probabilities for green turtles (involving 764 tagged turtles and 1579 re-captures) found that, within the reserve, mean annual survival probability was 0.89; this value fell to 0.76 once the turtles left the protection of the reserve (Bjorndal *et al.*, 2003). The cause for this discrepancy remains unclear, although exposure to human take cannot be ruled out.

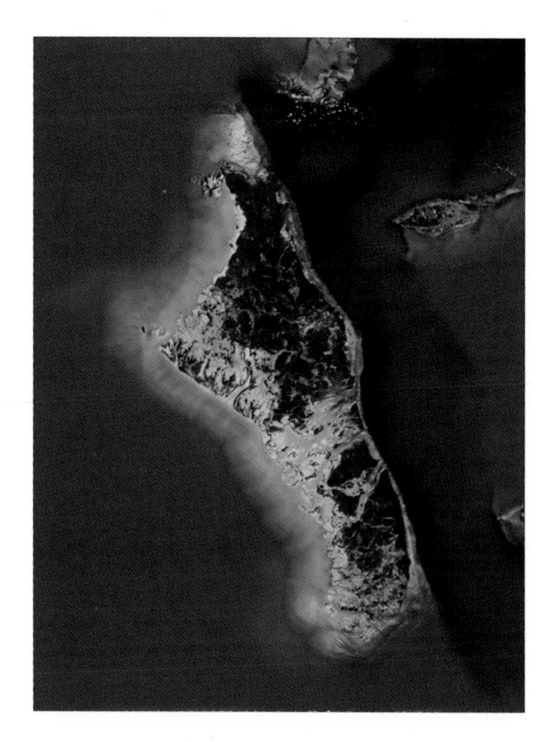

Fig. 8.18 Satellite image of Andros Island showing preponderance of wetlands (light brown). Dark green indicates coppice and woodlands. Landsat scene from the ESRI satellite library, courtesy of The Nature Conservancy, The Bahamas.

8.3.3 Wetland and creek habitat: The interface of land and sea

Craig A. Layman
Department of Applied Ecology, School of Agriculture, North Carolina State University, Raleigh, North Carolina, USA

8.3.3.1 The issue

Wetland ecosystems are ubiquitous and essential components of the coastal realm. According to the *Ramsar Convention of Wetlands* (Ch. 3), "wetlands" are defined as: "Areas of marsh, fen, peatland or water, whether natural or artificial, permanent or temporary, with water that is static or flowing, fresh, brackish or salt, including areas of marine water the depth of which at low tide does not exceed six meters" and including areas that "may incorporate riparian and coastal zones adjacent to the wetlands, and islands or bodies of marine water deeper than six meters at low tide lying within the wetlands." Accordingly, wetlands may be the most prevalent habitat type (by area) in The Bahamas. They provide a myriad ecological services, yet their value is often under-appreciated, placing them at risk of destruction or degradation.

8.3.3.2 Description and function

Marine-dominated tidal creeks are the most common wetland type in The Bahamas. Most tidal creeks receive little freshwater input because of small watershed size (essentially having no freshwater rivers or streams on any of the islands), little topographic relief, and porous calcium-carbonate geology. They typically consist of a main channel, formed from scouring of calcareous rock, that floods a broad, shallow, complex of mangrove, seagrass, and sandy substrate. Salinities are typically >30 ppt, with low nutrient concentrations (Allgeier *et al.*, 2010). Creeks range in size from less than a hectare to tens of thousands of hectares. The most expansive creeks and associated wetlands occur on the larger islands, primarily Andros, Grand Bahama, and The Abacos. Andros is most notable in this regard, with more than 250,000 ha of freshwater and marine wetlands—approximately 40% of the area of the entire island (Fig. 8.18).

Mangroves are critical components of tidal creeks. Red mangroves vary substantially in size (1–12 m in height; Fig. 8.4b). Smaller red "dwarf mangroves" can cover wide expanses of shallow portions of wetlands; growth in these areas is limited by nutrient-poor conditions, diminished tidal flushing, and karst rock that prevents root penetration. Taller red mangroves develop along main creek channels, where water is deeper and tidal flushing high. Black mangroves (Fig. 8.4c) are often interspersed at the landward edge of wetland systems and along the edges of isolated, high-salinity, wetlands. White mangroves and buttonwood occur at the transition from wetland to upland habitat, and also are common on rocky outcroppings interspersed among expanses of wetlands.

Mangrove-dominated wetlands and tidal creeks play a number of important functional roles in their ecosystems. They are critical nursery habitat for numerous ecologically and economically important species, including Nassau grouper, queen conch (*Strombus gigas*), and spiny lobster (*Panulirus argus*), by providing shelter from predators and/or food resources. Secondary production of snappers (Lutjanidae) and grunts (Haemulidae) may be especially high in mangroves (Valentine-Rose *et al.*, 2007). The red alga (e.g., *Laurencia* spp.) commonly occurs among seagrass beds and on hard bottoms of creeks, and serves as important habitat for newly settled fishes and invertebrates (Eggleston, 1995). Creeks, especially larger ones, are important feeding areas for larger organisms such as sharks, barracudas, and sea turtles. Shallow wetlands are also feeding areas for a diverse group of wading and shore birds. Mangroves play an important role in shoreline stabilization by trapping sediments, building up land areas, and preventing erosion during storms (Buchan, 2000). Creeks and wetlands also play critical roles in nutrient cycling and filtering runoff from adjacent terrestrial areas.

8.3.3.3 Conservation

Threats to wetlands are numerous and widespread. Because wetlands line coastal strands, and often occur in the vicinity of sought-after beaches and coastal views, they are among the first ecosystems to be affected by coastal development. The most severe threats are removal of mangroves, filling of wetlands for construction, and partial or complete blockage of creek flow (Fig. 8.19). Fragmentation by roads, i.e., those built with limited or no flow-conveyance structures, such as bridges or culverts, results in severe degradation. Fragmentation reduces, or completely eliminates, tidal exchange, effectively choking off the natural hydrological cycle that is essential to wetland function (Fig. 8.20; Layman *et al.*, 2007). More than 80% of the creeks along the east coast of Andros are fragmented to some degree by roads, and only a single unfragmented creek remains on New Providence (Layman *et al.*, 2004; Valentine-Rose *et al.*, 2007).

On 7 June 1997, The Bahamas signed the *Ramsar Convention on Wetlands*, an explicit call for national action through development of a framework for wetland protection, management, and conservation. The Bahamas National Wetland Policy, adopted in 2007, stated the goal: "to conserve, restore and manage wetlands wisely in conjunction with sustainable development practices," and established guidelines and a structure for the protection of Bahamian wetlands. The document clearly defines what a wetland is, why wetlands are important to the nation, and why their protection is a crucial part of national conservation strategy. The document includes well-defined guidelines for permitted activities, and an environmental impact assessment must also be conducted for all activities that are "likely to have an adverse impact on the wetlands." This broad stipulation is backed up by specific guidelines on how EIAs (environmental impact assessments) should be conducted. The EIA process can serve as a foundation for legislation, and also provides enforcement power.

Creek and wetland restoration projects are becoming a primary component of wetland management strategies in The Bahamas. The target is to restore a more-natural hydrology by installing culverts or bridges under roads (Fig. 8.21). Following restoration of tidal flow, recovery of ecological function can be rapid. Fish diversity can double in a period of weeks, and secondary production of fishes (i.e., accumulation of fish biomass over time) can increase more than twenty-fold in a year. Many restorations are conducted as community-based initiatives, serving as core educational tools for local students and the general public. From 2004–7, local NGOs, such as Friends of the Environment (Abaco) and the Andros Conservancy and Trust, have spearheaded a number of creek restoration projects, substantially improving the health and productivity of local creek systems.

Fig. 8.19 Somerset Creek, North Andros, mostly blocked by a road, showing effect of interrupted flow on turbidity. Photographs © Ray & McCormick-Ray.

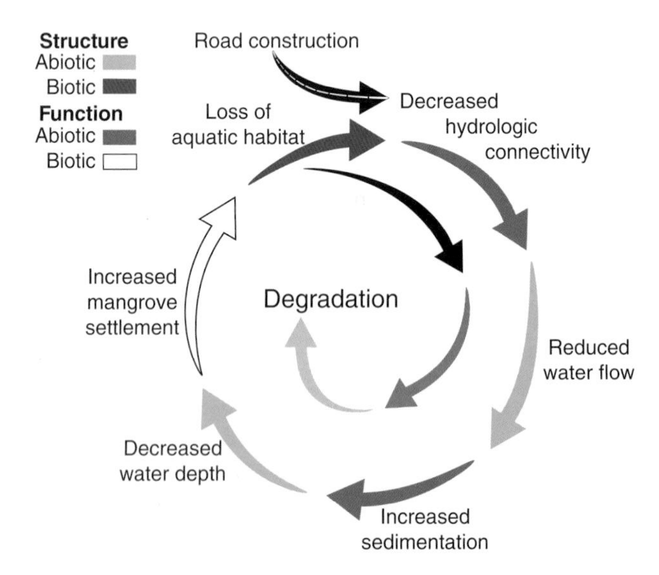

Fig. 8.20 Feedbacks involved in degradation of creeks by road construction, demonstrating the degradation feedback cycle that follows anthropogenic fragmentation of tidal creeks (Valentine and Layman, 2011). The shading of arrows represents the category of the effect, following the legend in the upper left-hand corner. Reproduced with permission of John Wiley & Sons.

Fig. 8.21 Installing culverts under roads increases hydrologic connectivity between wetlands and the ocean. These projects are a common form of ecosystem restoration in The Bahamas. Photograph courtesy C. R. Layman.

8.3.4 Invasion of Bahamian coral reefs by predatory Pacific red lionfish

Mark A. Hixon
University of Hawai'i at Manoa, Honolulu, Hawai'i, USA

Mark A. Albins
Marine Fish Laboratory, Auburn University, Fairhope, Alabama, USA

8.3.4.1 The issue

The threat of invasive species in marine ecosystems has recently been brought to the conservation forefront by the introduction, rapid spread, and expanding ecological effects of Pacific red lionfish (*Pterois volitans*) in The Bahamas and Greater Caribbean region (Fig. 8.22). Lionfish were accidentally or intentionally released from aquaria, and first noted off southeast Florida in the mid-1980s (Whitfield *et al.*, 2002; Schofield, 2009, 2010). Lionfish were first confirmed in The Bahamas in 2004 (Snyder and Burgess, 2007), underwent a population explosion there by 2007 (Green and Côté, 2008; Albins and Hixon, 2011), and have subsequently spread throughout most of the western tropical Atlantic and Caribbean region, including the Gulf of Mexico (Fig. 8.23; Freshwater *et al.*, 2009; Schofield, 2009, 2010). Lionfish now inhabit a broad variety of marine habitats in The Bahamas, ranging from coral reefs (Green and Côté, 2008), to nearby seagrass beds (Claydon *et al.*, 2012) and mangrove tidal creeks (Barbour *et al.*, 2010). Although visually appealing, lionfish are proving to be a dire threat to native fishes of The Bahamas (Albins and Hixon, 2008, 2011; Green *et al.*, 2012; Albins, 2013).

8.3.4.2 Natural history

Invasive red lionfish reach about 45 cm in length (Akins L, personal communication) and are effective predators of small fishes (Albins and Hixon, 2008; Morris and Akins, 2009; Albins, 2012). They are unique predators for several reasons. Their zebra-like color pattern, combined with their elaborately extended fin spines and rays, may breakup their silhouette, making it difficult for prey to identify them. When sitting still on the reef surface, a lionfish often resembles a clump of seaweed or the frills of a tubeworm, perhaps a form of mimicry (Albins and Hixon, 2011). When stalking prey, lionfish flare their large, fan-like pectoral fins, herd small fish into corners on the reef, and rapidly engulf them by suction feeding (Whitfield *et al.*, 2002). Lionfish also often blow a stream of water toward their prey just before striking, which results in the prey fish orienting in an easy-to-swallow, headfirst position (Albins and Lyons, 2012). Atlantic prey fishes have never before encountered lionfish, and native prey appear to take little if any evasive action. This unique combination of characteristics and behaviors may explain why lionfish are able to consume more prey and grow faster than similarly sized native predators (Albins, 2013). Additionally, comparative field observations show that invasive lionfish in The Bahamas and the Cayman Islands consume a broader variety of prey species, and eat larger prey on average, than native lionfish in the Pacific Ocean (Cure *et al.*, 2012).

Additionally, invasive lionfish are themselves resistant to predation. Atlantic predatory fishes have thus far largely ignored them. Multiple attempts have been made to feed live and speared lionfish to native sharks and groupers, both in the field and in large outdoor tanks, usually to no avail (Morris, 2009; Raymond *et al.*, in prep.), although Nassau groupers have been trained to eat speared lionfish (personal

Fig. 8.22 The invasive Pacific red lionfish (*Pterois volitans*) on a Bahamian coral reef near Great Exuma. Photograph © Mark Albins.

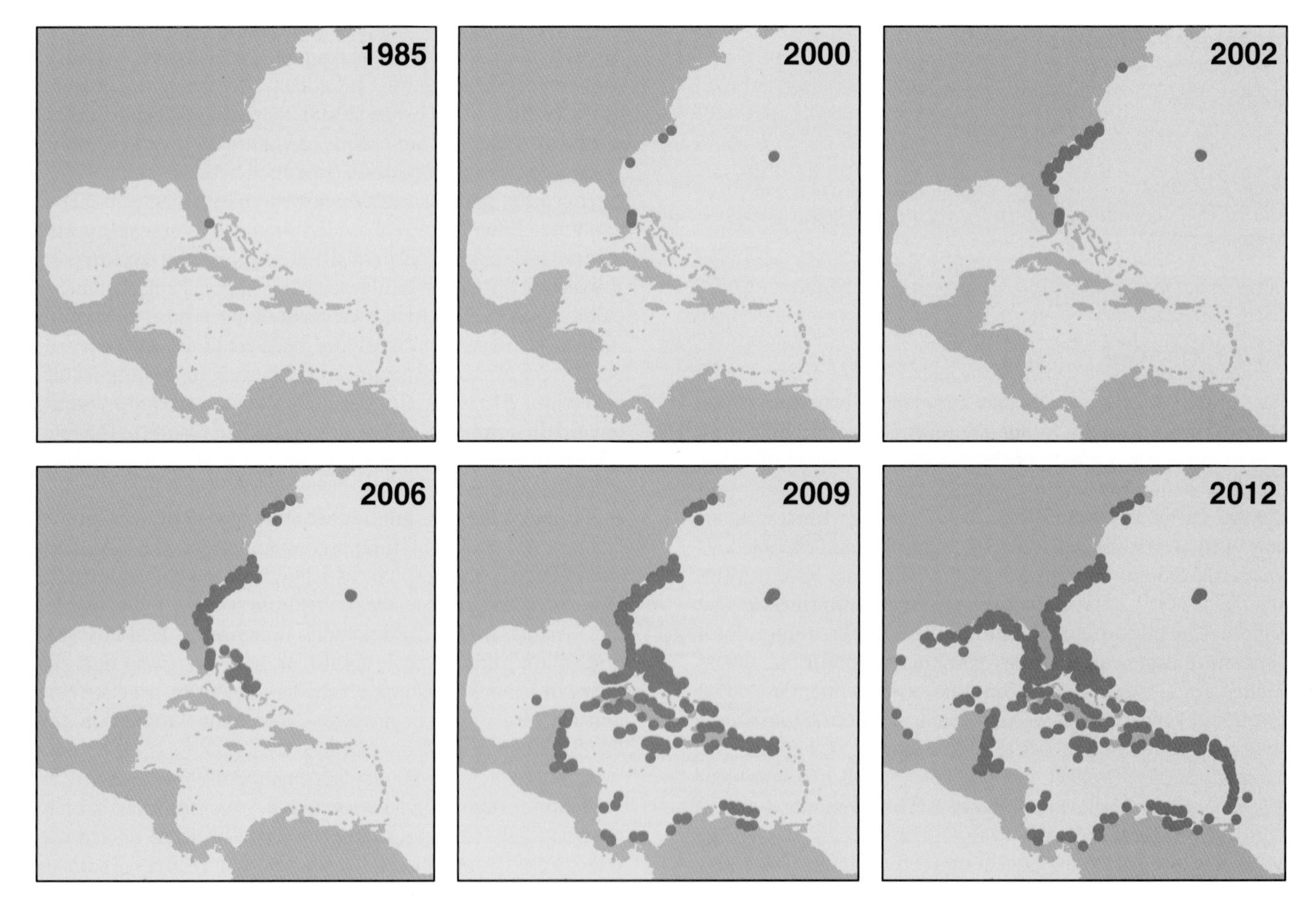

Fig. 8.23 Spread of the lionfish invasion from 1985 to 2012. Data courtesy of Dr. Pamela J. Schofield, USGS Nonindigenous Aquatic Species Database (2012).

observation). Perhaps the unique shape, color, and behavior of lionfish result in native predators not recognizing lionfish as prey. In any case, the numerous venomous spines of lionfish typically thwart those predators that do strike: 13 spines on the dorsal fin, three on the anal fin, and one on each of the paired pelvic fins. However, there are reports of tiger groupers (*Mycteroperca tigris*) and Nassau groupers (*Epinephelus striatus*) captured in The Bahamas with lionfish in their stomachs (Maljković *et al.*, 2008). Unfortunately, such larger predators are severely overfished through most of the Caribbean region (Sadovy and Eklund, 1999; Stallings, 2009). Additionally, lionfish in The Bahamas have very low levels of endo- and ecto-parasites that commonly infect native fishes inhabiting the same reefs (Sikkel *et al.*, in prep.; Tuttle *et al.*, in prep.).

The apparently low risk of predation and low parasite load, combined with high prey consumption rates, may explain the rapid individual growth, high reproductive rate, rapid population growth, and high density of invasive lionfish. Juveniles (<20 cm total length) add about 2 cm to their length monthly (Pusack *et al.*, in prep.). Adult females produce more than two million eggs annually (Morris, 2009). Fertilized eggs and developing larvae disperse in ocean currents for about 30 days (Ahrenholz and Morris, 2010). The estimated annual population growth rate throughout the Atlantic and Caribbean from 1992 through 2007 was 67% per year, and was even higher in Exuma Sound (Albins and Hixon, 2011). Densities in The Bahamas have reached more than 390 lionfish per hectare (Green and Côté, 2008)—far greater than in their native Pacific range (Kulbicki *et al.*, 2012)—and lionfish have been observed from submersibles at depths greater than 300 m in The Bahamas (R.G. Gilmore, personal communication).

8.3.4.3 Ecological effects

Invasive red lionfish now pose a substantial threat to not only native reef-fish communities (Albins and Hixon, 2008; Albins, 2013), but also indirectly the reefs themselves (Albins and Hixon, 2011). A single lionfish can reduce the number of small fish on a patch reef by about 80% in just a few weeks (Albins and Hixon, 2008). Lionfish consume a broad variety of reef fishes, and smaller lionfish also eat shrimp and other invertebrates (Morris and Akins, 2009). Victims include juveniles of important fishery species, such as groupers and snappers, as well as ecologically important species, such as parrotfishes, that help control algal cover on reefs (Mumby *et al.*, 2006). Divers have observed a single lionfish consume more than 20 juvenile reef fish in just 30 minutes (Albins and Hixon, 2008), and average consumption rates throughout the day are on the order of one to two prey per hour (Côté and Maljković, 2010).

A recent field experiment in The Bahamas determined that invasive lionfish might outcompete ecologically similar, native predators (Albins, 2013). The abundance of coney grouper (*Cephalopholis fulva*) and equal-sized lionfish were adjusted on patch reefs in three ways: one lionfish only, one coney only, and one lionfish and one coney together. After two months, coney alone had reduced the abundance of small fish on the reefs by an average of 36%, whereas lionfish alone had reduced prey fish by nearly 94%, over 2.5 times the effect of the native predator. The two species together reduced prey fish abundance to zero, indicating that one and/or both species suffered reduced consumption rates in the presence of the other. Lionfish in this experiment grew more than six times as rapidly as coney.

Ongoing research on the lionfish invasion focuses on four major themes. First, efforts to map and predict the geographic spread of the invasion quickly and accurately are reported in U.S. Geological Survey's Nonindigenous Aquatic Species Program (USGS, online). Oceanographic models predict future patterns of larval dispersal, while population genetics methods track actual patterns of dispersal (Freshwater *et al.*, 2009; Betancur *et al.*, 2011), and matrix models project future population growth (Morris *et al.*, 2011).

Second, studies continue to examine the ecological effects of this invasion on the structure and function of coral-reef ecosystems. Experimental studies in The Bahamas manipulate lionfish abundance relative to potential prey, predators, and competitors, and observational studies and models estimate how lionfish predation is altering biomass and energy flow in coastal food webs.

Third, efforts are underway to determine whether there are any mechanisms of natural resistance to this invasion. Biotic resistance appears to be low, although lionfish are rare on some reefs that tend to be buffeted by strong currents, which may reduce lionfish feeding rates. Perhaps native predators will eventually learn to consume lionfish more efficiently, or native parasites or diseases will ultimately control the invaders.

Fourth, natural mechanisms of population control are being explored. Within its native Indo-Pacific region, the red lionfish is typically rare, even at the geographic center of its range in the Philippines (Kulbicki *et al.*, 2012). Understanding whether and why this species is a relatively minor player on its native Pacific reefs may provide insight on how to control the invasion. One hypothesis is that lionfish in the Pacific are limited by early post-settlement predation, when newly settled lionfish (about 2 cm total length) have flexible spines and may be more palatable to predators. If such predation on small lionfish were confirmed in the Pacific, then effective management in the Atlantic would include conservation and enhancement of populations of ecologically similar native predators by means of fishing restrictions and marine reserves.

8.3.4.4 Conservation

The rapid and explosive invasion of red lionfish in The Bahamas has galvanized local and national responses. Joint initiatives among educational institutions, the government, and non-governmental organizations have been organized throughout the invaded region (Morris, 2012). Efforts include broad-reaching educational programs intended to teach citizens about basic lionfish biology, to explain why they are a threat to fisheries resources and the marine environment, and how they can be removed safely, effectively, and efficiently. Recreational divers assist marine scientists studying the invasion. Community events teach people to prepare and eat lionfish safely. A national "Eat Lionfish" campaign in The Bahamas seeks to assure citizens that lionfish are a viable alternative to more traditional food fishes. Lionfish are actually quite tasty, similar to their close relatives, the rockfishes and scorpionfishes (family Scorpaenidae). High-end restaurants have begun serving lionfish as a conservation dish (*The Economist*, online). However, as with native predatory reef fishes, eating lionfish caught in certain areas may cause ciguatera poisoning in humans. While many governments and NGOs are currently encouraging fisheries for lionfish, others caution that the economic benefits gained from such a fishery could encourage people to protect the invasive species, or even spread it to previously uninvaded areas, severely complicating management of the invasion (Nuñez *et al.*, 2012).

Control efforts notwithstanding, the rapid spread, high population sizes, and broad geographic range of invasive lionfish mean that complete eradication is highly unlikely. Until biotic control mechanisms are identified and fostered, there is only one effective means of controlling lionfish densities on shallow reefs: direct removal by divers. Such manual removals require extensive effort, which at high levels will come at a high cost. Lionfish inhabit deep waters that divers cannot reach and also occur on the most remote reefs of the Caribbean. Therefore, even the most comprehensive and effective program of direct removals will need to be sustained indefinitely. Otherwise, removals will be counteracted by reproductive lionfish at deep and remote locations continuing to generate larvae capable of continually re-colonizing more accessible locations.

In the long term, the invasion of Atlantic and Caribbean coral reefs by Pacific red lionfish will culminate in one of two outcomes. First, the invasion may worsen to the point where lionfish densities reach extreme levels and populations of many coral-reef fishes are decimated or even extirpated. Second, some combination of native species (competitors, predators, parasites, and/or diseases) may minimize the ecological effects of the invader (Albins and Hixon, 2011). Meanwhile, the best management strategies are likely to continue to involve organizing direct manual removal programs, fostering lionfish fisheries, and implementing marine reserves, the latter given that relatively intact ecosystems are often more resilient to invasions of exotic species. Only time will tell whether the population explosion of Pacific red lionfish in The Bahamas and Wider Caribbean Region becomes one of the most devastating marine invasions in history.

8.3.5 Social and systemic issues

The previous four issues are examples of specific bio-ecological issues facing The Bahamas. The following, in contrast, are anthropogenic and climate-related issues that tend to be systemic and emergent (Ch. 2), implying that future planning for conservation and management will necessarily be incremental.

8.3.5.1 Fisheries and fishing

The Nassau grouper exemplifies one of the most intractable, chronic challenges facing The Bahamas—fisheries—for which gaps in knowledge of natural history and critical habitats are major impediments. For commercial fisheries as a whole, small-scale foreign poaching and violations by domestic fishermen are difficult to track due to the very large size of the archipelago and remoteness of much of it. Harmful practices have included uses of chemicals, abandonment of traps, harvesting of juveniles, and catches during closed seasons. Some of these are no longer practiced (e.g., use of chemicals). However, regulatory measures are in place only for species of high commercial or sporting value and are handicapped by ineffective enforcement.

Other aspects of fishing represent significant gaps in management. For example, The Bahamas is one of the world's leading destinations for sportfishing, notably for bonefish (*Albula vulpes*), but little is known of the level of fishing pressure bonefish can withstand. The same may be said for open-ocean sport fishing for such large species as billfishes (marlins and sailfishes), even though it is generally agreed by scientists that several of these species are depleted throughout their ranges. Finally, the impacts of subsistence and sport fishing, mostly on reefs, can only be roughly estimated, although by general consensus among scientists and divers, reef fishes appear to be depleted, severely so near centers of population and recreation.

The Department of Marine Resources (DMR) is well aware of these problems, but has to concentrate almost all its efforts on commercial fisheries due to a lack of financial support, infrastructure, and personnel. Landings data highlight several features of management (Fig. 8.24). Lobsters (*Panulirus argus*; Fig. 8.25) are of greatest economic value, B$66.3 million out of a total of B$78 million for all fisheries in 2011; conch is second in value (B$ 5.1 million); third, snappers (B$3.1 million); fourth, Nassau grouper (B$0.86 million). Notably, Nassau grouper landings have declined from 0.79 million pounds in 1965 to 0.25 million pounds in 2011, representing a trend from abundance and overfishing to a need for restoration. Some fisheries species are lumped together, such as "snappers" (*Lutjanus* sp.) due to shortage of resources, symptomatic of DMR's inability to fully record levels of fishing on a biological species level. Similarly, a category of "other" includes many species of high ecological value, e.g., parrotfishes that have high ecological value—in their case, grazing on algae. Such lumping of species presents great difficulties for single species, data-based management. Furthermore, landings data alone usually fail to designate from which populations fish were actually caught. This "real-world" difficulty of species-by-species management is moving the DMR towards a broader approach, one aspect of which being establishment of fishery reserves as a means to restore and enhance fisheries and to maintain habitat and biodiversity (Section 8.4.1).

8.3.5.2 Water resources

Water is a critical resource problem for many island communities, including The Bahamas. Few surface-water streams occur in The Bahamas (Section 8.3.3). A notable exception is West Andros' large watersheds (Fig. 8.18). This largest of Bahamian islands is dominated by wetlands and numerous tidal creeks and has by far the largest groundwater aquifer of The Bahamas.

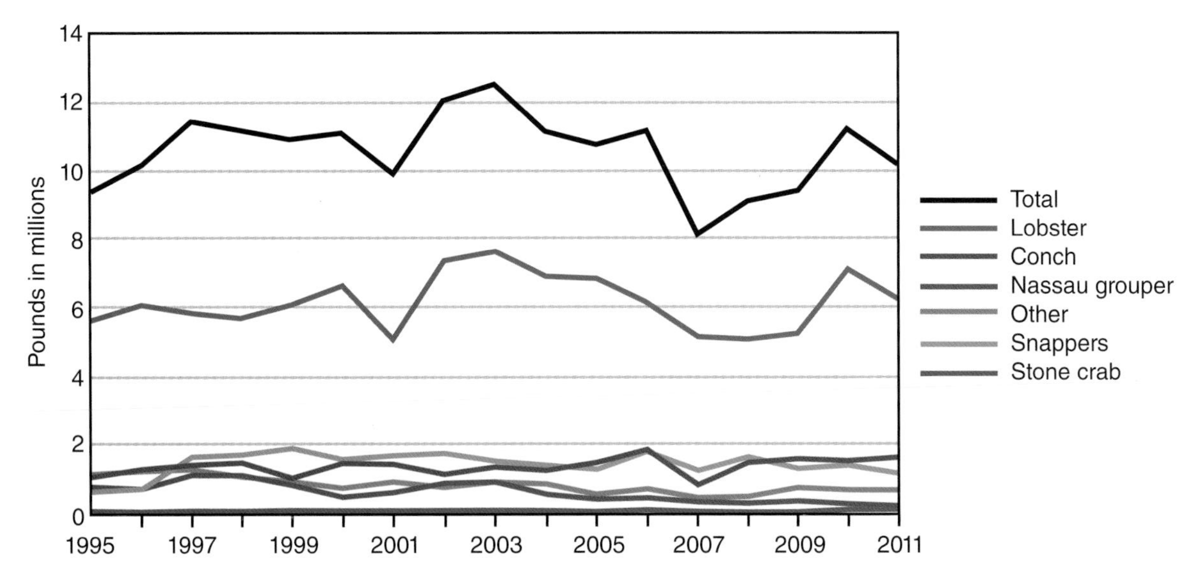

Fig. 8.24 Fisheries landings data. Total landings have been fairly consistent at ~10 million pounds annually from 1995 to 2011 (the last years of record at time of this writing). Note, however, decline for the Nassau grouper. See text for explanation. Data courtesy Bahamas Department of Fisheries.

Fig. 8.25 Spiny lobster, *Panulirus argus*, on coral reef. This species is by far the most economically valuable fisheries resource of The Bahamas. Photographs © Ray & McCormick-Ray.

Rainfall is the dominant source of freshwater, as influenced by the nature of the landscape and plant cover. The integrity of wetlands and landscape structure would seem integral to freshwater supply. However, the relationships among rainfall, wetlands, and freshwater supply are poorly understood. Groundwater resources are easy to exploit, and easily contaminated due to the porous nature of karst landscapes. In several Bahamian islands, dredging and channelization for development projects—e.g., in Grand Bahamas and New Providence—have cut through aquifers, resulting in massive losses of freshwater. The aquifer of New Providence, with approximately two-thirds of the country's human population, cannot supply the water needs of its people and the tourism industry, and has been seriously degraded by pollution and development.

Until recently, New Providence obtained up to 50% of its annual water supply by tanker barge from Andros' abundant supply, but water barging came to an end in 2011. The bulk of New Providence's water is now from reverse osmosis, an expensive operation dependent on fossil fuel energy.

8.3.5.3 Land-use policy and tourism

Many issues facing The Bahamas converge on the productive coastal strip of land and sea that includes creeks, wetlands, mangroves, seagrasses, and coral reefs—the same areas most desired for development and tourism—but which are essential habitat for many commercial and coral-reef fish (Nagelkerken *et al.*, 2000; Layman and Silliman, 2002). This situation sets up conflicts in cultural values, land use, and local and national economies.

Historically, much of The Bahamas was designated as Crown Land, including lands and waters held in trust by the English Crown on behalf of the Bahamian people. Presently, government holds all of the sea outward to the limits of the EEZ and 70% of the land in trust. In 1964, the *Promotion of Tourism Act* gave the Ministry of Tourism a mission "to make it increasingly easier to create, sell, and deliver a satisfying vacation product—satisfying to investors, employees and tourists." As a result, The Bahamas has now become the tourism leader of the Caribbean, supplying almost half The Bahamas' gross domestic product and about three-quarters of its gross foreign exchange (Cleare, 2007). Total annual arrivals from the mid-1980s (Fig. 8.26) now exceed the total Bahamian population (Fig. 8.10) by more than 1000%. Especially dramatic has been the growth in cruise ships (sea arrivals) that bring visitors for short stays. By 2010, cruise-ship arrivals brought 75% of all tourists. However, cruise-ship revenues to The Bahamas were only a tenth of revenues from visitors arriving by air who stayed for longer periods in local facilities. Nevertheless, Nassau's port has recently been dramatically expanded to accommodate the largest megacruise ships; even remote islands are not immune to becoming ports of call.

Related to the dramatic increase in tourism is immigration and growth of the general population. As a result of both population pressure and The Bahamas' overwhelming dependence on tourism, coastal lands near population centers have been over-developed, channels have been cut through reefs, marshes, and swamps, wetlands have been used as dump sites, and coastal waters have become increasingly contaminated, eutrophic, and turbid. Nearshore construction has initiated sedimentation that smothers coral. Compounding the effects of coastal development has been the introduction of the highly invasive, exotic Australian pine (*Casuarina equisetifolia*) that has facilitated erosion by inhibiting native vegetation (Fig. 8.27; Hammerton, 2001). And, through little fault of The Bahamas' own, many beaches are littered by refuse dumped at sea (Fig. 8.28). The Ministry of Tourism recognizes that these environmental impacts have affected tourists' perception of The Bahamas, and that rapid growth of large tourist facilities destroys the very habitats that support the resources upon which much of the tourism industry is based. Thus, government is seeking solutions involving land-use planning (see Section 8.4.3).

8.3.5.4 Climate change

Low-lying islands are especially at risk from rising sea level and a probable increase in storms (IPCC, 2007) due to increasing

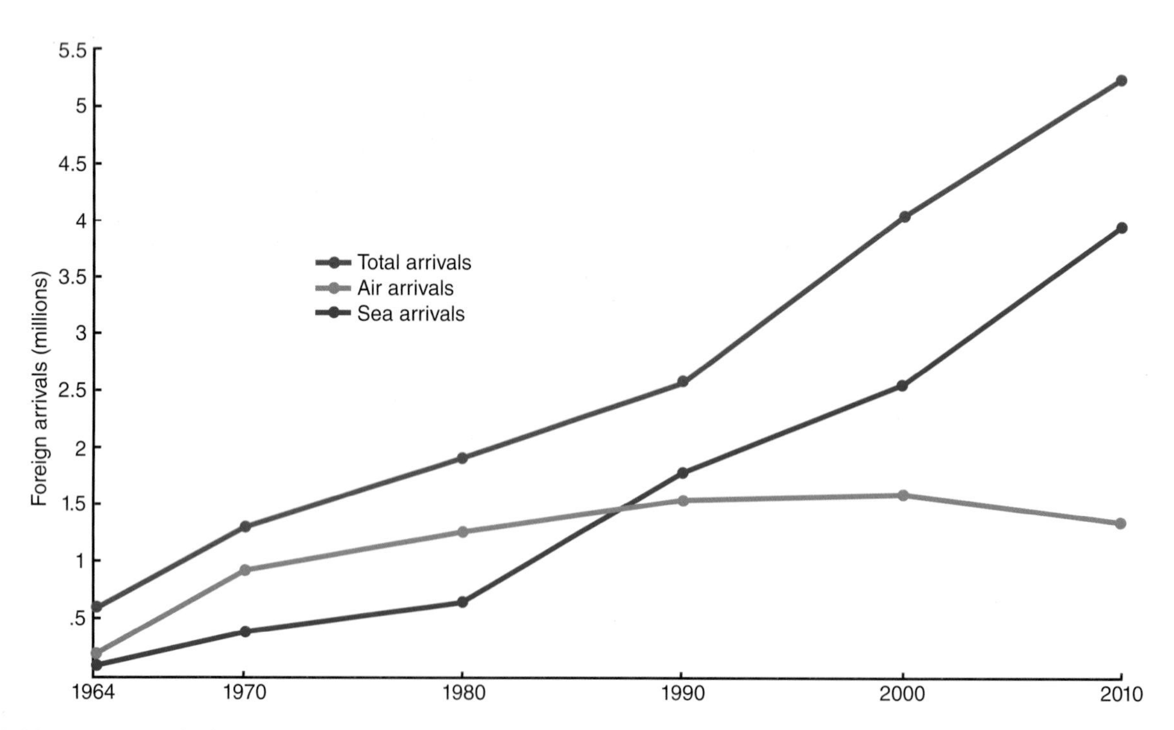

Fig. 8.26 Tourism growth 1964–2010. Data courtesy of Angela Cleare, Bahamas Ministry of Tourism.

Fig. 8.27 The invasive Australian "pine" (*Casuarina equisetifolia*) contributes to beach erosion. This species forms monospecific stands that shade and aggressively compete for resources with the native vegetation (Hammerton, 2001). Photograph © Ray & McCormick-Ray.

Fig. 8.28 Former green turtle nesting beach on Little Inagua. Litter from sea prevents further nesting there. Photograph © Ray & McCormick-Ray.

heat anomalies (Hansen *et al.*, 2012). The Bahamas' *National Policy for the Adaptation to Climate Change* (2005) accepts IPCC's findings that global temperatures are increasing due to human activities, and that this trend is likely to continue. As the *National Policy* states, "Many of the manifestations of global climate change are already occurring [bleaching, loss of coral reefs, submergence of low-lying islands, and increased frequency of cyclonic events]. Government recognizes that, while not all the processes relating to global climate change are fully understood. . .incomplete knowledge is not an acceptable basis for delay of taking no action." This statement reflects The Bahamas' adoption of the "precautionary principle" (Ch. 13). The Bahamas Policy also "proposes strategies for anticipating and ameliorating or avoiding the negative impacts. . .on: coastal and marine resource and fisheries, terrestrial biodiversity resources, agriculture and forestry, human settlements and human health, water resources, the energy and transportation sector, as well as on tourism and the finance and insurance sectors." As an initial step, the Department of Meteorology has undertaken a detailed study of storm surges that

Fig. 8.29 (a) Degraded Inagua reef, 1998, cause not known, but probably from multiple causes: e.g., El Niño events, overfishing, disease, etc. Photograph © Ray & McCormick-Ray.

are projected to have considerable potential to erode coasts, and recognizes the role that its extensive reef systems have played in attenuation of storms and waves.

8.3.5.5 Degraded coral reefs

Coral reefs have waxed and waned globally for millions of years, surviving the vicissitudes of climate and sea-level change. However, during the past few decades, the "coral-reef problem" (Box 2.3) has emerged as a major global conservation priority due to the sudden and obvious changes that have occurred during the past few decades. Notable for The Bahamas are coral bleaching (Fig. 2.5a) and the almost complete loss of staghorn and elkhorn corals (*Acropora cervicornis* and *A. palmata*, Fig. 8.7d,e); this situation has been linked to warming and El Niño events that occurred during the 1990s. Furthermore, overfishing has disrupted the reefs' complex food webs and ecological interactions with adjacent mangrove, seagrass, and creek environments (Mumby *et al.*, 2006). This, compounded with coastal development and sedimentation, has reduced both habitat and biodiversity. The result has been that many coral reefs have been seriously degraded (Fig. 8.29), thereby becoming less resilient to further environmental change. It is expected that declining reefs will leave less protected coastal environments. Increasing wave energy will reach the shore, allowing coastal erosion and overwash to further damage coastal structures and environments.

8.4 GOVERNANCE FOR SUSTAINABILITY

The Bahamas has emerged from a "boom-and-bust" history (Section 8.2.3) into a relatively prosperous nation today. Nevertheless, as the nation's population and economy have grown dramatically following World War II, land development has increasingly come into conflict with environmental and resource sustainability and social well-being. This situation has been exacerbated by economic dependency on tourism. For example, in 2001 following the "9/11" terrorism attacks on the U.S., tourist visitation to The Bahamas dropped sharply. Again, during the global economic crisis of 2008–9, The Bahamas suffered severe visitor decline. The present challenge is to manage tourism and economic development so that the benefits for investors, residents, and visitors are optimized, while social-environmental costs are minimized.

The Bahamas commitments to sustainability and a movement towards comprehensive planning and management have grown significantly stronger during the first decade of the 21st century. Three arenas of activity are notable: protected areas, public partnerships, and government policy.

8.4.1 Protected areas

Protected areas comprise The Bahamas' major, long-term, expression of conservation. Government policy obligates the nation, as a signatory to the 1992 *Convention on Biological Diversity* (CBD), to conserve biodiversity and to protect and preserve environments. In 2004, the government agreed to implement a Program of Work on Protected Areas, which was adopted at the Seventh Meeting of the Conference of Parties to the Convention; marine reserves were included in a further agreement in 2006. These actions commit The Bahamas to fulfill *The Bahamas 2020 Declaration*—i.e., to place 20% of its marine area under protection by 2020 and to effectively manage at least 50% of existing terrestrial and marine areas by that time.

The Bahamas National Trust (BNT) is The Bahamas' quasi-governmental agency for protected areas. Its vision is to "establish a comprehensive system of national parks and protected areas with every Bahamian embracing environmental stewardship." The Bahamian park system was initiated by an expedition to the Exuma Cays in 1968, which resulted in passage by Parliament of the *Bahamas National Trust Act* of 1959 (Ray,

1958, 1998). This action facilitated the establishment of the 456 km^2 Exuma Cays Land and Sea Park—a unified land-and-sea protected area under a single non-governmental jurisdiction, globally the first of its kind. A half-century later, the Exuma Park has become a premier tourist, yachting, and recreational destination, as well as a principal site for conservation-related research (Boxes 8.1, 8.2).

By the turn of the 21st century, protected areas included 12 national parks and protected areas totaling 131,077 ha—representing only 0.4% of all The Bahamas' territory, 7% of its land, and 0.1% of its marine environment. In 2008, the BNT in collaboration with The Nature Conservancy and others produced The Bahamas National Protected Area Master Plan. The government agreed to that Plan and, as a result, the park system has doubled in size by creating eight new parks, principally on Andros Island. At the time of this writing (2013), 25 National Parks covering 236,000 ha are managed by the BNT, housing an impressive array of ecosystems and serving as a tremendous source of pride to Bahamians (Bahamas National Trust, online). Nevertheless, The Bahamas park system covers only 0.7% of The Bahamas' total area, and remains far from the goal of 20% coverage. A serious impediment is that until 2008 the BNT had received less than 10% of its annual budget from government, and had been forced to raise the remainder of its support funds from private sources. That year, the government, recognizing the value of the BNT's protected-area system to its economy, pledged to enhance the budget by $1 million per year. This pledge has continued, albeit being subject to annual political decisions and availability of funds.

An important consideration for protected-area selection is application of science-based criteria. Historically, most protected areas worldwide have resulted from unplanned opportunities. Now, most are based on environmental data and information on species declines. Notably, the extensive protected-area system for Andros resulted from intensive rapid-assessment surveys, also involving consultation with local stakeholders. Nevertheless, the spatial scale for sustainable populations and ecosystems is often considerably larger than designated areas. For example, studies following creation of the Exuma Cays Land and Sea Park had indicated that abundances of lobster, conch, and certain fish had increased substantially. However, a recent study indicates that the Park is not sufficient in size to sustain conch, and that sustainable take of conch in the region will depend upon a network of MPAs along with other management measures to reduce fishing mortality (Stoner *et al.*, 2012). Additional studies of recruitment patterns in the Park support the conclusion that the scale for adequate sustainability of species encompasses the entire Exuma Sound basin (Boxes 8.1, 8.2). The inescapable

Box 8.1 Metapopulation dynamics and marine reserves: Caribbean spiny lobster in Exuma Sound, Bahamas

Romuald N. Lipcius
Department of Fisheries Science, Virginia Institute of Marine Science, College of William and Mary, Gloucester Point, Virginia, USA

William T. Stockhausen
Alaska Fisheries Science Center, National Marine Fisheries Service, NOAA, Seattle, Washington, USA

David B. Eggleston
Department of Marine, Earth and Atmospheric Sciences, North Carolina State University, Raleigh, North Carolina, USA

Marine reserves have been adopted worldwide to conserve the dwindling biodiversity of the world's oceans. One major postulated benefit of no-take marine reserves (i.e., Marine Protected Areas where exploitation of fishery species is prohibited) is that the number and biomass of breeding individuals (i.e., spawning stock) will increase within reserves and subsequently enhance larval production and recruitment to the reserves as well as exploited sites. The likelihood that this benefit will accrue depends on various factors, but particularly on the manner in which larvae produced in reserves are redistributed to reserve and exploited sites through hydrodynamic transport and recruitment processes, which can be conceptually subsumed under the theory of metapopulation dynamics.

Metapopulation dynamics, as used herein, deals with the dynamics of fragmented yet interconnected breeding populations. Populations are fragmented spatially, often by geographic barriers that restrict movements of breeding individuals between populations, such that interbreeding occurs predominantly between individuals within each population, rather than between individuals from different populations. These fragmented populations are, however, interconnected by dispersive stages in life history, which for marine species usually pertains to larvae and postlarvae dispersing via oceanic currents. In marine populations, dispersive stages in species with complex life cycles break the connection at the local scale between reproduction and recruitment; connectivity among subpopulations and metapopulation dynamics are emergent and critical properties of the system.

The degree to which marine reserves enhance recruitment under the influence of metapopulation dynamics is virtually unknown. In a recent study, which is summarized herein, we used field data on abundance of the

(*Continued*)

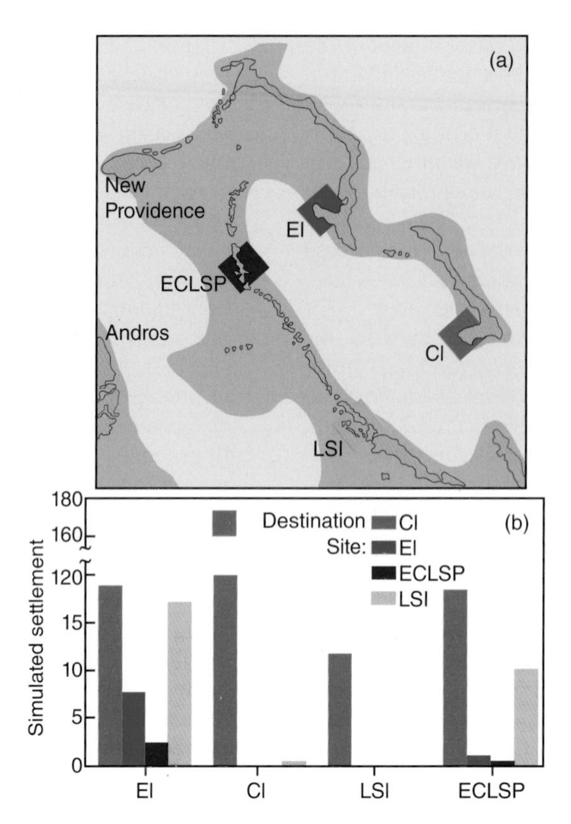

Fig. B8.1.1 (a) Caribbean spiny lobster and study sites in Exuma Sound, Bahamas—Exuma Cays Land and Sea Park (ECLSP), Eleuthera (EI), Cat Island (CI), and Lee Stocking Island (LSI). (b) Simulated larval transport and postlarval settlement in Exuma Sound. Release sites are on the abscissa; settlement sites are on the ordinate. Planktonic larvae were produced along the coast during the spring reproductive period at each of the four sites in equal numbers. In the simulations, larvae entered the offshore region and were transported by advection and diffusion as passive particles. After 180 d in the plankton, larvae metamorphosed to the postlarval stage, which actively migrated toward the nearest coast and settled.

Caribbean spiny lobster (*Panulirus argus*), habitat quality, and hydrodynamic transport patterns for an existing no-take reserve (Exuma Cays Land and Sea Park [ECLSP]) and three exploited, potential reserves (Cat Island [CI], Eleuthera [EI], and Lee Stocking Island [LSI]) in Exuma Sound, Bahamas, as a model system for assessing reserve success in enhancing metapopulation recruitment (i.e., spiny lobster postlarval influx to all sites encompassing the metapopulation).

Specifically, we assessed the likelihood of success of a reserve at each of the four Exuma Sound sites (= populations) in enhancing lobster recruitment, assuming that these sites comprise a semi-closed metapopulation. We assumed that the four sites in Exuma Sound harbor fragmented lobster populations because the sites are separated by either: (i) approximately 100 km of shoreline, which is much greater than the average benthic dispersal distance of most juveniles and adults in lobster populations; (ii) extensive shallow banks with little structural relief, food, or shelter, and which limit inter-site movements; or (iii) the deep, 2000 m basin of Exuma Sound (Fig. B8.1.1a). Furthermore, the major coral reefs, where adults reside and reproduce, are fragmented similarly across the four sites in Exuma Sound. The nominal populations are, however, interconnected due to larval and postlarval interchange between breeding populations in the sites.

Sites other than the existing marine reserve in the northwestern corner of the sound (ECLSP) were selected to provide broad spatial coverage of lobster habitats throughout Exuma Sound (Fig. B8.1.1a): Cat Island (CI) to the southeast, Eleuthera (EI) to the northeast, and Lee Stocking Island (LSI) to the southwest. The sites are characterized by coral reefs, patch corals, sand-covered hard-bottom, seagrass meadows, and fields of gorgonians and sponges. Coral reefs occur commonly to 30 m depth before the bottom plunges to 2000 m. Most small islands (Exuma Cays) lie on the western edge of Exuma Sound; tidal inlets connect the sound to the shallow (3–5 m depth) Great Bahama Bank. The Caribbean spiny lobster occurs commonly in reefs and inshore lagoons ringing Exuma Sound.

The ECLSP reserve, which extends 35.4 km from Wax Cay Cut southeast to Conch Cut in the northwestern quadrant of Exuma Sound, is a 456 km^2 no-take zone supervised by the Bahamas National Trust. The reserve, which was established in 1959 by the Bahamas National Trust, is the world's oldest land-and-marine park. Most islands within ECLSP are uninhabited, though recreational boats are common. Although ECLSP has been a no-take reserve since 1986, poaching by local fishers and visiting boaters is not uncommon.

We theoretically assessed the effectiveness of the existing reserve at ECLSP and single nominal reserves at each of the other three sites by estimating the degree to which larvae produced at a reserve were redistributed as recruits among all the sites. We did not address the relative merit of a single reserve and several small reserves, but instead emphasized the importance of spatial position upon reserve success. Enhancement of recruitment was deemed optimal when all four sites (i.e., the metapopulation) received recruits from the single functional or nominal reserve. Each nominal reserve was of a sufficient size (approximately 15% of coastal reefs) to meet the threshold requirement for enhancement of recruitment.

Using a circulation model, we theoretically assessed the effectiveness of the ECLSP reserve and nominal reserves at each of the exploited sites in enhancing metapopulation recruitment by estimating the degree to which larvae produced at a particular site were transported and redistributed to all the sites (Fig. B8.1.1b). Enhancement of recruitment was deemed optimal when all four sites (i.e., the metapopulation) received recruits from a particular site. Larvae discharged from ECLSP and EI recruited throughout Exuma Sound, with EI having the most equitable distribution of larvae. Larvae released from LSI and CI only recruited to CI and LSI due to selective hydrodynamic transport by gyres, which prevented larvae produced in southern Exuma Sound from reaching northern sites. Hence, a reserve at EI would be most suitable for enhancing metapopulation recruitment. A reserve at ECLSP functioned less effectively in enhancing metapopulation recruitment, though significantly more so than reserves at CI and LSI.

In assessing criteria useful in reserve designation for metapopulation enhancement, the use of information on habitat quality or adult density for the Caribbean spiny lobster in Exuma Sound was no more of a guarantee for success than if one were to determine the reserve site simply by chance. The only strategy that increased the likelihood of selecting an effective marine reserve for augmenting metapopulation recruitment was that which incorporated transport processes, resulting in the selection of either EI or ECLSP. Hence, the designation of a reserve location for exploited marine species requires careful attention to data on metapopulation dynamics and recruitment processes, when the opportunity to use such information presents itself.

Sources: Bohnsack (1994); Botsford *et al.* (2001); Crowder *et al.* (2000); Hanski (1998); Lipcius *et al.* (2001); Lipcius *et al.* (1997); Roberts (1997); Roberts and Polunin (1993); Stockhausen *et al.* (2000); Stockhausen and Lipcius (2001)

Box 8.2 Patterns of reef-fish larval dispersal in Exuma Sound, Bahamas

Mark R. Christie
Department of Zoology, Oregon State University, Corvallis, Oregon, USA
Mark A. Hixon
University of Hawai'i at Manoa, Honolulu, Hawai'i, USA

Most coral-reef fishes, like most marine animals, have a two-phase life cycle in which relatively sedentary adults in local populations release their gametes into the water column, where their developing larvae then disperse at sea before settling as new recruits to either nursery or adult habitats. Thus, offspring may settle at or near the same reef where they were spawned (self-recruitment) or disperse various distances to other local populations (connectivity). Documenting patterns of self-recruitment and connectivity is essential for understanding population dynamics, and consequently, for both managing and conserving marine populations effectively.

Determining geographic patterns of larval dispersal—where a fish was spawned and where it settled—is challenging due to the minuscule sizes of larvae and the vast areas of ocean through which they may travel. Directly determining patterns of larval dispersal involves the unequivocal identification of the birth and settlement locations of individual fish. Direct tagging of larval fish involves labeling the otoliths (ear stones) of developing embryos with radioisotopes before larval dispersal. Newly recruited juveniles can then be examined for distinctive traces of these chemical tags. Unfortunately, this method can be both expensive and logistically difficult.

One alternative means of directly documenting larval dispersal is genetic parentage analysis. Parentage analysis is a practical approach provided that the parent and offspring have not moved substantial distances before being sampled, as is true for most reef fishes. Furthermore, parentage analysis directly distinguishes among individuals such that numerous fish from many populations can be matched with their parents. Importantly, genetic sampling does not require a fish to be killed; a properly preserved fin clip is all that is needed.

In 2004 and 2005, we sampled 751 adult and recently settled bicolor damselfish (*Stegastes partitus*) from 11 reefs located around the perimeter of Exuma Sound, Bahamas (Fig. B8.2.1). Bicolors are abundant planktivores widely distributed throughout the Bahamas and Caribbean. After settlement to reef habitat, they rarely move more than a few meters. Males guard demersal eggs, which after hatching are planktonic for approximately 28 days. The Exuma

(*Continued*)

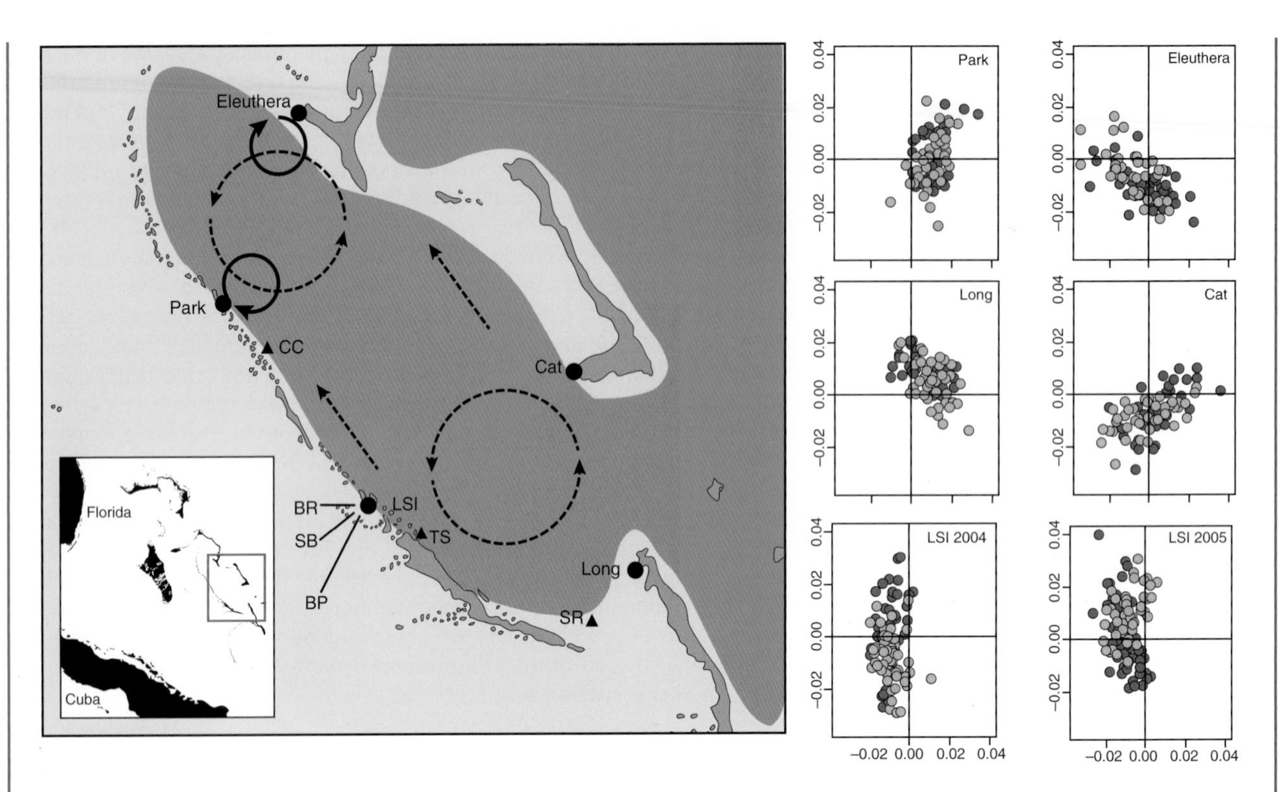

Fig. B8.2.1 Patterns of larval dispersal of bicolor damselfish in Exuma Sound, Bahamas. Parentage analysis identified two parent-offspring pairs (solid arrows), which directly documented self-recruitment at the two northern-most sites. Light seafloor indicates the shallow Great Bahama Bank, whereas dark seafloor indicates the Exuma Sound and nearby open Atlantic Ocean. Dashed arrows, including seasonal gyres, indicate prevailing surface currents. Triangles and straight arrows indicate 2004 sample sites, and filled circles indicate 2005 sample sites: Park = Exuma Cays Land and Sea Park, CC = Compass Cay, BR = Bock Rock, SB = String Bean Cay, BP = Barb's Point, LSI = Lee Stocking Island, TS = Three Sisters, SR = South Reef. The plots on the right represent principal coordinate analysis on all pair-wise relatedness values of sampled bicolor damselfish at five sites, with results separated by sampling location for clarity. Adults are represented by red circles and recruits are represented by yellow circles. Both axes combined explain 42% of the total variation.

Sound is a semi-enclosed basin with an average depth of 1500 m. To the west lies the Great Bahama Bank, a wide but shallow (<3 m deep) carbonate bank that has little suitable coral-reef habitat. To the east lies the open Atlantic Ocean. Within the Exuma Sound, seasonal mesoscale gyres could couple eastern and western reefs via larval dispersal. Furthermore, a general northwesterly flow derived from the Antilles Current could serve to transport larvae from southern to northern reefs.

All fish were genotyped at seven microsatellite loci, which are repeating, non-coding regions of nuclear DNA. These genetic markers are especially useful for parentage analysis because they are highly variable, which means that there are many different versions (alleles) of each microsatellite at each location (locus) on a chromosome. Pair-wise measures of genetic differentiation between sample sites (e.g., F_{ST}) were not significantly greater than zero, suggesting high levels of gene flow over evolutionary time periods.

In order to determine ecologically relevant patterns of larval dispersal, we applied novel Bayesian parentage analyses. These methods fully account for the hundreds of thousands of pair-wise comparisons that could result in pairs of fish sharing alleles by chance. Applying these methods, we identified two parent-offspring pairs at the two northern-most sampling locations: Exuma Cays Land and Sea Park (Park) and Eleuthera (Fig. B8.2.1). In both cases, the parent and offspring were sampled at the same site, indicating that self-recruitment occurred at both locations. Demographic estimates of adult population sizes at these sites were very large (>100,000 fish). These estimates, coupled with our relatively small sample sizes (Land and Sea Park: adults = 44, recruits = 45; Eleuthera: adults = 49, recruits = 37), suggest that self-recruitment must have occurred at high rates. The Exuma Cays Land and Sea Park is a long-established and effective marine reserve, and these parentage results indicate that larvae spawned within their borders may replenish marine reserves.

These estimates of self-recruitment were further bolstered by pair-wise relatedness analyses, which measure the number of shared alleles between all pairs of individuals. We used principal coordinates analyses to examine the relatedness values, an ordination procedure where individuals that share identical alleles would lie directly on top of one another, whereas genetically dissimilar individuals would be far apart in multivariate space. These analyses showed that recruits from all sampled sites shared more alleles with adults sampled from the same sites, which is further indicative of self-recruitment (Fig. B8.2.1).

These methods revealed that, for the years we sampled, self-recruitment is the relevant ecological pattern of larval dispersal of bicolor damselfish within Exuma Sound. We also detected patterns of sweepstakes reproduction, whereby only a small proportion of the total number of adults successfully contributed to the next generation. Indications of sweepstakes reproduction included greater levels of relatedness within recruit samples than within adult samples, greater genetic differentiation among recruit samples than among adult samples, and lower levels of genetic diversity within recruit samples than within adult samples. Further evidence of sweepstakes events included significant differences in allele frequencies between samples of recruits collected at Lee Stocking Island over the two years we sampled. These observations suggest that temporal variation may play a substantial role in shaping the genetic structure of populations of bicolor damselfish in the Exuma Sound.

Future work should focus in improving the resolution and ease of methods for detecting ecologically relevant patterns of larval dispersal. Future genetics methods will likely utilize thousands of markers that will undoubtedly help to elucidate subtle population structure. The combination of rapid methodological and theoretical advances will soon allow for near real-time estimates of population connectivity and self-recruitment. The challenge thereafter will be for scientists, policymakers, and other stakeholders to use such new information to design Marine Protected Areas and other spatially explicit management measures to sustain fisheries and conserve marine biodiversity more effectively than ever before.

Sources: Christie (2010); Christie *et al.* (2010); Hedgecock *et al.* (2007); Jones *et al.* (1991)

conclusion is that current parks are vulnerable to changes of their surroundings, and that the scale of management is considerably larger than protected areas themselves.

A related, important issue is the Department of Marine Resources' recent establishment of four marine reserves (Fig. 8.1), with the future intent of a network of marine reserves, i.e., "no-take" fishery areas that would also include conservation of biodiversity and essential habitats. Marine reserves differ substantially from parks, which emphasize biodiversity protection and education. Conversely, marine reserves are intended mainly for long-term preservation of essential habitats (reefs, seagrasses, mangroves, and creek nursery areas), enhancement of non-intrusive human activities, enhanced support for fisheries production, and scientific research. The complimentary purposes of parks and reserves were recognized in 1986 when the Exuma Cays Land and Sea Park was declared a fisheries "no-take" area in 1986, the first of its kind in the Caribbean.

8.4.2 Government policy

One of the most significant initiatives for environmental conservation in The Bahamas has been the establishment of the Ministry of the Environment in 2004, with a vision for future environmental and economic sustainability, accompanied by responsibilities for environmental policy, climate change, biodiversity, and coastal development (Box 8.3). One of the Ministry's most important accomplishments was passage by The Bahamas' Parliament in July 2010 of the *Planning and Subdivision Act: an Act to Combine, Consolidate and Revise the Law Relating to Town Planning and the Law Relating to the Development of Subdivisions and to Provide for Matters Connected Thereto.* Its purposes are to:

(a) provide for a land use planning based development control system led by policy, land use designations, and zoning;
(b) prevent indiscriminate division and development of land;
(c) ensure the efficient and orderly provision of infrastructure and services to the built environment;
(d) promote sustainable development in a healthy natural environment;
(e) maintain and improve the quality of the physical and natural environment;
(f) protect and conserve the natural and cultural heritage of The Bahamas;
(g) provide for planning processes that are fair by making them open, accessible, timely and efficient;
(h) recognize the decision making authority and accountability of the Government in land use planning; and
(i) plan for the development and maintenance of safe and viable communities, within the policies, and by the means, provided under this Act.

This Act thus seeks to address the root of the issues described above; that is, to promote sustainable development, to preserve and restore resources, and to avoid the rampant and unsustainable past development practices that have despoiled significant segments of The Bahamas' environment. In particular, it applies to the most valuable coastal strip for which comprehensive planning and zoning is fundamental—the coastal area where coral reefs, mangroves, seagrasses, beaches, and attractive seascapes predominate. The Act does not refer specifically to park or protected-area planning, or to fisheries, biodiversity, or other conservation priorities. Rather, it seeks to place development under a broad, consistent, fair, and depoliticized framework that has potential to control harmful effects on resources and to promote sustainability while maintaining and improving citizens' well-being. The Act also includes, for the first time in The Bahamas, formal environmental impact assessment (EIA) and statement (EIS) processes, intended to: (i) determine potential impacts, and the degree of such impacts of a proposed undertaking on the environment; and (ii) identify the measures to be established to mitigate against any potential adverse impacts that might occur as a result of the proposed undertaking. Should this groundbreaking Act be

Box 8.3 Environmental stewardship

Hon. Earl D. Deveaux
Former Minister of the Environment, The Bahamas, Nassau, Bahamas

The way of life for Bahamians and the economy of The Bahamas are inextricably linked to the environment of the archipelago. From earliest recorded history, humans in The Bahamas have defined their existence in relation to the environment. Columbus experienced this fundamental fact when he first encountered the Arawaks, Tainos, and Lucayans living off and, to some degree, in harmony with Bahamian land and seas. Later on, but still early in the history of the islands, when deprivation seized early European settlers on Eleuthera, sailors travelling to Boston saved them from starvation. The generosity of the sailors' gift of food was later repaid with timber and dyewood extracted from the forests so that the early settlers ". . .be not guilty of the foul sin of ingratitude. . ." (Michael Lightbourne, unpublished MS).

Through the years of the United States Civil War, the flight of Seminole Indians to the west coast of Andros, German U-boats marauding English merchant ships during World War II, sponging, and the U.S. Prohibition, the environment has been a central and constant feature of life and death in The Bahamas. As the population of the islands grew, more was demanded from island resources to sustain life and living. In fact, the failed experiments with sisal production in Andros and the depletion of soils through intensive cultivation of cotton in many parts of The Bahamas bear testimony to excessive resource demands. The demise of the sponging industry can largely be attributed to over-exploitation. The same is true of the timber resources of Abaco, Andros, and Grand Bahama, which were exploited in the same manner during the 1940s and 1950s, as were the hardwoods following the arrival of Columbus and the Spaniards.

The economic growth of The Bahamas is based on the unique appreciation by generations of Bahamians of the geography of The Bahamas and its most important resource, the environment. The twin pillars of the Bahamian economy, tourism and financial services, created the modern Bahamian economy and generate the third highest per capita income in the Western Hemisphere. From as early as 1958, visionary Bahamian policy-makers recognized the need to legislate protection for the environment.

As globalization reached all corners of the earth, it is the environment that feels its compelling urgency most acutely. The growth in the world's population, the spread of industrialization, and the connectedness of travel have, together, created a high level of pressure on the natural resources of the world. The pervasive extent of this intrusion has never before in history been witnessed to this degree. Global impacts have been categorized under the broad umbrella of climate change and its most insidious life-changing manifestation, rising sea levels, caused by increased temperatures and changes in weather patterns. The adverse impact of global warming and the need to foster a nationwide system of sustainable development informed the actions of governments of The Bahamas.

It is countries like The Bahamas—low lying coastal island nations with alluring displays of beautiful beaches, crystal-clear warm waters, and sublime winter climates—that face the adverse impact of climate change most acutely. Global warming in whatever form it eventually manifests (sea-level rise, coral-reef bleaching and death, depleted fish stocks, ecosystem collapse) will affect the islands of The Bahamas most directly. The terms are not abstract in relation to The Bahamas, which is most vulnerable to any worst-case scenarios. Because The Bahamas long recognized the environment as the basis for the continued wealth and well-being of its citizens, the country set out to create prescribed measures to sustain and protect it. The reality of life in The Bahamas underpins the policy objective behind the range of decisions taken to protect natural resources and set aside parts of national space in permanent protection under the stewardship of The Bahamas National Trust.

As challenges emerged and research informed clearer policy options, The Bahamas government responded in various far-reaching and consequential ways to protect and preserve its environment. In 1992, the country banned longline fishing, which has had a far-reaching, damaging impact on wild fish stocks. By banning this practice in The Bahamas, species such as tuna, sharks, and other migratory species were protected. In 1998, the country closed commercial fishing during critical periods of Nassau grouper spawning aggregations. In 1999, following a scientific panel review and completion of a detailed analysis, The Bahamas established the second in a series of permanent Marine Protected Areas, to preserve ecosystems, fish stocks, and replenishment and nursery areas These actions were followed by protecting sea turtles, banning shark fishing, and protecting marine mammals. Between 2007 and 2012, The Bahamas doubled the size of land and water areas under permanent protection.

The present vision of The Bahamas is to manage the country's natural resources to produce sustainable employment and prosperity through evidenced-based policies and practices. This vision has been formalized through legislation, specifically the *Planning and Subdivision Act*, the *Forestry Act*, and *The Bahamas National Trust Amendment Act*. Effectively, these Acts place all of The Bahamas under permanent legal environmental protection and scrutiny, and result in as much as 50% of the country's lands being set aside from commercial development. The *Planning and Subdivision Act* strengthens conservation of the physical landscape of The Bahamas, promotes viable communities by preventing the indiscriminate division of land, and pursues sustainable development of the built environment through the protection of natural and cultural heritage. The Act consolidates all aspects of Town Planning and Subdivision Development, expands public participation in approval, and mandates land-use plans for every island, based on National Land Development Policy. The requirement in the Act for an EIS and EIA for development have elevated the threshold of effective stewardship of sensitive areas. The *Forestry Act* will conserve Bahamian pine, hardwood, and mangrove forests while regulating the sustainable exploitation of commercial timber resources. The Act contains provisions for the declaration of Forest Reserves, Protected Forests, and Conservation Forests, with enhanced provision for the permanent protection of wetlands. *The Bahamas National Trust Act* was revised in 2010 for the first time since 1958, when the law establishing the BNT was first created. The new law gives expanded authority to the BNT to advise government and the public on development issues and policies.

The government pursued these measures because it recognized that environmental stakes are being inexorably raised. The cost of remedial actions is high and some degradation is irreversible with significant impact on the quality of life of Bahamian residents. Conservation in The Bahamas has benefitted from a unique confluence of factors, including heightened political maturity, increased environmental awareness, deliberate sponsorship of key partnerships, and a conscious application of lessons learned. The presence of scientists on BNT's board and the stewardship of scientists from around the world have greatly enhanced the environmental stock of The Bahamas.

sufficiently implemented and enforced, major conservation advancements will follow, to the benefit of the nation and its people.

8.4.3 Public partnerships and other mechanisms

Encouragement for protected areas has been accomplished through partnerships among NGOs and the government. The establishment of a Bahamas office of The Nature Conservancy (TNC) in 2002 was particularly significant (Box 8.4). TNC's cooperation with BNT and its influence with the government has proved invaluable for conservation planning, and for leading the team including the BNT, BEST, and the DMR that developed the BNT's Protected Area Master Plan. The Bahamas Reef Environmental Education Foundation (BREEF), an education leader, has also promoted The Bahamas' national park system, Nassau grouper conservation, and other initiatives.

Because many of the most pressing environmental problems in The Bahamas are local in nature, they also require local solutions. Thus, Friends of the Environment (FOE), established in 1988, is "dedicated to the preservation and protection of Abaco's marine and terrestrial environments in order to

Box 8.4 The end of paper parks—the Caribbean Challenge Initiative

Eleanor Phillips
The Nature Conservancy, Northern Caribbean Office, Nassau, The Bahamas

The Nature Conservancy (TNC) began working in The Bahamas in the late 1990s using a science-based approach for the protection of natural areas. TNC partners with local people, indigenous communities, businesses, governments, multilateral institutions, and other non-profits in pursuit of protecting ecologically important lands and waters.

The Conservancy recognized that there was a tremendous opportunity to advance its global marine goals by working in the Caribbean. The Bahamas, in particular, represents an important large marine ecosystem, straddling the Atlantic Ocean and the Caribbean Sea and containing 15% of the Caribbean region's reef inventory. The nearly 20,000 km^2 of geological reef structures in The Bahamas, of which an estimated 3150 km^2 is living coral reef, provide an important foundation for a healthy ocean environment, as well as shelter, nurseries, and feeding grounds for hundreds of marine species.

In May 2008, The Bahamas' government, alongside leaders from Grenada, Jamaica, the Dominican Republic, and St. Vincent and the Grenadines, launched the Caribbean Challenge Initiative, a region-wide campaign to protect 20% of the marine and coastal habitat of participating countries by 2020. Since 2008, St. Lucia, St. Kitts and Nevis, Antigua and Barbuda, The British Virgin Islands, and Puerto Rico have joined the Challenge. The 20% conservation goal may be achieved through the creation of new Marine Protected Areas and the expansion of existing ones. If accomplished, this will result in a wholesale transformation of countries' national park and reserve systems and will vastly increase the amount of marine and coastal habitat currently under protection by setting aside almost 8 million

(Continued)

ha of coral reefs, mangroves, creeks, seagrass beds, lagoons, banks, and ocean environments as essential habitat for sea turtles, whales, fishes, and other wildlife, as well as their support systems.

The three core components of the Challenge include:

- creating networks of Marine Protected Areas across 8 million ha of territorial coasts and waters;
- establishing protected-area trust funds to generate permanent, dedicated, and sustainable funding sources for the effective management, expansion, and scientific monitoring of all parks and protected areas; and
- developing national-level demonstration projects for climate change adaptation.

In order to implement the Program of Work on Protected Areas (PoWPA), the partner agencies of the PoWPA National Implementation Support Programme (NISP)—i.e., the Bahamas Environment, Science, and Technology (BEST) Commission, the Bahamas National Trust, Department of Marine Resources, and The Nature Conservancy Northern Caribbean Program—have worked together to complete the following tasks related to The Bahamas national system of protected areas: Ecological Gap Analysis, Rapid Assessment and Prioritization of Protected Area Management (RAPPAM), Capacity Action Plan, and Sustainable Finance Plan (SFP).

A key recommendation of the SFP was the need to establish a *Protected Area Fund* as a mechanism for sustained funding for The Bahamas National Protected Area System. This fund was envisioned as an endowment, with the interest generated from the capital investment being utilized for protected-area projects in the country. Subsequent to the development of the SFP, the Conservancy began to work on implementation of this fund as a priority. Its efforts resulted in the development of draft legislation for the establishment of the Bahamas Protected Area Fund (BPAF). Enactment of the BPAF bill will mean that a mechanism for sustainable financing of The Bahamas National Protected Area System has been established into perpetuity. It is envisioned that the BPAF will be an endowment fund, but will also have windows for revolving and sinking funds all to be utilized for protected-area projects in the country.

achieve sustainable living for the people and wildlife of Abaco." FOE places emphasis on "hands-on" learning experiences for students and community members. Similarly, the Andros Conservancy and Trust (ACT) has played a critical role in expansion of Andros' park system.

8.5 ISLAND SYSTEM AT A CROSSROADS

The variety of conservation issues presented in this chapter typifies problems faced by tropical island nations worldwide. The urgency for broader approaches to conservation is made clear by the case of The Bahamas' iconic Nassau grouper, which makes annual migrations of up to several hundred kilometers to breed; to sustain its populations, numerous habitat types and migration corridors are required. The conservation and restoration of wetlands are essential components of environmental sustainability on a far larger scale than can be accomplished by protected areas alone. Furthermore, The Bahamas' biodiversity and productivity result from the interactions among reefs, wetlands, seagrasses, and the land. These examples demonstrate that The Bahamas is a living system, wherein its banks, reefs, and fisheries are products of ecological processes, and its unique biodiversity depends on maintaining the full suite of interrelated environments. The invasion of the lionfish is a severe threat to this living system, and how this issue is to be resolved is not clear.

The recognition of inter-relationships among such issues as presented in this chapter is a clear demonstration of need for a shift from issue-by-issue management to comprehensive policy, planning, and management. In this context, protected areas are an essential, but not sufficient, component for meeting the goal of sustainable-resource and environmental conservation and use. Recent legislation such as the *Planning and Subdivision Act* is encouraging and, once again, illustrates The Bahamas' leadership in conservation among small-island states.

REFERENCES

Aguilar-Perera A, Aguilar-Davila W (1996) A spawning aggregation of Nassau grouper *Epinephelus striatus* (Pisces: Serranidae) in the Mexican Caribbean. *Environmental Biology of Fishes* **45**, 351–361.

Ahrenholz DW, Morris J (2010) Larval duration of the lionfish, *Pterois volitans* along the Bahamian Archipelago. *Environmental Biology of Fishes* **88**, 305–309.

Albins MA (2013) Effects of invasive Pacific red lionfish *Pterois volitans* versus a native predator on Bahamian coral-reef fish communities. *Biological Invasions* **15**, 29–43.

Albins MA, Hixon MA (2008) Invasive Indo-Pacific lionfish (*Pterois volitans*) reduce recruitment of Atlantic coral-reef fishes. *Marine Ecology Progress Series* **367**, 233–238.

Albins MA, Hixon MA (2011) Worst case scenario: potential long-term effects of invasive predatory lionfish (*Pterois volitans*) on Atlantic and Caribbean coral-reef communities. *Environmental Biology of Fishes*.

Albins MA, Lyons PJ (2012) Invasive red lionfish *Pterois volitans* blow directed jets of water at prey fish. *Marine Ecology Progress Series* **448**, 1–5.

Allgeier JA, Rosemond AD, Mehring AS, Layman CA (2010) Synergistic nutrient co-limitation across a gradient of ecosystem fragmentation in subtropical mangrove-dominated wetlands. *Limnology and Oceanography* **55**, 2660–2668.

Bahamas National Trust. www.bnt.bs.Bahamasa

Bahamas Department of Marine Resources. www.bahamas.gov.bs/marineresources

Barbour AB, Montgomery ML, Adamson AA, Díaz-Ferguson E, Silliman BR (2010) Mangrove use by the invasive lionfish *Pterois volitans*. *Marine Ecology Progress Series* **401**, 291–294.

BEST (1997) *Commonwealth of The Bahamas: National Biodiversity Strategy and Action Plan*. Prepared by Task Force: The Bahamas National Biodiversity Strategy and Action Plan for The Bahamas Environment, Science and Technology Commission (BEST), under contract from the United Nations Environment Programme. Nassau, The Bahamas.

Betancur-R R, Hines A, Acero A, Ortí G, Wilbur AE, Freshwater, DW (2011) Reconstructing the lionfish invasion: insights into Greater Caribbean biogeography. *Journal of Biogeography* **38**, 1281–1293.

Bjorndal KA (1980) Nutrition and grazing behavior of the green turtle, *Chelonia mydas*. *Marine Biology* **56**, 147–154.

Bjorndal KA, Bolten AB (2008) Annual variation in source contributions to a mixed stock: implications for quantifying connectivity. *Molecular Ecology* **17**, 2185–2193.

Bjorndal KA, Bolten AB, Chaloupka MY (2000) Green turtle somatic growth model: evidence for density dependence. *Ecological Applications* **10**, 269–282.

Bjorndal KA, Bolten AB, Chaloupka MY (2003) Survival probability estimates for immature green turtles, *Chelonia mydas*, in the Bahamas. *Marine Ecology Progress Series* **252**, 273–281.

Böhlke JE, Chaplin CCG (1968) *Fishes of the Bahamas*. Academy of Natural Sciences of Philadelphia. Livingston Publishing Company, Wynnewood, Pennsylvania.

Böhlke JE, Randall JE (1963) The fishes of the Western Atlantic genus *Gramma*. *Proceedings of the Academy of Natural Sciences of Philadelphia* **115**, 33–52.

Bohnsack JA (1994) How marine fishery reserves can improve reef fisheries. *Proceedings, Gulf and Caribbean Fisheries Institute* **43**, 217–241.

Bolten AB (2003) Variation in sea turtle life history patterns: neritic vs. oceanic developmental stages. In *The Biology of Sea Turtles*, **volume II** (eds Lutz PL, Music J, Weaken J). CRC Press, Boca Raton, FL, 243–257.

Bouchard SS, Bjorndal KA (2000) Sea turtles as biological transporters of nutrients and energy from marine to terrestrial ecosystems. *Ecology* **81**, 2305–2313.

Broecker WS, Takahashi T (1966) Calcium carbonate precipitation on the Bahama banks. *Journal of Geophysical Research* **71**, 1575–1662.

Buchan KC (2000) The Bahamas. *Marine Pollution Bulletin* **41**, 94–111.

Christie MR (2010) Parentage in natural populations: novel methods to detect parent-offspring pairs in large data sets. *Molecular Ecology Resources* **10**, 115–128.

Christie MR, Stallings CD, Johnson DW, Hixon MA (2010) Self-recruitment and sweepstakes reproduction amid extensive gene flow in a coral-reef fish. *Molecular Ecology* **19**(5), 1042–1057.

Claydon JAB, Calosso MC, Traiger SB (2012) Progression of invasive lionfish in seagrass, mangrove and reef habitats. *Marine Ecology Progress Series* **448**, 119–129.

Cleare A (2007) *History of Tourism in the Bahamas: A Global Perspective*. Xlibris Corporation.

Cohen DM, Robins CR (1970) A new ophidioid fish (Genus *Lucifuga*) from a limestone sink, New Providence Island, Bahamas. *Proceedings Biological Society Washington* **83**, 133–144.

Coleman FC, Koenig CC, Collins LA (1996) Reproductive styles of shallow-water groupers (Pisces: Serranidae) in the eastern Gulf of Mexico and the consequences of fishing spawning aggregations. *Environmental Biology of Fishes* **47**, 129–141.

Colin PL (1992) Reproduction of the Nassau grouper, *Epinephelus striatus* (Pisces: Serranidae) and its relationship to environmental conditions. *Environmental Biology of Fishes* **34**, 357–77.

Colin PL, Laroche WA, Brothers EB (1997) Ingress and settlement in the Nassau grouper, *Epinephelus striatus* (Pisces: Serranidae), with relationship to spawning occurrence. *Bulletin of Marine Science* **60**, 656–667.

Côté IM, Maljković A (2010) Predation rates of Indo-Pacific lionfish on Bahamian coral reefs. *Marine Ecology Progress Series* **404**, 219–225.

Crowder LB, Lyman SJ, Figueira WF, Priddy J (2000) Source-sink population dynamics and the problem of siting marine reserves. *Bulletin of Marine Science* **66**, 799–820.

Cure K, Benkwitt CE, Kindinger TL, Pickering EA, Pusack TJ, McIlwain JL, Hixon MA (2012) Comparative behavior of red lionfish *Pterois volitans* on native Pacific versus invaded Atlantic coral reefs. *Marine Ecology Progress Series* **467**, 181–192.

Dahlgren CP, Eggleston DB (2000) Ecological processes underlying ontogenetic habitat shifts in a coral reef fish. *Ecology* **81**, 222–2240.

Dietz RS, Holden JC, Sproll WP (1970) Geotectonic evolution and subsidence of the Bahama platform. *Geological Society of America Bulletin* **81**, 1915–1928.

Domeier ML, Colin PL (1997) Tropical reef-fish spawning aggregations defined and reviewed. *Bulletin of Marine Science* **60**(3), 698–726.

Economist, The. www.economist.com/sciencetechnology/displaystory.cfm?storyid=14637325

Eggleston DB (1995) Recruitment in Nassau grouper *Epinephelus striatus*: post settlement abundance, microhabitat features, and ontogenetic habitat shifts. *Marine Ecology Progress Series* **124**, 9–22.

Freshwater DW, Hines A, Parham S, Wilbur A, Sabaoun M, Woodhead J, Akins L, Purdy B, Whitfield PE, Paris CB (2009) Mitochondrial control region sequence analyses indicate dispersal from the US East Coast as the source of the invasive Indo-Pacific lionfish *Pterois volitans* in the Bahamas. *Marine Biology* **156**, 1213–1221.

Gould SJ (1983) Opus 100. *Natural History Magazine*, April, 10–19.

Green SJ, Côté IM (2008) Record densities of Indo-Pacific lionfish on Bahamian coral reefs. *Coral Reefs* **28**, 107.

Green SJ, Akins JL, Maljković A, Côté IM (2012) Invasive Lionfish Drive Atlantic Coral Reef Fish Declines. *PLoS ONE* **3**, e32596.

Hammerton JL (2001) Casuarinas in The Bahamas: a clear and present danger. *Bahamas Journal of Science* **9**, 2–14.

Hansen J, Sato M, Ruedy R (2012) Perception of climate change. *Proceedings of the National Academy of Sciences, Early Edition*. Online. www.pnas.org/cgi/doi/10.1073/pnas.1205276109.

Hanski I (1998) Metapopulation dynamics. *Nature* **396**, 41–49.

Hedgecock D, Barber PH, Edmands S (2007) Genetic approaches to measuring connectivity. *Oceanography* **20**, 70–79.

Huntsman GR, Potts J, Mays RW, Vaughan D (1999) Groupers (Serranidae, Epinephelinae): Endangered apex predators of reef communities. In *Life in the Slow Lane* (ed. Musick AJ), *Symposium* **23**. American Fisheries Society, Washington, D.C., 217–231.

IAC. www.iacseaturtle.org/

IPCC (2007) www.ipcc.ch/publications_and_data/ar4/wg1/en/contents.html

IUCN Red List. iucn-mtsg.org/about/structure-role/red-list/

Jackson JBC, Kirby MX, Berger WH, Bjorndal KA, Botsford LW, Bourque BJ, Bradbury RH, Cooke R, Erlandson J, Estes JA, Hughes TP, Kidwell S, Lange CB, Lenihan HS, Pandolfi JM, Peterson CH, Steneck RS, Tegner MJ, Warner RR (2001) Historical overfishing and the recent collapse of coastal ecosystems. *Science* **293**, 629–638.

Johannes RE (1978) Reproductive strategies of coastal fishes in the tropics. *Environmental Biology of Fishes* **3**, 65–84.

Jones GP (1991) Postrecruitment processes in the ecology of coral reef fish populations: a multifactorial perspective. In *The ecology of fishes on coral reefs* (ed. PF Sale). Academic Press, San Diego, CA.

Keegan WF (1997) *Bahamian Archeology*. Media Publishing, Nassau, Bahamas, 294–328.

Kulbicki M, Beets J, Chabanet P, Cure K, Darling E, Floeter SR, Galzin R, Green A, Harmelin-Vivien M, Hixon M, Letourneur Y, Lison de

Loma T, McClanahan T, McIlwain J, MouTham G, Myers R, O'Leary JK, Planes S, Vigliola L, Wantiez L (2012) Distributions of Indo-Pacific lionfishes (*Pterois* spp.) in their native ranges: implications for the Atlantic invasion. *Marine Ecology Progress Series* **446**, 189–205.

Lahanas PN, Bjorndal KA, Bolten AB, Encalada SE, Miyamoto MM, Valverde RA, Bowen BW (1998) Genetic composition of a green turtle (*Chelonia mydas*) feeding ground population: evidence for multiple origins. *Marine Biology* **130**, 345–352.

Layman CA, Arrington DA, Langerhans RB, Silliman BR (2004) Degree of fragmentation affects fish assemblage structure in Andros Island (Bahamas) estuaries. *Caribbean Journal of Science* **40**, 232–244.

Layman CA, Quattrochi JP, Peyer CM, Allgeier JE (2007) Niche width collapse in a resilient top predator following ecosystem fragmentation. *Ecology Letters* **10**, 937–944.

Layman CA, Silliman BR (2002) Preliminary survey and diet analysis of juvenile fishes of an estuarine crêche on Andros Island, Bahamas. *Bulletin of Marine Science* **70**, 199–210.

León YM, Bjorndal KA (2002) Selective feeding in the hawksbill turtle, an important predator in coral reef ecosystems. *Marine Ecology Progress Series* **245**, 249–258.

Lightbourne ME (n.d.) *The Primeval Forest*. Unpublished manuscript, Bahamas Historical Society, 1–10.

Lipcius RN, Stockhausen WT, Eggleston DB (2001). Marine reserves and Caribbean spiny lobster: empirical evaluation and theoretical metapopulation recruitment dynamics. *Marine and Freshwater Research* **52**, 1589–1598.

Lipcius RN, Stockhausen WT, Eggleston DB, Marshall Jr. LS, Hickey B (1997). Hydrodynamic decoupling of recruitment, habitat quality and adult abundance in the Caribbean spiny lobster: source-sink dynamics? *Marine and Freshwater Research* **48**, 807–815.

MacArthur RH, Wilson EO (2001) *The Theory of Island Biogeography*. New Edition, Princeton Landmarks in Biology, Princeton, New Jersey.

Maljković A, Van Leeuwen TE, Cove SN (2008) Predation on the invasive red lionfish, *Pterois volitans* (Pisces: Scorpaenidae), by native groupers in the Bahamas. *Coral Reefs* **27**, 501.

Moran KL, Bjorndal KA (2005). Simulated green turtle grazing affects structure and productivity of seagrass pastures. *Marine Ecology Progress Series* **305**, 235–247.

Moran KL, Bjorndal KA (2007) Simulated green turtle grazing affects nutrient composition of the seagrass *Thalassia testudinum*. *Marine Biology* **150**, 1083–1092.

Morris Jr. JA (2009) *The Biology and Ecology of the Invasive Indo-Pacific Lionfish*. Doctoral dissertation, North Carolina State University, 168pp.

Morris Jr. JA, ed. (2012) *Invasive lionfish: A guide to control and management*. Gulf and Caribbean Fisheries Institute Special Publication Series Number 1, Marathon, Florida.

Morris Jr. JA, Akins JL (2009) Feeding ecology of invasive lionfish (*Pterois volitans*) in the Bahamian archipelago. *Environmental Biology of Fishes* **86**, 389–398.

Morris Jr. JA, Shertzer KW, Rice JA (2011) A stage-based matrix population model of invasive lionfish with implications for control. *Biological Invasions* **13**, 7–12.

Mortimer JA, Donnelly M (2007) Marine Turtle Specialist Group 2007 IUCN Red List status assessment hawksbill turtle (*Eretmochelys imbricata*). www.iucnmtsg.org/red_list/ei/index.shtml

Mumby PJ, Dahlgren CP, Harborne AR, Kappel CV, Micheli F, Brumbaugh DR, Holmes KE, Mendes JM, Broad K, Sanchirico JN, Buch K, Box S, Stoffle RW, Gill AB (2006) Fishing, trophic cascades, and the process of grazing on coral reefs. *Science* **311**, 98–101.

Nagelkerken I, van der Velde G, Gorissen MW, Meijer GJ, Van't Hof T, den Hartog C (2000) Importance of mangroves, seagrass beds, and the shallow coral reef as a nursery for important reef fishes, using a visual census technique. *Estuarine, Coastal and Shelf Science* **51**, 31–44.

Nuñez MA, Kuebbing S, Dimarco RD, Simberloff D (2012) Invasive Species: To eat or not to eat, that is the question. *Conservation Letters*.

Pusack TJ, Kindinger TL, Benkwitt CE, Cure K (in preparation) Invasive lionfish (*Pterois volitans*) grow faster and larger in the Atlantic Ocean than in their native Pacific range.

Ray C, ed. (1958) *Report of the Exuma Cays Park Project*. Submitted to the Government of the Bahamas, May. Revised 1961, 1–39.

Ray GC (1998) Bahamian protected areas. Part 1: How it all began. *Bahamas Journal of Science* **6**, 2–11.

Ray GC, McCormick-Ray MG (1992) Functional coastal-marine biodiversity. *Transactions 57th Wildlife and Natural Resources Conference (1992)*, Wildlife Management Institute, Washington, D.C., 384–397.

Raymond WW, Albins MA, Pusack TJ (in preparation) Shelter competition between invasive Pacific red lionfish (*Pterois volitans*) and native Nassau grouper (*Epinephelus striatus*). Intended Journal: *Journal of Experimental Marine Biology and Ecology*.

Reich KJ, Bjorndal KA, Bolten AB (2007) The 'lost years' of green turtles: using stable isotopes to study cryptic lifestages. *Biology Letters* **3**, 712–714.

Reid RP, Macintyre IG, Steneck RS (1999) A microbialite/alga ridge fringing reef complex, Highbourne Cay, Bahamas. *Atoll Research Bulletin* **465**, 1–18.

Roberts CM (1997). Connectivity and management of Caribbean coral reefs. *Science* **278**, 1454–1457.

Roberts CM, Polunin NVC (1993). Marine reserves: simple solutions to managing complex fisheries. *Ambio* **22**, 363–368.

Sadovy Y (1993) The Nassau grouper, endangered or just unlucky? *Reef Encounter* **13**, 10–12.

Sadovy Y, Eklund AM (1999) Synopsis of biological data on the Nassau grouper, *Epinephelus striatus* (Bloch, 1792), and the jewfish, *E. itajara* (Lichtenstein, 1822). *NOAA Technical Report, NMFS* **146**, 1–68.

Schofield PJ (2009) Geographic extent and chronology of the invasion of non-native lionfish (*Pterois volitans* [Linnaeus, 1758] and *P. miles* [Bennett, 1828]) in the Western North Atlantic and Caribbean Sea. *Aquatic Invasions* **4**, 473–479.

Schofield PJ (2010) Update on geographic spread of invasive lionfishes (*Pterois volitans* [Linnaeus, 1758] and *P. miles* [Bennett, 1828]) in the Western North Atlantic Ocean, Caribbean Sea and Gulf of Mexico. *Aquatic Invasions* **5**, S117–S122.

Schlager W, Ginsburg RN (1981) Bahama carbonate platforms – The deep and the past, *Marine Geology* **44**, 1–24.

Sealey NE (1994) *Bahamian Landscapes: An Introduction to the Geography of the Bahamas*, Second edition. Media Publishing, Nassau, Bahamas.

Sealey NE, ed. (2001) *Caribbean Certificate Atlas*, Third edition. Macmillan Education Ltd., Oxford.

Sikkel PC, Tuttle LJ, Cure K, Dove AI, Passarelli J, McIlwain JT, Hixon MA (in preparation) Enemy release hypothesis tested: native Pacific lionfish (*Pterois volitans*) have more parasites than invasive Atlantic lionfish.

Smith CL (1972) A spawning aggregation of Nassau grouper, *Epinephelus striatus* (Bloch). *Transactions of the American Fisheries Society* **2**, 257–261.

Snyder DB, Burgess GH (2007) The Indo-Pacific red lionfish, *Pterois volitans* (Pisces: Scorpaenidae), new to Bahamian ichthyofauna. *Coral Reefs* **26**, 175.

Spalding MD, Ravilious C, Green EP (2001) *World Atlas of Coral Reefs*. Prepared at the UNEP World Conservation Monitoring Centre. University of California Press, Berkeley, California.

Spalding MD, Fox HE, Halpern BS, McManus MA, Molnar J, Allen GR, Davidson N, Jorge ZA, Lombana AL, Lourie SA, Martin KD, McManus E, Recchia CA, Robertson J (2007) Marine ecoregions of the world: A bioregionalization of coastal and shelf areas. *Bioscience* **57**, 573–583.

Stallings CD (2009) Fishery-independent data reveal negative effect of human population density on Caribbean predatory fish communities. *PLoS ONE* **4**, e5333.

Stockhausen WT, Lipcius RN, Hickey BH (2000). Joint effects of larval dispersal, population regulation, marine reserve design, and exploitation on production and recruitment in the Caribbean spiny lobster. *Bulletin of Marine Science* **66**, 957–990.

Stockhausen WT, Lipcius RN (2001). Single large or several small marine reserves for the Caribbean spiny lobster? *Marine and Freshwater Research* **52**, 1605–1614.

Stoner AW, Davis MH, Booker CJ (2012) Abundance and population structure of queen conch inside and outside a marine protected area: repeat surveys show significant declines. *Marine Ecology Progress Series* **460**, 101–114

Tuttle LJ, Sikkel PC, Williams EA, Bunkley-Williams L, Dove AI, Hixon MA (in preparation) Invasive lionfish (*Pterois volitans*) have fewer parasites than native piscivorous fishes found on the same Atlantic reefs.

USGS Nonindigenous Aquatic Species Database. nas.er.usgs.gov/queries/FactSheet.asp?speciesID=963

Valentine-Rose L, Layman CA, Arrington DA, Rypel AL (2007) Habitat fragmentation affects fish secondary production in Bahamian tidal creeks. *Bulletin of Marine Science* **80**, 863–878.

Valentine-Rose L, Layman CA (2011). Response of fish assemblage structure and function following restoration of two small Bahamian tidal creeks. *Restoration Ecology* **19**, 205–215.

Wallace AR (1880) *Island Life*. Re-published, with an Introduction by H. J. Birk. Prometheus Books, Great Minds Series. Amherst, New York.

Whitfield PE, Gardner T, Vives SP, Gilligan MR, Courtenay Jr WR, Ray GC, Hare JA (2002) Biological invasion of the Indo-Pacific lionfish *Pterois volitans* along the Atlantic coast of North America. *Marine Biological Progress Series* **235**, 289–297.

Wilson SM (1997) Introduction to the study of the indigenous people of the Caribbean. In *The Indigenous People of the Caribbean* (ed. Wilson SM). University Press of Florida, Gainesville, Florida, 1–8.

Witherington B, Kubilis P, Brost B, Meylan A (2009) Decreasing annual nest counts in a globally important loggerhead sea turtle population. *Ecological Applications* **19**, 30–54.

CHAPTER 9

THE ISLES OF SCILLY: SUSTAINING BIODIVERSITY

Richard M. Warwick

Plymouth Marine Laboratory, Plymouth, Devon, UK

When Nature addressed herself to the construction of this archipelago she brought to the task a light touch: at the moment she happened to be full of feeling for the great and artistic effects which may be produced by small elevations, especially in those places where the material is granite.
Sir Walter Besant *Armorel of Lyonesse* (1890)

The Zoology of Archipelagoes will be well worth examining.
Charles Darwin *Ornithological Notes on the Galapagos* (1835)

9.1 SETTING THE SCENE

Despite four thousand years of more or less continuous human occupation and dependency on marine resources, the marine environment of Scilly gives the impression of being undisturbed, natural, and biodiverse in comparison with other areas of Europe. Tourism is the main source of income, much of this falling under the broad heading of "ecotourism," which is dependent on sustaining a high environmental quality in the face of both global (climate change) and local (e.g., agriculture and fishing) environmental pressures. Thus, the main conservation issue for the Isles of Scilly is balancing of needs for tourism and other local sources of income with the imperative of sustaining biodiversity.

The biota has a number of special features. Because of the mild climate, several species from southern Europe and the Mediterranean occur on Scilly and nowhere else in Britain. The islands have more benthic species defined as "nationally rare" and "nationally scarce" than any other locality in southwest Britain, while many species that are normally sublittoral occur on the shore. Habitat diversity within the archipelago is high, and many sites have a complex array of habitat types in a small area. Wave exposure varies from extremely exposed to very sheltered, often within short distances. It is this complex array of habitat types, wave exposure, and the associated communities that is of significant marine biological value as all these features together contribute to the high conservation importance of the islands.

These features will be used in this case study as examples that address some general issues of biodiversity and sustainability and their management, and it is these two themes of biodiversity and sustainability that constitute the focus of this chapter.

9.2 PHYSICAL AND BIOGEOGRAPHIC SETTING

The Isles of Scilly are situated some 45 km southwest of Land's End, UK, and present the most southwesterly shores and shallow marine waters of Britain. The land area covers about 16 km^2 and consists of five relatively large inhabited islands—St. Mary's (654 ha), St. Martin's (238), Tresco (298), Bryher (134), and St. Agnes (145)—as well as innumerable smaller islets and rocks ranging in size from Samson (36) and Annet (22), through the Eastern Isles and Northern Rocks, all of which are less than a hectare, down to the rocky islets and reefs immediately offshore of the main islands and extending in a vast fringe southwest towards Bishop Rock from St. Agnes (Fig. 9.1).

By virtue of their oceanic position and southerly location, the Isles of Scilly enjoy the mildest climate in Britain, and have been described as "the only Lusitanian oceanic archipelago in Europe." Warm, predominantly southwest winds from the Atlantic combine with warm ocean currents generated by the tropical Gulf Stream, resulting in a clement oceanic climate with mild, frost-free winters (the average temperature in February being 7.6°C), but cooler summers than the mainland (average in August 16.2°C). The abnormally cold winter of 1963 that resulted in mortalities of many littoral species on the mainland coast of Cornwall had a much less devastating effect on Scilly. The small low-lying islands are not conducive to cloud formation from the moist southwesterly winds, so that Scilly receives on average an annual rainfall of only 850–900 mm, rather less than nearby mainland coastal areas. Sunshine hours, on the other hand, are higher than the English mainland (Fig. 9.2). Summer water temperatures are

Marine Conservation: Science, Policy, and Management, First Edition. G. Carleton Ray and Jerry McCormick-Ray.

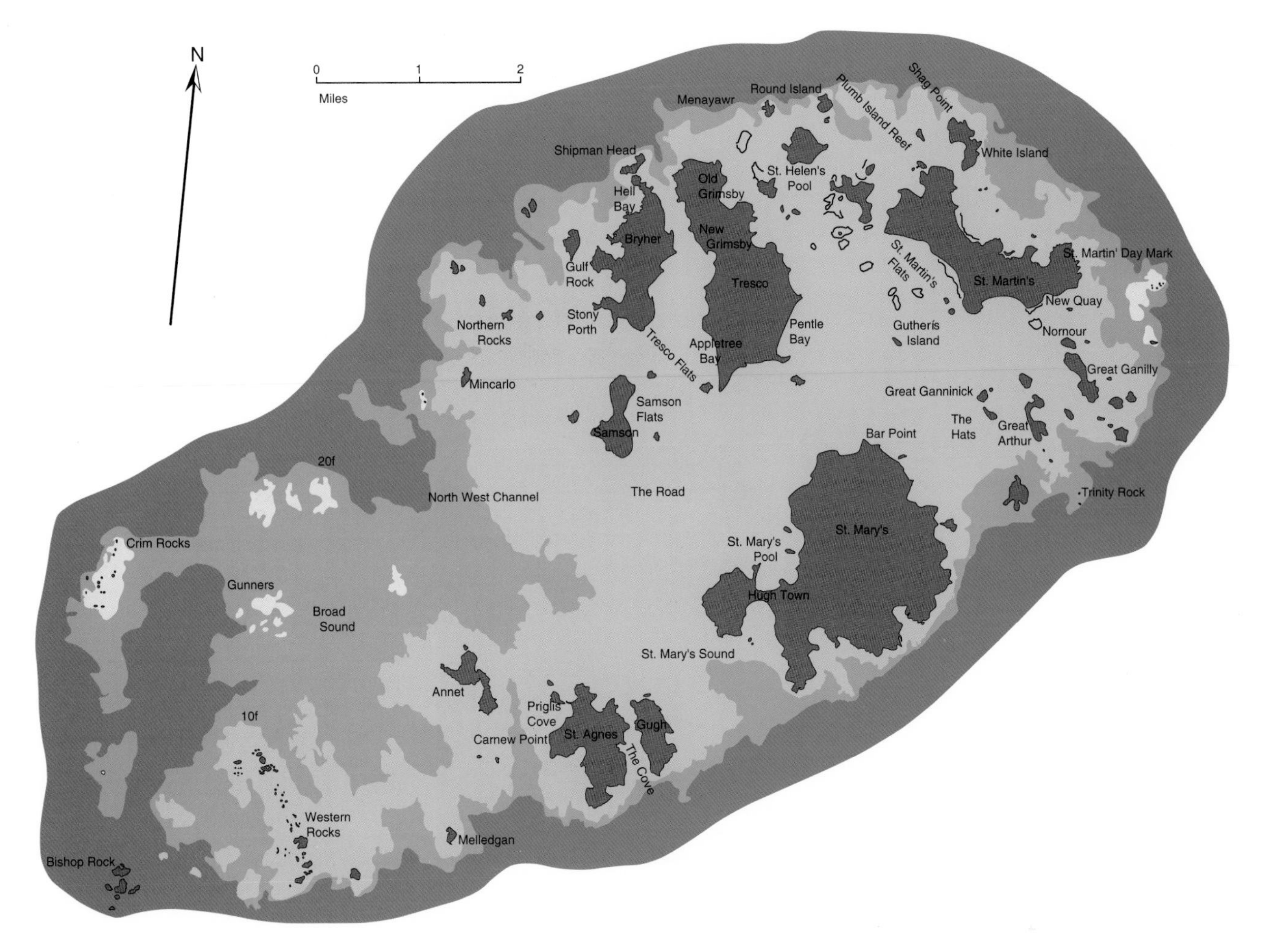

Fig. 9.1 Map of the Isles of Scilly, showing locations mentioned in the text.

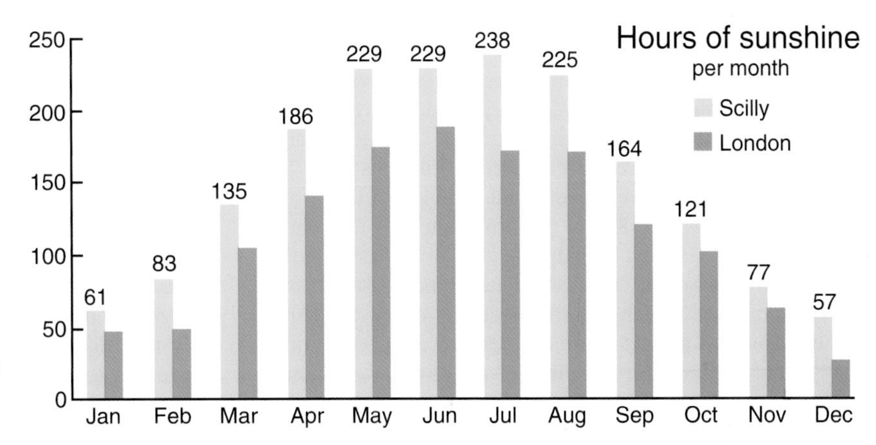

Fig. 9.2 Average monthly sunshine hours on Scilly compared with London. Based on data from Isles of Scilly Tourist Board.

much lower than for the adjacent mainland. In summer, coastal waters of mainland Britain become stratified and surface layers are warmed. Small islands such as Scilly, in a stratified region of the shelf sea, create conditions for a local increase in tidal mixing, bringing deeper cooler water to the surface, as is evident from satellite images (Fig. 9.3). Thus in summer, surface-water temperature rarely exceeds 15°C, and varies annually by only approximately 5°C. Long-term trends in water temperature are insignificant compared with seasonal cycles, but could be related to reported changes in pelagic species' distributions in the region. Although no records are available for Scilly itself, surface waters off Plymouth showed

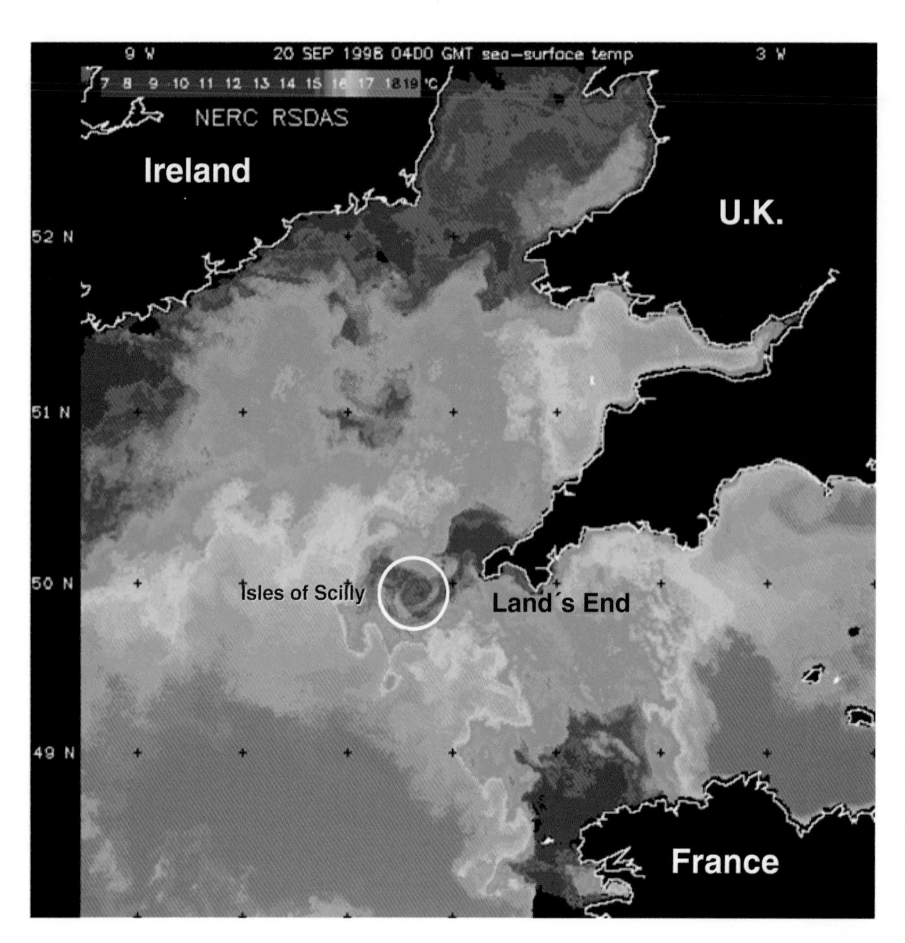

Fig. 9.3 Sea surface temperatures, September 1998. Note cold water in the vicinity of Scilly compared with much warmer adjacent stratified areas. NOAA Advanced Very High Resolution Radiometer image, processed at the Plymouth Marine Laboratory as part of the NERC Earth Observation Data Analysis and Acquisition Service (NEODAAS).

a rising trend of about 0.5°C from 1900 to 1970, followed by a period of cooling, possibly related to changes in the strength of the North Atlantic Oscillation (NAO).

Tidal mixing in the vicinity of the islands also results in enhanced nutrient concentrations. As a result, levels of phytoplankton biomass and productivity are enhanced by a factor of approximately five over a large region surrounding the islands, equivalent to approximately 20 times the island area. Predominantly west-to-east ocean currents and an almost total lack of freshwater and alluvial runoff from rivers or streams results in uniform salinity of around 35.3 psu, as well as low turbidity (Fig. 9.4a). Consequently, remarkable water clarity results in dense stands of kelp (*Laminaria ochroleuca*), which occurs in depths of up to 30 m. Strong tidal jets propagate in a clockwise direction round the islands, with a peak current flow of 175 cm s^{-1}. This has important ecological implications; for example, sewage from point sources and planktonic larvae or propagules of benthic animals and plants are likely to remain in coastal waters for relatively long periods. Also around Land's End, the nearest mainland to Scilly, net transport of water is in a northerly direction so that planktonic larvae and plant propagules released there would be rapidly transported northwards, away from the islands. That is, net flows in the area mid-way between the island and the mainland are relatively slight, so that overall exchanges of suspended particles between the islands and the mainland in either direction tend to be minimal.

Geologically, the islands are composed almost exclusively of granite with intrusions of many narrow veins of quartz. Erosion takes place more easily along these veins and cracks providing a network of crevices that afford refuges for many rocky-shore and subtidal animals, especially in more exposed localities (Fig. 9.4b). Eventually, erosion reduces the granite to sand, which is white due to the hard grains of feldspar and colorless to white quartz, and sparkling due to the small flat platelets of the colorless to light brown/black micas that reflect light.

In sum, the interplay between regional attributes of the physical environment (climate, tidal currents, geology) and the local distribution of a wide variety of different-sized islands gives rise to the unique character of Scillonian marine ecosystems. The mild climate gives the biota its unique regional character, and the mosaic of different habitats and resulting types of biotic organization determines its local character. Organisms have a range of morphological, physiological, and trophic adaptations allowing them to cope with a wide range of local conditions. At one end of the spectrum, animals will be physically robust filter feeders or grazers that can attach themselves firmly to the substratum, for example barnacles and limpets, and at the other end species tend to be more physically delicate motile deposit-feeders such as polychaete worms.

Fig. 9.4 (a) Clear blue waters of Scilly, conducive to luxuriant growth of macroalgae. Photograph courtesy of J. T. Davey, Plymouth Marine Laboratory. (b) Peninnis Inner Head, St. Mary's, showing vertical and horizontal fissures that provide a refuge for animals on this exposed section of coast. Photograph © R. M. Warwick.

9.3 MEASURING AND MEASURES OF BIODIVERSITY

Biodiversity is generally considered at regional, landscape, organismal, and genetic levels (Ch. 5). From a practical point of view, sustaining a range of high-quality habitats conserves organismal diversity, and sustaining organismal diversity conserves genetic diversity. This section gives a brief account of the range of habitats and their associated biota that occur on Scilly, based on a relatively long history of marine biological investigation (Box 9.1), followed by a more detailed account of approaches to the measurement of organismal diversity, an essential prerequisite to its conservation, with a detailed focus on the fauna of St. Martin's sandflat.

Box 9.1 History of marine biological studies

Richard M. Warwick
Plymouth Marine Laboratory, Plymouth, Devon, UK

Early observations

The first impetus for the serious study of marine natural history came in 1850 when Dr H. W. Acland, Lee's Reader in Anatomy at the University of Oxford, dispatched Dr. J. Victor Carus to the islands to collect specimens for the Anatomical Museum's collections. His rationale was that the islands had not been explored previously and, due to their geographical position, some curious and interesting species might be obtained. For marine fauna, Carus remarked that "it is not at all a dense one, although there are multitudes of zoophytes and hosts of fishes; there are only a few molluscs, some worms and a not very large but interesting number of echinoderms." For intertidal habitats he wrote that "it is perhaps worth remarking that I found some animals at low-water-mark or even higher, which commonly inhabit deeper water." Sporadic studies between the mid-1800s and early 1900s are summarized by Harvey (1969).

Recent studies

After a long period without marine biological work, a survey of the distributions of selected species along the coast of the English Channel (Crisp and Southward, 1958) noted the absence of certain species and the relative abundance of others compared to the mainland. Since then, the special nature of Scilly's marine environment has been a focus of interest. Professor L. A. Harvey, with students and staff of the Department of Zoology of the University of Exeter, made collections from intertidal habitats without break from 1956 to 2004. The University of London Sub-Aqua Club also organized six expeditions to Scilly from 1964 to 1968, which greatly increased knowledge of sublittoral fauna. These two initiatives prompted the *Journal of Natural History* to publish a paper on "The islands and their ecology" by Professor Harvey in 1969, followed by a series of papers on "Marine flora and fauna of the Isles of Scilly." To date, 20 papers on groups of marine animals have been published in the series, and more are on the way. Also, a few targeted studies have been conducted on inshore plankton and occasional mass strandings of pelagic species such as Scyphomedusae (jellyfish) and the by-the-wind sailor (*Velella velella*) have attracted attention (Fig. B9.1.1).

(Continued)

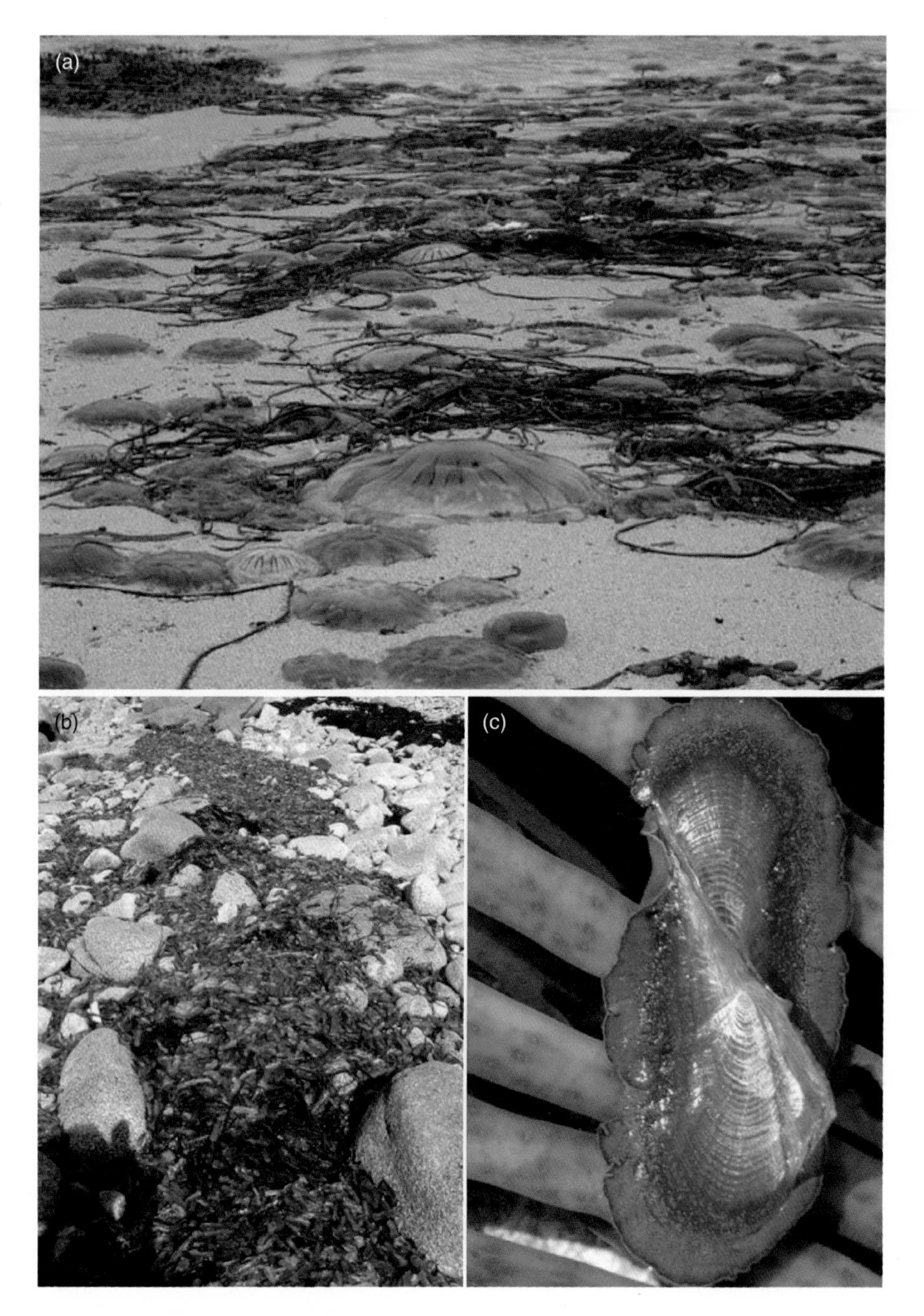

Fig. B9.1.1 A mass stranding of two species of jellyfish on Bryher (a), and millions of by-the-wind sailors (*Velella velella*) stranded on Scilly in September 2004 (b, c). Photographs © R. Pearce (a) and © R. M. Warwick (b, c).

The special nature of the Isles of Scilly attracted the attention of what was then the Nature Conservancy Council (NCC), and it commissioned a range of intertidal and subtidal survey and monitoring programs. Furthermore, under the *EU Habitats Directive*, the Isles of Scilly have been designated as a Special Area of Conservation based on the presence of three key habitats, subtidal rocky reefs, mudflats and sandflats not covered by water at low tide, and sandbanks which are slightly covered by seawater all the time (see Section 5.1). There is, therefore, an obligation on Natural England (part of the former NCC) to monitor these habitats, and this work is ongoing.

Another initiative, the European Concerted Action BIOMARE, had as a major objective to establish a network of marine coastal sites for comparative studies of marine biodiversity throughout Europe. Based on a number of strict criteria, a small subset of these sites approaching pristine conditions was selected for intensive comparative studies. Only six sites throughout Europe fulfilled these criteria, Scilly being the only one in the UK. An All Taxon Biodiversity Inventory (ATBI) is proceeding under national funding, and recent surveys have discovered a number of species that are new to science. As an example, Dr. Anno Faubel from the University of Hamburg sampled and studied marine

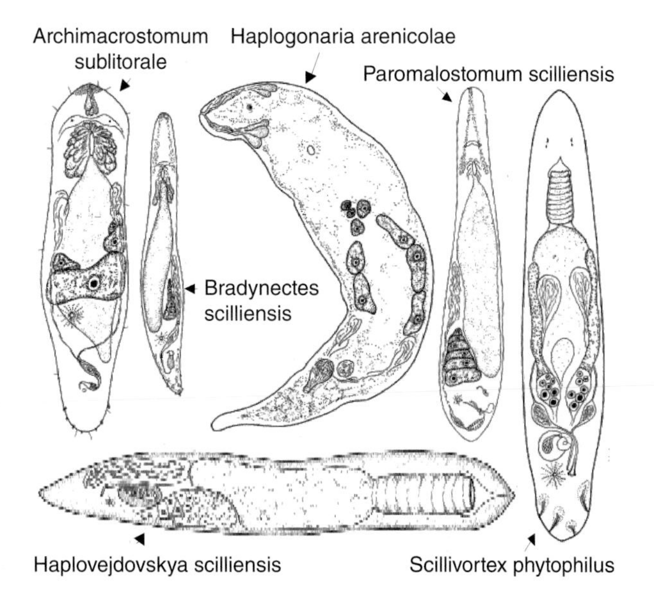

Fig. B9.1.2 New species of flatworms discovered by Anno Faubel on Scilly. From Faubel A, Warwick RM (2005) The marine flora and fauna of the Isles of Scilly: free-living Plathelminthes ("Turbellaria"). *Journal of Natural History* **39**, 1–45.

free-living flatworms and found 67 species, only three of which had been recorded from Scilly before, and six of the species were new to science, including one new genus (Fig. B9.1.2). Other groups for which new species have been found include free-living nematodes and harpacticoid copepods.

With respect to local initiatives, the Environmental Records Centre for Cornwall and the Isles of Scilly (ERCCIS) holds an extensive number of environmental and geological records for both mainland Cornwall and the Isles of Scilly. A great deal of information has been collated in an "Isles of Scilly Biodiversity Audit," produced collaboratively with the Wildlife Trusts of Cornwall, the Isles of Scilly, and Natural England. A comprehensive report assesses the Isles' biodiversity status and trends, the main issues affecting it, and priorities for conservation, and also makes recommendations for immediate action to conserve and enhance the Isles of Scilly's biodiversity.

Sources: Lewes (1858); Smart and Cooke (1885); Clarke (1909); Crisp and Southward (1958); Harvey (1969); Faubel and Warwick (2005); www.cornwallwildlifetrust.org.uk/conservation/livingseas/Isles_Scilly_Marine_Biodiversity_Project

9.3.1 Coastal-marine habitat diversity

Five major habitat types occur in the Isles of Scilly, each of which supports a distinct biota. Albeit diverse for a system of its size, biodiversity is not so great as in the rest of the UK. Probably, the most important factor preventing the spread of some animal species, or at least restricting their abundance, is the direction of the residual water currents (Section 9.2).

The *phenomenon of emergence* (not to be confused with ecological emergence, Chs. 2, 13) is also an important feature of Scilly's biota. Early naturalists noted that there are many species that occur intertidally in Scilly that are only found in deeper water elsewhere in Britain; for example, a number of bivalve mollusks, a brittle star (*Antedon bifida*), a conspicuous orange, seven-armed starfish (*Luidia ciliaris*, Fig. 9.9), and a lancelet (*Branchiostoma lanceolatum*). One possible cause for this is the scarcity of near-zero temperatures that may permit animals to emerge into littoral environments; another is the relatively high salinity that allows some species to enter littoral areas. Also, several small interstitial species (meiobenthos) occur in sandy beaches of Scilly that elsewhere only occur sublittorally. An additional possible explanation is the angularity of sand grains derived from granite, which are tightly packed and restrict drainage from the beach at low tide, resulting in an interstitial environment no different from the sublittoral.

9.3.1.1 Rocky shores

Rocky-shore communities of Scilly are distinctly southern in character, as indicated by the high abundance of several species of algae and invertebrates. Outer slopes of the island are dominated by rock and boulder substrates. The block-like structure of the granite bedrock gives a distinctive stepped appearance to many shores, with shallow, open crevices, but few surfaces eroded to form large or deep pools. The open nature of the few crevices and non-friable nature of the rock leads to the presence of poorly developed crevice faunas compared with other areas of southwest Britain.

Rocky intertidal areas occur in a full range of exposures to wave action, according to a biologically defined, eight-point exposure scale where 1 is very exposed to wave action and 8 is

Fig. 9.5 Grey Seals (*Halichoerus grypus*) hauled out on the rocks of the Eastern Isles. Photograph © R. M. Warwick.

very sheltered (Ballantine, 1961). For Scilly, a "super-exposed" category was added, indicating extremely severe wave action; for example, areas exposed to the open ocean used as haulouts by grey seals (Fig. 9.5). "Super-exposed shores" and "extremely exposed shores" are west facing and comprise steeply sloping bedrock on smaller islets and rocks to the west and north (Fig. 9.6). Rather few species can survive there, and shore zonation is in belts that are expanded in comparison with those of more sheltered shores.

As shelter from the prevailing westerly winds increases, shores become more gently sloping. "Very exposed shores" (Fig. 9.7) are richer in species, which need to be well adapted to resist dislodgement and damage. "Fairly sheltered shores" (Fig. 9.8) are generally of bedrock above mid-tide level and large boulders below, with a fairly gradual slope. "Very sheltered shores" occur in protected channels between the larger islands or in embayments protected from the west (Fig. 9.9) and are very species rich. The bedrock of the upper shore usually gives way to boulders, and the boulders to sand and gravel near the bottom of the shore. These shores often occur only a few kilometers from the super-exposed sites and a few hundred meters from the slightly less exposed sites, providing a mosaic of habitats in a relatively small geographic area.

Boulders small enough to be turned occur on fairly sheltered to very sheltered shores and are usually a feature of lower shores. Under-boulder animal communities are particularly rich, while those on very sheltered shores support fewer sponges, anthozoans, and echinoderms, but more ascidians than at more exposed sites. The fauna tends to be most diverse at the more exposed sites where boulders lie on a substratum of underlying boulders and cobbles that permits free flow of water.

9.3.1.2 Sandy shores

Intertidal sandflats of Scilly are extensive (Fig. 9.6), with highly diverse fauna. The sediment is derived from granite and is generally coarse grained, with low silt content and virtually no input of sediment from the land, due to Scilly's lack of significant streams (Section 9.2). The most stable areas are between islands, protected from wave action and bisected by narrow tidal channels. The largest of these areas is St. Martin's Flats (Fig. 9.10a); other important areas are east of Samson, between Bryher and Tresco, and adjacent to Old Grimsby on Tresco.

Although sandflats might appear to the casual observer to be rather barren, species diversity is actually very high, supporting a mobile fauna such as shrimps, crabs, and fish. Lower-shore sandflats are notable, for they include the fringes of the most extensive and diverse beds of eelgrass (*Zostera marina*) of southern England (Section 9.3.1.3), with an unusually species-rich biota of seaweeds, fish, and sediment communities of anemones, polychaete worms, bivalve mollusks, and burrowing echinoderms, including many species restricted to sublittoral areas elsewhere in the UK. As for rocky shores, many southern species are present, often in large numbers, including some that are recorded rarely in the UK.

The diversity of invertebrates is especially high due to large numbers of smaller invertebrates living between sand grains (the interstitial meiofauna), as well as larger species burrowing through the sediment. At a single location on St Martin's Flats, and on a single sampling occasion, 464 species of metazoans were found, the dominant taxa being nematodes with 207 species, and harpacticoid copepods with 75. On more open shores, sand beaches are more steeply sloping (Fig. 9.10b), but have been less thoroughly investigated as to species composition.

Substantial differences are evident between communities of each of the major sedimentary flats, and even within individual flats (Fig. 9.11), resulting from differences in physical and biotic factors. The heart urchin (*Echinocardium cordatum*) is probably the most conspicuous animal on these flats, occurring in a wide variety of sediment types, and in some areas

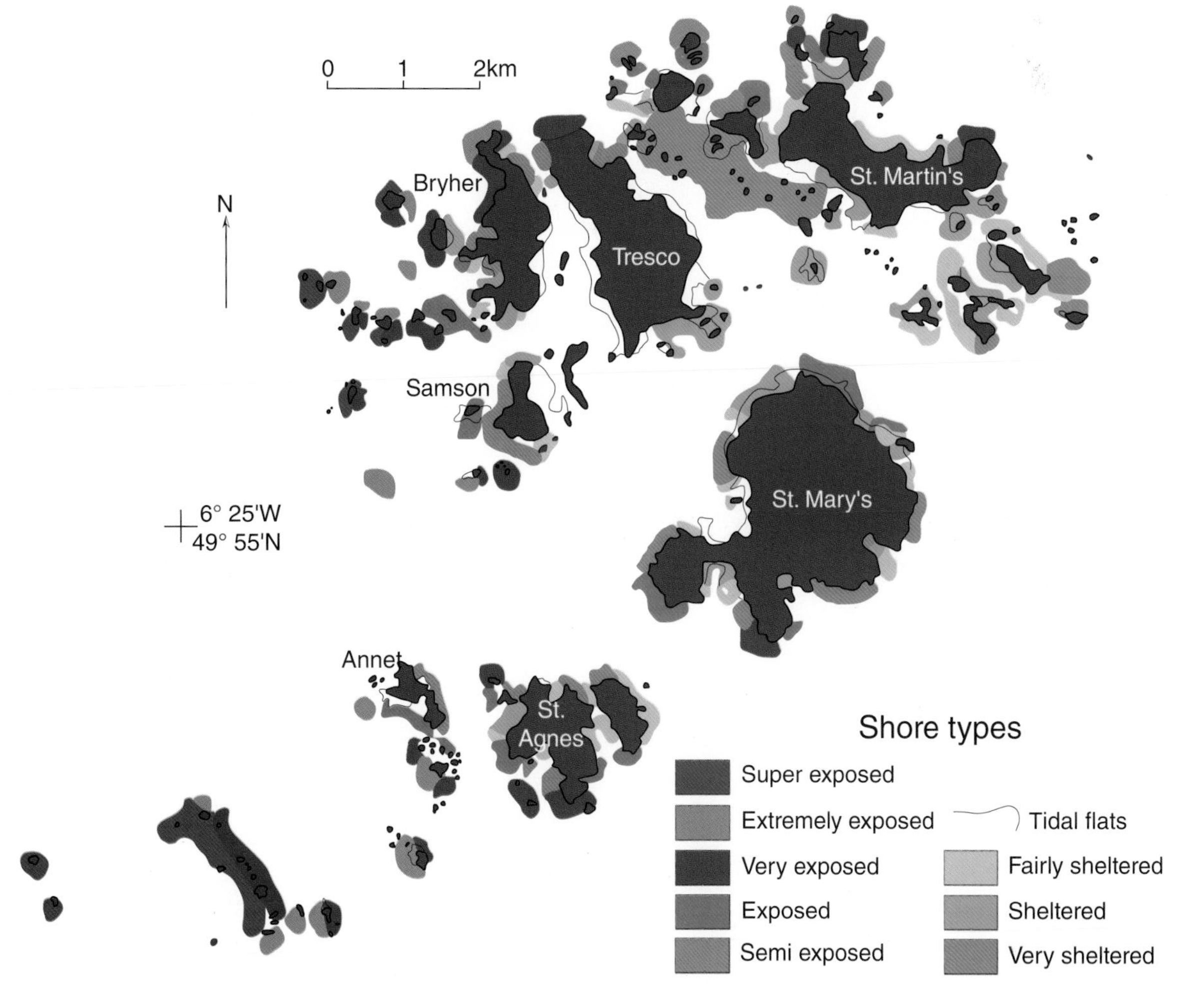

Fig. 9.6 Map showing distribution of rocky shores in the various exposure categories. The uncolored areas enclosed by lines are intertidal sandy flats. Based on data from Hiscock (1984).

may reach densities up to $25\,m^{-2}$. Where exposure and tidal influence are greater and sand is mixed with gravel, pebbles, and cobbles, infaunal communities live on stones, in addition to the subsurface; for example, nationally rare species such as red seaweeds. These stony areas may also support large populations of the turban top-shell (*Gibbula magus*) at population densities of up to $40\,m^{-2}$.

9.3.1.3 Seagrass beds

Much of the sandy sublittoral between islands is shallow and sheltered and the water is clear, providing an ideal habitat for seagrasses. Eelgrass (*Zostera marina*) is the most important seagrass in the UK. The only other UK seagrass is widgeon grass (*Ruppia maritima*), confined to the only area of sheltered brackish water on Scilly, Bryher Pool (Fig. 9.12a). Extensive meadows once extended over vast areas of sandflats and intertidal areas, but largely disappeared in the 1930s due to a wasting disease, caused by the slime mold *Labyrinthula*. Scilly did not suffer so badly as other areas, and has become one of the most important seagrass areas in the UK. Recovery has occurred to some extent, but seagrass is still very limited in intertidal areas (Fig. 9.12b). Symptoms of the wasting disease were observed in Scilly during the 1980s and 1990s and so the disease may be responsible for a recent, apparently declining trend in abundance. The disease is still present, but mainly affects only the outer leaves of the plants, not the inner, newly growing shoots.

Seagrass beds are of very high conservation value due to the richness of their associated flora and fauna. The extensive network of leaves, roots, and rhizomes provides structural complexity that offers a variety of ecological niches for other species. The most diverse faunal communities of any sedimentary area in Scilly are found among roots and rhizomes, including numerous species of algae, hydroids, anemones, mollusks, and fish. Eelgrass beds are particularly important as spawning and nursery areas for many fishes, including some that are commercially important. They also support nationally rare species, including an alga (*Asparagopsis armata*), a hydroid (*Laomedea angulata*), the spiny or long-snouted seahorse (*Hippocampus guttulatus*), and probably the short-snouted seahorse (*Hippocampus hippocampus*), thought to breed there.

Fig. 9.7 (a) Very exposed shore, White Island, St Martin's, at a low spring tide; (b) dog-whelk (*Nucella lapillus*) and beadlet anemone (*Actinia equina*) among barnacles and limpets mid-shore; (c) pink coralline alga (*Corallina officinalis*) and reproductive fronds of thong-weed (*Himanthalia elongata*) on the lower shore; (d) two kelps, dabberlocks (*Alaria esculenta*), and oarweed (*Laminaria digitata*) at extreme low water; (e) small blue mussels (*Mytilus edulis*) nestling among barnacles (*Chthamalus stellatus*) mid-shore; (f) bladderless form of bladder wrack (*Fucus vesiculosus f. linearis*) characteristic of high exposure in mid-upper shore; (g) variety of lichens in the splash zone. Photographs © R. M. Warwick.

Fig. 9.8 (a) Fairly sheltered shore south of Periglis, St. Agnes, at low spring tide; (b) lower shore festooned with reproductive fronds of thong-weed, with offshore kelps becoming exposed, including oarweed (*Laminaria digitata*) and furbelows (*Saccorhiza polyschides*); (c) algal-dominated lower mid-shore with serrated wrack (*Fucus serratus*), carrageen (*Chondrus crispus*), and vegetative buttons of thong-weed (*Himanthalia elongata*); (d) brown alga (*Fucus vesiculosus*) with flotation bladders, beadlet anemones (*Actinia equina*), grey top-shell (*Gibbula cineraria*), dog-whelk (*Nucella lapillus*), and limpet (*Patella vulgata*) on mid-shore; (e) brown seaweed (*Bifurcaria bifurcata*, a southern species), in upper shore pool; (f) black lichen (*Lichina pygmaea*) and channeled wrack (*Pelvetia canaliculata*), can survive almost complete desiccation during long periods of aerial exposure, with juvenile winkles (unidentified spp.) in crevice, on the upper shore. Photographs © R. M. Warwick.

Fig. 9.9 (a) Very sheltered shore at Porth Hellick, St Mary's, at a low spring tide, permanent pool is retained at low water by rocky sill; (b) seven-armed starfish (*Luidia ciliaris*, aberrant individual with eight arms) on gravel bed of low-water pool; (c) giant top-shell (*Gibbula magus*); (d) serrated wrack (*Fucus serratus*) and thong-weed (*Himanthalia elongata*) with developing reproductive fronds on lower shore; (e) bladder wrack (*Fucus vesiculosus*) and knotted wrack (*Ascophyllum nodosum*), with limpets and thick top-shells (*Monodonta lineata*), on mid-shore; (f) barnacle (*Chthamalus montagui*) on upper-mid shore; (g) channeled wrack, thick top-shell, and limpet (*Patella vulgata*) on upper-mid shore; (h) zonation of lichens with channeled wrack (*Pelvetia canaliculata*) below on upper shore. Photographs © R. M. Warwick.

Fig. 9.10 (a) Benthic sampling on St. Martin's Flats. (b) Steeply sloping sandy beach, Great Bay, north coast, St. Martin's. Photographs © R. M. Warwick.

Eelgrass beds are highly sensitive to substrate loss, smothering, turbidity, wave exposure, nutrients, and physical disturbance. Propeller wash can erode or smother seagrass beds. Anchoring and mooring can also cause direct physical damage, causing bare circular areas where mooring chains abrade the bed as boats move with the tide (Fig. 9.13a). Naturally occurring, ring-shaped areas of eelgrass ranging from 5 to 50 m in diameter, with bare sand in the center, are also found in some areas (Fig. 9.13b), possibly resulting from current action dislodging the center of the bed, forming a "blow out."

9.3.1.4 Sublittoral rocky reefs

Subtidal and intertidal rocky reefs attract particular attention on account of their luxuriant fauna and flora and water clarity. An unusual temperature regime has been invoked to account for the rich and varied algal communities (Section 9.2). Scuba divers have explored more than 100 reef sites; many others outside the shelter of the ring of islands remain unexplored. Thus, the full extent of this habitat is currently unknown. Granitic reefs and islets surrounding the main islands have walls covered with abundant, spectacularly colorful anemones and soft corals that are particularly attractive to divers (Fig. 9.14), including jewel anemones (*Corynactis viridis*), plumose anemones (*Metridium senile*), and dead man's fingers (*Alcyonium digitatum*). Fragile sponge communities occur particularly on the more sheltered eastern sides of the islands in >25 m of water.

9.3.1.5 Sublittoral soft sediments

Soft sediments have received less ecological attention than reefs. They occupy sheltered areas within the islands, and are complex due to a wide range of sediment types from clean and poorly sorted fine sands to gravels (Fig. 9.15); there are no muddy sediments, due to a paucity of streams (Section 9.1). Overall, species diversity appears to be high, but may be low at individual sites, and most species' abundances are low. Deeper fine and medium sand areas generally support a greater diversity of infauna than shallow sediment areas, those south of St. Agnes and at St Mary's being particularly rich. In sandy areas, there is little correlation between species composition and sediment type. There is great variability among species diversity and composition among individual samples, and because of this, assigning species associations to sediment types is considered inappropriate.

The conservation value of these habitats lies in the great variety of animal species and the rapid changes in species composition and habitat characteristics that occur over very short distances. These habitats appear to be unaffected by human activities at present, although they are potentially susceptible to physical disturbances such as gravel extraction and/or trawling. However, the latter is unlikely to occur close inshore because of shallow water and treacherous rock outcrops. These habitats are also susceptible to deposition of fine materials, which would result in increased numbers of common deposit-feeders at the expense of the more unusual species now present.

9.3.2 Organismal diversity

The marine fauna of Scilly has lesser representation of species for most major phyla compared with the nearest recorded region of the mainland (Marine Biological Association, 1957). That is, species richness is lower, implying that taxonomic distinctness is higher. One reason for lesser richness is that species with a long planktonic larval life have a greater chance of being swept away by currents from their parental breeding areas, with little probability of replenishment from the mainland. Such species are thus likely to be depleted in, or disappear from, Scilly waters. Another explanation may be the well-known species-area relationship, which predicts that smaller areas are capable of hosting fewer numbers of species than larger ones (MacArthur and Wilson, 1967). It would be interesting to compare species lists for various taxa from Scilly with

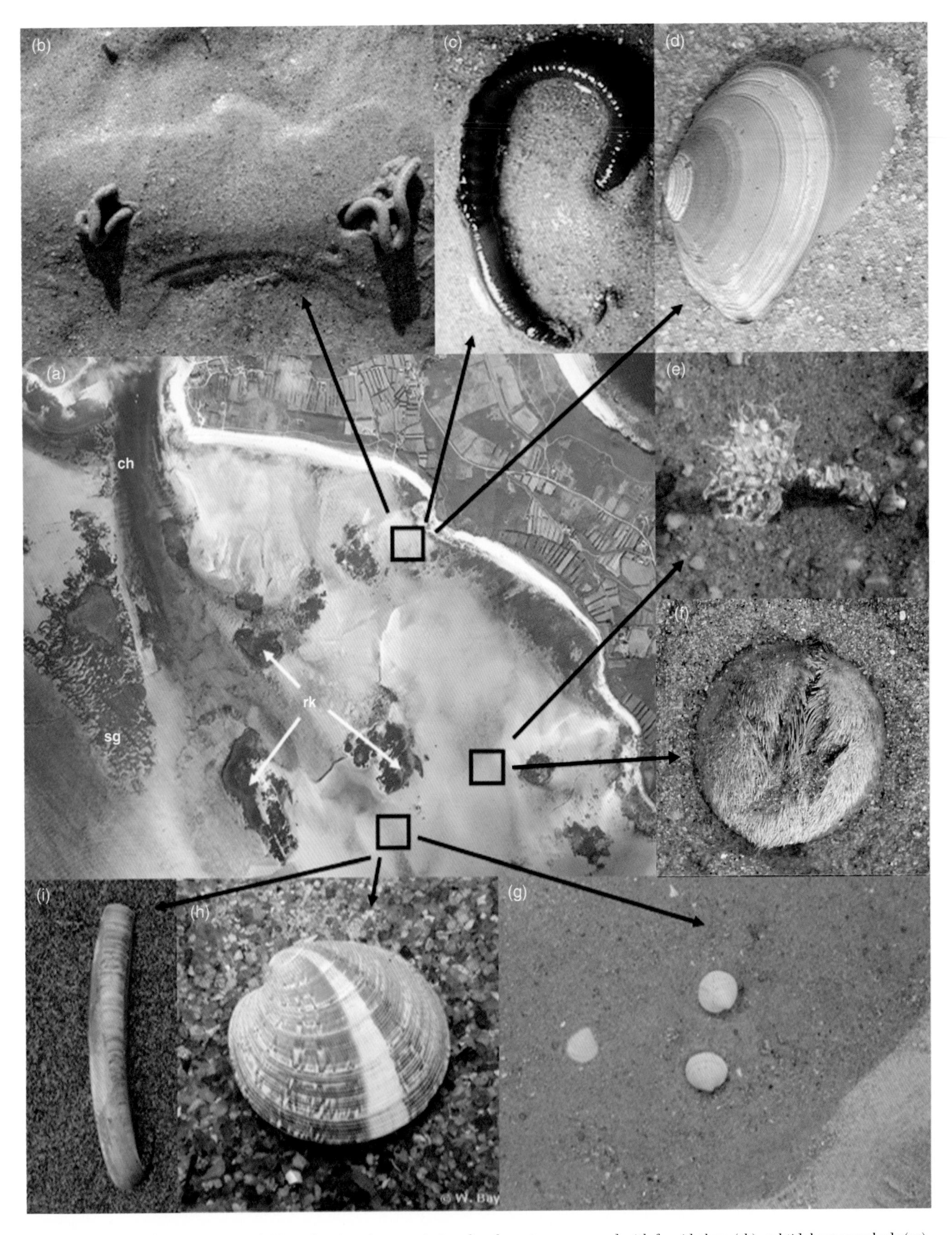

Fig. 9.11 (a) Aerial view, St. Martin's Flats, showing close proximity of rocky outcrops covered with fucoid algae (rk), subtidal seagrass beds (sg), and deep-water channel (ch). © Natural England 2013 material is reproduced with the permission of Natural England, http://www.naturalengland.org.uk/copyright. Three areas of flats with very different faunas: (b) fine rippled sand with sediment casts of polychaete worm (*Arenicola marina*); (c) polychaete worm (*Arenicola*); (d) bivalve mollusk (*Angulus tenuis*); (e) smooth sand with filtering fan of sand-mason worm (*Lanice conchilega*); (f) burrowing heart urchin ("sea potato," *Echinocardium cordatum*); (g) coarse hummocked sand with dead shells of bivalve mollusk (*Dosinia exoleta*); (h) clam (*Dosinia*); (i) burrowing razor-shell (*Ensis ensis*). Photographs (b–e, g–i) © R. M. Warwick. Photograph (f) © G. C. Ray.

Fig. 9.12 (a) Bryer Pool, southwest Bryer Island; (b) relatively extensive seagrass bed, Old Grimsby, Tresco, exposed at low spring tide. Photograph © R. M. Warwick.

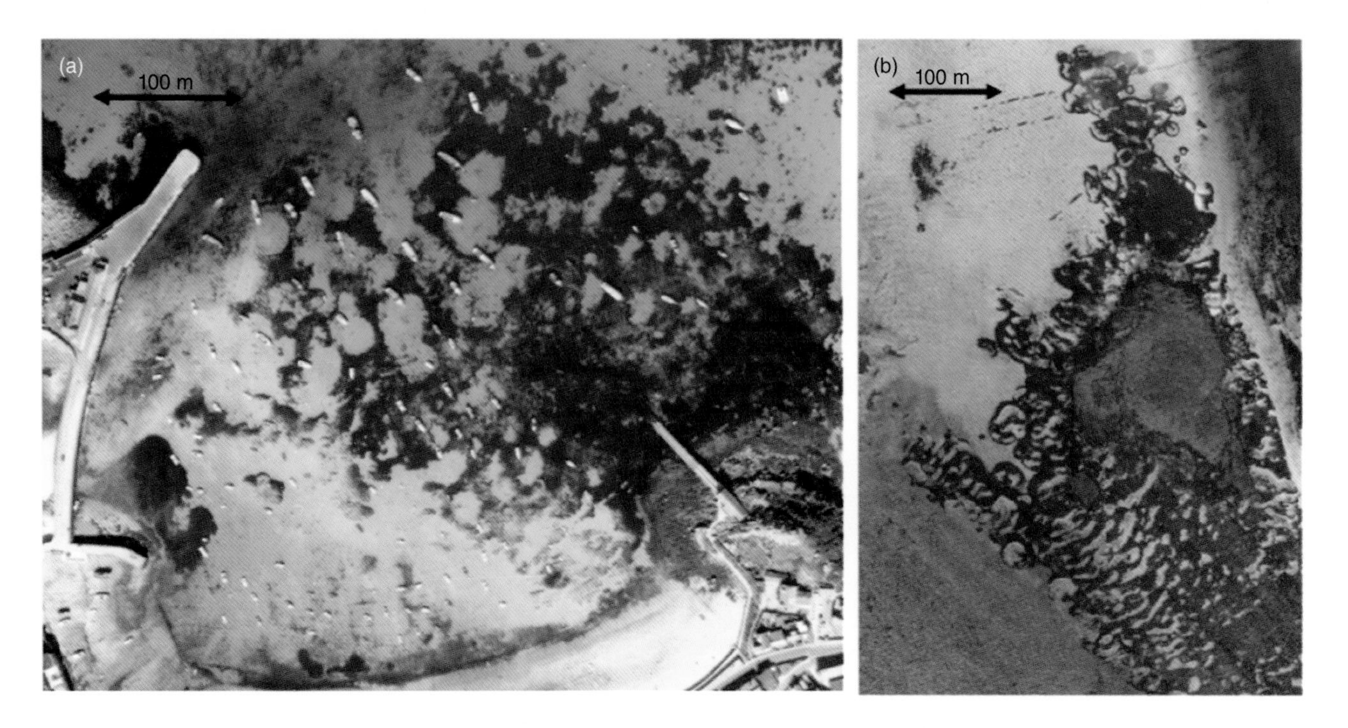

Fig. 9.13 Aerial photographs showing: (a) damage to seagrass beds caused by mooring chains—note bare circular patches associated with each boat; (b) natural ring-shaped patterns with bare sand in the center in a seagrass bed east of St. Martin's. © Natural England 2013 material is reproduced with the permission of Natural England, http://www.naturalengland.org.uk/copyright.

total numbers from the UK (Howson and Picton, 1997) to see whether the species-area relationship alone can explain the patterns observed.

In order to assess the effectiveness of measures to conserve biodiversity, an essential prerequisite is to define and measure it in practical and ecologically meaningful ways. At the organismal level, biodiversity is regarded as a measure of the number of species in an area, formulated in terms of species richness and/or evenness in a region (Ch. 5), as opposed to a small, defined area or sample. However, on large regional scales species richness is sometimes difficult to measure; the harder one looks the more species one finds. Apart from this dependence on sampling effort, species-richness measures have a number of other logistical and conceptual disadvantages. First, maximum species richness occurs at intermediate levels of disturbance and may either increase or decrease in response to changes in the level of environmental degradation (Ch. 5). Species richness is also dependent on habitat type and complexity, but lacks explicit links to functional diversity. Furthermore, biodiversity cannot be regarded as just the number of species in an area, and requires measures that reflect the phylogenetic divergence of the organisms present. To overcome some of these problems, the Isles of Scilly have been used as a testing ground for the development of biodiversity measures using simple species lists compiled from uncontrolled sampling effort.

Fig. 9.14 Colorful wall of sublittoral epifauna, very attractive to recreational divers. Photograph © courtesy F. Gloystein.

9.3.2.1 Average taxonomic diversity and distinctness

Assemblages of organisms subject to human (or other) impacts will exhibit reduced taxonomic spread of species; in extreme cases, very closely related species belonging to the same genus may occur (Fig. 9.16). Unimpacted assemblages, on the other hand, will have a wider taxonomic spread, with species belonging to many different genera, families, orders, classes, and phyla.

The method advanced here for developing biodiversity indices is based on tracing the average path length or taxonomic distance between every pair of individuals or species in

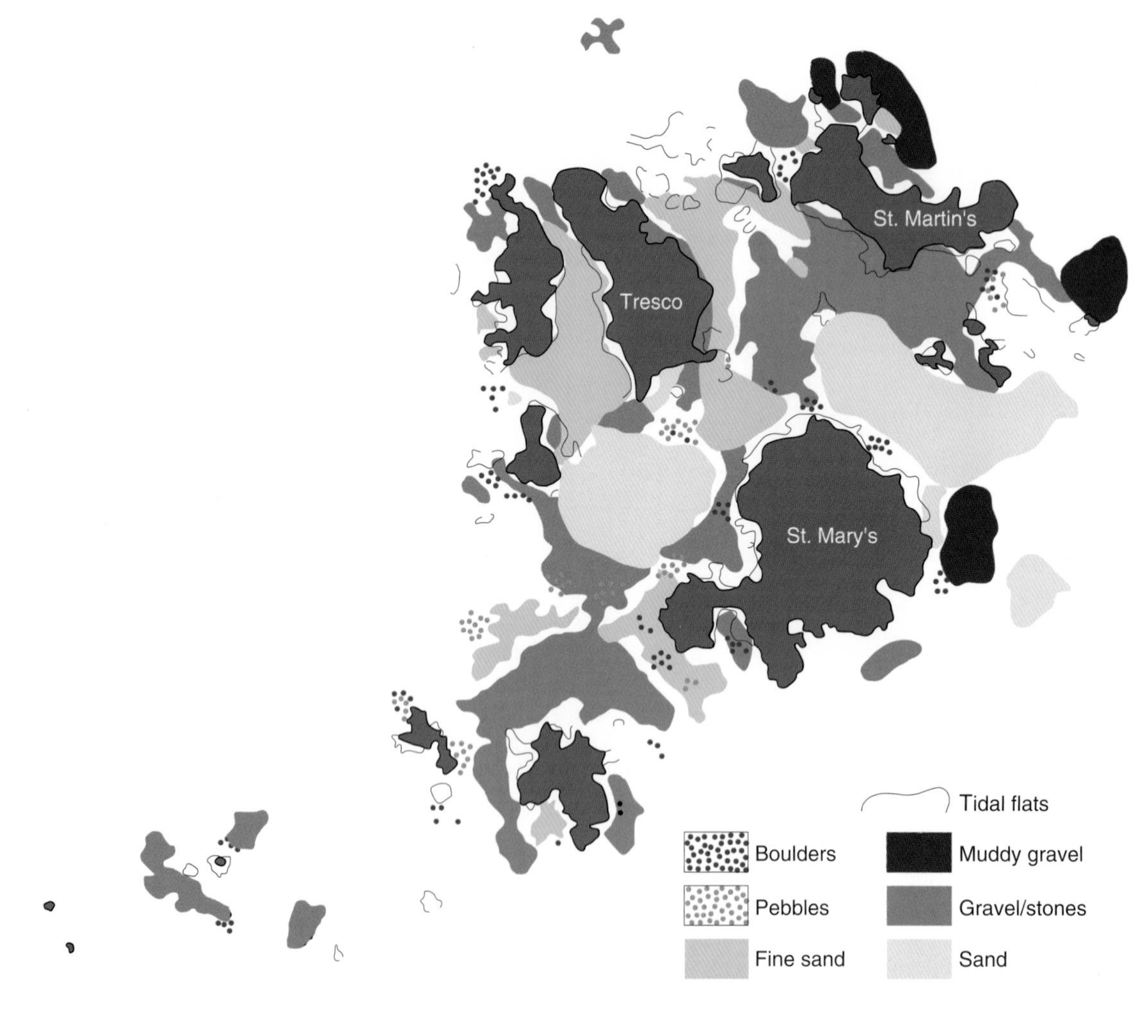

Fig. 9.15 Map showing distribution of sediment types. Based on data from Rostron (1989).

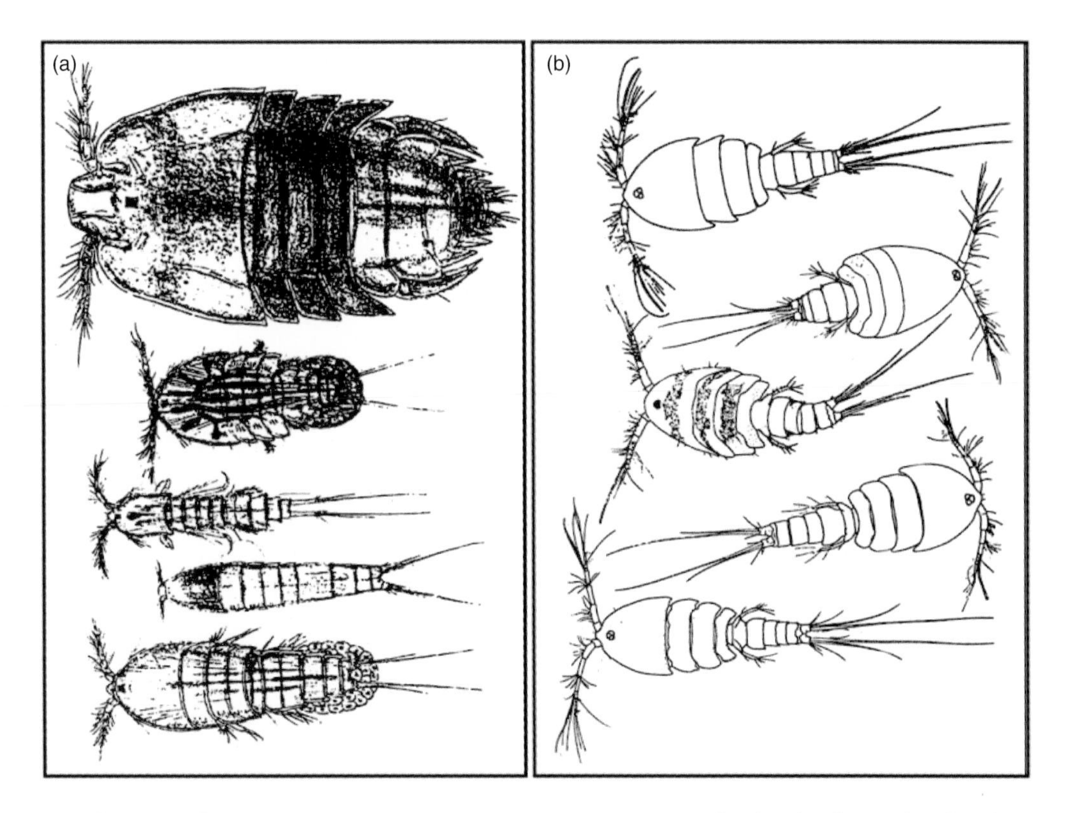

Fig. 9.16 Dominant harpacticoid copepods from: (a) a pristine environment, among seaweeds, Isles of Scilly, UK; (b) polluted environment, Belgian sluice dock. Those in (a) differ greatly in form and function and belong to various families, while those in (b) are sibling species of *Tisbe*.

a taxonomic classification tree, or measuring the variability in these path lengths (Fig. 9.17). These indices are independent of sample size or sampling effort, are little affected by small variations in habitat type, and can be used for data consisting simply of species lists and arising from unknown or uncontrolled sampling effort. The software package PRIMER 6 (Warwick and Clarke, 2001) can be used to calculate these indices. Average taxonomic diversity (Δ) of a sample is simply the average taxonomic distance apart of every pair of individuals in the sample. Average taxonomic distinctness (AvTD or Δ^*) is a measure that is more nearly a pure reflection of the taxonomic hierarchy with a similar formulation as Δ, but excludes paths between individuals belonging to the same species. Where quantitative data are not available and the sample consists simply of species' presence/absence data, the average taxonomic distinctness Δ^+ is the average taxonomic distance apart of all its pairs of species.

This method results in a highly intuitive definition of biodiversity, as it explains the average taxonomic breadth of a sample. In certain cases, biodiversity changes may involve situations in which some genera become highly species rich, while a range of other higher taxa are represented by only one (or a very few) species. Such is the case, for example, of the effects of demersal fishing on the macrobenthos of the North Sea, where polychaetes are increasing at the expense of such taxa as mollusks and echinoderms. Such changes can be summarized in a further statistic: the variance of the taxonomic distances between each pair of species about their mean value Δ^+, termed the variation in taxonomic distinctness (VarTD) denoted by Λ^+. The lack of dependence of both Δ^+ and Λ^+ on the number of species in the sample has far-reaching consequences for comparing historic data sets and other studies for which sampling effort is uncontrolled, unknown, or unequal.

This construction of taxonomic distinctness indices from simple species lists also makes it possible to address another desirable biodiversity measure: i.e., departures from expectation. This envisages an inventory of species encompassing an appropriate biogeographic region, from which the species found at one locality can be drawn. For example, Fig. 9.18 uses the entire British faunal "master" list of free-living marine nematodes (395 species identified to date) as a basis for comparison with the species complement for any particular locality and/or historic period within that region, to ask whether the observed subset of species is representative of the biodiversity expressed in the full species inventory. The value of Δ^+ for the full inventory is the expected value for average distinctness, and reductions from this level, at one place or time, can potentially be interpreted as loss of biodiversity. The reduction (or increase) can be regarded as statistically significant if the observed value falls outside the central 95% of a large number of simulated values, for that number of species drawn at random from the master inventory. For individual samples or locations, these values can be visualized as a histogram (Fig. 9.18), or if many values are to be compared then the 95% confidence limits for the simulated values can be visualized as a "probability funnel" for samples with varying species

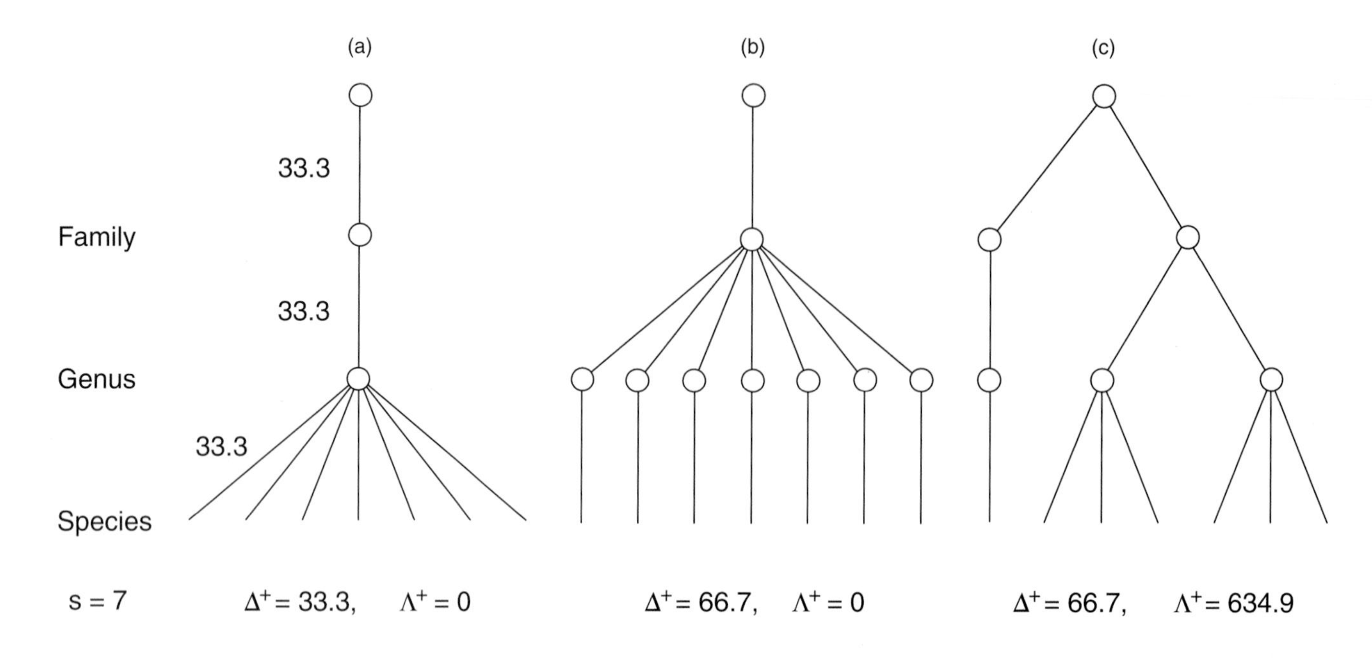

Fig. 9.17 Three taxonomic hierarchies for seven species with equal step lengths (distances) between each taxonomic level, using simple linear scaling whereby the largest number of steps in the tree (two species at greatest taxonomic distance apart) is set to 100. Thus, the path length (total distance) between individuals in species 1 and 2–7 = 100, between species 2–4 and 4–7 = 66.6, and between species 2, 3, 4 and 4, 5, 6 = 33.3. In (a) the average path length is 33.3 and in (b) 66.6, so these are the AvTD (Δ^+) values; in both figures there is no variation in step length, so that VarTD (Λ^+) is zero. In (c) the average path length, and hence Δ^+, is the same as in (b) but the path lengths are variable and the Λ^+ value is 634.9.

numbers (Fig. 9.19). In general, impacted assemblages are characterized by decreased AvTD and increased VarTD, although the latter may not always be the case.

It seems surprising that anything sensible can be said about diversity indices drawn from data consisting simply of species lists, and arising from unknown or uncontrolled sampling effort. Nevertheless, use of AvTD and VarTD has made possible the study of relatively large-scale patterns and anthropogenic impacts on biodiversity (Warwick and Clarke, 1998), and to make predictions about the effects of climate change (Warwick and Turk, 2002).

9.3.2.2 "Surrogacy" approaches to biodiversity estimation

Another important approach for measuring biodiversity is extrapolation from one taxonomic group to another, from site to site, and from sample to inventory, across spatial scales. Because of the difficulties related to changing baselines (Ch. 13) and of routinely undertaking comprehensive surveys, "surrogacy" methods will clearly become the norm in site assessment in the future, and the search for appropriate "indicators" will become an important research goal. Two possible methods have been developed on the Isles of Scilly: the use of death assemblages of mollusks as surrogates for the biodiversity of the regional living species pool, and the use of one size component as a surrogate for the overall biodiversity of the total size spectrum of the biota.

Molluskan death assemblages. On a regional basis, a census of all living species from the whole spectrum of habitat types would be time-consuming and costly. The concept of sampling a single "spatially averaged" death assemblage as a surrogate for regional biodiversity is therefore appealing, but depends on the death assemblage being fully representative of living biodiversity. On St Martin's Flats, a wide range of shells of species from habitats other than sand are present as a result of post-mortem transport by waves and currents (Fig. 9.20). The hypothesis is that this assemblage reflects the biodiversity of living molluskan fauna of the archipelago (Warwick and Light, 2002). Although relative numbers and abundances of species in a death assemblage would not be expected to be representative of living fauna, AvTD (Section 9.3.2.1) has proved to be a useful method for a rapid assessment for St. Martin's Flats. The Flats extend over the whole of the southwest shore of St. Martin's and outwards towards St. Mary's, and are in close proximity to outcrops of intertidal and subtidal rock, beds of seagrass, and deep sandy channels (Fig. 9.11).

The assessment method employed "parataxonomists" (persons capable of identifying organisms to known species from readily available literature) as well as taxonomic experts, who collected Gastropoda (snails) and Pelecypoda (bivalves) shells. Members of the Conchological Society of Great Britain

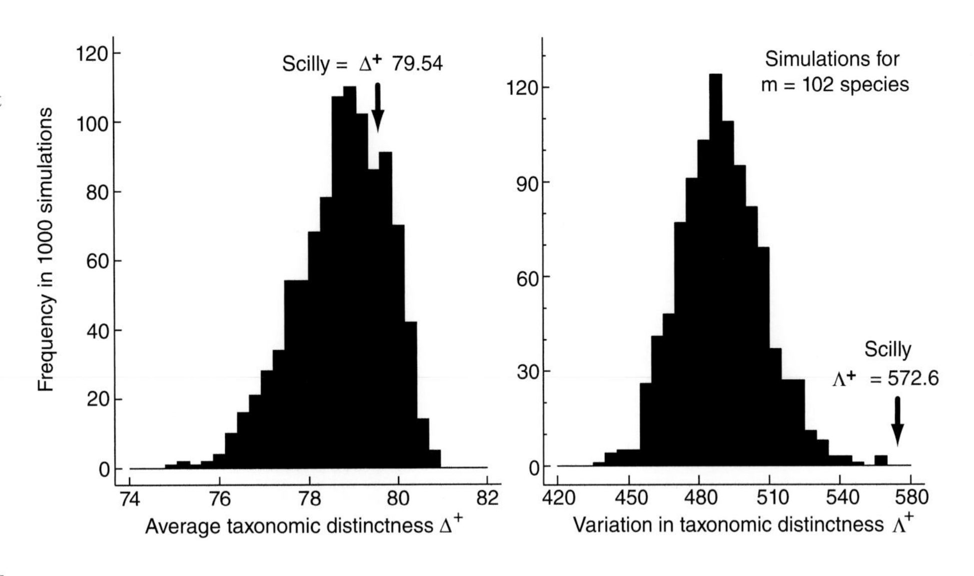

Fig. 9.18 Histograms of simulated AvTD and VarTD from 999 sublists drawn randomly from a UK master list of 395 species of free-living nematodes. A sublist of size 102 corresponds to the number of species recorded from the Isles of Scilly. Measures of Δ^+ and Λ^+ for Scilly are indicated by arrows: the Δ^+ value is central but the null hypothesis that VarTD equates to that for the UK list as a whole is rejected, falling well outside the 95% probability limits of the null distribution. This is thought to be due to the fact that muddy and low salinity habitats are absent from Scilly, and there are higher taxa of marine nematodes that characterized such habitats and are thus under-represented compared with the national average, giving a significantly higher VarTD.

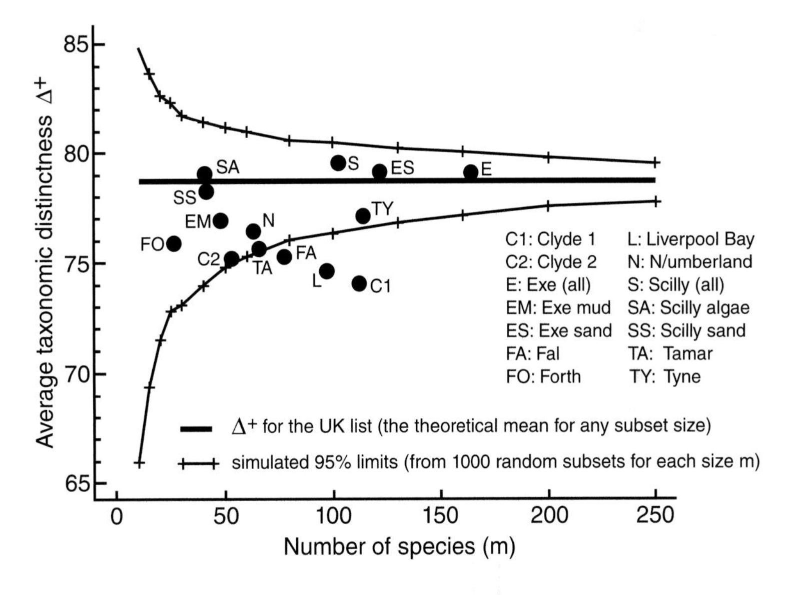

Fig. 9.19 UK regional study of free-living nematodes. Funnel plot for simulated AvTD for a range of sublist sizes (x-axis). Crosses and thin lines indicate limits within which 95% of simulated Δ^+ values lie. The thick line indicates mean Δ^+ (the AvTD for the master list), which is not a function of the number of species. Filled circles are the true AvTD (y-axis) for 14 location/habitat studies plotted against their sublist size (x-axis). Note that values for the polluted Clyde estuary, Liverpool Bay and Fal estuary fall below the 95% probability limits of the funnel, whereas values for Scilly (sand, algae, and all habitats combined) are amongst the highest and all close to the mean.

and Ireland made collections and processed sediment samples, comprising a high proportion of coarse gravel and broken shell, divided into four size fractions. The inventory of living mollusks from the Isles of Scilly is probably the most comprehensive of any group of animals. The total number of shelled mollusks found as a death assemblage on the Flats was slightly over half the total number of species recorded for the islands. For gastropods, average taxonomic distinctness (Δ^+) for all four sediment size fractions proved to be close to the mean for the full regional inventory, and within the 95% intervals of the probability funnel (Fig. 9.21). Thus, the data indicate that gastropods from the death assemblage are *fully representative* of the biodiversity of the regional fauna and can be regarded as a good surrogate for regional gastropod biodiversity. For bivalves, on the other hand, all four data points are well below the 95% probability intervals, indicating that they are significantly *unrepresentative* of the average taxonomic distinctness of regional fauna. The most obvious explanation is that bivalve shells in a death assemblage belong to species from the same habitat, and are not representative of the biodiversity of the region as a whole.

Thus, data for gastropods and bivalves yield quite different results. Of 42 species of gastropods in this study, only two (necklace shell, *Polinices polianus*, and wentletrap, *Epitonium clathrus*) are characteristic of shallow water and intertidal sand habitats. The remaining species are more typical of rocky, macroalgal, or seagrass habitats, or are eurytopic (i.e., found in a wide range of habitats). By contrast, only six of the 49 bivalve species sampled are characteristic of habitats *other than* shallow and intertidal sand, while families characterizing other habitats (e.g., mussels, family Mytilidae) are underrepresented. The key difference appears to lie in contrasting natural

Fig. 9.20 Shells from sediment surface. St. Martin's Flats, illustrating the concept of taxonomic distinctness. Both (a) and (b) have the same number of species, but in (a) the species have a wide range of morphology, feeding type (grazers, deposit feeders, suspension feeders), and belong to a variety of higher taxonomic groups, whereas in (b) the species are morphologically similar and all belong to a single family of suspension feeders. Species richness measures suggest that (a) and (b) have the same diversity, but (a) is obviously more "biodiverse," which taxonomic distinctness indices indicate. Photos © R. M. Warwick.

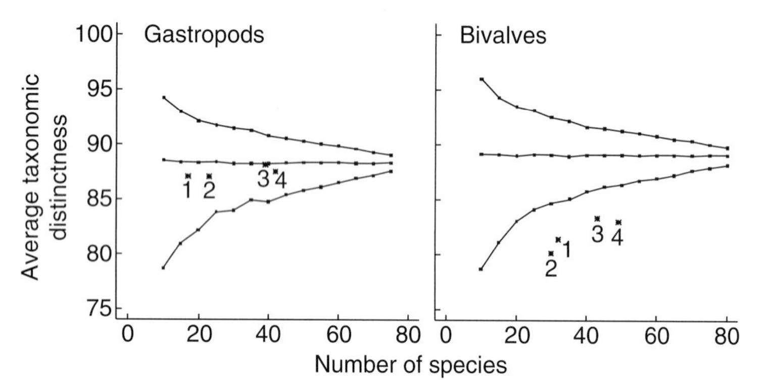

Fig. 9.21 Probability funnels (mean and 95% probability intervals) of average taxonomic distinctness (Δ^{+}) for 1000 random subsets of different numbers of species from the living regional species pool. Measured values from St. Martin's death assemblage (asterisks) are: (1), Conch. Soc. surface search; (2) RMW surface search; (3) Conch. Soc. all data including microscopic examination of sediment samples; (4) total for all studies.

histories. Bivalves are primarily infaunal, but gastropods are primarily epifaunal and more subject to transport from other habitats, either directly or via rafting on debris.

It is axiomatic that, in order to extrapolate regional biodiversity of any group of organisms from a death assemblage at one location or habitat, the assemblage sampled must have been constituted by methods and processes that randomize species composition from a wide range of habitats, and must not be biased by over-representation of species characterizing the habitat at one location. Thus we might expect, for example, that foraminiferan shells in a sediment core would be representative of regional planktonic biodiversity rather than of benthic taxa. Similarly, tree pollen in terrestrial soil cores would be representative of historical tree biodiversity, and marine beach litter would be representative of the sources of plastics pollution.

Diversity and species composition in different size components. A detailed study of a single site on St. Martin's Flats investigated the use of one faunal size component as a surrogate for the overall biodiversity of the total size spectrum of the biota (Warwick *et al.*, 2006). The objective was to determine the degree to which infaunal diversity and species composition varies with mesh size, sample size, and sample dispersion within an apparently homogeneous area of coarse intertidal sand (Fig. 9.22). Species diversity was found to be constant across the 63–125–250 µm size categories, with a stepwise reduction in the 500–1000 µm categories (Fig. 9.23). Similarly, diversity profiles visualized as *k*-dominance curves were remarkably constant within these same two size groupings (Fig. 9.24a). Non-metric multidimensional scaling (MDS) plots for the entire dataset also revealed two tight clusters corresponding with the same size groupings (Fig. 9.24b).

Although an analysis of each of these two clusters separately showed a gradual and regular change in community composition across the 63–250 μm size range, as well as a clear separation of the 500 and 1000 μm samples, the most dramatic change in species composition was evident between these two groupings. These conform to what have been traditionally regarded as meiofauna (<500 μm) and macrofauna (>500 μm), respectively. The constancy of species diversity, dominance profiles, and regular sequential change in community composition within these two domains is remarkable, at least for the meiofaunal domain.

Other studies of marine shallow-water, sediment-dwelling benthos from other latitudes have also shown bimodal species size distributions, with many meiofaunal and many macrofaunal species, but few of intermediate size. Comparative studies of non-marine and non-sedimentary habitats suggest that these patterns do not simply correspond to the physical scaling of habitat complexity, and must relate to some more universal scaling relationships that are not fractal-like. Warwick *et al.* (2006) suggested that important relationships are those between body size and various biological characteristics such as feeding behavior, reproduction, and life-history mode as they are affected by environmental spatial and temporal structure.

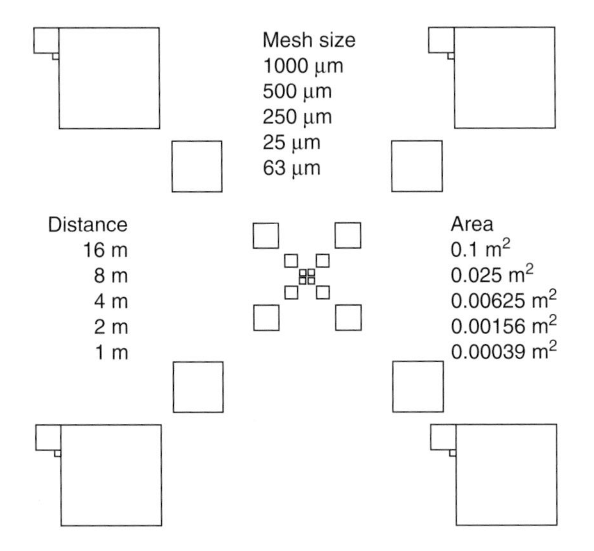

Fig. 9.22 Diagrammatic layout of sampling grid showing sample size, sample spacing, and sieve mesh sizes; standard range of mesh sizes (63, 125, 250, 500, 1000 mm) are scaled to sample areas and distances between samples. All metazoans were identified to species level.

9.4 SUSTAINING BIODIVERSITY FROM POSSIBLE THREATS

Environmental and social sustainability for the Isles of Scilly faces four major challenges.

9.4.1 Pollution

At present, the Isles of Scilly are relatively free from major sources of marine pollution and from the in situ industrial pollution that affects mainland Britain. There is no mining, and dumping and dredging are negligible. However, there is an ever-present threat of shipping accidents due to the many submerged reefs and rocks and the potential for stormy weather. Also, domestic sewage, refuse disposal, and agricultural runoff pose potential problems. Offshore sediments appear to be unaffected by human activities, although sediments are potentially susceptible to physical disturbances such as gravel extraction or trawling. The latter is unlikely to occur close inshore because of shallow water and treacherous rock outcrops. These habitats also are susceptible to increased deposition of fine materials, which would result in increased numbers of common deposit-feeding animals at the expense of the more unusual species now present.

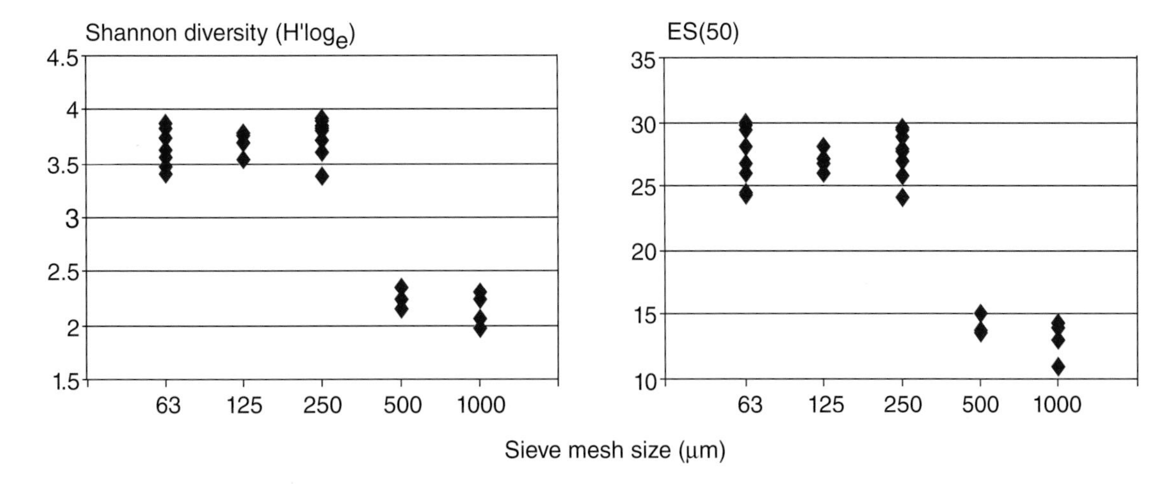

Fig. 9.23 Diversity indices calculated for all samples of each mesh size. Note constant species diversity across the 63–125–250 μm size categories, with a stepwise reduction to the 500 and 1000 μm categories, which also have equivalent values. See text for further explanation.

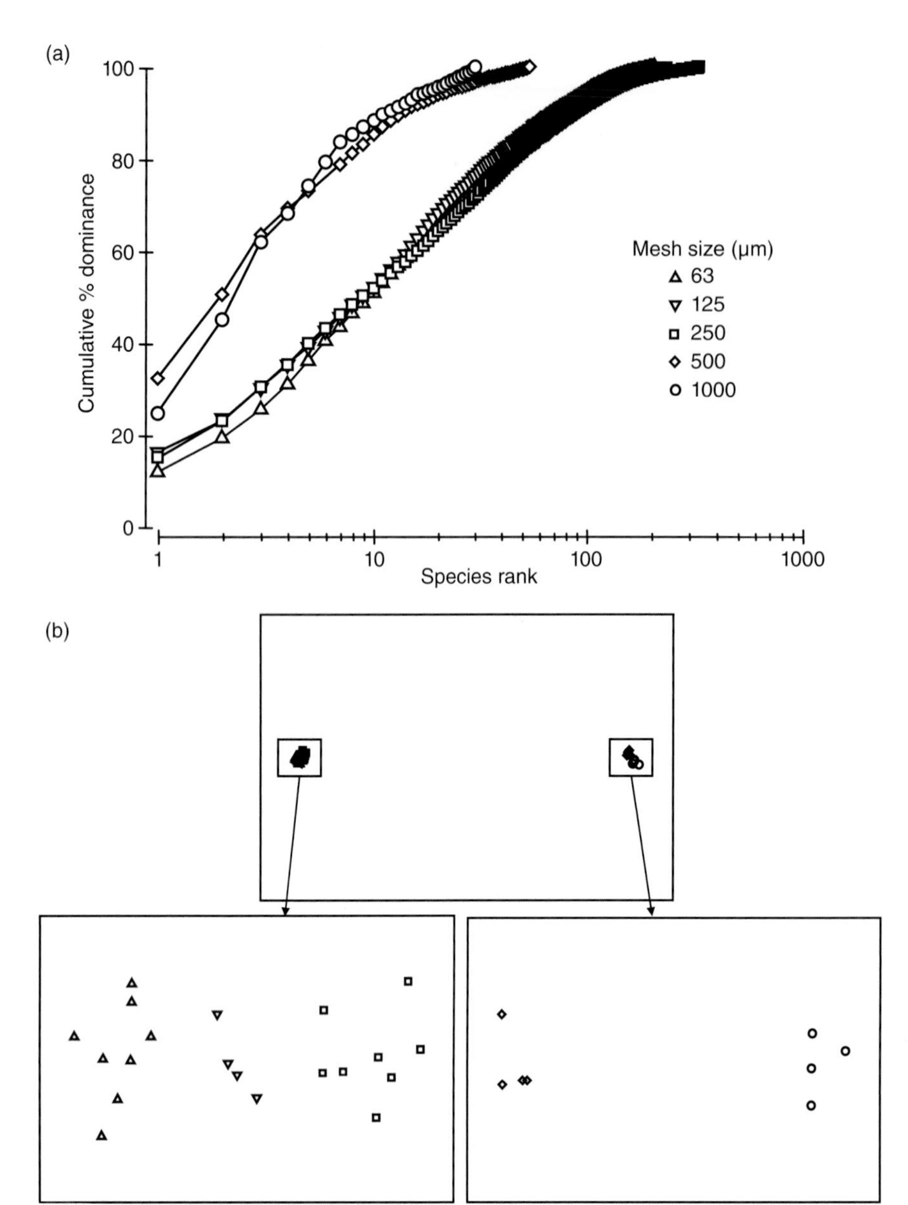

Fig. 9.24 (a) k-dominance curves averaged over all replicates for each mesh size. (b) MDS plots based on standardized √ transformed species abundance data and the Bray-Curtis similarity measure for samples of each mesh size. These plots can be thought of as a map in which the distances apart of the samples reflect their similarity in species composition (the closer together the more similar) rather than their geographical separation. Top: all samples. Bottom left: 63, 125, and 250 µm clusters analyzed separately. Bottom right: 500 and 1000 µm clusters analyzed separately.

One of the first and largest oil spills of modern times to affect the English coast occurred on the Seven Stones Reef, 8 km east of Scilly, when the *Torrey Canyon* oil tanker hit the reef in 1967, spilling 100,000 mt of crude oil. The oil damaged marine life on the mainland coast, and only the direction of prevailing winds prevented a major disaster. In March 1997 the *MV Cita*, a 3038 gross-tonne container vessel, foundered on the Newfoundland Rocks at Porth Hellick on St. Mary's, spilling 145 containers. Localized fuel spillage occurred and the contents of the containers were washed up on many beaches. The shipping company's insurers took no responsibility and the result was that the Isles of Scilly Council had to pay for clean-up. Rolls of polythene film, 3 ft wide and totaling 3740 miles in length, disintegrated and were distributed in the marine environment; despite huge efforts by volunteers, less than 1% was recovered. Spillage of oils, plastics, or chemicals demonstrates the potential for highly damaging marine pollution incidents that could seriously impact marine life, particularly sessile intertidal species. Clean-up and containment operations after oil spills are often more damaging to marine life than the oil spill itself, as they may involve the use of chemical dispersants, hot water, pressure-washing of the foreshore, and human trampling.

The sewage from Old Town, St. Mary's, is fully treated by an Accelerated Aerobic Digestion Plant ("Biobubble"), which includes ultraviolet treatment; the output (said to be drinkable) reaches the sea via a small stream. Septic tank products from outlying habitations are pumped out and pass through this treatment system. However, the output from the largest town, Hugh Town, St. Mary's, is macerated, but otherwise untreated, and reaches the sea via a main sewer serving about 1000 people. Dispersal is facilitated by strong tidal jets that propagate in a clockwise direction around the islands, with a peak current flow of 175 cm s^{-1}, which means that the sewage may remain in the vicinity of the islands for some time, although there is no evidence of it being washed ashore. Tests for the presence of *E. coli* in coastal waters of St Mary's have been undertaken sporadically, but have never been positive. Plans to build another Accelerated Aerobic Digestion Plant to process this effluent are on the table. There are no sewer mains for isolated buildings or on the four inhabited islands with much lower populations; buildings either use a tank and soak-away system or, on St Mary's, septic tanks that can be emptied and discharged to mains.

In the past, refuse from the islands was disposed of either by burning or dumping into the sea. For example on St Martin's, all the non-burnable rubbish was tipped over cliffs on the uninhabited north side of the island. Although this practice ceased in the late 1990s, there may still be impacts from chemicals and heavy metals that are likely to have accumulated in the sediments there.

The exceptionally mild climate of Scilly allows the production of early flowers and bulbs, mainly daffodils, and early potatoes, the main agriculture of the islands. Nitrates are no longer used as fertilizers, and the use of eelworm killers (for daffodils) has been banned. Potentially harmful runoff from fields is therefore negligible. However, herbicides are still used in flower farming, and runoff from newly ridged fields might possibly enter the marine environment. There are no slaughterhouses, and the few sheep and cattle present are taken to the mainland for slaughter. Runoff from grazing land probably has negligible impact on the marine environment.

9.4.2 Fishing

Fishing is arguably the main potential threat to Scilly's ecosystems. For many years, fishing was one of Scilly's major economic activities, but the industry has dwindled and fishing now accounts for only about 5% of the islands' gross domestic product (GDP). Only about 30 fishing boats are currently registered. Two trawlers fish year-round for a variety of demersal species. Other boats are mainly used for low-key seasonal (May to October) shellfish potting and small-scale netting. Between April and November, boats from the mainland may fish in the waters off Scilly during neap tides, returning to the mainland to land their catch before spring tides. Recreational angling is a popular sport and is important for tourism. Shore-based anglers and boat-fishers target similar species as do commercial fisheries. Sharks caught during these trips are tagged and released as part of a program run by Southampton University. During summer, artisanal shrimping by locals and visitors occurs at low-water spring tides on sandflats. The Area of Outstanding Natural Beauty Management Plan (2004, Section 9.5.2.1) expresses concern that the current level of shrimping activity may be too high, even though it is restricted by voluntary agreement to about 32 days each year.

In southwest Britain, despite management measures, some fish populations have declined due to overfishing. For example, the spawning stock of mackerel (*Scomber scombrus*) has fallen since the mid-1980s. While fishing effort can be controlled fairly easily within Scilly's territorial waters, the potential for overfishing of migratory species outside these waters is high. For example, non-migratory lobsters (*Homarus gammarus*) have maintained a fairly stable population, whereas migratory crawfish (*Palinurus elephas*) are in decline, probably due to overfishing by boats from other European Union (EU) nations. Although sharks are not targeted, they are globally threatened by the shark-fin trade; highly migratory species such as blue sharks (*Prionace glauca*) could be threatened locally by such global activities.

A problem associated with most types of fishing gear is bycatch of non-target species, which may include commercial or non-commercial fish, diving seabirds, and marine mammals. Deaths of stranded cetaceans can sometimes be attributed to fishing, and seals sometimes get entangled in fishing nets and lines, with harmful and even fatal results. In Scilly, the amount of discards produced by fishing non-target species is currently unknown. Potting is more discriminate and produces fewer discards than trawling. Small-scale netting produces very low discards, as the fish are not sold at market and are therefore not subject to quotas.

Scallop (*Pecten maximus*) dredging is damaging to seabed communities, in particular to fragile sessile species such as seafans, sponges, and corals, which are slow-growing and slow to recover. Although scalloping is restricted within 4 nmi of the islands, video-camera surveys of deeper-water hard-bottom areas, particularly east of St. Martin's, have provided evidence of physical damage resulting from this activity. Scallopers normally avoid areas of bedrock and large boulders that are preferred by many fragile sessile species, due to the risk of damaging their gear. Most dredging occurs on mixed bottoms of gravel, stones, and smaller rocks, where rocks are overturned and tracks and ridges on the seabed may be visible for long periods in areas of low currents. Also, potting in these areas easily damages fragile sponges.

Fishing activity is also one of the main sources of marine litter on UK beaches, accounting for 11.2% of the total. Fishing gear is usually made from durable materials and can persist for many years in marine waters. Lost gear, such as lobster pots that have lost their marker buoys, has the potential for "ghost fishing." Nets and other lost gear can entangle cetaceans, seals, turtles, and seabirds, and can become snagged on branching benthic species such as seafans, affecting their growth.

9.4.3 Climate change

Continuation of Scilly's clement climate is uncertain. Current climate models predict that if greenhouse gas emissions continue to increase, the component of the Gulf Stream driven by differences in water density is likely to decrease by 25% in the next 100 years. The loss of Arctic sea ice and the resulting influx of cold water into the North Atlantic also might have an effect in weakening the Gulf Stream, or even shutting it down completely in the future. A reduced Gulf Stream would mean that less heat would be brought to northwest Europe, resulting in harsher winters. However, current climate-model predictions are confident that the increase in temperatures resulting from increased greenhouse gas emissions is much greater than the potential cooling effect, so a cooling of the Scilly climate is

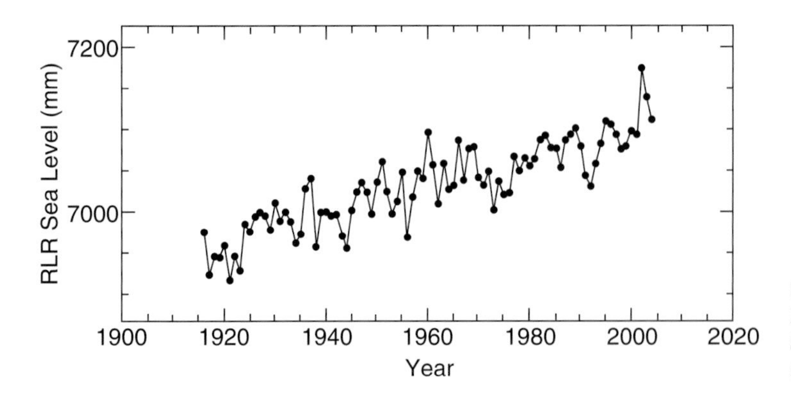

Fig. 9.25 Graph of sea-level rise since the early 1900s at Newlyn on the Cornish mainland, the nearest tide gauge to Scilly. Data from the Permanent Service for Mean Sea Level (PSMSL).

unlikely this century, unlike predictions for a colder mainland Europe.

Present sea level was established at Scilly about 6000 bp, with only minor fluctuations since then. Archaeological evidence supports the view that the islands (perhaps excluding St. Agnes) were united at some time in human history, only requiring a lowering of sea level by a few meters for this to happen again. A lowering of 10 m would unite most of the main islands, and a drop of 20 m would include Agnes and the Western Rocks. The vestiges of field walls and dwellings are still evident below the current high-tide mark in some places, and some recently excavated Romano-British buildings, such as those at Nornour in the Eastern Isles, now lie partially submerged at high tide. Accurate sea-level records for Scilly are only available from 1994 onwards, insufficient to establish more recent trends. However, tide-gauge records for Newlyn in southwest Cornwall began in 1915, and show a linear rise of nearly 20 cm during 90 years (Fig. 9.25). A continued linear rise of 20 cm per century does not in itself give immediate cause for concern, but predictions of increased storminess resulting from global warming in combination with this rise may have serious consequences for low-lying areas, including Hugh Town on the narrow isthmus between the main island of St Mary's and the Garrison.

9.4.4 Tourism

Tourism is the mainstay of Scilly's economy, generating in excess of 85% of the islands' GDP or about £62 million annually. The position of the islands attracts unusual and rare migrant birds both in early spring and late autumn, extending the visitor season from mid-March to mid-November. Recreational scuba diving is an extremely popular visitor activity, due to the clarity of the water, the richness of the fauna and flora, and the many shipwrecks. Many visitors are "ecotourists"; 90% of visitors surveyed by the Isles of Scilly Tourist Board felt that Scilly should promote a more environmentally sustainable environment. Many others cite the environment, the countryside, and beautiful views as their reason for visiting. Their respect for the environment minimizes their impact on it, although increased production of sewage and waste is inevitable.

In recent years, there has been a declining trend in the annual number of visitors from 124,000 in 2003 to 105,000 in 2007, mainly due to a reduction in the number of casual day-visitors. The number of longer-staying visitors has remained constant, reflecting the ceiling imposed by limited accommodation opportunities. Visiting boats possibly have the most impact on the marine environment, due to disposal of sewage and damage to the seabed caused by anchor chains, especially in seagrass areas (see Section 9.3.1.3).

9.5 CONSERVATION LEGISLATION, MECHANISMS, AND VOLUNTARY ACTIONS

In Scilly and the UK as a whole, conservation is affected by a range of statutory and non-statutory protection mechanisms operating at international, national, and local levels, including those administered by voluntary bodies and other organizations that own or control land. Statutory protected sites are those notified, designated, or authorized under European directives and/or implemented through British legislation by a statutory body (e.g., the 1981 *Wildlife and Countryside Act*), thereby conferring legal protection. Non-statutory sites are not directly protected by legislation, but are recognized by statutory bodies or owned, managed, or both by non-statutory organizations for their nature-conservation or aesthetic values.

9.5.1 International and national policies

The practice of marine conservation in Scilly has been directed by convoluted and continuingly evolving sets of national and European legislation. The Nature Conservancy was created by royal charter in 1949, and in 1973, the Nature Conservancy Council (NCC), a government agency, was established by act of Parliament. Its goals were designating and managing national nature reserves and other conservation areas, identifying Sites of Special Scientific Interest, advising government ministers on policies, providing advice and information, and commissioning or undertaking relevant scientific research. In 1992, the European Community adopted Council Directive 92/43/EEC on the *Conservation of Natural Habitats and of Wild Fauna and Flora*—the *EC Habitats Directive*. This is the means by which the Community meets its obligations as a signatory of the *Convention on the Conservation of European Wildlife and Natural Habitats* (Bern Convention). The Directive requires member states to introduce a range of measures, including the

Box 9.2 The UK Biodiversity Action Plan (UK BAP)

Richard M. Warwick
Plymouth Marine Laboratory, Plymouth, Devon, UK

In June 1992, the *Convention of Biological Diversity* was signed by 159 governments at the Earth Summit in Rio de Janeiro (see Ch. 3). In the UK this led in 1994 to the launch of *Biodiversity: the UK Action Plan* (UK BAP) for dealing with biodiversity conservation. A UK Biodiversity Steering Group was created in 1994 and published *Biodiversity: the UK Steering Group Report*, establishing a framework and criteria for identifying species and habitat types of conservation concern. As a result, action plans for 391 species and 45 habitats were eventually published. To measure progress on the 436 local action plans a three-to-five yearly reporting cycle was established. Responsibility for implementation of the BAP was subsequently devolved to the four separate countries of the UK (England, Wales, Scotland, and Northern Ireland), which have published country strategies to help guide the implementation of biodiversity conservation, sustainable development, and environmental concerns. England's strategy was published in *Working with the Grain of Nature* in 2002. Conservation work is guided by the England Biodiversity Group, with representation from a range of stakeholders. After the *UK List of Priority species and Habitats* was published in 2007, responsibility for implementing conservation of these priorities was also devolved to a country level. During 2008, the four countries set up a number of groups to reflect this responsibility and to start planning how best they could conserve these species and habitats. A complete review of the UK BAP priority species and habitats was published in August 2007 and the conservation approach for 1150 species and 65 habitats is being developed by both statutory and non-statutory sectors. In early 2008, a list of the priority conservation actions (signposting) was published. In addition, and partly in response to the publication of the country strategies and progress with the UK BAP, a refreshing of the UK BAP and of ways of delivering conservation was initiated, concluding with the publication of *Conserving Biodiversity—the UK Approach*.

Sources: www.ukbap.org.uk/library/UKSC/DEF-PB12772-ConBio-UK.pdf

protection of species listed in Annexes, to undertake surveillance of habitats and species, and to produce a report every six years on the implementation of the Directive. In response, the UK government launched the UK Biodiversity Action Plan (UK BAP) in 1994 (Box 9.2) responding to the *Convention of Biological Diversity* (Ch. 3), calling for the creation and enforcement of national strategies and action plans "to conserve, protect and enhance biological diversity." The *Lisbon Agenda* (March 2000) aimed at making the EU the most competitive economy in the world, and identified three pillars: economic, social, and environmental. The *EU Water Framework Directive* (December 2000) entered into UK law in 2003 and charged member states with reaching good chemical and ecological status in inland and coastal waters by 2015. The *EU Maritime Policy* (October 2007) provides a coherent framework to exploit synergies among different policies, to resolve potential conflicts among all elements of marine activity, and to provide a holistic and integrated approach for addressing economic and sustainable development. The *EU Marine Strategy* (June 2008) intends to protect the European marine environment more effectively, to achieve better environmental status of marine waters by 2021, and to protect the resource base upon which marine-related economic and social activities depend. The *UK Marine and Coastal Access Act* was passed through Parliament in November 2009 and is one of the most important pieces of environmental legislation to be passed recently as it creates a new, comprehensive management system for the UK's coasts and seas. Among other things, and of particular relevance to the conservation of biodiversity, it calls for creation of a network of **Marine Conservation Zones** to protect some of the UK's most important marine species and habitats, as well as improving fisheries management and public access to the English coast.

At the heart of this complex of policy drivers is recognition of the need to protect the marine environment and to preserve biodiversity, largely because of the socio-economic goods and services provided to marine users (Ch. 13). However, biodiversity per se has never been an explicit conservation objective in the UK. Action plans are strategically intended to conserve priority species and habitats, but lack tactical means for assessing their effectiveness in terms of the maintenance or enhancement of biodiversity.

9.5.2 Local applications

Many local organizations are involved with conservation in Scilly, including the Isles of Scilly Council, the Duchy of Cornwall, Natural England, the Isles of Scilly Wildlife Trust, the Royal Society for the Protection of Birds, the Isles of Scilly Bird Group, and the Tresco Estate.

9.5.2.1 Statutory designations (Fig. 9.26)

- *Special Areas of Conservation* (SACs) are a main mechanism for implementation of the 1992 *European Community Habitats and Species Directive*, and are outstanding examples of selected habitat types or areas important for the maintenance of

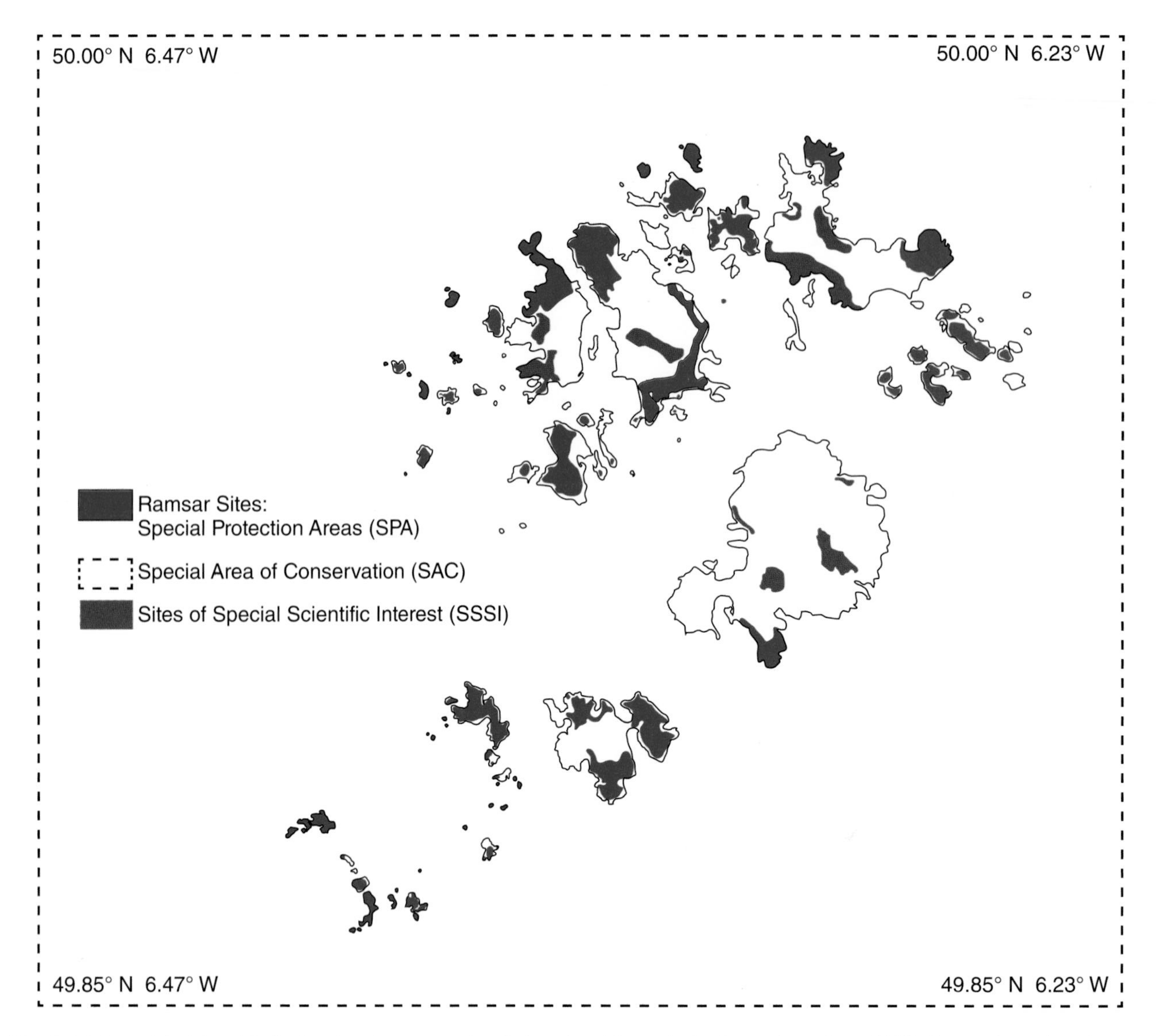

Fig. 9.26 Map of the Isles of Scilly showing site conservation designations. The entire land area shown is an AONB. See text for explanation.

selected species; each SAC covers the entire area down to mean low water. Annexes I and II list the habitats and species respectively that require the designation. In the UK, the Directive is implemented through 1994 Conservation Regulations. Some habitats have been designated for the whole of the Isles of Scilly (Fig. 9.26), the qualifying Annex I coastal-marine habitats being:

1110 "Sandbanks which are slightly covered by sea water all the time"

1140 "Mudflats and sandflats not covered by seawater at low tide"

1170 "Reefs" (defined as rocky marine habitats or biological concretions that rise from the seabed)

The total SAC coverage in Scilly is 26,851 ha, of which only 181 ha are terrestrial.

- *Sites of Special Scientific Interest* (SSSIs) are notified under the *Wildlife and Countryside Act* 1981 and constitute a national network of areas in Britain where natural features, especially those of greatest conservation value, are most highly concentrated or of highest quality. Provisions of the Act, and its 1985 amendments, aim to limit or prevent operations that are potentially damaging to wildlife of the area. In Scilly there are 26 SSSIs, five of which are geological, covering a total area of 554.98 ha. Most are reported to be in satisfactory condition, with little reported loss or degradation of terrestrial habitat; relatively little information exists for the marine environment against which to measure change.
- *Wetlands of International Importance* (Ramsar sites, Ch. 3) are designated for their waterfowl populations, their important plant and animal assemblages, their wetland interest, or a

combination of these. They must first be designated as SSSIs under national statute. All Ramsar sites on Scilly are coastal, covering an area of some 400 ha. A Ramsar site designation in 2001 was for a breeding population of lesser black-backed gulls (*Larus fuscus*).

- *Special Protection Areas* (SPAs) are implemented by government under the 1981 *Wildlife and Countryside Act*. All SPAs must first be designated as SSSIs, and on Scilly they are essentially the same as Ramsar sites, one being designated in 2001 for a population of European storm-petrels (*Hydrobates pelagicus*).
- *Areas of Outstanding Natural Beauty* (AONBs) are designated under the 1949 *National Parks and Access to the Countryside Act*, the primary purpose being to conserve natural beauty. The entire land mass of Isles of Scilly was designated in 1975, which is the smallest AONB in the UK. The Council of the Isles of Scilly ensures the delivery of statutory duties through an AONB Joint Advisory Committee (JAC), which recognizes that "The islands have a clean, safe, productive and biologically diverse marine and coastal environment that is valued and enjoyed. It is well recorded, understood and managed sensitively, enabling it to support a sustainable local fishery." One of its broad aims is to ensure that this situation is maintained. In the AONB context, the network of Marine Conservation Zones proposed in the government's *Marine Act* is designed to "protect the full range of marine biodiversity."

9.5.2.2 Non-Statutory designations

- *Heritage Coasts* are selected for their exceptionally fine scenic quality, exceeding 1 nmi in length, substantially undeveloped, and containing features of special significance and interest. This 1974 designation covers the entire coastline of Scilly (64 km) and is managed together with the AONB Joint Advisory Committee (see Section 9.5.2.1), since the objectives are complementary.
- *Voluntary Marine Nature Reserves* (also called "voluntary marine conservation areas" or "voluntary marine wildlife areas") may be established by representatives of users of a subtidal area or an area of shore in order to initiate management of that area. The Isles of Scilly Marine Park, established in 1989, comprises the whole archipelago down to the 50 m depth contour.
- *Sensitive Marine Areas* (SMAs) are non-statutory areas that are nationally important and notable for their marine animal and plant communities. They are identified by Natural England with the aim of raising awareness and disseminating information to be taken into account in management planning. Scilly was identified as such an area in 1994, relying on cooperation of local users for sustainable management, with the help of grant aid.
- *UK Wildlife Trusts* are established to promote nature conservation at a local level. In 1986 the Isles of Scilly Environmental Trust was formed as a registered charity, and in 2001 it became officially affiliated with the National Organization of Wildlife Trusts and has since become the Isles of Scilly Wildlife Trust—the 47th in the UK. The Trust is responsible for management of all land leased to it by the Duchy of Cornwall on the inhabited and uninhabited islands, islets, and rocks. Its primary commitments are conservation for public benefit of terrestrial and marine wildlife and their habitats, the landscape, and archaeological and historical remains. The Trust is also concerned with the furtherance of public education about the islands and promotion and coordination of research, information, and interpretative services. The Trust currently has three salaried staff and 12 Trustees, all being residents of Scilly. The full Trust Board is responsible for policy while a Management Team is responsible for day-to-day administration. Given its small income and small staff, the Trust recruits volunteers to help carry out its extensive program. While volunteers of all ages are welcomed, most are young people who are studying for qualifications in related fields of conservation, land management, etc.

9.5.2.3 Regulatory measures

The Isles of Scilly are also subject to national legislation regarding fishing gear, minimum species size limits, boat and engine size restrictions, closed seasons and quotas. Outside the 12 nmi limit of territorial waters, the *EU Common Fisheries Policy* applies, enabling other EU countries to fish within the 200 nmi Exclusive Economic Zone. The Isles of Scilly Sea Fisheries Committee (SFC) of the Isles of Scilly Council has the power to pass bylaws to regulate fisheries within 12 nmi of the islands under the *Sea Fisheries Regulation Act* of 1966. There are currently three bylaws:

1. Vessels greater than 10 gross mt or 11 m overall length are prohibited from fishing within the Sea Fisheries District out to 6 nmi.
2. Scallopers with more than four dredges are prohibited from fishing out to 4 nmi.
3. Removal of lobsters with a carapace length of less than 90 mm is prohibited.

The SFC enforces these bylaws with the assistance of the local Fishermen's Association, which acts voluntarily to improve sustainability and to provide a voice for local fishermen. Cornwall's SFC also enforces fisheries legislation. The new *Marine Act* will centralize fisheries management, in which case Scilly's SFC will lose responsibilities to the Cornwall County Council.

Environmental waste and pollution legislation is implemented by the Council, which replaces the responsibilities of the national Environment Agency on the islands. The islands have been given so-called "Area to be avoided" status by the International Maritime Organization (IMO) after the 1993 grounding of the *Braer* oil tanker off the Shetland Islands, but it has no statutory powers. Islanders claim that ships flout the present voluntary system. Therefore, the Council intends to introduce a four-point action plan to safeguard the environment, including a compulsory exclusion zone and radar surveillance. The Emergency Planning Department of the Council has produced a draft Oil Spill Contingency Plan that expands on the Maritime and Coastguard Agency's National Contingency Plan. In the event of an oil spill, government has the responsibility for managing pollution from 1 nmi offshore, whereas local authorities are responsible for shoreline clean-up. It is encouraging that single-hull tankers have now been phased out by the IMO.

The Isles of Scilly Council organizes cleaning of certain beaches for tourism, and the AONBJAC has been working with local schools to raise waste awareness among island children, with a focus on marine litter. Regular beach cleans are coordinated with schools during summer in conjunction with "Beachwatch," a project run by the national Marine Conservation Society involving volunteers in beach litter surveys and clean-ups. Beach clean-ups have been run in Scilly in 2005 and 2007 and the Marine Conservation Society hopes to continue this project.

Environmental protection is facilitated by limitations on the availability of accommodation and transport to the islands, hence on the number of tourists. Additionally, a Green Tourism Business Scheme has been promoted on Scilly through the AONB. Scilly has the greatest concentration of members of this scheme in the UK, and nine local businesses have received awards under this scheme.

9.6 THE CONSERVATION STATUS OF SCILLY

So far, the plethora of mandates and actions described above has been relatively successful in balancing the economic needs of the islands with sustaining biodiversity. Although they have all been working towards the same ends, their scope and number are quite mind-boggling, and it is difficult for environmental managers to see their way through this maze.

It is important that management measures are based on the best available scientific information. Academic research on biodiversity is well advanced in the Isles of Scilly (Box 9.1), but hitherto has been running a separate course with little dialogue between scientists and environmental managers. There are challenges on both sides: academics must provide the practical tools needed to study and measure biodiversity in ecologically meaningful ways on relevant spatial and temporal scales, while managers must find time to read journals in which this information is published. Face to face discussion is important, in order to break down the entrenched but erroneous views of environmentalists that academics are arrogant, and of academics that environmentalists are naive. Encouragingly, this dialogue is ongoing on Scilly.

In some respects there are similarities between Scilly and The Bahamas (Ch. 8), although Scilly is three orders of magnitude smaller in extent. Tourism, including ecotourism, is the mainstay of the economy of both these island groups, and maintenance of high environmental quality is therefore imperative for both. However in The Bahamas a host of social, economic, and environmental problems are evident, dramatic population growth and conflicts among conservation and developmental policies being among the most challenging. Conflicts among conservation goals and economic development are less apparent on Scilly. Part of the explanation for why Scilly currently has no serious identifiable environmental problems, and has maintained the well-being of its local people, is one of physical scale. As E. F. Schumacher wrote in his book *Small is Beautiful* (1973): "... greed and envy demand continuous and limitless economic growth of a material kind, without proper regard for conservation, and this type of growth cannot possibly fit into a finite environment." Firstly, planning regulations are under very strict local control of the landowner (the Duchy of Cornwall) and the Isles of Scilly Council, and the very limited development of dwellings has prevented any increase in population for over a century. It is still possible to walk along white sandy beaches, such as Great Bay on St Martin's, without a building in sight.

Although international, national, and local conservation legislation is in place and effective in Scilly, it is perhaps more than anything the attitude and actions of the local people that has contributed most to its present success. One has to get to know a cross-section of the resident population to realize that the beauty of the environment instills in them a general feeling of contentment that replaces "greed and envy." Most are content to earn a modest living and enjoy a high quality of life. They know when they are on to a good thing, whereas in more urban environments life's goals are very different.

With conservation agencies, local administrators, and local people working towards the same ends, we have grounds for cautious optimism. Nevertheless, social and political change, unexpected events, and climate change are beyond local control, and there is always a danger of complacency creeping in. For example, increasing demands for energy may result in the development of offshore wind farms and oil drilling. As Rosemary Parslow (2007) put it: "The Isles of Scilly are a classic case of the golden goose: it would be so easy to kill the goose."

REFERENCES

Ballantine WJ (1961) A biologically-defined exposure scale for the comparative description of rocky shores. *Field Studies* **1**, 1–19.

Besant W (1890) *Armorel of Lyonesse*. Chatto & Windus, London.

Clarke J (1909) Notes on Cornish Crustacea. *Zoologist* **13**, 281–308.

Crisp DJ, Southward AJ (1958) The distribution of intertidal organisms along the coasts of the English Channel. *Journal of the Marine Biological Association of the United Kingdom* **37**, 157–208.

Faubel A, Warwick RM (2005) The marine flora and fauna of the Isles of Scilly: free-living Plathelminthes ("Turbellaria"). *Journal of Natural History* **39**, 1–45.

Harvey LA (1969) The marine flora and fauna of the Isles of Scilly. The islands and their ecology. *Journal of Natural History* **3**, 3–18.

Hiscock K (1984) Rocky Shore Surveys of the Isles of Scilly March 27th to April 1st 1983 and July 7th to 15th 1983. *Nature Conservancy Council, CSD Report 509*.

Howson CM, Picton BE (1997) *The Species Directory of the Marine Fauna and Flora of the British Isles and Surrounding Seas*. Ulster Museum and Marine Conservation Society, Belfast and Ross-on-Wye.

Lewes GH (1858) Seaside Studies at Ilfracombe, Tenby, the Scilly Isles and Jersey. Wm. Beachwood & Sons, Edinburgh and London.

MacArthur RH, Wilson EO (1967) *The Theory of Island Biogeography*. Princeton University Press, Princeton, New Jersey.

Marine Biological Association (1957) *Plymouth Marine Fauna*. Marine Biological Association, Plymouth.

Parslow R (2007) *The Isles of Scilly*. Harper Collins, London.

Rostron DM (1989) Animal communities from sublittoral sediments in the Isles of Scilly. September 1988. *Nature Conservancy Council, CSD Report 918*.

Schumacher EF (1973) *Small is Beautiful: A Study of Economics as if People Mattered.* Frederick Muller Ltd., London.

Smart RWJ, Cooke AH (1885) The marine shells of Scilly. *Journal of the Conchological Society of London* **4**, 285–303.

Warwick RM, Clarke KR (1998) Taxonomic distinctness and environmental assessment. *Journal of Applied Ecology* **35**, 532–543.

Warwick RM, Clarke KR (2001) Practical measures of marine biodiversity based on relatedness of species. *Oceanography and Marine Biology Annual Review* **39**, 207–231.

Warwick RM, Dashfield SL, Somerfield PJ (2006) The integral structure of a benthic infaunal assemblage. *Journal of Experimental Marine Biology and Ecology* **330**, 12–18.

Warwick RM, Light J (2002) Death assemblages of molluscs on St. Martin's Flats, Isles of Scilly: a surrogate for regional biodiversity? *Biodiversity and Conservation* **11**, 99–112.

Warwick RM, Turk SM (2002) Predicting climate change effects on marine biodiversity: comparison of recent and fossil molluscan death assemblages. *Journal of the Marine Biological Association of the United Kingdom* **82**, 847–850.

SUGGESTED READINGS

Barne JH, Robson CF, Kaznowska SS, Doody JP, Davidson NC, Buck AL, eds (1996) *Coasts and seas of the United Kingdom. Region 11. The Western Approaches: Falmouth Bay to Kenfig.* Joint Nature Conservation Committee, Peterborough.

Fowler S, Laffoley D (1993) Stability in Mediterranean-Atlantic sessile epifaunal communities at the northern limits of their range. *Journal of Experimental Marine Biology and Ecology* **172**, 109–127.

Gaston KJ, ed. (1996) *Biodiversity. A Biology of Numbers and Difference.* Blackwell, Oxford.

Gaston KJ, Blackburn TM (2000) *Pattern and Process in Macroecology.* Blackwell, Oxford.

Gaston KJ, Spicer JI (2004) *Biodiversity: an Introduction.* Blackwell, Oxford.

Gee JM, Warwick RM (1994) Body-size distribution in a marine metazoan community and the fractal dimensions of macroalgae. *Journal of Experimental Marine Biology and Ecology* **178**, 247–259.

Gee JM, Warwick RM (1994) Metazoan community structure in relation to the fractal dimensions of marine macroalgae. *Marine Ecology Progress Series* **103**, 141–150.

Gibbs PE, Bryan GW, Pascoe PL, Burt GR (1987) The use of the dog-whelk, *Nucella lapillus*, as an indicator of tributyltin (TBT) contamination. *Journal of the Marine Biological Association of the United Kingdom* **67**, 507–523.

Gubbay S (1993) Management of marine protected areas in the UK: lessons from statuary and voluntary approaches. *Aquatic Conservation: Marine and Freshwater Ecosystems* **3**, 269–280.

Gubbay S, Welton S (1995) The voluntary approach to conservation of marine areas. In *Marine protected areas. Principles and techniques for management* (ed. Gubbay S). Chapman and Hall, London, 199–227.

Hawkins SJ, Southward AJ (1992) The Torrey Canyon oil spill: recovery of rocky shore communities. In *Restoring the nation's marine environment.* Proceedings of the NOAA Symposium on Habitat Restoration, held in Washington, D.C., September 25–26, 1990 (ed. Thayer GW). Maryland Sea Grant College, Maryland, 583–631.

Hiscock K, ed. (1998) *Marine Nature Conservation Review. Benthic marine ecosystems of Great Britain and the north-east Atlantic.* Joint Nature Conservation Committee, Peterborough.

Hughes TP, Bellwood R, Folke C, Steneck RS, Wilson J (2005) New paradigms for supporting the resilience of marine ecosystems. *Trends in Ecology and Evolution* **20**, 380–386.

John DM (1969) An ecological study on *Laminaria ochroleuca. Journal of the Marine Biological Association of the United Kingdom* **49**, 175–187.

Jones KJ, Gowen RJ (1990) Influence of stratification and irradiance regime on summer phytoplankton composition in coastal and shelf seas of the British Isles. *Estuarine, Coastal and Shelf Science* **30**, 557–567.

Kendall MA, Widdicombe S, Davey JT, Somerfield PJ, Austen MCV, Warwick RM (1996) The biogeography of islands: preliminary results from a comparative study of the Isles of Scilly and Cornwall. *Journal of the Marine Biological Association of the United Kingdom* **76**, 219–222.

Magurran AE (2004) *Measuring Biological Diversity.* Blackwell, Oxford.

Magurran AE, McGill BJ, eds (2011) *Biological Diversity.* Oxford University Press, Oxford.

Norton TA (1968) Underwater observations on the vertical distribution of algae at St. Mary's, Isles of Scilly. *British Phycological Bulletin* **3**, 585–588.

Pingree RD, Mardell GT (1986) Coastal tidal jets and tidal fringe development around the Isles of Scilly. *Estuarine, Coastal and Shelf Science* **23**, 581–594.

Simpson JH, Tett PB, Argote-Espinoza ML, Edwards A, Jones KJ, Savidge G (1982) Mixing and phytoplankton growth around an island in a stratified sea. *Continental Shelf Research* **1**, 15–31.

Summers CF (1974) The grey seal (*Halichoerus grypus*) in Cornwall and the Isles of Scilly. *Biological Conservation* **6**, 285–291.

Symes D, Phillipson J (1997) Inshore fisheries management in the UK: Sea Fisheries Committees and the challenge of marine environmental management. *Marine Policy* **21**, 207–224.

Tett P (1981) Modelling phytoplankton production at shelf-sea fronts. *Philosophical Transactions of the Royal Society of London, Series A* **302**, 605–615.

Warwick RM (1977) The structure and seasonal fluctuations of phytal marine nematode associations on the Isles of Scilly. In *Biology of benthic organisms* (eds Keegan BF, Ceidigh PO, Boaden PJS). Pergamon, Oxford, 577–585.

CHAPTER 10

GWAII HAANAS: FROM CONFLICT TO COOPERATIVE MANAGEMENT

N. A. Sloan

Gwaii Haanas National Park Reserve, National Marine Conservation Area Reserve, and Haida Heritage Site, Haida Gwaii, British Columbia, Canada

The Gwaii Haanas Marine Area now enjoys the highest standard of legal protection afforded by Canada.

Canada Gazette (official newspaper of the Government of Canada since 1841), June 17, 2010

10.1 NATION-TO-NATION PURSUIT OF LAND-SEA CONSERVATION

The creative flashpoint for the land-sea protected area called Gwaii Haanas was an act of civil disobedience in 1985 led by the Haida, the indigenous people of Haida Gwaii, northern British Columbia (Haida Gwaii, "islands of the people", was legally renamed from the Queen Charlotte Islands in 2010). A road blockade prevented access by loggers and resulted in mass arrests (Fig. 10.1). The heart of the conflict was a collapse in confidence among the Haida and others that the British Columbia (provincial) government would honor a promise to stop logging where steep island slopes and high rainfall exacerbated ruinous erosion from forest clear-cutting and attendant road building. The post-blockade legal tumult and intense lobbying with media coverage compelled the governments of Canada and British Columbia to agree in 1988 to create a linked land-sea conservation area associated with the southern 15% of the archipelago's lands. Thus, what the Haida had declared in 1985 as a Haida Heritage Site became recognized by other levels of government and was forged into a cooperatively managed entity.

Terrestrial conservation came first. Canada and the Haida created the Archipelago Management Board (AMB) through the 1993 *Gwaii Haanas Agreement*. The original AMB included

Fig. 10.1 The standoff on Lyell Island (Athlii Gwaii) in 1985. To the left, the Royal Canadian Mounted Police are reading legal instructions requesting dispersal of the logging road blockade to the seated Haida Elders in the middle in ceremonial dress. To the right, the Elders are backed by other Haida in the blockade. The blockade stood firm, leading to a mass arrest whose repercussions precipitated the *South Moresby Agreement* of 1988 that created Gwaii Haanas. Photograph courtesy of Skidegate Band Council.

Marine Conservation: Science, Policy, and Management, First Edition. G. Carleton Ray and Jerry McCormick-Ray.

two Council of the Haida Nation representatives and two Parks Canada Agency staff (representing Canada) for making consensus-based decisions for 1500 km^2 of conserved lands and 1700 km of shoreline of Gwaii Haanas National Park Reserve and Haida Heritage Site—mandated under the *Canada National Parks Act*. This Agreement implemented the first nation-to-nation cooperative protected-area management for Canada. There has been a federal Canadian terrestrial national park conservation mandate since the creation of Banff National Park in 1885. A national park system was created by 1911 with an attendant Act in 1930.

Marine and terrestrial conservation were linked by the 2010 *Gwaii Haanas Marine Agreement*, which added one Fisheries and Oceans Canada (DFO) representative on the Canada side of the AMB, as well as another Haida representative. Gwaii Haanas National Marine Conservation Area Reserve (hereafter called Gwaii Haanas Marine) was thereby established to manage 3400 km^2 of sea space, including the intertidal around the conserved lands. Parks Canada had acquired a marine-area conservation mandate through the 2002 *Canada National Marine Conservation Areas* (NMCA) *Act* that aspires eventually to have NMCAs represent all 29 of its designated marine natural regions. Five of these regions are in Pacific Canada, of which two are represented by Gwaii Haanas Marine—shallow eastern continental shelf waters and deep western North Pacific waters.

There are two great ironies to contemplate concerning such seemingly "wilderness" conditions. Gwaii Haanas' remoteness, with boat or airplane access only, and essentially undeveloped landscape, with virtually no year-round human occupation, means that it is relatively undisturbed. These largely unpolluted lands and waters host about 1800 "back-country" (self-sufficient camping and/or boating) visitors annually from May to September, before poor weather sets in. However, thousands of Haida likely subsisted year-round on the coast for millennia before post-contact disease epidemics and social dissolution depopulated the area by the late 19th century. Secondly, the ecological integrity of the lands and coast is deeply compromised by a human folly common to island ecosystems—introduced species (Gaston *et al.*, 2008). The introduction of ten mammals, including two species of rat (*Rattus* spp.), raccoon (*Procyon lotor*), and Sitka black-tailed deer (*Odocoileus hemionus sitkensis*), now hyper-abundant in the absence of predators, has had strong terrestrial and coastal ecological effects.

With Gwaii Haanas Marine linked to conserved lands in 2010, the evolution of cooperative, two-nation management for an integrated land-sea conservation continuum unfolds in this case study. An Interim Management Plan is in play until the formal Management Plan is developed by 2015. Developing Gwaii Haanas Marine comes at a time of changing national, provincial, regional, and First Nations' (indigenous peoples) approaches to spatial marine management and planning throughout the north coast of British Columbia. This is complimented by growing public awareness for much-needed reformed oceans management worldwide (Roberts, 2007, 2012). Elements driving change for Gwaii Haanas Marine include reconciliation with First Nations having unique constitutionally recognized resource access rights, new Canadian ocean law and policy infused with notions of sustainability, and ecosystem-based management. These new developments mobilize the natural and social sciences, and foster broader societal participation in spatial marine planning and management.

10.2 NATURAL HERITAGE

Biogeographic and cultural overviews presented in this section are distilled from a detailed multi-authored assessment of Gwaii Haanas Marine (Sloan, 2006). A conservation continuum spanning treeless alpine tundra to deep-sea (western) and continental shelf (eastern) ecosystems is featured. The extent of the combined area (5000 km^2 of land and sea) and the breadth and completeness of ecosystem coverage is globally unique for a protected temperate coastal rainforest area. Virtually all the conserved landscape's contiguous watersheds are intact and preserved in perpetuity.

The Haida Gwaii marine region includes the nearshore and adjacent waters of the eastern North Pacific (Fig. 10.2). Also shown are other regional spatial management entities to be discussed later. Gwaii Haanas comprises the southern portion of Moresby Island and associated islands 90 km off the northern British Columbia mainland. There are more than 200 islands exceeding 1.0 ha in area with a shoreline punctuated by streams draining hundreds of small watersheds—many hosting spawning salmon (*Oncorhynchus* spp.). The west coast is highly exposed to the North Pacific and has a narrow, <2 km, continental shelf that breaks into a continental slope descending rapidly to >1800 m depth within Gwaii Haanas Marine's footprint. The east coast faces across continental shelf waters, of mostly <200 m depth, of Hecate Strait (Fig. 10.3), and has a complex shoreline and many islands (Fig. 10.4).

10.2.1 Dynamic landscapes

The archipelago is part of British Columbia's land mass, composed of distinct tectonic belts, fused onto the North American Plate, forming mountains and controlling the major coastal alignment. Thus, Haida Gwaii graces the western margin of the North American Plate, featuring a mountainous terrain oriented northwest–southeast (Fig. 10.5). Plate tectonics render it one of the most seismically active areas of North America, with more than 11,000 mostly imperceptible earthquakes recorded between 1985 and 2005. The epicenter of the second largest earthquake ever recorded in Canada (magnitude Mw 7.8) occurred under Gwaii Haanas in October 2012. Seismic activity results from moving plates originating from spreading centers, and faults at plate interfaces—particularly the active Queen Charlotte Fault only 10 to 15 km offshore to the west.

Haida Gwaii has undergone great environmental fluctuations since the end of the last ice age. Between 20,000 to 16,000 BP, the ice-covered British Columbia mainland had glaciers reaching across Haida Gwaii to the continent's edge. The extent of ice cover is speculative, but minimally is known to include an ice-cap and seaward-flowing valley glaciers.

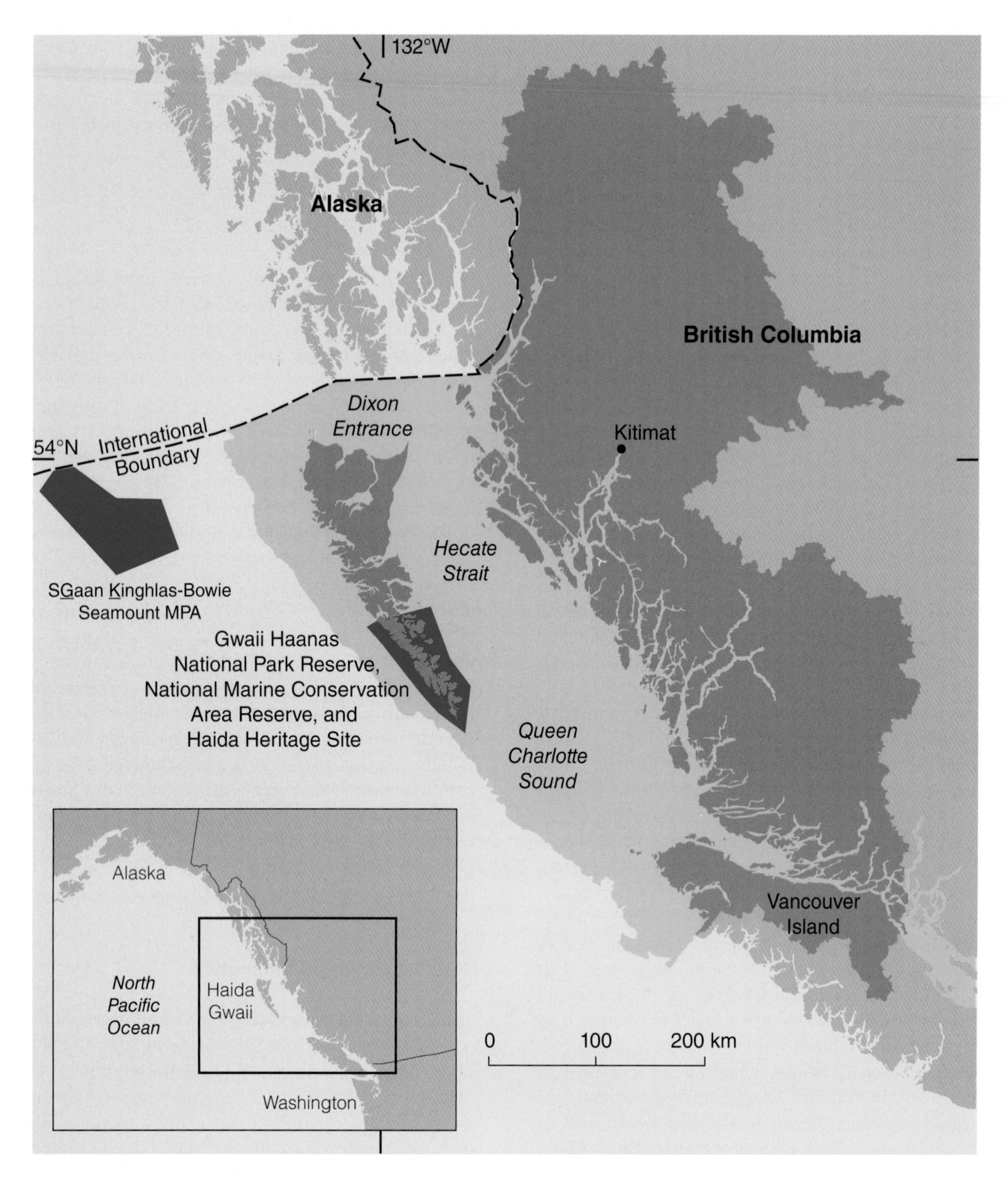

Fig. 10.2 The Haida Gwaii marine region including the footprint of Gwaii Haanas National Park Reserve, National Marine Conservation Area Reserve, and Haida Heritage Site. Enveloping Haida Gwaii is the shaded Fisheries and Oceans Canada large ocean management area (light blue) for the Pacific called the Pacific North Coast Integrated Management Area (PNCIMA). The shaded land area (dark brown) is the British Columbia mainland watersheds draining into PNCIMA. The seaward edge of PNCIMA is approximately the base of the continental slope. The shaded (dark blue) offshore area is SGaan Kinghlas-Bowie Seamount Marine Protected Area, established by Fisheries and Oceans Canada in partnership with the Haida Nation in 2008 under Canada's *Oceans Act*. The dark blue area surrounding the Gwaii Haanas lands is the representative area mandated under the NMCA.

Whether biological refugia persisted in the lowlands is uncertain. Ice disappeared from the archipelago's coast by about 15,000 BP, and by 14,500 BP receded from the Hecate Strait area. Between 16,000 to 13,000 BP, a cool tundra-like environment hosted the first pine (*Pinus* sp.) forests and these were replaced with open spruce forest by 11,000 BP. The climate again cooled from 10,600 to 10,000 BP when tree-lines were lower until warming returned and the closed-canopy western hemlock-spruce forests emerged to dominate. Cool, wet conditions have remained relatively stable since about 6000 BP.

Sea levels also fluctuated with tectonic activity and climate change. The shoreline in the early post-glacial period at around

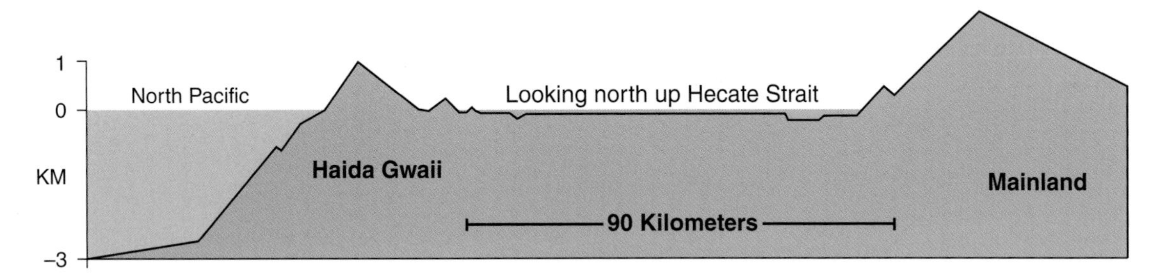

Fig. 10.3 Cross-section through Haida Gwaii and northern British Columbia mainland coast looking northward up Hecate Strait (scale exaggerated vertically).

Fig. 10.4 Thickly treed islands off the east coast of Gwaii Haanas Marine typical of the area's complex rocky shoreline. Photograph © Parks Canada/Andrew Wright.

12,500 BP was 150 m below current levels, and rolling meadowlands of the Hecate Strait area connected Haida Gwaii to the mainland. After 12,000 BP, sea-level rose, reaching modern levels by 9400 BP. From 9000 BP to about 5000 BP, sea levels reached their highest point at 15 m above current levels, then fell gradually to current levels about 3000 years ago.

This dynamic environmental history has given rise to the spatially complex ecological conditions of today. Gwaii Haanas' lands are characterized by rugged, high-relief mountains and incised shorelines, and although no point of land is more than 5 km from the sea, there is an east/west ecological distinction. The leeward eastern coast, which rises from hills to 700 m elevation with rounded tops, contrasts with the very steep windward ranges along the west coast rising to 1100 m elevation. The extremely rainy and windy west coast contains stunted trees, thin soils, and poorly drained, boggy woodlands, while the leeward eastern side is more typically a coastal temperate rainforest dominated by large conifers. On both coasts, powerful saline winds, clouds, and heavy rains greatly influence terrestrial ecosystems to the highest point of alpine tundra. Hence, the biogeoclimate typical of British Columbia and Haida Gwaii is characterized as "wet hyper-maritime" and supports coastal coniferous rainforests, dominated by western red cedar (*Thuja plicata*), Sitka spruce (*Picea sitchensis*), and western hemlock (*Tsuga heterophylla*). At higher elevations (>550 m), forests of mountain hemlock (*T. mertensiana*) give way to alpine tundra.

10.2.2 Climatic variability

Although Haida Gwaii straddles between 52 to 54°N latitude, the ocean moderates its seasonal "marine west coast cool" climate. Average daily air temperatures vary from 3.2°C in January to 15°C in August. There has been a coast-wide warming trend of +1.2°C over the period 1948 to 2004. As well, average annual sea-surface temperature has increased 0.38°C over the last 40 years.

The archipelago is wet and windy. Average annual precipitation is 1.4 m occurring about 230 d yr^{-1}, mostly as rain, with the driest months from June through September. Some short-lived (at sea level) snowfalls occur mostly from December through March, with up to 15 days of snowfall in amounts exceeding 1.0 cm. Annual rainfall increased 12.5% from 1950

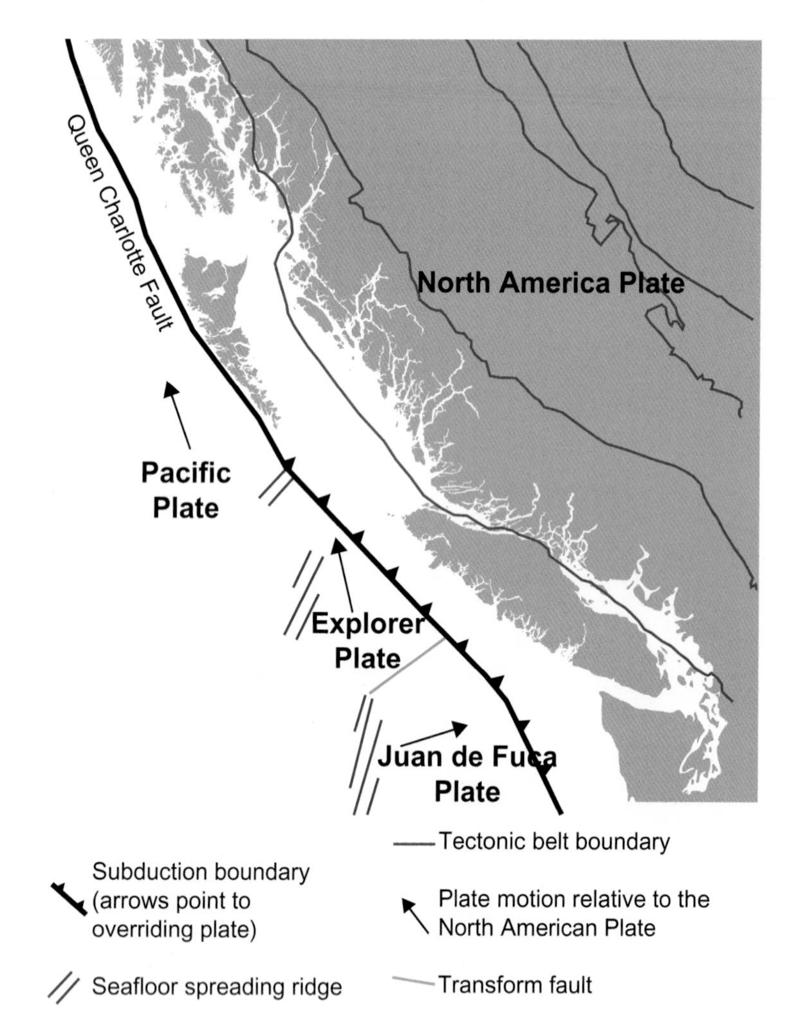

Fig. 10.5 Plate tectonic setting of the British Columbia coast, including Haida Gwaii. The tectonic belts' northwest–southeast orientation indicates general plate collision events. Modified from Sloan (2006).

to 2001, mainly in winter and spring. Local winds are some of the strongest and most persistent in Canada. For example, Hecate Strait winds average 8.5 ms^{-1}, mostly from the south and southeast, and there is <1% calm conditions. These winds attracted Canada's first proposed marine windfarm off the islands' northeast side. The project was environmentally approved, and had Haida Nation involvement, but costs proved to be too high, placing the project in abeyance.

Two seasonal systems influence local climate: the Aleutian Low (fall through winter) and the North Pacific High (spring though summer). These systems produce air pressure gradients associated with storms that seasonally dominate the northeast Pacific and control many aspects of climate. Winds are typically strongest in winter and cause topographic effects, such as valley-funneling and wind strengthening through mountains. The Aleutian Low develops and intensifies into a cyclonic system, producing counter-clockwise surface winds that deliver moisture from October through April. As the Aleutian Low diminishes and retreats in May, the North Pacific High expands and intensifies in summer, bringing clearer skies, less rain, and predominately westerly to northwesterly winds.

During longer time periods, Haida Gwaii has experienced significant inter-annual climatic variations. The quasi-periodic El Niño Southern Oscillation (ENSO) system brings warming and cooling periods known as El Niño and La Niña, respectively, to Haida Gwaii waters. Strong ENSO events have an average return period of three to seven years and events persist for 6 to 18 months. In El Niño years, water temperatures are raised in the eastern Pacific and upwelling along the northeast Pacific coast including Haida Gwaii is inhibited. Upwelling is a key oceanographic process enabling nutrient exchange that increases ecosystem productivity (Ch. 4). The other larger, inter-annual climatic system is the Pacific Interdecadal Oscillation (PDO). This poorly understood system is associated with warming or cooling of North Pacific waters on a cycle of about 20–30 years. Over the 20th century, there have been two full PDO cycles with the cool phases (1890–1924 and 1947–76) and warm phases (1925–46 and 1977–96).

Global and regional climate change is affecting Haida Gwaii. For example, sea level is predicted to rise by perhaps 90 cm worldwide by 2100 according to the Intergovernmental Panel on Climate Change, largely due to thermal expansion of warmer seawater and increased glacial meltwater. The current sea-level rise rate of 1.6 $mm\,yr^{-1}$ around Haida Gwaii suggests a possible rise of 11 to 22 cm by 2100, which could affect biogeographic patterns and human uses for which management planning must account.

10.2.3 Oceanographic variability

Haida Gwaii's surrounding ocean is characterized by three water-mass domains complicated by variable oceanic, weather, and local conditions. The *oceanic domain* off the west coast is the most uniformly marine with its prevailing offshore oceanic conditions. Typically, the water column is vertically stratified; the thermocline transition between 100 to 150 m depths abruptly separates the warmer, well-mixed surface layer from the deeper, cooler layer. Surface salinities vary little, usually 31.5 to 32.5 ppt, but sea-surface temperatures may range between 8 to 14°C. Below the thermocline, conditions are relatively stable, with deep-water salinities of 34.0 to 34.5 ppt and temperatures of 4 to 6°C. Occasionally, strong northwest winds upwell nutrient-rich waters to the surface, creating highly productive fronts. The *eastern coastal domain* encompasses Hecate Strait and Queen Charlotte Sound. Mainland river drainages influence this domain; less saline, surface water near the mainland transitions across Hecate Strait into more saline marine waters exceeding 31 ppt closer to Haida Gwaii. The *Dixon Entrance domain* is distinctly estuarine, being strongly influenced by outflow from large mainland rivers, e.g., the Skeena and Nass. This domain is dominated by seasonal low surface salinity water; the thin surface layer about 20 m deep has highly variable surface temperatures.

The major circulation patterns around Haida Gwaii shift seasonally (Fig. 10.6). In summer, the south-flowing shelf-break current caused by northwest winds originates mid-way along the west coast, merging with southward-flowing surface waters of Hecate Strait. During winter, dominant southerly winds create the north-flowing Davidson current, as well as north-flowing surface currents in Hecate Strait. These current systems create a mean northwestern flow around Haida Gwaii.

Poorly understood oceanic processes influence nearshore areas shallower than 30 m. Besides offshore influences, nearshore waters are also affected by local winds, tides, freshwater input, and the highly complex shoreline. These can interact to create unique local conditions. For example, the tidal range for Haida Gwaii is generally 4–5 m, but approaches 8 m in Hecate Strait and Skidegate Inlet, in contrast to less than 2 m in Masset Inlet (Graham Island) at the end of a long, narrow channel. Tidal currents can be strong, exceeding 2.5 ms^{-1} at Cape St. James at the southern tip. Summer northwesterly winds displace surface west coast waters offshore, upwelling typically colder, nutrient-rich waters that enhance biological productivity. Shallow east-coast waters lack upwelling, but are well mixed by winds and tides.

The wave climate around Haida Gwaii is the most energetic in Canada and has created exceptional shore exposures. Winter average wave heights measured at oceanographic buoys are typically twice those of the summer. Prevailing westerly winds of winter, huge open-water fetches, and intense low-pressure systems all contribute to waves >20 m height occurring off the east and west coasts; two are on record that exceeded 30 m.

10.2.4 Biodiversity

Canadian marine biodiversity studies are in their infancy (Archambault *et al.*, 2010). Nearly 4000 marine species have been recorded from Haida Gwaii (Table 10.1). The deep sea has been studied little, as more than 80% of sampling and observation has occurred nearshore or from shelf waters in <200 m depth. Also, plankton species are poorly documented. Marine invertebrate biodiversity (Fig. 10.7), although representing over 90% of animal species, may perhaps be a 10-fold underestimate. Regional species inventories on a geographic information system (GIS) are an early step towards linking local biodiversity and eventual marine ecosystem understanding (Sloan and Bartier, 2009). Gwaii Haanas Marine is a regional biodiversity reference location for which elements of biodiversity, in a spatial database, will aid describing ecosystem types and monitoring changes over time.

The islands are rich in charismatic species, including marine birds and mammals. More than 125 species of marine birds

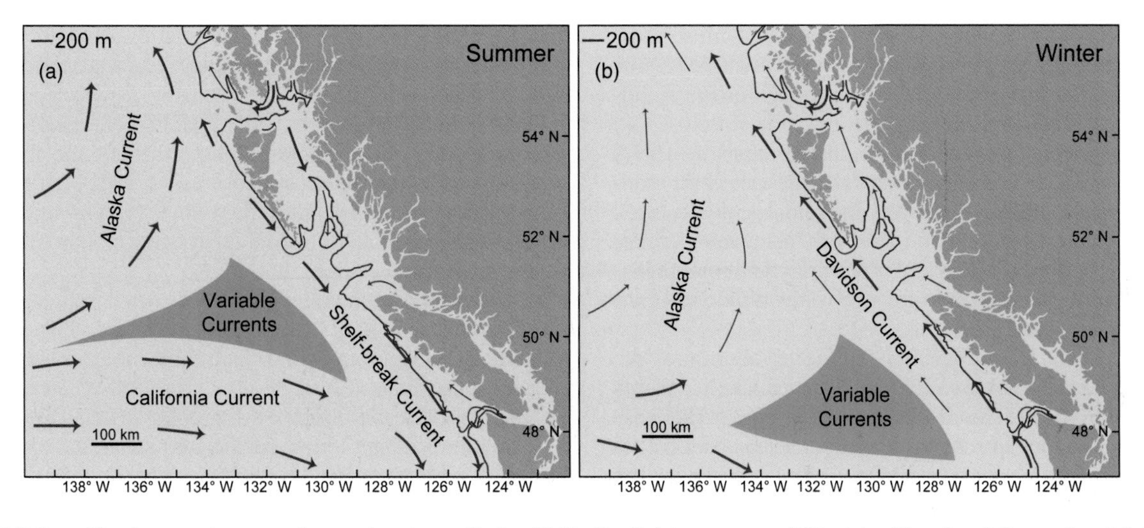

Fig. 10.6 Prevailing large-scale seasonal current systems affecting Haida Gwaii: (a) summer, and (b) winter. The edge of the continental shelf is indicated by the 200 m depth contour. Modified from Sloan (2006).

Table 10.1 Marine species diversity documented from around Haida Gwaii. Modified from Sloan (2006).

Marine group	Approximate number of species	Notes
Lichens (on rock)	43	Preliminary estimate
Seaweeds (algae)	360	Perhaps underestimated up to 40%
Seagrasses (angiosperms)	4	Likely close to the actual number
Invertebrates	3,000	Perhaps 10% of what is present
Fishes	400	Preliminary estimate
Reptiles	2	Leatherback and green sea turtles
Birds	125	Likely close to the actual number
Mammals	11 (maritime*) 26 (marine)	Likely close to the actual number
Total	3,970	Preliminary estimate

*"Maritime" mammals are terrestrial species that eat beached animal or plant species and transfer that food energy onto the land; some of these are introduced species such deer or raccoon.

Fig. 10.7 An example of the invertebrate species diversity. The large multi-armed predatory sea star, *Pycnopodia helianthoides*, and other predatory sea stars, are among many herbivorous turban snails, *Astraea gibberosa*, here exposed at extreme low tide. This photograph is from the wave-sheltered, strongly tidal Burnaby Narrows whose benthic community hosts many predators feasting on the enormous accumulation of biomass due to large amounts of suspended food particles provided by daily tidal flows. Photograph © Parks Canada/Graham Osborne.

(seabirds, marine waterfowl, shorebirds, marine raptors) straddle land and sea. Marine bird populations of national and international importance rely on nearshore and offshore areas for food when breeding, migrating, or wintering. While some marine birds breed locally, others breed elsewhere in the Northern Hemisphere and migrate along, or winter on, the Haida Gwaii coast and in offshore waters. Other species breed in the Southern Hemisphere and occur locally during the austral winter. Seabirds forage widely at sea, where oceanographic conditions concentrate their food, but must come ashore to breed in coastal colonies safe from predators. Some 1.5 million seabirds (13 species) breed colonially on the islands and disperse offshore during the non-breeding season. Trend data are available for some seabird colonies after the 1950s through surveys by the Canadian Wildlife Service of Environment Canada. Figure 10.8 provides an example of two species' breeding colony locations and sizes around Gwaii Haanas. Monitoring seabird colonies and protecting them from predation by introduced rats and raccoons will continue to be an important land-sea issue.

Mammals are also prominent. The term "maritime" describes terrestrial mammals that eat beached and live plants or animals in the nearshore, intertidal, or estuaries. There are five native and four introduced maritime species. Among the native maritime species, the river otter (*Lontra canadensis*) is the most marine-adapted and is common along coasts and into watersheds. Coastal river otters eat mostly nearshore marine fishes, although they can prey on land, including on adults and/or eggs of nesting seabirds. The endemic black bear subspecies (*Ursus americanus carlottae*) occurs from the alpine tundra to the intertidal. In the intertidal, bears (and raccoons) forage on invertebrates, especially crustaceans and mollusks. Bears transfer marine nutrients (especially nitrogen) to the land, during fall salmon spawning, by distributing carcasses into riparian (near stream or lake) forest ecosystems. This marine nutrient transfer into forests by salmon predators, or directly into waters from rotting salmon carcasses, is a key characteristic of temperate coasts throughout the North Pacific, including Haida Gwaii (Darimont *et al.*, 2010). Salmon themselves are important to the islands' ecology and culture, linking marine and terrestrial ecosystems perhaps more than any other marine group (Box 10.1).

The 26 marine-mammal species reported from around the islands (20 cetaceans, five pinnipeds, sea otter—*Enhydra lutris*) form an iconic group with a high public profile. Ten of these species are listed at some level of risk under Canada's 2003 *Species at Risk Act*. Species using onshore sites are the fish-eating harbor seal (*Phoca vitulina*) and Steller sea lion (*Eumetopias jubatus*). Harbor seals haul out and breed widely throughout Haida Gwaii whereas Steller sea lions haul out at only a few sites and have breeding rookeries at only two locations (Fig. 10.9).

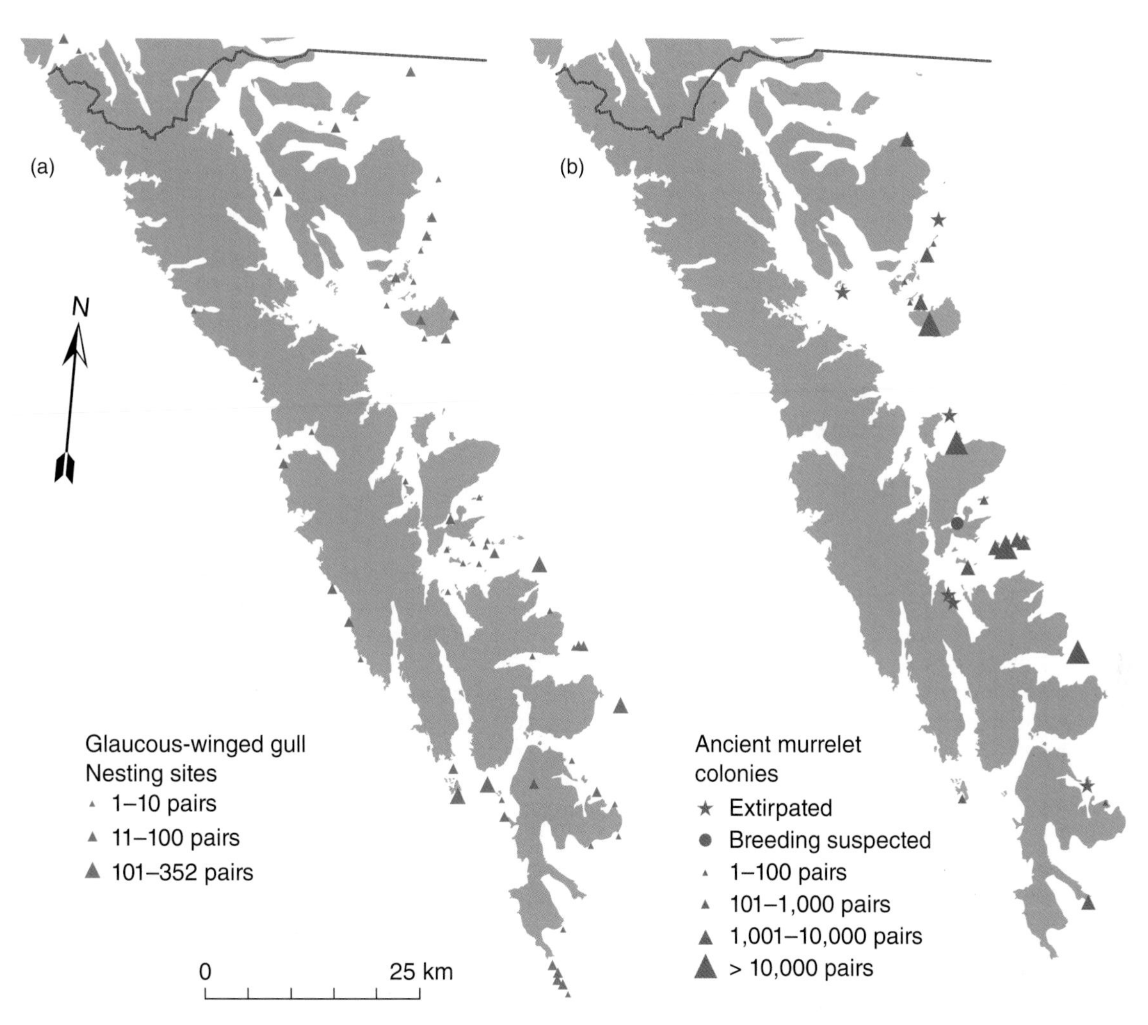

Fig. 10.8 Examples of locations and relative sizes of coastal seabird breeding colonies around Gwaii Haanas Marine: (a) glaucous-winged gull (*Larus glaucescens*), and (b) ancient murrelet (*Synthliboramphus antiquus*). Modified from Sloan (2006).

Introduced terrestrial species affect the biodiversity and ecosystem functioning of coastal lands. Predation by introduced rats and raccoons has caused major declines at some ground-nesting seabird colonies. Over-browsing by hyper-abundant deer has appreciably thinned riparian forest understory and estuarine wetlands of shrubs, grasses, and herbs, as well as many plants traditionally important to the Haida. Introductions of marine species along coastal North America have mainly been from overseas ships' ballast-water releases, species associated with imported aquaculture stock, and vessel hull fouling. Some introductions go unnoticed, but others can become "invasive" due to rapid local dispersal of larvae or propagules that overwhelm local species. At least three plant, 19 invertebrate, and two fish species have been introduced to Haida Gwaii, although none yet appears to be invasive. Perhaps the highest coast-wide profile relates to escape of Atlantic salmon (*Salmo salar*) from mainland-coast, floating net-pen aquaculture, but so far have not been recorded from Haida Gwaii freshwater systems. Some larval settlement away from locally raft-cultured Pacific oyster (*Crassostrea gigas*) has occurred. The Japanese seaweed *Sargassum muticum*, likely from imported oyster shell stock in southern British Columbia, is ubiquitous in the mid-intertidal along rocky shores.

10.2.5 Land-sea interface

The coastal zone is vital to managing the land-to-sea continuum of Gwaii Haanas. This transitional zone encompasses the maritime-influenced vegetated fringe, intertidal shores, estuarine salt marshes, kelp forests, seagrass meadows, and subtidal waters to about 20 m depth. This zone is defined by scales of time and space, and by the strength of interactions between its parts and adjacent systems. Due to these interactions, this zone is the most productive and biologically diverse of local ecosystems and supports the greatest levels of current and past human uses. This zone also falls between terrestrial and marine management, governance, and agency mandates (Bartier and Sloan, 2007). Linked land-sea conservation is an important and emerging field (Álvarez-Romero *et al.*, 2011; Samhouri and Levin, 2012; Ch. 4, Section 4.6).

Biophysical mapping provides the basis for spatial representation and conservation planning, including zoning and oil-spill response. Maps portray species and ecosystem spatial distributions by identifying patterns useful for depicting biological productivity and for marine spatial planning and management. A basic dataset for Gwaii Haanas Marine is a GIS-based, biophysical shoreline inventory. This enables integrating physical and biological characteristics that can be linked to other unmapped features. Complete aerial video imagery, verified by surveys at biological reference locations, enabled recognition of discrete "shore units" where topography and substrate within each unit could be mapped into shore types and where visible biological communities are represented as "biobands." The presence of a particular species assemblage can then be used to infer wave-exposure tolerance and substrate type and provides a general habitat class for other intertidal species. Thus, the integration of physical and biological data yields identification of species and/or community habitat for each shore unit.

The distributions, definitions, and proportions of substrate types along Gwaii Haanas' shoreline provide a first approxima-

Box 10.1 Salmon escapement and riparian forests

Salmon are creatures of the forest, they're born there and they die there.

Charles Frederick Bellis (Haida Elder)

Most Haida Gwaii salmonid species are anadromous (Table B10.1.1). The dominant species in Gwaii Haanas are chum, pink, and coho. Adults leave the ocean and enter river-lake systems in late summer-fall to spawn and die. Some species' young grow for different periods in freshwater and nearshore waters such as estuaries before going to sea, while other species emigrate directly. All species spend their adult lives in the open North Pacific before returning to spawn in their natal watersheds. Salmonids, therefore, feature highly in the ecology of coastal ecosystems through homing into their natal streams, providing a cyclical presence in forested watersheds, including nutrients from hundreds of thousands of carcasses into hundreds of local watersheds annually. The *Haida Gwaii Land Use Plan* of 2007 established riparian forests as protected landscape attributes. An example of the riparian fish-influenced forest on Gwaii Haanas lands reveals the potential extent of a marine nutrient shadow on the landscape (Fig. B10.1.1).

The number of adult salmon that survive to spawn is called "escapement." Escapement data underpin salmon stock assessment. Salmon escapement data are the oldest annual coastal biological time series in Gwaii Haanas. Annual escapement monitoring began in the 1930s for 73 streams of which a minority yielded annually consistent counts. Escapements have declined in recent decades.

Because so many streams host salmon, the Department of Fisheries and Oceans uses index streams of reliable producers as references for particular areas or species. Years of experience from walking streams yielded a focus on a sub-set of index streams called "key indicator" streams and are first priority for surveying. Returns from 1947 to 2005 for nine key indicator streams indicate low numbers of chum and coho since the 1970s (Fig. B10.1.2). Reasons for these declines are poorly understood. Stream habitat damage from logging, common elsewhere in the archipelago, is not an issue, but speculation includes overfishing, poor at-sea survival due to changing ocean conditions, climate change effects on watersheds, or effects of altered riparian vegetation due to intense browsing by deer. Monitoring escapement will be important for understanding the scale of salmon-mediated land-sea connectivity and for prospects of stock restoration.

Table B10.1.1 Salmonid species of Haida Gwaii. Life histories can range from full-time freshwater resident to anadromous.

Resident
Dolly Varden char* (*Salvelinus malma*)
Cutthroat trout† (*Oncorhynchus clarkii*)
Anadromous
Steelhead‡ (*O. mykiss*)
Sockeye salmon§ (*O. nerka*)
Coho salmon (*O. kisutch*)
Pink salmon (*O. gorbuscha*)
Chum salmon (*O. keta*)
Chinook salmon (*O. tshawytscha*)

*Dolly Varden is the most common year-round resident lake and stream salmonid, but some populations are anadromous; †can also be anadromous; ‡can also be permanent lake- and stream-resident rainbow trout; §can also be permanent lake-resident kokanee.

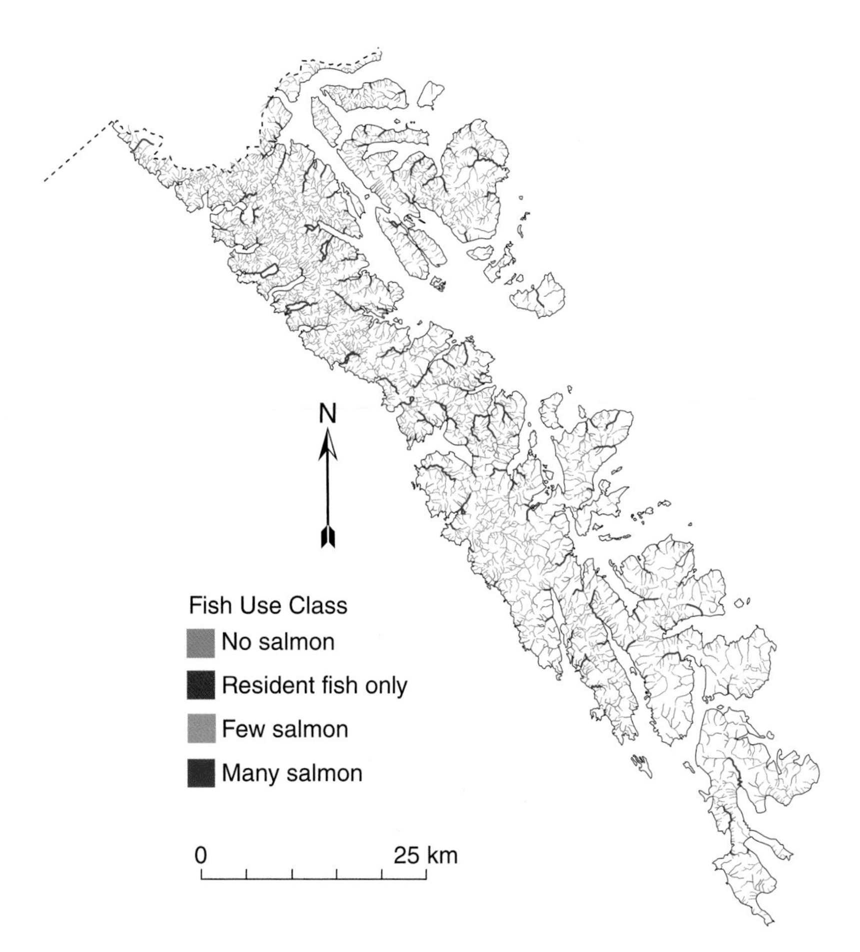

Fig. B10.1.1 Spatial network of the riparian fish-forest in Gwaii Haanas shown in different colors according to inferred levels of fish use. For example, upstream of watercourses with known impassable falls, the most fish use possible would be "resident fish only." Data from J. Broadhead in Sloan (2006).

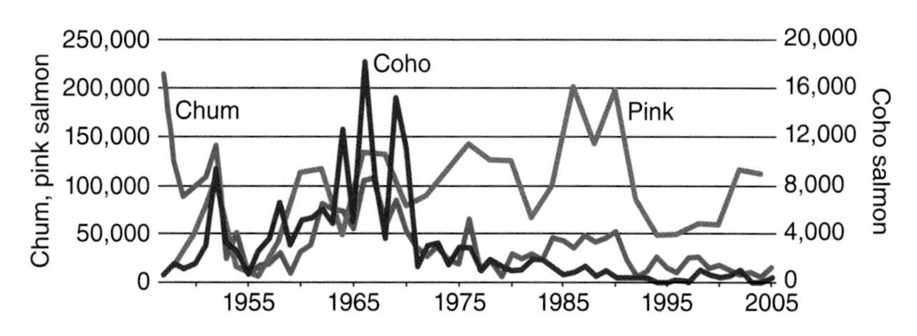

Fig. B10.1.2 Aggregate salmon escapement into the nine "key indicator" streams in Gwaii Haanas from 1947 to 2005. Data for chum (*Oncorhynchus keta*) and coho (*O. kisutch*) are annual while data for pink (*O. gorbuscha*) are even-year only. Note the much smaller scale for the numbers of coho.

tion of shore habitat (Fig. 10.10). Each of 1987 mapped shore units is one of the 34 shore types according to substrate, sediment type, beach width, and beach slope. The resolution of these classes is key to understanding shoreline ecology; for example, species associated with bare rock platforms differ from those associated with sediment-covered platforms where wave-borne sediments may scour off species. The presence of variable coarse sediments increases habitat complexity. Boulders and cobbles support a variety of attached species, as well as providing underlying crevice microhabitat for cryptic species. On sheltered beaches, species can encrust rocks to the size of boulders (>25.6 cm in diameter), whereas on exposed coasts boulders can be wave-shifted, abrading encrusted life. Estuarine salt marsh wetlands are generally small and widely distributed.

Wave exposure, in tandem with substrate type, determines shoreline habitat types. Physical wave exposure relies on "fetch"—the distance over open water that winds blow. The

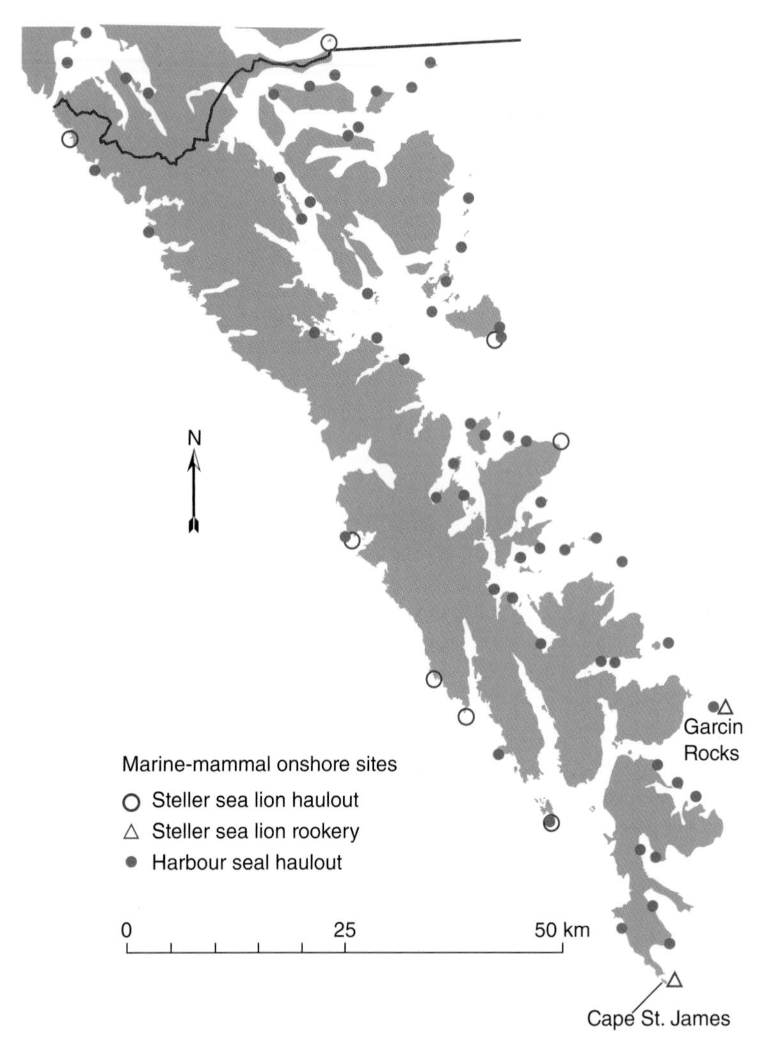

Fig. 10.9 Onshore sites for pinnipeds around Gwaii Haanas Marine. The haulout sites for harbor seal (*Phoca vitulina*) are from aerial surveys. The Steller sea lion (*Eumetopias jubatus*) haulouts also include breeding rookeries at Cape St. James and Garcin Rocks. Modified from Sloan (2006).

distribution, definitions, and proportions of wave-exposure classes range from very exposed to protected (Fig. 10.11). Very exposed shores occur only on the west coast of Moresby Island—the most exposed shoreline in all of British Columbia. The east coast exhibits greater shoreline diversity and protection.

Substrate type and physical exposure criteria support predictable bioband occurrences typifying shoreline biodiversity. Biological exposure of each shore unit can be inferred from species assemblages based on knowledge of species' exposure tolerances. Descriptions of selected biobands and their exposure tolerances (Table 10.2) link biodiversity and substrate with shore tidal height. Bioband distributions are exemplified in Figure 10.12 by relatively sheltered "eelgrass" in sediment vs. relatively exposed "California mussel" on rock.

10.3 CULTURAL AND COMMERCIAL HERITAGE

Protected areas are part of cultural heritage, and for Parks Canada the role of indigenous culture in protected-area management is growing (Dearden and Langdon, 2009; Langdon *et al.*, 2010). The Gwaii Haanas partnership is part of a major political shift towards greater inclusion of indigenous peoples, their culture, and traditional knowledge in resource management. Traditional knowledge is defined as: "A cumulative body of knowledge, practice, and belief, evolving by adaptive processes and handed down through generations by cultural transmission, about the relationship of living beings (including humans) with one another and with their environment" (Berkes, 2012). Canada's 1982 *Constitution Act*, in which aboriginal rights and title were entrenched, underlies the imperative for inclusion. Since this Act was passed, Supreme Court of Canada decisions have been cumulatively defining, in a still-unfolding, precedent-setting process, the responsibilities of government with respect to aboriginal rights, title, and appropriate consultation, such as for natural-resource uses within indigenous peoples' traditional territories. Within the spirit of Gwaii Haanas' cooperative management arrangement, therefore, traditional Haida knowledge will be respected and used along with natural and social science knowledge.

Humans have occupied Haida Gwaii for at least 12,000 years (Fedje and Mathewes, 2005). Further, the archipelago is

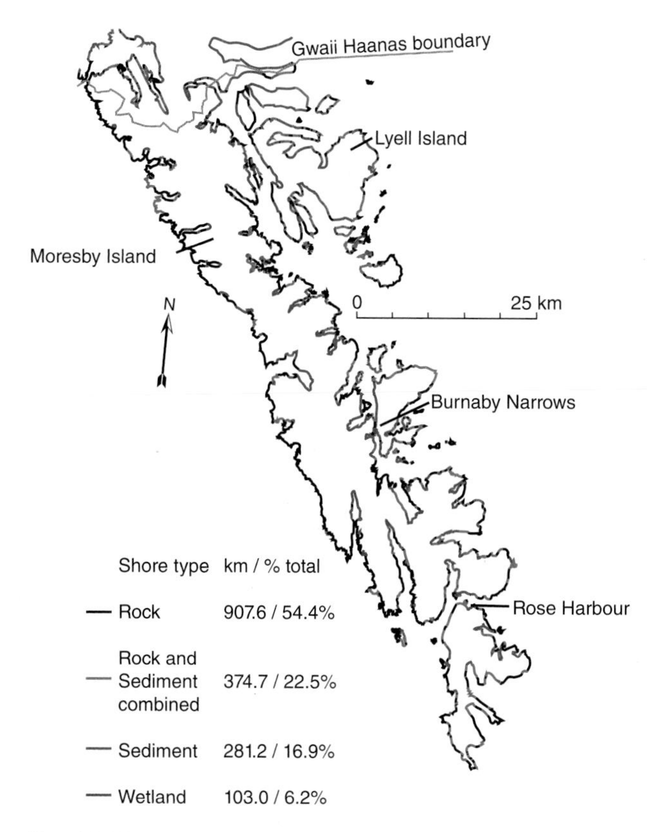

Fig. 10.10 Distribution, definitions, and proportions of shore substrate types along 1700 km of Gwaii Haanas Marine's shoreline. Also shown are some place names mentioned in the text. Shore types: *rock* is bedrock-dominated such as sea cliffs; *rock and sediment* includes bedrock ramps underlying a veneer of boulders or cobbles (typically associated with lower wave exposures); *sediment* ranges from boulders to muds; and *wetland* is typically estuarine with salt marsh vegetation rooted in sediments. Substrate particle diameters: boulder >256 mm, cobble 60 to 256 mm, gravel 2 to 60 mm, sand 0.06 to 2 mm, and mud <0.06 mm. Modified from Sloan (2006).

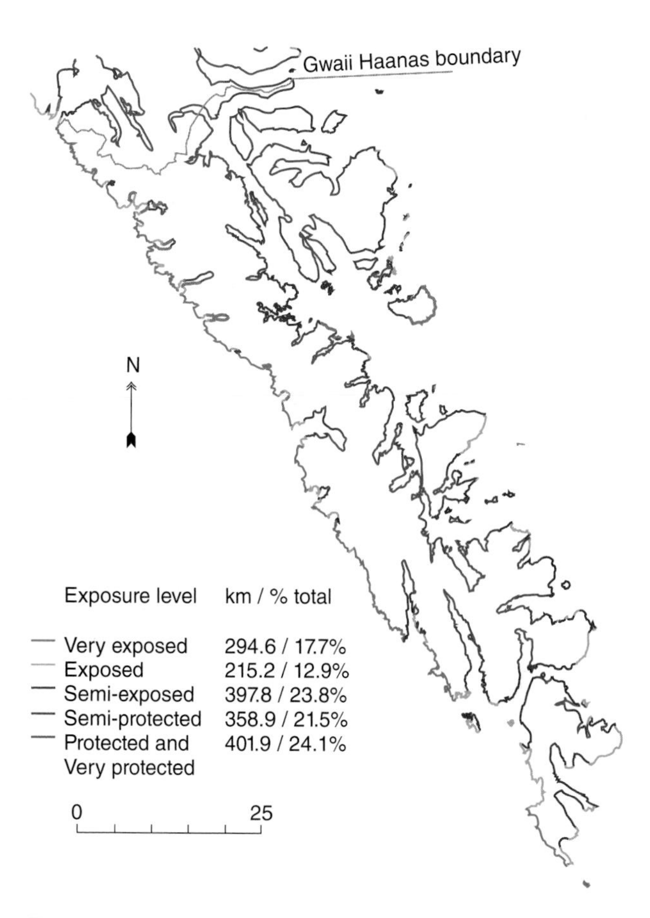

Fig. 10.11 Distribution, definitions, and proportions of the wave exposure classes, based upon observed biological assemblages, along Gwaii Haanas Marine's shoreline. Exposure levels: *very exposed* is the only extreme exposure level along the British Columbia coast, with little wave energy lost due to refraction or shoaling offshore, and has a very long fetch (distance of open water over which unobstructed winds blow); *exposed* experiences higher-energy swells (oceanic waves generated by offshore weather) and has a fetch of 500 to 1000 km; *semi-exposed* includes swells and locally generated waves with a fetch of 50 to 300 km; *semi-protected* is sufficiently exposed for waves to shift cobbles (rocks from ~60 to 256 mm diameter), and a fetch of 10 to 50 km; and *protected* and *very protected* (rare exposure level—virtually no wave action) are combined and sheltered enough that, at a maximum, gravels (2 to 60 mm diameter) can be shifted by waves and with a fetch from ~1 to 10 km. Modified from Sloan (2006).

important to theories about coastal human migration into North America, vs. over a land bridge and through the continent's interior, from northeastern Asia during about 14,000 to 10,000 BP. The archipelago is very archaeologically rich; Gwaii Haanas alone has more than 600 coastal archaeological sites. These include the oldest intact midden—a food waste heap that can contains stone tools, wood, bone, charcoal, and fiber—along coastal Pacific North America, dating from 10,800 BP. This ancient site reveals people who hunted both nearshore and offshore, including for marine birds and mammals. These were sea people with well-developed watercraft and pelagic hunting skills, not likely interior people newly migrated to the coast.

Interdisciplinary research is integrating the archaeological record with environmental history, for which prehistoric culture is divided into eras based upon stone tool technologies. After millennia of being a stone culture, the Haida began using metals, first copper, then iron. There are archaeological studies of the early (transitional) era of contact with Europeans on Haida Gwaii, during which Haida economic adaptations changed appreciably after trade with the "iron people" (Orchard, 2009). Further, faunal and floral remains from middens revealed a late pre-contact environmental baseline that aids understanding post-contact ecological changes.

First recorded European contact with the Spanish in 1774 was soon followed by a vigorous maritime fur trade creating a dramatic shift in Haida economy. Surviving ships' logs reveal that traders were mostly American and British who focused on

Table 10.2 Descriptions of selected biobands and their exposure criteria along the shoreline of Gwaii Haanas Marine. Biobands are ordered according to elevation, starting with the splash zone in the supratidal (*Verrucaria*) and downward into the shallow subtidal (bull kelp).

Bioband	Description	Wave exposure or habitat indicated
Verrucaria	Lichen (*Verrucaria* species) complex* visible as a black band on bare rock in the supratidal splash zone	Band width increases with exposure
Salicornia	Pickleweed (*Salicornia virginica*), salt marsh grasses, or other salt-tolerant herbaceous plants in muddy soils; appears in salt marsh wetlands around estuaries and along the log-line of sandy beaches with dune wildrye (*Leymus mollis*) grass	*Protected* to *semi-protected*
Rockweed	Dominated by rockweed (*Fucus* species) on rock	*Protected* to *semi-exposed*
California mussel	Dominated by large California mussel (*Mytilus californianus*) and thatched barnacle (*Semibalanus cariosus*) with scattered goose barnacle (*Pollicipes polymerus*) on rock at higher wave exposures	*Semi-exposed* to *very exposed*
Eelgrass	Eelgrass (*Zostera marina*) meadow rooted in sand or mud of the lower intertidal to shallow subtidal, including in estuaries	*Semi-protected* to *protected*
Urchin barrens	Pale pink, bleached-looking encrusting coralline algae on the shallow subtidal rock where red sea urchins (*Strongylocentrotus franciscanus*) have grazed away all the fleshy algae	*Semi-exposed* to *semi-protected* or current-dominated
Giant kelp	Nearshore canopy-forming giant kelp (*Macrocystis integrifolia*) forests on rocks; usually in more sheltered areas than bull kelp	*Protected* to *semi-exposed*
Bull kelp	Nearshore canopy-forming bull kelp (*Nereocystis luetkeana*) forests on rocks; often indicates current-affected areas if growing in semi-protected areas	*Semi-protected* to *exposed* or current-dominated

*May include up to nine *Verrucaria* species as well as other black lichen species.

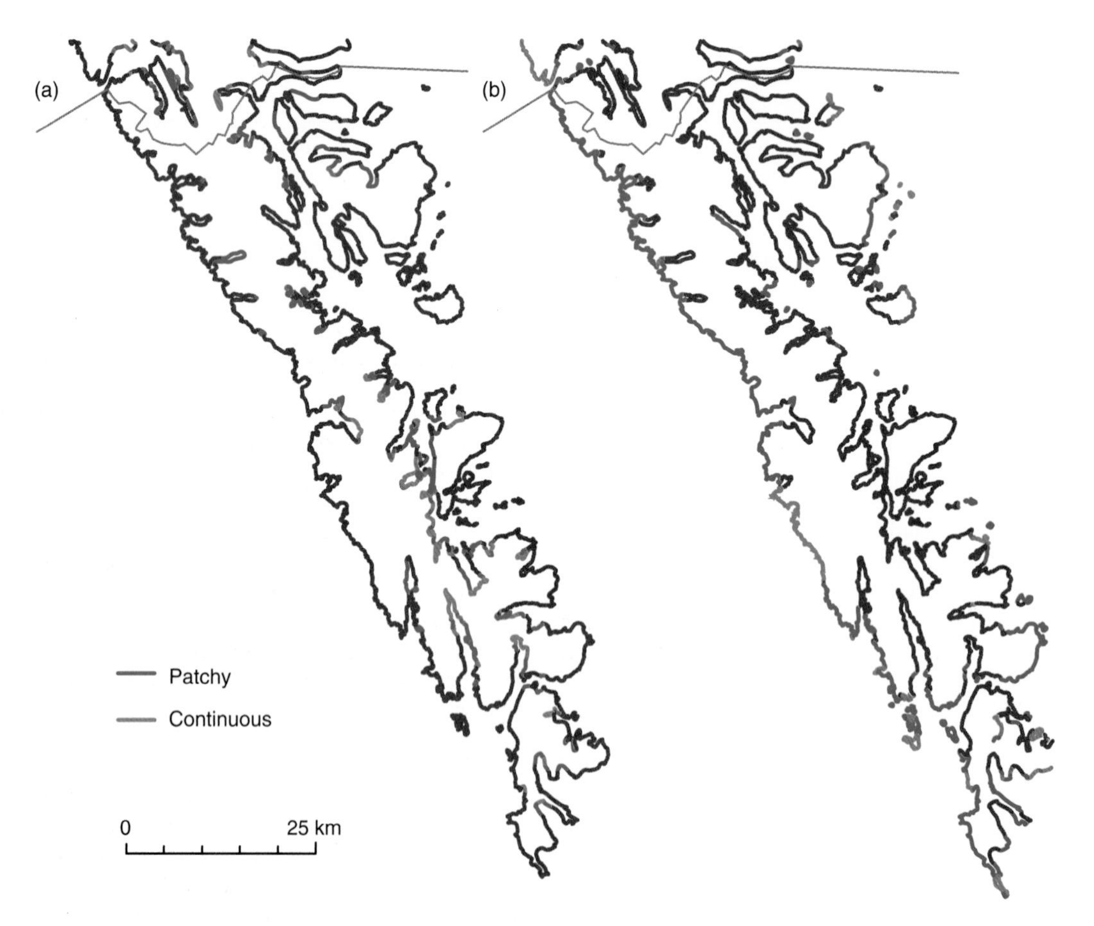

Fig. 10.12 Examples of distributions of biobands along strongly contrasting exposure regimes and substrate types around Gwaii Haanas Marine: (a) eelgrass (*Zostera marina*) bioband along *semi-protected* to *protected* sediment (sand and/or mud) shores, and (b) California mussel (*Mytilus californianus*) bioband along *semi-exposed* to *very exposed* rocky shores. Modified from Sloan (2006).

Fig. 10.13 Totem poles at SGang Gwaay that likely date from the mid-19th century, before most villagers left after the smallpox epidemic. SGang Gwaay contains the largest amount of in situ North Pacific coast monumental indigenous art in existence and is one of the most visited sites within Gwaii Haanas. SGang Gwaay has been a United Nations Educational, Scientific and Cultural Organization (UNESCO) World Heritage Site since 1981. Photograph © Parks Canada/Rebecca Cumming.

acquiring sea otter pelts, hunted by the Haida, for highly profitable exchange with China. Trading boomed from the 1790s to the 1830s around Haida Gwaii (Sloan and Dick, 2012). This was part of maritime fur trading along the northwest coast of North America and the prime focus of relations between Europeans and coastal aboriginal peoples for the first decades post-contact. By the mid-1800s, sea otter populations were heavily depleted coast-wide, including Haida Gwaii. Removal of sea otters, important predators in British Columbia kelp forest ecosystems, had important ecological effects still evident today (Watson and Estes, 2011).

The early contact era brought desired trade goods to the Haida such as iron tools, cloth, and firearms, but also a series of disease epidemics, particularly in the 19th century. Especially tragic was a smallpox epidemic in 1862–63, likely transmitted from mine workers, during which about 80% of the Haida perished. Estimates of pre-contact Haida population range from 15,000 to 30,000. After a low point of around 600 people by the 1910s, the Haida Nation currently has grown to about 4500, including its off-island diaspora. By the 1890s, the remaining Haida had moved from many original village sites (Fig. 10.13) to Skidegate and Old Massett, where they are mostly settled today. There is also post-contact industrial history in Gwaii Haanas from the 19th to the late-20th centuries in which mining, fish processing, and logging brought their own social and ecological changes, including temporary settlements and the first cash employment for the Haida. For example, at the Rose Harbour whaling station (active 1910 to 1943) over 5000 whales were processed. Whaling did not stop in Pacific Canada until 1968. Industrial logging, the largest land-based human activity within Gwaii Haanas, stopped in July 1987 with 6% (90 km^2) of the forest cover removed since first recorded logging in 1901.

The Haida of today take a wide variety of seaweeds and invertebrates (shellfish) archipelago-wide from intertidal and nearshore shallow waters (Turner, 2004; Sloan, 2006). The main Haida subsistence finfish are salmon, Pacific herring (*Clupea pallasi*), and Pacific halibut (*Hippoglossus stenolepis*). A major food fishery is for sockeye salmon (*O. nerka*) returning to fresh water to breed and gill-netted from various estuaries, none of which are in Gwaii Haanas. Another is for herring eggs deposited on kelp, taken nearshore during the early spring spawn. Halibut are longlined by baited hook mostly north of Gwaii Haanas.

Gwaii Haanas Marine heritage also includes commerce, particularly fisheries and tourism. By far the largest human effect on the Gwaii Haanas Marine is withdrawal of biomass by commercial fishing (Table 10.3; Sloan, 2006). The setting for commercial fisheries differs enormously. Listed shellfisheries all date from the 1980s, occur inshore and target geoduck clam (*Panopea abrupta*) and red sea urchin (*Strongylocentrotus franciscanus*) by diving, and prawn (*Pandalus platyceros*) by baited trap. Finfisheries target many species using a range of gear types from the shallow nearshore to deep offshore waters. Habitat damage by fisheries is not a major issue; for example, the groundfish trawl fishery expends relatively little effort within Gwaii Haanas Marine. Offshore, deep-water fisheries target sablefish (*Anoplopoma fimbria*), taken mostly by baited trap over the continental slope. A particularly important nearshore fishery has been for Pacific herring (Box 10.2). Herring stocks are much reduced from historical levels and there are expectations that management of Gwaii Haanas Marine will aid restoration. With the exception of herring, on-island representation among the license holders fishing in Gwaii Haanas Marine is low. This means that most economic benefits from commercial fisheries go off-island, particularly to southern British Columbia. For example, the geoduck industry is one of the most lucrative, taking about $10 million worth of catch every third year from Gwaii Haanas Marine, but there are no local license holders or crew, and catch processing occurs off-island.

Recreational fisheries are a major British Columbia marine industry and a vital activity around Haida Gwaii, where angling from small boats nearshore is one of the prime ways residents and tourists enjoy coastal waters. This mostly May-to-October fishery consists of independent local anglers and those guided by commercial operators of various types from small-boat daytrips to multi-day trips from luxury lodges. Target species are mostly salmon, Pacific halibut, and rockfish (*Sebastes* spp.) by hook and line generally around Graham Island and northern Moresby Island. Within Gwaii Haanas Marine, however, the recreational fishery is modest, due to the area's remoteness and lack of infrastructure, and is likely to remain so.

Gwaii Haanas offers a coastal wilderness ("backcountry") ecotourist experience enabled only by boat or floatplane: 95% and 5% respectively in 2011. The remote setting, with lack of human infrastructure and often challenging sea state, can create demanding conditions. Visitor experiences include scenic beauty, wildlife, solitude, and Haida culture. The management goal is safety and enjoyment without compromising

Table 10.3 The major commercial shellfisheries and finfisheries around Gwaii Haanas Marine. Each fishery is limited-entry, has its own Integrated Fishery Management Plan issued annually by Fisheries and Oceans Canada and is represented by its own industry association.

Fishery	Notes on fishing methods; licensing; seasons
Geoduck clam	By diver (<20 m depth) over sandy substrates using scuba or surface supply (HOOKAH); IVQ*; year-round, every third year only
Red sea urchin	By diver (<20 m depth) over rocky reefs with kelp forests, product is roe; IVQ*; September to July
Prawn	By baited trap (70 to 100 m depth) in sheltered inlets; mid-May to mid-July
Salmon	By purse seine or gill net in estuaries for returning spawners ("terminal") to Haida Gwaii targeting pink and chum; mid-August to October
	By purse seine or gill net at sea for passing fish ("interception") targeting sockeye and pink that mostly spawn elsewhere along the Canada/U.S. coast; July to August
	By troll (towed hook and line) at sea to intercept chinook and coho passing Haida Gwaii to spawn elsewhere; July to September
Pacific herring	By purse seine or gill net, product is roe from the body cavity; February to late March
	By purse seine, product is spawn-on-kelp from floating net frames in which kelp is suspended from lines on which captive herring deposit fertilized eggs and are then released; March to April
Groundfish trawl	By trawl (mid-water or bottom) for many species†; IVQ*; year-round
Groundfish hook and line‡	By long-line (on- or near-bottom with baited hooks), for certain species (lingcod, dogfish, skates, sole, flounder, Pacific cod); IVQ*; year-round or specific times for specific areas
Rockfish (outer coast)	By long-line, for rockfish (*Sebastes*) species (also thornyheads, greenlings, lingcod); IVQ*; year-round or specific times for specific areas; each licensee must choose one of four fishing options annually§
Pacific halibut	By long-line; IVQ*; February to November
Sablefish	Mostly by baited traps on a ground-line but also some longline (350 to 1100 m depth); IVQ*; year-round

*IVQ (individual vessel quota) in which each licensed vessel is allocated a portion of the total allowable catch annually, and multiple vessel-quotas can be batched and fished by one vessel—IVQs are freely transferable via the market; † there are over 75 species lumped together as "groundfish," of which 23 account for about 95% of the trawl landings; ‡ hook and line here includes longline, hand line, jig or troll; the three types of longline gear used are on-ground, off-bottom, and pelagic, among which groundfish gear is anchored directly at both ends; this gear can be modified to fish off-bottom or mid-water by adding floats along the mainline; use of longlines is not permitted for some species; § options by which fishers target species that satisfy different market sectors such as: live, fresh (iced), or frozen.

Box 10.2 Pacific herring

Pacific herring (*Clupea pallasi*) is a key species for Gwaii Haanas Marine. Herring are small, schooling pelagics whose populations naturally fluctuate widely, with unpredictable stock recruitment. Local herring migrate inshore during October to December from offshore feeding grounds in Hecate Strait to spawn in early spring. The species shows some site fidelity to natal spawning areas such as particular inlets. Herring are a key "forage" species, eaten by many fish, bird, and mammal predators, and transfer energy from the lower trophic levels of their plankton food to the higher trophic levels of their predators (Smith *et al*., 2011).

Haida subsistence and commercial herring fisheries are highly valued. Islanders have always been active in this fishery. Given that the Haida Gwaii stock is recognized as separate from all others coast-wide, and that stocks have declined, herring is symbolic in local concerns over fisheries management, restoration, and marine ecosystem health. The present era of stock decline, with consistently low recruitment for poorly understood reasons, heightens local concerns over sustainability. Moreover, Gwaii Haanas Marine could have a role, as a herring stronghold in years past, in ecosystem-related research and restoration.

Herring spawning is one of the year's most dramatic coastal events, causing nearshore waters to turn white with milt (sperm). Herring spawn in aggregations between 1.5 m above to perhaps 18 m depth below the zero tide line mostly on attached plants. Spawning occurs from February to July with a March to April peak. Spawning provides an enormous pulse of biomass and energy into nearshore food webs and attracts many predators to gorge on spawners and deposited eggs (Willson and Womble, 2006).

Herring spawn along the linear extent of shoreline, and this has historically occurred around Moresby Island (Fig. B10.2.1). The Haida Nation's fisheries program cooperates with DFO annually to complete diving surveys of spawn distribution and abundance. Given that spawning is related to depth, substrate, and vegetation characteristics, it provides a valuable shoreline reference dataset and a key stock assessment tool.

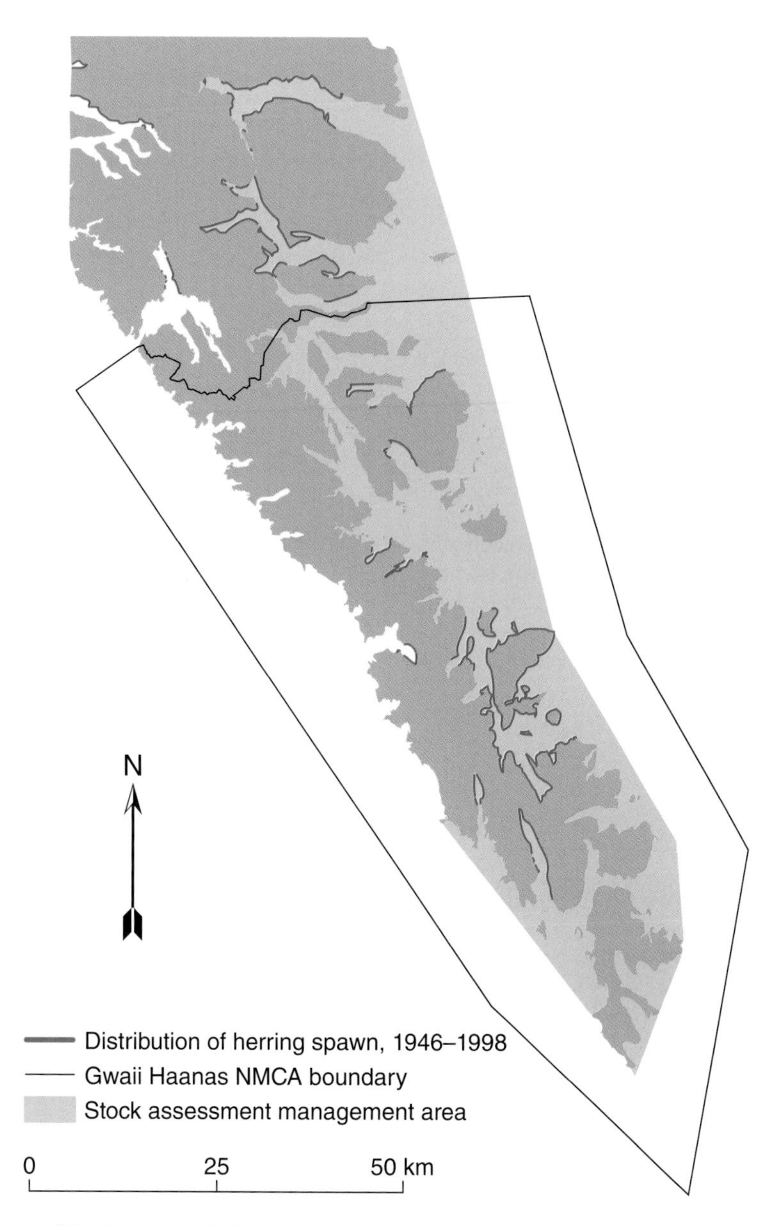

Fig. B10.2.1 Southern area of Haida Gwaii, including Gwaii Haanas, showing the main Fisheries and Oceans Canada Pacific herring stock assessment area and the distribution of herring spawn deposition based on combined data from 1946 to 1998. Modified from Sloan (2006).

This fishery experienced dramatic local catch variability from 77,500 tonnes in 1956 to some annual landings of less than 1000 tonnes, underscoring both high potential stock abundance and inherent stock instability—not uncommon among small schooling pelagic fishes. The modern limited entry (fixed number of licenses) fishery began in the early 1980s for high-value export products of herring roe and spawn-on-kelp. Within commercial allocations, spawn-on-kelp had precedence over roe; there have been no roe landings since 2002. Central to management is setting a fixed annual quota for the major stock area based on a take of 20% of the forecast surplus mature stock biomass determined by the previous year's spawn assessments; there must be a minimum spawning stock biomass ("cutoff"), below which the commercial fishery is closed, although the Haida subsistence fishery can occur.

Accordingly, herring management will be an important test for Gwaii Haanas Marine. The depressed fishery sparked concern among Haida Chiefs and political leadership about the effectiveness of marine resource management. For years, low herring stocks precipitated disagreements between DFO and the Haida. The Haida believe in managing smaller, more localized stocks, assert their traditional knowledge, and want to see a departure from single-species herring management towards a more multi-species, ecosystem-based approach in which the appropriate precaution considers the status and needs of other species while fostering herring recovery.

ecological and social carrying capacities underpinned by wilderness values, i.e., "the sensation of being the first person to set foot here." Important for Gwaii Haanas is that the AMB cut its teeth through developing visitor management in the 1999 Backcountry Management Plan (updated twice since), executing public consultations in that process, and working with industry (the Tour Operators' Association). Backcountry visitor management is recounted in Box 10.3.

10.4 INTEGRATING LAND-SEA CONSERVATION

Establishment of Gwaii Haanas Marine occurred during major changes in Canadian marine legislation, policy, and planning at national, regional, and local levels. The conservation and use vision for Gwaii Haanas Marine, with its terrestrial linkages, embodies sustainable uses while maintaining ecosystem structure and function, facilitating visitor experiences, and fostering Canadians' understanding of their marine heritage. This vision also champions the continuity of Haida culture and economic development while respecting Haida traditional knowledge, subsistence and spirituality.

Gwaii Haanas' terrestrial management plan (AMB, 2003) did not include marine waters, but, critically, stated: ". . . it cannot help but recognize the close relationship that exists between land and sea." Therefore, the inseparability of land and sea towards an integrated, long-term management approach is the area's cooperative management ethic. The essence of this ethic is: "Respect for the area developed through knowledge and understanding will be the surest means of protection for Gwaii Haanas" (AMB, 2003). Terrestrial management goals are: (i) preserving natural and Haida cultural heritage; (ii) managing human use (visitation); and (iii) informing citizens about conservation and heritage from a unique place-based perspective. The terrestrial area received the highest score in a survey of 55 national parks in the United States and Canada (Tourtellot, 2005). The survey focused on environmental quality and protected-area relations with gateway communities, including indigenous peoples.

10.4.1 Laws, accords, and policies set the stage

Gwaii Haanas Marine was established under the 2002 *Canada NMCA Act*. This Act is enabling, not prescriptive, legislation defining a process by which Parliament creates permanent NMCAs that are representative of Parks Canada's marine natural regions. Federal agencies cooperate with their individual mandates unchanged, such as those of DFO for

Box 10.3 Backcountry visitor management

Two hours by motorboat or two kayaking days are needed to reach Gwaii Haanas' northern boundary from the closest road-head. Since annual visitation was first recorded in 1996, it has never exceeded 2200 and was 1753 in 2011. More than 70% of visitors travel with licensed tour operators, the remainder being independent. Independents spend an average of seven nights while the stay of guided visitors averages four nights. There is little human infrastructure, with two Gwaii Haanas operations stations and four Haida Gwaii Watchmen sites within Gwaii Haanas' area. The Watchmen program was founded by the Haida Nation in 1981 and has been funded by Parks Canada since the early 1990s. This support has been important for strengthening Canada-Haida relations. Haida Watchmen are site guardians and interpretive hosts for visitors at historic village sites during the main visitor season of May to September.

Campers use mostly kayaks to access level beach areas. All visitors attend a mandatory orientation in which responsibility is stressed, such as no-trace, self-sufficient camping, and respect for the environment and Haida culture. Very high environmental quality and wilderness values (e.g., solitude and little infrastructure) are expected. Visitation occurs mostly along the east coast, especially around key attraction sites such as Rose Harbour and the Watchmen sites (Fig. B10.3.1). Few visitors go to the west coast because of its unpredictable weather and rugged shores with limited sheltered beach access.

Visitation is carefully controlled by the Archipelago Management Board (AMB), based on the notion of carrying capacity, assessed from tour operator and visitor experience feedback and monitoring of physical indicators, such as erosion, at campsites. The AMB uses an adaptive management feedback loop to refine evidence-based management. Annual visitor use patterns in time and space have been recorded on a GIS since 1996. Data are collected from the registration system, log books at Watchmen sites, and trip logs from tour operators and independent and tour visitors. Submission of tour operator trip logs is a mandatory business-licensing requirement (hence 100% returns). Tour guides must be certified for backcountry skills and wilderness first aid. Independent visitor registration is mandatory, but submission of tour visitor and independent visitor logs is voluntary. Accordingly, the spatial data from tour operators are complete, while the spatial and visitor experience data from tour and independent visitors are partial samples. Management actions include limiting numbers at any one place on shore to 12, giving specific allocations to each licensed operator and using a reservation system to control independent visitor use. The total annual allocation of 33,000 user-nights (visitor-nights plus guide-nights) was set by the AMB as a precautionary limit. There is room for growing visitation as annual use has averaged around 9700 user-nights (1996 to 2011).

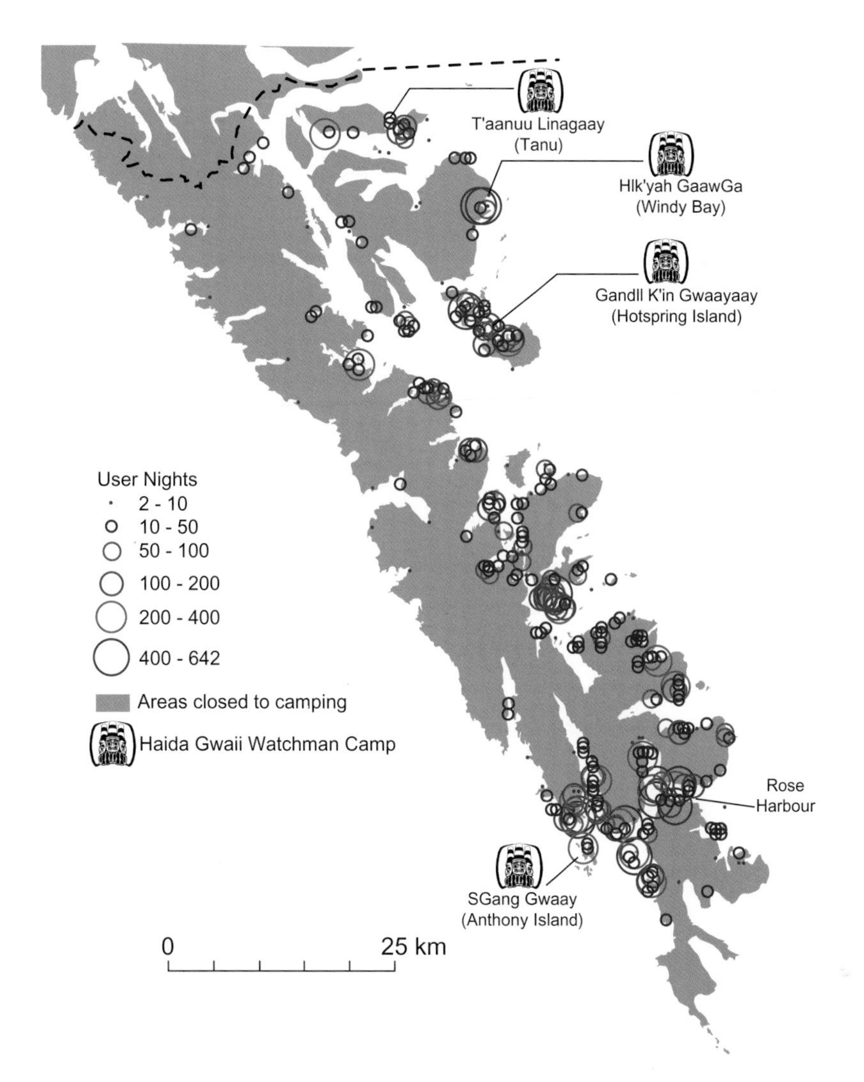

Fig. B10.3.1 Distribution of over 24,400 user-nights for camping by commercially guided visitors to Gwaii Haanas Marine, 1997 to 2004. User-nights include both visitors and tour operator staff. Data points close together can result in overlapping circles. T'aanuu Linagaay is an Indian Reserve and not legally part of Gwaii Haanas' lands. Modified from Sloan (2006).

fisheries, and Transport Canada for shipping. The NMCA mandate includes maintaining ecosystem structure and function as the first priority, while permitting multiple sustainable uses such as fisheries, tourism, or aquaculture. Other mandate aspects are informing the public on marine conservation, facilitating visitors' experiences, and cooperating with academia or NGOs for applied research. The only explicit prohibition is extraction of non-renewable resources (petroleum, minerals, sediments). For NMCA establishment, public consultations are tabled in Parliament along with an Interim Management Plan including a zoning strategy enabling coexistence of preservation with sustainable uses. Zoning is provided through two zone types—sustainable use and full protection—and there is a flexible stance towards establishing other zone types, depending on local conditions and consultations. In sharp contrast to the sea, Parks Canada's terrestrial mandate, under the *Canada National Parks Act*, is preservation with no commercial resource extraction. Hunting, fishing, gathering, and trapping on Gwaii Haanas' lands and waterways by the Haida for subsistence is permitted. Within Gwaii Haanas' footprint, therefore, Parks Canada has a preservationist terrestrial mandate in contrast to a sustainable-use marine mandate.

The need to improve Canadian oceans management led to the 1996 *Oceans Act* that mandated DFO to facilitate cooperative management, albeit after a slow start (OAGC, 2005, 2012; Ricketts and Hildebrand, 2011). Based on this Act, Canada's Oceans Strategy (DFO, 2002) guides actions grounded on principles of sustainable development, integrated management, and the precautionary approach. The Strategy promotes better ocean governance through broad societal collaboration (aboriginal, community, industry, agency, academia, NGO), public awareness of and care for oceans and coasts, and commits to implementing ecosystem-based management (EBM). Further, there is a national role for NMCAs in area conservation within Canada's Oceans Strategy, Marine Protected Areas Strategy (DFO, 2005a), and a Marine Protected

Areas network concept (DFO, 2011) in which a network is defined as: "... a set of complementary and ecologically linked marine protected areas in a particular region." Nevertheless, national progress in marine-area protection has been slow (Dearden and Canessa, 2009; Jessen, 2011).

Gwaii Haanas Marine will eventually be networked with other conservation areas. For Pacific marine-area protection, a 1998 Canada-British Columbia (BC) joint strategy was followed by a 2004 Memorandum of Understanding to implement Canada's Ocean Strategy and to coordinate protected-area establishment. A Canada-BC Oceans Coordinating Committee of 13 federal and provincial agencies was established in 2006 to administer that memorandum. This committee consulted with First Nations, communities, and the public and drafted a Pacific Marine Protected Area network strategy to improve on the ad hoc approach to area establishment and to begin networking area protection (C-BC, 2012). All of British Columbia's marine-related conservation areas, including Gwaii Haanas Marine, collectively represent 2.7% (12,400 km^2) of the 453,000 km^2 of Canada's Pacific Exclusive Economic Zone, and 161 of these 162 areas (11 federal, 150 provincial, one municipal) are open to some commercial fishing (Robb *et al.*, 2011).

Gwaii Haanas Marine has a role in emerging listed species-at-risk management under Canada's 2003 *Species at Risk Act*. This Act facilitates protection, e.g., for critical habitat, and recovery of species deemed at various levels of risk through independent scientific review. The Act is implemented federally by Environment Canada and DFO, with the latter being responsible for listed aquatic species and their habitats. Parks Canada is responsible for managing listed species within national parks, national historic sites, and NMCAs. Applying this Act has been particularly difficult for some listed marine species, such as those taken in targeted fisheries or as incidental catch within multispecies fisheries (Mooers *et al.*, 2007). Further, depending on the category of risk, some level of "incidental harm" to a listed species within a commercial fishery may be permitted. Within Gwaii Haanas Marine, listed species' "critical habitat" areas have been identified for northern abalone (*Haliotis kamtschatkana*) and humpback whales (*Megaptera novaeangliae*). As well, of the 26 marine-mammal species known locally, the conservation profile of 10 species of this charismatic group is enhanced because they are listed.

The *Gwaii Haanas Marine Agreement* (Box 10.4) is Gwaii Haanas Marine's political accord. Besides the Haida Nation, Parks Canada partners with DFO, the nation's leading marine agency. The DFO has a fisheries management mandate under the *Fisheries Act* whose roots pre-date Canadian confederation of 1867. This Act's first priority is fisheries (aboriginal, commercial, recreational) stock conservation. The second priority is for enabling aboriginal subsistence for food, social, and ceremonial purposes. This fundamental right of access, entrenched at the highest level of Canadian law (*Constitution Act*), is central to aboriginal culture. The Act's third priority is for provision of stock to commercial and recreational fishery sectors. In summary, the deeply entrenched and culturally powerful *Fisheries Act* was augmented by the 1996 *Oceans Act* that promotes DFO's lead in integrating oceans management, including EBM and marine-area conservation, towards broad societal involvement and greater collaboration with indigenous peoples.

Box 10.4 Gwaii Haanas Marine Agreement

This Agreement, signed January 2010, begins: "... the Government of Canada and the Council of the Haida Nation have a common desire that the Gwaii Haanas Marine Area shall be regarded with the highest degree of respect and will be managed in an ecologically sustainable manner that meets the needs of present and future generations, without compromising the structure and function of ecosystems." In the Agreement, Canada is represented by the Minister of Environment (responsible for Parks Canada Agency) and the Minister of Fisheries and Oceans (DFO); the Haida are represented by the Council of the Haida Nation. Management is through the Archipelago Management Board (AMB)—first established to cooperatively manage the conserved lands in 1993. In 2011, the AMB was expanded to six members with one more Haida (total three) and one DFO representative (added to two from Parks Canada). The AMB renders all decisions for the planning, operation, management, and uses of the integrated land-and-sea conservation area. The five objectives of the AMB are to: (i) maintain or restore healthy ecosystems; (ii) to be a benchmark for science and human understanding; (iii) to maintain the continuity of Haida culture; (iv) to facilitate ecologically sustainable resource uses respecting coastal community economic well-being; and (v) to promote understanding and appreciation of the marine environment along with opportunities for visitor experiences and awareness. Consensus is the goal in the AMB's decision-making; failing that, senior representatives of the signing parties will attempt agreement "in good faith," respecting the spirit of the Agreement's intent. Central for the AMB will be drafting the first *Gwaii Haanas Marine Management Plan*, due by 2015 and reviewed every five years thereafter. In the nearer-term, the AMB will operate under the *Interim Management Plan*, required for establishment, and formulate a *Marine Area Strategy* describing management approaches and actions to include zoning and fishing activities to meet ecosystem objectives. Yet, DFO independently manages commercial fisheries access (licensing) and allocation (of species and of quantities to particular fishing groups) according to its unimpeded *Fisheries Act* mandate. Meanwhile, the AMB oversees development of ecosystem, socio-economic, and cultural objectives (including for fishing) and monitors progress towards thresholds by which it will assess management progress, and subsequently adapt management as needed.

Table 10.4 Cooperative coastal and marine planning and management agreements for Haida Gwaii signed between provincial or federal levels of government and the Haida Nation, 1992 to 2010.

Management Agreement	Date	Notes
Framework Interim Measures Agreement	August 1992	Between Canada (Fisheries and Oceans Canada) and the Haida Nation, under the Aboriginal Fisheries Strategy, for future cooperative fisheries management arrangements
Gwaii Haanas Agreement for Gwaii Haanas National Park Reserve and Haida Heritage Site	January 1993	Between Canada (Parks Canada Agency) and the Haida Nation for cooperative management and planning of Gwaii Haanas' lands (1500 km²) and shoreline (1700 km)
Razor Clam Sub-agreement (under the Framework Agreement of 1992)	August 1994	Between Canada (Fisheries and Oceans Canada) and the Haida Nation for cooperative management of the intertidal digging for razor clam (*Siliqua patula*) fishery—the first such agreement with a First Nation in Pacific Canada
Cooperative management and planning of SGaan Kinghlas-Bowie Seamount Marine Protected Area	April 2007†	MOU* between Fisheries and Oceans Canada and the Haida Nation for cooperative management and planning of SGaan Kinghlas-Bowie Seamount MPA (6122 km²), 180 km west of Haida Gwaii in the North Pacific
Haida Gwaii Strategic Land Use Agreement	December 2007	Between the Province of British Columbia and the Haida Nation for collaborative land use (zoning and ecosystem-based management) and for compatible marine uses adjacent to new land conservancies on Haida Gwaii, north of Gwaii Haanas
Gwaii Haanas Marine Agreement for Gwaii Haanas National Marine Conservation Area Reserve	January 2010‡	MOU* between Canada (Parks Canada Agency and Fisheries and Oceans Canada) and the Haida Nation for cooperative management and planning of Gwaii Haanas marine area (3400 km²)

*MOU = a formal Memorandum of Understanding; †established as a Marine Protected Area under the *Oceans Act* in April, 2008; ‡established as a National Marine Conservation Area Reserve of Canada under the *Canada National Marine Conservation Areas Act* in June, 2010.

Formal cooperation between the Haida Nation and other levels of government is Haida Gwaii's resource planning and management model. Haida Gwaii marine and coastal areas have operated under cooperative agreements since the early 1990s (Table 10.4). Prominent is the *Haida Gwaii Strategic Land Use Agreement* signed in 2007 between British Columbia and the Haida Nation and implemented in 2010 (BCMFLNRO, 2012). A protocol Agreement to implement the Land Use Plan was signed in December 2009 and included officially renaming the province from Queen Charlotte Islands to Haida Gwaii, and the Haida Nation giving back the name Queen Charlotte Islands to the Crown. These arrangements became provincial law under the *Haida Gwaii Reconciliation Act* of April 2010 and were followed by a December 2010 *Land Use Objectives Order* guiding implementation. The Haida Gwaii Management Council implements the Land Use Plan, establishes forest practice objectives, oversees the annual forest cut, and approves management plans for the new terrestrial conservancies, including their adjacent marine areas. The Council has two Haida and two British Columbia representatives, and a fifth member agreed upon by both parties.

This Agreement protects: 48% of the total landscape as conserved areas, including federal Gwaii Haanas lands; another 4% as ecological reserves to protect nesting areas of bird species at risk or other forest areas of high cultural value; marine conservancies adjacent to the new land conservancies; and 74% of the archipelago's shoreline (3448 km). Further, the 48% of the Haida Gwaii landscape zoned for logging will be guided by a forest EBM regime. The 52% of lands conserved is high relative to the British Columbia mainland that has about 15% in conservancy—itself the highest for a province or territory in Canada. Finally, the key decision of the annual allowable cut of wood volume is now made on-island. Thus, the EBM ethos for Haida Gwaii land use influences societal expectations for on-island control of natural-resource management, including marine-use planning.

10.4.2 Interim management and zoning

Establishing Gwaii Haanas Marine required an Interim Management Plan, zoning strategy, record of public consultations, and legal descriptions of the area. These were developed by a marine project team of Haida Nation, federal, and provincial agency and fisheries liaison contractors, with input from a community advisory committee. The Interim Management Plan and the *Gwaii Haanas Marine Agreement* set the management tone. Interim management will guide operations during post-establishment until adoption of the formal Management Plan, due in 2015. All activities will be zoned and managed, including fishing, to meet ecosystem objectives that are being developed. The interim plan's guiding principles (Table 10.5) reveal the values underlying adaptive management. The explicit acknowledgement for the partnership to be innovative

Table 10.5 The five guiding principles from the 2010 Interim Management Plan required by Parliament for the establishment of Gwaii Haanas Marine.

Principle	Description
Showing respect	Successful management can only be achieved through respect, both for one another and for all living things
Working together	Inclusiveness acknowledging a diverse range of ideas, integrating programs and sharing responsibility for planning and management are fundamental
Balancing protection and sustainable use	Management will respect a range of environmental, social, economic, and cultural values to achieve protection and sustainable use
Fostering innovation	Due to the dynamism of marine ecosystems, adaptive management approaches are required to accommodate change through a combination of traditional, modified, and innovative methods
Demonstrating accountability	Accountability will be demonstrated through appropriate monitoring and transparent reporting

is particularly encouraging. The following management objectives ensure Gwaii Haanas Marine's consistency with those principles:

- implement effective collaboration for planning and management;
- protect, conserve, and restore marine biodiversity and ecosystems;
- sustain the continuity of Haida culture and protect features of spiritual and cultural importance;
- promote ecologically sustainable uses of marine resources;
- advance understanding of biodiversity and ecosystem functions, natural and social sciences, cultural resource values, and sustainable uses;
- enhance awareness and understanding, among local and national audiences, of the natural and cultural heritage; and
- foster meaningful connections with all Canadians and memorable visitors' experiences.

Broad approaches for AMB-sanctioned activities will also be articulated to achieve these management objectives, will aid development of annual work-plans, and be advanced through appropriate precaution, consultation, and adaptive management. A permanent committee selected from communities, industry sectors, NGOs, academia, and the public will advise the AMB. Planning and management must be transparent and inclusive, demonstrate considering stakeholders' advice, and adhere to high technical standards. Fostering strategic partnerships, such as with academia, will facilitate applied natural and social sciences research. Activities listed under one management objective may aid achievement of other related objectives; for example, fisheries management and zoning, when developed concurrently, can achieve conservation of biodiversity and culture under ecologically sustainable use objectives. This integrated approach will also help to implement education, awareness, and visitor experience objectives.

Critical to managing Gwaii Haanas Marine will be working with commercial fisheries. Indeed, increasing support and understanding for marine-area conservation within fishing industries is a worldwide challenge (Roberts, 2007, 2012). Local fisheries (Table 10.3) are sectored and have their own industry associations that are separately consulted towards drafting annual Integrated Fishery Management Plans negotiated with DFO. Further, all belong to the BC Seafood Alliance, a coast-wide and industry-wide lobby (BCSA, 2012) of 17 separate industry associations that collectively represents over 90% of the commercially fished, processed, marketed, and exported seafood. Although the Alliance supports marine-area conservation in principle, it questioned the establishment of Gwaii Haanas Marine. Despite the concerted pre-establishment efforts to minimize effects to the industry, and the modest size of Gwaii Haanas Marine's no-take zones, the Alliance objected to the proposed zoning. The Alliance demanded compensation for any loss of revenue arising from no-take zoning, questioned the social and natural science used for spatial analyses, objected to the prospect of Haida subsistence uses within proposed no-take zones, and criticized the industry consultation process as insufficient although there had been 70 meetings with particular industry sectors. Important aspects of ongoing commercial fisheries-Gwaii Haanas Marine relations include:

- coast-wide consultations with this well-organized industry;
- uncertainty on how the AMB will affect *Fisheries Act*-based management while also implementing *Oceans Act*-influenced EBM;
- the final extent of no-take zoning and accommodating unimpeded Haida subsistence rights within those no-take zones;
- demands for compensation for any fishing grounds lost through zoning;
- current low economic returns to island communities from most fisheries; and
- expectations for the area to affect management reform and stock restoration.

The development of interim zoning involved defining operating principles, collecting and mapping biophysical data, and identifying candidate areas of high biophysical value using a decision-support tool (Box 10.5). Overlap of information types yielded a first-cut spatial analysis sufficient to support the zoning requirement for establishing Gwaii Haanas Marine. An essential information source missing for the area, however, is fishers' experiential knowledge (Scholz *et al.*, 2011). Arguably, few know more about the distribution and behavior of local stocks over time than commercial fishers whose livelihoods depend on sound local knowledge.

Because of constitutionally protected aboriginal subsistence rights, the Haida can take for food, social, and ceremonial purposes within no-take zones, unless there is an overriding conservation concern. The outcome was six key areas representing 3.0% of the Gwaii Haanas Marine (Fig. 10.14). This is low, considering the recommended 20 to 40% range of no-take within marine conservation areas from the global literature

Box 10.5 Identifying areas of high biophysical importance using MARXAN

To identify areas of high biophysical value, 157 spatial information layers of features (e.g., oceanographic, prominent species [fish, bird, mammal], nearshore ecosystems) were compiled. The decision-support tool MARXAN (Ball *et al.*, 2009) was used to employ an optimization algorithm to select sets of areas achieving marine conservation targets from the information layers. Gwaii Haanas Marine was divided into about 60,000 6 ha hexagonal planning units. The amount of each feature within each planning unit was then recorded. MARXAN identified areas that most efficiently met the representation targets per feature. MARXAN was run 400 times to create a range of options for meeting representation goals. The operating principles included ensuring representation of all conservation elements such as ecosystems, unique places, and processes while not fragmenting the integrity of functional units. Other principles were minimizing negative socio-economic effects on stakeholders (commercial fisheries and tour operators) and maximizing economic, cultural, and spiritual benefits. Information from area users was also mapped based on some local experiential and Haida traditional knowledge. These included Haida Elders (with extensive childhood experience in Gwaii Haanas Marine during seasonal provisioning trips with their families), Gwaii Haanas' field staff, and the interim advisory committee. Results identified candidate areas that consistently and efficiently represented "candidate areas of high biophysical value." Conservation targets were used in the MARXAN analyses to capture a portion of each biophysical feature/value in the areas of high biophysical value. Targets for biological features and unique or special areas were set at 30%. This target was, in part, influenced by the international literature that recommends a range of 20 to 40% of conservation areas set aside as no-take zones (Roberts, 2007, p. 369). Further, features of highest ecological significance were targeted at between 45 to 60%. These included representative habitats or known biophysical or productivity hotspots, such as estuarine salt marshes, herring spawn areas, eelgrass meadows, kelp forests, and sea lion rookeries. Consultations with the Haida, commercial and recreational fishing sectors, the tour industry, and independent scientists were then held for feedback on these candidate areas. Weighing this feedback, plus an internal review by the DFO, enabled refining the candidate areas to ensure that zoning decisions for establishment minimized displacement of current users and took socio-economic considerations into account. These modifications lowered the footprint of the candidate areas from 30% to about 10% of the area. These modified candidate areas were further informed by independent traditional/experiential knowledge gathering by the Haida that identified ecologically and culturally significant areas and by additional consultations with the various commercial fishing sectors. In response to this second round of feedback, the six candidate areas were selected that represented 3% of Gwaii Haanas Marine as preservation (no-take) zones in the 2010 *Interim Management Plan*.

(Roberts, 2007) and the 30% no-take recommended by Jessen *et al.* (2011) for each Canadian marine bioregion.

In addition in 2004, DFO established in regulation under the *Fisheries Act* two Rockfish Conservation Areas (RCAs) covering 472 km^2 (13.6%) of Gwaii Haanas Marine. RCAs are year-round hook-and-line closures for commercial and recreational fisheries, although most net, diving, baited trap, and hand-picking fisheries are allowable. Also, all First Nations subsistence fisheries (including hook-and-line) are allowable in RCAs. RCAs were part of DFO's coast-wide rockfish conservation strategy launched in 2002 to rebuild stocks of five "inshore" (0–200 m depth) species. These species occur within the remarkable complex of 37 rockfish species (*Sebastes* spp., family Scorpaenidae) taken in fisheries from nearshore to continental slope depths. The long-term goal is to protect 20 to 30% of inshore rockfish habitat coast-wide, depending upon areas' histories of stock declines. Identifying RCAs was based upon DFO applying an inshore rockfish habitat (shallow rocky reef) model based on depth and substrate.

10.4.3 Gwaii Haanas Marine within regional marine planning and management

This is a dramatic moment in the islands' history, for cooperative marine-use planning and management is unfolding at multiple spatial scales simultaneously. Plans range from nearshore marine areas north of Gwaii Haanas to well offshore, such as SGaan Kinghlas-Bowie Seamount Marine Protected Area (Table 10.4). As well, since 2007 the Haida Nation has guided development of marine-use planning for the archipelago (Jones *et al.*, 2010; Haida Laas, 2012), with the exception of the AMB-guided Gwaii Haanas. The Haida held an on-island public oceans forum for community awareness (CHN, 2009) and formed a multi-sector Marine Advisory Committee (members both on- and off-island) in 2011 to advise on development of the Haida Gwaii marine-use plan due late 2013.

The Haida Gwaii marine planning process is also linked to a larger process for all of northern British Columbia. Canada's Oceans Action Plan (DFO, 2005b) outlined an integrated management framework including five large ocean management areas nationwide to pilot *Oceans Act* intentions of integrated management, while balancing human uses with sustaining marine ecosystems. Representing Pacific Canada is the Pacific North Coast Integrated Management Area (PNCIMA, 2012) whose 102,000 km^2 of sea space envelope Haida Gwaii (Fig. 10.2). A collaborative oceans governance agreement was signed in December 2008 between north-coast First Nations and DFO. The PNCIMA process was launched in a 2009 public forum co-hosted by DFO and First Nations. Regional community meetings were held early in 2010 to introduce the PNCIMA idea and to solicit citizens' opinions. At the first

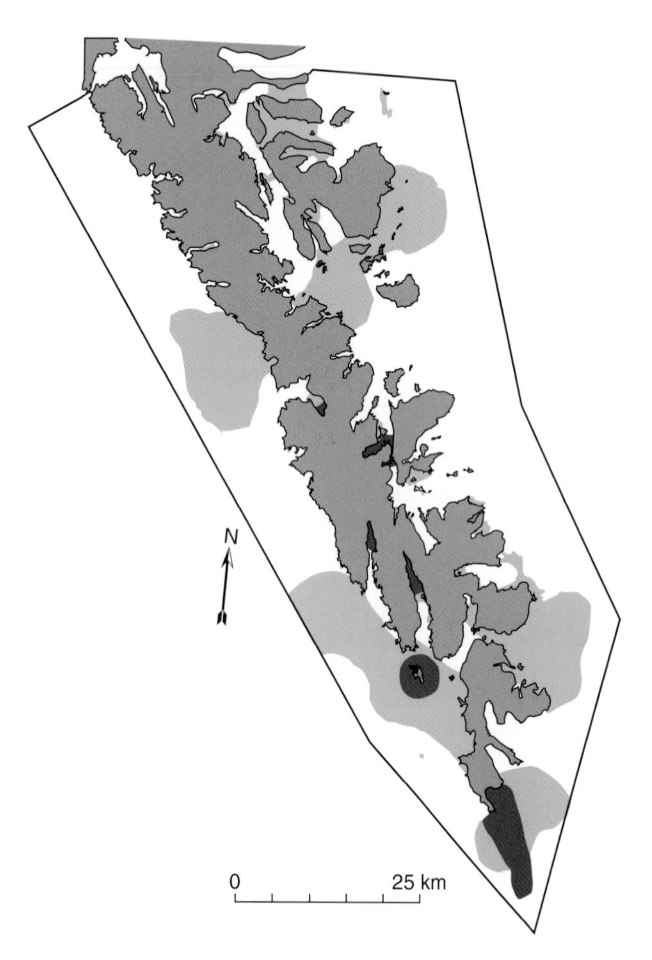

Fig. 10.14 Yellow areas of Gwaii Haanas Marine represent highest biophysical values resulting from the MARXAN analyses. Also shown are purple no-take areas (3% of Gwaii Haanas Marine) from the Interim Management Plan; www.pc.gc.ca/pn-np/bc/gwaiihaanas/index.aspx

meeting of its guiding body in June 2010, a public engagement strategy was launched and a December 2012 deadline was set for PNCIMA's Integrated Oceans Management Plan. The province of British Columbia, the lead agency for coastal lands management, signed the collaborative agreement in September 2010.

In November 2010, the Gordon and Betty Moore Foundation, a U.S. philanthropy, granted the PNCIMA process more than $8 million to be managed at arm's-length from the Foundation by a Canadian NGO (Tides Canada). An agreement was signed in January 2011 by Fisheries and Oceans Canada and north coast First Nations to activate this contribution. However, the Foundation's involvement came under attack as undue foreign influence on Canadian natural-resource development, and lobbying led to DFO announcing in September 2011 that, while retaining its deadline for the final PNCIMA Plan (now late 2013), Canada was withdrawing from the agreement to use Moore funding. Great disappointment was articulated by coastal First Nations at this unilateral decision, viewed as contrary to the PNCIMA's initial ethos of collaborative management. In November 2011 coastal First Nations and the province regrouped (with support of Moore funding) to form the Marine Planning Partnership (MaPP) for the North Pacific Coast to continue north-coast marine planning in parallel to the PNCIMA process, from which First Nations had withdrawn. The MaPP process will provide four sub-regional marine plans, including the one for Haida Gwaii, and an overarching regional plan due by the end of 2013. In June 2012, most of the MaPP First Nations rejoined the PNCIMA process. As of this writing, it remains unclear the extent to which First Nations will embrace the final PNCIMA process outcomes, or how PNCIMA and MaPP outcomes will mesh.

Prospects of energy megaprojects also complicate north-coast regional marine planning. In 2011, approvals (including First Nations support) were in place for a 5 million ton year^{-1} liquefied natural gas tanker terminal in Kitimat (directly across Hecate Strait from Haida Gwaii) that would be linked to a pre-existing gas pipeline. However, there is another Kitimat-based project proposal, the Enbridge Northern Gateway, that includes a new dual pipeline from the province of Alberta to a marine terminal. The terminal would load heavy oil (bitumen from Alberta's tar-sands) into Asia-bound supertankers (some exceeding 300,000 dead weight tons), with an outgoing daily pipeline capacity of 525,000 barrels outgoing and 190,000 barrels of condensate (diluting agent for bitumen) incoming. Proposed shipping lanes include confined mainland inlet waters with a northward option through Dixon Entrance or southward through Hecate Strait and Queen Charlotte Sound. A major public environmental and socio-economic assessment began in January 2012 that will report to the federal government by the end of 2013, after which a political decision will be made. From the outset, oil-spill concerns (*Exxon Valdez* redux) have led all north-coast First Nations and other coastal and inland communities to oppose this project. The current federal government views the project as a strategic national priority.

10.5 CRUCIBLE FOR ECOSYSTEM-BASED MANAGEMENT

Gwaii Haanas Marine presents intriguing prospects, among which realizing the full potential of the Canada-Haida cooperative management partnership is paramount. Besides this nation-to-nation partnership, the uniqueness of Gwaii Haanas Marine's conservation management for Canada is: the first NMCA to be established under the *Canada NMCA Act*, the first federal marine area for which Parks Canada and DFO partner, and the extent to which terrestrial and marine conservation will be integrated.

Gwaii Haanas Marine warrants an adaptive management approach, in essence to be treated as a policy experiment (Fox *et al.*, 2012). Key to success will be the partnership implementing EBM towards desired and measurable ecological, cultural, and economic outcomes based upon spatial delineation of ecosystems and their main properties. In addition to capacity building, the act of cooperative management yields "social capital" (Pretty and Smith, 2004), forging bonds within groups through which people invest in cooperation, knowing that others will also participate. The social capital emanating from cooperative management changes attitudes and expectations, creating further opportunities, and building new capacities. Gwaii Haanas Marine was established at a time of change in northern British Columbia, such as PNCIMA and MaPP, and it could be a reference location for management innovation and an environmental sentinel.

The PNCIMA process issued the following draft definition of EBM in 2011: "An adaptive approach to managing human activities that seeks to ensure the coexistence of healthy, fully functioning ecosystems and human communities." In comparison, the Haida marine-use planning process yielded a highly values-driven definition of EBM (Jones *et al.*, 2010): "Respect is the foundation of ecosystem-based management. It acknowledges that the land, sea, air and all living things, including the human community, are interconnected and that we have the responsibility to sustain and restore balance and harmony." Contemplating action on these definitions comes at a time of explosive growth of the marine EBM literature (e.g., McLeod and Leslie, 2009; Halpern *et al.*, 2010; Tallis *et al.*, 2010) in which there is widespread agreement that implementation will be challenging (Palumbi *et al.*, 2009).

Enabling EBM will be the vision of Gwaii Haanas Marine's future shared with the public. The AMB must first guide, after evaluating input from advisory processes and associated research, the development of ecosystem objectives, and validate indicators as a basis for establishing ecosystem status and trends. Each indicator will have a threshold for an agreed-upon level of well-being. The commitment to monitoring (e.g., Fancy and Bennetts, 2012) will use data in a feedback loop to adjust actions enabling adaptive management. Long-term data sets from this commitment will be a major technical legacy of Gwaii Haanas Marine. Part of the EBM approach is valuing ecosystem services—the resources and processes that nature provides to people. Emphasizing marine ecosystem services, such as provision of fish and a clean, healthy environment, underscores dependence on the sea and fosters social capital towards improved management (Granek *et al.*, 2010).

Underlying Gwaii Haanas Marine's cooperative management will be accommodating DFO's unimpeded *Fisheries Act* mandate over access and allocation, while the AMB ensures that commercial fishing also meets the partnership's overall ecosystem objectives. This compromise links back to the enabling, not prescriptive, nature of the *Canada NMCA Act*, in which agencies' legislated marine mandates remain intact within NMCAs. To add to this complexity, commercial fisheries are highly sectoral, and the AMB must engage each industry sector, with its own annual management plan, separately to achieve EBM. The ecosystem approach to fisheries (Rice, 2011), therefore, may help implement EBM in which separate fishing sectors retain their identities while their mandates and accountabilities are extended. With the AMB holding fisheries accountable for their effects on ecosystems, direct effects are likely manageable, but managing interactions among fisheries and between fisheries and other use sectors will be challenging. This framework of ideas for accommodating fisheries from an ecosystem perspective, and complemented by the parallel progress of socio-economic and cultural objectives, will propel innovation for Gwaii Haanas Marine.

Creating Gwaii Haanas Marine places into focus the values underlying decisions to conserve, restore, and sustainably use marine ecosystems. As people live within and can significantly affect marine ecosystems and species (Jackson *et al.*, 2001), ethical considerations are basic to human-ocean relations (Norton, 2003) and include notions of social justice. Social justice, not just resource allocation, is important to the Haida as they view participating fully in cooperative resource management within their traditional territory as appropriate progression in their nation's development. As part of this development, indigenous traditional ecological knowledge features respect as the correct attitude for both human-environment and human-human relations (Houde, 2007; Berkes, 2012). Moreover, their holistic, ecosystem-based view of human-environmental relations compliments the need to view fisheries as social-ecological enterprises (Bundy *et al.*, 2008). In proposing the Haida marine-use plan, ethics were central with respect (yah'guudang) and responsibility ('laagu-ugakanhlins) as key features underscoring human-ocean reciprocity (Jones and Williams-Davidson, 2000). These are early days for Gwaii Haanas Marine, but social capital has been generated in a cooperative management ethos that will aid EBM and provision of marine ecosystem services. A dramatic looming challenge that will draw upon this social capital will be sea otter restoration (Fig. 10.15).

As a postscript to this case study, it is interesting to reflect that, despite the decades of cooperation to create Gwaii Haanas, Canada, British Columbia, and the Haida Nation still agree to disagree over final ownership of the land and sea. The

Fig. 10.15 A sea otter eating a northern abalone (*Haliotis kamtschatkana*) off north-western Haida Gwaii, July 2013. The fur trade of sea otter pelts flourished until the 1830s and led to the species' extirpation by the early 1900s (Sloan and Dick, 2012). Since 1972, 17 confirmed sightings from Haida Gwaii indicate rapid expansion from their early 1970s reintroduction off northwest Vancouver Island. While sea otters re-establish, cascading ecosystem effects are expected as sea otters eat kelp-grazing red sea urchins (Watson and Estes, 2011), which could threaten the commercial sea urchin fishery. Both sea otters and their abalone prey are legally "listed" species-at-risk; the former special concern; the latter endangered. Thus, prospects could be poor for restoring the traditional Haida food of northern abalone that has been closed to all fishing (since 1990). Islanders will find this situation challenging, as it will compel people to share shellfish resources with these voracious predators (Sloan and Dick, 2012). Photograph © Parks Canada/Stef Olcen.

term "Reserve," used twice in Gwaii Haanas' formal name (Fig. 10.2), means that, although establishment proceeded, final title for lands and seas is unresolved (Langdon *et al.*, 2010). The Haida Nation filed a Statement of Claim in the Supreme Court of British Columbia in 2002 that aims to establish aboriginal rights and title to lands and seas around Haida Gwaii, within the meaning of the Canadian Constitution. This will be among the few aboriginal claims so far in Canada that includes both land and sea. The eventual settlement of maritime rights and title around Haida Gwaii will have important implications for marine management nationwide. Indeed, despite the various cooperative north-coast marine management processes underway, there is a regional backdrop of uncertainty in how First Nations' roles will unfold. This is because most coastal British Columbia First Nations have not yet reconciled land or sea treaty or title with other levels of government (Jones, 2006; Robinson *et al.*, 2007). Thus, Gwaii Haanas Marine is an evolving story in which civility nurtures cooperative management.

REFERENCES

Álvarez-Romero JG, Pressey RL, Ban NC, *et al.* (2011) Integrated land-sea conservation planning: the missing links. *Annual Review of Ecology, Evolution and Systematics* **42**, 381–409.

AMB (Archipelago Management Board) (2003) Gwaii Haanas management plan for the terrestrial area. Archipelago Management Board, Queen Charlotte, B.C.

Archambault P, Snelgrove PVR, Fisher JAD, *et al.* (2010) From sea to sea: Canada's three oceans of biodiversity. *PLoS ONE* **5**(8), e12182.

Ball IR, Possingham HP, Watts M (2009) Marxan and relatives: Software for spatial conservation prioritisation. In *Spatial conservation prioritisation: Quantitative methods and computational tools* (eds Moilanen A, Wilson KA, Possingham HP). Oxford University Press, Oxford, 185–195.

Bartier PM, Sloan NA (2007) Reconciling maps with charts towards harmonizing coastal zone base mapping: a case study from British Columbia. *Journal of Coastal Research* **23**, 75–86.

BCMFLNRO (British Columbia Ministry of Forests, Lands and Natural Resource Operations) (2012) Haida Gwaii Strategic Land Use Agreement implementation. Victoria, B.C. ilmbwww.gov.bc.ca/slrp/lrmp/nanaimo/haidagwaii/index.html (accessed April, 2012).

BCSA (BC Seafood Alliance) (2012) www.bcseafoolalliance.com/—also, BCSA position on marine area conservation: www.bcseafoodalliance.com/BCSA/BCSA_CONSERVATION.html (both accessed April, 2012).

Berkes F (2012) *Sacred ecology*. Third edition. Routledge, New York.

Bundy A, Chuenpagdee R, Jentoft S, *et al.* (2008) If science is not the answer, what is? An alternative governance model for the world's fisheries. *Frontiers in Ecology and Evolution* **6**, 152–155.

C-BC (Canada-British Columbia) (2012) *Canada-British Columbia marine protected area network strategy*; www.mccpacific.org/MPA-DraftCanadaBCStrat-May27.pdf (accessed March, 2013).

CHN (Council of the Haida Nation) (2009) Gaaysiigang—an ocean forum for Haida Gwaii. Haida Fisheries Program, Council of the Haida Nation, Old Massett, B.C. (Gaaysiigang means the-person-who-sizes-up-the-waves-so-the-canoe-can-go-out in Haida) www.haidanation.ca/Pages/Splash/Documents/Gaaysiigang_72.pdf (accessed April, 2012).

Darimont CT, Bryan HM, Carlson SM, *et al.* (2010) Salmon for terrestrial protected areas. *Conservation Letters* **3**, 379–389.

Dearden P, Canessa R (2009) Marine protected areas. In *Parks and protected areas in Canada—Planning and management*, Third edition (eds Dearden P, Rollins R). Oxford University Press, Don Mills, ON, 403–431.

Dearden P, Langdon S (2009) Aboriginal peoples and national parks. In *Parks and protected areas in Canada—Planning and management*. Third edition (eds Dearden P, Rollins R). Oxford University Press, Don Mills, ON, 373–402.

DFO (Fisheries and Oceans Canada) (2002) Canada's oceans strategy. Our oceans: our future – policy and operational framework for integrated management of estuarine, coastal and marine environments in Canada. Fisheries and Oceans Canada, Ottawa. www.dfo-mpo.gc.ca/oceans/publications/cos-soc/index-eng.asp (accessed April, 2012).

DFO (Fisheries and Oceans Canada) (2005a) Federal marine protected areas strategy. Fisheries and Oceans Canada, Ottawa. www.dfo-mpo.gc.ca/oceans/publications/fedmpa-zpmfed/index-eng.asp (accessed April, 2012).

DFO (Fisheries and Oceans Canada) (2005b) Canada's oceans action plan for present and future generations. Fisheries and Oceans Canada, Ottawa. www.dfo-mpo.gc.ca/oceans/publications/oap-pao/index-eng.asp (accessed April, 2012).

DFO (Fisheries and Oceans Canada) (2011) National framework for Canada's network of marine protected areas. Fisheries and Oceans Canada, Ottawa. www.dfo-mpo.gc.ca/oceans/publications/dmpaf-eczpm/docs/framework-cadre2011-eng.pdf (accessed July, 2012).

Fancy SG, Bennetts RE (2012) Institutionalizing an effective long-term monitoring program in the US National Park Service. In *Design and analysis of long-term ecological monitoring studies* (eds Gitzen RA, Millspaugh JJ, Cooper AB, *et al.*). Cambridge University Press, New York, 481–497.

Fedje DW, Mathewes RM, eds (2005) *Haida Gwaii: human history and environment from the time of Loon to the time of the Iron People*. University of British Columbia Press, Vancouver.

Fox HE, Mascia MB, Barsuto X, *et al.* (2012) Reexamining the science of marine protected areas: linking knowledge to action. *Conservation Letters* **5**, 1–10.

Gaston AJ, Golumbia TE, Martin J-L, *et al.*, eds (2008) *Lessons from the islands—Introduced species and what they tell us about how ecosystems work*. Proceedings from the Research Group on Introduced Species 2002 Symposium. Queen Charlotte, B.C. Special Publication, Canadian Wildlife Service, Environment Canada, Ottawa.

Granek EF, Polasky S, Kappel CV, *et al.* (2010) Ecosystem services as a common language for coastal ecosystem-based management. *Conservation Biology* **24**, 207–216.

Haida Laas (newsletter of the Haida Nation) (July 2012) www.haidanation.ca/Pages/Haida_Laas/Haida_Laas.htm (accessed July, 2012).

Halpern BS, Lester SE, McLeod KL (2010) Placing marine protected areas into the ecosystem-based management seascape. *Proceedings of the National Academy of Science* **107**, 18312–18317.

Houde N (2007) The six faces of traditional ecological knowledge: challenges and opportunities for Canadian co-management arrangements. *Ecology and Society* **12**(2), 34. www.ecologyandsociety.org/vol12/iss2/art34/

Jackson JBC, Kirby MX, Berger WH, *et al.* (2001) Historical overfishing and the recent collapse of coastal ecosystems. *Science* **293**, 629–639.

Jessen S (2011) A review of Canada's implementation of the Oceans Act since 1997 – from leader to follower? *Coastal Management* **39**, 20–56.

Jessen S, Chan K, Coté I, *et al.* (2011) *Science-based guidelines for Marine Protected Areas and MPA networks in Canada*. Canadian Parks and Wilderness Society, Vancouver. cpaws.org/uploads/mpa_guidelines.pdf (accessed April, 2012).

Jones R (2006) Canada's seas and her First Nations. In *Towards principled oceans governance – Australian and Canadian approaches and challenges* (eds Rothwell DR, VanderZwaag DL). Routledge, London, 299–314.

Jones R, Rigg C, Lee L (2010) Haida marine planning: First Nations as a partner in marine conservation. *Ecology and Society* **15**(1), 12. www.ecologyandsociety.org/vol15/iss1/art12/ (accessed April, 2012).

Jones R, Williams-Davidson T-L (2000) Applying Haida ethics in today's fishery. In *Just fish—ethics and Canadian marine fisheries* (eds Coward H, Ommer R, Pitcher T). Institute of Social and Economic Research, Memorial University of Newfoundland, St. John's, NL, 100–115.

Langdon S, Prosper R, Gagnon N (2010) Two paths one direction: Parks Canada and aboriginal peoples working together. *George Wright Forum* **27**, 222–233.

McLeod K, Leslie H (2009) *Ecosystem-based management for the oceans*. Island Press, Washington, D.C.

Mooers AØ, Prugh LR, Festa-Bianchet M, *et al.* (2007) Biases in legal listing under Canadian endangered species legislation. *Conservation Biology* **21**, 572–575.

Norton BG (2003) Marine environmental ethics—where we might start. In *Values at sea – ethics for the marine environment* (ed. Dallmeyer DG). University of Georgia Press, Athens, GA, 33–49.

OAGC (Office of the Auditor General of Canada) (2005) *Report of the Commissioner of the Environment and Sustainable Development to the House of Commons*. Chapter 1. Fisheries and Oceans Canada—Canada's Oceans Management Strategy. Public Works and Government Services Canada, Ottawa, ON. www.oag-bvg.gc.ca/internet/English/att_c20050900xe03_e_14094.html (accessed April, 2012).

OAGC (2012) *Report of the Commissioner of the Environment and Sustainable Development to the House of Commons*. Chapter 3. Marine Protected Areas. Public Works and Government Services Canada, Ottawa, ON. www.oag-bvg.gc.ca/internet/index.htm (accessed March, 2013).

Orchard TJ (2009) Otters and urchins: continuity and change in Haida economy during the late Holocene and Maritime Fur Trade periods. *British Archaeological Reports International Series* **2027**, Oxford.

Palumbi SR, Sandifer PA, Allan JD, *et al.* (2009) Managing for ocean biodiversity to sustain marine ecosystem services. *Frontiers in Ecology and the Environment* **7**, 204–211.

PNCIMA (Pacific North Coast Integrated Management Area) (2012) www.pncima.org (accessed April, 2012).

Pretty J, Smith D (2004) Social capital in biodiversity conservation and management. *Conservation Biology* **18**, 631–638.

Rice J (2011) Managing fisheries well: developing the promise of an ecosystem approach. *Fish and Fisheries* **12**, 209–231.

Ricketts PJ, Hildebrand L (2011) Coastal and ocean management in Canada: progress or paralysis? *Coastal Management* **39**, 4–19.

Robb CK, Bodtker KM, Wright K, *et al.* (2011) Commercial fisheries closures in marine protected areas on Canada's Pacific coast: the exception, not the rule. *Marine Policy* **35**, 309–316.

Roberts C (2007) *The unnatural history of the sea*. Island Press, Washington, D.C.

Roberts C (2012) *The ocean of life—the fate of man and the sea*. Viking, New York.

Robinson JL, Tindall DB, Seldat E, *et al.* (2007) Support for First Nations' land claims amongst members of the wilderness preservation movement: the potential for an environmental justice movement in British Columbia. *Local Environment* **12**, 579–598.

Samhouri JF, Levin PS (2012) Linking land- and sea-based activities to risk in coastal ecosystems. *Biological Conservation* **145**, 118–129.

Scholz AJ, Steinback C, Kruse SA, *et al.* (2011) Incorporation of spatial and economic analyses of human-use data in the design of marine protected areas. *Conservation Biology* **25**, 485–492.

Sloan NA, ed. (2006) Living marine legacy of Gwaii Haanas. V. Coastal zone values and management around Haida Gwaii. *Parks Canada Technical Reports in Ecosystem Science* **42**, Ottawa, 1–413.

Sloan NA, Bartier PM (2009) Historic marine invertebrate species inventory: case study of a science baseline towards establishing a marine conservation area. *Aquatic Conservation: Marine and Freshwater Ecosystems* **19**, 827–837.

Sloan NA, Dick L (2012) *Sea otters of Haida Gwaii: Icons in human-ocean relations*. Haida Gwaii Museum Press, Skidegate, B.C.

Smith ADM, Brown CJ, Bulman CM, *et al.* (2011) Impacts of fishing low-trophic level species on marine ecosystems. *Science* **333**, 1147–1150.

Tallis H, Levin PS, Ruckelshaus M, *et al.* (2010) The many faces of ecosystem-based management: making the process work today in real places. *Marine Policy* **34**, 340–348.

Tourtellot JB (2005) Traveler special report—destinations. *National Geographic Traveler* **22**(5), 80–92.

Turner NJ (2004) *Plants of Haida Gwaii*. Sono Nis Press, Winlaw, B.C.

Watson JC, Estes JA (2011) Stability, resilience and phase shifts in rocky subtidal communities along the west coast of Vancouver Island, Canada. *Ecological Monographs* **81**, 215–239.

Willson MF, Womble JN (2006) Vertebrate exploitation of pulsed marine prey: a review and the example of spawning herring. *Reviews of Fish Biology and Fisheries* **16**, 183–200.

CHAPTER 11

SOUTH AFRICA: COASTAL-MARINE CONSERVATION AND RESOURCE MANAGEMENT IN A DYNAMIC SOCIO-POLITICAL ENVIRONMENT

Barry Clark

Zoology Department, University of Cape Town, South Africa

Allan Heydorn

[Retired CEO, World Wildlife Fund, South Africa], Stellenbosch, South Africa

This Cape is a most stately thing and the fairest Cape we saw in the whole circumference of the Earth.

Sir Frances Drake upon sailing around the Cape Peninsula in the 16th century

11.1 A CHALLENGE FOR GOVERNANCE

South Africa's history of coastal development and marine resource exploitation highlights key successes and failures that challenge management for sustainable use. Easily accessible nearshore waters are frequently subjected to citizen demands in excess of what available natural resources can sustain. These demands for resources take many forms, including degradation of water catchments and consequent damage to estuary and river mouths, pollution, overfishing, and in some areas along the South African coast, blatant poaching. The effects of global climate change in the form of increasing sea-surface temperatures, altered and more variable weather patterns, rising sea levels, and increasing frequency and/or intensity of storm events are likely to exacerbate these concerns. But irrespective of which factors might be at play, sustainability of marine resources (a scientific/conservation recognition) and equitable access to them (a social/governance issue) remain of cardinal importance.

Loss of sustainability for key fisheries is obviously a huge social detriment, whether for commercial, recreational, or subsistence purposes, and hits hardest in the poorest communities, where fishing might be the only means of putting food on the table. Mistakes and inequalities in coastal and fisheries management that occurred under various government regimes intensified most unfortunately during the latter half of the 20th century when "apartheid"—a system of segregation and discrimination on grounds of race—became official national policy. In the early 20th century, the rights and privileges of people from different ethnic and racial origins began to be distinguished under law, and racially based policies denied entire sectors of society fair access to large sectors of the coast and its resources. This national policy of discrimination was revised dramatically with the emergence of the new South Africa in 1994, which brought a change of government that aimed to guarantee all South Africans equal rights previously held only by the white minority—essentially free movement, a healthy environment, and public participation in the political process. Social and political attempts have since been made to rectify injustices in terms of developing skills and ensuring equitable access to the coast and its resources. However, equity and sustainable development are still difficult to achieve for various reasons, including lack of competence due to inadequate training of officials at various levels since the change in government in 1994. Unemployment and poverty remain serious problems, forcing many inhabitants in disadvantaged settlements to turn to crime as a means of survival; it is from such communities that people are drawn into the illegal poaching industry.

Marine Conservation: Science, Policy, and Management, First Edition. G. Carleton Ray and Jerry McCormick-Ray.

While much research into the extent and sustainability of renewable marine resource harvesting was encouraged under the former South Africa regime, the management approach was bureaucratic and authoritative. Understandably, this left much bitterness and disillusionment in coastal communities dependent on fishing. Serious attempts have since been initiated, following the democratic elections of 1994, to implement more equitable and participatory policies, including a new Constitution and new environmental and fisheries policies spelling out frameworks for rights allocation and management of the coast. Hence, the environmental policies and legislation now in place are often cited as being among the world's most progressive. Nevertheless, effective implementation is fraught with difficulties, especially in attempts to meet demands of a large sector of poverty-stricken and unemployed people—demands that are often cast in socio-political rhetoric.

It is in this context of social and political change that South Africa embraced the biosphere-reserve concept as advocated by UNESCO and the IUCN, showcased here through the internationally recognized Kogelberg Biosphere Reserve (KBR) established in 1998. The intention was to resolve problems related to coastal habitats and their biota, which were (and still are) under severe threat in the KBR. Key threats include poorly regulated coastal development and resource exploitation, as well as the intensive, internationally orchestrated poaching for a marine mollusk—the abalone (perlemoen, *Haliotis midae*)—and to a lesser extent the rock lobster (*Jasus lalandii*). Overall, the major problem is one of ineffective governance, exacerbated by inadequate investment in staff and infrastructure at various governmental levels and by differences in interpretation over biosphere-reserve requirements for conservation and development. This situation has prompted hitherto unprecedented government and private-sector collaboration designed to improve management of the coast, including compilation and implementation of an Integrated Coastal Management Plan, more effective measures to establish Marine Protected Areas with "no-take" zones, and a proactive approach to curb poaching (Sections 11.5, 11.6). These initiatives have been subjected to rigorous debate and are not yet resolved, but hopefully will contribute to improved conservation and management strategies in future years, not just in the Kogelberg region but also elsewhere in South Africa.

11.2 SOUTH AFRICA'S COASTAL REALM: PHYSICAL, BIOTIC, AND HUMAN SETTING

The total length of South Africa's coastline is 3000 km, representing 10% of Africa's coastline. South Africa's Southern Hemisphere location in temperate to subtropical regions, adjacent to oceanic water masses, is a significant determinant of its coastal climate and its uniquely biodiverse and productive coastal system.

11.2.1 Physical-biotic characteristics

As described in more detail in Heydorn and Flemming (1983), South Africa's coast consists of four physically and biogeographically distinct ecoregions (Fig. 11.1). On either side of the southern African subcontinent, atmospheric influences are dominated by semi-permanent high-pressure systems—the South Atlantic anticyclone in the west and the South Indian anticyclone in the east—and by perturbations in the subtropical easterlies in the north and temperate westerlies in the south. Southerly seasonal displacement of these major pressure systems in summer and north in winter moderates the extent of these perturbations, establishing marked seasonal differences in climate across much of the country, particularly for the south and west coasts. Rainfall is a predominantly summer phenomenon on the east coast, bimodal along the south coast, and a predominantly winter phenomenon on the southwest and west coasts. The amount of rainfall received in these regions also differs markedly, e.g., Port Nolloth (west coast) and Durban (east coast) are located at similar latitudes but receive on average 50 vs. 760 mm per year, respectively. This is largely due to the influence of the warm, fast-flowing Agulhas Current that originates in the tropics and flows down the east coast, and the sluggish, northward-flowing cold Benguela Current on the west coast, where associated upwelling of cold water of Antarctic origin regularly occurs (Figs 11.1, 11.2). Tides are semi-diurnal, with a moderate difference between low and high tide of 1.6 m for spring tides and 0.5 m at neap.

Coastal-realm sectors are determined by a combination of climatic and geologic factors. Typical of the east coast are subtropical conditions, mangrove-lined estuaries, and beaches backed by dense dune forests. Sugarcane farming on floodplains of the lower reaches of rivers (Fig. 11.3) has resulted in serious erosion of riverbanks. The sediment thus released has in many cases blocked estuary mouths, damaged river banks, and has inhibited the utilization of estuarine feeding and breeding habitats by fish, prawns, and other marine life (Heydorn, 1989). The south coast region between Cape Padrone and Cape Agulhas is characterized by a varied topography and is spectacularly beautiful, with rugged coastal cliffs, extensive sandy beaches, deeply incised river mouths and wide, mangrove-lined estuaries. In the western portion of this region, a calcareous "karst" substrate holds valuable groundwater reserves (Heydorn and Tinley, 1980). Cape Agulhas, at this coast's western extremity, is Africa's southernmost point, but it is Cape Point (Cape of Good Hope) to the west—discovered by the Portuguese explorer Bartolomeu Dias in 1488—which is widely regarded as being the dividing point between the Atlantic and Indian oceans. There, powerful current systems generate high-energy climatic events, causing early sailors who had to endure northwesterly and southeasterly gales in wooden sailing vessels to poignantly name this region the "Cape of Storms." The southwest coast between Cape Agulhas and the Berg River is scenically spectacular, dominated by granite intrusives overlain by quartzitic and deeply folded sandstones, and contains extensive beaches and small estuaries. In contrast, the west coast between the Berg and Orange Rivers becomes progressively more arid and takes on semi-desert conditions northward; this area is rich in minerals, for copper mining inland, dredging for diamonds on the coast, and diving for them in shallow nearshore waters.

The dominant shoreline vegetation reflects South Africa's variable climatic and geomorphological coastal environment.

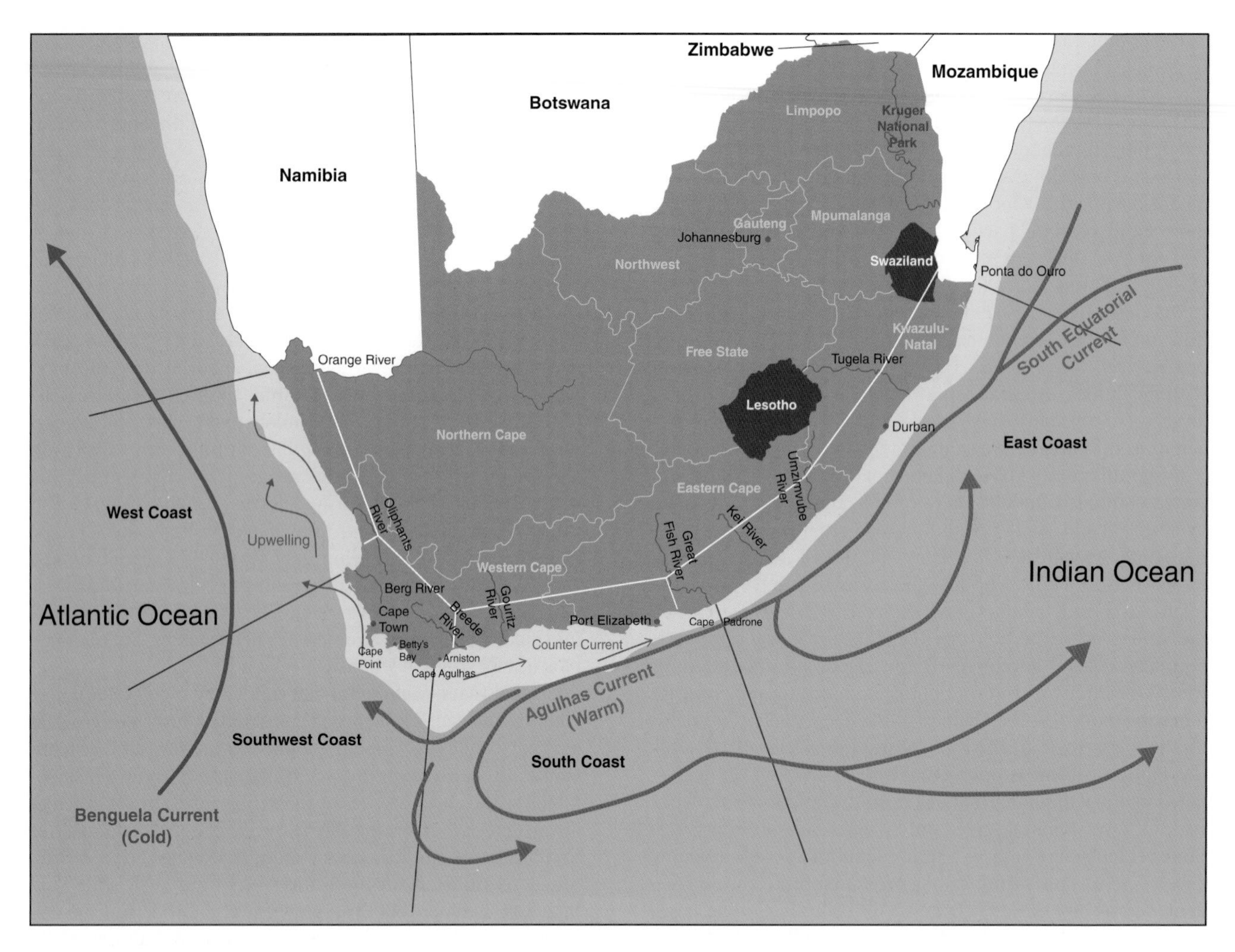

Fig. 11.1 Map of southern Africa showing key features mentioned in the text. Not all locations mentioned in text could be identified on this map; for those, consult atlases.

The east coast is characterized by subtropical vegetation, which is replaced on the south coast by grasslands that reach almost to the sea. Pioneer dune plants (e.g., *Scaevola thunbergii*, *Ipomoea pes-caprae*) are associated with warm Agulhas Current waters. These plants are replaced westward by coastal thickets and Afro-montane forests. Farther south, steep cliffs distinguish a raised coastal platform some 150–250 m above the sea, where drainage systems and river gorges lined by Afro-montane forests deeply incise the shore. Estuaries are prominent features. Mediterranean-type, fine-leaved "fynbos" vegetation characterizes the southwest coast, constituting the globally unique Cape Floral Kingdom (Box 11.1, Fig. B11.1.1), a portion of which is the Kogelberg Biosphere Reserve (see Section 11.5). By contrast, the arid west coast is characterized by hardy, shrub-like plant communities known as "strandveld."

In the nearshore marine environment, corals with a varied associated fauna of fishes, crustaceans, and mollusks are typical of the Agulhas Current system in the northeastern subtropical sector, while kelp communities (e.g., *Ecklonia maxima*, *Laminaria schinzii*) are dominant farther south and in the west, where upwelling of cold water of Antarctic origin occurs regularly within the realm of the Benguela Current. Kelp beds provide habitat for a less varied but abundant fauna, rendering commercial exploitation of organisms such as rock lobster, abalone, and linefish feasible. In deeper water beyond the kelp beds and beyond the narrow continental shelf of the west coast, pelagic and benthic fish support a substantial commercial fishery. Thus, in general, east coast waters are characterized by great biotic diversity, in contrast to the more productive nutrient-rich waters of the west and south coasts with fewer species in great abundance.

These distinct biogeographic zones are separated by indistinct physiographic boundaries equivalent to ecotones. The rate of species replacement with distance is generally low within the biogeographic zones, but is much higher at the boundaries between these areas, a feature with important consequences for establishing protected-area boundaries.

11.2.2 Cultural differences

Topographic and climatic features play key roles in human activities, and socio-political factors have profound influences

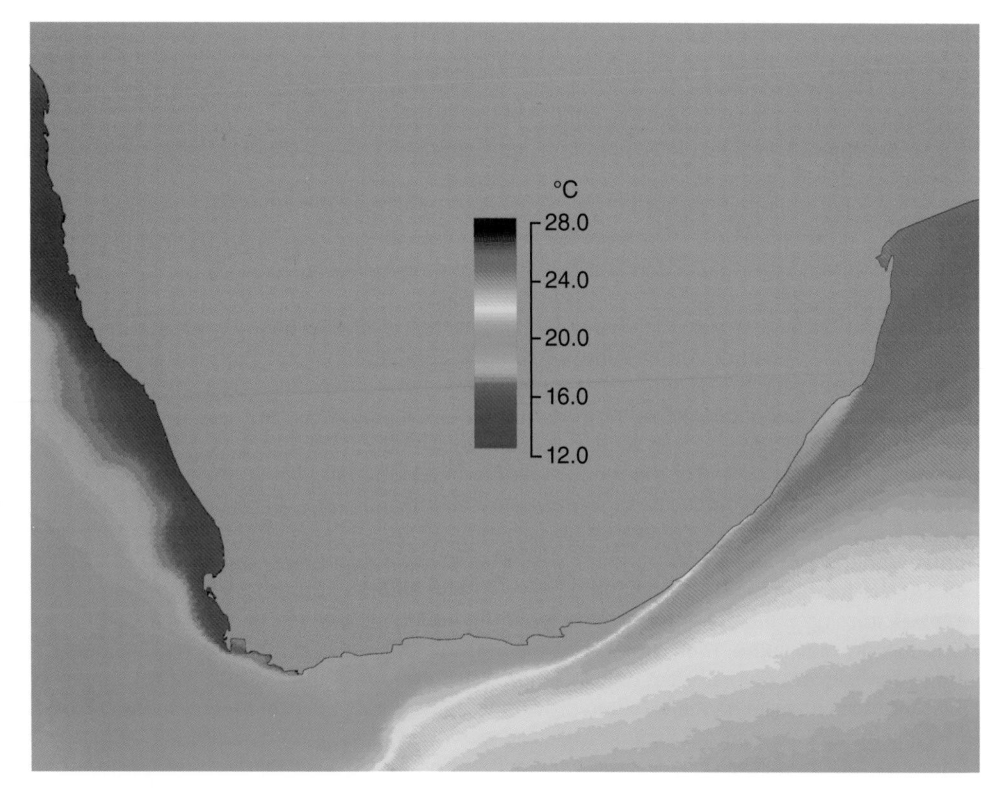

Fig. 11.2 Long-term (nine-year) averaged sea surface temperature based on Aqua MODIS satellite data, 4 km^2 resolution). Modified from Anchor Environmental Consultants 2012. Strategic Environmental Assessment to guide identification of potential marine aquaculture development zones for fin fish cage culture in South Africa. Unpublished report prepared for Directorate Sustainable Aquaculture Management: Aquaculture Animal Health and Environmental Interactions, Department of Agriculture, Forestry and Fisheries, South Africa.

Fig. 11.3 Tugela River estuary mouth. Extensive sugar cane lands are the lighter green vegetation. Sugar cane lands located in floodplains affect riverine vegetation and cause bank erosion, turbidity in lower river reaches, formation of sandbanks, and nearshore sediment deposition. These changes disrupt the natural interaction between river and sea and deprive fish and prawns of feeding and nursery grounds in the sheltered estuarine environment. Photograph © G.C.Ray.

on the utilization and sustainability of the coastal environment. The north portion of the east coast is known as KwaZulu-Natal, a popular tourism destination typified by cultural diversity, with beaches backed by dense dune forests that attract holiday-makers, surfers, and anglers from far and wide. Warm seas support a high diversity of marine life (corals, crustaceans, fish, sharks, and marine mammals). The metropolitan center of Durban, South Africa's third-largest city and busiest port, has high social and economic importance. Sugar cane farming is of substantial economic importance, but as

Box 11.1 The Kogelberg—heart of the Cape Floral Kingdom

Brian J. Huntley
Retired Director, Kirstenbosch National Botanical Garden, Capetown, South Africa

The great Swede, Carolus Linnaeus, writing in the 1770s in response to the endless stream of botanical marvels arriving in Uppsala from South Africa, described the Cape as "that Paradise on Earth, the Cape of Good Hope, which the beneficent Creator has enriched with his choicest wonders." Linnaeus was right. Today, over 20,456 higher plant species have been catalogued from South Africa, 9500 species of these in what became known as the Cape Floral Kingdom—the Cape Floristic Region (CFR) of recent literature (Fig. B11.1.1). With some 9500 species in 83,000 km^2, it is the smallest by far of the world's six Floral Kingdoms, just 1/500th of the area, for example, of the Holarctic Kingdom, which embraces most of Eurasia and North America. It has no fewer than eight endemic or near-endemic families. Over 20% of its nearly 1000 genera, and over 80% of its species, are endemic—found nowhere else on the planet.

At the heart of "that Paradise on Earth" we find the Kogelberg Biosphere Reserve. Part of a cluster of protected areas forming the Cape Floristic Region World Heritage Site (inscribed by UNESCO in 2002), the Kogelberg, of just 1000 km^2, has some 1880 species of plants, 132 of these endemic to the Reserve, and representing the hottest hotspot of plant diversity not only in Africa, but throughout the extra-tropical world.

The Kogelberg Biosphere Reserve owes its rich biodiversity to several factors, chief of these being its dramatic landscapes with high mountains sweeping down into the sea, coastal dunefields, wetlands, bogs, and patches of forest in deep valleys. The ancient sandstones and quartzites that were thrust up and folded to form its mountains some 250 million years ago are extremely poor in nutrients, especially nitrogen and phosphorus. Survival in such poor, acidic sands has led to complex adaptations in root structure and nutrient sequestration: i.e., eco-physiological diversity that has resulted in floristic diversity.

The mountains, which rise from the sea at Kogelbaai to 1250 m within 5 km of the sea (the steepest gradient anywhere on the African coast), catch the southeasterly winds of summer, bringing fog and rain and the flowering of ericas and proteas to the high peaks at the driest period of the region's Mediterranean-type climate. Altitudinal migration by sunbirds, some endemic to the area, ensures an all-year food source for these nectar-feeders, critical in the long, hot, and dry summers.

The evolution of the Cape's species richness has a long history, but with a recent rush of speciation. Old families such as Proteaceae, Restionaceae, and Bruniaceae provide traces of the ancient, tectonic, Gondwana origins of some of the flora, but most of the radiation of specious genera such as Ericaceae (with 760 of the world's 820

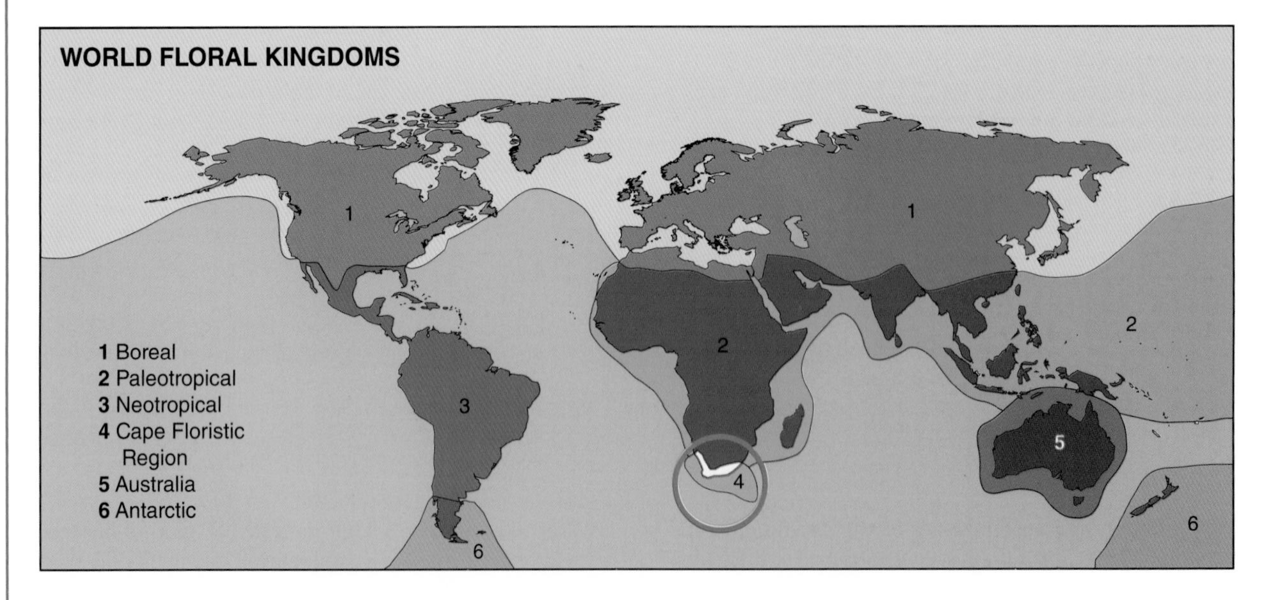

Fig. B11.1.1 World Floral Kingdoms. From Hyman (2006).

species in the Cape), *Aspalathus* with 245 species, *Ruschia* with 138 species, etc., has occurred since the onset of the Mediterranean winter-rainfall climate about three million years ago, with a final burst of speciation through the Pleistocene. Many endemics of the high mountains survived Pleistocene climatic oscillations by being able to migrate up and down the mountain slopes.

The vegetation of the Kogelberg is predominantly fynbos (fine-bush)—a Mediterranean-climate shrubland characterized by small-leaved, sclerophyllous species of three main plant families, Proteaceae, Ericaceae, and Restionaceae. The fynbos biome, defined on plant growth form and physiognomy, is roughly co-incident with the Cape Floristic Region—defined on the basis of the geographic distributions of families and species. Fynbos plants have long-lived, hard leaves, where carbohydrates are concentrated and the few available nutrients are tightly bound. In the low-nitrogen fynbos environment, leaf loss through predation is avoided by the production of poisonous secondary compounds such as tannins. The decomposition of these leaves gives the waters of Kogelberg streams a characteristic "tea" coloration, and coastal lakes a deep black gloominess.

Unique among the world's Mediterranean-climate ecosystems (of California, Chile, Australia, South Africa, and the Mediterranean Basin), fynbos lacks large trees. Short shrublands and heaths clothe the landscape, subject to recurrent and often scorching fires, at 10 to 25-year intervals. High winds, shallow soils, and fire reduce tree success, but have not prevented the survival of small patches of cool, temperate forests in sheltered valleys. These Afromontane forests link the Cape to similar forests along a chain of forested "islands"—notably the equatorial mountains of Kilimanjaro, Kenya, and Ruwenzori—where seed dispersal by frugivorous birds has maintained a remarkably homogenous forest flora. Only 32 tree species occur in Kogelberg forests, compared with over 1850 species in its fynbos shrublands. A third vegetation type "strandveld" (dune scrub) forms a narrow belt along the Kogelberg coast. This vegetation is adapted to extreme soil conditions—calcareous, hydrophobic sands—that are toxic to the growth of fynbos species, abundant just meters away on acid sands. Some strandveld species, such as *Chrysanthemoides monilifera*—have become major invasive pests in the dunes of the Australian coast, just as Australia's *Acacia cyclops* has become a plague on the Kogelberg coast.

The fascination of Kogelberg vegetation lies not only in its extreme diversity and the beauty of its flowers—proteas, ericas, gladioli, orchids, daisies, etc.—but also in the adaptations and interactions that they have evolved to survive challenging conditions. The transfer of nutrients from leached acid soils is facilitated by fine but abundant clouds of root hairs of specialized root systems of more than 20% of fynbos species, which are special adaptations related to fire. Bulb plants (geophytes) constitute 15% of the Kogelberg flora, their underground organs surviving fires, augmented by rapid flowering and seed-set before other plants shade them out in the spurt of growth that follows fire. The seeds of many protea species are equipped with fat-rich deposits on the seed surface that attracts ants to collect them and bury them under the soil, out of the fire's lethal heat pulse. Other proteas store their seeds in hard woody cones, which only release seeds after the passage of fires, usually in later summer, falling to the ground in the moist winter and autumn, an adaptation common in many Mediterranean ecosystems. The list of adaptations and interactions goes on—buzz-pollination by carpenter bees, pollination of long-tubed iris species by long-tongued flies, sweetly scented proteas pollinated by mice, bright red orchids attracting an endemic butterfly—enough to fascinate any visitor to the Kogelberg, making it a naturalist's paradise while its spectacular coastal scenery excites even the most casual tourist.

But all is not well in Eden. Invasion by non-native species of *Acacia* on the lowlands, and *Hakea* and *Pinus* in the mountains not only shade out natural fynbos plants, threatening many species with extinction, but, because of their much higher water use, result in desiccation of aquifers—a major concern in the Kogelberg's mountain catchments. Habitat loss, land transformation, invasive species, and over-harvesting are key threats to the survival of the rich flora of the Cape. A recent comprehensive analysis of South Africa's 20,456 plant species has indicated that 1831 fynbos species, 71% of species listed in South Africa's *Red Data Book*, are threatened with extinction. The importance of the Kogelberg Biosphere Reserve, offering long-term security for 1880 species of plants, many endemic animals, and an extensive marine environment, cannot be too strongly championed.

mentioned above (see Section 11.2.1), has deleterious consequences. Extensive trawling for fish and prawns occurs offshore in continental-shelf waters.

The east coast's southern portion is generally referred to as the "wild coast." The climate is subtropical, large-scale commercial farming is less prevalent, and the predominating local population of indigenous Xhosa relies heavily on cattle, subsistence crops, and harvesting seafood. Mussels and other intertidal organisms have traditional, economic, and subsistence importance, but their exploitation is frequently carried out indiscriminately with spades and crowbars, which injures the inshore ecosystem and the sustainability of important food sources.

Farther south, larger-scale cattle farming and forestry represent key industries, and extensive beaches and dune environments attract tourism. South Africa's second-oldest city, Port Elizabeth, renamed the Nelson Mandela Bay Metropolitan Area, is a major urban and port center. Farther westward, an area aptly named the "Garden Route," characterized by numerous coastal lakes, indigenous forests, and steep coastal cliffs, attracts local to international holiday-makers. Here, forestry remains an important industry. Even farther to the west is the region known as the "Overberg," where the coast begins to slope more gently towards the sea and a calcareous "karst" substrate holds valuable groundwater reserves; Cape Agulhas is the Overberg's most notable coastal feature. Extensive grain, sheep, and cattle farming are economic mainstays. Line fishing from boats and shore angling (see Section 11.3.2.2) are important to local people, and squid (*Loligo reynaudii*) fishing supports a significant commercial industry.

Cape Town (population approximately four million) is South Africa's oldest city and the economic epicenter of the southwest coast. This area supports diverse industries, including tourism, viticulture, and wheat and livestock farming that provide employment and are major attractors for job-seekers, as well as for coastal development for second homes and retirement. Industrial purse-net fishing for pelagic species such as pilchard (*Sardinops ocellata*), anchovy (*Engraulis japonicus*), and maasbanker (*Trachurus capensis*) is of considerable economic value, as is industrial deep-sea trawling for hake (*Merluccius capensis*) and other demersal and benthic organisms. Sport fishing is popular, but as already mentioned, poaching of abalone and rock lobster constitutes a major social problem.

On the west coast, the climate becomes progressively more arid northward. A small number of rivers bisect the coast and are centers of development, but are now suffering from diminishing flows due to excessive water extraction from catchments. Wine is produced in the south, and sheep, wheat, and vegetable farming are widely practiced. Spectacular displays of wild flowers in spring attract visitors, concurrently with those who visit the Garden Route. Iron ore deposits are mined on a large scale inland, and diamond-mining concessions cover much of the coastline, extending from several kilometers inland to 200 m depth offshore. Pelagic fishing, primarily for anchovy, has been a major commercial enterprise that has deteriorated significantly in recent times, probably from overexploitation exacerbated by changing oceanographic and climatic conditions.

11.3 MAJOR CONSERVATION ISSUES OF SOUTH AFRICAN COASTS

Human impacts on the coast differ in their nature and relative importance in different regions due to prevailing legislative and policy frameworks, stages of development, geological, geomorphological, climatic influences, and according to socio-economic, political, and ecological factors.

11.3.1 Population growth and coastal development

Human population densities along South African coasts tend to be much higher than in inland areas; approximately 40% of South Africa's population lives within 100 km of the coast. Principal reasons are socio-economic: quest for employment, basic livelihoods, the appeal of living near the coast, tourism, retirement opportunities, and availability of holiday homes, among others. Cumulative effects have led to over-loading of the carrying capacity of many coastal towns and, consequently, inadequacies in the provision of essential services such as water, electricity, sanitation, and health services. The United Nations estimates that of 45 million inhabitants of South Africa, 50% live in urban areas and 34% are less than 15 years of age (UN, 2011). Whereas the 1996 and 2010 census data reflect relatively small changes in population densities within municipalities, population growth and increased development continue to pose severe threats to resources in urban nodes. A prominent component of coastal development is new construction and expansion or upgrade of existing harbors and ports. South Africa currently hosts eight large commercial ports, the most westerly being Saldanha Bay and most easterly, Richards Bay. South Africa's newest, the Port of Ngqura immediately northeast of Port Elizabeth, completed in 2005, is a deepwater port with the longest breakwaters in South Africa. Such developments are attractive to people seeking employment and opportunities, but are highly disruptive to coastal ecosystems.

Modification of coastal ecosystems is especially damaging to estuaries. South Africa is well endowed with these highly productive systems (ca. 250, Whitfield, 2000), which include many species of fauna and flora found nowhere else. Estuaries are especially important as feeding and nursery grounds, as they contain much of the only suitably sheltered habitats along South Africa's highly exposed, almost linear coast (Beckley, 1984; Field and Griffiths, 1991). The contribution of estuaries to the national economy in terms of fisheries alone has been estimated to be on the order of R 952 million (ca. US$136 million) in 1997 (Lamberth and Turpie, 2003), which is highly significant when measured against the total estimated value of inshore fisheries (R 2441 million), and even more so when one considers that the contribution of fisheries represents only a fraction of the full value of wildlife habitat and scenic beauty for tourism.

Recent countrywide assessments of estuarine health have indicated that the majority of estuaries have been degraded to some extent, and that many are in a very poor state (Heydorn, 1986; Whitfield, 2000; Harrison *et al.*, 1994 and Colloty *et al.*, 2001). Arguably, the greatest threat is the abstraction of water from catchments that feed these systems. Reduction in flow, particularly in frequency or intensity of flooding, has several major consequences, including changes in erosional capacity and other sedimentary processes, altered depth profiles and mouth configurations, duration of open phases, and modification of tidal prisms. Changes in flow may also be accompanied by changes in nutrient levels, suspended particulate matter, temperature, conductivity, dissolved oxygen, and turbidity (Drinkwater and Frank, 1994), all of which play a role in structuring biological communities. The effects of freshwater deprivation vary according to estuary type and locality. Estuaries in regions of low rainfall and high evaporation rates tend to be more severely affected by a reduction in freshwater runoff than those in high rainfall regions. Hypersaline conditions (salinity >40 ppt) develop when evaporative water loss exceeds freshwater inflow, which usually occurs seasonally or in closed estuaries and is not uncommon in the upper reaches of permanently open estuaries. Hypersalinity is often accompanied by stagnant water unsuitable as habitat by most species. Furthermore, as the concentration of nutrients increases, dissolved oxygen levels decrease, further reducing estuarine capacity to support biological communities. Mass mortalities often result.

11.3.2 Extraction of natural resources

11.3.2.1 Non-living resources

South Africa's marine and coastal environment is mined in the northeast for heavy metals (titanium and zirconium), in the south for fossil fuel (gas and oil), and in the northwest for diamonds. Unavoidable consequences of mining are altered sediment transport, coastal erosion, and disruption of sedimentary and rocky groundwater-holding structures, all being detrimental to biotic communities and/or agricultural activities. Damage may range from extensive in the case of titanium mining to limited in cases where mining occurs only on a small scale, e.g., quarrying of building sand. Proposals to implement hydraulic fracturing ("fracking") for gas extraction from deep-lying shale formations are intensively controversial in South Africa at present. Fracking involves drilling long (3–5 km) core tunnels into subsurface sediment and rock structures, which are subjected to underground explosions to release gas. This activity can be particularly harmful, especially to precious groundwater reserves.

Mining has negative impacts on surface and subsurface water systems, particularly on wetlands and associated biological communities. New legislation governing mining in South Africa (*Mineral and Petroleum Resources Development Act*, 2002) requires coastal mining operations to rehabilitate mined areas. These areas may recover within a few years if rehabilitation is conducted in accordance with properly planned environmental procedures, which is often not the case. Coastal dune areas in northeast KwaZulu-Natal have been mined for heavy metals and successfully rehabilitated, but surrounding wetlands, estuaries, and water supplies have suffered long-term damage.

Inshore and offshore diamond mining involves the extraction and resuspension of benthic sediment, creating sediment plumes that carry heavy metals that may reach toxic concentrations (Lane and Carter, 1999), or that may settle on reefs and rocky shores to suffocate organisms that live there (Clark *et al.*, 1998). Marine mining, which mainly occurs along the arid west coast between Saldanha Bay and the Namibian border, is restricted to South Africa's Exclusive Economic Zone (EEZ). Less than 1% of this area is currently being mined for this purpose, suggesting that more mining may soon take place.

The negative consequences of oil and gas extraction in offshore marine environments are not as obvious as mining. The impact of oil spills has been vividly illustrated in various parts of the world (Ch. 2), and has consequences for South Africa as well (see Pollution, Section 11.3.2.3).

11.3.2.2 Living marine resources

Interactions of the Agulhas and Benguela current systems provide various oceanographic conditions for a wide diversity and abundance of life (Fig. 11.4) Many shore habitats and estuary and river mouths provide accessible sites for small-scale subsistence fishing by indigenous people (Sowman, 2006). Fishing had relatively little impact on fish populations and coastal ecosystems until technological development led to the establishment of formal fisheries. Now, increasingly severe impacts of industrial fishing have led to serious concern for the sustainability of these resources.

Human presence along the South African coast dates back to the Early Stone Age (1.0–0.5 million years BP). Systematic exploitation of marine resources can be traced to the Middle Stone Age (±150,000 years BP), but much of what is known about prehistoric exploitation derives from more numerous Late Stone Age sites, which had their origin about 12,000 years ago. Marine mollusks constituted the most abundant food in Middle and Late Stone Ages; some species found in middens (Fig. 11.5) include many that are still prized by fishers today. A large proportion of fish bones and crustacean skeletons in middens, especially of rock lobster, seem to indicate improved fishing skills towards the Late Stone Age. Tidal fish traps (Fig. 11.6) dating from about 6000 years BP consist of low walls of boulders built across gullies in the intertidal zone, and many are still in use (Avery, 1975). Other technology (e.g., bone hooks, baskets, and woven nets) also no doubt contributed to a greater percentage of fish in the diet of prehistoric South Africans. Human remains dated by isotopic measurements indicate that prehistoric exploitation of marine resources probably peaked about 3000 to 2000 years BP, but declined thereafter, mainly due to the advent of pastoralism.

The earliest reasonably accurate record of commercial marine resource harvesting coincides with the arrival of European settlers in South Africa in the 17th century, marking a major turning point in human impact on the coast and its associated resources, due to far more intense exploitation. Cape fur seals and seabirds were among the first species targeted, as they breed mostly on islands where they are easily accessible. Dutch, French, and English sailors aboard old sailing vessels plundered seal and seabird colonies in the vicinity of Cape Town and Saldanha Bay (Rand, 1972; Best, 1973); at least 23 colonies were destroyed by the end of the 19th century (Shaughnessy, 1984; David, 1989). Many fur seal colonies, also inhabited by penguins, did not recover, their breeding areas being taken over by seabirds. This in turn led to the development of manned stations on some islands to facilitate the harvesting of guano and collection of seabird eggs. Attempts to limit impacts of seal harvesting were introduced by the *Cape Fish Protection Act* (1893), followed by the *Sea Birds and Seals Protection Act* (1973). Human impacts on seabirds and seals were, however, not confined to direct exploitation, but were exacerbated by modification of habitats (e.g., construction of causeways linking islands to the mainland, oil spills) and through bycatch fishing mortality (Griffiths *et al.*, 2004). Until 1967, harvesting of penguin eggs on offshore islands was legally sanctioned. Populations of African penguins (*Spheniscus demersus*) reportedly decreased from more than 40,000 pairs in 1956 to only about 1000 pairs in 2000. Other species were also affected, for example, Cape gannet (*Sula capensis*) breeding areas decreased from 6.24 ha in 1956 to 0.63 ha in 1996 (Crawford *et al.*, 2001). This decline was attributed to the collapse of pelagic fish stocks and to oil pollution (Morant *et al.*, 1981).

Fig. 11.4 (a) African penguin (*Spheniscus demersus*) (photograph © G. Carleton Ray); (b) west coast rock lobster (*Jasus lalandii*); (c) abalone (perlemoen, *Haliotis midae*); (d) giant chiton (*Dinoplax gigas*); (e) granite limpet (*Scutellastra argenvillei*); (f) black musselcracker (*Cymatoceps nasutus*); (g) black mussel (*Chromytilus meridionalis*); (h) Cape fur seal (*Arctocephalus pusillus pusillus*). (b–h) Photographs © Richard Starke, with permission.

Whaling in South Africa began with the arrival of whaling ships from the United States, France, and Britain in the late 18th century (Griffiths *et al*., 2004). The southern right whale (*Eubalaena australis*) was the principal species targeted. The use of hand-thrown harpoons from open boats was dangerous, but nevertheless sufficient to reduce whale populations (Best and Ross, 1986). Early in the 1900s, steel-hulled and steam-driven catchers with canon-fired harpoons had devastating effects, leading to a population crash by 1915, when a wider range of whales then became targeted (Griffiths *et al*., 2004). After the International Whaling Commission banned the harvest of right whales in 1953, the population began to recover and continues to increase today.

Commercial line fishing developed after the British captured the Cape Colony in 1795. In South Africa, the term "linefishing" refers to a practice of fishing with hand-lines or rods from small boats (Fig. 11.7a). It is undertaken by both commercial and recreational fishers and is distinct from "shoreangling" that is generally practiced by recreational and subsistence fishers only, and refers to the practice in which line, sinker, and bait are cast from the shore, either from sandy beaches or from rocky, intertidal reefs (Fig. 11.7b). Technological developments facilitated the expansion of fishing, especially fast boats powered by outboard motors and fishing boats equipped with freezers, nylon lines, echo-sounders, and geographic positioning systems (GPS). Pelagic fisheries, mainly for sardines, anchovy, and maasbanker, which commenced in the early 1950s, rapidly expanded with catches peaking at almost 500,000 tons in 1962 that declined to 350,000 tons in 1987, thereafter declining to the level of stock collapse (Crawford *et al*., 1987). Smaller sectors of the fishing industry include net fisheries, where nets are laid in a

Fig. 11.5 A megamidden (prehistoric "rubbish dump") comprising mostly abalone shells in the Walker Bay Nature Reserve near the Kogelberg Biosphere Reserve. Photograph © Barry Clark.

Fig. 11.6 Operational fish traps near Arniston east of Cape Agulas. Low stone walls have been packed in the intertidal zone since Stone Age times and are maintained to this day. High-tide waters over-top the low stone walls, and when water levels recede during low tide, fish get trapped within easy reach of fishers. Photograph © Barry Clark.

u-shaped pattern from the beach by man-powered rowing dinghies, and the two ends are then drawn in by hand. Rock lobster are largely caught from dinghies with the aid of baited hoop nets. Deeper-water fishing is conducted from larger boats using baited traps.

The advent of diving heralded the advent of the abalone fishery in the late 1960s. The belief, especially in far-eastern countries, that abalone flesh is an aphrodisiac has driven black-market prices to astronomical heights, in excess of US$ 40/lb wet weight. The export, mainly to the Far East, takes place in a clandestine manner through black-market channels that deprive the national economy of income from this valuable resource. Rampant abalone poaching continues throughout South Africa's west and south coasts, including in marine reserve areas, where poachers operate openly from land or from fast, outboard-powered inflatable boats capable of carrying up to 12 divers (Fig. 11.8). Poachers are fully aware that excessive removal of organisms is detrimental to the fishery, but the lure of sky-high, black-market prices overrides wider interests. Since 1996, poaching incidents have fluctuated widely, with few signs of decreasing (Fig. 11.9). Sadly, abalone poaching is also closely linked to illegal drug trafficking (Hauck, 2001). Young people involved in poaching may become dependent on drugs and susceptible to other aspects of criminal activity, leading to a downward spiral in social justice within affected communities. The life history of abalone and its interdependence with other habitat members (e.g., sea urchins, fishes, and kelp) are important factors for conservation (Box 11.2, Fig. B11.2.1, and Fig. B11.2.2). Less well recognized is the ecological importance of abalone, and the

Fig. 11.7 (a) Traditional linefish boats at Arniston on the Cape south coast. Photograph © Barry Clark. (b) Shore anglers at Betty's Bay. Photograph © G. Carleton Ray. A shore-angling competition is depicted in the picture, in direct conflict with the objectives of biosphere reserves. During angling competitions the rock-living organisms in the inter-tidal zone are severely trampled and at times are utilized for bait.

impact that lobster and/or abalone depletion can have on other valued resources. Given the cycle of abalone depletion and habitat change, it is possible that this important species could become locally extirpated or even commercially or biologically extinct within its range (Fig. 11.10).

11.3.2.3 Coastal-marine pollution

Pollution of coastal waters originates from various sources, including land-based sources (industrial, municipal, agricultural runoff), shipping activity (accidental or deliberate discharges, garbage and dumping), and release of atmospheric gases (Attwood *et al.*, 2002). At least 67 licensed discharge points scattered along the coast release as much as 1.3 million m^3 of wastewater daily into the marine environment (DEAT, 2006), not including stormwater outfalls. Nevertheless, South Africa's coastal waters are relatively low in pollution (Brown, 1987; Griffiths *et al.*, 2004). Although the number of discharge points has decreased in recent times, the amount of wastewater discharge is increasing at around 10% annually, the majority being released mostly into the surf zone, with lesser amounts into estuaries and offshore. An alarming 275,000 m^3 of wastewater, 90% of which is domestic, is discharged daily into coastal waters off the Western Cape Province alone; offshore discharges along the KwaZulu-Natal coast amount to 500,000 $m^3 day^{-1}$, of which 61% is industrial. The Eastern

Fig. 11.8 Abalone poachers at Betty's Bay Marine Reserve. Poachers use scuba gear and concentrate on Betty's Bay Marine Reserve, as abalone stocks have been thinned out by poaching elsewhere. The abalone are often shucked before being brought ashore. Patrol vessels operating farther offshore are ineffective in kelp beds. However, the poachers are often apprehended and have their gear confiscated when they land to offload their catches. Photograph © Richard Starke, with permission.

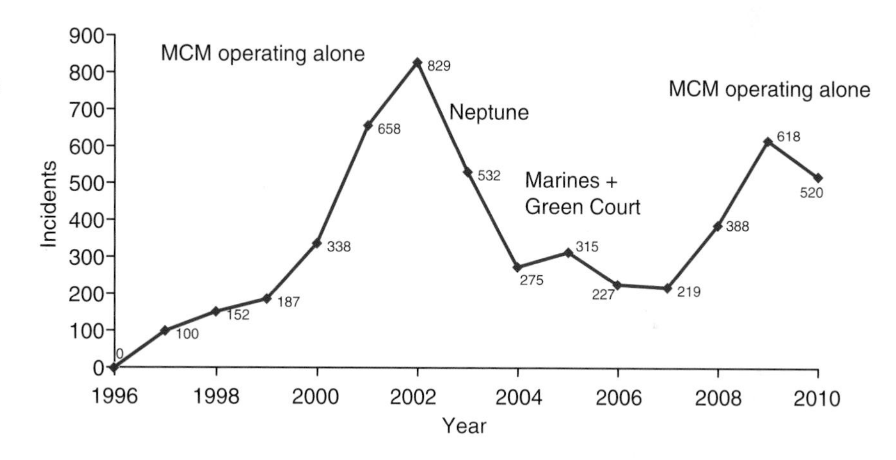

Fig. 11.9 Incidence of abalone poaching, 1996–2010. Various anti-poaching exercises are noted. Incidents were highest during 2002–10 when MCM (Marine and Coastal Management agents) unsuccessfully tried to curb poaching. During 2002–6, poaching declined substantially when special police units (Neptune and Marines) working collaboratively with local authorities were operational and after a special environmental court (the "Green Court") had been established. When these measures were discontinued, high-level poaching resumed. Figure prepared by Richard Starke of Seawatch.

Cape and Northern Cape provinces have considerably lower amounts of wastewater discharge than either of the other two coastal provinces. (Note: east coast, south coast, southwest- and west coast refer to actual geographic regions, each with its specific environmental characteristics (Section 11.2.1). These geographic designations are also aligned to a large extent to water catchment boundaries. Western Cape, Eastern Cape, and Northern Cape refer to politically assigned regional names and are not an accurate indication of ecoregions. Eutrophication is reported only for small semi-enclosed bays and estuaries, where it can have significant impact; causes include discharges of domestic wastewater (e.g., treated and untreated sewage), agricultural runoff (containing high concentrations of fertilizer), and industrial effluent (from fish processing plants). In open coastal environments, naturally high wave energy and strong prevailing currents are generally sufficient to ensure that nutrient concentrations do not reach problematic levels.

Oil pollution is a chronic problem worldwide. During the last decade, approximately 82,000 tonnes of oil have been accidentally or deliberately discharged into South African coastal waters (DEAT, 2006). The worst oil pollution incident in several decades occurred in 1992 when the *Katina-P* oil tanker sank off the coast of Mozambique, releasing >67,000 tonnes of oil; many lesser spills of 1500–3000 tonnes have also occurred. In 1976, South Africa acceded to the 1969 *Convention on Civil Liability for Oil Pollution Damage* ("Civil Liability Convention") by passing the *Marine Pollution (Control and Civil Liability) Act* of 1981, providing for compensation from the International Oil Pollution Compensation Fund in the event of an oil spill. There has been no update in the legislation since 1976, which means that any claim lodged by South Africa is pegged to 1976 levels. Ship-owners have a maximum liability of 14 million Special Drawing Rights—a monetary unit of international reserve assets defined and maintained by the International Monetary Fund—for any one oil pollution incident. In today's value, this is less than US$ 22 million, paltry in comparison to the likely real, long-term costs of an oil spill, as is illustrated by the case of the *Exxon Valdez* spill in Alaska in 1989 or the *Deepwater Horizon* in the Gulf of Mexico in 2010.

Box 11.2 Life history and ecological significance of the abalone

The abalone (*Haliotis midae*) is a gregarious species that lives in dense communities on sub-tidal rocky reefs in 1–20 m depths. Its life history is complex (Fig. B11.2.1). Dense clustering is part of its survival strategy. Ova and sperm are released into open water and fertilization is external. Lifestyle is to a very large extent sedentary. Feeding takes place by trapping kelp fronds and other algal material with the muscular foot and then rasping away the organic matter with the tongue-like radula.

Abalones occupy an important niche in energy-flow mechanisms of sub-tidal reefs through breakdown of free-floating organic material and conversion into nutrients. Movement by means of the muscular foot is possible, but is restricted. Re-colonization of denuded areas is, therefore, a slow process. Consequently, abalone are very vulnerable to over-exploitation. When thinned out too much, external fertilization and distribution of larvae by currents is no longer possible. Intensive poaching of the southern and southwestern Cape is therefore endangering survival of this valuable resource, especially as immature sub-adults are now being targeted.

Severe reduction of abalone may also have ecological consequences via feed-back loops. First, if abalone decline due to poaching, their reproductive potential is reduced. Second, a progressive shift southwards of oceanic conditions in the Benguela upwelling regime has probably triggered a shift of rock lobsters southwards. Third, rock lobsters eat juvenile abalone that have settled on reefs or within the protective spines of sea urchins (Fig. B11.2.2), but lobsters also eat sea urchins. Thus, young abalone become increasingly subject to lobster predation. Fourth, the habitat abalone once occupied on sub-tidal rock-faces is taken over by smaller algae and other life forms. Hence, a positive feed-back loop is created in which abalone decline towards local extirpation due to the combined effects of poaching and invading lobsters, with the dual consequences of progressive changes in reef ecology, as well as declining habitat conditions for abalone.

Source: Tarr (1989)

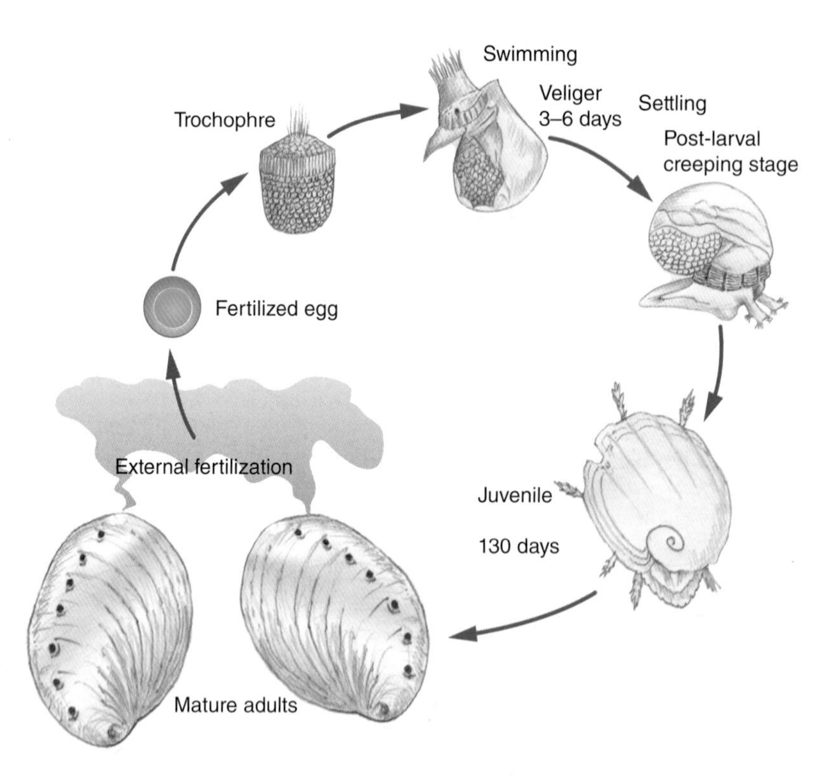

Fig. B11.2.1 Abalone life history. Adults live in dense groups on subtidal reefs along the south and west coast of South Africa. Fertilization is external, with eggs and larva being released directly into the water column by the adult individuals. Larvae drift in the water for a short period before settling on inshore reefs covered with coralline algae where their coloration closely matches that of the substratum and they feed by scraping microscopic algae from the reef. Within a few months they seek the shelter of spiny sea urchins and later, when they grow too large for this, they move into cracks in the reef and start to trap kelp fronds, which is the main adult diet. Adapted from Tarr RJQ (1989) Abalone. In *Oceans of Life off Southern Africa* (eds Payne AIL and Crawford RJM).

Fig. B11.2.2 Juvenile abalone sheltered beneath a sea urchin *Parechinus angulosus*. Photographs © Charles Griffiths.

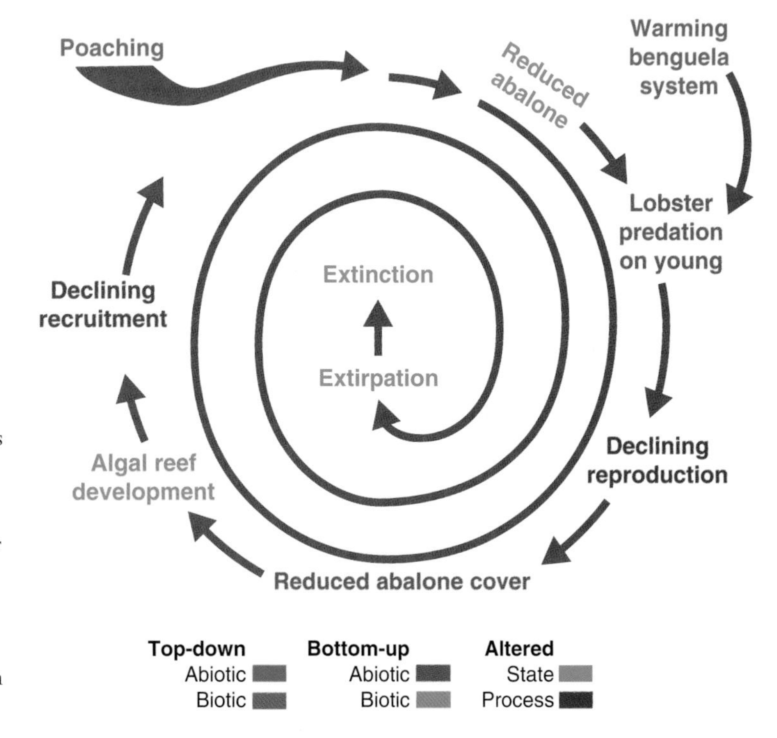

Fig. 11.10 Concept of dynamic interactions of poaching and habitat change leading to threatened abalone and a degraded altered state through inter-related processes. Poaching has drastically reduced adult abalone populations (and hence recruitment rates), while at the same time warming in the Benguela Current has caused a southward migration of rock lobster that predate on both urchins (under which juvenile abalone shelter from these and other predators) as well as the juvenile abalone themselves. Urchins also feed by scraping algae from the reef, which in turn helps to keeps these reefs free of sediment and foliose algae and hence promote the growth of encrusting coralline algae, which juvenile abalone use as a substratum for their initial settlement.

Fig. 11.11 Alien invasive marine species from the South African coastal and marine environment: (a) Pacific barnacle (*Balanus glandula*); (b) red-eyed wattle (rooikrans, *Acacia cyclops*); (c) European green crab (*Carcinus maenas*); (d) Mediterranean mussel (*Mytilus galloprovincialis*). Photographs © Charles Griffiths, with permission.

11.3.2.4 Alien invasive species

An estimated 85 marine species have been introduced to South African coastal waters (Mead *et al.*, 2011a). An additional 40 species are currently of uncertain origin (Robinson *et al.*, 2005; Griffiths *et al.*, 2010; Mead *et al.*, 2011b). These are most likely underestimates, given the dearth of taxonomic expertise and research in South Africa. Species originating from the Northern Hemisphere are highest on the cool-temperate west coast, whereas those originating mostly from the Southern Hemisphere mostly occur on the subtropical east coast.

Most introduced species occur in sheltered areas such as harbors. They may have been introduced through shipping activities, mostly ballast water releases, and/or from aquacultural activities. Ballast water tends to be loaded with species from sheltered harbors, and such species transported from these environments are probably poorly adapted to survive along South Africa's exposed coast. Similarly, although aquaculture has considerable potential to introduce alien species, it tends to be practiced mostly in sheltered bays or in onshore culture facilities and behind dams. For these reasons, only a small proportion of the alien species detected in South African waters to date have become invasive; those that are include the Mediterranean mussel (*Mytilus galloprovincialis*, Fig. 11.11d), the European green crab (*Carcinus maenas*, Griffiths *et al.*, 1992; Robinson *et al.*, 2005; Fig. 11.11c), and the recently detected barnacle (*Balanus glandula*, Laird and Griffiths, 2008; Fig. 11.11a).

Dune stabilization is a particular problem because it disrupts sediment transfer processes and leads to severe coastal erosion in some areas. Stabilization of coastal sand dunes by alien plants such as the red-eyed wattle (*Acacia cyclops*; Fig. 11.11b) was conducted by various government departments intensively between the 1940s and 1960s. At the time it was not foreseen that this alien plant, indigenous to Australia, would subsequently invade other habitats on a massive scale. This species is now considered to be the single biggest threat to plant and animal biodiversity for South Africa's coasts. Other invasive alien plants have now become established in more than 10 million hectares of land, much of it coastal (Richardson and van Wilgen, 2004). Primary concerns include reduced availability of freshwater due to high water consumption by these alien plants, reduced agricultural potential, increased flood and fire risk, increased erosion, destruction of riverine habitat, siltation of dams and estuaries, compromised water quality, and competition with indigenous plants and animal species. The fynbos biome (Box 11.1) is the most severely invaded biome, mostly by trees and shrubs of the genera *Acacia*, *Hakea*, and *Pinus*.

11.3.2.5 Climate change

Complex, uncertain relationships emerge from climate change (Table 11.1; Ch. 2). Impacts on South Africa's coastal-marine environments, biota, and fisheries are predictably pervasive, with highly deleterious cascading effects (Allison *et al.*, 2009). For example, increased atmospheric CO_2 results in ocean warming and acidification with potentially significant biological and ecosystem effects. Higher water temperatures may alter circulation and biotic patterns, resulting in phenological differences between spawning and recruitment for some species (Ch. 5). Extreme events are also likely, such as increased frequency of storms, floods, and droughts that can affect habitat structure, biodiversity, and commercial species. Other effects are related to changes in rainfall patterns, increased ultraviolet (UV) radiation, and sea-level rise.

Monitoring of sea-surface temperature, mean sea level, and rainfall has confirmed that South Africa's environmental changes mirror global patterns (IPCC, 2007; Tyson, 1990; Schumann *et al.*, 1995; Taunton-Clark and Shannon, 1998; Clark, 2005). Sea-surface temperatures off southern Africa have increased by about 0.25°C per decade for the last four decades (Schumann *et al.*, 1995; DEAT, 2006), or 1°C since World War II (Tyson, 1990). Warming of deeper oceanic waters (700–1100 m depth) has also been noted in the Southern Ocean, temperatures having risen by 0.17°C between the 1950s and the 1980s (Gille, 2002). Global circulation models (GCMs) predict that warming trends will continue, with mean surface air temperatures over southern Africa expected to rise between 1 and 3°C by 2050 (Hulme, 1996; Perks *et al.*, 2000; Ragab and Prudhomme, 2002). South Africa tide-gauge measurements indicate rising sea levels of approximately 1.2 mm/year during the last three decades, in close agreement with international estimates (Brundrit, 1995). The current trend of rising sea level is expected to accelerate in the future, with potentially catastrophic impacts on coastal environments and human settlements.

In keeping with international predictions, upwelling in the Benguela region appears to have intensified in recent decades (Shannon *et al.*, 1990). Contrary to expectations, primary production has declined (Brown and Cochrane, 1991) but zooplankton abundance appears to have increased over a similar period (Verheye *et al.*, 1998), accounting for an apparent reduction in phytoplankton biomass. Increases in zooplankton biomass may in turn be a function of reduced predation pressure by pelagic fish, whose populations were decimated by overfishing in the 1960s and possibly linked to changes in upwelling intensity.

11.4 COASTAL RESOURCE MANAGEMENT: PAST AND PRESENT

Resolution of the issues described above requires an integrated approach to management, i.e., a dynamic process that evolves through lessons learned from past experiences, and the understanding of the functioning of the coastal realm in its entirety. Integrated management seeks to coordinate and regulate various human activities and issues in order to achieve conservation and sustainable use (Celliers *et al.*, 2009). In this respect, one of the greatest challenges facing South Africa today is to find a balance between the requirements of development and conservation that will remain workable in the long term and that will provide optimal benefit to the country and its people. Development is essential if the needs of a growing population are to be catered for and if lack of employment opportunities and poverty are to be countered. By the same token, development at the expense of environmental integrity cannot be sustained. A sound economy and socio-political stability cannot be built upon a degraded environment. Most of all, equity among all citizens is paramount in the new South Africa.

Glavovic (2006) identified a number of coastal management eras in South Africa (Box 11.3) that are reflected in a variety of policies and laws adopted by the nation (Table 11.2). South Africa has a somewhat checkered history of coastal-marine management. Past management divorced land and sea resources, living and non-living, from one another. All land below the high water mark has always been retained as public property held in trust by the state, while land above the high water mark lies in the private or public domains. Land- and marine-based sources of pollution have also been managed through separate legislative Acts by separate government agencies, with the result that many pollution-related impacts are not adequately addressed. The same applies to the impacts of mining. Shortcomings of this situation have long been recognized, and collaborative efforts by government departments and coastal scientists since the 1970s have aimed at overcoming these problems.

11.4.1 Living marine resources management

Fisheries management is a particularly difficult issue. South Africa's efforts to control harvesting of marine living resources are among the oldest in the world, dating from the *Cape Fish Protection Act* of 1893. Concerted and coordinated efforts to manage living marine resources, however, have a short history. The first comprehensive legislation designed to protect marine resources was the *Sea Fisheries Act* of 1940. This Act and its subsequent revisions (1973 and 1988) were concerned primarily with the management and control of access to the larger commercial fish resources available within the territorial waters of South Africa (e.g., small pelagic fishes, demersal trawl species, and rock lobster). However, these Acts paid little heed to often more intensively exploited inshore resources, simply because these yielded smaller financial returns. Rights to harvest marine resources, allocated as a proportion of the

Table 11.1 Summary of potential pathways for climate change impacts on aquatic ecosystems, fish stocks, and fisheries. Based on data from Allison *et al.* (2009).

	Impacts on		
Climatic variable	**The environment**	**Biota**	**Fisheries**
Increased atmospheric CO_2	• Acidification of the oceans	• Changes in phytoplankton species composition, biomass, and productivity • Changes in coral reef species composition, biomass, and productivity	• Changes in production and availability of fished species • Reduced profitability
Global warming	• Higher water temperatures • Altered ocean circulation patterns, wind fields, and upwelling patterns • Changes in phenology (timing of natural phenomena)	• Changes in abundance, biomass, distribution of biota • Altered predator-prey and competitive interactions • Mismatch between spawning and recruitment events and natural processes that facilitate their success (e.g., food supply, transport processes, etc.) • Altered frequency and intensity of regime-scale and event-scale processes (e.g., ENSO events, monsoon processes) • Skewed sex ratios in some species with temperature dependent sex ratios • Spread of alien invasive species and pathogens	• Changes in production and availability of fished species; reduced catches • Altered catch composition; increased operating costs • Reduced spawning and recruitment success • Reduced profitability
Changes in rainfall	• Altered freshwater flows (timing, magnitude, frequency, intensity of base flows and flood events) • Altered lake levels, and changes in lake level variability	• Changes in species composition abundance, biomass, distribution of biota • Altered predator-prey and competitive interactions • Mismatch between spawning and recruitment events and natural processes that facilitate their success (e.g., food supply, transport processes, etc.)	• Changes in production and availability of fished species; reduced catches; altered catch composition; increased operating costs • Reduced spawning and recruitment success • Reduced profitability
Increased UV radiation	• Increased UV-B exposure & damage	• Inhibition of photosynthesis and nutrient uptake DNA in aquatic plants	• Changes in production and availability of fished species
Sea-level rise	• Loss and modification of coastal habitat Saline intrusion into freshwater and estuarine habitats	• Changes in coastal profiles • Changes in abundance, biomass, distribution of biota	• Reduced productivity of coastal-marine ecosystems • Loss of harbor, fishery and aquaculture, and infrastructure
Extreme events	• Increased frequency of storm events, floods and droughts	• Increased mortality, reduction in spawning and recruitment success	• Reduced catches • Loss and damage to infrastructure and equipment (harbors, vessels, etc.)

Box 11.3 Eras in coastal management in South Africa as defined by Glavovic (2006)

"Ad hoc sector-based management in the 1970s"—a period in which concerns began to grow over the impacts of uncoordinated interventions by a range of management agencies on the environment and coastal communities.

"Introduction of coastal zone management procedures and regulations based on expert ecological inputs in the 1980s"—where coastal management became a national priority, where more research into impacts of development on coastal processes was initiated, as were efforts to regulate physical development in the coastal zone. Environmental impact assessments for any developments in the coastal zone became mandatory.

"Establishment of participatory policy formulation procedures in the 1990s"—a period where managers were required to come to grips with the transition to a new political dispensation, the end of apartheid, the need to re-address imbalances of the past and recognition of the coast as a specific entity. All of this was compressed into an extensive consultation process, which finally culminated in the development of a new coastal policy or white paper.

"Start of the drafting of a new people-centered and pro-poor *Integrated Coastal Management Act* in 2000." This Act was to include legislation to put into effect a new coastal policy.

"A people-centered and pro-poor ICM in 2000"—during which time legislation was drafted that could give effect to the new coastal policy (a new coastal management act, 10 years in the making, finally approved only in 2010) and a number of programmes were initiated to leverage the value of the coast in alleviating poverty in coastal communities (notably the Working for the Coast Programme and the Sustainable Coastal Livelihoods Programme).

Table 11.2 Summary of key legislation governing management of the coast in South Africa.

Constitution of the Republic of South Africa (1996): Requires the state to ensure that "everyone has the right (a) to an environment that is not harmful to their health or well-being; and (b) to have the environment protected, for the benefit of present and future generations, through reasonable legislative and other measures that (i) prevent pollution and ecological degradation; (ii) promote conservation; and (iii) secure ecologically sustainable development and use of natural resources while promoting justifiable economic and social development."

National Environmental Management Act (1998): This Act promotes co-operative environmental governance. It establishes principles for decision-making on matters affecting the environment and provides for environmental institutions that promote co-operative governance. It also creates procedures for coordinating the actions of different government bodies that impact on the environment. The Act includes protocols for conducting environmental impact assessments and strategic environmental assessments for listed activities specified in regulations published under this act.

National Environmental Management: Integrated Coastal Management Act (2008): Defines the components of the coastal zone in South Africa, provides for the efficient and coordinated management of estuaries, defines institutional arrangements for the management of the coast, governs all development in the coastal zone, provides measures for protection of coastal resources from detrimental activities, and establishes procedures for regulating disposal of waste into estuaries and the sea.

National Environmental Management: Protected Areas Act (2003): Provides for the protection and conservation of ecologically viable areas representative of South Africa's biological diversity and its natural landscapes and seascapes; and for establishment of a national register of national, provincial, and local protected areas, describes the different types of protected areas that can be declared.

National Environmental Management: Biodiversity Act (2004): Provides for the conservation of biological diversity, and regulates sustainable use of biological resources in South Africa.

Marine Living Resources Act (1998): Regulates living resource use within marine and estuarine areas, mainly through licensing; provides for establishment of Marine Protected Areas.

National Water Act (1998): Defines the environmental reserve in terms of quantity and quality of water; provides for national, catchment, and local management of water.

Marine Pollution (Control and Civil Liability) Act (1981): Provides for the protection of the marine environment from pollution by oil and other harmful substances, the prevention and combating of such pollution, and the determination of liability in certain respects for loss or damage caused by the discharge of oil from ships, tankers, and offshore installations.

Local Government: Municipal Systems Act (2000): Requires each local authority to adopt a single, inclusive plan for the development of the municipality intended to encompass and harmonize planning over a range of sectors such as water, transport, land use, and environmental management.

Mineral and Petroleum Resources Development Act (2002): Deals with environmental protection and management of mining impacts, including coastal and offshore mining.

Total Allowable Catch (TAC) or Total Allowable Effort (TAE), were mainly allocated to a small number of large, mainly white operators. Recreational fisheries received scant attention and subsistence-style fisheries were hardly recognized.

This situation changed abruptly after the first democratic elections of 1994 that ended apartheid rule. A new fisheries policy was released in 1997 (*White Paper on Marine Fisheries Policy*) and new fisheries management legislation was introduced in 1998 (*Marine Living Resources Act*). To rectify historical imbalances, one of the key objectives of the new fisheries policy was to restructure the fishery, which proved far more challenging than the government had imagined. Most marine resources were already fully exploited or, in fact, over-exploited, and the personnel, vessels, and equipment available to exploit these resources were already in oversupply. For this reason, it was not possible to increase limits on catch (TAC) or effort (TAE). The solution was to open the fishery to new operators and drastically reduce the shares of existing operators. But many new operators came with their own vessels, equipment, and crew, which exacerbated fishing over-capacity—i.e., too many vessels seeking too few fish. Predictably by the turn of the 21st century, stocks of many marine fish species had declined further, fisheries compliance withered, and poaching and under-reporting of catches began to spiral out of control. This was particularly evident for valued inshore resources such as abalone, rock lobster, and linefish, for which access was more difficult to control (Sauer *et al.*, 1997; Hauck and Sweijd, 1999; Hauck, 2001; Groeneveld, 2003).

11.4.2 Recent developments

The most recent step in South Africa's attempt to develop sustainable coastal resource management was the adoption of the *Integrated Coastal Management Act* of 2008. Drastically revised policies and legislation represent long-overdue steps forward, including: recognition of the coast as a specific entity that functions as an interconnected system, including human activities; overlapping, expanded, and/or merged laws governing terrestrial, marine, living and non-living portions of the coastal-marine environment; and devolution of powers for hierarchical management of the coast, with much responsibility given to local government. Veto powers, however, remain vested at the provincial level, and implementation of the Act is still far from satisfactory. Government departments at all levels are struggling to come to grips with new responsibilities, made difficult by a politically inspired split of interrelated coastal functions originally housed within a single department (the Department of Environmental Affairs and Tourism—DEAT). The new Department of Agriculture, Forestry and Fisheries (DAFF) was given responsibility for managing coastal fisheries, while the Department of Environmental Affairs (DEA) retained responsibility for coastal management. Nevertheless, inextricably linked living renewable resources and the environment upon which they depend cannot be managed independently. Predictably, the splitting of these functions into two departments has led to immense confusion. Government officials are no longer certain about their specific fields of responsibility, thus hindering interdepartmental collaboration.

In spite of concerted and collaborative efforts during the past few decades, much remains to be done. The most important aim now is to ensure the sustainability of renewable resources and ecosystem functioning, especially in nearshore areas. This goal seeks effective implementation of existing protective measures for marine resource protection (e.g., fish size and bag limits, and closed seasons) and establishment of Marine Protected Areas (MPAs) with no-take zones to ensure adequate recruitment into adjacent areas open to fishing. Over and above these, equitable access to the coast and its resources are essential, accompanied by a more participatory approach to coastal resource management.

It is against this background that the marine portion of the Kogelberg Biosphere Reserve arose, including the establishment of a no-take Marine Protected Area.

11.5 IN PURSUIT OF THE KOGELBERG BIOSPHERE RESERVE

South Africa's internationally acclaimed Kogelberg Biosphere Reserve (KBR; Fig. 11.12) provides a perfect opportunity to coordinate protection of exceptional biodiversity and marine productivity with expanding urbanization and development under the biosphere-reserve concept (Ch. 3). This unique land/seascape of great scenic beauty and outstanding diversity is named after the imposing mountains that fringe the coast (Fig. 11.13), and includes expansive coastal plains, freshwater aquatic systems, sand dunes and beaches, rocky headlands, and subtidal seascapes. This land- and seascape hosts more than 1800 species of plants, 70 mammals, 43 reptiles, 22 amphibians, innumerable invertebrates, a rich and varied avifauna including an expanding colony of African penguins, and a diverse range of marine mammals and other forms of marine life—all supported by the exceptional marine productivity at the confluence of the Agulhas and Benguela currents. Ironically, coastal and marine resource management problems, especially coastal development, manifest themselves most intensively at the foot of the KBR.

South Africa has a long history of wildlife and resource conservation, including a spectacular array of terrestrial national parks. However, in conservation terms, the marine portion remains poorly protected and hardly managed. Nineteen mostly small, disconnected MPAs have been established, and a few more are currently under consideration (Sink and Attwood, 2008). This series of isolated MPAs cannot fulfill conservation needs due to the large sizes of marine ecosystems and their interconnectedness. The goal of a biosphere reserve is to address this problem. Within the larger land- and seascape, marine core areas are designated to provide meaningful protection to all components of the marine ecosystem against undue human inroads, mainly by fishing and poaching. Implementation of this concept has proved difficult, as the case of the KBR illustrates.

11.5.1 The biosphere-reserve concept

During the early 1970s, UNESCO's Man and the Biosphere Programme emerged as an offshoot of the scientifically ori-

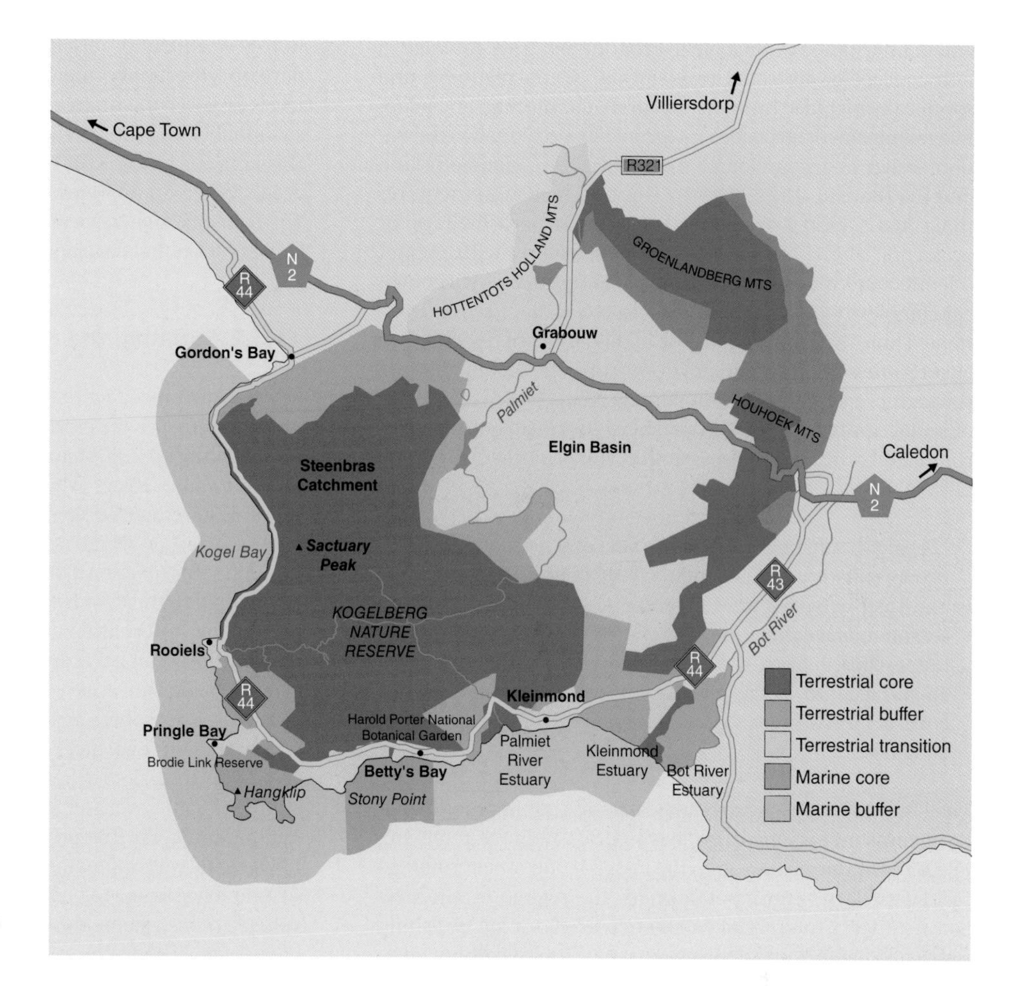

Fig. 11.12 Kogelberg Biosphere Reserve (KBR) zonation. The light-blue area indicates the proposed Marine Protected Area (MPA) at the foot of the KBR, designated as "buffer zone"; the darker blue area is the already proclaimed Marine Reserve at Betty's Bay. From Johns and Johns (2001), with permission.

Fig. 11.13 Foot of the Kogelberg Biosphere Reserve between Gordon's Bay and Betty's Bay. The mountains (±1100 m) form part of the core zone of the KBR, while the coastal road and associated camping and picnic areas are located in the buffer zone. Photograph © G. Carleton Ray.

ented International Biological Programme. This initiated a new concept for protected areas through spatial planning, and specifically included humans, research, monitoring, and adaptive management strategies integrated in a regional conservation design to accommodate man and nature, and both land and sea (Batisse, 1986, 1990). This holistic concept requires that conservation problems not be addressed parochially or for protection alone, but within a regional setting with greater scientific and management depth and encompassing ecological complexity along with consideration of land/sea interactions. Thus, the primary purpose of all biosphere reserves is to ensure integration of the environment with sustainable use through proper conservation management. Clear recognition of social equity with the cardinal need to protect natural environmental processes is also required. Importantly, as UNESCO (1996) made clear:

> "People should be considered as part of a biosphere reserve. People constitute an essential component of the landscape and their activities are fundamental for its long-term conservation and compatible use. People and their activities are not excluded from a biosphere reserve, rather they are encouraged to participate in its management and this ensures a stronger social acceptance of conservation activities."

The latter principle, in particular, stresses that improved environmental management in South Africa requires rectifying injustices of the past through the introduction of fair management practices, as promoted under the regime of the new South Africa. Thus, social issues such as allocation of fishing rights and low-income housing must be resolved without compromising the sustainability of the resources upon which society depends. In this context, efforts to establish and manage the KBR have been augmented by local organizations dedicated to conservation, albeit challenged by urbanization and development due to Cape Town's expanding proximity.

To obtain UNESCO's approval of the KBR as a biosphere reserve, several requirements needed to be met. First was designation of a zonation scheme in accord with the BR concept. This required identification of: *core areas* consisting of all state-owned lands to be managed strictly as protected areas for resource conservation; *buffer areas* where activities compatible with core conservation activities take place; and *transition areas* where sustainable management practices can be developed, including agricultural lands, townships, and industrial areas to be managed largely for human needs and (Batisse, 1990; Fig. 11.12). Theoretically, buffer and transition areas are compatibly managed to protect the core. The second requirement is the integration of three statutory functions: *conservation* of ecosystems, landscapes, and genetic resources (species); *encouragement* of sustainable development that included all members of society; and a *logistic* function consisting of research, monitoring, and education. These requirements can be extraordinarily difficult to achieve, but the area cannot be regarded a true biosphere reserve if these requirements are not met. No funds are allocated by UNESCO to meet such requirements. Instead, the concept relies on the commitment of local communities, farmers, conservation agencies, non-governmental organizations, and municipalities to develop governance mechanisms, in collaboration with all levels of government, to assure that the landscape is managed sustainably and biodiversity is conserved. The application of this concept to the KBR's biophysically and socially complex area seemed an appropriate mechanism to encompass different cultures and harness rapid growth while meeting both conservation and development needs.

11.5.2 Initiating the Kogelberg Biosphere Reserve

Seven years preceding the official establishment of the KBR, a desire among private citizens to establish a protected area was motivated by opposition to building a dam in the lower reaches of the Palmiet River, which was to meet about 10% of the water requirements of greater Cape Town (Hyman, 2006). In 1995, a non-profit citizen group became established as the Kogelberg Biosphere Association (KOBIO) that adopted a strategy to oppose the dam by designating the Kogelberg area as a prestigious, international biosphere reserve. First, however, the South African government had to apply for UNESCO membership before an area could be eligible to apply for official status as a biosphere reserve. Membership was achieved in 1996, and in December 1998, the KBR came into being according to a signed agreement between UNESCO and the South African government—the first biosphere reserve in the nation among six that presently exist.

The KBR was originally established in principle to provide a sustainable management mechanism for protecting this area's valued fynbos against the expected growth and water needs of Cape Town. The KBR's terrestrial 103,629 ha encompasses a part of the globally significant Cape Floristic Region (Box 11.1), and falls under the jurisdiction of several municipalities. Terrestrial protected core areas, of which the Kogelberg Nature Reserve forms the largest part, are state-owned, mostly lightly inhabited, and managed by a semi-governmental agency, the Western Cape Nature Conservation Board (Cape Nature), which also manages several other Western Cape nature reserves. The marine component covers about 24,500 ha and includes a 70 km buffer strip of coastline seaward to 2 nmi, excluding the nearshore waters of Gordon's Bay (Fig. 11.12), and is intended to be managed jointly by DEA and DAFF (see Section 11.4.2), including the proposed core area of Betty's Bay Marine Reserve, which had been proclaimed a marine reserve in the early 1960s. While the initial objectives for the establishment of the reserve may not have been the most noble (and potentially even motivated by personal gain), over half a century of protection of what are, perhaps coincidentally, very important biodiversity features means that this area is deserving of the highest level of protection that it can be assigned.

A major requirement of UNESCO is designation of some organization to manage the KBR, from the bottom up, that is by the people living there. This requirement led to the formation of the voluntary KBR Committee (KBRC) composed of 58 member stakeholders united by the main objective, "conservation and protection of the environment." KBRC membership included KOBIO, Cape Nature, various interest groups, civic associations, developers, skilled labor, property owners, and interested citizens. KRBC's unity of purpose, however, was not to last. Some KBRC members interpreted the KBR as a resource to be conserved (protectionists, including KOBIO), while others saw it as a restraint on development (developers and some land-owners defending their property values). The protectionists opposed development per se, taking positions of defending property rights and opposing low-cost housing, thereby forming an anti-development agenda masked as conservation. This extreme position was in marked contrast to the principles of biosphere reserves as established at the International Conference on Biosphere Reserves in Seville, Spain, in 1995—the Seville Strategy (UNESCO, 1996)—that biosphere reserves are intended to reconcile conservation of biodiversity with sustainable resource use. Furthermore, sustainable resource activities would be intended to benefit the region's least advantaged residents: that is, poverty alleviation that can complement conservation. An understandable reaction to KBRC's philosophical divide emerged from the disadvantaged sector, that the KBR was designated with the concealed purpose of denying access to parts of the country.

Despite being approved in principle in 1998, the KBR did not exist in law, had no budget, no office, nor a full-time administrator. To make matters worse, the KBRC's growing membership soon proved unwieldy. To resolve this situation, the KBRC formed a company in 2001 composed of 12 board members and a smaller Business Plan Committee of eight members who agreed to actions that were considered a watershed for resolving the contentious dispute. The composition of these two committees was to ensure that all statutory management bodies and key non-governmental organizations would be positively involved in the management process. By 2002, the company became registered, completed a Business Plan, identified seven portfolios, and proposed a R3.4 million budget that dwarfed KBRC's budget of only R200 thousand. KBRC then submitted a grant proposal to South Africa's Critical Ecosystem Partnership Fund for much-needed support. The proposal was rejected, as it lacked clarity about what the KBRC would do and how it would accomplish its goals. Meanwhile, the controversy continued as to the KBR's purpose, while poaching and coastal development intensified and the human landscape was dramatically changing (Fig. 11.14).

By 2003, a philosophical, managerial, and financial crisis became clearly evident. Local government had not been involved in KBR's planning, being caught up in spiraling development, and most low-income residents hardly knew of KBR's

Fig. 11.14 Historic and current mountain views of the eastern portion of Betty's Bay: (a) 1950s, (b) 2011. Note rapid and intense development in (b). Note the substantial increase in residential areas. The arrows on (a) and (b) indicate location of the African penguin colony, and are for orientation between the two photos. Photographs © Richard Starke, with permission.

existence. Conservation professionals became cynical, albeit recognizing the potential benefits of a functioning KBR. The Provincial Premier then stepped in, obliging the Overberg District Municipality that held the majority of the KBR's area, to revive the KBRC. This led all municipalities within the KBR area to sign a declaration of commitment. A Technical Advisory Committee was formed and a Steering Committee advised that KBRC required a full-time professional. The KBRC's main objective became:

> "The main business of the company is integration of the conservation of biological diversity, critical values, and economic development in the area designated as the Kogelberg Reserve."

Up to this point, living renewable resources and the coastal-marine environment were independently managed. Therefore, the next steps were to include the marine portion into an overall planning process, to create a strategic management plan overall, and to seek funding.

11.5.3 Implementing marine conservation: the Kogelberg Marine Working Group

With respect to the marine sector of the KBR, urgent attention was required to fulfill a priority for adequate marine conservation and resource protection. Poaching was out of control and shore angling was still permitted throughout the KBR, despite the requirement for "strict protection" within core areas. A proposal (Attwood, 2000) had suggested the establishment of closed areas for key exploited species (abalone, rock lobster, linefish) and the appointment of a designated authority responsible for management. Attwood proposed several areas located in the buffer areas of the KBR, one of which corresponded to the existing Betty's Bay Marine Reserve.

Betty's Bay Marine Reserve occupies a 3 km section of coastline between a rocky promontory known as Stony Point and a small eastward-facing pocket bay known as Jock's Bay located about in the middle of the coastal town of Betty's Bay (Fig. 11.14). This environmentally diverse strip of coast is surrounded by mountains, forested ravines, waterfalls, streams, and lakes. When streams flood, they deliver brown, nutrient-rich, tannin-stained freshwater into the bay. The coast, typified by high dunes, sandy beaches, sheltered lagoon-like areas, rocky shores, nearshore kelp-dominated habitat, and high- and low-energy wave-action zones, has an exceptional potential for the dispersal of marine larvae (Newman, 1964). Although Attwood's proposal represented a major step forward towards improved marine resource conservation and management in the KBR, limited national government interest resulted in no further action for its protection.

At about this time of KBRC's crisis, the Cape Action Plan for People and the Environment (CAPE) began to have significant influence on marine conservation in the KBR. CAPE was one of a number of ecoregion-based initiatives worldwide that arose from the need to find ways to operate at a scale large enough to achieve ecologically viable conservation results that would maintain healthy ecosystems (Younge and Fowkes, 2003). As a government program, CAPE was originally formed to ensure effective conservation of the Cape Floristic Region through partnerships between NGOs, and local, provincial, and national governments. It was supported by Global Environment Facility (GEF, Ch. 3) and coordinated through the CAPE Coordination Unit located in the South African National Biodiversity Institute (SANBI, Kirstenbosch National Botanical Garden). From 1998 to 2000, CAPE developed a strategy to conserve the entire Cape Floristic Region, and later expanded to include freshwater, estuarine, and marine components. Included among its recommendations were: effective management and conservation of entire catchments, and upgrading the capacity of marine areas to conserve resources and to improve enforcement capabilities (Younge and Fowkes, 2003). The KBR was selected as one of CAPE's several initiatives.

One of CAPE's first interventions, jointly with the World Wildlife Fund (WWF-SA), was the establishment of the Kogelberg Marine Working Group (KMWG) in 2007. This Working Group's primary purpose was to provide a platform for different levels of government to interact. WWF-SA provided support in terms of administration, funding, and coordination, while NGOs such as Seawatch (a local vigilant organization) provided reports on poaching and other intrusions in the nearshore zone. Academics and scientists assisted by compiling scientifically based reports, and environmental education specialists consulted with affected fishing communities, especially in previously disadvantaged sectors. The work of the KMWG was aided by representatives of the Kogelberg Biosphere Reserve Company and other interested and affected parties. From the outset, CAPE and the KMWG took a goal-oriented approach, focusing on key ecological, economic, and social issues, the premise being that sustainable exploitation of living renewable marine resources depends on a recruitment rate equal to or greater than the combined effects of natural and human-induced mortality. If excessive exploitation occurs, it will eventually have deleterious impacts on the ecosystem, with inevitable economic and socio-political consequences.

A number of key tasks have subsequently been initiated by the KMWG. These included development of a Management Plan for the Betty's Bay Marine Protected Area (Du Toit and Attwood, 2008), an assessment of the recreational use and value of the entire Kogelberg Coast (Turpie and de Wet, 2009), and an assessment of the ecology, value, and management of the Kogelberg Coast (Turpie *et al.*, 2009). These studies provided valuable, new insights into the geographic and socio-economic context of the Kogelberg region and associated coastal-marine ecosystems, the use and value of the coast and its resources, management issues and legislation, and opportunities for enhancing coastal integrity and value. Additionally, the government's *Integrated Coastal Management Act* of 2008 provided new opportunities for development of a Kogelberg Coast Integrated Management Plan (KCIMP), in which key elements of significance for the Kogelberg region were: recognizing the value of the coast, the need for establishing development setback lines to protect against inappropriate developments and activities, and in the case of small-scale fisheries policy, that local coastal communities should be the primary beneficiaries of resources available within their area of abode, as per the Territorial Use Rights in Fisheries (TURF) concept (Christy, 1982).

The KCIMP aimed to convey proper understanding of the importance of rational management of living marine resources upon which the fishing communities and the recreational fishing sector ultimately depend. Public participation accompanied KCIMP development and publication, but implementation of the Plan again proved controversial. Local fishers were naturally very quick to embrace the TURF concept as a system for management of inshore marine resources (abalone, rock lobster, linefish, etc.) in the Kogelberg, especially as one of their main complaints had always been that outsiders—very often illegal fishers—were the primary beneficiaries of marine resources. (Note: offshore resources, which can be much more effectively and efficiently harvested using larger vessels based at centralized landing sites, were specifically excluded from these discussions.) The fishers were less eager, however, to accept responsibility for future sustainability of these resources. Although they did not initially support the establishment and enforcement of no-take sanctuary zones within the same area, they soon became convinced this was in their own self-interest, especially if their rights of access to these resources could be secured into the future. Other complicating factors were more difficult to resolve: inadequate communication between responsible officials at all governmental levels; interference by politicians seeking voter support; resistance by local shore-angling interests who did not want to sacrifice access to favorite angling spots and for whom selfish short-term benefit appears to be more important than long-term sustainability; and impoverished individuals or families who need to put food on the table on a day-to-day basis. The persistent problem of poaching remained unresolved.

The final proposal currently (2012) under consideration by national government is that subsistence and commercial exploitation of all inshore living marine resources (abalone, rock lobster, kelp, etc.) within the marine component of the KBR be set aside for designated locally resident fishers only. The existing Betty's Bay Marine Protected Area, together with sufficient additional habitat areas (equivalent to 20% of the total) within the existing KBR buffer, would be proclaimed as no-take sanctuary zones ("core" areas according to UNESCO guidelines) where no fishing of any form would be permitted. Recreational fishing by outsiders would still be permitted in the non-sanctuary (buffer) zones of the KBR, but existing or improved protective measures—e.g., size and bag limits and fishing seasons—would be rigorously enforced. Of great importance to local fishing communities is fair and non-prejudicial zoning and allocation of commercial and subsistence fishing rights; i.e., fishing licenses would be granted to area residents but not to parties from outside. Financial benefits of this proposal for local communities could be substantial, with the total value of harvestable living marine resources in the KBR estimated to be almost R100 million per year in 2004 (Clark *et al.*, 2004), discounting the black-market value of poached abalone, which is quite possibly higher than the total value of all other fisheries combined. At the time of writing (2012), there are strong indications that abalone are so depleted that their full recovery will take many years, if at all.

The above actions are intended to motivate the current proposal to declare Betty's Bay a "no-take zone" within the KBR. But because shoreangling was not prohibited when the area was first proclaimed as a marine reserve, a no-take zone becomes especially difficult to implement even though progressive declines in resource yields indicate that ongoing fishing pressure has been too high (Attwood and Farquhar, 1999), and no doubt remains so. Furthermore, removal of organisms through poaching or angling has potential negative cascading effects for other species and other locations, including the kelp-bed ecosystem beyond KBR boundaries. Ironically, declining fish yields can lead directly to increasing demands to relax protective measures. Whether such relaxation would ameliorate hardship or exacerbate declining trends is a matter requiring serious debate and active research.

11.6 THE FUTURE OF COASTAL MANAGEMENT IN SOUTH AFRICA

A number of conclusions can be drawn from this case study. Foremost is the failure of governance to manage growth and protect natural resources. Government priorities promote economic development and socio-economic benefits, leaving protection and management of large state-owned areas to local, minimally funded agencies composed of citizens pulling from different directions. Integrating conservation measures to protect the resource base and carry out a management plan requires coordinated leadership directed by the highest levels of government. Lack of a national political-will to coordinate and protect nature and sustainably manage development, particularly in coastal-marine environments, creates unanticipated discordance in meeting socio-political priorities.

South Africa is blessed with 3000 km of highly diverse and productive coastal systems. Its oceanographic configuration and climate determine the productivity of terrestrial and marine ecosystems and the all-important land-sea interactions between them. Effective coastal management must therefore be based on recognition of natural processes and their protection. Sadly, this has not yet been recognized adequately in South Africa. For example, sugar cane farming on riverine floodplains is creating extreme habitat degradation. Estuaries are being degraded, despite their essential role as feeding and nursery areas for commercially valued species, so critical to local communities. Furthermore, the apparent inability or unwillingness of governmental authorities to plan for sustainable resource use, to regulate and enforce line fishing, or to curb blatant and large-scale poaching of abalone, will predictably contribute to ecological collapse. Of equal importance, especially with respect to the biosphere-reserve concept, is effective planning of development in which equal access is given to all members of society.

This KBR case study broadly outlines efforts being carried out through private-sector collaboration with government to overcome these problems. The development of an Integrated Coastal Management Plan, which enjoyed wide acceptance during the course of public participation, has yet to be implemented due to lack of political and financial support and lack of coordination among government departments at national to local levels that share common goals. South African authorities at all governmental levels, and all who have effective coastal zone management at heart, must confront issues of poaching, over- or unplanned land-based development, and establishment of fisheries no-take zones, or otherwise fail to meet internationally acceptable standards for the management of the Kogelberg Biosphere Reserve. As such, the basic biosphere-reserve functions of *conservation, sustainable and equitable development, and logistical support for research, monitoring, and education* simply cannot be met. Furthermore, if government is perceived as being incapable of facing these problems, recruiting support from coastal communities for conservation, including the biosphere-reserve concept, will also be problematical—to the detriment of the economy and the tourist industry.

The beginning of this chapter stated that South Africa's coastal management regime is relatively advanced among the world's nations. Add to this the country's magnificent and diverse coastal realm and its wide range of terrestrial and marine coastal resources, and it becomes clear that superb opportunities arise, among which is a fully operational Kogelberg Biosphere Reserve. South Africa's coast and its natural resources cannot, and will not, deliver their full potential until wise management, based on scientific information, overrides political expediency, local indecision, and short-sightedness. Essential prerequisites must be the government's determination to support the efforts of affected coastal communities towards sustainability, and to integrate coastal-marine conservation science and policy throughout South Africa. Honest assessment of the shortcomings of existing management strategies and meaningful efforts to overcome them can place the management of the coast—on land as in the sea—on a sound footing to make the Kogelberg Biosphere Reserve a functional reality. This task remains a huge challenge, both here and for coastal regions throughout the world.

REFERENCES

Allison EH, Perry AL, Badjeck, *et al.* (2009) Vulnerability of national economies to the impacts of climate change on fisheries. *Fish and Fisheries.*

Anchor Environmental Consultants 2012. Strategic Environmental Assessment to guide identification of potential marine aquaculture development zones for fin fish cage culture in South Africa. Unpublished report prepared for Directorate Sustainable Aquaculture Management: Aquaculture Animal Health and Environmental Interactions, Department of Agriculture, Forestry and Fisheries, South Africa.

Attwood CG, Farquhar M (1999) Collapse of linefish stocks between Cape Hangklip and Walker Bay, South Africa. *South African Journal of Marine Science* **21**, 415–431.

Attwood C, Moloney CL, Stenton-Dozey, *et al.* (2002) Conservation of Marine Biodiversity in South Africa, National Research Foundation Publication. In *Summary Marine Biodiversity Status Report* (eds Durham BD, Pauw JC). 68–83.

Attwood C (2000) Proposal for a Kogelberg Marine Park. Unpublished report, Marine and Coastal Management, DEAT, 14pp.

Avery G (1975) Discussion on the age and use of tidal fish-traps (visvywers). *South African Archaeological Bulletin* **30** (119/120), 105–113.

Batisse M (1986) Developing and focusing the biosphere reserve concept. *Nature and Resources (UNESCO)* **22**, 1–10.

Batisse M (1990) Development and Implementation of the Biosphere Reserve Concept and its Applicability to Coastal Regions. *Environmental Conservation* **17**, 111–116.

Beckley LE (1984) The ichthyofauna of the Sundays River estuary with particular reference to the juvenile marine component. *Estuaries* **7**, 248–250.

Best PB, Ross GJB (1986) Catches of right whales from shore-based establishments in southern Africa, 1792–1975. *Report of the International Whaling Commission*, Special Issue **10**, 275–289.

Best PB (1973) Seals and sealing in South and South West Africa. *South African Shipping News and Fishing Industry Review* **28**, 49–57.

Brown PC, Cochrane KL (1991) Chlorophyll a distribution in the southern Benguela, possible effects of global warming on phytoplankton and its implications for pelagic fish. *South African Journal of Science* **87**, 233–242.

Brown AC (1987) Marine pollution and health in South Africa. *South African Medical Journal* **71**, 244–248.

Brundrit GB (1995). Trends in southern Africa sea level: statistical analysis and interpretation. *South African Journal of Marine Science* **16**, 9–17.

Celliers L, Breetzke T, Moore, Malan D (2009) A User-friendly Guide to South Africa's Integrated Coastal Management Act. The Department of Environmental Affairs and SSI Engineers and Environmental Consultants. Cape Town, South Africa. Printed and distributed free of charge by The Department of Environmental Affairs, South Africa and SSI Engineers and Environmental Consultants.

Christy Jr. FT (1982) Territorial use rights in marine fisheries: definitions and conditions. *FAO Fisheries Technical Paper* **227**, 1–10.

Clark BM (2005) Climate change: A looming challenge for fisheries management in southern Africa. *Marine Policy* **30**, 84–95.

Clark BM, Atkinson L, Attwood C, *et al.* (2004) *Business Plan for the proposed Kogelberg Marine Park.* Prepared for: Kogelberg Biosphere Reserve Company and the World Wide Fund for Nature – South Africa (WWF-SA), 41pp.

Clark BM, Meyer WF, Smith CE, *et al.* (1998) *Synthesis and assessment of information on the Benguela Current Large Marine Ecosystem: Integrated overview of diamond mining in the Benguela Current region.* Report prepared for the Beguela Current Large Marine Ecosystem Programme, 72pp.

Colloty BM, Adams JB, Bate GC (2001) *The Botanical Importance Rating of the estuaries in former Ciskei/Transkei.* WRC Report no TT 160/01, 135pp.

Crawford RJM, David JHM, Shannon LJ, *et al.* (2001) African penguins as predators and prey: coping (or not) with change. *South African Journal of Marine Science* **23**, 435–447.

Crawford RJM, Shannon LV, Pollock DE (1987) The Benguela ecosystem. Part IV. The major fish and invertebrate resources. *Oceanography and Marine Biology: An Annual Review* **25**, 353–505.

David JHM (1989) Seals. In *Oceans of Life off Southern Africa* (eds Payne AIL, Crawford RJM). Cape Town, Vlaeberg, 288–302.

DEAT (Department of Environmental Affairs and Tourism) (2006) *South Africa Environment Outlook. A report on the state of the environment.* Department of Environment Affairs and Tourism, Pretoria, 371pp.

Drinkwater KF, Frank KT (1994) Effects of river regulation and diversion on marine fish and invertebrates. *Aquatic Conservation: Marine & Freshwater Ecosystems* **34**, 135–151.

Du Toit J and Attwood C (2008) *Management plan for the Betty's Bay Marine Protected Area.* Prepared for: World Wildlife Fund & the Department of Environmental Affairs and Tourism, Marine and Coastal Management. Coastal and Marine Eco-Tourism Corporation, Roggebaai, South Africa.

Field JG, Griffiths CL (1991) Littoral and sublittoral ecosystems of southern Africa. In *Ecosystems of the World 24: Intertidal and Littoral Ecosystems* (eds Mathieson AC, Niehuis PH). Elsevier, Amsterdam, 323–346.

Gille ST (2002) Warming of the Southern Ocean since the 1950s. *Science* **295**, 1275–1277.

Glavovic BC (2006) Coastal sustainability – an elusive pursuit?: Reflections on South Africa's coastal policy experience. *Coastal Management* **34**, 111–132.

Griffiths CL, Hockey PAR, van Erkom Schurink C, le Roux PJ (1992) Marine invasive aliens on South African shores – implications for community structure and trophic functioning. *South African Journal of Marine Science* **12**, 713–722.

Griffiths Cl, Mead A, Robinson TB (2010) A brief history of marine bio-invasions in South Africa. *African Zoology* **44**, 241–247.

Griffiths CL, van Sittert L, Best PB, *et al.* (2004) Impacts of human activities on marine animal life in the Benguela – An historical overview. *Oceanography and Marine Biology Annual Review* **42**, 303–392.

Groeneveld JR (2003) Under-reporting of catch of South coast rock lobster *Palinurus gilchristi*, with implications for the assessment and management of the fishery. *African Journal of Marine Science* **25**, 407–411.

Harrison TD, Cooper JAG, Ramm AEL, Singh RA (1994) *Health of South African Estuaries, Orange River-Buffels (Oos)*. Executive Report. CSIR Catchment and Coastal Environmental Programme, Congella.

Hauck M, Sweijd NA (1999) A case study of abalone poaching in South Africa and its impact on fisheries management. *ICES Journal of Marine Science* **56**, 1024–1032.

Hauck M (2001) An overview of state and non-state responses to abalone poaching in South Africa. *Journal of Shellfish Research* **19**, 518–519.

Heydorn AEF, ed. (1986) *An assessment of the state of the estuaries of the Cape and Natal in 1985/86*. South African National Scientific Programmes Report No. 130, 39pp.

Heydorn AEF (1989) The conservation status of southern African estuaries. In *Biotic Diversity in Southern Africa – Concepts and Conservation* (ed. Huntley BJ). 290–297.

Heydorn AEF, Flemming BW (1983) South Africa. In *The World's Coastline* (by Bird EC, Schwartz ML). 653–667.

Heydorn AEF, Tinley KL (1980) *Estuaries of the Cape. Part 1: Synopsis of the Cape Coast: Natural Features, Dynamics & Utilization*. Stellenbosch, CSIR Research Report 380, 1–96.

Hulme M (1996) Climate change and southern Africa: an exploration of some potential impacts and implications in the SADC region. Norwich: University of East Anglia, Climatic Research Unit.

Hyman MG (2006) How a Powerful Minority has Exploited UNESCO Biosphere Reserve Status: A Case Study of the Kogelberg Biosphere Reserve. Institut d'études politiques de Paris, Université de Paris-Sorbonne, 30 October 2006, 100 + xi.

IPCC (2007) Climate Change 2007: The Physical Science Basis. *Contribution of Working Group I to the Fourth Assessment Report of the Intergovernmental Panel on Climate Change* (eds Solomon S, Qin D, Manning M, Chen Z, Marquis M, Avery KB, Tignor M, Miller HL). Cambridge University Press, Cambridge, United Kingdom and New York, NY, USA, 996pp.

Johns A, Johns M (2001) *Kogelberg Biosphere Reserve – Heart of the Cape Flora*. Struik Publishers, Cape Town, 1–56.

Laird MC, Griffiths CL (2008) Present distribution and abundance of the introduced barnacle Balanus glandula in South Africa. *African Journal of Marine Science* **30**, 93–100.

Lamberth SJ, Turpie JK (2003) The role of estuaries in South African fisheries: economic importance and economic implications. *African Journal of Marine Science* **25**, 131–157.

Lane SB, Carter RA (1999) Generic environmental management programme for marine diamond mining off the west coast of South Africa. Marine Diamond Miners Association, Cape Town, South Africa. Six volumes.

Mead A, Carlton JC, Griffiths CL, Ruis M (2011a) Revealing the scale of marine bioinvasions in developing regions: the South African example. *Biological Invasions* **13**, 1991–2008.

Mead A, Carlton JT, Griffiths CL, Rius M (2011b) Introduced and cryptogenic marine and estuarine species of South Africa. *Journal of Natural History* **45**, 2463–2524.

Morant PD, Cooper J, Randall RM (1981) The rehabilitation of oiled jackass penguins *Spheniscus demersus*, 1970–1980. In *Proceedings of the Symposium on Birds of the Sea and Shore* (ed. Cooper JE). Cape Town, African Seabird Group, 267–301.

Newman CG (1964). *Movements of the South African Abalone Haliotis midae*. Investigational Report, Division of Sea Fisheries, Cape Town, 1–19.

Perks LA, Schulze RE, Kiker GA, *et al.* (2000) *Preparation of climate data and information for application in impact studies of climate change over southern Africa*: a report to the "South African Country Studies on Climate Change Programme". University of Natal, Pietermaritzburg School of Bioresources, Engineering and Environmental Hydrology.

Ragab R, Prudhomme C (2002) Climate and water resources management in arid and semi-arid regions: prospective and challenges for the 21st century. *Biosystems Engineering* **81**, 3–4.

Rand RW (1972) *The Cape fur-seal Arctocephalus pusillus. 4. Estimates of population size*. Investigational Report of the Sea Fisheries Research Institute, South Africa **89**, 1–28.

Richardson DM, van Wilgen BW (2004) Invasive alien plants in South Africa: how well do we understand the ecological impacts? *South African Journal of Science* **100**, 45–52.

Robinson TA, Griffiths CL, McQuaid CD, Rius M (2005) Marine alien species of South Africa – status and impacts. *African Journal of Marine Science* **27**, 297–306.

Sauer WHH, Penney AJ, Erasmus C, Mann BQ, *et al.* (1997) An evaluation of attitudes and responses to monitoring and management measures for the South African boat-based line fishery. *South African Journal of Marine Science* **18**, 147–163.

Schumann EH, Cohen AL, Jury MR (1995) Coastal sea surface temperature variability along the south coast of South Africa and the relationship to regional and global climate. *Journal of Marine Research* **53**, 231–248.

Shannon LV, Lutjeharms JRE, Nelson G (1990) Causative mechanisms for intra- and interannual variability in the marine environment around southern Africa. *South African Journal of Science* **86**, 356–373.

Shaughnessy PD (1984) *Historical population levels of seals and seabirds on islands off southern Africa, with special reference to Seal Island, False Bay*. Investigational Report of the Sea Fisheries Research Institute, South Africa 127, 61pp.

Sink K, Attwood C (2008) Guidelines for Offshore Marine Protected Areas in South Africa. *SANBI Biodiversity Series* **9**, South African Biodiversity Institute, *Pretoria*.

Sowman S (2006) Subsistence and small-scale fisheries in South Africa: A ten year review. *Marine Policy* **30**, 60–73.

Tarr RJQ (1989) Abalone. In *Oceans of Life off Southern Africa* (eds Payne AIL, Crawford RJM). Cape Town, Vlaeberg, 62–69.

Taunton-Clark J, Shannon LV (1998) Annual and interannual variability in the southeast Atlantic during the 20th century. *South African Journal of Marine Science* **6**, 97–106.

Turpie JK, de Wet J (2009) The recreational value of the Kogelberg Coast. Unpublished report to WWF-SA. Anchor Environmental Consultants, Cape Town.

Turpie JK, Clark BM, Hutchings K, Orr KK, de Wet J (2009) *Ecology, value and management of the Kogelberg coast*. Anchor Environmental Consultants, Cape Town.

Tyson PD (1990) Modelling climate change in southern Africa: a review of available methods. *South African Journal of Science* **86**, 318–330.

United Nations (UN) (2011) World Population Prospects: The 2011 Revision, Volume I, Comprehensive Tables (United Nations publication, ST/ESA/SER.A/313) and Volume II: Demographic Profiles (United Nations publication, ST/ESA/SER.A/317).

UNESCO (1996) *Biosphere Reserves: The Seville Strategy and the Statutory Framework of the World Network*. UNESCO, Paris.

Verheye HM, Richardson AJ, Hutchings L, *et al.* (1998) Long-term trends in the abundance and community structure of coastal zooplankton in the southern Benguela system, 1951–present. In: *Benguela Dynamics. Impacts of Variability on Shelf-Sea Environments and their Living Resources* (eds Pillar SC, Moloney CL, Payne AIL, Shillington FA). *South African Journal of Marine Science* **19**, 317–332.

Whitfield AK (2000) *Available scientific information on individual estuarine systems*. Water Research Commission Report No. 577/3/00.

Younge A, Fowkes S (2003) The Cape Action Plan for the Environment: overview of an ecoregional planning process. *Biological Conservation* **112**, 15–28.

CHAPTER 12

SPECIES-DRIVEN CONSERVATION OF PATAGONIAN SEASCAPES

Claudio Campagna, Valeria Falabella, and Victoria Zavattieri

Wildlife Conservation Society, Buenos Aires, Argentina

> Yet if in any country a forest was destroyed, I do not believe nearly so many species of animals would perish as would here, from the destruction of the kelp. Amidst the leaves of this plant numerous species of fish live, which nowhere else could find food or shelter; with their destruction the many cormorants and other fishing birds, the otters, seals, and porpoises, would soon perish also; and lastly, the Fuegian savage, the miserable lord of this miserable land, would redouble his cannibal feast, decrease in numbers, and perhaps cease to exist.
>
> Charles Darwin *The Voyage of the Beagle* (1839)

12.1 DARWIN'S PATAGONIA

Darwin's words are a very early example of integrated ecosystem thinking about interdependence, community ecology, food webs, life histories, biodiversity, and ecosystem services at a time when some of these concepts had not yet been forged. Darwin reasoned about the consequences of damaging the support system of a biological community depending on the kelp forest, with cascading costs all the way up to humans. But what forces would have been in Darwin's mind that could have destroyed a keystone species? Did he have in mind the idea of a human footprint strong enough to threaten an ecosystem? Would Darwin have proposed creation of marine reserves if threats had been evident?

This case study is about a vision for the Patagonian seascape and its wildlife, and urges creation of a large marine reserve as a tool for open-ocean conservation. It is about a process to advance candidate areas worthy of special management considerations, while also confronting practical solutions and the chronic paucity of scientific data. Consistent with Darwin's thinking, protecting species today requires a large-scale spatial approach, one that considers entire ecosystems, rather than small patches of habitat. We use the word "seascape" to refer to extensive ocean space that sustains foraging requirements of pelagic top predators and other functionally important species. We focus on one of the world's most productive large ocean provinces, the Patagonian Sea (PS; Fig. 12.1), which is a functional seascape with a large and shallow continental shelf, where iconic resident, migratory, and transboundary predators migrate and forage, and are, per se, targets of conservation; i.e., whales, seals, penguins, albatrosses, among others (Conway, 2005; Fig. 12.2a). Those that are large pelagic predators are effective in conveying esthetic and conservation values, in addition to playing essential roles in ecosystem health, structure, and function (Boyd *et al.*, 2006). Their role in the system is the equivalent to an ambassador representing the requirements and risks of many others. In this context, we define seascapes to include those species habitats and to reflect the heterogeneity resulting from the integrated functions of environmental variables, such as currents, depth profiles, temperature, and nutrients.

This case study has the purpose of illustrating processes and needs of large ocean conservation, coupled with sustainable use, such as Darwin illustrated with the example of the kelp forest. It also reflects current thinking in community ecology. We have a goal in mind: to set aside large areas of the Patagonian seascape to protect the foraging grounds of charismatic, and other species. The concept of charismatic species is critical, as such species have differential values for promoting conservation. Clearly, marine mammals and large marine birds are both charismatic and functionally important as predators. But important species for the PS that are not charismatic are equally important, particularly squids and fishes such as anchovies and sardines that sustain the trophic web. In Darwin's terms, kelp may not be charismatic, but the otters that depend on it are.

Marine Conservation: Science, Policy, and Management, First Edition. G. Carleton Ray and Jerry McCormick-Ray.

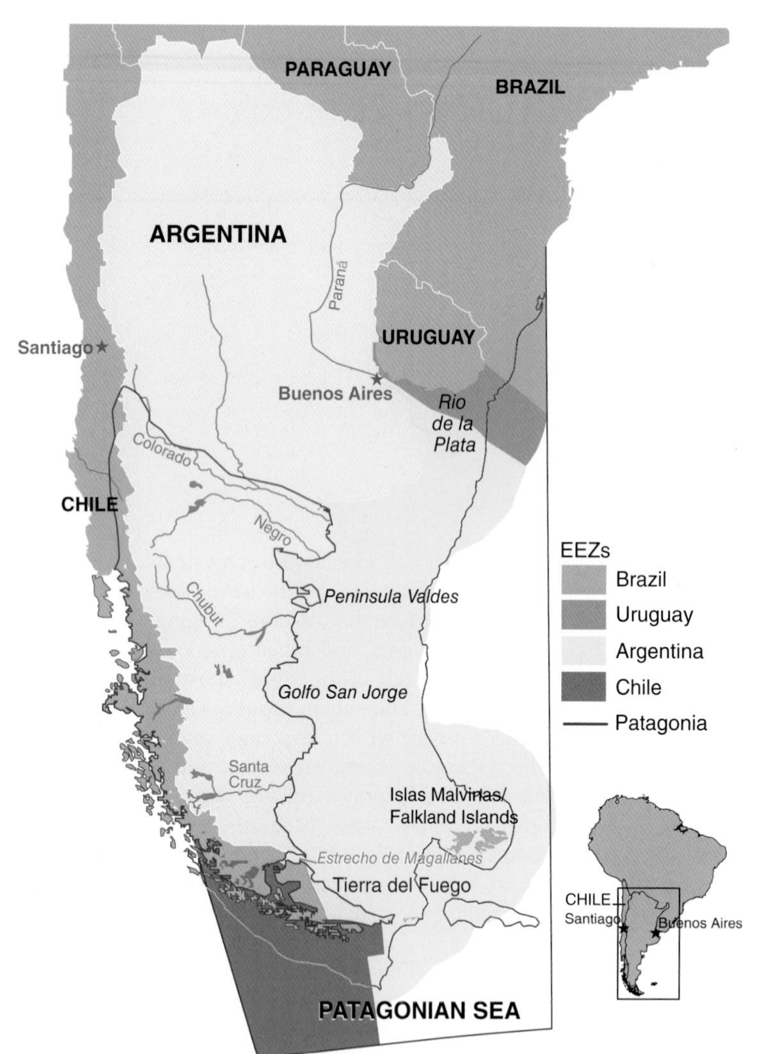

Fig. 12.1 The Patagonian Sea is a critical seascape of the Southwest Atlantic and Southeast Pacific. The extended, shallow continental shelf off the coasts of Argentina and Uruguay is a unique habitat, as also are the estuary of the Río de la Plata, the Chilean fiords, and the Strait of Magellan. National jurisdictions illustrate the transboundary nature of the system. Patagonia is a terrestrial landscape region of both Argentina and Chile, delineated by a black line, from near the Colorado R. southward.

12.2 A CONSERVATION DILEMMA

Darwin experienced the PS up-close, as he traversed it from north to south and east to west. In fact, a significant proportion of the Beagle's voyage occurred in our target seascape. The Beagle's voyagers must have experienced an extended area of a million km^2 that supported the most wonderful wildlife. Darwin was not as much inclined to describe marine wildlife as he was for terrestrial species and habitats, perhaps because of the uncomfortable consequences of seasickness. Yet, he described dolphins and penguins at the entrance of the Río de la Plata, a rare occurrence today. He may have missed breeding aggregations of penguins and other unique birds, such as albatrosses and petrels, sea lions and fur seal rookeries, and whales, all dependent on the same seascape (Fig. 12.2b). Ironically, it may be that subsequent "visitors," including today's human inhabitants of the continent, are also missing the importance of the PS and its dependent species and their roles in providing "services" (Ch. 13), as is the case today when fisheries pose a threat to the PS under the predominant vision of exploitation at any cost.

Conserving the ocean is about the art of integrating visions and needs, goals and options, approaches, methods, concepts, and interventions into an evolving perspective. But how does a conservationist decide what to protect in an ocean fully exploited by fisheries? Should the focus be on saving species as biodiversity or as resources? Or, should it be about protecting spaces? These questions are not negligible, considering that present approaches to ocean governance do not optimize any of them, despite the assumption that all are compatible.

Today, sustaining fisheries is the predominant concern of decision-makers for ocean policy and for managers that must decide on fishing grounds, targets, quotas, seasons, and effort through permit procedures and other management tools. Quotas are often determined species-by-species, rather than by the properties of ecosystems or the state of their habitats. Effective conservation, on the other hand, requires focusing on the diversity of life and habitats, on ecological functions, natural cycles, and species' life histories. Despite alleged differences in priorities, a portion of the conservation community shares the concerns of managers, and devotes efforts to ensure, for example, human livelihoods over protecting the diversity of

Fig. 12.2a Top predators and iconic species of differential conservation value. (i) Southern elephant seal (*Mirounga leonina*; photograph © Jim Large). (ii) Black-browed albatross (*Thalassarche melanophrys*; photograph © Jim Large). (iii) Magellanic penguin (*Spheniscus magellanicus*; photograph © Leonardo Campagna). (iv) Commerson's dolphin (*Cephalorhynchus commersonii*; photograph © Valeria Falabella). (v) Southern right whale (*Eubalaena australis*; photograph © J. McCormick-Ray). (vi) South American sea lion (*Otaria flavescens*; photograph © Claudio Campagna). (vii) Southern rockhopper penguin (*Eudyptes chrysocome*; photograph © Jim Large). (viii) Antarctic fur seal (*Arctophoca gazella*; photograph © Jim Large).

species. Others, these authors among them, suggest that conservation of biodiversity and ecosystem function are the principal priorities, even if the costs of eliminating threats to species means compromising short-term economic gain, the rationale being that without appropriate abundance of species and their healthy habitats, the chances for sustained human use are negligible.

Thus, the practice of conservation faces dilemmas about what a priority is and what the essential interventions are. Targeting a system to maintain its functionality may be the ideal end-goal of conservation efforts, yet available tools are more readily prepared to reflect and implement actions about species, spaces, or uses, than about underlying processes. Additionally, although species and spaces conservation are compatible, they often progress through different paths. Species conservation, for example, may require managing the resources of a fishery, while habitat conservation could be about creating temporary Marine Protected Areas. Both may convey that a no-take protected area, where fisheries may not operate, might be helpful for recovery of depleted

Fig. 12.2b Wildlife spectacles: colonies of iconic species. (i) Southern elephant seal (*Mirounga leonina*; photograph © Claudio Campagna). (ii) Imperial shag (*Phalacrocorax atriceps*; photograph © Victoria Zavattieri). (iii) Gentoo penguin (*Pygoscelis papua*; photograph © Valeria Falabella). (iv) Magellanic penguin (*Spheniscus magellanicus*; photograph © Guillermo Harris—WCS). (v) South American sea lion (*Otaria flavescens*; photograph © Claudio Campagna). (vi) King penguin (*Aptenodytes patagonicus*; photograph © Valeria Falabella). (vii) Black-browed albatross (*Thalassarche melanophrys*; photograph © Guillermo Harris). (viii) South American fur seal (*Arctophoca australis*; photograph © Daniel Costa).

species and fisheries. But should species conservation guide the protection of spaces or vice versa? And how should the ecological integrity of ocean systems be maintained, considering that integrity is about functional links of species and spaces?

A great deal of confusion and discussion among conservationists, managers, users, and scientists is reminiscent of a tug-of-war in which the ocean is least benefitted. More species are being placed at risk of depletion or extinction every year, more habitats are degraded, more fisheries are over-exploited or collapsing (e.g., Worm *et al.*, 2006; Polidoro *et al.*, 2008), and more marine habitats are globally placed at risk (Halpern *et al.*, 2008). It is also a fact that the proportion of the ocean presently protected as reserves, despite national and international commitments taken by governments, is negligible. It is finally a fact that without protected areas there will be always gaps in biodiversity conservation when attempted only via management of human activities. Increasing the proportion of the ocean protected or managed for sustainable use then becomes an undisputed urgency.

The tendency among some people, even some conservationists, is to believe that resources would be in better shape though simplistic approaches. Yet, despite often-conflicting perspectives, many conservationists and managers have in common the need for large spatial scale in their visions and interventions on species and habitats. The small scale of the great majority of protected areas that have a marine component rarely matches the natural history of species, especially large pelagic predators. But how could a reserve be effectively designated to cover hundreds of thousands of km^2 of open-ocean space of three dimensions, in which species are patchy, movable, and often inconstant? How could a reserve cope with relatively sudden changes in productivity, such as El Niño/La Niña cycles? Scenarios of conservation decisions are mined by such dilemmas.

This case study is, by no means, the only way to go about protecting large seascapes, but is one that begins with species, mostly marine birds and mammals for which we have satellite-tracking data of their movements at sea, and ends with spaces to promote conservation at a large scale in order to counteract the threats of unsustainable use. We here propose examples of areas (spaces) of the PS that, if protected, could benefit species and even resource use. Our rationale follows these steps:

1. Define the seascape (Sections 12.3 and 12.4) by stating the background geographic and biological information that support the relevance of the PS and its conservation status.
2. Define the relevant species selected as priority conservation targets on the assumption that if they are fully protected, many others will also be benefitted. In this process we discuss "candidate seascape species" and the resulting "seascape species" as the best ambassadors (Coppolillo *et al.*, 2004; Section 12.5).
3. Show how some of these important species relate to spaces, that is: describe areas within the target ecosystem where they forage or migrate (Section 12.6).
4. Propose examples of spaces that could serve as pilot sites towards the long-term goal of a Large Ocean Reserve (LOR), compatible with a large ecosystem perspective and the natural-history needs of seascape species (Sections 12.7 and 12.8).

12.3 OCEANOGRAPHIC AND BIOGEOGRAPHIC SETTINGS

The PS is a sector of the Southwestern Atlantic, lying mostly within Argentina's Exclusive Economic Zone (Fig. 12.1), in the order of the one million km^2. It is one of the most productive seas in the Southern Hemisphere, sustaining a variety of top predators, particularly marine mammals and seabirds that forage in pelagic habitats. Some species that find food in the PS come from distant places, as far away as New Zealand and South Georgia. The name PS is intended to promote a bio-centered perspective in contrast to jurisdictional, use-driven denominations that have been established by international agencies such as "Argentine Sea" or "FAO Area 41."

For conservation purposes, it is important to justify boundary conditions. The PS is not an enclosed system, so its distinction as a "sea" is functional rather than geographical, being defined by bathymetric and water circulation parameters rather than the limits set by the coastal realm (Ch. 4). Its physical profile defines three distinct environments: the extended, relatively shallow, flat continental shelf; the steep continental margin; and part of the deep Patagonian basin (Fig. 12.3a). Functionally, the PS depends on the circulation of the Malvinas Current (Piola and Matano, 2001; Boersma *et al.*, 2004), a northward-flowing branch of the Antarctic Circumpolar Current that carries cold, nutrient-rich water that meets the warm, nutrient-poor waters of the southward-flowing Brazilian Current located between 40° and 47° S, depending on season. The confluence of these two important currents occurs over a major bathymetric feature, the Patagonian continental slope that descends from 100 m to nearly 4000 m depth, and is commonly demarcated by the 200 m contour. As water circulation is significantly determined by the topography of the seabed, the spatial pattern of productivity is predictable in this system. The interaction of the Malvinas Current with the 1500 km-long shelf-break drives a seasonally variable, nutrient upwelling zone of biological productivity that spills up and across the Patagonian coastal shelf, making the shelf-break front the most productive feature of the PS seascape (Fig. 12.3b; Acha *et al.*, 2004). This spatial predictability and seasonal consistency of biological production is a functional signature of the PS that favorably influences its abundance of life.

Biogeographically, the PS links two marine ecoregions of the world, the Magellanic and Southwest Atlantic Warm Temperate provinces (Spalding *et al.*, 2007; Fig. 12.4). However, the PS is not the only template that has been suggested to integrate form (geology), function, and biodiversity in the Southwest Atlantic. Several alternatives differ in priorities and approaches. The closest template to the PS within a global framework of shelf ecosystems, both in terms of regional characteristics and large-scale ecological functions such as production processes, is the Patagonian Large Marine Ecosystem (LME). This LME is one of 64 LMEs defined by US/NOAA (Ch. 3; Sherman and Alexander, 1989). LMEs work within the framework of ecosystem approaches rooted in shelf systems, but represents only a subset of the larger PS as we have defined it, as it omits deep international waters important for biodiversity. Thus, the PS covers oceanographic regimes, including the high seas and parts of the Economic Exclusive Zones of Argentina and Uruguay that are excluded by the LME approach. Likewise, some iconic species of the PS, such as some pinnipeds, wandering albatross, and *Loligo* squid, are distributed beyond the limits of the LME, but are included in the PS.

Other templates for the region include marine provinces, ecoregions, and biomes. The only global regions that are consistently based on a standardized set of empirical data are the biogeographical coastal provinces and oceanic biomes of Longhurst (1998; Fig. 12.4). The Southwest Atlantic Shelf Province, for example, is based on environmental profiles and ocean-color satellite data, which identify an area similar in extent to the southern portion of the PS. In contrast, the warm temperate Southwest Atlantic and Magellanic provinces, as defined by Spalding *et al.* (2007), encompass coastal and near-shore pelagic waters, and are classified according to ocean circulation, biogeographic boundaries, and expert consultation, but do not strictly follow a functional perspective. Another classification for South America was based on the work of

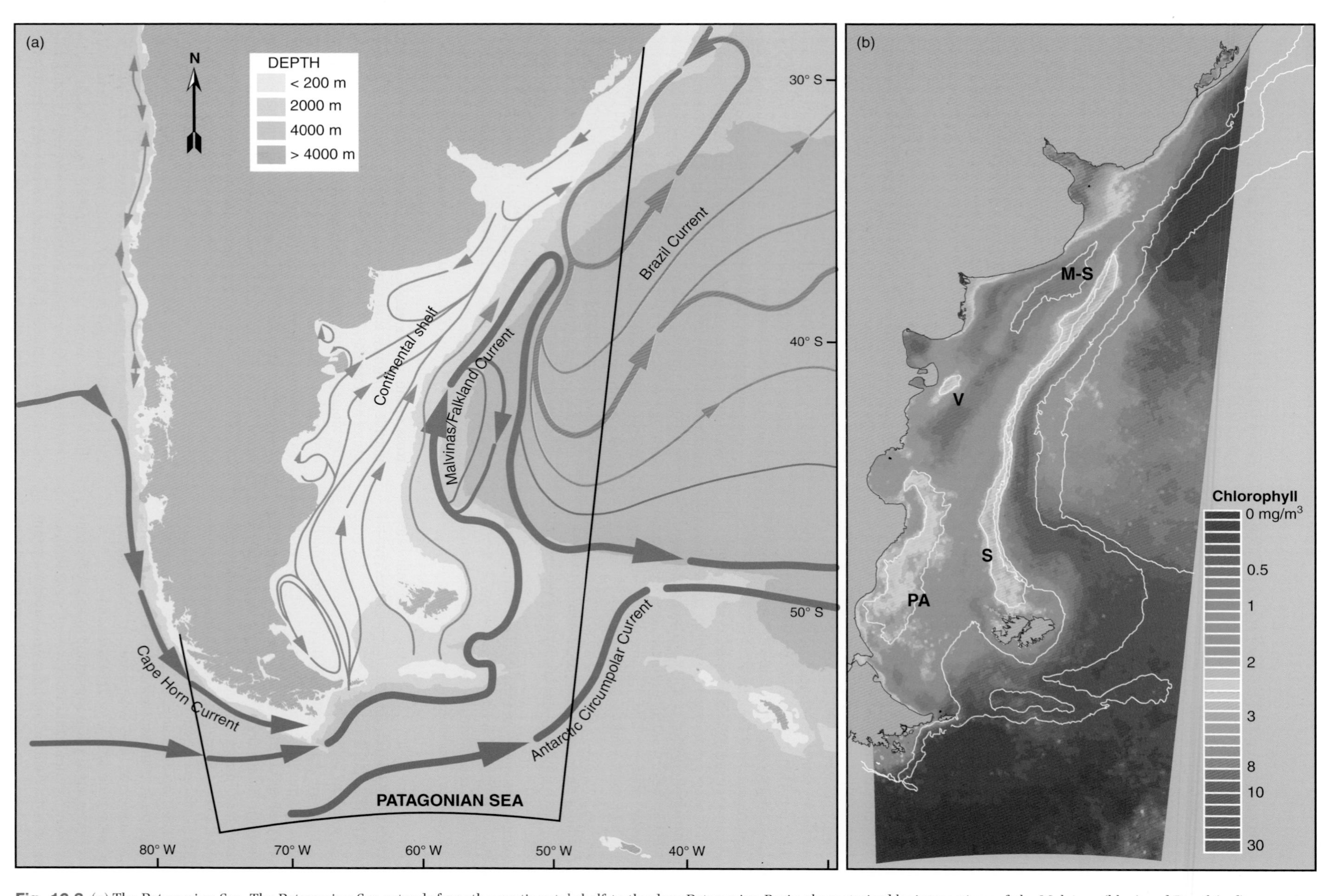

Fig. 12.3 (a) The Patagonian Sea. The Patagonian Sea extends from the continental shelf to the deep Patagonian Basin characterized by interactions of the Malvinas (blue) and Brazil (red) currents. The latitude and longitude lines that circumscribe the PS distinguish it from well-defined, neighboring environments such as the subantarctic system to the south (Matano *et al.*, 2010). (b) Ocean fronts. Average chlorophyll data for the Southern Hemisphere peak summer (January 1998–2009; values in mg/m^3). The $3\,mg/m^3$ chlorophyll contour marks the position of ocean fronts that are predictable in space and time: slope (S), mid-shelf (M-S), Valdés (V), and Patagonia Austral (PA). SeaWiFS data supplied by M Carranza, S Romero, and A Piola.

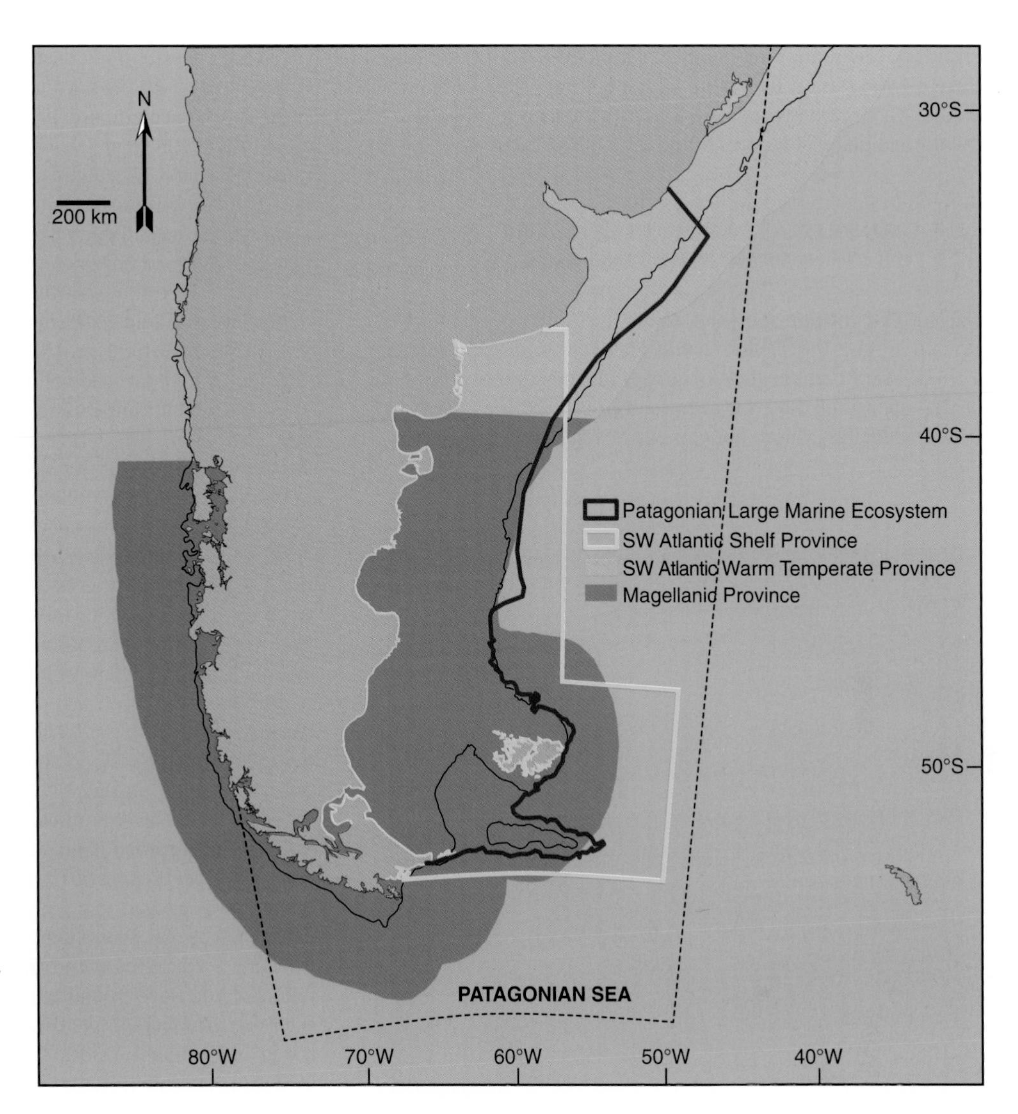

Fig. 12.4 Ecoregions and other comparative ocean templates. Compare with Fig. 12.3 (a), the Patagonian Sea. See text for references.

Sullivan and Bustamante (1999), who defined and delineated ecoregions according to patterns of ocean circulation, coastal geomorphology, and distribution of major faunal populations, and covers coastal areas and continental shelves, but not the high seas.

Regarding diversity, at least 700 species of vertebrates, a variety of invertebrates (mollusks alone account for over 900 species), and some 1400 species of zooplankton at the base of the food web have been described for PS waters, including the Brazil and Malvinas currents and associated pelagic and benthic habitats. As for most temperate systems, the PS is not as biodiverse as tropical seas. Rather, it is the abundance, very high productivity, and high biomass of fewer species that make temperate and cold-water systems unparalleled in marine environments; these features also make the PS ecosystem globally important as a source of food for resident and migratory species of fish, turtles, birds, and marine mammals (FORO, 2008), as well as for humans. Among the 83 species of seabirds, there are seven penguins (five residents) and 33 albatrosses and petrels (eight residents). Of the 47 species of marine mammals that reproduce or visit the area, four dolphins are endemic or have a limited world distribution and require urgent protection: La Plata river dolphin (*Pontoporia blainvillei*), Peale's dolphin (*Lagenorhynchus australis*), Chilean dolphin (*Cephalorhynchus epotropia*), and Commerson's dolphin (*Cephalorhynchus commersonii*, Fig. 12.2a,iv). The southern right whale (*Eubalaena australis*, Fig. 12.2a,v) breeds off the same coast and is now recovering from the threat of extinction. The only continental (and growing) breeding population of the southern elephant seal (*Mirounga leonina*; Fig. 12.2a,i; Fig. 12.2b,i) occurs on Patagonian shores. About 75% of the world population of black-browed albatrosses (*Thalassarche melanophrys*, some 400,000 pairs: Fig. 12.2a,ii; Fig. 12.2b,vii) breed and feed in the region. And, more than a million pairs of Magellanic penguins (*Spheniscus magellanicus*; Fig. 12.2a,iii) breed annually in colonies on the shores of the mainland and on islands. What makes all this possible is the high biomass of some forage species, such as Argentine anchovy (*Engraulis anchoita*), Fuegian sprat (*Sprattus fueguensis*), and Argentine hake (*Merluccius hubbsi*), all of which are important members of the PS ecosystem food web and provide food for the aforementioned marine mammals and birds (FORO, 2008).

12.4 CONSERVATION SETTING: THE STATUS OF A NON-PRISTINE OCEAN

The intense human footprint that affects all oceans is also evident in the PS (FORO, 2008; Halpern *et al.*, 2008; Fig. 12.5). A very conservative estimate indicates that at least 65 of 223 species of the PS that have been categorized following IUCN red listing criteria are globally or regionally threatened,

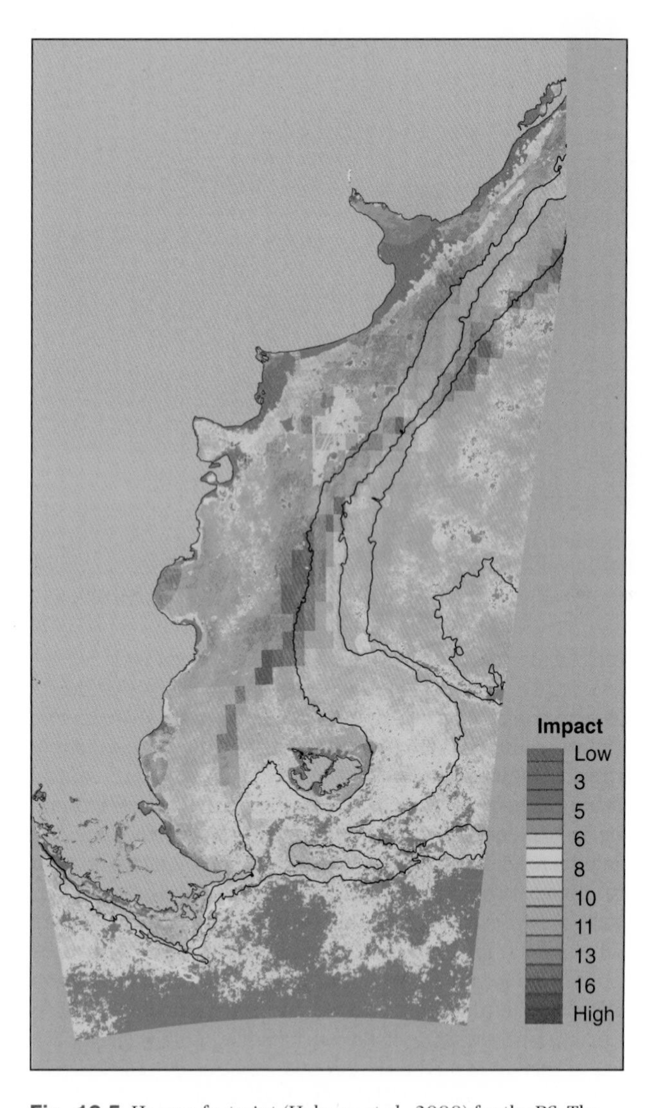

Fig. 12.5 Human footprint (Halpern *et al.*, 2008) for the PS. The analysis synthesizes 17 global data sets of anthropogenic drivers of ecological change, such as fisheries, introduced species, commercial shipping, and habitat modification, among others. Higher scores mean stronger ecological impacts of anthropogenic origin for the region relative to the rest of the ocean.

including invertebrates, cartilaginous and bony fishes, marine turtles, birds, and mammals. Most of the species that are targets of large commercial fisheries have not been categorized as in the IUCN *Red List of Threatened Species*. Threats are many—disease, oil, solid waste, pollution, introduced species, etc. (Ch. 2)—but most important is unsustainable fisheries. In addition to excess fishing capacity, threats of fisheries include illegal fishing, intentional discard of species of lower value (including an unknown number of invertebrates), and incidental bycatch and entanglement of seabirds, mammals, and sea turtles in fishing gear. Fishing boats are also sources of chronic oil and garbage pollution.

Garbage pollution has not been quantified at sea, but is apparent along the entire coast of Patagonia, and in the stomach contents of seabirds and turtles. The Magellanic penguin supports significant nature tourism activities in coastal areas and is the species most affected by accidental oil spills and chronic hydrocarbon pollution. It has suffered the most mass mortality events of any other species for which we have data and is also exposed to the greatest number of infectious diseases; fortunately, it probably has a high resistance to common avian pathogens. Other seabirds and marine mammals have been exposed to the agents of epidemic diseases (FORO, 2008). Seabirds have been exposed to paramyxoviruses (types 1, 2 and 3), avian smallpox virus, and non-pathogenic strains of the avian influenza virus. Two mass mortality events affecting 100,000–200,000 seabirds have been recorded since 2002 in the Malvinas Islands. Seals have been exposed to morbillivirus and herpes viruses, which have caused mass mortalities in other places of the world. Effects on ecosystems, such as harmful algal blooms (red tides) adjacent to the shelf-break front, have also been reported and have spread in time and space. Each year a considerable number of dead right whales is observed on the coasts of Península Valdés. In 2007, a record 83 right whales died from unknown causes, and 93% of these were juveniles.

Unsustainable fisheries are, beyond argument, the most insidious and detrimental of all present threats to marine biodiversity. "Unsustainable" is defined as extraction beyond replenishment; it means paying no attention to the precautionary principle and to scientific data that show high impact, even when provided by advisory bodies to managers. All species of marine turtles and a growing number of shark and ray populations face a high probability of local extinction if overfishing is not mitigated. The three pinnipeds resident in the PS have not been officially designated as "threatened" in IUCN's *Red List of Threatened Species*, but are susceptible to threats that have caused population declines in other parts of the world, for example by entanglement and accidental catch in fishing gear (FORO, 2008).

The fishing industry in the PS has expanded during the last two decades, and is incurring environmental costs that are not subtracted from the benefits. In 1997, Argentine fisheries had their highest level of success, with well over a 1 million mt of capture and >US$ 1 billion of exports. This is an underestimate for the total catch; for only Argentine hake, the 1997 take (including an estimate of bycatch and unreported data) was estimated at >800,000 tons, or 111% higher than the allowable catch for that year. Again in 1998 and for the first decade of the 21st century, income from fishing exports has been typically over the US$ 1 billion per annum level. Fishery declines are also apparent. In 2010, the reported total catch in the PS reached only 715,000 mt, the lowest level in a decade; nevertheless, exports amounted to US$ 1.3 billion, a record. Thus, 2010 may be remembered not only for its record income, but also for its declining catch, which also seems to be the case nationally. Some Patagonian provinces are landing 70% less biomass in fish than a half-decade before. For Santa Cruz province, the catch was almost 206,000 mt in 2006, but 46,000 in 2010. For Chubut province, second in importance after Buenos Aires, the catch decreased by 50% during the same period. Overfishing may also be measured by other means; for some time, the fishery of the Argentine-Uruguayan Common Fishing Zone has extracted 49% of the maximum photosynthetic energy of the ecosystem, a larger proportion than in other intensely exploited temperate ecosystems around the world. The conclusion is that this fishery has reduced the yield for the most-prized species.

The conflict between fisheries and marine predators centers on spaces where they overlap. PS fisheries primarily target species that depend on the shelf and shelf-break fronts. The primary productivity and biomass of these areas sustains fisheries that supply high-quality products to wealthy markets (e.g., shrimp, squid, and scallops to Spain and North America, for example)—at the risk of depletions and loss of ecological sustainability. In this regard, the fishery for anchovy and other forage fishes, largely for fish meal, is of special concern. These species provide a crucial intermediate step in the flow of energy through the food web between plankton and wildlife. In 2004 and 2005, Argentine catches of anchovy exceeded 30,000 tons of anchovy for the first time in 30 years, and demand is rising. Changes in forage-fish populations can alter populations of both predator and prey, with consequences for penguins, seals, and whales that are of major value to the tourist industry (Skewgar *et al.*, 2007; Smith *et al.*, 2011).

Other current uses have also affected PS diversity and abundance of marine life to a point that some populations, such as albatrosses and marine turtles, are rapidly decreasing in numbers. The black-browed albatross, for example, is a widespread, endangered species of the region; the majority of its world population inhabits offshore and pelagic systems of the PS, where it interacts heavily with fishing vessels and is part of the incidental bycatch of longliners and trawlers. Breeding colonies are localized, and although not all are declining at the same rate, the main overall trend indicates a substantial decrease, largely due to fisheries interactions (FORO, 2008).

12.5 SEASCAPE SPECIES: A FIRST APPROACH TO SETTING CONSERVATION PRIORITIES

Up to this point, we have observed that production and biodiversity of large predators make the PS functionally unique, what some of the best representatives of its biodiversity might be, and what threatens biodiversity and habitat quality. The purpose of this case study is to illustrate approaches to species and spaces conservation for the PS in order to identify areas appropriate for conservation, to address threats, and to achieve sustainability. Here, we describe an approach for selecting differentially important seascape species that together represent targets as a first step for identifying important, large-scale areas of the total PS seascape for conservation. Critical areas, once conserved, have the potential to help secure the sustainability of the ecosystem, as well as the viability of its species. A detailed application of the seascape species approach for the PS is reported in Campagna *et al.* (2008).

The PS, as mentioned above, is a heterogeneous region comprised of a number of mesoscale seascapes that supports hundreds of vertebrate and thousands of invertebrate species, which are distributed according to its oceanographic, bathymetric, water circulation, and productivity patterns. How may conservation action be guided, then, by differentially targeting some species based on clues provided by their life histories? How could species, drawn from the PS's total species pool, serve to identify areas in the coastal and open ocean as potential areas for protection of that biodiversity, as well as for maintenance of system function? One way to get started is to select from the total diversity of species those that are known to have habitat requirements that make them particularly susceptible to human alteration and use of the natural system. Fig. 12.6 lists 33 "candidate seascape species" important to the region and for which natural histories are reasonably well known; the figure also identifies ocean regimes and habitats in which they may occur. The "seascape species approach" is the selection process for selecting "seascape species" from the candidates by considering the following five criteria:

- area requirements: use of many habitats within the seascape;
- dependence on large-scale, heterogeneous areas;
- performance of critical ecological functions;
- vulnerability: sensitivity to human use, and;
- socio-economic significance.

These selection criteria allow reduction of potential candidates to a manageable number of mostly large-bodied, predatory vertebrates, and a few invertebrates, which collectively occupy the full range of habitats of the PS. Data for these species are then entered into decision-support software that reports summary scores and relative ranks for each species.

The application of the seascape species approach for the PS was first inspired by studies of the foraging behavior of pelagic PS predators, such as southern elephant seals. Dispersal of this species' adults and juveniles at sea is better understood when the seascape is seen functionally and all habitats are understood as a combination of benthic topography, circulation patterns, and food supply (Campagna *et al.*, 2006, 2008). Additional studies on other pelagic predators that migrate by water or air, and cover hundreds of thousands to millions of square kilometers, e.g., penguins and albatrosses, also support the scenario illustrated by elephant seals (BirdLife International, 2004). The seascape species selected in previous work for the PS are listed in Table 12.1, and include marine mammals, birds, other vertebrates, and some invertebrates that collectively affect the trophic web at different levels, exhibit extended foraging ranges, depend on the productivity and diversity of many habitats, and forage in pelagic and benthic habitats.

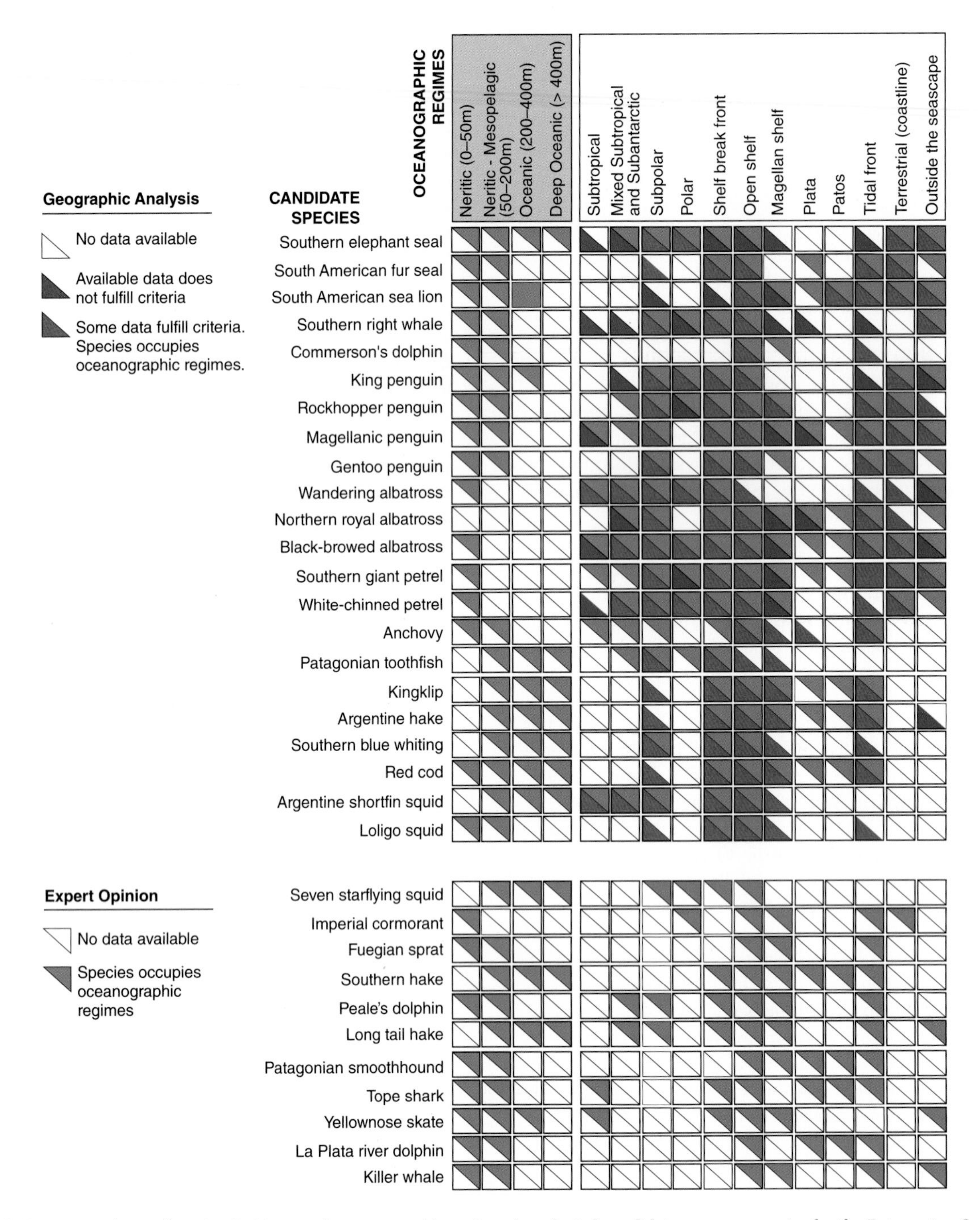

Fig. 12.6 Summary of use of marine habitats and oceanographic regimes by selected candidate seascape species for the Patagonian Sea, based on analysis of available spatially explicit data and expert opinion (Campagna *et al.*, 2008). Reproduced with permission of John Wiley & Sons.

The suite of seascape species listed in Table 12.1 captures the full range of habitat types, management units, and threats in the PS. The table also points out particular regions of the ocean that are differentially relevant for conservation as foraging grounds. However, the selection of species that are widely distributed and depend on heterogeneous habitats does not alone address specific ocean spaces that are differentially relevant for them. For that, we need to integrate seascape species with other sources of data; we use satellite tracking for this purpose.

12.6 FROM SEASCAPE SPACES TO IMPORTANT FORAGING AREAS

The seascape species approach is useful to secure biodiversity conservation. The method is complemented by identification of foraging areas for seascape species by deploying state-of-the-art satellite-tracking techniques (tracked species, Table 12.1). We propose that the foraging distribution of species that fulfill the criteria listed above (Section 12.5) helps identify important, large-scale areas essential for those species. The

Table 12.1 Critical species for the Patagonian Sea. For definition of "seascape species" see text. "Other predators" refers to species that were identified as candidate seascape species in Campagna *et al.* (2008), except for the gray-headed and light-mantled albatrosses that did not fulfill all criteria. Seascape species and candidates fulfill similar criteria, but the former have an ecological profile that best represents all the rest. In bold: 16 species that have been satellite tracked (see Fig. 12.7 for distribution of some of them at sea). Not all seascape species have been satellite tracked, and some satellite-tracked species did not make the list of seascape species, despite being candidates. "Areas" link to Fig. 12.8 and identify which of the important regions of the seascape are particularly relevant for each species. Analytical methods and threats to these species are also described in detailed in Campagna *et al.* (2008). Resident is a species that spends the entire annual cycle in the Patagonian Sea; visitor is a species that forages in the Patagonian Sea but reproduces elsewhere. IUCN status is the *Red List* category for each species (NT: Near Threatened; EN: Endangered; VU: Vulnerable; LC: Least Concern). See *Red List* criteria in: www.iucnredlist.org/technical-documents/categories-and-criteria.

Common name	Spanish name	Scientific name	Areas	Use of the system	IUCN status
SEASCAPE SPECIES					
Magellanic Penguin	Pingüino de Magallanes	*Spheniscus magellanicus*	3, 5	Resident	NT
Black-browed Albatross	Albatros ceja negra	*Thalassarche melanophrys*	1, 2, 3, 4, 5	Resident	EN
Southern Rockhopper Penguin	Pingüino de penacho amarillo del sur	*Eudyptes chrysocome*	5	Resident	VU
Yellownose skate	Raya picuda	*Dipturus chilensis*			
Argentine shortfin squid	Calamar argentino	*Illex argentinus*			
Southern right whale	Ballena franca del sur	*Eubalaena australis*	The species reproduces in the seascape but forages beyond its limits.		LC
Southern blue whiting	Polaca	*Micromesistius australis*			
OTHER PREDATORS					
Wandering Albatross	Albatros errante	*Diomedea exulans*	1, 2, 4, 5	Visitor	VU
Northern Royal Albatross	Albatros real del norte	*Diomedea sanfordi*	1, 2, 4, 5	Visitor	EN
Grey-headed Albatross	Albatros cabeza gris	*Thalassarche chrysostoma*	5	Visitor	VU
Light-mantled Albatross	Albatros manto blanco	*Phoebetria palpebrata*		Visitor	NT
Southern Giant Petrel	Petrel gigante del sur	*Macronectes giganteus*	1, 4, 5	Resident	LC
Northern Giant Petrel	Petrel gigante del norte	*Macronectes halli*	1, 5	Visitor	LC
White-chinned Petrel	Petrel negro	*Procellaria aequinoctialis*	1, 3, 4, 5	Resident	VU
King Penguin	Pingüino rey	*Aptenodytes patagonicus*	4, 5	Resident	LC
Gentoo Penguin	Pingüino Papúa	*Pygoscelis papua*	5	Resident	NT
Southern Elephant Seal	Elefante marino del sur	*Mirounga leonina*	1, 2, 3, 4, 5	Resident	LC
South American Sea Lion	Lobo marino de un pelo sudamericano	*Otaria flavescens*	3, 5	Resident	LC
Antarctic Fur Seal	Lobo marino de dos pelos antártico	*Arctophoca gazella*	4, 5	Visitor	LC
South American Fur Seal	Lobo marino de dos pelos sudamericano	*Arctophoca australis*	5	Resident	LC

purpose is to guide procedures about where to concentrate conservation efforts, as the choice of conservation areas must necessarily integrate ecological relevance, use, and natural-history knowledge, and will result from a process of selecting seascape species and understanding how and why they are distributed at sea.

All seascape species, by definition, use large areas of the ocean, interact with many other species, and overlap with human activities. The seascape species approach makes clear that although no single species may inhabit more than a small proportion of the hundreds of thousands of km^2 of the PS, only a few highly mobile representatives range over the entire area. This approach exposes the weaknesses of protecting only very small coastal areas, while attempting to fulfill the requirements of conservation of pelagic predators. Some seascape species of the PS, for example, have extensive breeding distributions along the coast. Others have annual migratory cycles divided into breeding and foraging periods, during which they make intense use of the vertical and horizontal dimensions of the seascape. Still others are strictly coastal, with overlapping breeding and foraging areas. Most migrate through a variety of jurisdictions, thus have transboundary distributions. Most colonies of land-breeding species use small coastal areas susceptible to protection; these breeding spaces contrast with the huge ranges of some species' dispersion at sea of up to hundreds of thousands or even millions of km^2 (Falabella *et al.*, 2009). Some individuals travel thousands of kilometers away from the coast and may remain at sea for weeks or even months (Campagna *et al.*, 2006, 2007; BirdLife International, 2004). This is not to say that small, protected areas are to be neglected; strict protection of reproduction colonies (rookeries) can effectively restrict human impacts such as extensive disturbance by tourism or coastal development. Nevertheless, they are insufficient to protect a species throughout its annual cycle.

Reproduction areas may be readily identified, at least for land-breeding birds and mammals, but equally important foraging areas are more obscure. Therefore, to increase protection, it is necessary to locate these areas and to determine if a proportion of them can be secured against threats, such as competition or overlap with unsustainable fisheries, bycatch, waste, or entanglement. Satellite-tracking studies are available for 16 species of large predators of the PS (Table 12.1; Falabella *et al.*, 2009), including residents and four species that visit the PS: grey-headed, northern royal, and light-mantled albatrosses, and Antarctic fur seals.

The results of tracking clearly illustrate the extraordinarily large oceanic distributions of seascape species, for example pinnipeds, although some are more restricted than others (Fig. 12.7a,b,c). Other tracked species reveal similar results. Satellite-tracking and natural-history data also demonstrate that critical areas include land and nearshore environments, oceanographic fronts associated with the continental shelf and slope, and areas adjacent to the Malvinas Islands. The areas occupied by these species encompass many different habitats, and are connected by migratory corridors that species such as penguins and seals use to travel from breeding colonies to their foraging areas. By integrating satellite-tracking data from these seascape species, a few places appear to be most important and susceptible to spatial conservation (Fig. 12.8):

1. habitats influenced by the outflow of the Río de la Plata;
2. the slope area of the continental shelf or shelf-break front;
3. the corridor that links Península Valdés with the continental slope;
4. the shelf slope at the latitude of Golfo San Jorge; and
5. habitats adjacent to the Malvinas-Falkland Islands.

Satellite and natural-history data show that no one species may suffice to identify the most relevant large-scale regions of the ocean for conservation. Rather, the seascape species approach searches for a suite of species that together represent the ecological relevance of species and spaces and the need to reconcile conservation of biodiversity on a large scale with human demands. The distributions of 33 candidate species (Fig. 12.6) and the distributions of a few selected seascape species (Fig. 12.7) are in agreement that conservation of the PS must be conducted at a large regional scale. Satellite studies suggest, in addition, that strict area protection (e.g. MPAs) may not be practical for some species due to their extraordinarily large dispersal. Even a large pelagic protected area, for example, may not cover a significant proportion of the home range of the southern elephant seal that is distributed very widely over the ocean surface and in the water column (Fig. 12.7a). Indeed, almost the entire PS would need to be turned into a protected area to cover the total requirements of that one species. Such a large protected area may arguably not be even desirable, and is certainly neither practical nor feasible because jurisdictional complexities would become almost insurmountable transboundary obstacles. Contrary to elephant seals, some otariids are more susceptible to spatial conservation tools as they are distributed over smaller areas (Fig. 12.7b). Female South American sea lions in the first stages of the nursing period when they are most vulnerable could be used to estimate size and shape of protected pelagic areas. Magellanic penguins complement sea lions as another example of species susceptible to protection by ocean reserves of reasonable size.

In sum, effective conservation or management for large, ecologically important, ocean predators cannot be successful on a species-by-species level, but requires a multispecies approach that includes both terrestrial breeding rookeries (Fig. 12.7c) and marine feeding areas. It is the integration of species and spaces that defines important areas of the seascape. Once distributions are made comparable for a suite of species, and are pooled in a common database, a few large areas, at seascape scales in the order of several thousand km^2, arise as preeminent for use by pelagic predators and for conservation.

12.7 THE CONCEPT OF "LARGE OCEAN RESERVES"

Estimating the optimal dimension of an ocean conservation area is fundamental to long-term sustainability. We have shown that the PS is a productive system, where ocean fronts are predictable in space and time. We have suggested that oceanic productivity is predictable and an asset for top predators that reside or arrive in the PS from distant breeding areas. Also, we have estimated that some of these species are better ambassadors for biodiversity than others, and have listed a

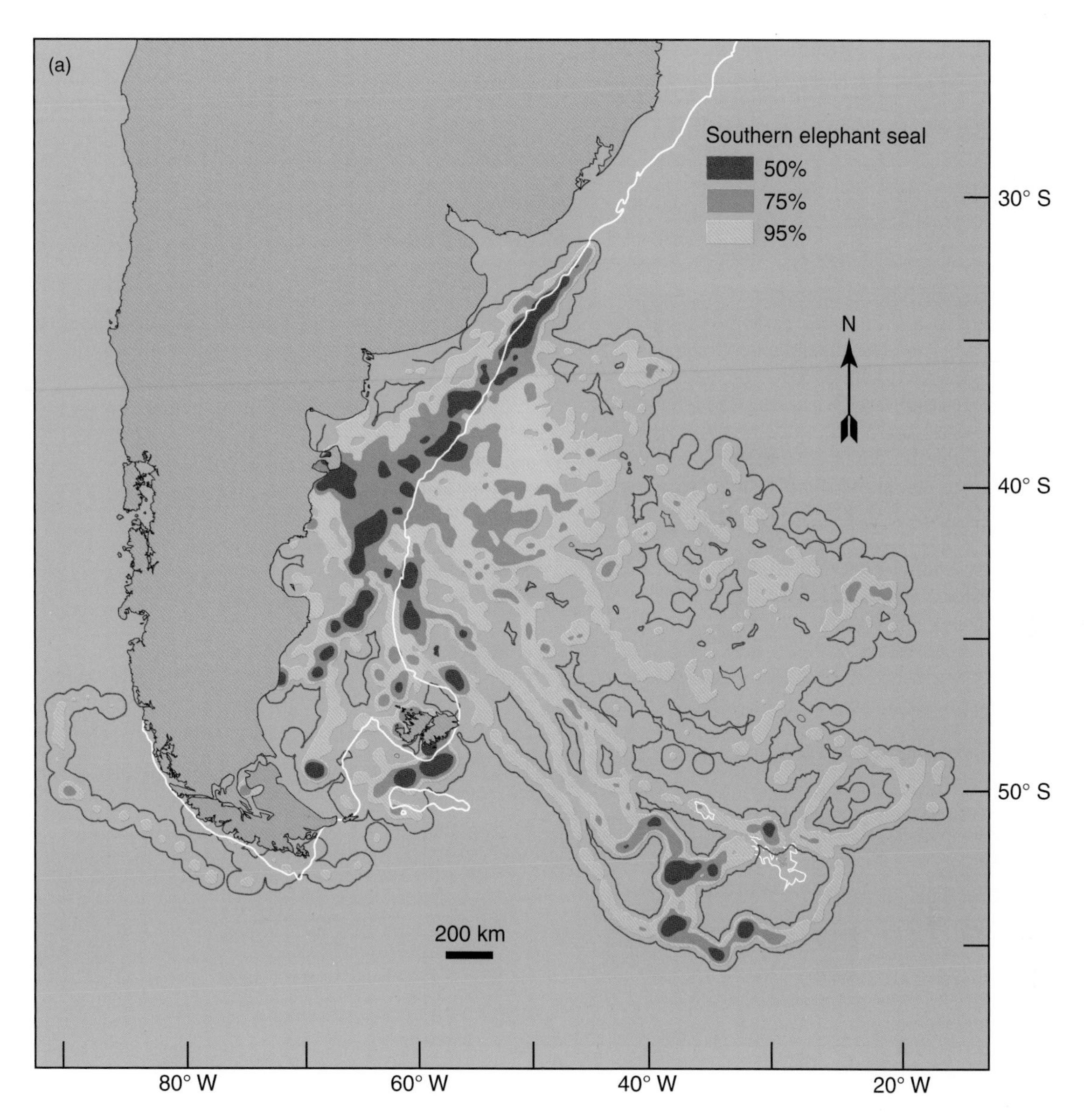

Fig. 12.7 Pinnipeds and the Patagonian Sea; ocean use and breeding colonies. Distribution contours indicate utilization areas within which satellite-tracked individuals spent 50, 75, and 95% of their time at sea. (a) Utilization distribution contours for southern elephant seals from Península Valdés. Elephant seals disperse over millions of km^2. (b) Utilization distribution contours for South American sea lion, South American fur seal, and Antarctic fur seal. The fur seal is distributed widely; the other two are exclusive or predominant shelf foragers. (c) Pinniped breeding colonies (Dans *et al.*, 2004; Reyes *et al.*, 1999; Schiavini *et al.*, 2004).

small group of them as candidate and/or seascape species. We have finally determined that the distribution of individuals that were satellite tracked during the foraging phase of their annual cycles indicates that pelagic predators use the ocean at a large scale, as do fisheries. Yet, Marine Protected Areas for the PS are small and coastal (see Section 12.8; Fig. 12.10a). This mismatch between use and conservation requires fixing. We thus propose a means to advance the application of large-scale conservation areas that could have a positive impact on reducing the human footprint, and in the long-run, enhancing sustainable species and spaces in the PS. Our perspective is one possible way to advance the goal of ocean conservation; we acknowledge that there may be complementary or overlapping procedures. Progress will require use of all relevant information, and expert opinion should be considered feasible for future guidance.

A report by the Ocean Governance Council of the World Economic Forum Global Agenda concludes that a priority vision for ocean conservation should be the creation of Large Ocean Reserves (LORs), particularly High Seas Large Ocean Reserves (WEF, 2012, online). LORs are areas that encompass hundreds of thousands of km^2 where jurisdictions overlap, including those between land and sea and high seas waters beyond national control. The report states:

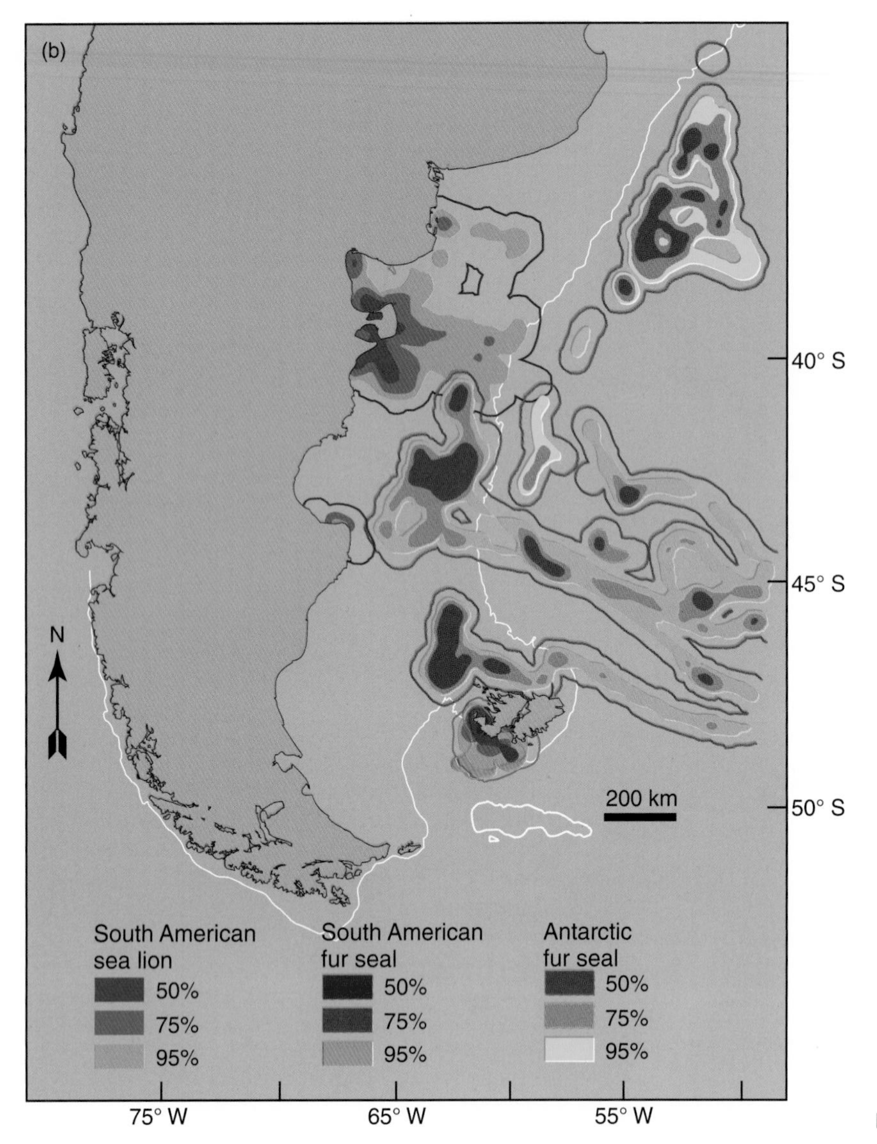

Fig. 12.7 (*Continued*)

"While it is clearly desirable to increase the number of coastal protected areas, they will always constitute a very small fraction of the world's ocean. For protected areas in the ocean to play a major role in the long-term sustainability of biodiversity, biological production and ecological services, very large protected regions, preferably no-take zones that encompass significant habitat space in the coastal ocean, the open ocean and the deep sea, will need to be established."

This statement recognizes that major threats to ocean species operate at large scales and that human impacts on very wide-ranging species can span entire biogeographic regions, in some cases extending globally. The recent Gulf of Mexico oil spill *Deepwater Horizon*; Table 2.2), for example, affected species on a regional scale, some of which were already globally threatened (Campagna *et al.*, 2011). Likewise, commercial and international fisheries are virtually all large-scale operations. Even the small fleet of a few boats targeting scallops on the continental shelf of the PS covers huge areas. This fleet operates in three sectors covering about 11,250 km^2 of shelf habitat. Every day, all year round, about 5 km^2 of shelf bottom are trawled. A much larger operation is the jigging fleet targeting Argentine squid (Waluda *et al.*, 2002); boats deploy high-intensity, incandescent lights in their operations, which can be detected by satellite, reminiscent of a large city spread over hundreds of kilometers of shelf edge (Fig. 12.9). The Argentine shortfin squid (*Illex argentinus*) supports one of the most selective fisheries, and represents the highest catches in the region. In 1987, night fishing began with the use of automatic machines and squid jigs. This method of selective fishing by bright lights placed on deck concentrates the squid during the night. The population of this species has fluctuated and it is highly possible that it will be damaged by the intensity of the fishing effort. Other ocean species are even more widespread. Recent integration of satellite-tracking data for 16 species of birds and mammals showed that it is possible to identify putative foraging areas in the open ocean where several species, such as albatrosses and seals, converge (Falabella *et al.*, 2009).

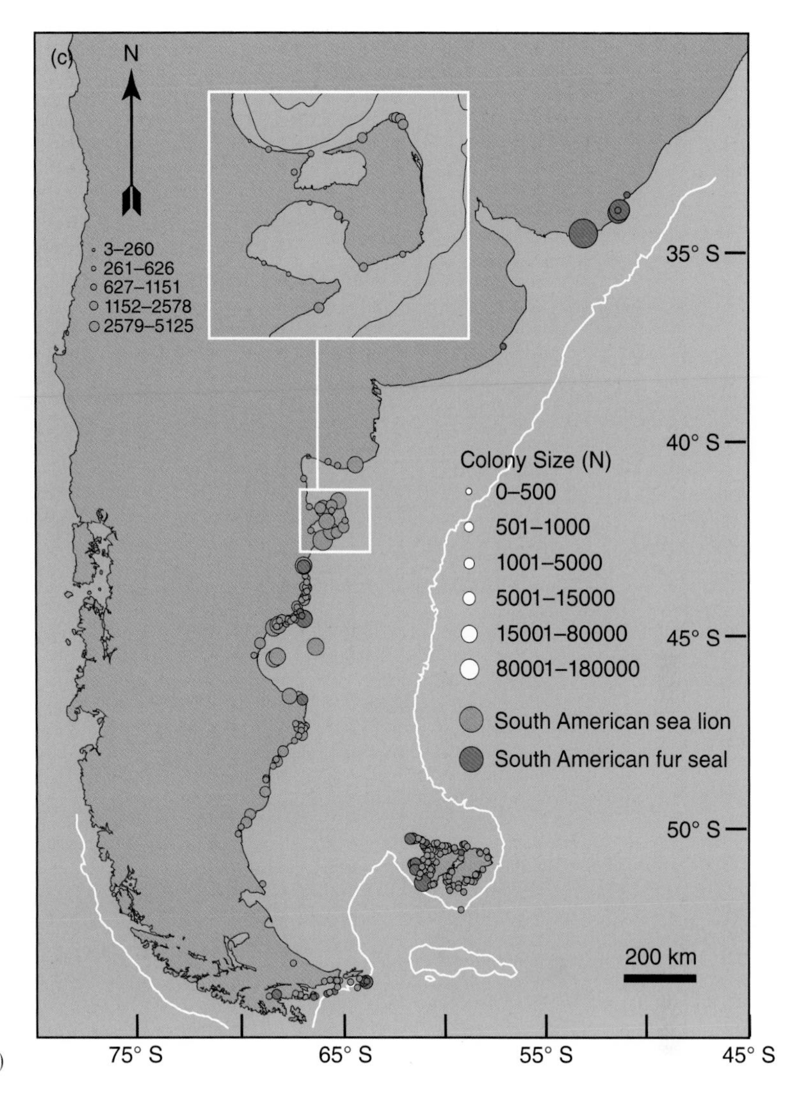

Fig. 12.7 *(Continued)*

Threats and foraging distributions of seascape and candidate seascape species illustrate generally how fisheries and pelagic species may converge over hundreds of thousands of km^2 of ocean space. What then are the conservation and/or management mechanisms that can sustain commercially important resources, biodiversity, and functionally critical processes occurring within the same area? The question may be appropriate, but answers are controversial. Paucity of data about species-ecosystem relationships and lack of precaution in management decisions slow conservation interventions. Institutional and jurisdictional frameworks provide little guidance for decision-making as they are most often directed towards economic development, spread over many governance levels, and require regional or international consensus that is always difficult to achieve. Therefore, it is not surprising that all but a few marine environments are poorly represented in global conservation efforts, a notable exception being coral reefs. Offshore, pelagic protected areas are a particular challenge in concept and in practice, as they need to be dynamic in space and time (seasonal or mobile), requiring complicated monitoring, enforcement, logistics, and legal agreements.

Are LORs, then, an idealistic vision with little chance of implementation? Despite the difficulties for effective open-ocean conservation to protect biodiversity, it has been possible to create some LORs that include large expanses of open waters. Several examples exist, one of the oldest and best known being the Great Barrier Reef Marine Park (1975; 344,000 km^2) and one of the most recent the Papahānaumokuākea marine area (2006; 360,000 km^2) in Hawaiian waters (see Toropova *et al.*, 2010 for a complete list). Some of these initiatives may not solve problems where the most pervasive threats occur. As a matter of fact, ocean conservation is still guided by the "low-hanging fruit" effect; i.e., easily achievable goals that are more opportunistic than scientifically guided. The high seas beyond national EEZ jurisdictions is especially in need of large-scale spatial conservation tools. Important initiatives such as for the Sargasso's Sea seascape and the South Orkney Marine Protected Area encourage more thinking in a direction that expands beyond biodiversity protection to ecological function. The Sargasso's Sea seascape is an initiative supported by the Government of Bermuda in collaboration with scientists, international marine conservation groups, and private donors.

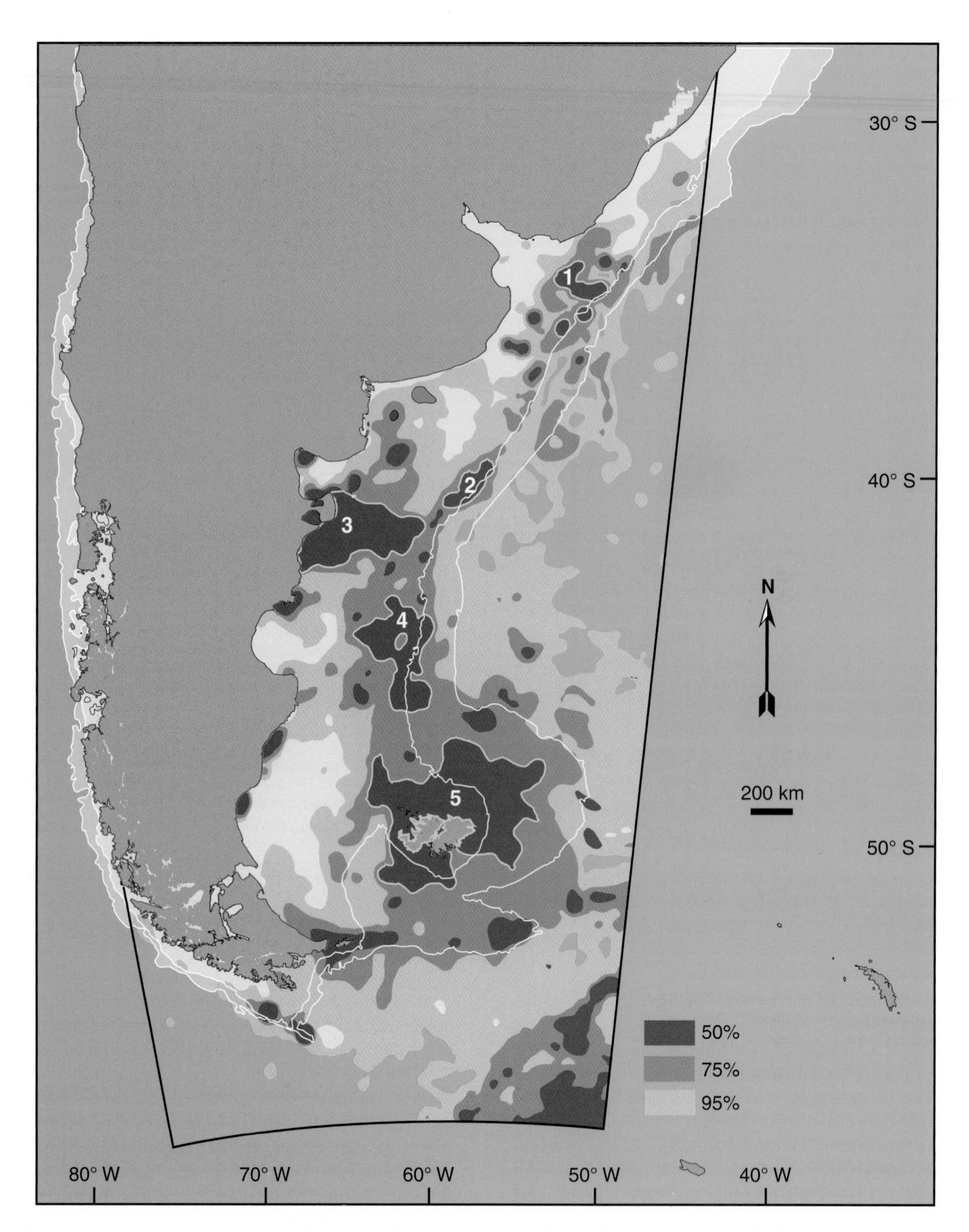

Fig. 12.8 Important areas for foraging top predators. Data for 16 species of satellite-tracked, foraging marine birds or mammals. Distribution at sea is represented by the 50, 75, and 95% utilization distribution contours, indicating areas within which individuals spent 50, 75, and 95% of their time at sea. Most relevant migratory and foraging areas based on the available information are: (1) habitats influenced by the outflow of the Río de la Plata; (2) the slope area of the continental shelf or shelf-break front; (3) the corridor that links Península Valdés with the continental slope; (4) the shelf-slope at the latitude of Golfo San Jorge; (5) the habitats adjacent to the Malvinas Archipelago, and extended southward towards the Burdwood Bank. See also Table 12.1 (column for "Areas").

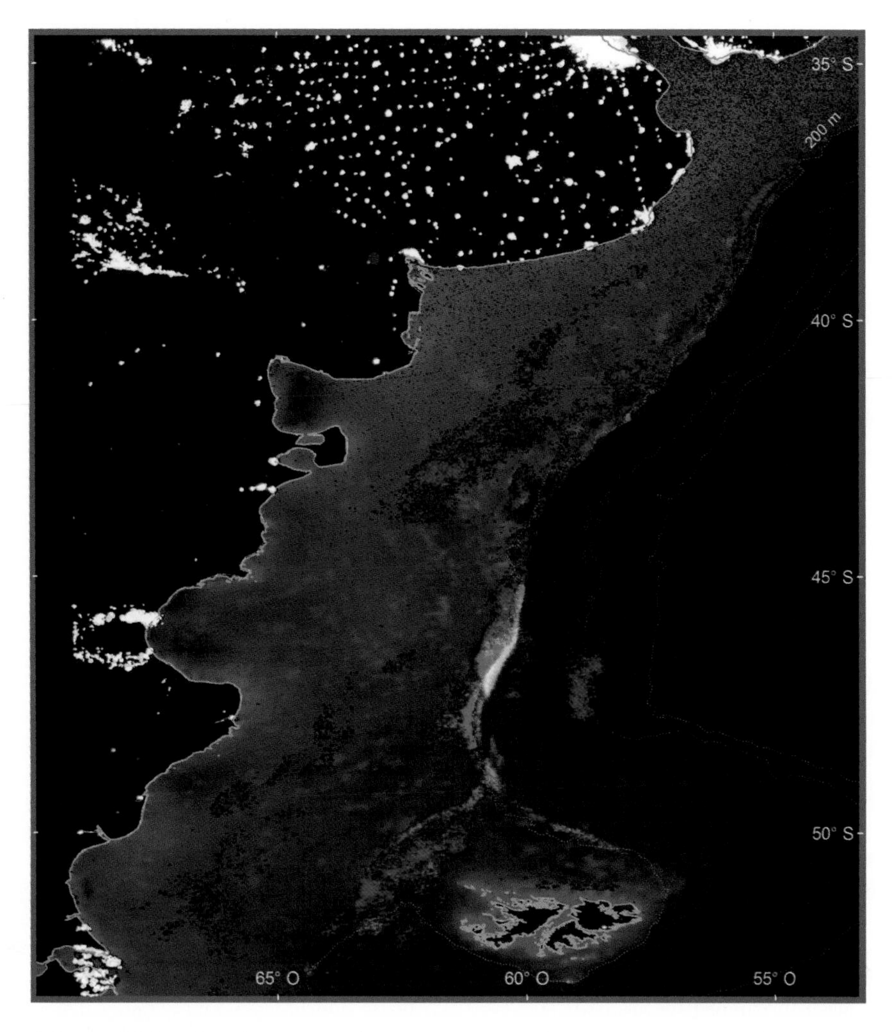

Fig. 12.9 Distribution of the squid fishing fleet (See Fig.12.1 for orientation). Squid fishing takes place mainly from squid jiggers. The fishery operates day and night and uses lights so powerful that they can be detected by satellites. It is thus possible to identify the movement and distribution of the fleet using satellite images. Map shows the distribution of the combined squid fishing fleets in 1995 (blue), 2000 (green), 2008 (red), and when two or more years are combined, no matter where located (yellow). The areas where the fishing boats concentrate coincide with the most relevant areas identified for top predators, namely: the waters adjacent to the Malvinas Islands; the slope front, mainly on the latitude of the Gulf of San Jorge and the Río de la Plata; as well as the shelf and slope waters opposite the Valdés Peninsula. The concentration of the squid fishing fleet on the slope, opposite the Gulf of San Jorge, was particularly noticeable in the three years shown on the map. Low light imaging data collected by the US Air Force Defense Meteorological Satellite Program. Data processing by the NOAA National Geophysical Data Center. Courtesy Christopher Elvidge.

The South Orkney MPA is the first within the jurisdiction of the CCAMLR. It was declared at the 28th annual meeting of CCAMLR and came into force in May 2010 and is the first step towards the development of a representative network of MPAs in the Southern Ocean.

12.8 A FIRST STEP TOWARDS A PATAGONIAN SEA LOR: CANDIDATE AREAS FOR CONSERVATION

We have seen that scale is important in ocean conservation, and that the present status for protected spaces in the PS is not fulfilling the profile for an LOR. Coastal MPAs are of limited conservation effectiveness in that they are typically small, generally undermanaged, do not effectively restrain fisheries, and fail fully to protect pelagic fishes, marine mammals, and birds that are distributed over expanses of ocean space. Coastal breeding colonies such as those of the Magellanic penguin, South American sea lion, and elephant seal are rarely overlooked as conservation targets, as they represent wildlife spectacles of enormous esthetic and economic potential for ecotourism (Conway, 2005; FORO, 2008; Falabella *et al.*, 2009). Conversely, distant seascapes at sea, where the same colonial species forage, are out of sight and consequently can be easily disregarded as conservation priorities. Yet, it is in these same oceanic foraging areas that bycatch in fishing gear and competition for food between fisheries and marine mammals inflict mortality and cause decreases in population sizes (Croxall and Prince, 1996; García-Borboroglu *et al.*, 2006).

We are now at the stage of advancing an integrated, science-supported rationale for large-scale conservation of the PS seascape. Today, a step towards integrated LORs is possible for the PS and could be supported by the following available tools:

• Application of a focal-species approach (seascape species approach, Section 12.4) that promotes an integrated, multi-species understanding of the target seascape.
• Integration of satellite-tracking data to indicate putative foraging areas (Section 12.5).
• Analysis of threats with a diagnosis of the conservation status of the target system (Section 12.3).
• Bio-zoning of the seascape, with prioritization of sites important for human use and biodiversity (to be developed).

Use of these tools points toward a long-term, sustainable integration of Patagonian coastal-marine development and biodiversity conservation, in which a prominent role will initially be played by a mesoscale network of coastal and oceanic protected areas. Hopefully, these can eventually evolve into an integrated Patagonian Sea LOR as their advantages become clear. Most threatened species respond to conservation measures only if the areas where they reproduce, forage, or migrate are placed under integrated management decisions that consider multiple species at large scales. But what is the scenario that needs to be adopted by means of large-scale spatial and species conservation tools?

Presently (2012), the system of MPAs for the Argentinian coast of the PS, including the intertidal zone, covers an insignificant proportion of the estimated total area of Argentina's marine environment (FORO, 2008), as is typical for marine conservation efforts worldwide. These MPAs have not necessarily been created to target threatened species (with exceptions, such as for the southern right whale), but to protect wildlife aggregations, such as breeding colonies. Of the 52 protected areas reported for the Argentine continental coast (Fig. 12.10a), only eleven include more than 100 km^2 of ocean space. The majority (42) are under provincial or municipal jurisdictions, thus are limited to inshore waters and gulfs. Some of them are located in important tourist attractions (e.g., Península Valdés). Some suffer direct threats such as impacts of overfishing, and some indirect threats such as unlimited tourist exploitation. As a rule, enforcement is deficient.

In summary, present conservation and institutional contexts for the PS are insufficient to guarantee the protection of virtually any species that ventures into the open ocean, or any marine habitat threatened by fishing operations, oil exploration, or other forms of exploitation, waste disposal, or pollution related to urban development. To improve on the present protected-area scenario for the PS (Fig. 12.10a), and as a step towards an LOR, it would be advantageous to advance pilot initiatives dedicated to large-scale conservation of spaces and sustainable use, guided by available scientific knowledge. The distributions of candidate seascape and/or seascape species and their uses of ocean space offer a means to identify candidate protected areas. Based on that rationale, four types of areas, in addition to improvements in existing protected areas, are proposed here (Fig. 12.10b; Table 12.2): (i) Blue Hole (EEZ and high seas); (ii) canyons of the slope (BRW, AMG, etc.); (iii) Burdwood Bank; and (iv) corridors and marine extensions of coastal protected areas. These areas represent a first approach to spatial conservation, as guided by seascape species and their uses of the PS environment and by existing ecological knowledge of the PS. As such, they constitute a pilot project for a network of conservation areas directed towards the inception of an LOR for the PS. All areas have been selected because they offer opportunities to conserve unique habitats, whose values are yet to be fully understood. Extension of present coastal protected areas and protection of corridors (Figs 12.10a,b) serve two additional purposes: first to protect essential breeding habitats and, second, to allow population interchange. These initiatives would be particularly effective for species such as penguins and sea lions, whose foraging occurs close to breeding areas and also occur as metapopulations (Ch. 5).

12.8.1 Blue Hole

This is a transboundary portion of the central Patagonian continental shelf. At the Blue Hole, 200 miles of the EEZ intercepts the 200 m bathymetric contour. Where the shallow shelf exceeds EEZ limits, a "high seas" area of ca. 6600 km^2—physically part of the shelf-break ocean front—is created; hence the name "Blue Hole." Remote sensing studies and satellite tracking have shown that the Blue Hole is heavily utilized by seascape species (Fig. 12.8, area 4), due to its position within the very productive continental shelf slope. The area is also subject to an intensive squid fishery (Waluda *et al.*, 2002). Estimates by FAO (1994) indicate that 11–35% of the biomass of Argentine squid (*Illex argentinus*) concentrates in this area. Argentine squid migrate along the slope where fishing boats concentrate and where they are heavily fished (see Section 12.7).

The complexities of enforcing regulations in waters beyond national jurisdiction make this area particularly susceptible to unregulated and unreported fisheries, due to lack of any fisheries management organization that might administer such fisheries. The problem of regulating fisheries in the Blue Hole is reminiscent of several other areas, such as the Bering Sea's Donut Hole (Ch. 7), and the Challenger Plateau near New Zealand. Such areas would benefit from regulations on highly migratory species. From a species perspective, a protected Blue Hole could benefit transboundary species targeted by commercial fisheries other than Argentine squid, such as longtail hake (*Macruronus magellanicus*), southern blue whiting, grenadiers, *Loligo* squid, and seascape species such as the wandering albatross, northern royal albatross, southern giant petrel, black-browed albatross, king penguin, and southern elephant seal (Falabella *et al.*, 2009). The Blue Hole has been listed by UNEP-WCMC (2006) as a candidate area for high seas protection (Corrigan and Kershaw, 2008) and the area has been explored in some detail in recent surveys of the continental edge of Argentina (Portela *et al.*, 2010).

The national/international jurisdictional nature of the Blue Hole poses great difficulties for governance. Although these complexities are typical of other areas where fisheries occur, the *United Nations Convention on Law of the Sea* (Ch. 3) provides

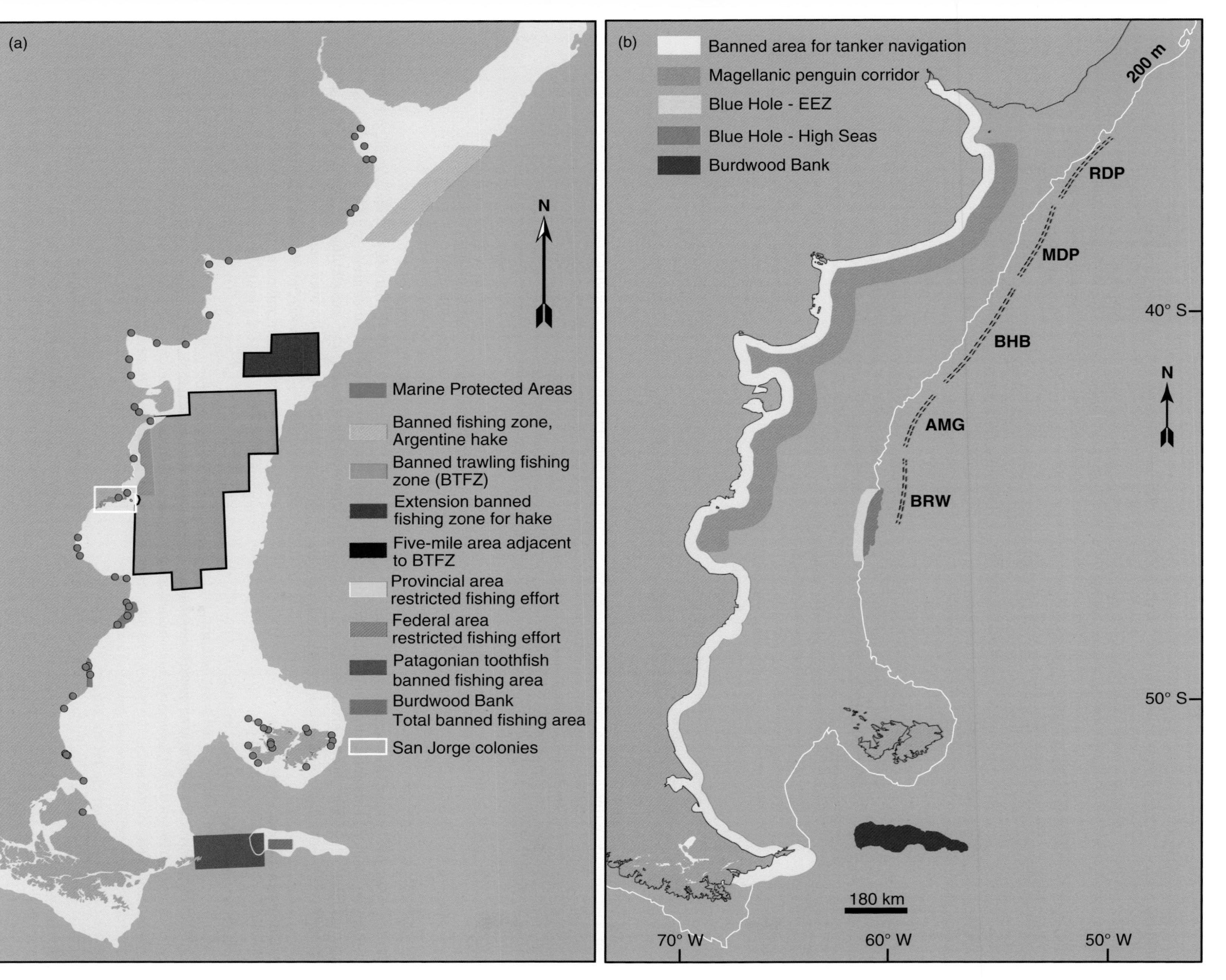

Fig. 12.10 Examples of spaces that could protect species. (a) Present protected areas have little or no associated marine area. Many banned fishing areas are declared for a target species but may serve as templates for biodiversity conservation. (b) Candidate areas for conservation (see Table 12.2). Submarine canyons have been described on the slope of the continental shelf: RDP, Río de la Plata; MDP, Mar del Plata; BHB, Bahía Blanca; AMG, Ameghino; BRW, Almirante Brown (adapted from Hernández-Molina *et al.*, 2009). The Burdwood Bank is a submerged plateau rich in endemic species. The Blue Hole (BH) is a transboundary area that concentrates effort of the Argentine squid fleet. The proposed corridor for penguins is a tentative area to minimize risk for migrating animals to encounter oil spills.

Table 12.2 Candidate areas for the conservation of pelagic environments, also important for top predators, seascape species, and iconic megafauna sustained by the Patagonian Sea.

Candidate area	Ecological relevance for biodiversity	Human activities	Main threats
Blue Hole	Located in the most extensive frontal zone of the PS. Foraging area for top predators (some endangered). Traversed during migration of transboundary species such as Argentine hake.	Located in the most extensive frontal zone of the PS	Overfishing, bycatch, accidental catch of seabirds, entanglement, unreported, unregulated fishing, discards.
Canyons of the Shelf Slope	The canyons have been poorly explored. Data-deficient areas. Possibly differentially relevant for benthic fauna, particularly endemic invertebrates.	Fisheries in the water column and benthic fisheries in nearby areas.	Fisheries in the water column and benthic fisheries in nearby areas.
Burdwood Bank	An area of abundance and diversity of species, some endemic (e.g., Southern king crab, cold water corals, sponges, bivalves, isopods). Species particularly vulnerable to trawl fishing. Spawning area for Fuegian sprat and Southern blue whiting. Nursery area for fishes and invertebrates. Feeding area of albatrosses, petrels, penguins, sea lions and fur seals.	Demersal and pelagic trawling. Longlining.	Bycatch of fishes, seabirds and invertebrates. Bottom trawling effect on the sea bed. A small area of the Bank has been protected by the Fisheries authority of Argentina. It is the only area under some form of protection for biodiversity in the EEZ.
Extension of Coastal MPAs	Expand conservation efforts to cover foraging areas for species that breed in colonies protected by coastal MPAs.	Artisanal and commercial fisheries.	Accidental catch, entanglement, competition for food, disease transmission.
Seasonal Corridors	Extend protection efforts to migratory path of species whose breeding colonies on land are protected. The Magellanic penguin is a candidate for a pilot effort.	Oil transportation, fisheries.	Oil spills, accidental catch in fishing gear.

an institutional and policy basis that has potential for advancing conservation initiatives in such circumstances. What is most lacking is the will of governments to do so.

12.8.2 Canyons of the shelf slope

The Patagonian continental slope is the location of some of the most productive ocean fronts of the Southwest Atlantic, and is important for pelagic predators such as the black-browed albatross and southern elephant seal. The ecological connections between benthic habitats of the slope, the water column, and resources that support satellite-tracked pelagic predators to this area are unknown. However, recent studies (Bremec and Schejter, 2010) have shown that the diversity of some benthic areas where submarine canyons occur is particularly important. These canyons may soon become targets of deep-sea fisheries. Therefore, the creation of protected areas that include slope canyons would be a precautionary pilot initiative for the PS, even though their biological importance is insufficiently known (Parker *et al.*, 1997; Bertolino *et al.*, 2007). A recent survey described seven submarine canyons in the middle slope, transversally crossing the continental rise, between 44° S and 48° S (Portela *et al.*, 2010). Adult males of southern elephant seal, for example, are shown by satellite tracking to forage along the edge of the shelf and within these canyons, suggesting sufficient abundance of prey to provide for rapid replenishment of depleted fat reserves. Beyond being foraging areas, other values of these canyons need to be recognized for fisheries and for biodiversity conservation. One canyon was recently described in the vicinity of commercial scallop beds; 86 taxa of mega- and macro-invertebrates were also found in the area, including sponges, hydroids, bryozoans, and ascidiaceans; and nine species of demosponges were identified, two of them new species for the PS (Bertolino *et al.*, 2007; Bremec and Schejter, 2010). Some of these species may represent range extensions, rare species, or of species dispersed by the Malvinas Current from subantarctic waters. Nevertheless, they are indicators of potentially high biodiversity. Preliminary evidence has indicated that benthic species common in South Georgia habitats or in the Magellanic region of Southern Chile range into the area of the canyons (Bremec and Schejter, 2010). Due to complex topography, fishery vessels do not yet trawl canyons and surrounding areas, but the bathymetry of the canyons may not prevent future oil exploration and exploitation in the area.

Lack of detailed information prevents delineation of specific protected areas for slope canyons. Only with more physical and biological information we will be able to design a conservation strategy for them. Nevertheless, the precautionary approach suggests their protection pending further information.

12.8.3 Burdwood Bank

Burdwood Bank is an isolated, relatively shallow-water habitat, influenced by the productivity of the shelf-edge front. It is included here for its unique community of benthic invertebrates, as a foraging area for many candidate species, and as a breeding ground for ecologically important fishes. Analysis of seascape species to date does not suggest this area, but expert knowledge of the area's diverse species and their natural histories, and its ecology, provide substantial bases for conservation.

Burdwood Bank is an underwater plateau of ca. 28,000 km^2; depths range from 50 m on the submerged plateau to 3000 m in surrounding waters. The Bank is under the influence of the Malvinas Current, which brings nutrients that drive high productivity, particularly during spring and summer months. Several species of seabirds and marine mammals forage extensively on the Bank (Falabella *et al.*, 2009). The area is also a breeding ground for two ecologically important fishes, the southern blue whiting and Fuegian sardine. The Bank's benthic environment seems to host endemic species, as would be expected from its geography. Some benthic invertebrate groups are thought to be well-represented and diverse, such as cnidarians, gastropods, bivalves, and seastars. The Bank is a hot spot for anthozoans, especially gorgonians (Arntz and Brey, 2003), including hard and soft corals. The biomass of these species appears to be similar to the Weddell Sea, but of even greater diversity, represented by six families of gorgonians and some habitat-forming, cold-water endemic coral species. Among the crustaceans, the squat lobster is most abundant, and endemic species of decapods (family Campylonotidae) have also been described.

The important conservation relevance of the Bank was acknowledged in 2008, when the federal fishing authority of Argentina (Consejo Federal Pesquero) banned commercial fishing in a small sector (1800 km^2) of this submerged habitat (ACTA Consejo Federal Pesquero N° 18/2008). Commercial fishing occurs in the Bank (outside the banned area) and on the neighboring shelf (Crespo *et al.*, 1997). Mid-water and bottom trawlers and longliners target Patagonian toothfish, Patagonian grenadier (*Macruronus magellanicus*), and southern blue whiting (*Micromesistius australis*). Under the criteria of vulnerability and differential relevance for endemic species, we suggest that the existing protected area be extended to cover a larger proportion of the Bank, While this chapter was in press, the Argentine Government passed a National Law declaring the Burdwood Bank as the first Oceanic Protected Area of Argentina (Buenos Aires, July 3rd 2013).

12.8.4 Corridors and marine extensions of coastal areas

Some seascape species migrate from coastal colonies along the Patagonian coast to foraging grounds in more northern latitudes, or to the shelf edge and the Malvinas Islands. Corridors are also important areas for migration and dispersal and for maintenance of metapopulations (Ch. 5). Most corridors are better defined by one or a few species than by a suite of candidate seascape species. For example, one corridor proposed for the PS is a transit area for elephant seals and sea lions to the edge of the continental margin that links Península Valdés to the shelf slope (Fig. 12.8, area 3). Another corridor involves migratory paths of Magellanic penguins along the coast (Fig. 12.10b), which is a priority seascape species that would be adversely affected by oil pollution. It migrates from colonies along coastal Patagonia to foraging grounds extending to the Brazilian coast (Boersma and Parrish, 1999; Boersma, 2008). Adults travel hundreds of kilometers from their breeding colonies to forage, even during the breeding season, mostly in shelf waters: in autumn to the north, up to 9°40'S, and in spring, to return south, for a total of about 4000 km, traversing the jurisdictions of Argentina, Uruguay, and Brazil.

The paths of migratory penguins and other species partially overlap with fisheries and with the routes of oil tankers transporting crude oil from Patagonian land sources to the refineries in the north. Spills from loading and transporting oil result in chronic pollution that impacts penguin populations (García-Borboroglu *et al.*, 2006). In 1997, tanker trajectories were moved to offshore waters and measures were adopted to decrease the impact of oil pollution on penguin mortality. This decision was only partially effective and would be improved by establishing a Magellanic penguin corridor that would help create a route for oil tankers that would minimize the interaction time between tankers and migratory penguins, and that could be seasonal. The corridor and tanker routes could be improved as more information becomes available, and/or further adapted to responses of penguins to annual variation in the sources and abundance of food. Most importantly, corridor establishment represents a precautionary approach to decrease risks in case of accidents.

12.9 MAKING SLOW PROGRESS

This case study on open-ocean conservation is guided by species' natural histories and suggests establishment of LORs for the PS, consistent in scale with both fisheries and wildlife uses of the seascape. The PS is unique in the wide extent of its shelf, mostly under Argentine jurisdiction, and in the predictability of its bathymetry- and circulation-dependent fronts. This system supports and attracts top predators from distant places, such as the Antarctic Peninsula, South Georgia, and New Zealand, thereby elevating its global importance. Many of the iconic species of the PS, such as penguins, albatrosses, right whales, elephant seals, and others, are threatened or are declining due to direct human impact. A combination of species knowledge and spatial conservation can result in better management at an ecosystem level.

The Eighth Conference of the Parties to the Convention on Biological Diversity held in Brazil, March 2006, defined criteria for Ecologically and Biologically Significant Areas. Criteria relevant for identification of potential sites for protection that include areas of the open ocean and the deep sea (CBD, 2006) are: (i) uniqueness or rarity; (ii) special importance for the life history of species; (iii) importance for threatened, endangered or declining species and/or habitats; (iv) vulnerability, fragility, sensitivity, slow recovery; (v) biological productivity; (vi) biological diversity; and (vii) naturalness. Similar criteria have been suggested under the *Convention on Biological Diversity*, as well as for regional seas (Ch. 3). How might these apply to the PS?

We here propose a set of candidate areas for large-scale, open-ocean conservation guided primarily by the distribution, needs, functions, and threats for iconic marine mammals and seabirds, some of whose foraging distributions have been determined by satellite-tracking techniques. Initial conservation options could be used as "place-holders," while at the same time nurturing the concept of large, seasonal, movable, high seas, protected, and multiple-use areas. These candidate areas would be especially benefitted by the extension of existing coastal MPAs into the ocean and the expansion of no-take fishery areas as partial templates for biodiversity conservation, as well as for resource replenishment and sustainable use. For conservation of species that restrict their movements relatively close to land, protected coastal areas are required to shelter them from land-based human threats or from overfishing, whereas those that venture farther from land require additional measures to protect their foraging grounds. It would seem feasible, in either case, to designate temporal buffer areas that would encompass foraging grounds during vulnerable breeding periods of critical species that reproduce on protected islands; for example, initial protection of foraging areas for Magellanic penguins could guide future steps towards open-ocean areas designed to guard against potential deleterious uses, especially overfishing. Extension of coastal protected areas into the sea can also be guided by following the criteria of Marine Important Bird Areas (IBAs; BirdLife International, 2009) to identify globally relevant sites supporting the breeding and foraging activity of seabirds on the Patagonian shelf. Seabirds are important oceanic predators and can add to the process of setting geographical priorities for conservation in the absence of detailed information on other taxa. Marine IBAs are particularly relevant as candidate areas for achieving a representative network of Marine Protected Areas that take account of annual life cycles, life-history stages, and migration routes of many representative species. Such protected areas are necessary for sustainable use of the total seascape in which species protection and human resource uses could be made compatible.

Another statement of the World Economic Forum's Ocean Governance Council on LORs is relevant to the conservation of the PS at a large spatial scale:

> "We recommend also that Large Ocean Reserves should be established even in the absence of prior characterization if political, economic and environmental conditions are favorable, on the presumption that biodiversity, productivity and ecological services will be best maintained under a fully protected condition."

Coastal and ocean protection are often delayed by requirements for more information on species and environmental conditions. These are real concerns, yet there is an urgent need to reconcile the necessary with the possible in order to achieve sustainable coastal-ocean conservation and use. One thing that should not be lost in the process of conserving ocean habitats is an integrated, large-scale vision, based on emergent conservation science. Species cannot be managed in isolation of other species or if their total life-history needs are not met. Furthermore, the diversity, distribution, and abundance of predatory species are important elements in the ecological function of the PS system as a whole. Failure to reconcile conservation with human use risks further ecosystem dysfunction. Thus, the creation of networks of large protected areas in the PS as a stepping stone towards an LOR of international significance is strongly justified. The seascape concept as applied to species and areas should guide interventions and use coherent with ecological requirements, as well as to reconcile conservation with human needs and well-being.

REFERENCES

Acha EM, Mianzan HW, Guerrero RA, Favero M, Bava J (2004) Marine fronts at the continental shelves of austral South America Physical and ecological processes. *Journal of Marine Systems* **44**, 83–105.

Arntz E, Brey T (2003) Expedition ANTARKTIS XIX/5 (LAMPOS) of RV "Polarstern" in 2002. Ber. Polarforsch. *Meeresforsch* **462**, 1618–3193.

Bertolino M, Schejter L, Calcinai B, Cerrano C, Bremec C (2007) Sponges from a submarine canyon on the Argentine Sea. In *Porifera Research: Biodiversity, Innovation and Sustainability. Série Livros 28* (eds Custódio MR, Lôbo-Hajdu G, Hajdu E, Muricy). Museu Nacional, Rio de Janeiro, 189–201.

BirdLife International (2004) *Tracking ocean wanderers: the global distribution of albatrosses and petrels.* Results from the Global Procellariiform Tracking Workshop, September, 2003, Gordon's Bay, South Africa. Cambridge, UK: BirdLife International.

BirdLife International (2009) *Using seabird satellite tracking data to identify marine IBAs.* Report of a workshop 1–3 July 2009. CNRS, Chize, France. Cambridge, UK: BirdLife International (internal report).

Boersma D, Parrish JK (1999) Limiting abuse: marine protected areas, a limited solution. *Ecological Economics* **31**, 287–304.

Boersma D, Ogden J, Branch G, Bustamante R, Campagna C, Harris G, Pikitch EK (2004) Lines on the water: ocean use planning within large marine ecosystems. In *Defying Ocean's End* (eds Glover LK, Earle SA). Island Press, Washington, D.C., 125–138.

Boersma D (2008) Penguins as marine sentinels. *Bioscience* **58**, 597–607.

Boyd IL, Wanless S, Camphuysen CJ, eds (2006) *Top predators in Marine Ecosystems.* Cambridge University Press, Cambridge.

Bremec C, Schejter L (2010) Benthic diversity in a submarine canyon in the Argentine Sea. *Revista Chilena de Historia Natural* **83**, 453–457.

CBD Secretariat of the Convention on Biological Diversity (2006) *Global Biodiversity Outlook 2.* Montreal, 81 + vii pages.

Campagna C, Piola AR, Marin MR, Lewis M, Fernández T (2006) Southern elephant seal trajectories, fronts and eddies in the Brazil/ Malvinas Confluence. *Deep-Sea Research, Part I – Oceanographic Research Papers* **53**, 1907–1924.

Campagna C, Piola AR, Marin MR, Lewis M, Zajaczkovski U, Fernández T (2007) Deep divers in shallow seas: Southern elephant seals on the Patagonian shelf. *Deep-Sea Research, Part I – Oceanographic Research Papers* **54**, 1792–1814.

Campagna C, Sanderson EW, Coppolillo PB, Falabella V, Piola AR, Strindberg S, Croxall JP (2008) A species approach to marine ecosystem conservation. *Aquatic Conservation (Marine and Freshwater Ecosystems)* **17**, S122–S147.

Campagna C, Short FT, Polidoro BA, McManus R, Collette BB, Pilcher NJ, Sadovy Y, Stuart SN, Carpenter KE (2011) Gulf of Mexico oil blowout increases risks to globally threatened species. *BioScience* **61**, 393–397.

Conway W (2005) *Act III in Patagonia. People and Wildlife*. Island Press, Washington, D.C.

Coppolillo P, Gomez H, Maisels F, Wallace R (2004) Selection criteria for suites of landscape species as a basis for site-based conservation. *Biological Conservation* **115**, 419–430.

Corrigan C, Kershaw F (2008) *Working Toward High Seas Marine Protected Areas: An Assessment of Progress Made and Recommendations for Collaboration*. UNEPWCMC, Cambridge, UK.

Crespo E, Pedraza SN, Dans SL, Koen Alonso M, Reyes LM, García NA, Coscarella MA, Schiavini A (1997) Direct and indirect effects of the highseas fisheries on the marine mammal populations in the northern and central Patagonian coast. *Journal of the Northwest Atlantic Fisheries Science* **22**, 189–207.

Croxall JP, Prince PA (1996) Potential interactions between Wandering Albatrosses and longline fisheries for Patagonian toothfish at South Georgia. *CCAMLR Science* **3**, 101–110.

Dans SL, Crespo EA, Pedraza SN, Alonso MK (2004) Recovery of the South American sea lion (Otaria flavescens) population in northern Patagonia. *Canadian Journal of Fisheries and Aquatic Sciences* **61**(9), 1681–1690.

Darwin C (1839) *Voyage of the Beagle*, Ch. XI: Strait of Magellan, Climate of the Southern Coasts. Republished (1962) The Natural History Library, Anchor Books, Doubleday & Company, Inc., Garden City, New York.

Falabella V, Campagna C, Croxall J, eds (2009) *Atlas del Mar Patagónico. Especies y espacios*. Wildlife Conservation Society y BirdLife International, Buenos Aires. Online. www.atlas-marpatagonico.org

FAO (1994) *Examen de la situación mundial de las especies altamente migratorias y las poblaciones transzonales*. FAO Fisheries Technical Paper, Organización de las Naciones Unidas para la Agricultura y la Alimentación, Roma.

Foro para la Conservación del Mar Patagónico y Áreas de Influencia, ed. (2008) *Síntesis del estado de conservación del Mar Patagónico y áreas de influencia*. Puerto Madryn, Argentina.

García-Borboroglu P, Boersma PD, Ruoppolo V, Reyes L, Rebstock GA, Griot K, Heredia SR, Adornes AC, da Silva RP (2006) Chronic oil pollution harms Magellanic penguins in the Southwest Atlantic. *Marine Pollution Bulletin* **52**, 193–198.

Hernández-Molina EJ, Paterlini CM, Violante RA, Marshall P, de Isasi M, Somoza L, Rebesco M (2009) Contourite depositional system on the Argentine Slope: An exceptional record of the influence of Antarctic water masses. *Geology* **37**, 3.

Halpern BS, Walbridge S, Selkoe KA, Kappel CV, Micheli F, D'Agrosa C, Bruno JF, Casey KS, Ebert C, Fox HE, Fujita R, Heinemann D, Lenihan HS, Madin EMP, Perry MT, Selig ER, Spalding M, Steneck R, Watson R (2008) A global map of human impact on marine ecosystems. *Science* **319**, 948–952.

Longhurst AR (1998) *Ecological Geography of the Sea*. Academic Press, San Diego, CA.

Matano RP, Palma ED, Piola AR (2010) The influence of the Brazil and Malvinas Currents on the Southwestern Atlantic Shelf circulation. *Ocean Science* **6**(4), 983–995.

Parker G, Paterlini CM, Violante RA (1997) El Fondo Marino. In *El Mar Argentino y sus Recursos Pesqueros* (ed. Boschi EE). INIDEP, Mar del Plata, Argentina, 65–87.

Piola AR, Matano RP (2001) Brazil and Falklands (Malvinas) currents. In *Encyclopedia of Ocean Sciences* (eds Steele JH, Thorpe SA, Turekian KK). Academic Press, London, UK, 340–349.

Polidoro BA, Livingstone SR, Carpenter KE, Hutchinson B, Mast RB, Pilcher N, Sadovy de Mitcheson Y, Valenti S (2008) Status of the world's marine species. In *The 2008 Review of The IUCN Red List of Threatened Species* (eds Vié JC, Hilton-Taylor C, Stuart SN). IUCN, Gland, Switzerland.

Portela JM, Pierce GJ, del Rio JL, Sacau M, Patrocinio T, Vilela R (2010) Preliminary description of the overlap between squid fisheries and VMEs on the high seas of the Patagonian Shelf. *Fisheries Research* **106**, 229–238.

Reyes LM, Crespo EA, Szapkievich V (1999) Distribution and population size of the southern sea lion (*Otaria flavescens*) in central and southern Chubut, Patagonia, Argentina. *Marine Mammal Science* **15**(2), 478–493.

Schiavini ACM, Crespo EA, Szapkievich V (2004) Status of the population of South American sea lion (*Otaria flavescens* Shaw, 1800) in southern Argentina. *Mammalian Biology* **69**(2), 108–118.

Sherman K, Alexander LM, eds (1989) *Biomass yields and geography of large marine ecosystems*. AAAS Selected Symposium 111. Westview Press Inc., Boulder, CO.

Skewgar E, Boersma D, Harris G, Caille G (2007) Anchovy fishery threat to Patagonian ecosystem. *Science* **315**, 45.

Smith ADM, Brown CJ, Bulman CM, *et al.* (2011) Impacts of fishing low-level species on marine systems. *Science* **333**, 1147–1150.

Spalding MD, Fox HE, Halpern BS, McManus MA, Molnar J, Allen GR, Davidson N, Jorge ZA, Lombana AL, Lourie SA, Martin KD, McManus E, Recchia CA, Robertson J (2007) Marine ecoregions of the world: A bioregionalization of coastal and shelf areas. *Bioscience* **57**, 573–583.

Sullivan Sealey K, Bustamante G (1999) *Setting geographic priorities for marine conservation in Latin American and the Caribbean* (ed. T.N. Conservancy). The Nature Conservancy, Arlington, Virginia.

Toropova C, Meliane I, Laffoley D, Matthews E, Spalding M, eds (2010) *Global Ocean Protection: Present Status and Future Possibilities*. Brest, France: Agence des aires marines protégées, Gland, Switzerland, Washington, D.C. and New York, USA: IUCN WCPA, Cambridge, UK: UNEP-WCMC, Arlington, USA: TNC, Tokyo, Japan: UNU, New York, USA: WCS.

UNEP (2006) *Ecosystems and Biodiversity in Deep Waters and High Seas*. UNEP Regional Seas Reports and Studies No. 178. UNEP/IUCN, Switzerland.

Waluda CM, Trathan PN, Elvidge CD, Hubson VF, Rodhouse PG (2002) Throwing light on straddling stocks of *Illex argentinus*: assesing fishing intensity with satellite imagery. *Canadian Journal of Fisheries and Aquatic Sciences* **59**, 592–596.

WEF (World Economic Forum) (2012) Online. members.weforum.org/pdf/GAC09/council/ocean_governance/proposal.htm

Worm B, Barbier EB, Beaumont N, Duffy JE, Folke C, Halpern BS, Jackson JBC, Lotze HK, Micheli F, Palumbi SR, Sala E, Selkoe KA, Stachowicz JJ, Watson R (2006) Impacts of biodiversity loss on ocean ecosystem services. *Science* **314**, 787–790.

SUGGESTED READINGS

Costello MJ, Coll M, Danovaro R, *et al.* (2010) A Census of Marine Biodiversity Knowledge, Resources, and Future Challenges. *PLoS ONE* **5**(8), 1–15.

IUCN-Species Survival Commission (2010) *Global Marine Species Assessment*. IUCN Annual Report.

Miloslavich P, Klein E, Díaz J, *et al.* (2011) Marine Biodiversity in the Atlantic and Pacific Coasts of South America: Knowledge and Gaps. *PLoS ONE* **6**(1), 1–43.

Schipper J, Chanson JS, Chiozza F, *et al.* (2008) The Status of the World's Land and Marine Mammals: Diversity, Threat, and Knowledge. *Science* **322**(5899), 225–230.

Tittensor DP, Mora C, Jetz W, *et al.* (2010) Global patterns and predictors of marine biodiversity across taxa. *Nature* **466**, 1098–1101.

Wu J, Hobbs RJ, eds (2007) *Key topics in Landscape* Ecology. Cambridge University Press, Cambridge, UK.

CHAPTER 13

FROM BEING TO BECOMING: A FUTURE VISION

> . . . in practice, possibilities for national action are now constrained by the increasing power of international forces, the declining influence of federal agencies over state and local government actions, an unmanageable workload, and an aging system of laws and regulations sometimes out of sync with new science and new issues.
>
> Graham (1999) *The Morning After Earth Day*.

13.1 THE NEW NORMAL

The 21st century entered an era of limits for marine systems—from managing abundance to managing increasing scarcity. Marine policies that arose in the era of the Marine Revolution (Ch. 1) became instituted under social and environmental conditions different from today. Institutions were established to carry out fisheries management, species protection, regulation of shipping, oil, gas, and mining, and coastal zone management to resolve internal conflicts separately. This sector-based decision-making lacked clear authority to resolve cross-sector conflict and cumulative effects on coastal and marine systems.

Human civilization thrived during the last 10,000 years within a relatively narrow range of regular temperatures, freshwater availability, and biogeochemical flows, in a period when Earth was unusually stable (Dansgaard *et al.*, 1993). With the global human population expected to reach 8 billion in 2024 and 9 billion in 2045, human activity may soon exceed the "safe operating space" for humanity (Rockström *et al.*, 2009), especially in that narrow coastal fringe covering only 7.6% of the Earth's total land area that presently holds approximately 40% of the world's population, dependent in varying degrees on coastal resources for its livelihood.

Now, population growth, technological achievements, and improved social standards encompass a world society intimately connected to oceans, climate, and biogeochemical cycles. Twenty-six of the 33 world's megacities are located on the coast (Table 2.4), and have expansive ecological footprints (Folke *et al.*, 1997). Through increasing use of fossil fuel energy and natural resources, and institutional capacity to resolve issues, the social system is transforming coastlines, marine ecosystems, and the global climate. The accelerating scale of human activity intertwines social and ecological crises (Walker *et al.*, 2009) and increases synergistic effects that lead to mass ocean extinctions (Jackson, 2008).

The recognition of increasing frequency of deadly and costly natural disasters is alerting scientists, decision-makers, and the public to take precaution. New geospatial tools focus on large regional scales to foretell potential climate-related disasters. Scientists are now able to forecast the impact of climate change on major cities and urban land susceptible to high-risk coastal flooding (Strauss *et al.*, 2012). Yet, fragmented policies and management directives continue to fail holistically to address the interconnected coastal-ocean system undergoing change. Under this "New Normal," marine conservation remains stubbornly dependent on past policy instruments and soft-policy agreements.

13.2 FROM BEING . . .

Ilya Prigogine, in his classic book *From Being to Becoming* (1980), explained from theoretical non-equilibrium thermodynamics how systems evolve from low-level chaos to high-level order, and that change is always accompanied by degradation into a more dispersed, chaotic state of greater disorder (Ch. 4). Panarchy theory describes four stages of ecosystem change: expansion, maturity, senescence, resilience, and back to expansion (Fig. 4.21). These explanations of system behavior become metaphors for the transition of marine conservation, from the necessary but insufficient protection of species and spaces and narrowly based regulations (fisheries, pollution, etc.) into the broader context of complex eco- and social-system behavior that ultimately governs the success of conservation effort.

Marine conservation is rooted in both social and environmental conditions, with expectations of building on the past. Our case studies illustrate that conservation issues extend throughout whole systems, evolving with the social-ecological system from past history that is moving into a dramatically different 21st century. As such, marine conservation remains a work in progress, encompassing present mechanisms to preserve biodiversity, ecosystem resilience, and human well-being, while working within the bounds of a changing socio-ecological system. That solutions are not yet "complete" has less to do with lack of scientific information than with social, economic, and political forces and ecosystem adjustments. Failure to treat conservation goals systemically, while socio-political forces exacerbate the global environmental deficit,

Marine Conservation: Science, Policy, and Management, First Edition. G. Carleton Ray and Jerry McCormick-Ray.

risks the loss of socially and environmentally valued assets. According to Bormann (1990), "This deficit results from the collective and often unanticipated impact of our alteration of the Earth's atmosphere, water, soil, biota, ecological systems, and entire landscapes [and seascapes]. The social and economic costs to human welfare are ultimately greater than the short-term benefits that flow from this activity."

Especially in the vital area of the biosphere we call the coastal realm, societies at all governing levels require integrative conservation action to protect and restore species and environments. This narrow land-sea region of Earth, where social and economic uses impinge most strongly on ecosystem integrity, is increasingly exposed to intense droughts, floods, earthquakes, and storms from human-induced climate change. "Natural disasters" are increasingly affecting urbanized coasts and industrialized seas. Most significant was the 2004 Indian Ocean earthquake and its tsunami that shocked the world when it caused the deadliest natural disaster in recorded history. In 2011, Japan experienced the most costly powerful earthquake on record, revealing the vulnerability of offshore waters to coastal nuclear power and to tons of debris that were added to the North Pacific "garbage patch" (Ch. 2, Section 2.2.2.4). In 2012, Hurricane Sandy became the costliest natural disaster in U.S. history when it struck the mid-Atlantic states, surpassing the horrific damage of Hurricane Katrina in the northern Gulf of Mexico in 2005. Under these new conditions, sector-based management continues to follow institutional frameworks that were created under different social and environmental conditions.

13.3 . . . TO BECOMING

Three historic arenas most prominent for marine conservation are fisheries, coastal management, and Marine Protected Areas. The following three sections describe their present states. The succeeding Section 13.4 offers arenas for change. New conditions suggest new strategies.

13.3.1 Managing fish, fisheries, and fishermen

Fisheries have long been acknowledged to be the most ubiquitous of human interventions into marine environments (NRC, 1995). Fish, fisheries, and fishermen are managed through established institutions influenced by the industrialization of capture fisheries, which generated enormous wealth in the 19th to 20th centuries. Modern fishing now supports about 20 million fishermen worldwide, 90% of which are small-scale fishers that remove an estimated 25% of the global catch (FAO, 2008). Fisheries support a global fleet of about 4.3 million vessels: 59% powered by engines and 41% by traditional craft operated by sails and oars, primarily in Asia (77%) and Africa (20%). About 40% of the total global fisheries catch is for human consumption in an expanding market that supports an industry that employs almost ten times as many persons as fishermen, including processors, shippers, and marketers. Fishers, along with aqua culturists and those supplying services and goods to them, sustain the livelihoods of about 540 million people, or 8.0% of the world population (FAO, 2010). Small-scale, nearshore fisheries in the tropical Pacific are for subsistence, social and cultural purposes, and for food, trade, and recreational resources (e.g., Dalzell *et al.*, 1996). Among South Pacific islanders, coastal fisheries target reef fishes and coastal pelagic fishes, with a total removal of 100,000 tons per year and a value of $262 million dollars, 80% harvested for subsistence (Dalzell *et al.*, 1996). But as Ludwig *et al.* (1993) observed, ". . . the larger and the more immediate are prospects for gain, the greater the political power that is used to facilitate unlimited exploitation."

The principal, direct impact of fishing is that it reduces the abundance of target species, especially the largest, most reproductive, and ecologically valuable species (Pauly *et al.*, 2002). This process of "fishing down the food web" (Pauly *et al.*, 1998) selectively reduces economically valued top predators, such as the endangered Atlantic bluefin tuna (Ch. 2), Nassau groupers (Ch. 8), and others now in very reduced numbers and even threatened with extinction. Removal of top predators has strong effects on ecosystems and biodiversity (NRC, 1995); e.g., prey numbers increase to force an ecological shift toward short-lived species on lower trophic level species, thereby affecting biodiversity (Branch *et al.*, 2010) and ecosystem function (Choi *et al.*, 2004). Meanwhile, these species attract new fisheries, furthering a cascade of decline, appropriately termed a "March of Folly" (Sumaila and Pauly, 2011).

Fisheries management to date has most often been ineffective (Pikitch *et al.*, 2004). It has focused, until recently, on single-species, "maximum sustainable yield" management in which the natural histories of fish and habitat have only lately been included (Box 3.5). Management effectiveness within national jurisdictions varies greatly among nations, but high seas international management is in a state of crisis. This crisis relates to Garrett Hardin's (1968) resounding essay "Tragedy of the Commons" in which environmental disaster occurs in the public commons from individual decisions and too many people exploiting limited common-pool resources. This tragedy is particularly devastating for fisheries, which remains the last, large-scale "hunter-gatherer" activity of humans, affecting whole ecosystems, in many cases irreversibly, and also the economic prospects of numerous dependent fishermen worldwide. Fisheries management is also affected by investment policies of multilateral development banks for developing countries that aim to generate foreign exchange through expanding fisheries export capacity, as well as other environmentally destructive practices; e.g., aquaculture, larger vessels for increased capacity, and encouraging foreign access to fisheries. The management of wide-ranging, anadromous fish is especially contentious, for example high-profile icons such as Pacific salmon, because these fish are affected by nearly everything local people do (Box 13.1). Therefore, potential remedies impact traditional ways in which people extract water, generate electricity, transport goods, harvest fish, develop industrial, commercial, and private properties, and conduct their daily lives (Ruckelshaus *et al.*, 2002).

In 2006, United Nations General Assembly Resolution 61/105 called "upon States to take action immediately to sustainably manage fish stocks and protect vulnerable marine ecosystems (VMEs) (Ch. 3; Section 3.5.2.3), including

Box 13.1 Pacific salmon: science policy

James H. Pipkin
[Retired, U.S. Department of State Special Negotiator for Pacific Salmon (1994 to 2001); U.S. federal Commissioner on the bilateral Pacific Salmon Commission (1999 to 2002); Counselor to the U.S. Secretary of the Interior (1993 to 1998); and Director of the Interior Department's Office of Policy Analysis (1998 to 2001)], Bethesda, Maryland, USA.

Historically, more than ten million salmon and steelhead trout (all *Oncorhynchus* species) returned to the Columbia River basin each year to spawn. Now there are less than two million, and most of those originate from hatcheries, not from the wild. Some salmon populations are already extinct and others are headed that way.

Few people would deny that many stocks of wild salmon in the Pacific northwest are in trouble, or that an ecosystem approach must be taken if the problem is to be corrected. However, the application of ecosystem principles to the Pacific salmon issue is enormously complicated and requires the accommodation of a number of important, and often competing, interests. This is true from many standpoints—economic, ecological, social, cultural, institutional, and political.

Severe declines of wild salmon in the Pacific northwest are the product of many factors. The major contributing factors are sometimes referred to as the four "Hs": harvests, hydropower, hatcheries, and habitat. The first two require little explanation. The term "harvests" simply refers to the fact that salmon are caught by commercial, recreational, and tribal fishermen, both in the ocean and when they return to the stream in which they were born. "Hydropower" relates to the huge dams that have been constructed on important salmon rivers as part of our quest for cheap electricity. Those dams form barriers that make migration more difficult, as well as altering water temperatures and flow rates. The main issue concerning "hatcheries" is that salmon from hatcheries can reduce the genetic diversity of wild salmon or make them more susceptible to disease and predation. The fourth factor, "habitat," is complicated, since a wide variety of human activities have an adverse impact on salmon habitat. Clear-cutting of forests removes tree cover and increases siltation, which increases turbidity and raises water temperatures. Drawing water from rivers to irrigate agriculture lowers water levels in the river. Urban and agricultural runoff contains fertilizers, pesticides, herbicides, and other materials that result in habitat degradation in rivers and streams. Wastes produced by industries find their way into aquatic habitat. The building of homes along rivers removes tree cover and woody debris, as well as increasing runoff. Widespread diking of rivers to protect development in floodplains also affects salmon habitat adversely.

Because there are so many "culprits," the norm has been for each group to defend its actions and point fingers at the others. Fishermen say that they have sacrificed more than other sectors and that further cuts in their catch will not reverse the real problems, which are rooted in habitat degradation. Indian fishermen also point to their treaties (under which they ceded most of their land to the United States), which reserved to them the right to continue to fish in their "usual and accustomed" fishing places. Farmers say that they have relied on irrigation water for decades and cannot give it up without going out of business. Land developers say that their contribution to the problem is minuscule when compared to the other factors and that the economic costs of non-development are excessive. Some landowners also argue that the right to make maximum economic use of their properties is protected under the Fifth Amendment to the Constitution.

Many of the arguments about salmon restoration boil down to economics. Unfortunately, the economic consequences of various restoration strategies are rarely clear, and gaps in the science make the implications even more controversial. Beyond economics, in the Pacific northwest, salmon are cultural icons. This is especially true of Indian tribes for which the salmon has both cultural and spiritual significance. Even for non-Indian residents who have no intention of ever catching a fish, salmon represent part of their heritage. To many, salmon are like the canary in the coal mine: the serious decline in the abundance of wild salmon is an indication that streams are no longer healthy, that riparian habitat has been degraded, and that desirable recreational opportunities have been lost.

Politically, the salmon situation is also complex. On one level, an elected politician (state or federal) representing farmers who irrigate their fields will have a different perspective on the salmon issue than a politician whose constituents prize wilderness and sport-fishing opportunities. Both will differ sharply with a politician from another region who questions the large amounts of money being spent to restore salmon in an area that means little to his constituents. The approach of each will make it virtually impossible to find common ground, and in the end the political debate will likely be resolved through seniority and relative political power, rather than by a thoughtful weighing of the merits or a quest for a compromise solution. The same difficulty of achieving consensus exists in the international arena as well.

Jurisdictionally, the problem is equally complicated. No single governmental or other entity controls any one of the four Hs referred to above. Take harvests as an example. In general, each state regulates fishing within its borders

(Continued)

and off its coast to a distance of three miles. But unfortunately, salmon are not very good at respecting state borders. They swim across state lines, as well as into Canadian waters where harvests are regulated by the Canadian government, and into international waters where their principal protection is the *North Pacific Anadromous Fisheries Convention*. Beyond that, tribal governments have treaty rights that affect permissible harvests. The U.S. federal government can step in to regulate salmon harvests only when they affect species "listed" as threatened or endangered under the *Endangered Species Act* or when other federal statutes come into play. Similar dispersions of authority exist with respect to the other Hs.

Institutions that play a role in the Pacific salmon controversy include federal and state legislatures, state and federal fish and wildlife agencies, the courts, the Bonneville Power Administration, the Northwest Power Planning Council, the U.S. Army Corps of Engineers, the Environmental Protection Agency, federal and state land management agencies, harvest management organizations such as the Pacific Fishery Management Council and the U.S./Canada Pacific Salmon Commission, and tribal governments. Many institutions at different levels focus primarily on isolated issues that bear on salmon recovery—for example, how many fish can be caught and by whom, how timber will be harvested on national forest land, or how a hydropower facility will be managed to reduce the impact on fish. No existing institution has the mandate to determine answers to broader ecosystem-based questions, such as how a watershed will be managed. Current institutional arrangements are poorly suited to development of an ecosystem (or bioregional) perspective on the salmon problem. Indeed, the U.S. National Research Council concluded that the current set of institutional arrangements "contributes to the decline of salmon and cannot halt that decline."

In sum, it is hard to imagine a more complex problem than what to do about Pacific salmon. Does that mean we should throw in the towel and schedule funeral services for the remaining wild salmon? For all but the hardened pessimist, the answer to that question has to be "no." The sharp declines in salmon abundance, the "listings" of salmon populations under the *Endangered Species Act*, and the economic dislocations to commercial fishermen and to the recreational infrastructure of the Pacific northwest, have dramatically increased the general awareness of the problem and the willingness of many residents of the region to make economic sacrifices to ensure the survival of remaining salmon populations.

A few encouraging signs have appeared. Among them, plans being developed at the federal, state, and tribal levels attempt to take an ecosystem approach to salmon recovery. Long-term arrangements for ocean salmon fisheries put into place by the United States and Canada, together with harvest limits being imposed under the *Endangered Species Act*, provide some confidence that harvests will be under control. Moreover, restrictions have been imposed on timber activities in the northwest for the benefit of salmon and other anadromous fish.

Although some reasons for optimism exist, for that optimism to come to fruition will require a huge commitment, by many interested parties, during many years, to fill gaps in scientific information, to coordinate management, to provide funding, and to make and implement hard choices among painful alternatives that have serious and interrelated ecologic, economic, social, and cultural implications. We should not delude ourselves into thinking that the effort to restore wild salmon is anything other than an uphill struggle. Success will no doubt require a willingness on the parts of federal, state, and tribal governments to conclude that current institutional arrangements are not adequate and to establish a new decision-making framework that can address broad issues such as watershed management and fish governance.

These tasks are formidable, similar to the daunting obstacles faced by the juvenile salmon making the difficult and challenging journey to the sea. The salmon has no option but to begin the journey and to commit fully to completing its task. For that matter, if we are to bring salmon and back from the brink of extinction, neither do we.

seamounts, hydrothermal vents and cold water corals, from destructive fishing practices . . ." individually and through regional fisheries management organizations and arrangements consistent with precautionary and ecosystem approaches (UNGA, 2006). Nations and regional fishery management organizations are to manage fisheries so as to prevent significant adverse impacts to areas identified as VMEs. The European Union and the Northeast Atlantic Fisheries Commission have taken measures for action, including closing bottom fisheries in large areas of the high seas identified as VMEs, e.g., on the Mid-Atlantic Ridge. And in March 2011, North Pacific nations agreed to interim measures to curb expansion of destructive bottom trawling for more than 16.1 million mi^2 of seafloor habitat in accordance with the North Pacific Fisheries Commission (NPFC, 2012).

To further remedy fisheries problems, NGOs, national agencies, and international institutions are forming collaborative partnerships. And formally separate fisheries science and ecosystem science are seeking common ground to ensure that fishing remains a viable occupation and that ecosystems are viable enough to maintain productive fisheries (see EBFM, Section 13.4.4.1). For rebuilding fisheries in many poorer regions, co-management involves collaboration of local communities with government or non-governmental organizations. And because of the impacts of international fleets, lack of fishing alternatives, and poor documentation that complicate recovery, a global perspective is required; 63% of assessed stocks worldwide still require rebuilding (Worm *et al.*, 2009). Large Marine Ecosystems (LMEs, Ch. 3, Section 3.5.1) lend particular opportunities for nations to integrate fisheries and ecosystem principles into effective fisheries management (Pauly *et al.*, 2008). LMEs occur in nations' Exclusive Economic Zones where overfishing is most severe, marine pollution is concentrated, and eutrophication and anoxia are increasing (Sherman *et al.*, 2009).

13.3.2 Managing uses along congested coasts

The coastal economy is big business, in which expanding marine transportation and ports, coastal fisheries, coastal forestry (mangroves, etc.), aquaculture, offshore energy, minerals, tourism, and recreation are in conflict with human health, biodiversity, ecosystem function, coastal real estate values, and other interests, further exacerbated by exponential population growth, climate change, and rising sea levels. Coastal management is being undertaken in multiple dimensions, first in the full extent of the coastal realm, including the nation's Exclusive Economic Zone (Ch. 4), as in the case of anadromous fishes, and second in the narrower land-sea boundary zone (Fig. 4.7). In both cases, management and conservation efforts are confronted by jurisdictional, social, and economic conflicts. Additionally, all coastal nations are facing significant global pressures, escalating risks, and issues of private ownership. Ports that were once considered public entities are today either privately owned or operated by maritime shipping companies and port terminals (Rodrigue, 2010). Any attempt to maintain marine biodiversity and protect natural habitats must address issues of natural disasters, coastal degradation, and destruction of natural landforms (marshes, beaches) and seascapes (reefs, seagrasses) to resolve multiple-use conflicts brought on by intensifying population growth.

Changes of ecosystems, demography, institutions, and high costs of ecosystem restoration are forcing coastal management to undergo a major transformation (Olsen, 2000). Integrated Coastal Zone Management (ICZM) is being undertaken by many nations through an approach that attempts to incorporate natural-resource management, conservation of biodiversity, maximization of socio-economic benefits, and protection of life and property from natural hazards within existing institutional and organizational frameworks of individual nations' particular characteristics (Clark, 1996). Nations are mostly undertaking ICZM through the power of their constitutions or laws without a comprehensive context, delegating such authority to subnational levels. As the practice of ICZM moves from developed to developing nations, it is also being promoted by international assistance; at least 142 ICZM efforts had been established by about 57 sovereign and semi-sovereign states by the 1990s (Sorensen, 1993). The accomplishments to date are modest, having only modulated some localized impacts of anthropogenic change. The institutions involved often protect and promote jurisdictional interests without cross-sector collaboration, while weak public constituencies strive for better equity and accountability. These institutions often succumb to private interests. For example, some owners on eroding shores have gained court permission to build expensive barriers to protect their threatened homes from the advancing ocean, defying state officials and scientific advice that such activities accelerate erosion, thus burdening their neighbors with the erosion problem (Dean, 2005). Olsen (2000) suggests that to promote sustainable development through a governing process that is rooted in participatory democracy, ICZM needs to integrate scientific principles within a curriculum that includes a diverse mixture of knowledge and skills, and where public education plays an increased role.

Land-based activities are major drivers of coastal-marine change. Nevertheless, agencies and institutions that manage land and sea traditionally operate under separate legislative mandates and generally lack adequate communication among them. This situation is well recognized, bolstered by cases where only changes in land management will solve coastal-marine issues; e.g., for estuaries (Ch. 6), pollution (*Deepwater Horizon*, Ch. 2, Section 2.2.2.2), and dead zones (Box 2.5). Thus, heightened awareness and economic incentives are inspiring hope for significant change in governance at national and local levels. The international community has also taken significant action (albeit without regulatory power) by agreeing on the *Global Programme of Action for the Protection of the Marine Environment from Land-based Activities* (GPA) in 1995. The Program is designed to assist nations to preserve and protect the marine environment by preventing, reducing, controlling, and/or eliminating degradation, and to recover from land-based impacts, with assistance from UNEP's Regional Seas Program (Ch. 3). UNCLOS obligates signatories to protect and preserve the marine environment through regional and global cooperation by adopting laws and regulations to deal with land-based sources of marine pollution. Also, the World Bank in 2001 launched a new global partnership for the oceans, *Oceana*, intended to become the largest international organization on ocean conservation, involving the world's top conservation organizations, and private sectors (Oceana.org).

In sum, these activities offer ways to confront the socio-economic and ecological dilemma of protecting coastal-marine systems while also maintaining healthy economies. Solutions are bound to be case-dependent due to the huge array of differing mixes of social, legal, and ecological conditions.

13.3.3 Protecting species and spaces

Iconic species—whales, walruses, sea turtles, manatees, penguins, great white sharks, tuna, marlins, groupers, oysters, corals, seabirds, and many other marine and coastal species—are seen by the public at large as the primary symbols of ocean well-being, and many NGOs promote their protection. Indeed, without that recognition, species' protection, even their survival, would be immeasurably more difficult, as the public would be less involved. Consequently, major efforts are devoted to listing depleted species as threatened, endangered, or vulnerable to extinction by international, national, provincial, and in a few special cases, by local and native institutions and cultures (Fig. 3.1). It is less well recognized that these icons of the seas are also often the fish that fishermen target, that depend on habitats coveted by developers (seagrass beds, coastal marshlands, reefs, mangroves, kelp forests, etc.), or that are most threatened by human activity (pollution, invasive species, disease, global climate change, ocean acidification, and ocean warming). The case studies highlight the vulnerability of these icons to human pressures in a changing ocean system, where increasing noise, disturbance, pollutants, and anthropogenic change threaten their ability to make a living. For recovery, these species depend on increasing understanding of their geographic distributions, their natural histories, their sensitivities to disturbance, and their resilience. All have coevolved during decadal to century-long periods of interconnected ecosystems, characterized by various degrees of ecological complexity and biomass production. Many are top-predator, "keystone" species (Ch. 5) that exert significant effects on their ecosystems, while others are forced to adapt to change, if possible.

Designated Marine Protected Areas (MPAs; Ch. 3) constitute a major form of protection for species and spaces. They have formed the foundation for marine conservation in many areas of the world, being scientifically proven for effective fishery management and biodiversity protection and vital for protecting habitat, maintaining food-web integrity, and enhancing biodiversity when appropriately managed. Within designated locations, MPAs can address both specific and multiple objectives, ranging from strict nature conservation and scenic beauty, to protecting specific life-history stages and biodiversity, controlling fishing mortality, and facilitating fish recruitment outside their boundaries. MPAs can also serve as a hedge against the uncertain outcomes of environmental change. And because of their value, signatory nations to the *Convention on Biological Diversity* (Ch. 3) agree to establish national and regional systems of protected areas that are comprehensive, effectively managed, and ecologically representative, leaving to nations the establishment of target goals. This Convention brought incentives to support systematic conservation planning in which to develop ecologically representative MPA networks (UNEP-WCMC, 2008). Among the first general models proposed for a MPA network design was by Sala *et al.* (2002) for the Gulf of California.

Some MPAs are included in networks to help achieve regional or target goals. Worldwide, there are at least 30 national and 35 subnational ecological MPA network initiatives, most still under development and none yet fully managed (UNEP-WCMC, 2008). Networks of MPAs help achieve regional or target goals through a range of different types of MPAs that include both no-take areas (NTAs) and multiple-use sites (U.S. marine sanctuaries). Some involve hierarchical approaches, with small networks nested within larger national networks, as in Mexico, Indonesia, Australia, and the U.S. However, implementing a coherent network of MPAs can prove challenging. In California, for example, the *Marine Life Protection Act* of 1999 aimed to develop and apply the best available science in resource management and decision-making to improve ecosystem protection (Carr *et al.*, 2010). The task took seven years with an investment of about $38 million (Gleason *et al.*, 2012), but aroused much controversy. Thus, despite growing awareness of the need to curb biodiversity loss and protect ecosystem function at regional scales, MPA management plans must often be accompanied by complex governing structures that accommodate strong differences of opinion.

13.4 EMERGING CONCEPTS FOR MARINE CONSERVATION

The global ecological crisis stems from a political-economic organization of societies increasingly out of balance with ecological systems. Thus, global societies are being propelled into a future of intractable ocean and global change. The nexus of system behavior (Ch. 4) with species natural history (Ch. 5) makes clear the complex nature of marine systems. As the case studies illustrate, conservation principles are being applied in the attempt to transition into a coevolving, interdependent social/ecological system. And when such principles are incorporated into a spatial matrix of integrative planning, conservation targets are more likely to succeed. Thus, new ocean policies are being advocated that can be based on scientific models, spatial tools, system science, and a better-informed public (Crowder *et al.*, 2006).

Three emergent conceptual applications offer high potential for addressing this current situation: ecosystem-based management, marine spatial planning, and resilience thinking.

13.4.1 Ecosystem-based management (EBM)

EBM recognizes that all forms of resource use are interrelated and that humans form an integral part of ecosystems (McLeod and Leslie, 2009a). EBM also recognizes the utility of sound ecological models to address complexity, connectivity, and the dynamic, heterogeneous nature of ecosystems (Thrush and Dayton, 2010). Thus, the application of EBM has the potential to establish a commitment to adaptability, accountability, and inclusive decision-making, in which MPAs and marine spatial planning can play important roles and resilience practice offers adaptation to changing conditions.

Essential to EBM is recognizing the hierarchal scales in which both ecosystems and societies are organized. The self-organizing capacity of coastal-marine ecosystems operates over a range of different spatial and temporal scales, not always easy to identify. Each scale moves through its own adaptive cycle, building capacity to absorb disturbance and to maintain function (Walker and Salt, 2006, 2012). For a particular conservation target, choice of spatial and temporal dimension defines the scale, and multiple scales define the target holistically in a broader context of change and availability of conservation tools (Swaney *et al.*, 2012). Conserving coastal seagrass, for example, requires management not only of the aquatic area it occupies, but also the patterns and regulation of water flows and substrate quality that affect it. Thus, the target (i.e., a seagrass bed) is the smallest spatial unit (grain, patch) set within a larger region that defines the total area occupied ("extent") (O'Neill *et al.*, 1996). This consideration of multi-scale hierarchy also encompasses the concept of "panarchy" (Fig. 4.21). Panarchy describes the adaptive cycles that define ecosystems across scales at various stages of development (Holling, 2001) and how, through self-organizing capacity, a healthy system can "invent" and "experiment," and create opportunity while being kept safe from destabilization or "excessive exuberance." This is a concept applicable to sustainable development in which multi-scale, panarchy concepts can help foster approaches for adaptive management by using fine-scale, mechanistic understanding to screen hypotheses to be tested at large scales (Hobbs, 2003).

It is no small coincidence that social-economic systems operate in much the same cyclical way as ecosystems in reacting to environmental or other conditions on which both society and ecosystems depend. Thus, EBM embraces human activities and explicitly deals with trade-offs among human activities and ecosystem services (McLeod and Leslie, 2009b). Yet, current management strategies have rarely been designed to incorporate sectors of human activity *holistically with the ecosystem*. Hence, a necessary condition for implementing EBM is moving from sector-specific mandates to comprehensive planning, to be carried out by managers under comprehensive and clear legal mandates, in a forum for comprehensive ocean planning (Rosenberg and Sandifer, 2009). No better examples of this transition may be found than for fisheries and MPAs in their quests for sustainable resource management.

13.4.1.1 Ecosystem-based fisheries management

The paradigm for sustainable fisheries is becoming ecosystem-based fisheries management (EFBM; Pikitch *et al.*, 2004; Box 3.5). This approach has evolved from the uncertain results of single or multispecies management and the acknowledged interactions between fisheries and ecosystems (FAO, 2011). EBFM goes beyond traditional management of fish, based on simplistic, yield-based models, to involve interactions with the human system and ecosystem. Most significantly EBFM aims to ensure that total biomass removed by all fisheries in an ecosystem does not exceed a total amount of system productivity, *after* accounting for the requirements of other ecosystem components (e.g., non-target species, community interactions, habitat dependencies, etc.). For example, sessile, estuarine-dependent shellfish species, e.g., oysters, requires that sustainable management involves recognition of an historical legacy of overfishing and present depletions related to habitat loss, pollution, and disease in an estuarine system undergoing change (Ch. 6). A management plan for a single species that meets both fisheries-use needs and ecological restoration involves cooperation among numerous agencies in order to safeguard a mutually shared resource. Rebuilt oyster beds in a semi-enclosed, tidally controlled system require appropriately designated locations protected from harvest. Identification of such locations could provide a "free service" in a regional setting, ensuring natural recruitment through life-history dispersals, and for establishing restoration sites selected for a marine reserve network (Box 13.2). Ecosystem-based shellfish management is entering U.S. national and state-agency programs in which to enhance commercial, ecological, and social benefits. NOAA, a federal agency, has established National Shellfish Initiatives to build partnerships with states that will increase shellfish populations and encourage EBM to optimize economic and ecologic benefits. This comprehensive approach is embraced by Washington State, whose oyster protection plan involves global concerns for ocean acidification and climate change while also focusing on local management action.

Protected-area designation, fisheries initiatives, and EBFM, together, provide potential means for restoring fisheries into ecologically and economically sustainable enterprises, while also sustaining biodiversity and ecosystem function. This requires a significant paradigm shift in fisheries management. To stimulate discussion and bridge the gap between principles and methodology, Francis *et al.* (2007) propose "ten commandments" for EBF scientists that include: "characterize and maintain viable fish habitats"; "account for ecosystem change through time"; and "characterize and maintain ecosystem resilience." The intention is to integrate EBFM with natural and social sciences. Application is challenging, requiring "an expanded empirical basis as well as novel approaches to modeling." Under present circumstances, there appears to be no viable alternative, as scientific understanding, social support, and public understanding, *operating together*, are key factors for EBFM success.

13.4.1.2 Ecosystem-based Marine Protected Area management

EBM is also being extended to apply to MPAs, but how MPAs fit into the "ecosystem," where they have been established and for what purpose, is rarely apparent. In almost all cases, MPAs are islands in a sea of economic use, therefore highly vulnerable to outside influences. Thus, their designation requires "agonies of choice" (Ray, 1999) and careful consideration (Agardy *et al.*, 2011). MPAs designed solely around the distribution and apparent resource needs of particular species are likely to fall short of their intended purpose (Box 13.3). And unless they serve the public good and are locally acceptable to the user community (stakeholder), they may also fail to be viable entities for conservation. Justification for their selection involves opportunities, funding, planning, and community involvement and acceptance. As criteria for selection, biogeography of "key" species and biological associations can help address selection issues about which and how many protected areas are required to meet a conservation goal, where they should be placed, how big they should be, and what legal, social, and economic constraints affect their success. Nevertheless, exclusively protected habitats will lose species and fail to protect valued social functions, such as sustainable fishing and attraction of tourism, if not ecologically selected, established, and managed *within a wider ecosystem context*. By incorporating ecosystem principles, MPAs have the potential to fit well within an ecosystem-based approach (Lubchenco *et al.*, 2003).

MPAs, at present, have proved to be limited in their applications. First, only rarely are their boundaries based on ecological characteristics, and many have been designated opportunistically. Furthermore, many are "set aside" from outside, human-caused, deleterious influences, in which case sustainability is questionable. Despite such problems, there are signs among the scientific community and conservation biologists especially, that perspectives are changing towards broader applications, as the case studies exemplify. A fundamental part of this is a paradigm shift in both policy and management towards more comprehensive ways to protect biodiversity and marine space—a move towards EBM and marine spatial planning, with special attention given to

Box 13.2 Demography and connectivity: metapopulation dynamics guide the design of a marine reserve network

Brandon J. Puckett and David B. Eggleston
Department of Marine, Earth and Atmospheric Sciences, North Carolina State University, Raleigh, North Carolina, USA

Application of the metapopulation concept and identification of source and sink populations is key to designing marine reserve networks—multiple Marine Protected Areas connected by dispersal. Many marine populations are open, whereby larval settlement is uncoupled from local reproduction. At sufficiently large spatial sales, local populations characterized by demographic rates that vary in space and time are connected to one another via larval dispersal forming a closed metapopulation. Within a metapopulation, local populations can be classified as sources, which contribute more births than deaths to the metapopulation, or sinks if the converse is true. Metapopulation analyses require data on local demographic rates—growth, survival, and reproduction—and connectivity, i.e., the probability of local populations dispersing offspring among them.

Herein, we illustrate how metapopulation dynamics can guide reserve network design and evaluation using the eastern oyster (*Crassostrea virginica*) and Pamlico Sound, North Carolina, USA, as the model system. Specifically, we conducted mark-recapture studies and fecundity analyses to measure demographics and used a three-dimensional water circulation model to estimate larval connectivity among a network of no-take oyster reserves. Ultimately, we were interested in (i) identifying source and sink reserves; (ii) assessing the potential for reserves to function as a self-sustaining network (i.e., positive metapopulation growth, $\lambda \geq 1$); and (iii) comparing network sustainability among reserve size-increase design scenarios. The eastern oyster is particularly amenable to metapopulation analyses because adults are permanently attached benthic organisms representing discrete local populations only capable of dispersing via weakly swimming planktonic larvae. Unfortunately, populations of this commercially and ecologically important bivalve are at historic population lows along the Atlantic and Gulf coasts of the USA.

An increasingly popular strategy (and one used in North Carolina) for oyster restoration is to establish reserves that contain artificial reefs to provide hard settlement substrate for oyster larvae. In Pamlico Sound, which is a shallow, well-mixed, predominately wind-driven estuary, there are 10 oyster reserves that range in size from 0.03 to 0.2 km^2 and are spaced anywhere from c. 20 to 105 km apart. Our field measurements indicated that oyster growth, survival, and reproduction varied from 30 to 90% among reserves, and not in the same manner, such that certain reserves could be classified as the "growers" (i.e., reserves with fastest growth), others the "survivors", and yet others the "spawners" (Fig. B13.2.1a). Inter-reserve connections were rare—only 4 of a possible 90 connections were present, unidirectional, and relatively low in magnitude (<5%; Fig. B13.2.1a). Self-recruitment, while also rare (three of the 10 reserves self-recruited), was relatively high in magnitude (20–50%) when present (Fig. B13.2.1a). As a result of limited network connectivity via larval dispersal, only c. 6% of larvae released from reserves were subsequently retained within the reserve network. Limited connectivity and reserve-specific demographics resulted in only three of the 10 reserves functioning as sources (Fig. B13.2.1a) and an exponential decline in metapopulation size over time. Thus, we concluded that the network of oyster reserves in Pamlico Sound, as currently configured, is not self-sustaining or capable of persisting through time.

How can network connectivity be improved? One way is by increasing the size of reserves within the network. As reserve size is increased, a larger target for dispersing larvae is provided which should increase connectivity. However, socioeconomic and resource (e.g., artificial reef material or personnel) constraints prevent managers from indiscriminately increasing reserve size, so how should one prioritize reserves for expansion? To answer this question, we conducted computer simulations that increased reserve size under three scenarios: (i) uniform 10% size increase of all 10 reserves; (ii) size increase of three source reserves by 33.3%; and (iii) size increase of three worst sink reserves (i.e., lowest λ_C) by 33.3%. Model results for all three scenarios suggested that relatively small increases in reserve size led to rapid increases in larval retention within the network (i.e., increased network connectivity), as well as metapopulation growth rate (Fig. B13.2.1b). Of the three scenarios, increasing all reserves uniformly in size led to the largest improvements in larval retention and metapopulation growth rate, and thus was deemed the best strategy. Counter intuitively, increasing the size of the three source reserves ultimately resulted in the lowest larval retention and metapopulation growth rate (Fig. B13.2.1b), suggesting that a reserve's source strength is dependent on the location and size of surrounding reserves within the network (Fig. B13.2.1b). Given the current number, spatial configuration, and a uniform reserve size-increase strategy, approximately 5% of Pamlico Sound needs to be designated as reserves—an increase of two orders of magnitude from current size—for the network to be self-sustaining.

Spatial variation in oyster demographic rates in Pamlico Sound, as well as rare, but non-trivial larval connections, provided proof of the metapopulation concept in this reserve system, which enabled us to develop a metapopulation model to evaluate and guide reserve network design. Based on this analysis, we offer three concluding remarks: (i)

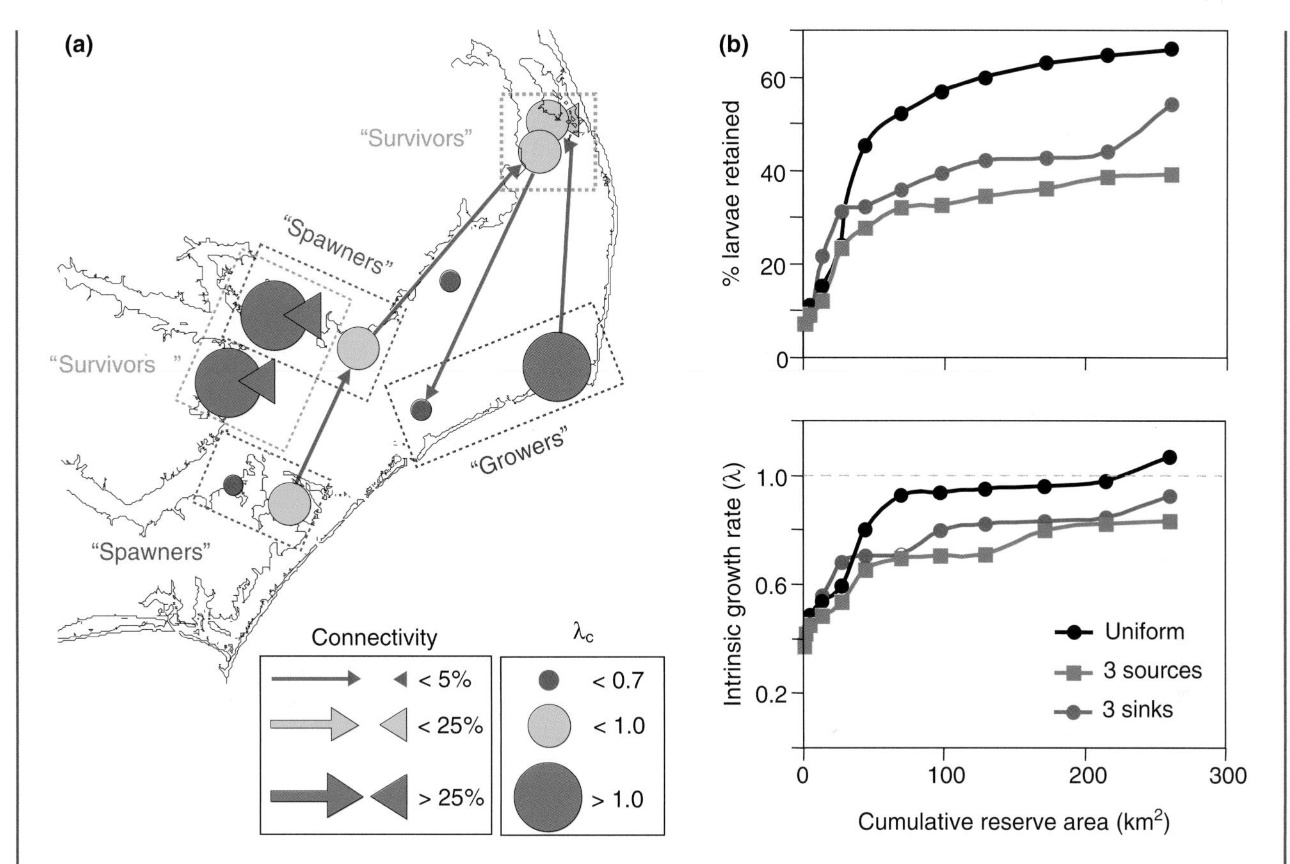

Fig. B13.2.1 (a) Map of no-take oyster reserves (circles) in Pamlico Sound, North Carolina, USA. Oyster demographic superlatives (e.g., fastest-growing oysters) are indicated by the dashed boxes. Larval connectivity is proportional to the thickness of the arrows and triangles, which depict inter-reserve dispersal and self-recruitment, respectively. Metapopulation source (largest circle, $\lambda c > 1$) and sink reserves ($\lambda c < 1$) are denoted by the circle size at each reserve. (b) Results of model simulations to determine the impact of increasing reserve size (measured as cumulative reserve area) on the percentage of larvae retained within the metapopulation (top panel) and growth rate of the metapopulation (bottom panel) for three scenarios whereby size increases were allocated (i) uniformly among all reserves (black); (ii) among the three source reserves (green); and (iii) among the three worst sinks (red). The initial points on each panel represent the current state of the reserve network.

not all habitats, local populations, or reserves function equally (i.e., source-sink dynamics); (ii) marine reserves and reserve networks do not guarantee restoration or conservation success and only when they are evaluated and designed at appropriate scales can they achieve management objectives; and (iii) metapopulation-based analyses that integrate demographic rates and population connectivity provide decision support tools that are essential for effective conservation and management.

Sources: Caley *et al.* (1996); Figueira and Crowder (2006); Gaines *et al.* (2010); Hanski and Gilpin (1991); Lester *et al.* (2009)

ecosystem resilience (see Section 13.4.3). That is, as MPAs are an essential part of marine conservation and must increase in number and magnitude, they also need to be implemented and managed in the context of the ecological and social dynamics of their surroundings at larger scales. Thus, to protect entire biotic communities, MPA strategies must seek a nested, hierarchical approach, from species and habitat protection at local scales to regulatory procedures for communities and ecosystems at regional scales.

13.4.2 Marine spatial planning

Marine spatial planning (MSP) is a recent attempt to move from sector-specific area protection (e.g., MPAs) and human uses of the coasts and oceans towards integrated multiple use through ocean zoning over large spatial scales. Sustainable use is a fundamental guiding principle (Douvere and Ehler, 2008; Gilliland and Laffoley, 2008). Mapping and analysis of biophysical features, socio-economic uses and values, and

Box 13.3 How not to design a sea otter reserve

James A. Estes
Center for Ocean Health, University of California, Santa Cruz, California, USA

The Bering Sea/North Pacific Ecosystem (BSNPE, south of the shelf, Ch. 7) is in a state of spectacular change, characterized in particular by precipitous population declines of various marine birds and mammals. Sea otters are the most recent addition to the list of dwindling species, having declined by >90% or more across large regions of western Alaska since about 1990. In contrast to the situation with most of the region's other declining species, the immediate cause of sea otter decline is reasonably well understood. But for the moment, let us imagine that we know little more than its pattern, other than to suspect that the cause is in some important way anthropogenic. These conditions define our present understanding for most of the BSNPE's declining species.

Given this state of affairs, how might one design a marine reserve for sea otters? We may begin by defining the goal, which is to protect a sufficiently large segment of habitat from human impacts to provide for a viable sea otter population. One approach is to identify an area capable of supporting an effective population size (*N*) of 500 individuals at an ecologically effective population density. Recent analyses indicate that about 6.5 sea otters km^{-1} is the minimum density required to prevent the coastal ecosystem from undergoing a phase shift from healthy kelp forests to deforested sea urchin barrens. Based on known features of the sea otter's mating system, sex ratio, and demography, this population size translates to an actual population (*N*) of about 1850 individuals. Armed with this figure and knowing the minimum effective population density of sea otters (*D*), the area (*A*) needed to support a viable population is estimated by *N/D*. These figures lead to values of A in the minimum range of about 300 km^2, roughly the size of the larger-sized islands in the Aleutian archipelago.

Although the reserve size specified by this scenario is large by existing average sizes of coastal-marine protected areas, in this case it would almost surely fail. The problem is that the ultimate cause of the sea otter decline has little or nothing to do with events in the protected area, but rather with those in far-removed oceanic waters. Tagging studies show that the demographic cause of the decline is elevated mortality, not reduced fertility or redistribution. Increased killer whale predation is one apparent reason for elevated mortality.

What may have caused killer whales to do this is less clear. Perhaps killer whales have hunted otters for years and killer whale numbers are on the rise. Or maybe a few killer whales fortuitously discovered sea otters as easy prey. The most likely explanation, in my view, is that killer whales turned to consuming sea otters as their preferred prey, as harbor seals and Steller sea lions have declined. The apparent fact that the otter declines occurred on the heels of the pinniped declines fits this explanation although it leaves the question of why pinniped populations have declined unanswered. Some suspect that pinniped declines were caused by negative interactions with local fisheries (Ch. 7), others favor oceanic regime shifts, and still others suspect that, as with the sea otter, predation had a lot to do with it.

This example serves to make the general points that marine reserves, even when very large, may not serve their intended purposes of when adjoining ecosystems are not considered. Marine ecologists recognize that functionally important linkages occur through such processes as larval dispersal, physical and biological transport of food and nutrients, and the movement of apex predators. However, the fact that marine reserves must be large enough to capture these diverse linkages remains underappreciated. Although one can argue the details, the take-home message is that marine reserves designed solely around the distribution and apparent resource needs of species targeted for protection are likely to fall short of their intended purpose.

Can marine reserves stem the tide of species decline in the BSNPE? Probably, but only *if* human impacts are instrumental in causing the declines and *if* the reserves are large enough to buffer populations against these impacts. The problem is that the burden of proof traditionally falls on scientists and resource managers to first demonstrate human cause and then to engineer the details of an explicit solution. Time is too short and the system is too complex for that approach to work in most cases. For the BSNPE, we might begin by establishing one or more very large Marine Protected Areas together with the time and resources to demonstrate their effects. These areas must be large because of the demographic characters and ecosystem linkages described above. If overfishing is responsible, either wholly or in part, for the declines of marine birds and mammals in the BSNPE, this approach has the potential to both demonstrate and begin to mitigate those effects. If killer whale predation turns out to be a factor, however, a management dilemma is raised that may be beyond human control.

Sources: Anderson and Piatt (1999); Estes *et al.* (1998); Estes *et al.* (2010); Ralls *et al.* (1983)

jurisdictional arrangements provide a first step toward producing meaningful mosaics of spaces suitable for place-based, marine ecosystem protection and resource management (Ray *et al.*, 1979, 1980; NOAA/NOS, 1988; Crowder and Norse, 2008). Principles of land- and seascape ecology have the potential to be applied as useful frameworks to guide managers in addressing multiple goals and to communicate with major stakeholders on shared responsibilities. MSP has entered national and international policy development, being one among nine priority objectives of the U.S. Ocean Policy Task Force (CEQ, 2010). The goal is to reconcile overlapping and conflicting management practices carried out by numerous agencies with conflicting mandates, rules, and management. MSP provides a transparent communication and application tool to address coastal and ocean-space zoning according to uses—e.g., fisheries, shipping, recreation, species protection, mineral extraction, and maintenance of biodiversity, among others. While MSP goals may be explicit, constraints lie in how boundaries may be set in highly dynamic ocean space. From a management point of view, a strategy is required that involves a balancing act: i.e., understanding the management cycle that links strategy with operations and the tools that apply at each stage of cycle (Kaplan and Norton, 2008). Furthermore from an ecosystem point of view, natural cycles of change (Section 4.8) need accounting. Yet, recognizing cumulative effects of development and application of zoning for use is moving marine conservation in a more realistic direction.

Any newly proposed ocean zoning will have to accommodate the past with future jurisdictions. It is not yet clear how existing zonation of the coasts and oceans (Fig. 3.4), or land-sea interactions, can be accommodated. Virtually the entire open ocean and coastal realms are captured in some sort of regulation or agreement, nationally and/or internationally. Ocean zones have been designated for a host of purposes, exclusively for specific uses and economic benefits, or to permit multiple uses within designated boundaries under varying sets of rules. To date, these zones have been almost exclusively directed towards human uses and endeavors, but overlapping uses have led to the recognition that more than jurisdictional zoning is required to resolve increasing conflicts, as between renewable and non-renewable resource extraction, marine transportation impacts, sensitivity of established protected areas, fisheries, mariculture, recreation and tourism values, and recovery of endangered species and habitats. More often than not, such zoning does little to address such emergent issues as climate change, sea-ice diminishment, acidification, anoxia, etc., all of which require changes in geopolitical thinking and improved administrative performance. Although jurisdictional boundary designations too often lack ecological context, thereby often hindering and complicating efforts for natural-resource sustainability, the question is not that zoning is not necessary, but how it can accommodate ever-changing and dynamic coastal and ocean systems (Ray, 2010; Spalding, 2011). For solutions to become workable, central government and intergovernmental policies need to be in place to guide decision-making in resolving competing goals among all stakeholders, addressing mismatches in ocean governance, and protecting common-pool resources.

The biosphere reserve is a spatial planning concept that addresses the above issues (Ch. 11). It was proposed in the mid-1970s, with coastal-marine conservation, conflict resolution the dynamic nature of marine systems, and boundary limitations of MPAs in mind (Ray and McCormick-Ray, 1987; Batisse, 1990). The first step in considering a biosphere reserve is to identify a "core" protected area that has legal protection, and the purpose for which it has been, or will be, established. Surrounding a core is a "buffer" area for the purpose of maintaining ecological and social support. Outside the buffer zone is a zone of "transition" that inhibits intensive, potentially deleterious human activities. Thus, the area a biosphere reserve potentially encompasses is large enough to maintain the identity and resilience of the designated core protected area. This "idealistic" solution has proved difficult to implement (Ch. 11). Nevertheless, the biosphere reserve concept offers an excellent model for the development of ecosystem-based marine spatial planning, particularly as it incorporates land-sea interactions, which, as yet, MSP does not.

13.4.3 Achieving resilience practice

The concept of "resilience practice" arises from the fact that resilience is an emergent property of ecosystems (Fig. 4.21). Walker and Salt (2006) clarified resilience as: "the capacity of the system to absorb disturbance and reorganize so as to retain essentially the same function, structure, and feedbacks—to have the same identity." Put more simply, resilience is "the ability to cope with shocks and keep functioning in much the same kind of way," being analogous to how healthy humans maintain health or recover from disease. Thus, given that ecosystems naturally change, and that human activities can disrupt this process, resilience is a necessary concept for conservation. Regeneration, or restoration following disturbance, depends on sources of resilience that operate at multiple scales, wherein ecosystems can absorb recurrent perturbations and cope with uncertainty and risk (Hughes *et al.*, 2005). Walker and Salt (2012) describe three broad activities for incorporating resilience into practice: describe the system, assess its resilience, and manage for resilience.

In the present situation of biodiversity loss and perturbed ecosystems, a fundamental aspect of resilience is restoration. Keeping in mind the above definition, this means restoring ecosystems to states that *have the same identity*, i.e., that processes are intact. As ecosystems progress through cycles of change, the likelihood is that community composition will shift along a trajectory of change. Jackson *et al.* (2011) call attention to "shifting baselines" as "a truly fundamental and revolutionary idea, but the revolution has not yet happened because the challenges are enormous." Which is to say that restoration to any precondition along a continuum of change may not be realistic or even possible absent knowledge of its history. Therefore, conservation for ecosystem resilience means keeping ecosystem function in mind when undertaking EBM, MPAs, and/or MSP. There are a number of ways this can be approached. A first step lies in recognizing that the system is self-organizing, moves through adaptive cycles in constant states of change, reaches limits (thresholds), and is linked to social, economic, and biophysical domains. In this context, one needs to consider what A.N. Whitehead termed the "fallacy of misplaced concreteness" in which attention is diverted from the underlying causes that drive change (Daly and Cobb, 1989).

Resilience practice requires flexibility and adaptability. Feedbacks between the manager and what is being managed may offer surprises and unexpected outcomes. Most importantly maintaining or restoring ecosystem resilience involves *assessing the sources of resilience in the system* (e.g., biodiversity, biophysical properties, evolutionary processes, etc.), monitoring them to determine a trend towards any new state, determining the extent and cause of change, and assessing costs and benefits socially, economically, and environmentally. Fundamentally, conservation for resilience allows engaging with the issue of complexity while focusing on what is important (Walker *et al.*, 2012). Goals to conserve biodiversity and valued species and species diversity, and to maintain or restore ecological services depend on the resilience of the interdependent socio-ecological system.

13.4.4 Engaging the social system

The human-ecological system is coupled to interactions and feedbacks that promote change in time, space, and organization; insights into this complexity cannot be gained through ecological or social research alone (Liu *et al.*, 2007). Different, interacting sectors of the human system play key roles in advancing integrative associations among coastal and marine systems, species, and their habitats that are needed to optimize goals of sustainable development and ecosystem resilience and also to promote social well-being (Box 13.4). This situation emphasizes the underlying role of governance, defined as the act or process of governing, taking account of social, economic, and ecological inputs, outputs, and functions.

13.4.4.1 The role of good environmental governance

There is a need for improved collective consciousness for making better decisions that impact ocean systems. The lifeboat metaphor (Hardin, 1974) exemplifies how human activities might be better managed to conserve ecosystem resilience for the common good. The globalization of opportunistic species that are homogenizing the planet into simpler forms of biodiversity and the disempowerment of people trying to manage their own affairs locally result from human-accelerated environmental change and eroding social, political, and jurisdictional boundaries that make up traditional governing practice (Swaney *et al.*, 2012).

The tools of good governance depend on law, science, economics, and public education entering into a process of decision-making, and of making decisions that affect outcomes of marine conservation. This process occurs where political authority facilitates resource management through a clear and legitimate process applied at all levels of governance, including the way the nation is governed or not. Marine and coastal protection, conservation, and restoration can only take place within a framework of good governance, respect for the rule of law, an environmentally sensitive legal system, and consent of people affected.

13.4.4.2 Roles of law, science, economics, and universities

Environmental law and policy are presently plagued by policy and jurisdictional mismatches (Alder, 2005). A mismatch is also evident between the nature and scope of coastal-marine issues and those of the institutions charged with solving them (Crowder *et al.*, 2006). Division of authority and responsibility for environmental protection among federal, provincial, and local governments has led to lack of cohesive rationale or justification for conservation action. Inevitably, these mismatches produce suboptimal levels of environmental protection, waste resources, discourage innovation, and inhibit the adoption and evolution of more effective measures for environmental protection.

Classic economics provides yet another critical mismatch, supporting a "growth economy" that runs into limits of finitude, ecological interdependence, and system entropy. The first and second laws of thermodynamics play key roles in economics (Georgescu-Roegen, 1971; Ch. 4). As the economist Herman Daly (1996) explained: "Finitude would not be so limiting if everything could be recycled, but entropy prevents complete recycling. Entropy would not be so limiting if environmental sources and sinks were infinite, but both are finite. That both are finite, plus the entropy law, means that the ordered structures of the economic subsystem are maintained at the expense of creating a more than offset amount of disorder in the rest of the system." This is to say that current economic processes and practices serve to degrade natural systems, deplete resources, and pollute the environment. As water and air are priced at zero to benefit markets, natural resources priced at market values become undervalued economic commodities and expose coastal and marine systems to exploitation, alteration, and degradation. As the economic system has advanced, the Earth has moved toward greater disorder (entropy), and as resources become scarce, their values rise and illegal trade advances. Therefore, market indicators are not good indicators of long-term social benefit. Robert Johnson, Executive Director of the Institute for New Economic Thinking recognized: "there's something amiss in a theory of value that doesn't value these common resources, the common pool on which we all base our lives" (Institute for New Economic Thinking, online).

Environmental degradation, pollution, climate change, and energy extractions cumulatively reduce the social values of coastal recreation, tourism, and renewable, common resources that most developing nations and individuals depend on. Furthermore, scientific evidence makes clear that coastal and marine systems cannot be managed for stakeholder benefits while ignoring ecosystem behavior. Present uses of these systems, as observed in the expanding human ecological footprint, are unsustainable (Jackson, 2010). Strauss *et al.* (2012) show that global warming has raised sea level about 20 cm since 1880, and the rate of rise is accelerating. How society may decide on reducing heat-trapping pollution will ultimately determine the fate of millions of people that live only 1 m above

Box 13.4 Well-being assessment

Robert Prescott-Allen
[Retired, International Union for Conservation of Nature and Natural Resources, author of *The Wellbeing of Nations*, co-author of *Blueprint for Survival, World Conservation Strategy*, and *Caring for the Earth: a Strategy for Sustainable Living*], Victoria, British Columbia, Canada

Well-being means a good condition or quality of life, both human life and the rest of the living world. Assessment means measurement and evaluation. Hence well-being assessment is the measurement and evaluation of human and ecosystem conditions and of the factors that produce them. Coastal-marine conservation needs to be guided by periodic assessments of human and ecosystem well-being to determine the socio-ecological state of a particular coastal or marine area, priority issues for action, and the effectiveness of the actions. This box reviews the dimensions of human and ecosystem well-being and how to assess them.

Human well-being

Depending on how it is defined, human well-being more or less equals human development or human happiness. Perhaps the simplest yet most encompassing definition is "human flourishing in its fullest sense" (Alkire, 2002). But no definition gives more than an inkling of human well-being. We need to know its dimensions—that is, the basic values or ends to be fulfilled for people to be well.

Over millennia, philosophers and spiritual leaders have proposed many sets of dimensions, addressing "being good" rather than "being well" or the proposition that "well being" comes from "well behaving" or "right conduct." In the 20th and 21st centuries, these have been joined by economists, sociologists, and psychologists with a focus on "well being" only, split into objective well-being and subjective well-being (or "well feeling"). The former tends to be examined through a process of argument. The latter is discovered through surveys of populations. Ends shared by many studies of objective well-being (e.g., Grisez *et al.*, 1987; Max-Neef *et al.*, 1991; Centre for Bhutan Studies, 2008) and of subjective well-being (e.g., Narayan *et al.*, 2000; Diener and Biswas-Diener, 2008; Rath and Harter, 2010) include health, wealth, rich use of time (work and play), inner well-being, and good relationships. Less common are freedom and rights, knowledge, aesthetics, and relationships with the supra-human. These broad ends give plenty of room for interpretation; different cultures value some aspects more than others.

Ecosystem well-being

Interest in human well-being is as old as people. The idea of its ecological equivalent is new, originating in the 20th century. It's not an exact equivalent. Ecologists are uncomfortable with "ecosystem well-being" (or "ecosystem health"), wary of its apparently teleological, Gaian suggestions. They prefer the more neutral "condition" or "state" or no label at all. Regardless of label, review of this book and the Millennium Ecosystem Assessment (2005) suggests that five essential features—or dimensions—need to be examined to determine the state of any ecosystem: (i) biophysical structure (structure of land, shore, and seafloor forms and of fixed biota); (ii) energy transformation (food chains and webs and production and productivity); (iii) biogeochemical processes (biogeochemical recycling and retention); (iv) diversity (species diversity, variability within species, and community diversity); and (v) ecosystem services (provisioning services such as food, biochemicals, ornamentals, and regulating services such as climate regulation, erosion regulation, waste treatment, natural hazard regulation, and other services such as heritage, recreation, aesthetic, spiritual, and intellectual).

Assessment

To be a reliable and useful guide for policy-making and action, a well-being assessment of a coastal and/or marine area has to follow some rules. First, combine measurement and narrative. *Measurement* provides a common language, an unambiguous account, enabling analysis of change over time and space. But not everything is measureable; and much that is measureable may not matter. *Narrative* is needed to tell the whole story, to interpret what is measured and put it in context; and to cover important aspects that cannot be measured without distortion.

Second, assess separately states (outcomes) and influences (inputs). One cannot tell how much, or in what ways, an influence such as economic performance or governance or education affects human well-being if it is included as a component of well-being. Nor can one learn the impact of energy use or resource extraction or emission rates on an ecosystem if it is part of the measurement of the ecosystem's condition. The four domains of well-being assessment are, therefore, human well-being, ecosystem condition, influences on human well-being, and influences on ecosystem condition (Fig. B13.4.1).

Third, develop an assessment protocol (*what* to assess) before designing the assessment (*how* to assess). This asks the questions: For the various groups of people in our area, what are the dimensions of human well-being?

(*Continued*)

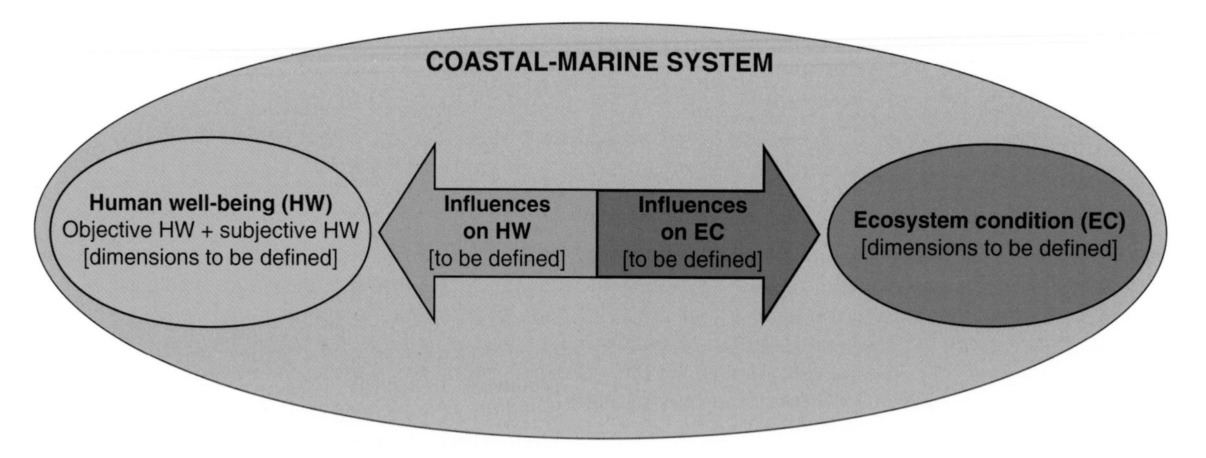

Fig. B13.4.1 The four domains of well-being assessment. Assessment participants decide the dimensions and influences to be assessed.

What features must we assess to determine the state of the ecosystem? What do we think are the main influences on human well-being and on ecosystem condition?

Fourth, construct a hierarchy of components of each dimension and influence, going from the general and unmeasurable (health, ecosystem services, economic performance, etc.) to the particular and measurable (indicators). For example, if level 1 is ecosystem services, level 2 might be provisioning services, regulating services, and other services, level 3 of provisioning services might be fisheries and others, and level 4 (the indicators) of fisheries might be the state of particular fishery stocks. Such systematic selection of indicators ensures they are representative of their dimension or influence and clearly reveals gaps in representation. In addition, the hierarchy of components provides a structure for the orderly aggregation of indicators should you wish to construct a human well-being index, ecosystem condition index, human influences index, and ecosystem influences index, to facilitate communication of the assessment's findings.

Sources: Alkire (2002); Centre for Bhutan Studies (2008); Diener and Biswas-Diener (2008); Grisez *et al.* (1987); Max-Neef *et al.* (1991); Millennium Ecosystem Assessment (2005); Narayan *et al.* (2000); Rath and Harter (2010)

high tide, saying nothing about the many lives that will be affected by increased storms, flooding, droughts, loss of food resources, and disease. Thus, managing the effects of such changes requires not only good governance, but also good science that brings to society an understanding of system behavior and natural history. Environmental change may be better understood through monitoring, creative synthesis of information, and continued research, all of which will ultimately determine the quality of coastal life in the 21st century and beyond (Swaney *et al.*, 2012). In this context, universities are centers of learning that have the capacity to integrate reliable knowledge into democratic principles that address coastal and marine issues. Universities can also help generate thinking about integrating good science with good governance, and bring awareness to the need to integrate land and sea into ecosystem-based goals, e.g., fisheries, pollution abatement, and ecosystem resilience.

Social values play strong roles in achieving marine conservation goals. The ethics of protecting marine and coastal biological diversity presents a struggle between environmental beliefs and values held in competing worldviews for institutional recognition, including law. Because translating biodiversity into legally enforceable obligations requires major changes in traditional habits, marine conservation reflects a cultural divide in human values: those that see the world through value of individual autonomy and mastery of the physical world, and those whose values urge individual restraint in order to preserve a common good that fits within the "balance of nature" (Cannon, 2007). As political will reflects the values of society at the present time, these values intertwine the social and ecological system into a future of environmental risk, chance, change, or opportunity for which there is no turning back, and society must make adjustments. However, interpretations of the law and implementing regulatory authority reflect cultural value judgments (Fig. 13.1). In the coastal realm where economic and social issues are especially difficult for politicians to resolve, implementation of treaties and national programs remain unfulfilled, lagging behind scientific understanding and public perception.

Good governance is essential for improving resource management. At present, numerous agencies operating under traditional institutional structures cannot adequately address new challenges or move from single-sector, issue-by-issue management and toward EBM. The combined effect of too many agencies seeking prescribed mandates, often in conflict with others, can exacerbate rather than solve environmental

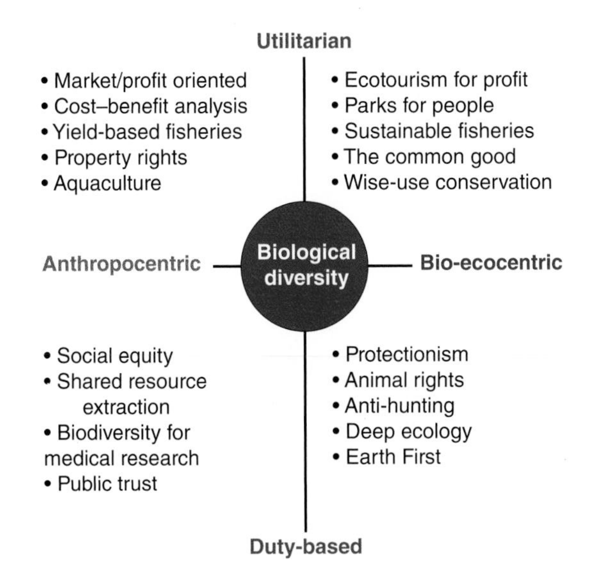

Fig. 13.1 Perceptions about values (e.g., biological diversity) are conceived in a decision matrix on two axes: from utilitarian to duty-based and from anthropocentric to bio- or ecocentric. Holders of various positions are exemplified in four segments, suggesting perceptual differences that are often difficult to resolve. Data from Beatley (1994).

problems. For example, the impact of many small decisions being made locally can accrete into big outcomes that no one intended, a phenomenon known as the "tyranny of small decisions" (Odum, 1982). Conservation goals confined to narrowly focused legal mandates are inhibited by lack of tools needed to meet intended goals, e.g., staff, funding, equipment, scientific data, etc. Management practices intended to do public good, such as government subsidies, may become "perverse" and need reform (Steenblik, 1998). In the case of fisheries, unintended and perverse consequences of governmental regulation of fishery resources that lack a uniform imposition of central regulations, or that ignore or contradict local regulations, exacerbate Hardin's "tragedy of the commons," where each person is trapped in a system that compels him to increase his profits without limits (Hardin, 1968; Davis, 1984).

13.5 LOOK TO THE FUTURE

F. Herbert Bormann (1996) observed, after long acquaintance with forestry management and acidification: "Never again would I be so naive as to think that natural resource decisions are based only on good scientific evidence. Social and economic factors will override science . . . if the public is not educated to understand the relationship of that science to their own long-term welfare."

Similarly, marine conservation faces its own set of social-scientific dilemmas. The ocean is a vast and powerful system that has entered into the human psyche for millennia. Its energy and productivity have nurtured the planet and have delivered benefits to humans that have seemed inexhaustible. Today, human activity no longer stops at the seashore, but encompasses huge arterial watersheds that deliver chemicals into the air and water to affect coastal and open-ocean systems. Intensifying human activities continue to alter the structure, timing, and recycling of marine ecosystems, systems that have self-organized into patterns of production over eons of time to sustain themselves and to provide resources to humans. As scientific evidence reveals, the diversity of life in the ocean is shifting toward smaller forms, undermining the leviathans that once roamed freely and that today must dodge human traffic, noise, and garbage in order to seek a decent meal.

Humans, as stewards of the oceans, are collectively pondering the need for marine protection and conservation. However, traditional beliefs that the oceans are too big to pollute, too productive to deplete, too out-of-sight to justify expensive actions, and too valued for their economic worth continue to hinder effective action. Our 20th century achievements have been enormous, having sent men to walk on the moon, created highways of information and communication, and provided nourishment for more people than Malthus ever thought possible. We have also achieved a social-political order through economic globalization, while also capturing fossil fuel energy by going deeper into the oceans, expanding global trade, and homogenizing a heterogeneous globe. Now, the extended human presence in the oceans, on land, and in the atmosphere faces challenges never before experienced in human history.

Conservation in general is a highly complex issue. Natural resources—water, food, air, and biota—are what the global society ultimately depends upon. The oceans, whether territorial or not, consist of a common pool of limited resources that require shared responsibility. This means that conservation is a social-scientific enterprise that needs to be put squarely on political and economic agendas. Addressing the ecological debt is as much a priority as the economic debt, as the latter is affected by the former. Thus, marine conservation can no longer be ignored without costing nations and the global society dearly. Marine conservation is no longer an option—it is the mandate for the 21st century.

The case studies exemplify the central role of governance, and the social tug-of-war for marine conservation. We may ask if we now know enough to proceed towards sustainable and resilient coastal and marine systems; the sole response has to be "possibly," but only if the coupled socio-economic-environmental system is addressed, and not only its components. And given the need for restoration of depleted species, disturbed spaces, and altered ecosystems under conditions of uncertainty, precaution is required. The "Earth Summit" (Rio Conference, 1992) suggested that the precautionary principle become operational as policy: "In order to protect the environment, the precautionary approach shall be widely applied by States according to their capabilities. Where there are threats of serious or irreversible damage, lack of full scientific certainty shall not be used as a reason for postponing cost-effective measures to prevent environmental degradation."

REFERENCES

Agardy T, Notarbartolo di Sciara G, Christie P (2011) Mind the gap: Addressing the shortcomings of marine protected areas through large scale marine spatial planning. *Marine Policy* **35**, 226–232.

Alder JH (2005) Jurisdictional mismatch in environmental federalism. *New York University Environmental Law Journal* **14**, 130–178.
Alkire S (2002) Dimensions of human development. *World Development* **30**, 181–205.
Anderson PJ, Piatt JF (1999) Community reorganization in the Gulf of Alaska following ocean climate regime shift. *Marine Ecology Progress Series* **189**, 117–123.
Batisse M (1990) Development and implementation of the biosphere reserve concept and its applicability to Coastal Regions. *Environmental Conservation* **17**, 111–116.
Beatley T (1994) *Ethical land use*. Johns Hopkins University Press, Baltimore, Maryland.
Bormann FH (1990) The global environmental deficit. *BioScience* **40**, 74.
Bormann FH (1996) Ecology: a personal history. *Annual Review of Energy and the Environment* **21**, 1–29.
Branch TA, Watson R, Fulton EA, Jennings S, McGilliard CR, Pablico GT, Ricard D, SR Tracey (2010) The trophic fingerprint of marine fisheries. *Nature* **468**, 431–435.
Caley MJ, Carr MH, Hixon MA, Hughes TP, Jones GP, Menge BA (1996) Recruitment and the local dynamics of open marine populations. *Annual Review of Ecology, Evolution and Systematics* **27**, 477–500.
Cannon J (2007) Words and Worlds: The Supreme Court in *Rapanos* and *Carabell*. *Virginia Environmental Law Journal* **25**(277), 1–44.
Carr MH, Saarman E, Caldwell MR (2010) The role of "rules of thumb" in science-based environmental policy: California's Marine Life Protection Act as a case study. *Stanford Journal of Law, Science and Policy* **2**, published online, March 2010.
Centre for Bhutan Studies (2008) *Gross National Happiness*. www.grossnationalhappiness.com/
CEQ (2010) *Final recommendations of the interagency ocean policy task force July 19, 2010*. The White House Council On Environmental Quality, Washington D.C., online.
Choi JS, Frank KT, Leggett WC, Drinkwater K (2004) Transition to an alternate state in a continental shelf ecosystem. *Canadian Journal of fisheries and Aquatic Science* **61**, 501–510.
Clark JR (1996) *Coastal zone management handbook*. CRC Press, Boca Raton, Florida.
Crowder LB, Osherenko G, Young OR, Airamé S, Norse EA, Baron N, *et al.* (2006) Resolving mismatches in U.S. ocean governance. *Science* **313**, 617–618.
Crowder L, Norse E (2008) Essential ecological insights for marine ecosystem-based management and marine spatial planning. *Marine Policy* **32**, 772–778.
Daly HE (1996) *The economics of sustainable development*. Beacon Press, Boston, Massachusetts.
Daly HE, Cobb Jr. JB (1989) *For the Common Good: Redirecting the Economy toward Community, the Environment, and a Sustainable Future*. Beacon Press, Boston, Massachusetts.
Dalzell P, Adams TJH, Polunin NVC (1996) Coastal fisheries in the Pacific islands. *Oceanography and Marine Biology Annual Review* **34**, 395–531.
Dansgaard W, Johnsen SJ, Clausen HB, Dahl-Jensen DD, Gunndestrup NS, Hammer CU, Hvidberg CS, *et al.* (1993) Evidence for general instability of past climate from a 250-kyr ice-core record. *Nature* **364**, 218–220.
Davis A (1984) Property rights and access management in the small boat fishery: a case study from southwest Nova Scotia. In *Atlantic Fisheries and Coastal Communities: Fisheries Decision-Making Case Studies* (eds Lamson C, Hanson AJ). Dalhousie Ocean Studies Programme, Halifax, N.S., 133–164.
Dean C (2005) *Calling the ocean's bluff. New York Times*, November 3.
Diener E, Biswas-Diener R (2008) *Happiness: Unlocking the Mysteries of Psychological wealth*. Blackwell Publishing, Malden, Massachusetts, and Oxford, UK.
Douvere F, Ehler E (2008) Introduction. *Marine Policy* **32**, 759–761.
Estes JA, Tinker MT, Williams TM, Doak DF (1998) Killer whale predation on sea otters linking costal with oceanic systems. *Science* **282**, 473–476.
Estes JA, Tinker MT, Bodkin JL (2010) Using ecological function to develop recovery criteria for depleted species: sea otters and kelp forests in the Aleutian Archipelago. *Conservation Biology* **24**, 852–860.
FAO (2008) *Fishery and aquaculture statistics*. Food and Agriculture Organization of the United Nations, Rome.
FAO (2010) *The state of world fisheries and aquaculture*. Food and Agriculture Organization of the United Nations, Rome.
FAO (2011) *A world overview of species of interest to fisheries*. In FAO Fisheries and Aquaculture Department [online]. Rome. Updated. [Cited 14 June 2011]. www.fao.org/fishery/topic/2017/en
Figueira WF, Crowder LB (2006) Defining patch contribution in source-sink metapopulations: the importance of including dispersal and its relevance to marine systems. *Population Ecology* **48**, 215–224.
Folke C, Jansson Å, Jonas Larsson J, Costanza R (1997) Ecosystem appropriation by cities. *Ambio* **26**, 167–172.
Francis RC, Hixon MA, Clarke ME, Murawski SA, Ralston S (2007) Ten commandments for ecosystem-based fisheries scientists. *Fisheries* **32**, 217–233.
Gaines SD, White C, Carr MH, Palumbi SR (2010) Designing marine reserve networks for both conservation and fisheries management. *Proceedings of the National Academy of Science USA* **107**, 18286–18293.
Georgescu-Roegen N (1971) *The entropy law and the economic process*. Harvard University Press, Massachusetts.
Gilliland PM, Laffoley D (2008) Key elements and steps in the process of developing ecosystem-based marine spatial planning. *Marine Policy* **32**, 787–796.
Gleason M, Fox E, Ashcraft S, Vasques J, Whiteman E, *et al.* (2012) Designing a network of marine protected areas in California: achievements, costs, lessons learned, and challenges ahead. *Ocean and Coastal Management*.
Graham M (1999) *The Morning After Earth Day*. The Brookings Institution, Washington DC, p.9.
Grisez G, Boyle J, Finnis J (1987) Practical principles, moral truth and ultimate ends. *The American Journal of Jurisprudence* **32**, 98–151.
Hanski I, Gilpin M (1991) Metapopulation dynamics: brief history and conceptual domain. *Biological Journal of the Linnean Society* **42**, 3–16.
Hardin G (1968) The tragedy of the commons. *Science* **162**, 1243–1248.
Hardin G (1974) Commentary: Living on a Lifeboat. *BioScience* **24**, 561–568.
Hobbs NT (2003) Challenges and opportunities in integrating ecological knowledge across scales. *Forest Ecology and Management* **181**, 223–238.
Holling CS (2001) Understanding the complexity of economic, ecological, and social systems. *Ecosystems* **4**, 390–405.
Hughes TP, Bellwood DR, Folke C, Steneck, Wilson J (2005) New paradigms for supporting the resilience of marine ecosystems. *TRENDS in Ecology and Evolution* **20**, 380–386.
Institute for New Economic Thinking. ineteconomics.org/about
Jackson JBC (2008) Ecological extinction and evolution in the brave new ocean. *Proceedings of the National Academy of Sciences* **105** (suppl. 1), 11458–11465.

Jackson JBC (2010) Review: the future of the oceans past. *Philosophical Transactions of the Royal Society B* **365**, 3765–3778.

Jackson JBC, Alexander KE, Sala E, eds (2011) *Shifting baselines*. Island Press, Washington.

Kaplan RS, Norton DP (2008) Mastering the management system. *Harvard Business Review* January, 63–77.

Lester SE, Halpern BS, Grorud-Colvert K, Lubchenco J, Ruttenberg BI, Gaines SD, Airamé S, Warner RR (2009) Biological effects within no-take marine reserves: a global synthesis. *Marine Ecology Progress Series* **384**, 33–46.

Lubchenco J, Palumbi SR, Gaines SD, Andelman S (2003) Plugging a hole in the ocean: the emerging science of marine reserves. *Ecological Applications* **13**(1), Supplement, S3–S7.

Liu J, Dietz T, Carpenter SR, Alberti M, Folke C, Moran E, *et al.* (2007) *Science* **317**, 1513–1516.

Ludwig D, Hilborn R, Walters C (1993) Uncertainty, resource exploitation, and conservation: lessons from history. *Science* **260**, 17, 36.

Max-Neef M, Elizalde A, Hopenhayn M (1991) Development and human needs. In *Human scale development: conception, application and further reflections* (ed. Max-Neef M). The Apex Press, New York and London, 13–54.

McLeod KL, Leslie HM (2009a) Why ecosystem-based management? In *Ecosystem-based management for the oceans* (eds McLeod KL, Leslie HM). Island Press, Washington, D.C., 3–12.

McLeod KL, Leslie HM (2009b) Ways forward. In *Ecosystem-based management for the oceans* (eds McLeod KL, Leslie HM). Island Press, Washington, D.C., 341–351.

Millennium Ecosystem Assessment (2005) *Ecosystems and human well-being: Synthesis*. Island Press, Washington, D.C.

Narayan D, Chambers R, Shah MK, Petesch P (2000) *Voices of the Poor. Crying out for Change*. Oxford University Press for the World Bank, Oxford and New York.

NOAA/NOS (1988) *Bering, Chukchi, and Beaufort Seas coastal and ocean zones strategic assessment: data atlas*. U.S. Department of Commerce, Strategic Assessment Branch, NOAA National Ocean Service, Government Printing Office, Washington, D.C.

NPFC (2012) Multilateral Meeting on Management of High Sea fisheries in the North Pacific Ocean. Interim measures. North Pacific Fisheries Commission online: nwpbfo.nomaki.jp/

NRC (1995) *Understanding Marine Biodiversity: A Research Agenda for the Nation*. National Research Council. National Academy Press, Washington, D.C.

Odum WE (1982) Environmental degradation and the tyranny of small decisions. *BioScience* **32**, 728–729.

Olsen SB (2000) Educating for the governance of coastal ecosystems: the dimensions of the challenge. *Ocean and Coastal Management* **43**, 331–341.

O'Neill RV, Hunsaker CT, Timmins SP, Jackson BL, Jones KB, Riitters KH, Wickham JD (1996) Scale problems in reporting landscape pattern at the regional scale. *Landscape Ecology* **11**(3), 169–180.

Pauly D, Christensen V, Guénette S, Pitcher TJ, Sumaila UR, Walters CJ, *et al.* (2002) Towards sustainability in world fisheries. *Nature* **418**, 689–695.

Pauly D, Alder J, Booth S, Cheung WWL, Christensen V, Close C, *et al.* (2008) Fisheries in Large Marine Ecosystems: descriptions and diagnoses. In *A Perspective on Changing Conditions in LMEs of the World's Regional Seas, the UNEP Large Marine Ecosystems Report* (eds Sherman K, Hempel G). Nairobi, Kenya, UNEP.

Pauly D, Christenson V, Dalsgaard J, Froese R, Torrres Jr. F (1998) Fishing down marine food webs. *Science* **279**, 860–863.

Pikitch EK, Santora C, Babcock EA, Bakun A, Bonfil R, Conover DO, *et al.* (2004) Ecosystem-Based Fishery Management. *Science* **305**, 346–347.

Prigogine I (1980) *From being to becoming*. W. H. Freeman and Company, New York, 272pp.

Ralls K, Brownell RL, Ballon J (1983) Genetic diversity in California sea otters: theoretical considerations and management implications. *Biological Conservation* **25**, 209–232.

Rath T, Harter J (2010) *Well being: the five essential elements*. Gallup Press, New York.

Ray GC (1999) Coastal-marine protected areas: agonies of choice. *Aquatic Conservation* **9**, 607–614.

Ray GC (2010) Editorial: coastal and marine spatial planning: a policy waiting to happen. *Aquatic Conservation: Marine and Freshwater Ecosystems* **20**, 363–364.

Ray GC, McCormick-Ray MG, Salm RV, Campbell DG, Dobbin NE, Smith VE, Miller KR, Putney AD (1979) *Preliminary data atlas: planning a marine conservation strategy for the Caribbean Region*. International Union for the Conservation of Nature, Gland, Switzerland.

Ray GC, McCormick-Ray MG, Dobbin JA, Ehler CN, Basta DJ (1980) *Eastern United States Coastal and Ocean Zones Data Atlas*. U.S. Council on Environmental Quality and the National Oceanic and Atmospheric Administration, Office of Coastal Zone Management, Washington, D.C.

Ray GC, McCormick-Ray MG (1987) Coastal and marine biosphere reserves. In *World Wilderness Congress Worldwide conservation. Proceedings of the Symposium on Biosphere Reserves*. MAB (Man and the Biosphere), Estes Park, Colorado, Sept. 14–17, 68–78.

Rockström J, Steffen W, Noone K, Persson Å, Chapin III FS, Lambin EF, *et al.* (2009) A safe operating space for humanity. *Nature* **461**, 472–475.

Rodrigue J-P (2010) Ports and maritime trade. In *The Encyclopedia of Geography* (ed. Warf B). SAGE Publications, University of Kansas.

Rosenberg AA, Sandifer PA (2009) What do managers need? In *Ecosystem-based management for the oceans* (eds McLeod KL, Leslie HM). Island Press, Washington, D.C., 13–29.

Ruckelshaus MH, Levin P, Johnson JB, Kareiva PM (2002) The Pacific salmon wars: what science brings to the challenge of recovering species. *Annual Review of Ecology, Evolution and Systematics* **33**, 665–706.

Sala E, Aberto-Gropeza O, Paredes G, Para, I, Barrera JC, Dayton PK (2002) A general model for designing networks of marine reserves. *Science* **298**, 1991–1993.

Sherman K, Belkin I, Seitzinger S, Hoagland P, Jin D (2009) Indicators of changing states of large marine ecosystems. In *Sustaining the World's Large Marine Ecosystems* (eds Sherman K, Aquarone MC, Adams S). Office of Marine Ecosystem Studies, NOAA, Narragansett, Rhode Island, 13–49.

Sorensen J (1993) The international proliferation of integrated coastal zone management efforts. *Ocean and Coastal Management* **21**, 45–80.

Spalding MJ (2011) Making lemonade. *The Environmental Forum* **28**(1), 30–33.

Steenblik R (1998) *Subsidy reform: doing more to help the environment by spending less on activities that harm it*. Presented at a workshop on "Doing More With Less," IUCN's 50th Anniversary Fontainebleau, France, November 1998. economics.iucn.org

Strauss B, Tebaldi C, Ziemlinski R (2012) *Surging seas: sea level rise, storms and global warming's threat to the US coast*. A Climate Central Report March 14, 2012. www.climatecentral.org/; sealevel.climatecentral.org/research/papers/

Sumaila UR, Pauly D (2011) The "March of Folly" in global fisheries. In *Shifting Baselines: The Past and the Future of Ocean Fisheries* (eds Jackson JBC, Alexander KE, Sala E). Island Press, Washington, D.C., 21–32.

Swaney DP, Humborg C, Emeis K , Kannen A, Silvert W, Tett P, *et al.* (2012) Five critical questions of scale for the coastal zone. *Estuarine, Coastal and Shelf Science* **96**, 9–21.

Thrush SF, Dayton PK (2010) What can ecology contribute to ecosystem-based management? *Annual Review of Marine Science* **2**, 419–441.

UNEP-WCMC (2008) *National and regional networks of marine protected areas: a review of progress.* UNEP-WCMC, Cambridge, UK.

UNGA (2006) *Sustainable Fisheries.* United Nations General Assembly A/RES/61/105.

Walker B, Barrett S, Polasky S, Galaz V, Folke C, Engström G, Ackerman F, *et al.* (2009) Looming global-scale failures and missing institutions. *Science* **325**, 1345–1346.

Walker B, Salt D (2006) *Resilience Thinking.* Island Press, Washington, D.C.

Walker B, Salt D (2012) *Resilience Practice: Building Capacity to Absorb Disturbance and Maintain Function.* Island Press, Washington, D.C.

Worm B, Hilborn R, Baum JK, Branch TA, Collie JS, Costello C, *et al.* (2009) Rebuilding global fisheries. *Science* **325**, 578–585.

SPECIES INDEX

Notes: Page numbers in *italics* refer to figures; those in bold refer to **tables**, and bold and *italics* are ***boxes***.
Species listed are those emphasized in text and accompanied by scientific names. Species are grouped in major subdivisions (microorganisms, plants, invertebrates, and vertebrates). When two or more species share a common name (e.g., barnacles, albatrosses), they are grouped together. For each species, vernacular ("common") names are given first; in some cases, the scientific name serves as the common name. Vernacular names may not indicate evolutionary relationships, e.g., "prawn" and "shrimp" may commonly be used interchangeably and "rockfish" is used for several unrelated species.

Microorganisms

Plants

Invertebrates

Vertebrates

SUBJECT INDEX

Notes: Page numbers in *italics* refer to figures; those in bold refer to **tables**, and bold and *italics* are ***boxes***. Organisms are listed under their common name, when given.

Marine Conservation: Science, Policy, and Management, First Edition. G. Carleton Ray and Jerry McCormick-Ray.